Your Learning Style

To discover what kind of learner you are, complete the Learning Styles Inventory on page xiv (or in MyMathLab). Then, in the textbook, watch for the Learning Strategy boxes and the accompanying icons that provide ideas for maximizing your own learning style.

_____ 1. I remember information better if I write it down or draw a picture of it.

_____ 2. I remember things better when I hear them instead of just reading or seeing them.

_____ 3. When I receive something that has to be assembled, I just start doing it. I don't read the directions.

_____ 4. If I am taking a test, I can visualize the page of text or lecture notes where the answer is located.

_____ 5. I would rather have the professor explain a graph, chart, or diagram to me instead of just showing it to me.

_____ 6. When learning new things, I want to do it rather than hear about it.

_____ 7. I would rather have the instructor write the information on the board or overhead instead of just lecturing.

_____ 8. I would rather listen to a book on tape than read it.

_____ 9. I enjoy making things, putting things together, and working with my hands.

_____ 10. I am able to conceptualize quickly and visualize information.

_____ 11. I learn best by hearing words.

_____ 12. I have been called hyperactive by my parents, spouse, partner, or professor.

_____ 13. I have no trouble reading maps, charts, or diagrams.

_____ 14. I can usually pick up on small sounds like bells, crickets, or frogs, or distant sounds like train whistles.

_____ 15. I use my hands and I gesture a lot when I speak to others.

Your Learning Strategies

Learning Strategy

Developing a good study system and understanding how you best learn is essential to academic success. Make sure you familiarize your-self with the study system outlined in the To the Student section at the beginning of the text. Also take a moment to complete the Learning Styles Inventory found at the end of that section to discover your per-sonal learning style. In these Learning Strategy boxes, we offer tips and suggestions on how to connect the study system and your learn-ing style to help you be successful in the course.

Learning Strategy

In the To the Student section, we suggest that when taking notes, you use a red pen for definitions and a blue pen for rules and pro-cedures. Notice that we have used those colors in the design of the text to connect with your notes.

Definition **Additive inverses:** Two numbers whose sum in zero.

Examples of additive inverses:

15 and -15 because $15 + (-15) = 0$

-9 and 9   because $-9 + 9 = 0$

0 and 0   because $0 + 0 = 0$

Notice that 0 is its own additive inverse and that for numbers other than 0, the additive inverses have the same absolute value but *opposite* signs.

Rules

The sum of two additive inverses is zero.

The additive inverse of 0 is 0.

The additive inverse of a nonzero number has the same absolute value but opposite sign.

The Math Study System

The Carson Math Study System is designed to help you succeed in your math course. You will discover your own learning style, and you will use study strategies that match the way you learn best. You will also learn how to organize your course materials, manage your time efficiently, and study and review effectively.

Your Math Notebook

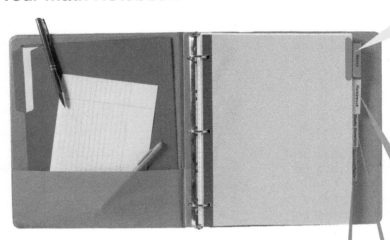

Notes
(see pages xvi–xvii)

Section #   9/20

We can simplify an expression by combining like terms.

p. 37 def. Like terms: constant terms or variable terms that have the
same variable(s) raised to the same powers.

Ex 1) 2x and 3x are like terms.
Ex 2) 5x and 7y are not like terms.

Homework
(see pages xvii–xix)

Section # Homework   9/21

#1 – 15 odd

1. $5^2 + 3 \cdot 4 - 7$
$= 25 + 3 \cdot 4 - 7$
$= 25 + 12 - 7$
$= 37 - 7$
$= 30$

Quizzes/Tests
(see page xxi)

Chapter # Quiz   10/10

For 1-4, simplify.

1. $|-8|$ $= 8$ ✓   4/4 = 100%
Nice
work!

2. $|9|$ $= 9$ ✓

3. $-15 + 5$ $= -10$ ✓

4. $-8 + -6$ $= -14$ ✓

Study Materials
(see pages xix–xx)

Chapter # Study Sheets   9/10

To graph a number on a   ex) Graph -2.
number line, draw a dot on   -4 -3 -2 -1   0   1
the mark for the number.

The absolute value of a positive
number is positive.   ex) $|7| = 7$

The absolute value of a negative
number is positive.   ex) $|-12| = 12$

Fourth Edition

Prealgebra

Tom Carson

PEARSON

Boston   Columbus   Indianapolis   New York   San Francisco   Upper Saddle River
Amsterdam   Cape Town   Dubai   London   Madrid   Milan   Munich   Paris   Montréal   Toronto
Delhi   Mexico City   Sao Paulo   Sydney   Hong Kong   Seoul   Singapore   Taipei   Tokyo

Editorial Director: Chris Hoag
Editor in Chief: Maureen O'Connor
Executive Editor: Cathy Cantin
Executive Content Editor: Kari Heen
Associate Content Editor: Christine Whitlock
Assistant Editor: Jonathan Wooding
Editorial Assistant: Kerianne Okie
Senior Development Editor: Elaine Page
Senior Managing Editor: Karen Wernholm
Associate Managing Editor: Tamela Ambush
Digital Assets Manager: Marianne Groth
Supplements Production Coordinator: Kerri Consalvo
Associate Producer: Jonathan Wooding
Content Development Manager: Rebecca Williams
Senior Content Developer: Mary Durnwald
Marketing Manager: Rachel Ross
Associate Marketing Manager: Alicia Frankel
Marketing Assistant: Ashley Bryan
Senior Author Support/Technology Specialist: Joe Vetere
Rights and Permissions Advisor: Michael Joyce
Image Manager: Rachel Youdelman
Procurement Manager/Boston: Evelyn Beaton
Procurement Specialist: Linda Cox
Associate Director of Design, USHE North and West: Andrea Nix
Senior Designer: Heather Scott
Text Design: Tamara Newnam
Production Coordination: Laura Hakala
Composition: PreMediaGlobal
Illustrations: PreMediaGlobal
Cover Designer: Tamara Newnam
Cover Image: F1 Online/SuperStock

For permission to use copyrighted material, grateful acknowledgment is made to the copyright holders on page P-1, which is hereby made part of this copyright page.

Many of the designations used by manufacturers and sellers to distinguish their products are claimed as trademarks. Where those designations appear in this book, and Pearson Education was aware of a trademark claim, the designations have been printed in initial caps or all caps.

Library of Congress Cataloging-in-Publication Data
Carson, Tom, 1967–
    Prealgebra/Tom Carson. —4th ed.
        p. cm.
    Includes index.
    ISBN 978-0-321-75695-4 (alk. paper)
1. Mathematics—Textbooks.   I. Title.

QA39.3.C39 2013
510—dc22   2011007280

18

ISBN 13: 978-0-321-75695-4
ISBN 10: 0-321-75695-9

Contents

Preface

Prealgebra, Fourth Edition, provides a fresh approach for the student who needs a brush-up in arithmetic and basic algebra concepts and for those who are encountering algebra for the first time. Written in a relaxed, nonthreatening style, this text takes great care to ensure that students who have struggled with math in the past will be comfortable with the subject matter. Explanations are carefully developed to provide a sense of *why* a mathematical process works the way it does, instead of just *how* to follow the process.

Problems pulled from science, engineering, accounting, health fields, the arts, and everyday life link mathematics to the real world. The link to real-world problem solving is further developed through the Project Portfolio Workbook (see page ix for a complete description of the online *Project Portfolio Workbook*).

Part of the freshness of *Prealgebra*'s approach is the sequencing of ideas. For example, students learn the basics of polynomials early in this text to give them a patient introduction and ample practice of this fundamental algebra topic.

In addition, a complete study system with a Learning Styles Inventory and learning strategies integrated throughout the text provides further guidance for students (see pages xvi–xxi).

Upon completing the material in this text, a student should be able to proceed successfully to an introductory algebra course or a survey of math course. This text is designed to be versatile enough for use in a standard lecture format, a hybrid course, a self-paced lab, and even an independent-study format. A strong ancillary package provides a wealth of supplemental resources for both instructors and students (see pages ix–xi for complete descriptions).

What's New in This Edition?

In addition to revised explanations and a new design, the following changes were made to create the fourth edition.

Features

The **Math Study System** has been expanded with sample student notes, homework, and study materials to guide students in developing their own math notebooks to complement their personal learning styles. In addition, learning strategies written by students and recent college graduates have been added throughout the text, providing helpful tips from successful students. These can be identified by the student's name at the end of the strategy.

Chapter Openers now provide a brief topical overview of the chapter. In the instructor's edition, each opener also now includes teaching suggestions for the chapter.

Each section now begins with **Warm-up** Exercises that review previously learned material helpful to understanding the concepts in the section.

Section Exercises are now grouped by objective, to make it easier for students to connect examples with related exercises. In addition, Prep Exercises help students focus on the terminology, rules, and processes corresponding to the exercises that immediately follow.

Chapter Summary and Review The Chapter Summary is now interactive with Review Exercises integrated within the summary to encourage active learning and review. For each key topic, the student is prompted to complete the corresponding definitions, rules, and procedures. A key example is provided for reference. Review exercises complete each section of summarized material, so the student can apply the skills and concepts for immediate reinforcement.

Updates

- Fresh data: Real-data examples and exercises have been revised with more current data as needed to keep the applications relevant and contemporary.
- New examples and exercises: Many new examples and over 650 new exercises have been added throughout the text.

Content Changes

To improve readability, lengthy explanations were simplified and, where possible, visuals were added to show rather than say.

- The problem-solving outline is used in the more complicated application examples as needed.
- Discussion boxes have been turned into discussion suggestions to the instructor that now appear only in the annotated instructor's edition. This reduces some distractions to the student while still giving the instructor ideas for enriching class discussion.

In addition, the following specific content changes have been made to improve student comprehension and retention of material:

Chapter 1

In Section 1.1, number lines are now taught before inequalities and rounding. Also, expanded notation is now shown with words instead of multiplication to strengthen the connection with place value names.

Square roots and the order of operations agreement were moved to Chapter 2, improving the pacing of Chapter 1.

Mean, median, and mode are now introduced in Section 1.5 and revisited throughout the text.

Chapter 2

Additive inverse is now taught in Section 2.2 after the addition of integers is introduced. This simplifies the topic and places it closer to Section 2.3, where it is used to write subtractions as equivalent additions.

Chapter 3

Section 3.1 now covers translating word phrases to mathematical expressions.

For clarity, combining like terms was moved from Section 3.3 to Section 3.2.

Chapter 6

Estimating was added to make decimal coverage consistent with fraction coverage.

Chapter 7

A more visual approach was used to explain how to translate application problems to a proportion to help students see the correct relationships between the numerators and denominators.

Chapter 8

Sections 8.2 and 8.3 were combined so that word-for-word translation to an equation and the proportion method are now taught side-by-side in one section. Teaching both methods for each example is a more efficient use of class time and makes it easier to discuss which method is best for the situation.

Chapter 9

Congruent triangles have been added to Section 9.1.

Hallmark Features

Real, relevant, and interesting applications   Nearly every application problem is a real situation taken from science, engineering, health, finance, the arts, or everyday life. The real-world applications not only illustrate the uses of basic arithmetic and algebra concepts, but they also expose students to the wonders of the world in which we live. Often the problems follow up with open-ended discussion questions where there is no "correct" answer. These questions help students to think beyond just getting a numeric answer by encouraging them to apply mathematical results to solve problems (see pages 72, 234, and 547).

The how and the why   This text explains not only how to do the math but also why the math works the way it does, where it comes from, and how it is relevant to students' everyday lives. Knowing all of this helps the students remember the concepts.

A Study System is presented in the To the Student section on pages xvi-xxii and is reinforced throughout the text. The system recommends using color codes for taking notes (red for definitions; blue for rules and procedures; and black for notes and examples) which are reinforced throughout the text. In addition, students are encouraged to use the Learning Styles Inventory (see p. xiv).

A Learning Styles Inventory is presented on page xiv to help students assess their particular style of learning and use it effectively in the course.

Learning Strategy boxes appear where appropriate in the text to offer advice on how to effectively use the study system and how to study specific topics based on a student's learning style (see pages 19, 282, 401, 536, and 605).

Connection boxes appear where appropriate in the text to help students see relationships between topics to develop a deeper understanding of the math (see pages 107, 377, and 603).

Problem-solving process   On page 62 of Section 1.6, a problem-solving process is introduced:

1. Understand
2. Plan
3. Execute
4. Answer
5. Check

This process is revisited throughout the text to reinforce the use of effective problem solving strategies (see pages 298, 439, and 513).

Student Supplements	Instructor Supplements

Student's Solutions Manual

ISBN-13: 9780321782915

ISBN-10: 0321782917

Complete solutions to the odd-numbered section exercises and solutions to all of the section-level review exercises, chapter review exercises, practice tests, and cumulative review exercises.

MyWorkBook

ISBN-13: 9780321782939

ISBN-10: 0321782933

MyWorkBook can be packaged with the textbook or with the MyMathLab access kit and includes the following resources for each section of the text:

- Key vocabulary terms and vocabulary practice problems.
- Guided examples with stepped-out solutions and similar practice exercises, keyed to the text by learning objective.
- References to textbook examples and section lecture videos for additional help.
- Additional exercises with ample space for students to show their work, keyed to the text by learning objective.

Chapter Test Prep Videos

Chapter Tests can serve as practice tests to help students study. Watch instructors work through step-by-step solutions to all the Chapter Test exercises from the textbook. These videos are available on YouTube (search CarsonPrealgebra) and in MyMathLab.

Video Resources

- Series of lectures correlated directly to each section of the text.
- Video lectures include English captions.
- Ideal for distance learning or supplemental instruction.
- Available in MyMathlab.

Annotated Instructor's Edition

ISBN-13: 9780321782854

ISBN-10: 0321782852

- Includes answers to all exercises, including puzzle problems, printed in blue near the corresponding problem.
- Useful instructor notes are printed in the margin.

Project Portfolio Workbook

- Activities and problems related to the theme of building a house are keyed to the relevant sections of the text. As they work through each project over the course of the semester, students apply the skills and concepts they've learned in a realistic and integrated context.
- Available in MyMathLab and from www.pearsonhighered.com/irc, this newly interactive feature is now assignable in MyMathLab.
- *Note to instructors*: The *Instructor's Resource Manual* includes tips and strategies for incorporating the project portfolio into your course.

Instructor's Solutions Manual (Download only)

ISBN-13: 9780321782861

ISBN-10: 0321782860

- Contains complete solutions to all even-numbered section exercises and puzzle problems.

Instructor's Resource Manual with Printable Test Forms (Download only)

ISBN-13: 9780321782922

ISBN-10: 0321782925

- A mini-lecture for each section of the text, organized by objective, includes key examples and teaching tips.
- Designed to help both new and adjunct faculty with course preparation and classroom management.
- Offers helpful teaching tips correlated to the sections of the text.
- Contains one diagnostic test per chapter; four free-response test forms per chapter, one of which contains higher-level questions; one multiple-choice test per chapter; one free-response midterm exam; two free-response final exams; and one multiple-choice final exam.

Additional Media Supplements

MyMathLab® Online Course (access code required)

MyMathLab delivers **proven results** in helping individual students succeed. It provides **engaging experiences** that personalize, stimulate, and measure learning for each student. And, it comes from a **trusted partner** with educational expertise and an eye on the future.

To learn more about how MyMathLab combines proven learning applications with powerful assessment, visit **www.mymathlab.com** or contact your Pearson representative.

MyMathLab Standard allows you to build your course your way, offering maximum flexibility and complete control over all aspects of assignment creation. Starting with a clean slate lets you choose the exact quantity and type of problems you want to include for your students. You can also select from pre-built assignments to give you a starting point.

Ready-to-Go MyMathLab comes with assignments pre-built and pre-assigned, reducing start-up time. You can always edit individual assignments as needed throughout the semester.

MyMathLab®Plus/MyStatLab™Plus

MyLabsPlus combines proven results and engaging experiences from MyMathLab® and MyStatLab™ with convenient management tools and a dedicated services team. Designed to support growing math and statistics programs, it includes additional features such as:

- **Batch Enrollment:** Your school can create the login name and password for every student and instructor, so everyone can be ready to start class on the first day. Automation of this process is also possible through integration with your school's Student Information System.
- **Login from your campus portal:** You and your students can link directly from your campus portal into your MyLabsPlus courses. A Pearson service team works with your institution to create a single sign-on experience for instructors and students.
- **Advanced Reporting:** MyLabsPlus's advanced reporting allows instructors to review and analyze students' strengths and weaknesses by tracking their performance on tests, assignments, and tutorials. Administrators can review grades and assignments across all courses on your MyLabsPlus campus for a broad overview of program performance.
- **24/7 Support:** Students and instructors receive 24/7 support, 365 days a year, by phone, email, or online chat.

MyLabsPlus is available to qualified adopters. For more information, visit our website at www.mylabsplus.com or contact your Pearson representative.

MathXL® Online Course (access code required)

MathXL® is the homework and assessment engine that runs MyMathLab. (MyMathLab is MathXL plus a learning management system.) With MathXL, instructors can:

- Create, edit, and assign online homework and tests using algorithmically generated exercises correlated at the objective level to the textbook.
- Create and assign their own online exercises and import TestGen tests for added flexibility.
- Maintain records of all student work tracked in MathXL's online gradebook.

With MathXL, students can:

- Take chapter tests in MathXL and receive personalized study plans and/or personalized homework assignments based on their test results.
- Use the study plan and/or the homework to link directly to tutorial exercises for the objectives they need to study.
- Access supplemental animations and video clips directly from selected exercises.

MathXL is available to qualified adopters. For more information, visit our website at www.mathxl.com, or contact your Pearson representative.

TestGen®

TestGen® (www.pearsoned.com/testgen) enables instructors to build, edit, print, and administer tests using a computerized bank of questions developed to cover all the objectives of the text. TestGen is algorithmically based, allowing instructors to create multiple but equivalent versions of the same question or test with the click of a button. Instructors can also modify test bank questions or add new questions. The software and test bank are available for download from Pearson Education's online catalog.

PowerPoint® Lecture Slides and Active Learning Questions

- PowerPoint Lecture Slides are written and designed specifically for this text, including figures and examples from the text.
- Active Learning Questions for use with classroom response systems include multiple choice questions to review lecture material.
- Both are available within MyMathLab® or from the Instructor Resource Center at www.pearsonhighered.com/irc.

Acknowledgments

So many people have helped me in so many ways that I could write another book in saying thank you. Though the words of thanks to follow may be few, no amount of space can contain the genuine gratitude that I feel toward each and every person that gave of themselves to make this work the best that it can be.

I would like to thank the following people who gave of their time in reviewing the text. Their thoughtful input was vital to the development of the text through multiple editions.

Lynn Beckett-Lemus, *El Camino College*
Terry Cheng, *Irvine Valley College*
Ines Figueiras, *Essex County College*
Matthew Flacche, *Camden County College*
Anne Franklin, *Pima Community College–Downtown*
Christian Glardon, *Valencia Community College–East*
Richard Halkyard, *Gateway Community College*
Mark Hunter, *Cosumnes River College*
Eric Kaljumagi, *Mount San Antonio College*
Patrick Kent, *North Central State College*
Iris Magee, *Los Angeles City College*
Brent Monte, *Irvine Valley College*
Evelyn Porter, *Utah Valley University*
Laurence Small, *Pierce College*
Joseph Spadaro, *Gateway Community College*
Susan Tummers Stocum, *El Camino College*
Marjorie Szoke, *Valencia Community College–East*
Joe Terreri, *Mount San Antonio College*
Marcia Wallace, *South Suburban College*
Dave Williamson, *Pima Community College–Downtown*
Dmitriy Zhiv, *Berkeley City College*

In addition, I would like to thank the students who provided me with their successful learning strategies for inclusion in the book. I hope their tips will help other students succeed as well.

Alison, *Denison University*
Deanna, *Seton Hall University*
Cara, *Drew University*
Dasle, *Goizueta Business School, Emory University*
Linda, *California Polytechnic State University, San Luis Obispo*
Stefanie, *Providence College*
Hannah, *Emmanuel College*
Alex, *Bentley University*
Sindhu, *Bentley University*
Evan, *Boston University*
Maria, *St. Lawrence University*
Arbora, *Stonehill College*
Elizabeth, *Boston College*
Morgan, *Capital University Law School*
Sarah, *University of Massachusetts, Amherst*
Maria, *The Ohio State University*
Carlin, *Drew University*
Bryon, *San Francisco State University*
Briana, *San Francisco State University*

I would like to extend special thanks to Aimee Tait, who encouraged me to write and put me in touch with Jennifer Crum and Jason Jordan, who gave me the opportunity. I would be remiss without also thanking John Hornsby, who answered my questions about the business. Very special thanks to Pamela Watkins and Carol Nessmith for their inspiration and contribution to the development of the study system.

My heartfelt thanks go to Cathy Cantin, Jonathan Wooding, Christine Whitlock, Tamela Ambush, Michelle Renda, Rachel Ross, Alicia Frankel, Ashley Bryan, Kerianne Okie, Elaine Page, Kari Heen, Greg Tobin, and all of the wonderful people at Pearson Arts & Sciences who helped so much in the revision of the text. Special thanks to Janis Cimperman, Beverly Fusfield, Deana Richmand, and Becky Troutman for checking the manuscript for accuracy.

To Laura Hakala, Tracy Duff, and all of the folks at Pre-Press PMG, thank you for your attention to detail in putting together the finished product.

To Louise Capulli and her team GEX, thank you for the exceptional quality of the solutions manuals, MyWorkBook and the Instructor's Resource Manual with Printable Test Forms.

Finally, I'd like to dedicate this work in thanks to my parents, Tom and Janice, who gave me all that I am, and to my wife Laura for her unwavering support and love.

Tom Carson

To the Student

Why Do I Have to Take This Course?

Often, this is one of the first questions students ask when they find out they must take an algebra course. What a great question! But why focus on math alone? What about English, history, psychology, or science? Does anyone really use *every* topic of *every* course in the curriculum? Most jobs do not require that we write essays on Shakespeare, discuss the difference between various psychological theories, or analyze the cell structure of a frog's liver. So what's the point? The issue comes down to recognizing that general education courses are not job training. The purpose of those courses is to stretch and exercise the mind so that the educated person can better communicate, analyze situations, and solve problems, which are all valuable skills in life and any job.

Professional athletes offer a good analogy. A professional athlete usually has an exercise routine apart from their sport designed to build and improve their body. They may seek a trainer to design exercises intended to improve strength, stamina, or balance and then push them in ways they would not normally push themselves. That trainer may have absolutely no experience with their client's sport, but can still be quite effective in designing an exercise program because the trainer is focused on building basic skills useful for any athlete in any sport. Education is similar: it is exercise for the mind. A teacher's job is like that of a physical trainer. A teacher develops exercises intended to improve communication skills, critical-thinking skills, and problem-solving skills. Different courses are like different types of fitness equipment. Some courses may focus more on communication through writing papers, discussion, and debate. Other courses, such as mathematics, focus more on critical thinking and problem solving.

Another similarity is that physical exercise must be challenging for your body to improve. Similarly, mental exercise must be challenging for the mind to improve. Expect course assignments to challenge you and push you mentally in ways you wouldn't push yourself. That's the best way to grow. So as you think about the courses you are taking and the assignments in those courses, remember the bigger picture of what you are developing: your mind. When you are writing papers, responding to questions, analyzing data, and solving problems, you are developing skills important to life and any career out there.

What Do I Need to Do to Succeed?

☑ **Adequate Time** To succeed, you must have adequate time and be willing to use that time to perform whatever is necessary. To determine if you have adequate time, use the following guide.

Step 1. Calculate your work hours per week.
Step 2. Calculate the number of hours in class each week.
Step 3. Calculate the number of hours required for study by doubling the number of hours you spend in class.
Step 4. Add the number of hours from steps 1–3 together.

Adequate time: If the total number of hours is below 60, then you have adequate time.

Inadequate time: If the total number of hours is 60 or more, then you do not have adequate time. You may be able to hang in there for a while, but eventually, you will find yourself overwhelmed and unable to fulfill all of your obligations. Remember, the above calculations do not consider other likely elements of life such as commuting, family, recreation, and so on. The wise thing to do is cut back on work hours or drop some courses.

Assuming you have adequate time available, choosing to use that time to perform whatever is necessary depends on your attitude, commitment, and self-discipline.

☑ **Positive Attitude** We do not always get to choose our circumstances, but we do get to choose our reaction and behavior. A positive attitude is choosing to be cheerful, hopeful, and encouraging no matter the situation. A benefit of a positive attitude is that it tends to encourage people around you. As a result, they are more likely to want to help you achieve your goal. A negative attitude, on the other hand, tends to discourage people around you. As a result, they are less likely to want to help you. The best way to maintain a positive attitude is to keep life in perspective, recognizing that difficulties and setbacks are temporary.

☑ **Commitment** Commitment means binding yourself to a course of action. Remember, expect difficulties and setbacks, but don't give up. That's why a positive attitude is important, it helps you stay committed in the face of difficulty.

☑ **Self-Discipline** Self-discipline is choosing to do what needs to be done—even when you don't feel like it. In pursuing a goal, it is normal to get distracted or tired. It is at those times that your positive attitude and commitment to the goal help you discipline yourself to stay on task.

Thomas Edison, inventor of the lightbulb, provides an excellent example of all of these principles. Edison tried over 2000 different combinations of materials for the filament before he found a successful combination. When asked about all his failed attempts, Edison replied, "I didn't fail once, I invented the lightbulb. It was just a 2000-step process." He also said, "Our greatest weakness lies in giving up. The most certain way to succeed is always to try just one more time." In those two quotes, we see a person who obviously had time to try 2000 experiments, had a positive attitude about the setbacks, never gave up, and had the self-discipline to keep working.

Behaviors of Strong Students and Weak Students

The four requirements for success can be translated into behaviors. The following table compares the typical behaviors of strong students with the typical behaviors of weak students.

Strong Students . . .	Weak Students . . .
• are relaxed, patient, and work carefully.	• are rushed, impatient, and hurry through work.
• almost always arrive on time and leave the classroom only in an emergency.	• often arrive late and often leave class to "take a break."
• sit as close to the front as possible.	• sit as far away from the front as possible.
• pay attention to instruction.	• ignore instruction, chit-chat, draw, fidget, etc.
• use courteous and respectful language, encourage others, make positive comments, are cheerful and friendly.	• use disrespectful language, discourage others, make negative comments, are grumpy and unfriendly. Examples of unacceptable language include: "I hate this stuff!" (or even worse!) "Are we doing anything important today?" "Can we leave early?"
• ask appropriate questions and answer instructor's questions during class.	• avoid asking questions and rarely answer instructor's questions in class.
• take lots of notes, have organized notebooks, seek out and use study strategies	• take few notes, have disorganized notebooks, do not use study strategies
• begin assignments promptly and manage time wisely, and almost always complete assignments on time.	• procrastinate, manage time poorly, and often complete assignments late.
• label assignments properly and show all work neatly.	• show little or no work and write sloppily.
• read and work ahead.	• rarely read or work ahead.
• contact instructors outside of class for help, and use additional resources such as study guides, solutions manuals, computer aids, videos, and tutorial services.	• avoid contacting instructors outside of class, and rarely use additional resources available.

Assuming you have the prerequisites for success and understand the behaviors of a good student, our next step is to develop two major tools for success:

1. **Learning Style:** complete the Learning Styles Inventory to determine how you tend to learn.
2. **The Study System:** this system describes a way to organize your notebook, take notes, and create study tools to complement your learning style. We've seen students transform their mathematics grades from D's and F's to A's and B's by using the Study System that follows.

Learning Styles Inventory

What Is Your Personal Learning Style?

A learning style is the way in which a person processes new information. Knowing your learning style can help you make choices in the way you focus on and study new material. Below are fifteen statements that will help you assess your learning style. After reading each statement, rate your response to the statement using the scale below. There are no right or wrong answers.

3 = Often applies   **2** = Sometimes applies   **1** = Never or almost never applies

_____ 1. I remember information better if I write it down or draw a picture of it.

_____ 2. I remember things better when I hear them instead of just reading or seeing them.

_____ 3. When I receive something that has to be assembled, I just start doing it. I don't read the directions.

_____ 4. If I am taking a test, I can visualize the page of text or lecture notes where the answer is located.

_____ 5. I would rather have the professor explain a graph, chart, or diagram to me instead of just showing it to me.

_____ 6. When learning new things, I want to do it rather than hear about it.

_____ 7. I would rather have the instructor write the information on the board or overhead instead of just lecturing.

_____ 8. I would rather listen to a book on tape than read it.

_____ 9. I enjoy making things, putting things together, and working with my hands.

_____ 10. I am able to conceptualize quickly and visualize information.

_____ 11. I learn best by hearing words.

_____ 12. I have been called hyperactive by my parents, spouse, partner, or professor.

_____ 13. I have no trouble reading maps, charts, or diagrams.

_____ 14. I can usually pick up on small sounds like bells, crickets, or frogs, or distant sounds like train whistles.

_____ 15. I use my hands and I gesture a lot when I speak to others.

Write your score for each statement beside the appropriate statement number below. Then add the scores in each column to get a total score for that column.

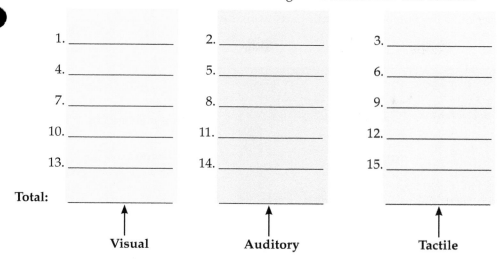

1. _____	2. _____	3. _____
4. _____	5. _____	6. _____
7. _____	8. _____	9. _____
10. _____	11. _____	12. _____
13. _____	14. _____	15. _____

Total: _____

Visual   **Auditory**   **Tactile**

The largest total of the three columns indicates your dominant learning style.

VISUAL

Visual learners learn best by seeing. If this is your dominant learning style, then you should focus on learning strategies that involve seeing. The color coding in the study system (see page xvii–xviii) will be especially important. The same color coding is used in the text. Draw lots of diagrams, arrows, and pictures in your notes to help you see what is happening. Reading your notes, study sheets, and text repeatedly will be an important strategy.

AUDITORY

Auditory learners learn best by hearing. If this is your dominant learning style, then you should use learning strategies that involve hearing. After getting permission from your instructor, bring a tape recorder to class to record the discussion. When you study your notes, play back the tape. Also, when you learn rules, say the rule over and over. As you work problems, say the rule before you do the problem. You may also find the videotapes to be beneficial because you can hear explanations of problems taken from the text.

TACTILE

Tactile (also known as kinesthetic) learners learn best by touching or doing. If this is your dominant learning style, you should use learning strategies that involve doing. Doing lots of practice problems will be important. Make use of the Your Turn exercises in the text. These are designed to give you an opportunity to do problems that are similar to the examples as soon as a topic is discussed. Writing out your study sheets and doing your practice tests repeatedly will be important strategies for you.

Note that the study system developed in this text is for all learners. Your learning style will help you decide what aspects and strategies in the study system to focus on, but being predominantly an auditory learner does not mean that you shouldn't read the textbook, do lots of practice problems, or use the color-coding system in your notes. Auditory learners can benefit from seeing and doing, and tactile learners can benefit from seeing and hearing. In other words, do not use your dominant learning style as a reason for not doing things that are beneficial to the learning process. Also, remember that the Learning Strategy boxes presented throughout the text provide tips to help you use your personal learning style to your advantage.

The Math Study System

Organize the notebook into four parts using dividers as shown:

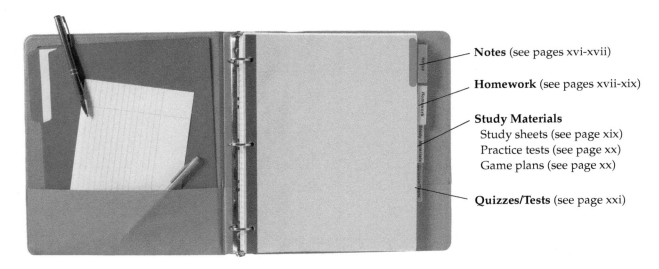

Notes (see pages xvi-xvii)

Homework (see pages xvii-xix)

Study Materials
  Study sheets (see page xix)
  Practice tests (see page xx)
  Game plans (see page xx)

Quizzes/Tests (see page xxi)

Notes

- Use a color code: red for definitions, blue for rules or procedures, and pencil for all examples and other notes.
- Begin notes for each class on a new page (front and back for that day is okay). Include a topic title or section number and the date on each page.
- Try to write your instructor's spoken explanations along with the things he or she writes on the board.
- Mark examples your instructor emphasizes in some way to give them a higher priority. These problems often appear on quizzes and tests.
- Write warnings your instructor discusses about a particular situation.
- Include common errors that your instructor illustrates, but mark them clearly as errors so that you do not mistake them for correct.
- To speed note taking, eliminate unnecessary words like "the" and use codes for common words like + for "and" and $\therefore$ for "therefore." Also, instead of writing complete definitions, rules, or procedures, write the first few words and place the page reference from the text so that you can copy from the text later.

Sample Notes with Color Code

Include title.

Include date.

Section #   9/20

We can simplify an expression by combining like terms.

Definition in red with textbook page reference.

p. #   def. Like terms: constant terms or variable terms that have the same variable(s) raised to the same powers.

Ex 1) 2x and 3x are like terms.

Ex 2) 5x and 7y are not like terms.

Consider 2x + 3x

2x means two x's are added together.

3x means three x's are added together.

$$2x \quad + \quad 3x$$

$$= x + x + x + x + x$$

We have a total of five x's added together.

$$= \qquad 5x$$

We can just add the coefficients.

> **Procedure in blue with textbook page reference.**

p. # Procedure: To combine like terms, add or subtract the coefficients and keep the variables and their exponents the same.

Ex 1) $7x + 5x = 12x$

Ex 2) $4y^2 - 10y^2 = -6y^2$

Homework

This section of the notebook contains all homework. Use the following guidelines whether your assignments are from the textbook, a handout, or a computer program like MyMathLab or MathXL.

- Use pencil so that mistakes can be erased (scratching through mistakes is messy and should be avoided).
- Label according to your instructor's requirements. Usually, at least include your name, the date, and assignment title. It is also wise to write the assigned problems at the top as they were given. For example, if your instructor writes "Section 1.5 #1–15 odd," write it that way at the top. Labeling each assignment with this much detail shows that you take the assignment seriously and leaves no doubt about what you interpreted the assignment to be.
- For each problem you solve, write the problem number and show all solution steps neatly.

Why do I need to show work and write all the steps? Isn't the right answer all that's needed?

- Mathematics is not just about getting correct answers. You really learn mathematics when you organize your thoughts and present those thoughts clearly using mathematical language.
- You can arrive at correct answers with incorrect thinking. Showing your work allows your instructor to verify you are using correct procedures to arrive at your answers.
- Having a labeled, well-organized, and neat hard copy is a good study tool for exams.

What if I submit my answers in MyMathLab or MathXL? Do I still need to show work?

Think of MyMathlab or MathXL as a personal tutor who provides the exercises, offers guided assistance, and checks your answers before you submit the assignment. For the same reasons as those listed above, you should still create a neatly written hardcopy of your solutions, even if your instructor does not check the work. Following are some additional reasons to show your work when submitting answers in MyMathlab or MathXL.

- If you have difficulties that are unresolved by the program, you can show your instructor. Without the written work, your instructor cannot see your thinking.
- If you have a correct answer but have difficulty entering that correct answer, you have record of it and can show your instructor. If correct, your instructor can override the score.

Sample Homework: Simplifying Expressions or Solving Equations

Suppose you are given the following exercise:

For Exercises 1–30 simplify.

1. $5^2 + 3 \cdot 4 - 7$

Your homework should look something like the following:

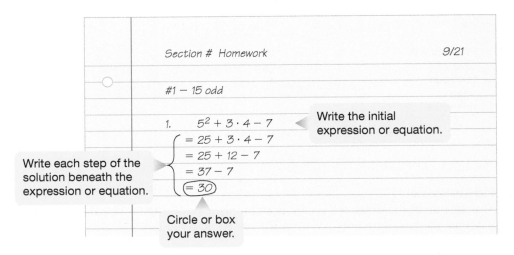

Section # Homework   9/21

#1 – 15 odd

1. $5^2 + 3 \cdot 4 - 7$ **Write the initial expression or equation.**
   $= 25 + 3 \cdot 4 - 7$
   $= 25 + 12 - 7$
Write each step of the solution beneath the expression or equation.
   $= 37 - 7$
   $= 30$

Circle or box your answer.

Sample Homework: Solving Application Problems

Example: Suppose you are given the following two problems.

For Exercises 1 and 2, solve.

1. Find the area of a circle with a diameter of 10 feet.
2. Two cars are traveling toward each other on the same highway. One car is traveling 65 miles per hour and the other is traveling at 60 miles per hour. If the two cars are 20 miles apart, how long will it be until they meet?

Your homework should look something like the following:

Section # Homework   10/6

If applicable, draw a picture or table.

#1 – 15 odd

If the solution requires a formula, write the formula.

1. $A = \pi r^2$

10 ft. $r = 1/2 \ (10 \text{ ft.})$
          $r = 5 \text{ ft.}$

	rate	time	distance
car 1	65 mph	t	$65t$
car 2	60 mph	t	$60t$

car 1 distance	+	car 2 distance	=	total distance
$65t$	+	$60t$	=	20
		$125t$	=	20
		$\dfrac{125t}{125}$	=	$\dfrac{20}{125}$
		t	=	0.16

$A = \pi(5 \text{ ft.})^2$
$A = \pi(25 \text{ ft.}^2)$ **Translate to an equation. Then show all solution steps.**
$A = 25\pi \text{ ft.}^2$
$A \approx 25 \ (3.14) \text{ ft.}^2$
$A \approx 78.5 \text{ ft.}^2$

The cars meet in 0.16 hours.

Answer the question.

Sample Homework: Graphing

Example: Suppose you are given the following two problems.

For Exercises 1 and 2, graph the equation.

  1. $y = 2x - 3$ 2. $y = -2x + 2$

Your homework should look something like the following:

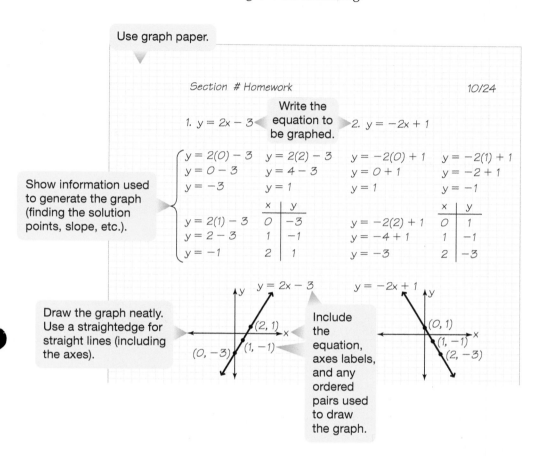

Use graph paper.

Write the equation to be graphed.

Show information used to generate the graph (finding the solution points, slope, etc.).

Draw the graph neatly. Use a straightedge for straight lines (including the axes).

Include the equation, axes labels, and any ordered pairs used to draw the graph.

Study Materials

This section of the notebook contains three types of study materials for each chapter.

   Study Material 1: The Study Sheet A study sheet contains *every* rule or procedure in the current chapter.

Use the chapter summary at the end of each chapter as a guide.

Write each rule or procedure studied. They are in blue in your notes and in the text.

If you are a visual or tactile learner, include a key example to illustrate what the rule or procedure says.

Include anything that helps you remember the procedures and rules. For example, auditory learners might write poems, rhymes, or jingles, as shown here.

To add same sign,

Add and keep the same sign	ex) $7 + 5 = 12$
With different signs, subtract	ex) $-7 + (-5) = -12$
Keep the greater value's sign	ex) $-7 + 5 = -2$
Add this, subtract that	ex) $7 + (-5) = 2$

Can't you read the signs?

Can't you read the signs?

Study Material 2: The Practice Test If your instructor gives you a practice test, proceed to the discussion of creating a game plan.

If your instructor does not give you a practice test, use your notes to create your own practice test out of the examples given in class. Include only the instructions and the problem, not the solutions. The following sample practice test was created from examples in the notes for Chapter 2 in a prealgebra course.

Chapter # Practice Test   10/10

For # 1 and 2, graph on a number line.

1. 4

2. −3

For each example in your notes, write the directions and the problem but not the solution.

For #3 and 4, simplify.

3. $|-8|$

4. $|9|$

After working through the practice test, use your notes to check your solutions.

Study Material 3: The Game Plan The game plan refines the study process further. It is your plan for the test based on the practice test. For each problem on your practice test, write the definition, rule, or procedure used to solve the problem.

The sample shown gives the rule or procedure used to solve each problem on the sample practice test on the previous page. The rules and procedures came from the sample study sheet.

Chapter # Game Plan   9/10

Write the rule or procedure used to solve the problems on the practice test.

Multiple problems that use the same rule or procedure can be grouped together.

#1 and 2: Draw a dot on the mark for the number.

#3 and 4: The absolute value of a positive number is a positive number.
The absolute value of a negative number is a positive number.
The absolute value of 0 is 0.

Quizzes/Tests

Archive all returned quizzes and tests in this section of the notebook.

- Midterm and final exams questions are often taken from the quizzes and tests, so they make excellent study tools for those cumulative exams.
- Keeping all graded quizzes and tests offers a backup system in the unlikely event your instructor should lose any of your scores.
- If a dispute arises about a particular score, you have the graded test to show your instructor.

Chapter # Quiz   10/10

For 1-4, simplify.

1. $|-8|$ $= 8$ ✓   4/4 = 100%
                                    Nice
                                    work!

2. $|9|$ $= 9$ ✓

3. $-15 + 5$ $= -10$ ✓

4. $-8 + -6$ $= -14$ ✓

Whole Numbers

Chapter Overview

To be successful in studying mathematics, we must build a solid foundation. Chapter 1 is the foundation of this course. In it, we explore:

- ► Arithmetic with whole numbers.
- ► Key geometry concepts: perimeter, area, and volume.
- ► Key statistical concepts: mean, median, and mode.

The skills developed here will be used in every chapter; so talk with your instructor if you have difficulty in this chapter.

1.1 Introduction to Numbers, Notation, and Rounding

Objectives

1 Name the digit in a specified place.

2 Write whole numbers in standard and expanded form.

3 Write the word name for a whole number.

4 Graph a whole number on a number line.

5 Use <, >, or = to write a true statement.

6 Round numbers.

7 Interpret bar graphs and line graphs.

Babylonian

Egyptian

Greek

Roman

Chinese

Hindu

Arabic

Warm-up Refer to the *To the Student Section* on p. xx.

1. What are the four sections of the notebook?
2. What is the color code for notes?
3. Complete the Learning Style Inventory. What is your learning style?

Learning Strategy

Every day after class, read over your notes and the related parts in the chapter. By looking at the information while it's still fresh in your mind, it will be easy to remember.

—Sarah S.

Numbers form the foundation of mathematics. To write numbers, we use ten *numerals*, or *digits*: 0, 1, 2, 3, 4, 5, 6, 7, 8, 9. Because our numeral system has ten numerals, it is a *base-10* system. To write numbers beyond 9, we use combinations of numerals in a place-value system, which we will explore in a moment.

Of Interest

Numerals have been written many different ways throughout history. Our modern numerals are derived from Hindu-Arabic forms. Numerals from various cultures are shown in the margin.

Numbers are classified into groups called **sets.** In this chapter, we focus on the set of **whole numbers,** which contains 0 and all of the counting numbers: 1, 2, 3, and so on. The counting numbers make up the set of **natural numbers.** Because every natural number is in the set of whole numbers, we say that the set of natural numbers is a **subset** of the set of whole numbers.

Definitions **Set:** A group of elements.

Subset: A set within a set.

Natural numbers: The natural numbers are 1, 2, 3, . . .

Note The three periods are an *ellipsis*, indicating that the numbers continue forever. ◄

Whole numbers: The whole numbers are 0, 1, 2, 3, . . .

The following figure shows the set of natural numbers contained within the set of whole numbers.

Note 0 is the only whole number ► that is not a natural number.

Whole Numbers: 0, 1, 2, 3, . . .

Natural Numbers: 1, 2, 3, . . .

Objective 1 Name the digit in a specified place.

Figure 1-1 shows how the place values are arranged.

Answers to Warm-up
1. notes, homework, study materials, graded work
2. red = definitions; blue = rules/procedures; pencil = all other notes
3. Answers may vary.

Figure 1-1 Place Values

Place Values														
Trillions period			Billions period			Millions period			Thousands period			Ones period		
Hundred trillions	Ten trillions	Trillions	Hundred billions	Ten billions	Billions	Hundred millions	Ten millions	Millions	Hundred thousands	Ten thousands	Thousands	Hundreds	Tens	Ones

◀···

The dashed arrow to the left of the table indicates that the place values continue to the left indefinitely. Notice that the place values are arranged in groups of three, called *periods*. Numbers written using the place-value system are said to be in *standard form*. When writing numbers in standard form, we separate the periods with commas, as in 3,243,972.

Example 1 What digit is in the thousands place in 209,812?

Answer: 9

Explanation: The digit in the thousands place is the fourth digit from the right.

▶ Do Your Turn 1

Objective 2 Write whole numbers in standard and expanded form.

Another way to write numbers is in *expanded form*, which shows a number's meaning using its digits and place values. For example, 430 is the sum of 4 hundreds, 3 tens, and 0 ones.

4	3	0
hundreds	tens	ones

4 hundreds + 3 tens + 0 ones

Note Adding 0 does not affect a sum; so we do not have to write 0 ones and can ◀ write 4 hundreds + 3 tens.

Procedure

To write a number in expanded form, write each digit followed by its place-value name and a plus sign until you reach the last place, which is not followed by a plus sign.

Example 2 Write the number in expanded form.

a. 57,483

Answer: 5 ten thousands + 7 thousands + 4 hundreds + 8 tens + 3 ones

b. 4,705,208

Answer: 4 millions + 7 hundred thousands + 5 thousands + 2 hundreds + 8 ones

▶ Do Your Turn 2

Connection The place-value table is like the money tray in a cash register. Writing a digit in a particular place is like putting that many bills in the designated tray of the register. For example, writing a 7 in the hundreds place is like putting seven $100 bills in the hundreds tray of the register, which is worth a total of $700.

Note The comma is optional in a four-digit number. For example, we can write 4,538 or 4538.

▶ **Your Turn 1**

a. What digit is in the hundred thousands place in 62,407,981?

b. What digit is in the ten millions place in 417,290,006?

▶ **Your Turn 2**

Write the number in expanded form.

a. 82,469
b. 7,082,049
c. 410,159,200

Answers to Your Turn 1
a. 4   b. 1

Answers to Your Turn 2
a. 8 ten thousands + 2 thousands + 4 hundreds + 6 tens + 9 ones
b. 7 millions + 8 ten thousands + 2 thousands + 4 tens + 9 ones
c. 4 hundred millions + 1 ten million + 1 hundred thousand + 5 ten thousands + 9 thousands + 2 hundreds

Learning Strategy

If you are using the note-taking system from the *To the Student* section of this text, remember to use a red pen for definitions and a blue pen for procedures and rules.

▶ **Your Turn 3**

Write the number in standard form.

a. 2 hundred thousands + 3 ten thousands + 1 thousand + 5 hundreds + 9 tens + 8 ones

b. 9 hundred millions + 1 ten thousand + 4 thousands + 7 hundreds

Let's convert from expanded form to standard form.

Procedure

To convert a number from expanded form to standard form, write each digit in the place indicated by the corresponding place value. If the expanded form is missing a place value, write a 0 for that place value in the standard form.

Example 3 Write the number in standard form.

a. 6 ten thousands + 9 thousands + 2 hundreds + 5 tens + 3 ones

Answer: 69,253

b. 9 millions + 2 ten thousands + 7 hundreds + 9 tens + 3 ones

Answer: 9,020,793 ◀ **Note** There are no hundred thousands and no thousands in the expanded form. So we write 0s in those places in the standard form.

◀ **Do Your Turn 3**

Objective 3 Write the word name for a whole number.

The word name for a number is the way we say the number.

Procedure

To write the word name for a whole number, work from left to right through the periods.

1. Write the word name of the number formed by the digits in the period.
2. Write the period name followed by a comma.
3. Repeat steps 1 and 2 for each period except the ones period. For the ones period, write only the word name formed by the digits. Do not follow the ones period with its name.

▶ **Your Turn 4**

Write the word name.

a. 49,777 (2009 median family income; *Source:* U.S. Bureau of the Census)

b. 14,256,000,000,000 (2009 Gross Domestic Product; *Source:* U.S. Bureau of the Census)

c. 847,716 (Diameter of the sun in miles)

Warning Do not write the word *and* in the word name for a whole number. As we will see in Chapter 6, the word *and* takes the place of a decimal point.

Example 4 Write the word name.

a. 2160 (diameter of the moon in miles)

Answer: two thousand, one hundred sixty

Explanation:

Thousands **Period**		Ones **Period**		
	2	1	6	0
two thousand		one hundred sixty		

b. 92,958,349 (average distance from the earth to the sun in miles)

Answer: ninety-two million, nine hundred fifty-eight thousand, three hundred forty-nine

Note All two-digit numbers from 21 to 99 whose names contain two words, such as forty-nine, fifty-eight, or ninety-two, require a hyphen between the two words. ◀

Explanation:

Millions **period**		Thousands **period**			Ones **period**		
9	2	9	5	8	3	4	9
ninety-two million		nine hundred fifty-eight thousand			three hundred forty-nine		

◀ **Do Your Turn 4**

Answers to Your Turn 3
a. 231,598
b. 900,014,700

Answers to Your Turn 4
a. forty-nine thousand, seven hundred seventy-seven
b. fourteen trillion, two hundred fifty-six billion
c. eight hundred forty-seven thousand, seven hundred sixteen

Objective 4 Graph a whole number on a number line.

We sometimes use a number line as a visual tool. On the following number line, we have marked the whole numbers through 7.

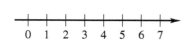

> **Note** The arrow pointing to the right indicates that the numbers increase indefinitely. When we discuss integers in Chapter 2, we will see that the number line continues indefinitely to the left as well.

Procedure

To graph a whole number on a number line, draw a dot on the mark for that number.

Example 5   Graph 5 on a number line.

Solution: 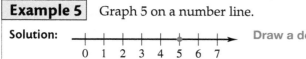   Draw a dot on the mark for 5.

▶ **Do Your Turn 5**

▶ **Your Turn 5**

Graph the number on a number line.

a. 3   **b.** 6

Objective 5 Use <, >, or = to write a true statement.

Now let's compare two numbers. A statement that contains an equal sign, such as $12 = 12$, is an **equation**. A statement that contains an inequality symbol, such as $6 < 10$, is an **inequality**.

Definitions **Equation:** A mathematical relationship that contains an equal sign ($=$).

Inequality: A mathematical relationship that contains an inequality symbol ($<$, $>$).

The inequality symbol $>$ means "greater than," whereas $<$ means "less than." Equations and inequalities can be true or false. The following table lists some equations and inequalities, shows how they are read, and whether they are true or false.

Statement	Translation to words	True or False
$12 = 12$	"twelve is equal to twelve"	True
$15 = 7$	"fifteen is equal to seven"	False
$6 < 10$	"six is less than ten"	True
$9 < 4$	"nine is less than four"	False
$11 > 8$	"eleven is greater than eight"	True
$0 > 5$	"zero is greater than five"	False

Procedure

To use $<$, $>$, or $=$ to make an incomplete statement true:

- If the two numbers are the same, use $=$.
- If the two numbers are different, choose $<$ or $>$ so that the inequality symbol opens toward the greater of the two numbers.

Answers to Your Turn 5

Learning Strategy

If you are a visual learner, imagine < and > as mouths that open to eat the bigger meal.

▶ **Your Turn 6**

Use <, >, or = to write a true statement.

a. 15,907 ? 15,906

b. 1,291,304 ? 1,291,309

c. 64,108 ? 64,108

d. 24,300 ? 25,300

Example 6   Use <, >, or = to write a true statement.

a. 208 ? 205

Answer: 208 > 205

Explanation: On a number line, 208 is farther to the right than 205; so 208 is greater. We use the greater than symbol to open toward 208.

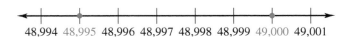

201 202 203 204 205 206 207 208 209 210

b. 48,995 ? 49,000

Answer: 48,995 < 49,000

Explanation: On a number line, 49,000 is farther to the right than 48,995; so 49,000 is greater. We use the less than symbol to open toward 49,000.

48,994 48,995 48,996 48,997 48,998 48,999 49,000 49,001

◀ **Do Your Turn 6**

Objective 6   Round numbers.

In Example 4b, we saw that the average distance from the earth to the sun is 92,958,349 miles. Because numbers such as 92,958,349 are rather tedious to say and work with, we often round such numbers so they are easier to communicate.

When we round a number, we must identify a place value to round to; often the place value will be specified. For example, let's round 92,958,349 to the nearest million. First, we must determine whether 92,958,349 is closer to the exact million greater than 92,958,349, which is 93,000,000, or the exact million less than 92,958,349, which is 92,000,000. A number line can be helpful.

Warning   Make sure you include all 0s after the rounded place. Instead of 93, write 93,000,000. Think about money. $93 is very different from $93,000,000.

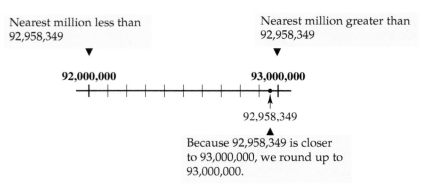

Nearest million less than 92,958,349

Nearest million greater than 92,958,349

92,000,000

93,000,000

92,958,349

Because 92,958,349 is closer to 93,000,000, we round up to 93,000,000.

Although we could use a number line every time we round, the digit in the place to the right of the place to be rounded can tell us whether to round up or down. If that digit is greater than 5, the number is closer to the greater number; so we round up. If that digit is less than 5, the number is closer to the lesser number; so we round down. If it is 5, we agree to round up. Our example suggests the following procedure.

Answers to Your Turn 6
a. 15,907 > 15,906
b. 1,291,304 < 1,291,309
c. 64,108 = 64,108
d. 24,300 < 25,300

Procedure

To round a number to a given place value:
1. Identify the digit in the given place value.
2. If the digit to the right of the given place value is 5 or greater, round up by increasing the digit in the given place value by 1. If the digit to the right of the given place value is 4 or less, round down by keeping the digit in the given place value the same.
3. Change all of the digits to the right of the rounded place value to zeros.

Example 7 Round 43,572,991 to the specified place.

a. hundred thousands

Solution: 43,572,991

Hundred thousands — The digit to the right of 5 is 7, which is greater than 5; so we round up.

Answer: 43,600,000

b. ten thousands

Solution: 43,572,991

Ten thousands — The digit to the right of 7 is 2, which is less than 4; so we round down.

Answer: 43,570,000

c. millions

Solution: 43,572,991

Millions — The digit to the right of 3 is 5; so we round up.

Answer: 44,000,000

d. hundreds

Solution: 43,572,991

Hundreds — The digit to the right of 9 is 9, which is greater than 5; so we round up.

Answer: 43,573,000 ◄ **Note** When the 9 in the hundreds place is rounded up, it becomes 10. Because 10 hundreds is 1000, we must add 1 to the digit in the thousands place so that the 2 becomes a 3. This process of adding to the next place to the left is sometimes called *regrouping* or *carrying*.

► Do Your Turn 7

In the real world, we are seldom told what place to round to; so we round to a place that makes sense for the situation. Some questions to consider are as follows:

- How precise must the numbers be?
- How accurately can the amounts involved be measured?
- Are others depending on what we do?
- Are lives at stake?

For rough estimation purposes, we can round so that there is only one digit that is not 0. This means to round to the place value farthest to the left.

► **Your Turn 7**

Round 602,549,961 to the specified place.
a. ten thousands
b. tens
c. millions
d. thousands
e. hundreds

Answers to Your Turn 7
a. 602,550,000
b. 602,549,960
c. 603,000,000
d. 602,550,000
e. 602,550,000

▶ **Your Turn 8**

Round each number so that there is only one nonzero digit.

a. 27,502,341

b. 6,128,200

c. 453,219

Example 8   Round each number so that there is only one nonzero digit.

a. 36,568

Solution: 36,568

Farthest left   The digit to the right of 3 is 6, which is greater than 5; so we round up.

Answer: 40,000

b. 621,905

Solution: 621,905

Farthest left   The digit to the right of 6 is 2, which is less than 5; so we round down.

Answer: 600,000

Note Remember, when rounding down, the digit in the rounded place remains the same.

◀ **Do Your Turn 8**

Objective 7 Interpret bar graphs and line graphs.

Numbers are also common in graphs, such as bar graphs, which are usually used to compare amounts.

▶ **Your Turn 9**

Use the graph in Example 9.

a. What revenue did New York generate?

b. Which state had the least revenue?

c. Round New York's revenue to the nearest hundred thousand.

Example 9   Use the following graph, which shows the five states with the highest revenues for public education in 2006–2007.

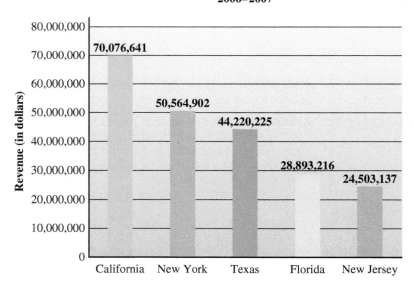

Revenue for Public Elementary and Secondary Schools, by State 2006–2007

Answers to Your Turn 8

a. 30,000,000

b. 6,000,000

c. 500,000

Answers to Your Turn 9

a. $50,564,902

b. New Jersey

c. $50,600,000

a. What revenue did Florida generate?

Answer: $28,893,216

b. Which state had the greatest revenue?

Answer: California

c. Round Texas' revenue to the nearest million.

Answer: $44,000,000

▶ **Do Your Turn 9**

We may also encounter line graphs, which are usually used to show how amounts have changed over time.

Example 10   Use the following graph, which shows average tuition and fees at two-year colleges.

▶ **Your Turn 10**

Use the graph in Example 10.

a. What were the average tuition and fees per semester in '99–'00?

b. Round the average tuition and fees per semester in '09–'10 so that there is only one nonzero digit.

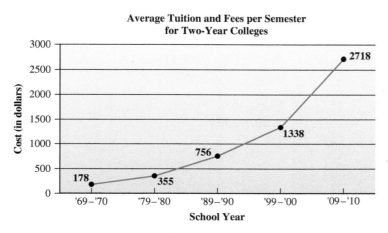

a. What were the average tuition and fees per semester in '89–'90?

Answer: $756

b. Round the average tuition and fees per semester in '79–'80 to the nearest hundred.

Answer: $400

c. What general trend does the graph indicate?

Answer: The cost of tuition has increased over the years (nearly doubling every decade).

▶ **Do Your Turn 10**

Answers to Your Turn 10
a. $1338   b. $3000

1.1 Exercises For Extra Help MyMathLab®

Objective 1

Prep Exercise 1 List the first four numbers in the set of whole numbers.

Prep Exercise 2 Is every whole number a natural number? Explain.

Prep Exercise 3 Complete the following table of place values.

Billions period		Millions period		Thousands period			Ones period	
Hundred Billions	Billion	Ten Millions	Millions	Hundred Thousands	Ten Thousands		Hundreds	Ones

For Exercises 1–4, name the digit in the requested place in the number 56,324,092.

1. hundreds
2. thousands
3. ten millions
4. hundred thousands

For Exercises 5–8, name the place held by the 7.

5. 2,457,502
6. 57,414
7. 9,706,541
8. 70,412,581

Objective 2

Prep Exercise 4 Explain how to write a number in expanded form.

For Exercises 9–14, write the number in expanded form.

9. 24,319
10. 78,625
11. 5,213,304
12. 2,410,512
13. 93,014,008
14. 86,300,905

For Exercises 15–20, write the number in standard form.

15. 8 thousands + 7 hundreds + 9 tens + 2 ones
16. 9 ten thousands + 2 thousands + 5 hundreds + 8 tens + 1 one
17. 6 millions + 3 ten thousands + 9 thousands + 2 tens
18. 8 hundred millions + 7 thousands + 2 hundreds + 3 ones
19. 4 ten millions + 9 hundred thousands + 8 ten thousands + 1 hundred + 9 ones
20. 7 hundred millions + 2 millions + 5 ten thousands + 8 tens

Objective 3

Prep Exercise 5 Explain how to write the word name of a number.

For Exercises 21–26, write the word name.

21. 7768 (diameter of the earth in miles)
22. 29,035 (height of the peak of Mount Everest in feet)

23. 310,464,734 (estimated resident population of the United States in 2010 according to the U.S. Bureau of the Census)

24. 13,615,674,949,267 (U.S. gross outstanding national debt in 2010)

25. 186,171 (speed of light in miles per second)

26. 299,792,458 (speed of light in meters per second)

Of Interest

The sun is 92,958,349 miles from the earth. The light leaving the sun travels at a speed of 186,171 miles per second. Even at this speed, it takes the light from the sun a little over 8 minutes to reach the earth.

Objective 4

Prep Exercise 6 Draw a number line showing the first six whole numbers.

For Exercises 27–32, graph the number on a number line.

27. 4

28. 2

29. 7

30. 9

31. 0

32. 1

Objective 5

Prep Exercise 7 When comparing two different numbers, choose $<$ or $>$ so that the inequality opens toward the _____ of the two numbers.

For Exercises 33–38, use $<$, $>$, or $=$ to write a true statement.

33. 599 ? 899

34. 89,900 ? 89,902

35. 609,001 ? 609,001

36. 88,332 ? 88,332

37. 4,299,308 ? 4,298,308

38. 9911 ? 9199

Objective 6

Prep Exercise 8 Explain how to round a number.

For Exercises 39–46, round 5,652,992,481 to the specified place.

39. thousands

40. hundred millions

41. millions

42. ten millions

43. billions

44. hundred thousands

45. hundreds

46. ten thousands

For Exercises 47–52, round each number so that there is only one nonzero digit.

47. 32,607

48. 281,506

49. 851,220

50. 4,513,541

51. 8723

52. 54,298

53. The average distance from Mars to the sun is 141,639,520 miles. Round the distance to a reasonable place. Why did you round to the place you chose?

54. The U.S. gross national debt was $13,615,674,949,267 in 2010. Round the debt to a reasonable place. Why did you round to the place you chose?

Objective 7

55. Use the graph to answer the following questions.

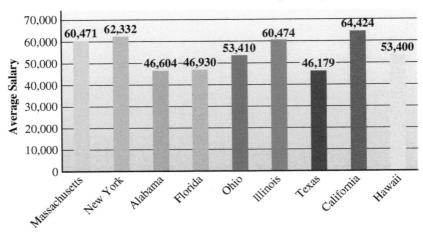

Source: National Education Association.

a. What was the greatest average salary among the states listed?

c. List the four highest-paying states of those listed in order from greatest average salary to least average salary.

e. Round the average salary for public school teachers in Hawaii to the nearest thousand.

b. What was the least average salary among the states listed?

d. Round the average salary for public school teachers in Florida to the nearest thousand.

f. Although salaries may vary greatly from state to state, what factor may "equalize" the income?

56. Use the graph to answer the following questions. Write numeric answers in standard form.

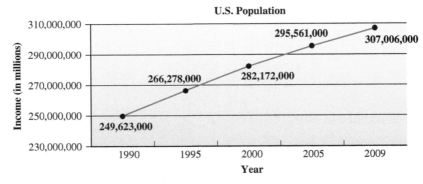

Source: U.S. Census Bureau.

a. What was the U.S. population in 2000?

b. What was the U.S. population in 2009?

c. Round the population in 1990 to the nearest million.

d. Round the population in 2005 so that it has one nonzero digit.

e. What does the graph indicate about the U.S. population?

1.2 Adding and Subtracting Whole Numbers; Solving Equations

Warm-up

[1.1] 1. Write 43,075 in expanded form.

[1.1] 2. Use $<$, $>$, or $=$ to write a true statement.

$$9871 \quad ? \quad 9786$$

Objective 1 Add whole numbers.

Let's examine the operation **addition.**

Definition **Addition:** The arithmetic operation that combines amounts.

When we write an addition sentence, the *addends* are the numbers that are added and the *sum* is the answer.

$$3 + 5 = 8$$
$$\uparrow \quad \uparrow \quad \uparrow$$
$$\text{Addends} \quad \text{Sum}$$

Notice how the order of addends can be changed without affecting the sum.

$$5 + 3 \text{ and } 3 + 5 \text{ both make a sum of 8.}$$

This fact about addition is called the *commutative property of addition*. The word *commutative* comes from the root word *commute*, which means "to move from one place to another." That's what happens when we move the addends—they commute.

Commutative Property of Addition

Changing the order of the addends does not affect the sum.

In math language: $a + b = b + a$, where a and b are any numbers.

We add whole numbers according to place value. Most people prefer to stack the numbers, aligning the like place values. Sometimes when we add digits, the sum is greater than 9; so we have to add the extra to the next column.

Procedure

To add whole numbers:

1. Stack the numbers so that the corresponding place values align.
2. Add the digits. If the sum of the digits in any place value (or column) is greater than 9, write the ones digit of that sum in that place-value column and add the remainder of the sum to the digits in the next place-value column. This process is sometimes called regrouping or carrying.

Example 1 Add.

a. $423 + 64$

Solution:
$$
\begin{array}{r}
423 \\
+64 \\
\hline
487
\end{array}
\quad \text{or} \quad
\begin{array}{r}
64 \\
+423 \\
\hline
487
\end{array}
$$
▲   ▲

Note The commutative property allows us to stack either way.

Connection When we add the digits 2 and 6, we add 2 tens and 6 tens, which gives 8 tens. In expanded form:

$$
\begin{array}{l}
4 \text{ hundreds} + 2 \text{ tens} + 3 \text{ ones} \\
\underline{\phantom{4 \text{ hundreds} +} + 6 \text{ tens} + 4 \text{ ones}} \\
4 \text{ hundreds} + 8 \text{ tens} + 7 \text{ ones} = 487
\end{array}
$$

Objectives

1 Add whole numbers.

2 Estimate sums.

3 Solve applications involving addition.

4 Subtract whole numbers.

5 Solve equations containing an unknown addend.

6 Solve applications involving subtraction.

7 Solve applications involving addition and subtraction.

Of Interest

In the 1400s, the Latin word *et*, which means "and," was used to indicate addition. The writers of the time would generally write the word so that the *e* and *t* would run together, creating our modern + sign.

Source: D. E. Smith, *History of Mathematics.*

Answers to Warm-up
1. 4 ten thousands + 3 thousands + 7 tens + 5 ones
2. 9871 > 9786

Add.

a. $527 + 41$

b. $5802 + 549$

c. $59,481 + 8574$

Explanation: We stacked the numbers with the ones and tens places aligned, then added the corresponding digits.

b. $5408 + 916$

Solution: $\overset{1\ 1}{5408}$
$\underline{+916}$
6324

Explanation: Notice that when we add the 8 and 6, we get 14. The 4 is placed in the ones place. The extra 10 in 14 is expressed as a 1 over the tens column of digits and is added with the other tens digits. The same thing happens with the 9 and 4 in the hundreds column. Finally, we added the 1 and 5 in the thousands column.

◀ **Do Your Turn 1**

When we have three or more addends, we can group them any way we wish. Consider $2 + 3 + 4$. We can add the 2 and 3 first, which is 5, and then add the 4 for a total of 9. Or we can add the 3 and 4 first, which is 7, and then add the 2 to total 9 again.

This property of addition is called the *associative property* of addition because to associate is to group. To write the associative property symbolically, we use parentheses to indicate the different ways we can group the addends.

$(2 + 3) + 4$ indicates add the $2 + 3$ first, then add 4 to the result.

$2 + (3 + 4)$ indicates add the $3 + 4$ first, then add that result to 2.

Either way, we get 9.

Associative Property of Addition

Grouping three or more addends differently does not affect the sum.

In math language: $(a + b) + c = a + (b + c)$, where $a, b,$ and c are any numbers.

The commutative and associative properties allow for flexibility in the way we add. We can change the order of the addends or group the addends any way we wish without affecting the sum.

Add.

a. $48 + 70 + 62$

b. $114 + 85 + 30 + 79$

c. $35,604 + 907 + 3215 + 42,008$

Example 2 Add. $35 + 60 + 18$

Solution: $\overset{1}{35}$
60
$\underline{+18}$
113

Connection The addition properties tell us that we can rearrange or group the addends differently in $35 + 60 + 18$ and the sum will still be 113.

Here are a couple of examples:

$35 + \underbrace{60 + 18}$ or $\underbrace{35 + 18} + 60$

$= 35 + 78$ $= 53 + 60$

$= 113$ $= 113$

Explanation: We stacked the numbers by place value and added the digits.

◀ **Do Your Turn 2**

Objective 2 Estimate sums.

In some situations, a quick approximation for a calculation, or an *estimate,* is preferable to finding the actual answer. An estimate can also serve as a rough check of the actual calculation. In general, to estimate a calculation, we round the numbers so that the calculation is easy to perform mentally. In this text, we will estimate sums (and differences) of whole numbers using the following procedure.

Procedure
To estimate a sum of whole numbers:
1. Round the smallest number to the highest possible place value so that it has only one nonzero digit.
2. Round all of the other numbers to that same place value.
3. Add the rounded numbers.

Note Estimates are meant to be quick and easy to perform mentally. The procedures given in this text are merely suggestions.

Example 3 Estimate each sum by rounding. Then find the actual sum.

a. 68,214 + 4318

Solution: To estimate, we first round the smallest number, 4318, so that it has one nonzero digit, which is 4000. Because we rounded 4318 to the nearest thousand, we round 68,214 to the nearest thousand as well, which is 68,000. The estimate is the sum of 68,000 and 4000.

$$
\begin{array}{ll}
\textbf{Estimate: } \overset{1}{6}8,000 & \textbf{Actual: } \overset{1}{6}8,\overset{1}{2}14 \\
\quad\quad +4,000 & \quad\quad +4,318 \\
\hline
\quad\quad 72,000 & \quad\quad 72,532
\end{array}
$$

Note The estimate is quite close to the actual sum, which reassures us that the actual ◀ sum is reasonable.

b. 5896 + 539 + 72

Solution: To estimate, we round the smallest number, 72, so that it has one nonzero digit, which is 70. Because we rounded 72 to the tens place, we round the other numbers to the tens place as well, which gives 540 and 5900. Then we add all three rounded numbers.

$$
\begin{array}{ll}
\textbf{Estimate: } \overset{11}{5}900 & \textbf{Actual: } \overset{121}{5}896 \\
\quad\quad 540 & \quad\quad 539 \\
\quad\quad +70 & \quad\quad + 72 \\
\hline
\quad\quad 6510 & \quad\quad 6507
\end{array}
$$

▶ **Do Your Turn 3**

Objective 3 Solve applications involving addition.

Often certain words, called *key words,* in a problem statement can help us determine how to solve the problem. Below are some key words and key questions that help us recognize addition.

Key Words for Addition
add, plus, sum, total, increased by, more than, in all, altogether, perimeter

Key Questions   How much in all?
How much altogether?
What is the total?

▶ **Your Turn 3**
Estimate each sum by rounding. Then find the actual sum.
a. 67,482 + 8190
b. 4586 + 62 + 871

Answers to Your Turn 3
a. Estimate: 75,000
   Actual: 75,672
b. Estimate: 5520
   Actual: 5519

▶ **Your Turn 4**

Below are the base price and prices for upgrades. What will be the total cost of the house with the upgrades? Check by estimating.

House base price = $225,480

Deck = $2280

Fireplace = $620

Kitchen upgrade = $1875

Example 4   Bob and Kim find out that the basic cost of building the house they have chosen is $212,500. They decide that they want to add some extra features. A wood-burning fireplace costs an additional $680. They also want to upgrade the fixtures and appliances at a cost of $3158. What will be the total price of the house?

Solution: Notice the word *total* in the question. Because the upgrades are costs in *addition* to the base price, we must add.

$$\begin{array}{r} \overset{1}{2}\overset{1}{1}2,500 \\ 3,158 \\ +680 \\ \hline 216,338 \end{array}$$ **Check by estimating:** $$\begin{array}{r} \overset{1}{2}12,500 \\ 3,200 \\ +700 \\ \hline 216,400 \end{array}$$ **This is reasonably close to the actual sum.**

Answer: The total price of the house is $216,338.

◀ **Do Your Turn 4**

Suppose we want to construct a fence around some property. To determine how much fence we need, we measure or calculate the total distance around the region to be fenced. This distance is called **perimeter.**

Definition Perimeter: The total distance around a figure.

Procedure

To find the perimeter of a figure, add the lengths of all sides of the figure.

Example 5   The Jensons want to put a border on the walls of their son's room. The room is a 10-foot-wide by 12-foot-long rectangle. How much border material must they buy?

Solution: To find the total amount of material, we need the total distance around the room. It is helpful to draw a picture.

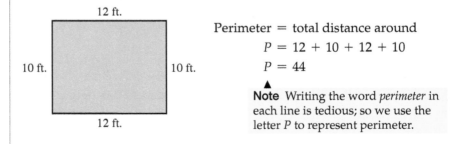

Perimeter = total distance around

$P = 12 + 10 + 12 + 10$

$P = 44$

▲
Note Writing the word *perimeter* in each line is tedious; so we use the letter P to represent perimeter.

Answer: The Jensons will need 44 ft. of border material.

Check by estimating: $10 + 10 + 10 + 10 = 40$ ft.

In Example 5, we used the letter P to represent the perimeter. In other problems involving perimeter, the perimeter value may be different; so P could represent a different amount altogether. Because P can vary in value, we call it a **variable.** We use variables to represent unknown amounts.

Definition Variable: A symbol that can vary or change in value.

Answers to Your Turn 4
Actual: $230,255
Estimate: $230,300

Some symbols do not vary in value. We call these symbols **constants.**

Definition Constant: A symbol that does not vary in value.

All numerals are constants. For example, the symbol 3 always means three; so it is a constant.

▶ **Do Your Turn 5**

Objective 4 Subtract whole numbers.

Subtraction is the *inverse* of addition. Addition and subtraction are inverse operations because they *undo* each other. If $5 + 4 = 9$, we can undo the addition with subtraction, as in $9 - 4 = 5$. There are three ways of interpreting subtraction.

Definition Subtraction: An operation of arithmetic that can be interpreted as follows:

1. Take away
2. Difference
3. Unknown addend

We interpret subtraction to mean *take away* when we remove an amount from another amount.

We interpret subtraction as *difference* when we must find the distance between two positions. For example, if you are at highway mile marker 42 and later you see that you are at mile marker 60, you can use subtraction to find the difference between the two positions and thus determine how far you have traveled.

We interpret subtraction as an *unknown addend* when an addend, which is part of an addition statement, is unknown. For example, in $(?) + 4 = 9$, the first addend is unknown. This corresponds to having \$4 and finding how much more is needed to bring the total up to \$9. We use subtraction to find that unknown addend.

When we write a subtraction sentence, the *subtrahend* is subtracted from the *minuend* and the *difference* is the answer.

$$9 - 4 = 5$$

Minuend Subtrahend Difference

Connection Because addition and subtraction are inverse operations, we can check the accuracy of $9 - 4 = 5$ by adding $5 + 4 = 9$.

Let's look at how we subtract. We align place values vertically, just like in addition. However, subtraction is *not* commutative, which means we cannot change the order of the numbers in the minuend and subtrahend positions. If we did, we would get a different result.

$7 - 4$ is 3, but $4 - 7$ is not 3. When we discuss integers in Chapter 2, we will learn that $4 - 7$ is -3.

Conclusion: To find the difference of two whole numbers, we stack the greater number on top, align the place values, and subtract the corresponding digits. For example, consider $795 - 32$.

Note Because 795 is the greater number, it goes on top, and we subtract the digits.

$$\begin{array}{r} 795 \\ -32 \\ \hline 763 \end{array}$$

Check:

$$\begin{array}{r} 763 \\ +32 \\ \hline 795 \end{array}$$

◀ **Note** We can use the inverse operation to check. Adding the difference, 763, to the subtrahend, 32, should give the minuend, 795.

▶ **Your Turn 5**

Mr. Marcus needs to fence a space in his yard for a garden. It is a rectangular space 45 feet wide by 60 feet long. How much fence will he need?

Of Interest

Fifteenth-century writers generally used a lowercase m for minus. Writing the m quickly usually produced a dash, so the m was simplified to a $-$ sign.

Source: D. E. Smith, *History of Mathematics.*

Connection When we study negative numbers in Chapter 2, we will see that the greater number is not always the minuend.

Answer to Your Turn 5
210 ft.

When subtracting, if a digit in the top number is less than the digit directly below, we must rename the top number. Consider $8982 - 765$.

$$\begin{array}{r} \overset{7\,12}{898\overset{}{2}} \\ -765 \\ \hline 8217 \end{array}$$

◄ **Note** Because the digit 2 is less than the 5 beneath it, we rename 2 by taking 1 ten from the 8 tens and adding that 10 to 2 to make 12. This process is sometimes called *regrouping* or *borrowing*. Taking 1 ten from the 8 tens leaves 7 tens. Notice that this does not change the value of 8982. We merely moved 1 ten from the tens place over to the ones place so that we could carry out the subtraction.

Connection Think about the cash register analogy. Suppose there are eight $10 bills and two $1 bills in the register, which amount to $82. To give more than two $1 bills in change, the cashier must break one of the $10 bills. The manager gives the cashier ten $1 bills in exchange for one $10 bill. There are now seven $10 bills and twelve $1 bills in the register, which is still a total of $82.

Procedures

To subtract whole numbers:

1. Stack the greater number on top of the smaller number so that the corresponding place values align.
2. Subtract the digits in the bottom number from the digits directly above. If a digit in the top number is less than the digit beneath it, rename the top digit.

To estimate a difference of whole numbers:

1. Round the smallest number to the highest possible place value so that it has only one nonzero digit.
2. Round all of the other numbers to that same place value.
3. Subtract the rounded numbers.

▶ **Your Turn 6**

Estimate each difference by rounding. Then find the actual difference.

  a. $48,975 - 6241$

  b. $5941 - 218$

  c. $520,048 - 63,793$

Example 6   Estimate $45,002 - 8473$ by rounding. Then find the actual difference.

$$\text{Estimate:} \quad \begin{array}{r} \overset{3\,15}{4\overset{}{5},000} \\ -8,000 \\ \hline 37,000 \end{array} \qquad \text{Actual:} \quad \begin{array}{r} \overset{14\;\;99}{\underset{}{3\,4}\;\;\overset{}{10}\overset{}{10}\,12} \\ 45,0\,0\,2 \\ -8,4\,7\,3 \\ \hline 36,52\,9 \end{array} \qquad \text{Check by adding:} \quad \begin{array}{r} \overset{11\;11}{36,529} \\ +8,473 \\ \hline 45,002 \end{array}$$

Explanation: The tens and hundreds places contain 0s; so when we renamed the 2, we had to take 1 thousand from the 5 thousands to make 10 hundreds. Then we took 1 hundred from the 10 hundreds to make 10 tens. This, in turn, allowed us to take 1 ten from the 10 tens to make 12 in the ones place.

◄ **Do Your Turn 6**

Answers to Your Turn 6
a. Estimate: 43,000
    Actual: 42,734
b. Estimate: 5700
    Actual: 5723
c. Estimate: 460,000
    Actual: 456,255

Objective 5 Solve equations containing an unknown addend.

Suppose we want to buy an appliance that costs $350 but we have only $200. How much more do we need to have enough to make the purchase? This problem translates to an equation with an unknown addend.

$$200 + (?) = 350$$

Because the additional amount needed is the difference between the $350 price and the $200 we have, we can write a **related equation** involving subtraction to find the needed amount.

Definition Related equation: A true equation that relates the same pieces of a given equation using the inverse operation.

We write the related subtraction equation for $200 + (?) = 350$ below.

$(?) = 350 - 200$ ◄ **Note** In the related subtraction equation, we subtract
$(?) = 150$ the known addend, 200, from the sum, 350.

Answer: You will need $150.

Procedure

To solve for an unknown addend, write a related subtraction equation in which the known addend is subtracted from the sum.

Earlier we defined a variable as a symbol or letter that can vary in value, and we said that we use variables to represent unknown amounts. Because (?) indicates an unknown amount, it is a variable. However, it is more common to use a letter such as x, y, or t for a variable. From here on, we will use letter variables.

Instead of: $200 + (?) = 350$ we write: $200 + x = 350$
$\qquad\qquad (?) = 350 - 200$ $\qquad\qquad\qquad\qquad\quad x = 350 - 200$
$\qquad\qquad (?) = 150$ $\qquad\qquad\qquad\qquad\qquad\quad x = 150$

The number 150 is the **solution** to the equation because the equation is true when the variable is replaced with 150. The act of finding this solution is called *solving the equation.*

Definition Solution: A number that makes an equation true when it replaces the variable in the equation.

The definition for *solution* suggests a method for checking. If 150 is correct, it will make the equation true when it replaces the x in $200 + x = 350$.

Check: $200 + x = 350$
$\qquad\quad 200 + 150 \stackrel{?}{=} 350$ ◄ **Note** The symbol $\stackrel{?}{=}$ is used to indicate that we
$\qquad\qquad\quad 350 = 350$ $\qquad$ are asking whether $200 + 150$ is equal to 350.

Because $200 + 150 = 350$ is true, 150 is the solution to $200 + x = 350$.

Example 7 Solve and check. $45 + x = 73$

Solution: $x = 73 - 45$ Write a related subtraction equation in which
$\qquad\qquad\qquad\qquad$ the known addend, 45, is subtracted from the sum, 73.

$\qquad\qquad x = 28$

Check: $45 + x = 73$
$\qquad\quad 45 + 28 \stackrel{?}{=} 73$ In $45 + x = 73$, replace x with 28.
$\qquad\qquad\quad 73 = 73$ Verify that the equation is true: $45 + 28 = 73$ is true.

Explanation: Think about solving the unknown addend statement in money terms. If we have $45 and we want to end up with $73, we subtract 45 from 73 to find how much more is needed.

► **Do Your Turn 7**

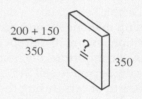

VISUAL

Learning Strategy

If you are a visual learner, imagine the equal sign in an equation as a wall separating two amounts. When we check a solution, we calculate the amount on each side of the wall to determine whether the results are identical. If they are, the equation is true, indicating that the solution checks.

► **Your Turn 7**

Solve and check.
 a. $14 + x = 20$
 b. $y + 32 = 60$
 c. $140 + t = 216$
 d. $m + 89 = 191$

Answers to Your Turn 7
 a. $x = 6$
 b. $y = 28$
 c. $t = 76$
 d. $m = 102$

Objective 6 Solve applications involving subtraction.

As in addition, there are key words and key questions that help us recognize subtraction situations.

> ## Key Words for Subtraction
>
> subtract, minus, difference, remove, decreased by, take away, left, less than
>
> **Key Questions**   How much is left? (take away)
>   How much more/higher/warmer/colder/shorter? (difference)
>   How much more is needed? (missing addend)

▶ **Your Turn 8**

Solve.

a. A food bank has 3452 boxes of food. Volunteers distribute 928 boxes in one neighborhood. How many boxes do they have left?

b. A plane is flying at an altitude of 24,500 feet. The plane experiences a downdraft, which causes it to drop abruptly to an altitude of 22,750 feet. How much altitude did the plane lose?

c. Debbie sells cars at a dealership that encourages competition among its salespeople. As an incentive, the first salesperson to sell $150,000 worth of cars during the year will win a cruise to the Caribbean. Debbie has sold $78,520 worth so far. How much more does she need to sell to win?

Example 8   Solve.

a. A small computer business has $5678 in the bank. During one month, $2985 is spent on parts for production. How much is left?

Solution: The word *spent*, indicating *to remove*, and the key question *How much is left?* indicate that we must subtract.

$$\begin{array}{r} \overset{15}{}\\ \overset{4\,5\,17}{5678}\\ -2985\\ \hline 2693 \end{array} \qquad \textbf{Check:}\quad \begin{array}{r} \overset{1\,1}{2693}\\ +2985\\ \hline 5678 \end{array}$$

Answer: $2693 is left.

b. A room-temperature (68°F) mixture of water and sugar is heated to boiling (212°F) so that more sugar may be added to supersaturate the mixture. How much did the temperature change?

Solution: Because we want to know the amount of change, we are looking for the difference; so we must subtract.

$$\begin{array}{r} \overset{10}{}\\ \overset{1\,0\,12}{212}\\ -\ 68\\ \hline 144 \end{array} \qquad \textbf{Check:}\quad \begin{array}{r} \overset{1\,1}{144}\\ +68\\ \hline 212 \end{array}$$

Answer: 144°F

c. A charity organization has collected $43,587 through a fund-raiser. The goal for the year is to raise $125,000. How much more is needed?

Solution: Because we must increase the original amount of $43,587 by an unknown amount to end up with $125,000, we can translate to an equation with an unknown addend.

$$\$43,587 + x = \$125,000 \qquad \text{Translate to an equation.}$$

$$x = 125,000 - 43,587 \qquad \text{Write a related equation in which the known addend is subtracted from the sum.}$$

$$x = 81,413$$

$$\begin{array}{r} \overset{9\ \ 9}{}\\ \overset{0\ 124\ \ 10\,10\,10}{125,000}\\ -43,587\\ \hline 81,413 \end{array} \qquad \textbf{Check:}\quad \begin{array}{r} \overset{1\ 11}{81,413}\\ +43,587\\ \hline 125,000 \end{array}$$

Answer: $81,413 is needed.

◀ **Do Your Turn 8**

Answers to Your Turn 8
a. 2524 boxes
b. 1750 ft.
c. $71,480

Objective 7 Solve applications involving addition and subtraction.

Often the problems we encounter in life require more than one step to find a solution. Let's consider some problems that involve addition and subtraction in the same problem.

Example 9

a. In circuits, a wire connection is often referred to as a *node*. Current is a measure of electricity moving through a wire and is measured in amperes (or amps, A). One property of circuits is that all of the current entering a node must equal all of the current exiting the node. In the circuit diagram shown, how much must the unknown current be?

Solution: We can calculate the total current entering the node and then subtract the known current that is leaving the node.

Entering current: $9 + 15 = 24$

If 24 amps are entering and we know that 13 amps are leaving, we can subtract 13 from 24 to find the unknown current.

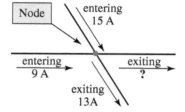

Unknown current: $24 - 13 = 11$

Answer: The unknown current is 11 A.

b. An accountant is given the following spreadsheet of expenses and income for a company. What is the final balance?

Description of expenses	Amount	Description of assets/income	Amount
Payroll	$10,548	Checking account	$5,849
Utilities	$329	Income	$19,538
Water	$87	Donations	$1,200
Waste disposal	$285		
New inventory	$12,243		

Solution: We must calculate the total expenses and total assets/income, then find the difference of the amounts.

Expenses	Assets/Income
$10,548	$ 5,849
329	19,538
87	+ 1,200
285	Total $26,587
+ 12,243	
Total $23,492	

Now that we have the total expenses and the total assets/income, we need to find the difference of those totals to determine the final balance.

Balance

$26,587
− 23,492
‾‾‾‾‾‾‾
3,095

Answer: The final balance is $3095.

► Do Your Turn 9

► **Your Turn 9**

Solve.

a. Find the unknown current in the circuit shown.

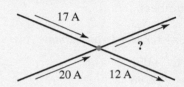

b. Lynda is a checkout clerk. At the end of her shift, the store manager makes sure the amount of money in the register equals Lynda's sales. The printout breaks down the sales into three categories: food, clothing, and nonperishable items. Lynda must count the money in the register and her credit card sales. Finally, $200 always stays in the register. The following chart shows the amounts. Based on the information in the chart, how did Lynda do? Is her register in balance? Does she have too little or too much? If so, how much?

Category	Amount
Food	$1587
Clothing	$2768
Nonperishables	$988
Totals	**Amount**
Register count	$4192
Credit cards	$1339

Answers to Your Turn 9
a. 25 A
b. There should be $12 more in the register.

1.2 Exercises For Extra Help MyMathLab®

Objectives 1 and 2

Prep Exercise 1 In an addition sentence, the numbers that are added are the _____ and the answer is the _____.

Prep Exercise 2 What does the commutative property of addition say?

Prep Exercise 3 Explain how to estimate a sum of whole numbers.

For Exercises 1–10, estimate each sum by rounding. Then find the actual sum.

1. 6051
    +2798

2. 52,407
    +31,596

3. 91,512
    +8,756

4. 82,098
    +7,971

5. 10,516
    782
    +4,516

6. 150,412
    258
    +6,239

7. $9319 + 519 + 5408$

8. $6809 + 398 + 2087$

9. $43,210 + 135,569 + 2088 + 516$

10. $128,402 + 4480 + 93,095 + 298$

Objective 4

Prep Exercise 4 In a subtraction sentence, the first number is the _____ , the number subtracted is the _____ , and the answer is the _____ .

For Exercises 11–20, estimate each difference by rounding. Then find the actual difference.

11. 5873
     −521

12. 9478
     −253

13. 40,302
     −6,141

14. 510,304
     −42,183

15. 50,016
     −4,682

16. 42,003
     −23,567

17. $51,980 − 25,461$

18. $413,609 − 20,724$

19. $210,007 − 43,519$

20. $6,005,002 − 258,496$

Objective 5

Prep Exercise 5 Explain the difference between a constant and a variable.

Prep Exercise 6 What is a solution to an equation?

Prep Exercise 7 Explain how to solve for an unknown addend.

For Exercises 21–30, solve and check.

21. $8 + x = 12$

22. $12 + y = 19$

23. $t + 9 = 30$

24. $m + 16 = 28$

25. $17 + n = 35$

26. $r + 27 = 70$

27. $125 + u = 280$

28. $88 + a = 115$

29. $b + 76 = 301$

30. $470 + c = 611$

Objectives 3, 6, and 7

Prep Exercise 8 In the following table, list four key words and one key question that indicate addition and four key words and one key question that indicate subtraction.

Addition key words/key question	Subtraction key words/key question

For Exercises 31–48, solve.

31. Tamika has to keep track of total ticket sales at a box office. Four attendants sell tickets. The first attendant sold 548 tickets; the second, 354; the third, 481; and the fourth, 427. How many tickets were sold in all?

32. Juan's monthly expenses are shown below. Find his total monthly expenses.
rent = $925   car payment = $380
utilities = $465   groceries/food = $600

33. Crown molding is a decorative strip of wood attached to the wall at the ceiling. How many feet of crown molding are needed to go around the perimeter of a rectangular room that measures 16 feet by 18 feet?

34. Mr. Grayson has some land that he wants to enclose with a fence. The figure below shows the dimensions of the land. How many feet of fence will he need in all?

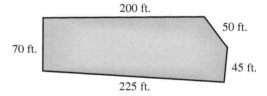

35. In the morning, Serene had $47 in her wallet. She bought coffee for $2, bought lunch for $8, and filled her car's gas tank for $21. How much money did Serene have left when she got home?

36. Jesse's student loan and scholarship totaled $5976. He paid $4612 in tuition and fees, $676 for textbooks, and $77 for other school supplies. How much money does Jesse have left?

37. A family is purchasing a home in Florida that is said to be 2400 square feet. The family later finds out that this figure includes an unfinished bonus room over the garage and the garage itself. The garage is 440 square feet, and the bonus room is 240 square feet. What is the area of the actual living space?

38. An animal rescue group had 37 cats in its shelter at the end of April. In May, 12 cats were returned to their owners, 21 cats went to new homes, and 26 new cats were brought to the shelter. How many cats were in the shelter at the end of May?

39. Margo Incorporated produces a motor that turns the fan in air conditioners. The annual cost of materials last year was $75,348. The labor costs were $284,568. Utilities and operations costs for the plant were $54,214. The company received $548,780 in revenue from sales of the motors. What was the profit? (*Note:* Profit can be found by subtracting all costs from revenue.)

40. Jose is raising funds to run for a seat in his state's senate. Currently, he has raised $280,540, and he intends to add $28,580 of his own money. His competitor has raised a total of $485,300. How much more does Jose need to raise to match his competitor's funds?

41. The photosphere, which is the outer layer of the sun, has an average temperature of about 5500 K. A typical atomic blast at ground zero has a temperature of about 3400 K. How much warmer is the sun's temperature?

42. The sun is 92,958,349 miles from the earth. The moon is 615,042 miles from the earth. During a solar eclipse, the moon is directly between the sun and the earth. How far is the moon from the sun during a solar eclipse?

Of Interest

The symbol K represents Kelvin temperature units. The scale was proposed by William Thomson Kelvin in 1848 and is used extensively in sciences. Increments on the Kelvin scale are the same as the Celsius scale. However, 0 K is equivalent to −273°C.

43. Find the unknown current in the circuit shown.

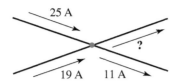

44. Find the unknown current in the circuit shown.

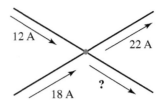

45. An accountant is given the following spreadsheet of expenses and income for a small company. What is the final balance?

Description of expenses	Amount	Description of assets/income	Amount
Payroll	$18,708	Checking account	$10,489
Utilities	$678	Income	$32,500
Water	$126		
Waste disposal	$1,245		
New inventory	$15,621		

46. An accountant is given the following spreadsheet of expenses and income for a small company. What is the final balance?

Description of expenses	Amount	Description of assets/income	Amount
Payroll	$22,512	Checking account	$15,489
Utilities	$1,265	Income	$82,569
Water	$240		
Waste disposal	$1,350		
New inventory	$24,882		

47. Use the graph to answer the following questions. Write answers in standard form.

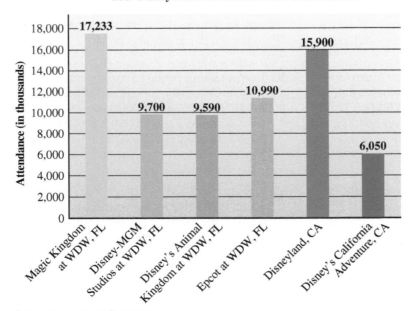

Source: Amusement Business.

   a. Find the total number of people who attended Disney parks in the United States in 2009.

   b. Find the total number of people who attended Florida's Disney parks.

   c. Estimate the total number of people who attended California's Disney parks.

   d. How many more people attended Disney-MGM Studios than Disney's Animal Kingdom?

   e. Estimate how many more people attended Epcot than Disney's California Adventure.

   f. How many more people attended Florida's Disney parks than California's Disney parks?

48. Use the graph to answer the following questions.

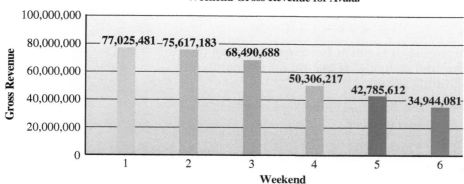

Weekend Gross Revenue for *Avatar*

   a. Find the total revenue the film earned over the first two weekends.

   b. Find the total revenue the film earned over weekends 3 and 4.

   c. Estimate the total revenue earned by the film during the first six weekends.

   d. How much more revenue did the film earn the first weekend compared with the second weekend?

   e. Find the difference in revenue between the second and third weekends.

   f. Estimate the difference in revenue between the first weekend and the sixth weekend.

Puzzle Problem Eight people are at a group therapy session. Everyone hugs everyone once. How many hugs take place?

Review Exercises

[1.1] **1.** Write 307,491,024 in expanded form.

[1.1] **2.** Write the word name for 1,472,359.

[1.1] **3.** Graph 5 on a number line.

[1.1] **4.** Use $<$, $>$, or $=$ to write a true statement.

           12,305 ? 12,350

[1.1] **5.** Round 23,405,172 to the nearest ten thousand.

1.3 Multiplying Whole Numbers; Exponents

Objectives

1 Multiply whole numbers.
2 Estimate products.
3 Solve applications involving multiplication.
4 Evaluate numbers in exponential form.
5 Write repeated factors in exponential form.
6 Solve applications.

Warm-up

[1.2] 1. Add. $5672 + 453 + 29{,}841$
[1.2] 2. Subtract. $36{,}004 - 9876$

Objective 1 Multiply whole numbers.

Suppose we need to install five kitchen cabinet doors, each requiring four screws. Note that we could find the total number of screws needed by adding five 4s.

$$4 + 4 + 4 + 4 + 4 = 20$$

However, it is much faster to memorize that the sum of five 4s is 20. This is the nature of **multiplication.**

Definition Multiplication: Repeated addition of the same number.

When we write a multiplication sentence, the *factors* are the parts that are multiplied and the product is the answer.

$$4 \times 5 = 20$$
$$\uparrow \quad \uparrow \qquad \uparrow$$
$$\text{Factors} \quad \text{Product}$$

We can also write:

$$4 \cdot 5 = 20$$
$$(4)(5) = 20$$
$$4(5) = 20$$
$$(4)5 = 20$$

To make multiplication easier, you should memorize a certain number of multiplication facts. The more facts you memorize, the faster you will be at multiplying. Figure 1-2 has multiplication facts through 12.

Figure 1-2 Multiplication Table

3	0	1	2	3	4	5	6	7	8	9	10	11	12
0	0	0	0	0	0	0	0	0	0	0	0	0	0
1	0	1	2	3	4	5	6	7	8	9	10	11	12
2	0	2	4	6	8	10	12	14	16	18	20	22	24
3	0	3	6	9	12	15	18	21	24	27	30	33	36
4	0	4	8	12	16	20	24	28	32	36	40	44	48
5	0	5	10	15	20	25	30	35	40	45	50	55	60
6	0	6	12	18	24	30	36	42	48	54	60	66	72
7	0	7	14	21	28	35	42	49	56	63	70	77	84
8	0	8	16	24	32	40	48	56	64	72	80	88	96
9	0	9	18	27	36	45	54	63	72	81	90	99	108
10	0	10	20	30	40	50	60	70	80	90	100	110	120
11	0	11	22	33	44	55	66	77	88	99	110	121	132
12	0	12	24	36	48	60	72	84	96	108	120	132	144

Answers to Warm-up
1. 35,966
2. 26,128

Like addition, multiplication is commutative. The *commutative property of multiplication* says that changing the order of factors does not affect the product.

$$(2)(5) = 10 \quad \text{and} \quad (5)(2) = 10$$

Commutative Property of Multiplication

Changing the order of the factors does not affect the product.

In math language: $a \cdot b = b \cdot a$, where a and b are any numbers.

Note When writing multiplication with variables, we can eliminate the dot symbol. For example, $a \cdot b = b \cdot a$ can be written as $ab = ba$.

Also like addition, multiplication is associative. The *associative property of multiplication* says that if three or more factors occur, grouping them differently does not affect the product.

$$(2 \cdot 3) \cdot 5 = 30 \quad \text{and} \quad 2 \cdot (3 \cdot 5) = 30$$

Associative Property of Multiplication

Grouping three or more factors differently does not affect the product.

In math language: $(ab)c = a(bc)$, where a, b, and c are any numbers.

Note that the product of 0 and a number is always 0.

$$0 \cdot 9 = 0 \quad \text{and} \quad 9 \cdot 0 = 0$$

Multiplicative Property of 0

The product of 0 and a number is 0.

In math language: $0 \cdot n = 0$ and $n \cdot 0 = 0$, where n is any number.

Also, the product of 1 and a number is always the number.

$$1 \cdot 7 = 7 \quad \text{and} \quad 7 \cdot 1 = 7$$

Multiplicative Property of 1

The product of 1 and a number is the number.

In math language: $1 \cdot n = n$ and $n \cdot 1 = n$, where n is any number.

Another property, called the *distributive property,* can be used when a sum or difference is multiplied by a number, as in $2(3 + 5)$. Think about the word *distribute*. When tests are distributed to students in a class, every student receives a copy. Similarly, when the distributive property is applied, each number inside the parentheses is multiplied by the number outside the parentheses.

$$2(3 + 5) = 2 \cdot 3 + 2 \cdot 5 \qquad \textbf{Multiply 3 and 5 by 2.}$$

If we calculate $2(3 + 5)$ and $2 \cdot 3 + 2 \cdot 5$, we can see that they are equal.

$2(3 + 5)$		$2 \cdot 3 + 2 \cdot 5$	
$= 2 \cdot 8$	**Find the sum of 3 and 5.**	$= 6 \ + \ 10$	**Find each product.**
$= 16$	**Multiply the sum by 2.**	$= 16$	**Add.**

Distributive Property

If a sum or difference is multiplied by a number, each number inside the parentheses may be multiplied by the number outside the parentheses.

In math language: $a(b + c) = ab + ac$ and $a(b - c) = ab - ac$, where a, b, and c are any numbers.

We use the distributive property to multiply numbers with more than one digit. Let's see how this works.

▶ **Your Turn 1**

Multiply.
  a. 18×0
  b. $248 \cdot 1$
  c. $(5)(31)$
  d. $3(402)$

Example 1   Multiply. 2×34

Solution: We use the distributive property to multiply each digit in 34 by 2.

$$\begin{array}{r} 34 \\ \times\, 2 \\ \hline 68 \end{array}$$ Multiply 2 times 4, then 2 times 3.

Explanation: Because 34 is $30 + 4$ in expanded form, we could write 2×34 as $2(30 + 4)$, then use the distributive property:

$$2(30 + 4) = 2 \cdot 30 + 2 \cdot 4$$
$$= 60 + 8$$
$$= 68$$

Because it is a bit tedious to write the multiplication in sentence form, we usually stack the numbers and use the distributive property to multiply the digits.

$$2 \times 30 = 60 \qquad \begin{array}{r} 3\ 4 \\ \times\ 2 \\ \hline 6\ 8 \end{array} \qquad 2 \times 4 = 8$$

◀ **Do Your Turn 1**

▶ **Your Turn 2**

Multiply.
  a. 32×41
  b. $(51)(23)$
  c. $802 \cdot 74$
  d. $512(61)$

Example 2   Multiply. $503 \cdot 62$

Solution: Use the distributive property to multiply each digit in 503 by each digit in 62; then add the two products.

$$\begin{array}{r} \overset{1}{5}03 \\ \times\ 62 \\ \hline 1\ 006 \\ +30\ 18 \\ \hline 31,186 \end{array}$$

Multiply 2 times 503.
Multiply 6 (tens) times 503.
Add.

Note When we multiply 6 tens times 3 ones, we get 180. The 8 in the tens place is the 80 part of 180. The 100 part of 180 is expressed as a small 1 over the 0 in 503. When we multiply 6 tens times 0 tens, we get 0 hundreds, but we have to add the extra 1 hundred. We write this 1 hundred to the left of the 8.

◀ **Do Your Turn 2**

Answers to Your Turn 1
a. 0   b. 248   c. 155   d. 1206

Answers to Your Turn 2
a. 1312   b. 1173   c. 59,348
d. 31,232

Procedure

To multiply two whole numbers, stack them and then apply the distributive property.

Objective 2 Estimate products.

To make estimating a product easy to perform mentally, we modify the procedure we used for estimating a sum or difference so that we round *each* factor to the highest possible place value.

Procedure

To estimate a product:
1. Round each factor to the highest possible place value so that each has only one nonzero digit.
2. Multiply the rounded factors.

Example 3 Estimate $42,109 \cdot 7104$ by rounding. Then find the actual product.

Estimate: To estimate, we round both factors to the highest possible place so that each has one nonzero digit. Rounding 42,109 to the highest possible place gives 40,000. Rounding 7104 gives 7000. The estimate is the product of 40,000 and 7000.

$$
\begin{array}{r}
40,000 \leftarrow \text{Four 0s} \\
\times\ 7,000 \leftarrow \text{Three 0s} \\
\hline
280,000,000
\end{array}
$$

Multiply 7×4 Total of seven 0s

Note Because the factors contain many 0s, we can multiply the 7 times 4 mentally and write the total number of 0s from both factors after the product.

Actual:

$$
\begin{array}{r}
\overset{1}{}\overset{6}{}\overset{3}{} \\
42,109 \\
\times\ 7,104 \\
\hline
168\,436 \\
000\,00 \\
42\,109 \\
+294\,763 \\
\hline
299,142,336
\end{array}
$$

Multiply $4 \times 42,109$.
Multiply $0 \times 42,109$.
Multiply $1 \times 42,109$.
Multiply $7 \times 42,109$.

Notice how the estimate and the actual answer are reasonably close.

▶ **Do Your Turn 3**

Objective 3 Solve applications involving multiplication.

There are certain key words you can look for that indicate multiplication.

Key Words for Multiplication

multiply, times, product, each, of, by

Example 4 The human heart averages about 70 beats each minute.

a. How many times does the heart beat in an hour?

Solution: Because there are 60 minutes in an hour and the heart averages 70 beats each minute, we multiply $60 \cdot 70$ to find the number of heartbeats in an hour. Let's multiply mentally as we learned in Example 3.

$$60 \cdot 70 = 4200$$
$$\qquad\quad 6 \cdot 7$$

Multiply $6 \cdot 7$ and write each 0 from each factor after the product.

Answer: The heart beats about 4200 times per hour.

▶ **Your Turn 3**

Estimate each product by rounding. Then find the actual product.
a. $56,045 \cdot 6714$
b. $24 \cdot 365 \cdot 12$

Connection If we use the same procedure to estimate a product that we use to estimate a sum or difference, the estimate is more accurate but it is more difficult to perform mentally.

$$
\begin{array}{r}
42,000 \\
\times\ 7,000 \\
\hline
294,000,000
\end{array}
$$

The estimate is more accurate, but 42×7 is more difficult to perform mentally.

Answers to Your Turn 3
a. Estimate: 420,000,000
    Actual: 376,286,130
b. Estimate: 80,000
    Actual: 105,120

▶ **Your Turn 4**

Solve.

a. If the heart averages about 70 beats per minute, how many times does the heart beat in a 30-day month?

b. A manager decides to monitor copy machine use in her office. She finds that the people in her area are using two boxes of paper each week. Each box contains 10 reams of paper, and each ream contains 500 sheets of paper. How many pages are being copied each week?

b. How many times does the heart beat in a week?

Solution: For the number of heartbeats in a week, we first need to figure out how many hours are in a week. Because there are 24 hours in a day and 7 days in a week, we multiply 24×7 to get the number of hours in a week.

$$\begin{array}{r} \overset{2}{2}4 \\ \times\ 7 \\ \hline 168 \end{array}$$

Connection We can use the distributive property to do this calculation in our head.

$$7(24) = 7(20 + 4) = 7 \cdot 20 + 7 \cdot 4 = 140 + 28 = 168$$

There are 168 hours in a week. From part a, we found that there are 4200 beats in an hour. To calculate the number of beats in a week, we multiply 168×4200.

$$\begin{array}{r} 4200 \\ \times\ 168 \\ \hline 33\,600 \\ 252\,00 \\ +\,420\,0 \\ \hline 705{,}600 \end{array}$$

Answer: The heart beats about 705,600 times in a week (a truly amazing muscle).

◀ **Do Your Turn 4**

We can use multiplication to count objects arranged in a **rectangular array**.

Definition   Rectangular array: A rectangle formed by a pattern of neatly arranged rows and columns.

The buttons on a cell phone form a rectangular array.

Four rows with three buttons in each row is a rectangular array. Because each row contains the same number of buttons, we can multiply to find the total number of buttons.

$$4 \cdot 3 = 12$$

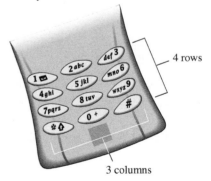

Conclusion: To calculate the total number of items in a rectangular array, multiply the number of rows by the number of columns.

Answers to Your Turn 4
a. 3,024,000 times
b. 10,000 each week

Example 5 A section in the upper deck of a football stadium has 42 rows, each containing 40 seats. There are 8 sections in the upper deck. How many people can be seated in the upper deck altogether?

Solution: The 8 sections of 42 rows with 40 seats in each row, form a large rectangular array; so we multiply 42 × 40 × 8.

```
       42              ⁵⁶
      ×40            1̇680
      ───             × 8
       00          ──────
                   13,440
    +168                ↑
    ──────      Number of seats
     1680       in all 8 sections
       ↑
 Number of seats
 in one section
```

Connection We could have applied the commutative and associative properties and multiplied any two factors first.

$$(42 \times 8) \times 40 \quad \text{or} \quad 42 \times (40 \times 8)$$
$$= 336 \times 40 \qquad\qquad = 42 \times 320$$
$$= 13,440 \qquad\qquad\quad = 13,440$$

Answer: The upper deck can seat 13,440 people.

Explanation: Notice that when the example says "42 rows, *each* containing 40 seats," the key word *each* tells us that we have a repeated addition situation; so we multiply. Because there are 8 identical sections, we are once again repeatedly adding; so again we multiply.

▶ **Do Your Turn 5**

A rectangular array can be used to determine how many different combinations are possible in a code containing two or more digits, as in Example 6.

Example 6 To get into a certain restricted building, a person must enter a code. The code box has two digital windows (see illustration). The first window can contain any number from 1 to 5. The second window can contain any letter from A to Z (only uppercase letters appear). How many possible codes are there in all?

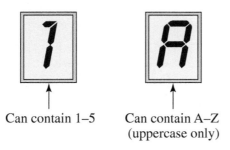

Can contain 1–5   Can contain A–Z
                     (uppercase only)

Solution: Each number in the first window could be paired with all 26 letters of the alphabet. In other words, the 1 could be paired with all 26 letters, then 2 with all 26 letters, and so on (see next page). That means there could be a total of 5 sets of 26. Notice how we can make a rectangular array out of the combinations. Therefore, we multiply 5 times 26.

▶ **Your Turn 5**

For a concert, chairs are arranged on the floor of an arena in 4 groups. Each group has 20 rows with 15 chairs in each row. How many chairs are on the floor of the arena?

Answer to Your Turn 5
1200 chairs

▶ **Your Turn 6**

A certain combination lock has three dials. The first dial has the letters A–F inscribed. The second dial has the numbers 0–9. The third dial has the names of the planets. How many combinations are possible?

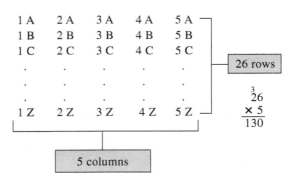

Answer: There are 130 possible codes.

Another way to analyze this problem is to think of each window as a place value or a *slot*, like on a slot machine. Determine the number of possible values each slot can contain and multiply.

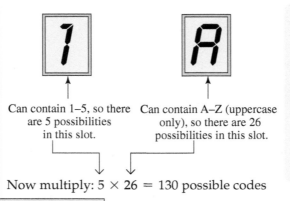

Can contain 1–5, so there are 5 possibilities in this slot.

Can contain A–Z (uppercase only), so there are 26 possibilities in this slot.

Now multiply: 5 × 26 = 130 possible codes

◀ **Do Your Turn 6**

Connection How many two-digit numbers can be made if each digit can be 0–9? 0–9 means that each of the two places has 10 different possibilities.

$$10 \cdot 10 = 100$$

That's why 00–99 represents 100 different two-digit numbers.

Of Interest

The branch of mathematics that deals with counting total combinations and arrangements of items is called *combinatorics*.

Procedure

To find the total number of possible combinations in a problem with multiple slots to fill, multiply the number of items that can fill the first slot by the number of items that can fill the next slot, and so on.

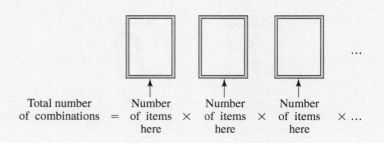

Objective 4   Evaluate numbers in exponential form.

Every operation performed by a computer is translated to numbers expressed in *binary* code. Each place value of a number in binary code can contain one of two possible digits, 0 or 1. In computer lingo, the place values are called *bits*. Let's determine how many numbers can be represented by using a 7-bit code.

Answer to Your Turn 6
480 combinations

Suppose each box shown represents a bit, which can contain a 0 or a 1.

0 or 1   0 or 1   0 or 1   0 or 1   0 or 1   0 or 1   0 or 1

Of Interest

In computer terminology, the place values are called *bits*. *Bit* is short for *binary digit*.

Each of the 7 boxes can have 2 different values, so we multiply seven 2s.

$$2 \cdot 2 \cdot 2 \cdot 2 \cdot 2 \cdot 2 \cdot 2 = 128$$

Notice that we multiplied the same number repeatedly. We can use **exponential form** to indicate repeated multiplication. For example, we can write 7 factors of 2 in exponential form where 2 is the *base* and 7 is the *exponent* as follows.

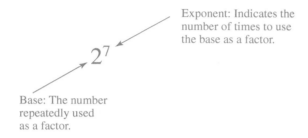

2^7

Exponent: Indicates the number of times to use the base as a factor.

Base: The number repeatedly used as a factor.

Definition **Exponential form:** A notation having the form b^n, where b is the base and n is the exponent.

The notation 2^7 is read "two to the seventh power," or "two to the seventh." To evaluate it, we multiply seven 2s.

Seven 2s

$$2^7 = \overbrace{2 \cdot 2 \cdot 2 \cdot 2 \cdot 2 \cdot 2 \cdot 2} = 128$$ Warning 2^7 does not mean $2 \cdot 7$.

Procedure

To evaluate an exponential form with a natural-number exponent, write the base as a factor the number of times indicated by the exponent, then multiply.

Example 7   Evaluate. 3^4

Solution: $3^4 = 3 \cdot 3 \cdot 3 \cdot 3$ Write 3 as a factor four times.

$\quad\quad\quad = 9 \cdot 3 \cdot 3$ Multiply the first two 3s.

$\quad\quad\quad = 27 \cdot 3$ Multiply 9 by 3.

$\quad\quad\quad = 81$ Multiply 27 by 3.

▶ **Do Your Turn 7**

▶ **Your Turn 7**

Evaluate.

a. 7^2 b. 10^4

c. 1^9 d. 0^4

e. 2^{12} f. 5^7

Answers to Your Turn 7

a. 49   b. 10,000   c. 1   d. 0

e. 4096   f. 78,125

Objective 5 Write repeated factors in exponential form.

► **Your Turn 8**

Write in exponential form.

a. $14 \cdot 14 \cdot 14 \cdot 14$

b. $2 \cdot 2 \cdot 2 \cdot 2 \cdot 2 \cdot 2 \cdot 2 \cdot 2 \cdot 2 \cdot 2$

c. $9 \cdot 9 \cdot 9 \cdot 9 \cdot 9 \cdot 9 \cdot 9$

d. How many different values can be represented by a 16-bit computer chip? Write your answer in exponential form; then use a calculator to evaluate.

Example 8 Write $5 \cdot 5 \cdot 5 \cdot 5 \cdot 5 \cdot 5$ in exponential form.

Answer: 5^6

Explanation: Because six 5s are multiplied, the exponent is 6.

◄ Do Your Turn 8

When an exponential form has a base of 10, we say that it is a *power* of 10. Consider the following pattern:

$10^2 = 10 \cdot 10 = 100$

$10^3 = 10 \cdot 10 \cdot 10 = 1,000$

$10^4 = 10 \cdot 10 \cdot 10 \cdot 10 = 10,000$

$10^5 = 10 \cdot 10 \cdot 10 \cdot 10 \cdot 10 = 100,000$

$10^6 = 10 \cdot 10 \cdot 10 \cdot 10 \cdot 10 \cdot 10 = 1,000,000$

Conclusion: Because the exponent indicates the number of 10s to be multiplied and each 10 contributes an additional 0 after the 1, the number of 0s in the product is equal to the exponent.

We can write expanded notation using powers of 10. Instead of writing the place value name, we write a product with the power of 10 that matches the place value.

► **Your Turn 9**

Write each number in expanded form using powers of 10.

a. 9,210,984

b. 46,009

c. 1,204,918,205

Connection To see why $10^0 = 1$, look at the pattern of exponents. Start with 10^3. If $10^3 = 1000$, $10^2 = 100$, and $10^1 = 10$, 10^0 has to be 1 for the pattern to hold true.

Example 9 Write 1,239,405 in expanded notation using powers of 10.

Solution:

| 1 million | + | 2 hundred thousands | + | 3 ten thousands | + | 9 thousands | + | 4 hundreds | + | 5 ones |

$1 \times 1,000,000 + 2 \times 100,000 + 3 \times 10,000 + 9 \times 1,000 + 4 \times 100 + 5 \times 1$

$= 1 \times 10^6 + 2 \times 10^5 + 3 \times 10^4 + 9 \times 10^3 + 4 \times 10^2 + 5 \times 10^0$

$= 1 \times 10^6 + 2 \times 10^5 + 3 \times 10^4 + 9 \times 10^3 + 4 \times 10^2 + 5 \times 1$

Note We did not write the tens digit in this expanded notation because it was 0.

◄ Do Your Turn 9

Each power of 10 represents a different place value. Remember that the period names change every three places. So every third power of 10 will have a new name, as Figure 1-3 illustrates.

Figure 1-3 Powers of 10 and Period Names

Power	Standard form	Period name
10^3	= 1,000	= one thousand
10^6	= 1,000,000	= one million
10^9	= 1,000,000,000	= one billion
10^{12}	= 1,000,000,000,000	= one trillion
10^{15}	= (1 with 15 zeros)	= one quadrillion
10^{100}	= (1 with 100 zeros)	= googol
10^{googol}	= (1 with a googol of zeros)	= googolplex

Note Some names are rather colorful.

Answers to Your Turn 8

a. 14^4 b. 2^{10} c. 9^7

d. $2^{16} = 5\,65,536$

Answers to Your Turn 9

a. $9 \times 10^6 + 2 \times 10^5 + 1 \times 10^4 + 9 \times 10^2 + 8 \times 10 + 4 \times 1$

b. $4 \times 10^4 + 6 \times 10^3 + 9 \times 1$

c. $1 \times 10^9 + 2 \times 10^8 + 4 \times 10^6 + 9 \times 10^5 + 1 \times 10^4 + 8 \times 10^3 + 2 \times 10^2 + 5 \times 1$

Objective 6 Solve applications.

Suppose we need to know how much surface space is within the borders, or edges, of a shape. We measure the space in **square units** and call the measurement **area.**

Definitions Square unit: A 1 × 1 square.

Area: The total number of square units that completely fill a shape.

Examples of square units:

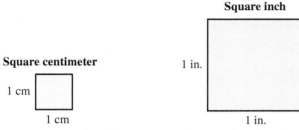

Square centimeter

1 cm

1 cm

A square centimeter is a 1 cm by 1 cm square and is represented mathematically by cm^2.

Square inch

1 in.

1 in.

A square inch is a 1 in. by 1 in. square and is represented mathematically by in.^{2}.

To *measure* the area of a rectangular room, we could lay square tiles until we completely cover the floor with tiles and then count the total number of tiles. In the following figure, we illustrate this approach for an 8-foot × 4-foot bathroom.

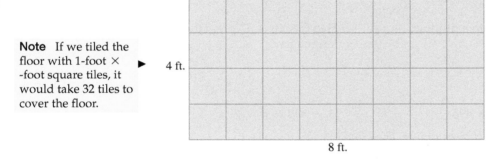

Note If we tiled the floor with 1-foot × -foot square tiles, it would take 32 tiles to cover the floor.

4 ft.

8 ft.

Notice that the square units form a rectangular array; so we can *calculate* the area by multiplying the length of the rectangle by the width, rather than measuring the area with square tiles. Calculating the area, we have $8 \cdot 4 = 32$ square feet, which agrees with our measurement.

Conclusion: Area of a rectangle = length × width, or, using variables, $A = lw$

The equation $A = lw$ is called a **formula.**

Definition   Formula: An equation that describes a procedure.

Procedure

To use a formula:

1. Replace the variables with the corresponding known values.
2. Solve for the unknown variable.

► **Your Turn 10**

A room measures 15 feet wide by 24 feet long. How many square feet of carpet will be needed to cover the floor of the room?

Example 10 A standard football field measures 100 yards by 50 yards, not counting the end zones. Find the area of a standard football field.

Solution: Because the field is a rectangle, to calculate the area, we multiply length times width.

$$A = lw$$
$$A = 100 \text{ yd.} \cdot 50 \text{ yd.}$$
$$A = 5000 \text{ yd.}^2$$

Note This unit of area is read "yards squared" or "square yards." All area calculations involve multiplying two distance measurements. Area units are ◄ expressed as the distance unit with an exponent of 2.

Connection The unit yd.^{2} means "square yards." Similarly, when we raise a number to the second power, we say that we are *squaring* that number, or the number is *squared*. The result is said to be the *square* of the number.

Answer to Your Turn 10
360 ft.^{2}

◄ **Do Your Turn 10**

1.3 Exercises
For Extra Help   MyMathLab®

Objectives 1 and 2

Prep Exercise 1 In a multiplication sentence, the numbers that are multiplied are the _____ and the answer is the _____.

Prep Exercise 2 What does the commutative property of multiplication say?

Prep Exercise 3 Explain how to estimate a product.

Prep Exercise 4 Express the distributive property in math language.

For Exercises 1–4, multiply.

1. 352×1 **2.** 0×495 **3.** $8 \times 9 \times 0 \times 7$ **4.** $12 \times 3 \times 1$

For Exercises 5–10, estimate each product by rounding. Then find the actual product.

5. 24×8 **6.** 63×4

7. $54 \cdot 91$ **8.** $47 \cdot 86$

9. $83(272)$ **10.** $68(826)$

For Exercises 11–20, multiply.

11. 642×70 **12.** 408×92 **13.** $2065 \cdot 482$

14. $9402 \cdot 608$ **15.** $207(41{,}308)$ **16.** $361(80{,}290)$

17. $(60{,}309)(4002)$ **18.** $(40{,}058)(7090)$ **19.** $205 \cdot 23 \cdot 70$

20. $14 \cdot 1 \cdot 289 \cdot 307$

Objective 4

Prep Exercise 5 Explain how to evaluate 3^4.

For Exercises 21–28, evaluate.

21. 2^4 **22.** 5^3 **23.** 1^6 **24.** 13^2

25. 3^5 **26.** 10^5 **27.** 10^7 **28.** 0^7

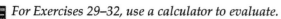

For Exercises 29–32, use a calculator to evaluate.

29. 6536^2 **30.** 279^3 **31.** 42^5 **32.** 34^6

Objective 5

For Exercises 33–38, write in exponential form.

33. $9 \cdot 9 \cdot 9 \cdot 9$ **34.** $12 \cdot 12 \cdot 12$ **35.** $7 \cdot 7 \cdot 7 \cdot 7 \cdot 7$

36. $10 \cdot 10 \cdot 10 \cdot 10 \cdot 10 \cdot 10 \cdot 10$ **37.** $14 \cdot 14 \cdot 14 \cdot 14 \cdot 14 \cdot 14$ **38.** $2 \cdot 2 \cdot 2 \cdot 2 \cdot 2 \cdot 2 \cdot 2 \cdot 2$

Prep Exercise 6 List three key words that indicate multiplication.

Prep Exercise 7 What is a rectangular array?

Prep Exercise 8 What is the area of a shape?

For Exercises 39–44, write each number in expanded form using powers of 10.

39. 24,902   **40.** 604,057

41. 9,128,020   **42.** 10,945

43. 407,210,925   **44.** 3,029,408

Objectives 3 and 6

For Exercises 45–64, solve.

45. In a certain city, avenues run north/south and streets run east/west. If there are 97 streets and 82 avenues, how many intersections are there? If the city must put up 8 traffic lights at each intersection, how many traffic lights are required in all?

46. A parking lot for a department store has 23 rows of spaces. Each row can hold 36 cars. What is the maximum number of cars that can be parked in the lot? Now suppose the lot is full. Estimate the number of people in the store assuming that there are two people in the store for each car in the lot.

47. On average, the human heart beats 70 times in 1 minute. How many beats is this in a year? (Use 365 days in a year.) If the average male in the United States lives to be 75 years old, estimate the number of heartbeats in an average male's lifetime. (*Source: National Vital Statistics Reports*, Vol. 52, No. 14, February 18, 2004, p. 33)

48. The average female life span in the United States is about 80 years. Using the information in Exercise 47, estimate the number of heartbeats in an average female's lifetime. (*Source: National Vital Statistics Reports*, Vol. 52, No. 14, February 18, 2004, p. 33)

49. A prescription indicates that a patient is to receive 10 milligrams of aminophylline per kilogram of weight. If the person weighs 74 kilograms, how many milligrams should the patient receive?

50. A nurse sets the drip rate for an IV at 26 drops each minute. How many drops does the patient receive in 1 hour?

51. An advertisement indicates that a person can make up to $800 per week as a telemarketer. How much would a telemarketer earning $800 per week earn in one year?

52. Michael works 5 days every week and buys lunch each workday. On average, he spends $9 each day. How much does he spend on lunch in a week? If he takes 10 vacation days each year, how much does he spend on lunch in a year?

53. An article recommends that a person drink eight 8-ounce glasses of water every day. If every one of the 300 million Americans drank eight 8-ounce glasses of water on the same day, how many ounces of water would be consumed that day?

54. On the evening news, you hear a conversation between the anchor and a commentator. The commentator says that a computer company sold nearly 800,000 copies of a new software package at $40 each. The anchor then says, "So that's about $320 billion. Wow." Is the anchor's response accurate? Explain.

55. A combination lock is opened by a combination of three things. You must turn a key to one of two positions, left or right; set a dial to a day of the week; and set another dial to a letter from A to H. How many possible combinations are there?

56. If a license plate consists of three capital letters and a three-digit number, how many license plates are possible? How many are possible in the entire United States if each state uses the same system of three letters and a three-digit number?

57. A certain computer chip processes 5-digit binary numbers, meaning that each of the five places could contain a 1 or 0. How many different binary numbers can it process? If each letter, A–Z, and each numeral, 0–9, corresponds to a different binary number, would the 5-digit chip be able to recognize them all? Explain.

58. How many codes are possible using a 12-bit binary coder?

59. MIDI stands for musical instrument data interface and is the common coding for digital instruments such as synthesizers. Most MIDI devices have a 7-bit binary memory chip. How many memory locations are possible?

60. A psychology quiz has 10 true-false questions. How many different answer keys are possible? Express the answer in exponential form; then calculate the actual number.

61. A man walks into a carpet store and says that he has a 24-foot-wide by 26-foot-long room and therefore needs about 4000 square feet of carpet. Exactly how many square feet does he need? Is his estimate reasonable?

62. You want to tile the kitchen floor in your home. The rectangular kitchen is 16 feet long by 12 feet wide. If you used 1 foot by 1 foot square tiles, how many would you need?

63. Delia plans to stain a new wood deck. The deck is 15 feet wide by 20 feet long. What is the area of the deck?

64. Taylor plans to paint a wall in his living room. The wall is 18 feet long by 9 feet high. What is the area of the wall?

Puzzle Problem A certain bacterium reproduces by dividing itself every minute. This causes the total population to double every minute. If we begin with a single bacterium on a petri dish, how many bacteria will exist after 5 minutes? 10 minutes? Write a formula that describes the pattern.

Review Exercises

[1.1] **1.** Write the word name for 16,507,309.

[1.1] **2.** Write 23,506 in expanded form.

[1.2] **3.** Add. 54,391 + 2079 + 518

[1.2] **4.** Subtract. 901,042 − 69,318

[1.1] **5.** Solve and check. $n + 19 = 32$

1.4 Dividing Whole Numbers; Solving Equations

Warm-up

[1.3] 1. Multiply. 345×67

[1.3] 2. Evaluate. 4^3

Objective 1 Divide whole numbers.

Like addition and subtraction, multiplication and division are inverse operations. Because multiplication means to add repeatedly, it follows that **division** means to subtract repeatedly.

Definition Division: Repeated subtraction of the same number.

When we write a division sentence, the *dividend* is divided by the *divisor* and the *quotient* is the answer.

Notation: $20 \div 5 = 4$

      ↑   ↑   ↑

   Dividend Divisor Quotient

Like multiplication, there are several ways to indicate division. We can write $20 \div 5 = 4$ using a slash, a fraction bar, or long division.

Slash: $20/5 = 4$

Fraction bar: $\dfrac{20}{5} = 4$

Long division: $5\overline{)20}$ with quotient 4

At one time, people divided by repeatedly subtracting the same number until they couldn't subtract anymore.

Let's calculate $20 \div 5$ by repeated subtraction. The quotient is the number of times we have to subtract 5 to get to 0 or a number between 0 and 5.

```
  20
  -5    1st subtraction
  ---
  15
  -5    2nd subtraction
  ---
  10
  -5    3rd subtraction
  ---
   5
  -5    4th subtraction
  ---
   0
```

Note Because we subtracted 5 from 20 four times to get 0, the quotient is 4.

The amount left after dividing is called the **remainder.**

Definition Remainder: The amount left over after dividing two whole numbers.

In the case of $20 \div 5$, there are exactly four 5s in 20; so the remainder is 0. Sometimes the remainder may not be 0. For example, if we split $21 into $5 bills, we would have four $5 bills with $1 left over. The $1 left over is the remainder, and we would write $21 \div 5 = 4\,r1$, using the letter r to indicate the remainder.

Objectives

1 Divide whole numbers.

2 Solve equations containing an unknown factor.

3 Solve applications involving division.

Connection Because multiplication and division are inverse operations, we can check the accuracy of $20 \div 5 = 4$ by multiplying $4 \cdot 5 = 20$.

Let's explore a few properties of division. Like subtraction, division is neither commutative nor associative.

Note $\neq$ means "is not equal to."

Division is not commutative: $20 \div 5 = 4$ while $5 \div 20 \neq 4$

Division is not associative: $(32 \div 8) \div 2$ $32 \div (8 \div 2)$

$$= 4 \div 2 \qquad \text{while} \qquad = 32 \div 4$$

$$= 2 \leftarrow \text{Different results} \rightarrow = 8$$

What if the divisor is 1, as in $\dfrac{24}{1}$? To check the quotient, we must be able to multiply the quotient by the divisor to get the dividend.

$$? \cdot 1 = 24$$

In Section 1.3, we learned that the product of 1 and a number is the number. Therefore, $24 \cdot 1 = 24$, which means $\dfrac{24}{1} = 24$.

Conclusion: Any number divided by 1 is that number.

Division Property

When any number is divided by 1, the quotient is the number (the dividend).

In math language: $n \div 1 = \dfrac{n}{1} = n$ when n is any number.

What if the divisor is 0, as in $\dfrac{14}{0}$? To check the quotient, we must be able to multiply the quotient by the divisor to get the dividend.

$$? \cdot 0 = 14$$

Because the product of any number and 0 is 0, the equation $? \cdot 0 = 14$ has no solution. Therefore, $\dfrac{14}{0}$ has no numerical quotient. We say that $\dfrac{14}{0}$ is undefined.

Conclusion: It is impossible to divide by 0.

Division Property

If the divisor is 0 and the dividend is any number other than 0, the quotient is undefined.

In math language: $n \div 0$, or $\dfrac{n}{0}$, is undefined when $n \neq 0$.

What if the dividend is 0, as in $\dfrac{0}{12}$? Again, we must be able to multiply the quotient by the divisor to get the dividend.

$$? \cdot 12 = 0$$

Notice that the missing factor must be 0; so $\dfrac{0}{12} = 0$. In general, the quotient of 0 and a nonzero number is 0, but before we formalize this rule, we must consider 0 divided by 0.

$$\dfrac{0}{0} = ?$$

To check, we must be able to multiply the quotient by the divisor to get the dividend.

$$? \cdot 0 = 0$$

It is tempting to say that the missing factor is still 0, but actually, the product of any number and 0 is 0; so we cannot determine a value for $\frac{0}{0}$. We say that $\frac{0}{0}$ is *indeterminate*.

Conclusion: When 0 is divided by any number other than 0, the quotient is 0. When 0 is divided by itself, the quotient is indeterminate.

Division Properties

When 0 is divided by any number other than 0, the quotient is 0.

In math language: $0 \div n = \dfrac{0}{n} = 0$ when $n \neq 0$.

When the dividend and divisor are both 0, the quotient is indeterminate.

In math language: $0 \div 0$, or $\dfrac{0}{0}$, is indeterminate.

What if we divide a number other than 0 by itself, as in $\frac{65}{65}$? Again, we must be able to check by multiplying the quotient times the divisor to get the dividend.

$$? \cdot 65 = 65$$

The only number that works here is 1 because $1 \cdot 65 = 65$; so $\dfrac{65}{65} = 1$.

Division Property

When a number (other than 0) is divided by itself, the quotient is 1.

In math language: $n \div n = \dfrac{n}{n} = 1$ when $n \neq 0$.

Example 1   Determine the quotient and explain your answer.

a. $34 \div 1$

Answer: 34 because $34 \cdot 1 = 34$

b. $27 \div 0$

Answer: undefined because there is no number that can make $? \cdot 0 = 27$ true

c. $0 \div 16$

Answer: 0 because $0 \cdot 16 = 0$

d. $45 \div 45$

Answer: 1 because $1 \cdot 45 = 45$

▶ **Do Your Turn 1**

▶ **Your Turn 1**

Determine the quotient and explain your answer.
a. $17 \div 1$
b. $0 \div 0$
c. $0 \div 42$
d. $38 \div 38$

Answers to Your Turn 1
a. 17 because $17 \cdot 1 = 17$
b. indeterminate because $x \cdot 0 = 0$ is true for any number
c. 0 because $0 \cdot 42 = 0$
d. 1 because $1 \cdot 38 = 38$

In division, when the remainder is 0, we say that the divisor is an *exact divisor*. For example, because $20 \div 5 = 4$, we say that 5 is an exact divisor of 20 or that 20 is divisible by 5. We can use divisibility rules to determine whether a number is an exact divisor for a given number without dividing the numbers. There are many divisibility rules, but we will consider rules for only 2, 3, and 5.

Divisibility Rules

2 is an exact divisor for any even number. Even numbers have 0, 2, 4, 6, or 8 in the ones place.

> Example: 3596 is divisible by 2 because it is an even number. 3596 has the digit 6 in the ones place.

3 is an exact divisor for a number if the sum of the digits in the number is divisible by 3.

> Example: 58,014 is divisible by 3 because the sum of its digits is $5 + 8 + 0 + 1 + 4 = 18$, which is divisible by 3.

5 is an exact divisor for any number with a 0 or 5 in the ones place.

> Example: 91,285 is divisible by 5 because it has 5 in the ones place.

▶ Your Turn 2

Use divisibility rules to determine whether the given number is divisible by 2.

a. 4,519,362

b. 5385

Example 2   Use divisibility rules to determine whether the given number is divisible by 2.

a. 45,091

Answer: 45,091 is not divisible by 2 because it is not even. It does not have 0, 2, 4, 6, or 8 in the ones place.

b. 691,134

Answer: 691,134 is divisible by 2 because it is an even number. It has 4 in the ones place.

◀ Do Your Turn 2

▶ Your Turn 3

Use divisibility rules to determine whether the given number is divisible by 3.

a. 93,481

b. 412,935

Example 3   Use divisibility rules to determine whether the given number is divisible by 3.

a. 76,413

Answer: 76,413 is divisible by 3 because the sum of its digits is $7 + 6 + 4 + 1 + 3 = 21$, which is divisible by 3.

b. 4256

Answer: 4256 is not divisible by 3 because the sum of its digits is $4 + 2 + 5 + 6 = 17$, which is not divisible by 3.

◀ Do Your Turn 3

Answers to Your Turn 2
a. yes
b. no

Answers to Your Turn 3
a. no
b. yes

Example 4   Use divisibility rules to determine whether the given number is divisible by 5.

a. 2,380,956

Answer: 2,380,956 is not divisible by 5 because it does not have 0 or 5 in the ones place.

b. 49,360

Answer: 49,360 is divisible by 5 because it has 0 in the ones place.

▶ **Do Your Turn 4**

▶ **Your Turn 4**

Use divisibility rules to determine whether the given number is divisible by 5.

  a. 6945

  b. 42,371

Now that we have learned some properties and rules of division, we can discuss more complex division. Because repeated subtraction is tedious, the common method for dividing is *long division.*

In a completed long division, it can be difficult to comprehend everything that happens. So in our first example of long division, we will break the process into steps. Thereafter, we will show only the completed long division.

Example 5   Divide. 6408 ÷ 7

Solution: $7\overline{)6408}$

Because 6 is less than 7, we include the 4 with the 6. Because 64 is greater than 7, we are ready to divide.

$$
\begin{array}{r}
9 \\
7\overline{)6408} \\
-63 \\
\hline
1
\end{array}
$$

$$
\begin{array}{r}
9 \\
7\overline{)6408} \\
-63\downarrow \\
\hline
10
\end{array}
$$

$$
\begin{array}{r}
91 \\
7\overline{)6408} \\
-63 \\
\hline
10 \\
-7 \\
\hline
3
\end{array}
$$

$$
\begin{array}{r}
915 \\
7\overline{)6498} \\
-63| \\
\hline
10| \\
-7\downarrow \\
\hline
38 \\
-35 \\
\hline
3
\end{array}
$$

◀ 1. Write the problem in long division form.

    2. Compare the leftmost digit in the dividend with the divisor. If it is less than the divisor, include the next digit. Continue in this way until the digits name a number that is greater than the divisor.

◀ 3. Divide 7 into 64 and write the quotient, 9, over the 4 in the dividend.

    4. Multiply 9 times 7 to get 63. Subtract 63 from 64 to get 1.

◀ 5. Write the next digit in the dividend, 0, beside the 1 remainder. It is common to say that you are *bringing down* the 0.

◀ 6. Divide 7 into 10 and write the quotient, 1, over the 0 in the dividend.

    7. Multiply 1 times 7 to get 7. Subtract 7 from 10 to get 3.

    8. Write the last digit in the dividend, 8, beside the 3 to make the number 38.

    9. Divide 7 into 38 and write the quotient, 5, over the 8 in the dividend.

◀ 10. Multiply 5 times 7 to get 35. Subtract 35 from 38 to get a remainder of 3.

Answer: 915 r3

Check: Multiply the quotient by the divisor and then add the remainder. The result should be the dividend.

$$
\begin{array}{r}
\overset{13}{91}5 \\
\times\ 7 \\
\hline
6405 \\
+3 \\
\hline
6408
\end{array}
$$

  Quotient

  Divisor

  Remainder

  Dividend

▶ **Do Your Turn 5**

▶ **Your Turn 5**

Divide.

  a. 5028 ÷ 3

  b. 6409 ÷ 5

  c. 12,059 ÷ 23

Warning If a remainder is greater than the divisor, the divisor can divide the digits in the dividend more times. For example, in our first step in Example 5, suppose we mistakenly thought that 7 divides into 64 only 6 times.

$$
\begin{array}{r}
6 \\
7\overline{)6408} \\
-42 \\
\hline
22
\end{array}
$$

Because the remainder, 22, is greater than the divisor, 7, the digit, 6, in the quotient is not big enough. Further, because 7 goes into 22 three times, adding 3 to the incorrect digit, 6, gives the correct digit, 9.

Answers to Your Turn 4
a. yes
b. no

Answers to Your Turn 5
a. 1676
b. 1281 r4
c. 524 r7

Procedures

To divide whole numbers, use long division.

To estimate a quotient of whole numbers:

1. Round each number to the highest possible place value so that each has one nonzero digit.
2. Divide the rounded numbers. If the calculation is difficult to perform mentally, try rounding the dividend so that it has two nonzero digits.

▶ **Your Turn 6**

Estimate each quotient by rounding. Then find the actual quotient.

  a. $2617 \div 6$

  b. $10,091 \div 97$

  c. $425,231 \div 705$

Note With division, rounding the dividend to two nonzero digits sometimes makes the division easier to perform mentally than does rounding to one nonzero digit. For example, estimating $44,602 \div 28$ by rounding both numbers to one nonzero digit gives $40,000 \div 30$, which is not so easy. However, rounding 44,602 so that it has two nonzero digits gives $45,000 \div 30$, which is easier to calculate mentally. We can think $45 \div 3 = 15$ so that the estimate is 1500.

Example 6 Estimate $42,017 \div 41$ by rounding. Then find the actual quotient.

Estimate: Round 42,017 and 41 to their highest place. Rounding 42,017 gives 40,000. Rounding 41 gives 40. The estimate is the quotient of 40,000 and 40.

$$
\begin{array}{r}
1\,000 \\
40\overline{)40,000} \\
-40 \\
\hline
0\,0 \\
-0 \\
\hline
00 \\
-0 \\
\hline
00 \\
-0 \\
\hline
0
\end{array}
$$

◀ **Note** To perform the division mentally, first eliminate the common zeros: $\dfrac{4000\cancel{0}}{4\cancel{0}} = \dfrac{4000}{4}$.

Then divide the nonzero digits, $4 \div 4 = 1$, and attach the three remaining zeros to the result so that you have 1000.

Actual:

$$
\begin{array}{r}
1,024 \\
41\overline{)42,017} \\
-41 \\
\hline
1\,0 \\
-0 \\
\hline
1\,01 \\
-82 \\
\hline
197 \\
-164 \\
\hline
33
\end{array}
$$

◀ **Note** Because 41 goes into 10 zero times, we write a 0 in the quotient.

Answer: 1024 r33

Check: Multiply the quotient by the divisor, and then add the remainder. You should get the dividend.

$$
\begin{array}{r}
1\,0\overset{1}{2}4 \quad \text{Quotient} \\
\times 41 \quad \text{Divisor} \\
\hline
1\,024 \\
40\,96 \\
\hline
41,984 \\
+33 \quad \text{Remainder} \\
\hline
42,017 \quad \text{Dividend}
\end{array}
$$

◀ **Do Your Turn 6**

Answers to Your Turn 6

a. Estimate: 500; Actual: 436 r1

b. Estimate: 100; Actual: 104 r3

c. Estimate: 600; Actual: 603 r116

Objective 2 Solve equations containing an unknown factor.

Suppose we have to design a room in a house, and we know that we want the area to be 150 square feet and the length to be 15 feet. What must the width be?

We know that the area of a rectangle is found by the formula $A = l \cdot w$; so we can write an equation with the width, w, as an unknown factor.

$$15 \cdot w = 150$$

When we solved equations with an unknown addend in Section 1.2, we used a related subtraction sentence. Similarly, to solve an equation with an unknown factor, we use a related division equation.

$$\textbf{Related division: } w = 150 \div 15 \quad \text{or} \quad w = \frac{150}{15}$$

$$w = 10 \qquad\qquad w = 10$$

The width needs to be 10 feet. We can return to the formula for area to check.

$$A = 15 \cdot 10 = 150$$

Our example suggests the following procedure.

Procedure

To solve for an unknown factor, write a related division equation in which the product is divided by the known factor.

Example 7 Solve and check.

a. $x \cdot 16 = 208$

Solution: $x = 208 \div 16$ Write a related division equation in which the product, 208, is divided by the known factor, 16.

$$x = 13$$

$$\begin{array}{r} 13 \\ 16\overline{)208} \\ -16 \\ \hline 48 \\ -48 \\ \hline 0 \end{array}$$

Check: $x \cdot 16 = 208$

$13 \cdot 16 \overset{?}{=} 208$ In $x \cdot 16 = 208$, replace x with 13.

$208 = 208$ Verify that the equation is true: $13 \cdot 16 = 208$ is true.

$$\begin{array}{r} \overset{1}{16} \\ \times 13 \\ \hline 48 \\ +16 \\ \hline 208 \end{array}$$

b. $14 \cdot n = 0$

Solution: $n = \dfrac{0}{14}$ Write a related division in which the product, 0, is divided by the known factor, 14.

$$n = 0$$

Check: $14 \cdot n = 0$

$14 \cdot 0 \overset{?}{=} 0$ In $14 \cdot n = 0$, replace n with 0.

$0 = 0$ Verify that the equation is true: $14 \cdot 0 = 0$ is true.

▶ **Your Turn 7**

Solve and check.

a. $6 \cdot x = 54$
b. $n \cdot 18 = 378$
c. $28 \cdot a = 0$
d. $r \cdot 0 = 37$

c. $y \cdot 0 = 5$

Solution: $y = \dfrac{5}{0}$, which is undefined.   Write a related division equation in which the product, 5, is divided by the known factor, 0.

Because $\dfrac{5}{0}$ is undefined, $y \cdot 0 = 5$ has no solution.

Check: We cannot check because no number can replace y to make the equation true.

◀ **Do Your Turn 7**

Objective 3 Solve applications involving division.

Like addition, subtraction, and multiplication, there are key words or phrases that indicate division.

Key Words for Division

divide, distribute evenly, each, split, quotient, into, per, over

▶ **Your Turn 8**

Solve.

a. Alicia wants to place a wooden fence along the back of her property. She measures the distance to be 180 feet. She decides to buy prefabricated 8-foot sections of fencing. How many sections must she purchase?

b. Juan is to set an IV drip so that a patient receives 300 units of heparin, an antico-agulant medication, in D/W solution for 1 hour. How many units should the patient receive each minute?

Of Interest

D/W is an abbreviation for dextrose in water. Dextrose is a type of sugar in animal and plant tissues.

Example 8   An egg farmer has 4394 eggs to distribute into packages of a dozen each. How many packages can the farmer make? How many eggs will be left?

Solution: We must split the total number of eggs into groups of a dozen.

$$
\begin{array}{r}
366 \\
12\overline{)4394} \\
-36 \\
\hline
79 \\
-72 \\
\hline
74 \\
-72 \\
\hline
2
\end{array}
$$

Answer: 366 packages, each holding a dozen eggs, can be made with 2 eggs left over.

Check:
$$
\begin{array}{r}
\overset{1\,1}{366} \\
\times 12 \\
\hline
732 \\
+366 \\
\hline
4392 \\
+2 \\
\hline
4394
\end{array}
$$

Quotient
Divisor

Remainder
Dividend

◀ **Do Your Turn 8**

Answers to Your Turn 7
a. $x = 9$ b. $n = 21$ c. $a = 0$
d. no solution

Answers to Your Turn 8
a. 23 whole sections
b. 5 units

1.4 Exercises For Extra Help MyMathLab®

Objective 1

Prep Exercise 1 In $40 \div 5 = 8$, 40 is the _____, 5 is the _____ , and 8 is the _____.

Prep Exercise 2 $n \div 1 = \dfrac{n}{1} =$ _____ when n is any number.

Prep Exercise 3 $0 \div n = \dfrac{0}{n} =$ _____ when $n \neq 0$.

Prep Exercise 4 $n \div 0$, or $\dfrac{n}{0}$, is _____ when $n \neq 0$.

Prep Exercise 5 $0 \div 0$, or $\dfrac{0}{0}$, is _____ .

Prep Exercise 6 $n \div n = \dfrac{n}{n} =$ _____ when $n \neq 0$.

For Exercises 1–8, determine the quotient and explain your answer.

1. $26 \div 26$

2. $381 \div 1$

3. $0 \div 49$

4. $29 \div 0$

5. $0 \div 0$

6. $462 \div 462$

7. $22 \div 0$

8. $0 \div 95$

Prep Exercise 7 2 is an exact divisor for all _____ numbers.

For Exercises 9–14, use divisibility rules to determine whether the given number is divisible by 2.

9. 19,761

10. 24,978

11. 143,706

12. 801,907

13. 431,970

14. 8445

Prep Exercise 8 3 is an exact divisor for a number if _____.

For Exercises 15–20, use divisibility rules to determine whether the given number is divisible by 3.

15. 19,704

16. 1101

17. 26,093

18. 450,917

19. 98,757

20. 241,080

Prep Exercise 9 5 is an exact divisor for numbers that _____.

For Exercises 21–26, use divisibility rules to determine whether the given number is divisible by 5.

21. 27,005

22. 148,070

23. 64,320

24. 704,995

25. 4,195,786

26. 319,424

Prep Exercise 10 When estimating a quotient, explain what to do if rounding the numbers to one nonzero digit produces a division that is difficult to perform mentally.

For Exercises 27–44, estimate. Then find the actual quotient.

27. $3627 \div 9$

28. $2364 \div 6$

29. $1038 \div 5$

30. $24,083 \div 4$

31. $3472 \div 16$

32. $9979 \div 17$

33. $6399 \div 26$

34. $9770 \div 19$

35. $\dfrac{12,464}{41}$

36. $\dfrac{15,542}{38}$

37. $\dfrac{24,400}{610}$

38. $\dfrac{20,800}{260}$

39. $\dfrac{235,613}{124}$ **40.** $\dfrac{403,218}{96}$ **41.** $\dfrac{331,419}{207}$

42. $\dfrac{510,631}{204}$ **43.** $\dfrac{725,480}{0}$ **44.** $\dfrac{174,699}{0}$

Objective 2

Prep Exercise 11 Explain how to solve for an unknown factor.

For Exercises 45–56, solve and check.

45. $9 \cdot x = 54$ **46.** $y \cdot 4 = 40$ **47.** $m \cdot 11 = 88$

48. $15 \cdot a = 45$ **49.** $17 \cdot n = 119$ **50.** $b \cdot 21 = 105$

51. $24 \cdot t = 0$ **52.** $u \cdot 0 = 409$ **53.** $v \cdot 2 \cdot 13 = 1092$

54. $16 \cdot 5 \cdot k = 560$ **55.** $29 \cdot 6 \cdot h = 3480$ **56.** $18 \cdot c \cdot 41 = 6642$

Objective 3

Prep Exercise 12 List three key words that indicate division.

For Exercises 57–70, solve.

57. Dedra's gross annual salary is $34,248. She gets paid once a month. What is her gross monthly salary?

58. A financial planner is asked to split $16,800 evenly among 7 investments. How much does she put in each investment?

59. Carl is to set an IV drip so that a patient receives 840 milliliters of 5% D/W solution in an hour. How many milliliters should the patient receive each minute?

60. The federal government grants a state $5,473,000 to be distributed equally among the 13 technical colleges in the state. How much does each college receive?

61. An employee of a copying company needs to make small flyers for a client. The employee can make 4 flyers out of each piece of paper. The client needs 500 flyers. How many pieces of paper must the employee use?

62. A printing company is asked to make 800 business cards for a client. It can print 12 cards on each sheet of card stock. How many sheets will it use?

63. How many 44¢ stamps can be bought with $10? Explain.

64. A long-distance company charges 12¢ per minute. How long can a person talk for $5? Explain.

65. A cereal factory produces 153,600 ounces of cereal each day. Each box contains 16 ounces. The boxes are packaged in bundles of 8 boxes per bundle, then stacked on pallets, 24 bundles to a pallet. Finally, the pallets are loaded onto trucks, 28 pallets to each truck.

  a. How many boxes of cereal are produced each day?

  b. How many bundles are produced?
  c. How many pallets are needed?
  d. How many trucks are needed? Explain.

66. A bottling company produces 48,000 bottles each day. The bottles are packaged in six-packs, then bundled into cases of 4 six-packs per case. The cases are then loaded onto pallets, 36 cases to a pallet.

  a. How many six-packs are produced?
  b. How many cases are produced?
  c. How many pallets are needed? Explain.

67. In the design of a room for a house, the desired length is 24 feet. The desired area is 432 square feet. What must the width be?

68. A landscaper needs to cover a 4000-square foot yard with sod. Sod is delivered in pallets. Each pallet holds enough sod to cover 504 square feet. How many pallets must the landscaper purchase if the company that produces the pallets will sell only whole pallets?

69. Barry decides to use prefabricated sections of wooden fence to fence his backyard. Each section is 8 feet in length. How many sections must he purchase to complete 170 feet?

70. The building code says that electrical outlets must be spaced no more than 8 feet apart along the wall perimeter in each room. If a room has a perimeter of 60 feet, how many electrical outlets will be needed to meet the code requirements?

Review Exercises

[1.2] **1.** Solve and check. $184 + t = 361$

[1.2] **2.** Connie is placing a decorative border near the ceiling in her bedroom. The room is 14 feet wide by 16 feet long. How much border will she need?

[1.3] **3.** Estimate 42×367; then find the actual product.

[1.3] **4.** Evaluate. 5^4

[1.3] **5.** Write 49,602 in expanded notation with powers of 10.

1.5 Order of Operations; Mean, Median, and Mode

Objectives

1 Simplify numerical expressions by following the order of operations agreement.

2 Find the mean, median, and mode of a list of values.

Warm-up

[1.4] 1. Use divisibility rules to determine whether 414 is divisible by 3.

[1.4] 2. Divide. 7052 ÷ 12

Objective 1 Simplify numerical expressions by following the order of operations agreement.

Now let's simplify expressions such as $10 + 3 \cdot 2$ that contain a mixture of operations. Notice that we get different answers depending on the order in which we perform the operations.

If we add first, we get:	If we multiply first, we get:
$10 + 3 \cdot 2$	$10 + 3 \cdot 2$
$= 13 \cdot 2$	$= 10 + 6$
$= 26$	$= 16$

The correct answer is 16 because we must multiply before adding. The correct order in which operations are to be performed is determined by an agreement. Because the specific order of operations that we follow is an agreement, it cannot be derived or proven.

Order of Operations Agreement

Perform operations in the following order:
1. Grouping symbols, which include parentheses ( ), brackets [ ], braces { }, and fraction bars ——
2. Exponents
3. Multiplication or division in order as these operations occur from left to right
4. Addition or subtraction in order as these operations occur from left to right

Note We will learn about other grouping symbols and square roots in Chapter 2.

Warning Some people make the mistake of thinking that multiplication always comes before division and addition always comes before subtraction. However, proper order is to multiply or divide from *left to right* in the order these operations occur. Add or subtract from *left to right* in the order these operations occur.

▶ **Your Turn 1**

Simplify. Show your steps.
  a. $15 + 4 \cdot 6$
  b. $20 - 3 \cdot 6 \div 2$
  c. $12 + 14 \div 2 \cdot 5$

Example 1   Simplify. $18 - 24 \div 6 \cdot 2$

Solution:
$$18 - 24 \div 6 \cdot 2$$
$$= 18 - 4 \cdot 2 \quad \text{Divide. } 24 \div 6 = 4$$
$$= 18 - 8 \quad \text{Multiply. } 4 \cdot 2 = 8$$
$$= 10 \quad \text{Subtract. } 18 - 8 = 10$$

Note We multiply or divide from left to right before adding or subtracting. Reading $24 \div 6 \cdot 2$ from left to right, we see $24 \div 6$; first; so we divide first.

◀ **Do Your Turn 1**

Answers to Your Turn 1
a. 39   b. 11   c. 47

Answers to Warm-up
1. yes
2. 587 r8

Example 2   Simplify. $\dfrac{8}{2^2} + 5$

Solution: $\dfrac{8}{2^2} + 5$

$\quad = \dfrac{8}{4} + 5$ Evaluate the exponential form. $2^2 = 4$

$\quad = 2 + 5$ Divide. $\dfrac{8}{4} = 2$

$\quad = 7$ Add. $2 + 5 = 7$

▶ Do Your Turn 2

Example 3   Simplify. $16 - 3^2 + 4 \cdot 10 \div 5$

Solution: $16 - 3^2 + 4 \cdot 10 \div 5$

$\quad = 16 - 9 + 4 \cdot 10 \div 5$ Evaluate the exponential form. $3^2 = 9$

$\quad = 16 - 9 + 40 \div 5$ Multiply. $4 \cdot 10 = 40$

$\quad = 16 - 9 + 8$ Divide. $40 \div 5 = 8$

$\quad = 7 + 8$ Subtract. $16 - 9 = 7$ ◀ **Note** We add or subtract from left to right; so in $16 - 9 + 8$ we subtract first.

$\quad = 15$ Add. $7 + 8 = 15$

▶ Do Your Turn 3

Example 4   Simplify. $3(12 - 9) + 5^2 - (18 + 2) \div 5$

Solution:
$3(12 - 9) + 5^2 - (18 + 2) \div 5$

A number next to parentheses means multiply. ↗

$\quad = 3 \cdot 3 + 5^2 - 20 \div 5$ Perform operations in parentheses.

$\quad = 3 \cdot 3 + 25 - 20 \div 5$ Evaluate the exponential form. $5^2 = 25$

$\quad = 9 + 25 - 4$ Multiply and divide from left to right. $3 \cdot 3 = 9$ and $20 \div 5 = 4$

$\quad = 30$ Add and subtract from left to right. $9 + 25 = 34$ and $34 - 4 = 30$

▶ Do Your Turn 4

When grouping symbols appear within other grouping symbols, we say that they are *embedded*. When we have embedded grouping symbols, we must work from the innermost symbols outward, performing the inner computations first.

Example 5   Simplify. $28 - 4[30 \div (10 + 5)] + (7 - 3)^2$

Solution: $28 - 4[30 \div (10 + 5)] + (7 - 3)^2$

$\quad = 28 - 4[30 \div 15] + 4^2$ Perform operations in parentheses.

$\quad = 28 - 4 \cdot 2 + 4^2$ Divide in the brackets. $30 \div 15 = 2$

$\quad = 28 - 4 \cdot 2 + 16$ Evaluate the exponential form. $4^2 = 16$

$\quad = 28 - 8 + 16$ Multiply. $4 \cdot 2 = 8$

$\quad = 20 + 16$ Subtract. $28 - 8 = 20$

$\quad = 36$ Add. $20 + 16 = 36$

▶ Do Your Turn 5

▶ **Your Turn 2**

Simplify. Show your steps.

a. $7 - \dfrac{3^2}{9}$

b. $2^2 - \dfrac{12}{4}$

c. $5^2 \cdot 3 - 10$

▶ **Your Turn 3**

Simplify. Show your steps.

a. $12 + 2^4 - 9 \cdot 2$
b. $2 \cdot 6 \div 4 + 5^2$
c. $3^3 - 5 \cdot 3 + 24 \div 3$

Note Remove the parentheses, brackets, or braces after completing the calculations in them. Some operations can be completed in the same step. For example, the operations in the parentheses and 5^2 can be calculated in the same step. ◀

▶ **Your Turn 4**

Simplify. Show your steps.

a. $7(9 - 5) + 16 \div (3 + 5)$
b. $3^4 - 5(8 - 2) \div 6 + (7 - 3)^2$

▶ **Your Turn 5**

Simplify. Show your steps.

a. $13 + [23 + (16 - 7)] \div 4$
b. $7^2 - \{5 + 3[18 \div (2 + 7)]\}$

Answers to Your Turn 2
a. 6   b. 1   c. 65

Answers to Your Turn 3
a. 10   b. 28   c. 20

Answers to Your Turn 4
a. 30   b. 92

Answers to Your Turn 5
a. 21   b. 38

The fraction bar can be used as a grouping symbol. For example, in $\dfrac{7^2 - 4}{29 - 3(8 - 6)^3}$, the fraction line means to divide the result of the top expression by the result of the bottom expression. In fraction terminology, the top is called the *numerator* and the bottom is called the *denominator*. Think of $\dfrac{7^2 - 4}{29 - 3(8 - 6)^3}$ as having implied parentheses in the numerator and denominator. In other words, the order of operations is the same as $(7^2 - 4) \div [29 - 3(8 - 6)^3]$.

> ▶ **Your Turn 6**
>
> Simplify. Show your steps.
>
> a. $\dfrac{12^2 - 24}{42 - 3(9 - 7)^2}$
>
> b. $\dfrac{10 + 2(12 - 5)}{(3 + 5)2 - (7 - 5)^3}$

Example 6   Simplify. $\dfrac{7^2 - 4}{29 - 3(8 - 6)^3}$

Solution: Simplify the numerator and denominator separately; then divide.

$$\frac{7^2 - 4}{29 - 3(8 - 6)^3}$$

$$= \frac{49 - 4}{29 - 3(2)^3} \qquad \text{Evaluate the exponential form in the numerator. } 7^2 = 49$$
$$\text{Subtract in the denominator. } 8 - 6 = 2$$

$$= \frac{45}{29 - 3 \cdot 8} \qquad \text{Subtract in the numerator. } 49 - 4 = 45$$
$$\text{Evaluate the cube in the denominator. } 2^3 = 8$$

$$= \frac{45}{29 - 24} \qquad \text{Multiply in the denominator. } 3 \cdot 8 = 24$$

$$= \frac{45}{5} \qquad \text{Subtract in the denominator. } 29 - 24 = 5$$

$$= 9 \qquad \text{Divide. } 45/5 = 9$$

◀ **Do Your Turn 6**

Objective 2 Find the mean, median, and mode of a list of values.

Mean

We can use a number called a **statistic** to describe data. Some statistics describe the central tendency, or middle, of a list of values. One such statistic is the **mean**, also known as the *arithmetic mean* or *average*.

Definitions Statistic: A number used to describe some characteristic of a set of data.

Mean: The quotient of the sum of the given values and the number of values.

Suppose a student has the following test scores: 94, 82, 88, and 96. To find the mean, we add the scores and then divide by the number of scores, which is 4.

$$\text{Mean} = \frac{94 + 82 + 88 + 96}{4} = \frac{360}{4} = 90$$ ◀ **Note** The mean, 90, describes a central tendency of the four scores.

> **Procedure**
>
> To find the mean, or average, of a set of data:
> 1. Find the sum of the values.
> 2. Divide the sum by the number of values.

Answers to Your Turn 6
a. 4   b. 3

Median

Another statistic that describes the central tendency of a set of numbers is the **median**.

Definition **Median:** The middle number in an ordered list of numbers.

Look at the following list of numbers arranged in order from least to greatest.

1, 2, 3, 5, <u>6</u>, 7, 9, 9, 12   ◄ **Note** This list has an odd number of values. The
              median is 6 because it is the middle number.
↑
median

Notice that the median was easy to find in the previous set of numbers because it had an odd number of values. Now look at a set of numbers containing an even number of values.

2, 3, 4,   8, 9, 10   ◄ **Note** This list has an even number of
       ↑   values. The median must be between 4
     median   and 8; so it is the mean of 4 and 8.

$$\text{median} = \text{mean of 4 and 8} = \frac{4 + 8}{2} = \frac{12}{2} = 6$$

Procedure

To find the median of a set of data:
1. List the values in order from least to greatest.
2. If the list has an odd number of values, the median is the middle value. If the list has an even number of values, the median is the mean of the middle two values.

Mode

The third statistic we will learn is **mode**.

Definition **Mode:** The value that occurs most often.

Look at the following list of numbers.

1, 2, 3, 5, 6, 7, <u>9, 9</u>, 12   **Note** Because 9 occurs most often,
            ↑   it is the mode.
          mode

Procedure

Mode: To find the mode of a set of data, count the number of repetitions of each score. The score with the most repetitions is the mode.

Note If no value is repeated, we say that the data has no mode. If there is a tie for the most repetitions, list each of those values as the mode.

▶ **Your Turn 7**

Find the mean, median, and mode.

a. Height, in inches, of softball players: 67, 70, 66, 67, 68, 68, 65, 68, 64

b. Salaries: $42,000, $37,000, $54,000, $70,000, $35,000, $41,000

c. Age of students in a class: 20, 19, 32, 18, 20, 18, 20, 19, 18, 19, 18, 19

Example 7   Find the mean, median, and mode.

a. Weight of basketball players in pounds: 192, 235, 225, 206, 187

Solution:

Mean: Divide the sum of the values by the number of values.

$$\text{mean} = \frac{192 + 235 + 225 + 206 + 187}{5} = \frac{1045}{5} = 209$$

Median: Arrange the weights in order from least to greatest. The list has an odd number of values; so the median is the middle value.

$$187, 192, \underset{\underset{\text{median}}{\uparrow}}{206}, 225, 235$$

Mode: Because no number is repeated, there is no mode.

b. Exam scores: 95, 79, 86, 76, 93, 95, 80, 76

Solution:

Mean: Divide the sum of the values by the number of values.

$$\text{mean} = \frac{95 + 79 + 86 + 76 + 93 + 95 + 80 + 76}{8} = \frac{680}{8} = 85$$

Median: Arrange the scores from the least to greatest. The list has an even number of values; so the median is the mean of the middle two values.

$$76, 76, 79, 80, \underset{\uparrow}{} 86, 93, 95, 95$$

$$\text{median} = \text{mean of 80 and 86} = \frac{80 + 86}{2} = \frac{166}{2} = 83$$

Mode: The list has two numbers tied for the most repetitions; so we list them both.

$$\underset{\underset{\text{mode}}{\uparrow}}{76, 76}, 79, 80, 86, 93, \underset{\underset{\text{mode}}{\uparrow}}{95, 95}$$

The modes are 76 and 95.

Answers to Your Turn 7

a. mean: 67; median: 67; mode: 68
b. mean: $46,500; median: $41,500; no mode
c. mean: 20; median: 19; modes: 18 and 19

◀ **Do Your Turn 7**

1.5 Exercises   For Extra Help   MyMathLab®

Objective 1

Prep Exercise 1 List the four stages of the order of operations.

Prep Exercise 2 What is the first step in simplifying the expression $15 - 3 \cdot 4$? Explain.

Prep Exercise 3 What is the first step in simplifying the expression $13 + 20 \div 5 \cdot 2$? Explain.

Prep Exercise 4 What is the first step in simplifying the expression $8 + 2(10 - 4)$? Explain.

For Exercises 1–46, simplify. Show your steps.

1. $7 + 5 \cdot 3$

2. $35 - 3 \cdot 7$

3. $18 + 36 \div 4 \cdot 3$

4. $20 \div 2 \cdot 5 + 6$

5. $12^2 - 6 \cdot 4 \div 8$

6. $7^2 + 5 \cdot 6 \div 10$

7. $36 - 3^2 \cdot 4 + 14$

8. $16 - 48 \div 4^2 + 5$

9. $\dfrac{39}{13} + 6 - 2^3$

10. $29 - 4^2 + \dfrac{21}{7}$

11. $25 + 2(14 - 9)$

12. $43 - 2(2 + 4)$

13. $2 \cdot 5^2 - \dfrac{24}{4}$

14. $3 \cdot 6^2 - \dfrac{15}{3}$

15. $7 \cdot 8 - 40 \div 5 \cdot 2 + 9$

16. $9 \cdot 4 - 24 \div 4 \cdot 2 - 6$

17. $2^6 - 18 \div 3 \cdot 5 - \dfrac{60}{6}$

18. $5^2 - 24 \div 8 \cdot 6 + \dfrac{44}{4}$

19. $3^2 \cdot 4 \div 6 - 5 + 2^2$

20. $5^2 \cdot 3 \div 15 + 4 - 3^2$

21. $2(14 + 3) - 8 \cdot 2 + (9 - 2)$

22. $5(13 - 7) + 4 \cdot 3 - (4 + 2)$

23. $(2 + 9)(19 - 16)^2$

24. $(5 - 3)(2 + 4)^2$

25. $7(23 - 12) + \dfrac{2^5}{(14 - 6)}$

26. $2(7 - 4) + 8^2 - \dfrac{(16 + 5)}{7}$

27. $(13 - 8)3^2 - \dfrac{48}{(15 - 9)}$

28. $(3 + 1)5^2 - \dfrac{(19 - 11)}{4}$

29. $9(2 \cdot 7 - 6 \cdot 2)$

30. $2(4 \cdot 5 - 3 \cdot 5)$

31. $(3 + 4)11 + 24 \div 8(6 - 4)$

32. $(19 - 16)13 + 54 \div 9(7 - 5)$

33. $12 \div (4 \cdot 3) + 5(9 - 6)^2$

34. $36 \div (2 \cdot 6) + 3(5 - 1)^2$

35. $[14 - (3 + 2)] \div 3 + 4(17 - 6)$

36. $[20 - (7 + 5)] \div 2 + 3(13 - 4)$

37. $36 - 5[(28 - 4) \div 2^3] + 3^3$

38. $31 - 3[(56 - 6) \div 5^2] + 2^4$

39. $\{18 - 4[21 \div (3 + 4)]\} + 48 \div 2$

40. $54 \div 2 + \{15 - 3[20 \div (4 + 6)]\}$

41. $\dfrac{9^2 - 21}{56 - 2(4 + 1)^2}$

42. $\dfrac{22 + 3^3}{3(14 - 6) - (8 + 9)}$

43. $\dfrac{38 - 4(15 - 12)}{(3 + 5)^2 - 2(39 - 8)}$

44. $\dfrac{(12 - 5)^2 + 2^3}{10 \div 2 - (11 - 9)}$

45. $[485 - (68 + 39)] + 4^5 - 24 \cdot 16 \div 8$

46. $(25 - 9)^3 + 420 \div [3 \cdot 5 + (19 - 13)]$

For Exercises 47–50, explain the mistake and work the problem correctly.

47. $48 - 6(9 - 4)$

$= 48 - 6(5)$

$= 42(5)$

$= 210$

48. $24 \div 6 \cdot 2 + 5$

$= 24 \div 12 + 5$

$= 2 + 5$

$= 7$

49. $(3 + 5)^2 - 2 \cdot 7$

$= 9 + 25 - 2 \cdot 7$

$= 9 + 25 - 14$

$= 34 - 14$

$= 20$

50. $[12 - 2 \cdot 3] + 4(3)^2$

$= [12 - 6] + 4(3)^2$

$= 6 + 4(3)^2$

$= 6 + 144$

$= 150$

Objective 2

Prep Exercise 5 A statistic is a number used to _____ of a set of data.

Prep Exercise 6 Explain how to find the mean of a list of values.

Prep Exercise 7 If a list contains an even number of values, after it is arranged in order from least to greatest, explain how to find the median.

Prep Exercise 8 What is the mode of a set of data?

For Exercises 51–58, find the mean, median, and mode.

51. 5, 7, 3, 13, 9, 6, 13   **52.** 9, 6, 15, 4, 12, 18, 6   **53.** 10, 4, 7, 17, 9, 14, 4, 8, 17   **54.** 18, 5, 8, 22, 16, 8, 15, 22, 12

55. 22, 35, 45, 46   **56.** 81, 83, 91, 97   **57.** 83, 37, 68, 54, 83, 47   **58.** 89, 19, 98, 19, 108, 87

For Exercises 59 and 60, use the following table, which shows the height and weight of the starting offense of a college football team.

Player	Height (in inches)	Weight (in pounds)	Player	Height (in inches)	Weight (in pounds)
M. Adams	72	210	S. Roberts	71	185
J. Berry	74	285	T. Roberts	73	210
K. Evans	73	200	R. Smith	75	300
T. Graves	74	205	J. Stephens	74	245
C. Johnson	74	207	T. Williams	78	310
B. Morris	76	305			

59. Find the mean, median, and mode height.

60. Find the mean, median, and mode weight.

61. The following table shows the ACT scores of a group of students applying to a college. Find the mean, median, and mode score.

Student	ACT score	Student	ACT score
Barnes, K.	31	Johnson, E.	28
Compton, S.	28	Milton, C.	25
Donaldson, R.	34	Nelson, D.	24
Garcia, J.	29	Robinson, W.	21
Hamilton, S.	25	Stevenson, L.	25
James, T.	21	Wilson, B.	21

62. The following table contains the test grades for the students in a chemistry class. Find the mean, median, and mode.

Student	Score	Student	Score
Allen, G.	84	Harris, A.	76
Baker, S.	75	Little, M.	71
Barren, T.	64	Moore, S.	93
Carter, J.	93	Payne, J.	60
Donaldson, R.	82	Reiche, D.	78
Green, R.	60	Rizvi, Z.	100

63. The following bar graph shows the total compensation of the 10 highest-paid CEOs of 2009 in the United States. Find the mean, median, and mode of their salaries.

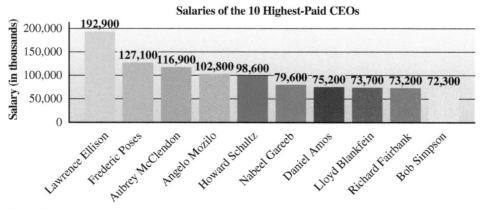

Source: Forbes 500.

64. The following bar graph shows the gross earnings in North America of the 10 top-grossing tours in 2009. Find the mean, median, and mode of their gross income.

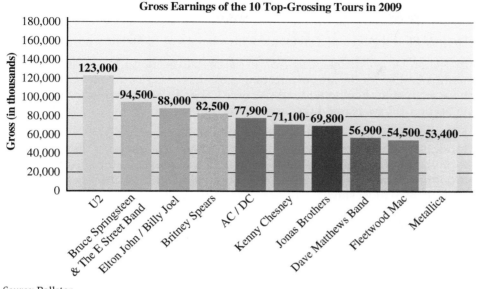

Gross Earnings of the 10 Top-Grossing Tours in 2009

Source: Pollstar.

Review Exercises

[1.2] 1. Estimate $42{,}320 + 25{,}015$ by rounding to the nearest ten thousand.

For Exercises 2 and 3, perform the indicated operation.

[1.3] 2. $498{,}503 \cdot 209$ **[1.4] 3.** $21{,}253 \div 17$

[1.3] 4. A 12-foot-wide by 16-foot-long room is to have crown molding installed. How much molding is needed? If the carpenter charges \$4 per foot to install the molding, how much will it cost?

[1.3] 5. The floor of the same 12-foot-wide by 16-foot-long room is to be covered with square 1 foot by 1 foot tiles. How many tiles must be purchased?

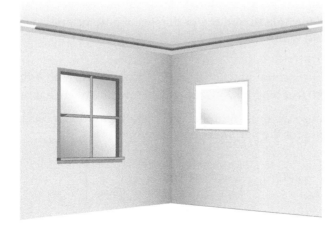

1.6 More with Formulas

Objectives

1 Use the formula $P = 2l + 2w$ to find the perimeter of a rectangle.

2 Use the formula $A = bh$ to find the area of a parallelogram.

3 Use the formula $V = lwh$ to find the volume of a box.

4 Solve for an unknown number in a formula.

5 Use a problem-solving process to solve problems requiring more than one formula.

Warm-up

[1.5] 1. $(8 - 3)^2 + 30 \div 3 \cdot 5$

[1.3] 2. Find the area of a rectangle with a length of 16 feet and a width of 9 feet.

In Section 1.2, we defined *variable* and *constant*. In Section 1.3, we used variables to write a *formula* for the area of a rectangle. We have also learned how to solve equations with an unknown addend (Section 1.2) or an unknown factor (Section 1.4). In Section 1.5, we learned the order of operations.

In this section, we will bundle all of those topics together. We will derive more formulas and use them to solve problems. When we are solving the problems, our formulas may contain numerical expressions that must be simplified using proper order of operations or may lead to unknown addend or unknown factor equations that we must solve.

Objective 1 Use the formula $P = 2l + 2w$ to find the perimeter of a rectangle.

First, let's derive a formula for the perimeter of a rectangle. Look at the rectangle below. We use the variable l to represent length and w to represent width. In Section 1.2, we learned that to find perimeter, we add the lengths of all the sides, which suggests the following formula:

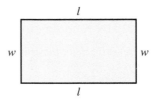

$$\text{Perimeter of a rectangle} = \text{length} + \text{width} + \text{length} + \text{width}$$
$$P = \quad l \quad + \quad w \quad + \quad l \quad + \quad w$$

Because two lengths and two widths are added, we can multiply both the length and the width by 2 and then add.

$$P = 2l + 2w$$

We can use the formula to find the perimeter of any rectangle when given values for its length and width.

▶ Your Turn 1

Solve.

a. A rectangle has a length of 42 meters and a width of 19 meters. Find the perimeter.

b. A rectangle has a length of 7 miles and a width of 6 miles. Find the perimeter.

Answers to Your Turn 1
a. 122 m   b. 26 mi.

Answers to Warm-up
1. 75
2. 144 ft.²

| **Example 1** | A school practice field is to be enclosed with a chain-link fence. |

Example 1   A school practice field is to be enclosed with a chain-link fence. The field is 400 feet long by 280 feet wide. How much fence is needed?

Solution: Because we must find the total distance around a rectangular field, we use the formula for the perimeter of a rectangle.

$P = 2l + 2w$
$P = 2(400) + 2(280)$ **Replace *l* with 400 and *w* with 280.**
$P = 800 + 560$ **Multiply.**
$P = 1360$ **Add.**

Answer: 1360 ft. of fence is needed.

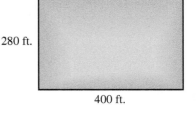

280 ft.

400 ft.

◀ **Do Your Turn 1**

Objective 2 Use the formula $A = bh$ to find the area of a parallelogram.

In Section 1.3, we discussed that area is measured in square units and developed a formula for calculating the area of a rectangle, $A = lw$. Rectangles are special forms of a more general class of figures called **parallelograms,** which have two pairs of **parallel** sides.

Definitions **Parallel lines:** Straight lines lying in the same plane that do not intersect.

Parallelogram: A four-sided figure with two pairs of parallel sides.

Examples of parallelograms:

To find the area of a parallelogram, we need the measure of the base and the height. The height is measured along a line that makes a 90° angle with the base. An angle that measures 90° is called a **right angle.**

Definition **Right angle:** An angle that measures 90°.

We will use the letter b for the measure of the base and the letter h for the height.

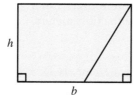

Note The right angle is indicated by a small
◄ square where the height line meets the base.

Notice that if we cut along the height line, we cut off a triangle. If we slide that triangle around to the right side and place the parallel sides together, the resulting figure is a rectangle.

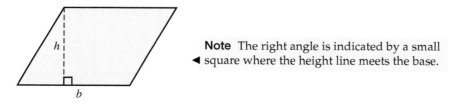

The rectangle we have made has the same area as the original parallelogram. The length of this rectangle is b, and the width is h. So the formula to calculate the area of a parallelogram is $A = bh$.

Note Because squares and rectangles are special parallelograms, we
◄ can use $A = bh$ for them as well.

▶ **Your Turn 2**

Find the area.

a.

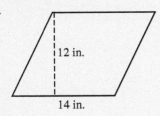

12 in.

14 in.

b.

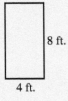

8 ft.

4 ft.

Example 2   Find the area of the parallelogram.

Solution: Use the formula for the area of a parallelogram, $A = bh$.

$A = bh$

$A = (15)(9)$ In $A = bh$, replace b with 15 and h with 9.

$A = 135$ Multiply. Because the base and height are measured in feet, the area unit is square feet.

9 ft.

15 ft.

Answer: The area is 135 ft.^{2}

◀ **Do Your Turn 2**

Objective 3 Use the formula $V = lwh$ to find the volume of a box.

Now let's consider **volume,** which is a measure of the amount of space inside a three-dimensional object. We measure volume using **cubic units.**

Definitions Cubic unit: A $1 \times 1 \times 1$ cube.

Volume: The total num ber of cubic units that completely fill an object.

Examples of cubic units:

Cubic centimeter

1 cm

1 cm

1 cm

A cubic centimeter is a 1 cm by 1 cm by 1 cm cube and is represented mathematically by cm^3.

Cubic inch

1 in.

1 in.

1 in.

A cubic inch is a 1 in. by 1 in. by 1 in. cube and is represented mathematically by in.^{3}.

Of Interest

In mathematics, we write cubic centimeters as cm^3. Another com mon abbreviation is cc, which is used in medical dosage or in a description of engine size.

Let's develop a formula for calculating the vol ume of a box. Consider a box that is 3 meters long by 2 meters wide by 4 meters high. Let's *measure* its volume by filling it with cubic meters.

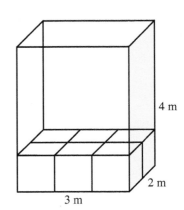

4 m

2 m

3 m

Note Six cubic meters form a layer at the bottom. Because each layer is 1 meter high, four of those layers fill the box so that it holds a total of 24 cubic meters.

▶

Answers to Your Turn 2

a. 168 in.^{2}

b. 32 ft.^{2}

Notice that we can *calculate* the volume by multiplying the length, width, and height.

Volume of a box = length · width · height
Volume = 3 m · 2 m · 4 m = 24 m^3

So the formula for calculating the volume of a box is $V = lwh$.

Note The unit m^3 is read *cubic meters* or *meters cubed*. All volume calculations involve multiplying three distance measurements. Therefore, volume units are always a distance unit with an exponent of 3.

Connection When a number is raised to the third power, we say that we are *cubing* that number, or the number is *cubed*. The result is the *cube* of the number.

Example 3 A computer tower is 30 centimeters long by 10 centimeters wide by 35 centimeters high. What is the volume of the tower?

Solution: The tower is a box; so use the formula for the volume of a box, $V = lwh$.

$V = lwh$
$V = (30)(10)(35)$ In $V = lwh$, replace l with 30, w with 10, and h with 35.
$V = 10{,}500$ Multiply. Because the length, width, and height are in centimeters, the volume is in cubic centimeters.

Answer: The volume is 10,500 cm^3.

▶ **Do Your Turn 3**

▶ **Your Turn 3**

Find the volume.

a.
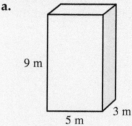
9 m   5 m   3 m

b.
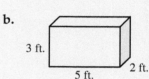
3 ft.   5 ft.   2 ft.

Objective 4 Solve for an unknown number in a formula.

Sometimes when we substitute numbers into a formula, the resulting equation contains an unknown factor.

Example 4 The area of a parallelogram is 40 square feet. If the base is 8 feet, find the height.

Solution: Use the formula $A = bh$.

$A = bh$
$40 = 8h$ Replace A with 40 and b with 8.
$40 \div 8 = h$ Write a related division statement.
$5 = h$ Divide 40 by 8.

Answer: The height is 5 ft.

Check: Verify that a parallelogram with a base of 8 feet and height of 5 feet has an area of 40 square feet.

$A = bh$
$A = (8)(5)$ Replace b with 8 and h with 5.
$A = 40$ Multiply. The height is correct.

▶ **Do Your Turn 4**

▶ **Your Turn 4**

Solve.

a. The area of a parallelogram is 72 square inches. If the height is 9 inches, find the base.

b. The area of a rectangle is 60 square meters. If the width is 5 meters, find the length.

Connection $40 = 8h$ is an unknown factor equation. In Section 1.4, we learned to solve unknown factor equations by dividing the product by the known factor.

Answers to Your Turn 3
a. 135 m^3 **b.** 30 ft.^{3}

Answers to Your Turn 4
a. 8 in.   **b.** 12 m

▶ **Your Turn 5**

Solve.

a. The volume of a box is 336 cubic centimeters. If the length is 12 centimeters and the width is 7 centimeters, find the height.

b. The volume of a box is 810 cubic inches. If the width is 9 inches and the height is 5 inches, find the length.

Example 5 The volume of a box is 480 cubic centimeters. If the length is 10 centimeters and the width is 8 centimeters, find the height.

Solution: Use the formula $V = lwh$.

$$V = lwh$$
$$480 = (10)(8)h \qquad \text{Replace } V \text{ with 480, } l \text{ with 10, and } w \text{ with 8.}$$
$$480 = 80h \qquad \text{Multiply 8 by 10.}$$
$$480 \div 80 = h \qquad \text{Write a related division statement.}$$
$$6 = h \qquad \text{Divide 480 by 80.}$$

Answer: The height is 6 cm.

Check: Verify that a box with a length of 10 centimeters, width of 8 centimeters, and height of 6 centimeters has a volume of 480 cubic centimeters.

$$V = (10)(8)(6) \qquad \text{Replace } l \text{ with 10, } w \text{ with 8, and } h \text{ with 6.}$$
$$V = 480 \qquad \text{Multiply. The height is correct.}$$

◀ **Do Your Turn 5**

Objective 5 Use a problem-solving process to solve problems requiring more than one formula.

As the problems we solve get more complicated, it is helpful to follow a problem-solving outline. The purpose of the outline is to suggest strategies for solving problems and to offer insight into the process of problem solving. Look for the outline in some of the more complicated examples.

Problem-Solving Outline

1. **Understand** the problem.
   a. Read the question(s), underline key words, and note what you are to find.
   b. If useful, draw a picture, make a list or table, simulate the situation, or search for a related example problem.
2. **Plan** your solution strategy. Search for a formula or translate key words to an equation.
3. **Execute** the plan. Solve the formula or equation.
4. **Answer** the question. Include appropriate units.
5. **Check** the results.
   a. Make sure the answer is reasonable for the situation.
   b. Check calculations by solving the problem a different way, reversing the process, or estimating the answer and verifying that your estimate and actual answer are reasonably close.

 **Learning Strategy**

If you are a visual learner, focus on visual strategies such as drawing pictures and making lists. If you are an auditory learner, try talking through the situation, expressing your ideas verbally. If you are a tactile learner, try to simulate or model the situation using pencil, paper, or any item nearby.

Answers to Your Turn 5
a. 4 cm
b. 18 in.

Example 6 Crown molding is to be installed in a room that is 20 feet long by 15 feet wide. The molding costs $6 per foot to install. How much will it cost to install the molding?

Understand We must find the total cost of installing crown molding around a 20-foot by 15-foot room. A picture of the rectangular room is shown below.

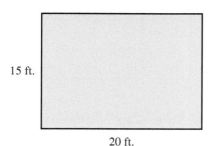

15 ft.

20 ft.

We are told it costs $6 per foot; so the total cost will be the length of molding required multiplied by the price per foot. Crown molding is installed on each wall flush with the ceiling of the room; so the total length required will equal the perimeter of the room.

Plan Find the perimeter of the room, which will equal the length of molding required; then multiply that length by $6 per foot.

Execute To find the perimeter of the room, we use $P = 2l + 2w$, replacing l with 20 and w with 15.

$$P = 2(20) + 2(15)$$ Replace *l* with 20 and *w* with 15.
$$P = 40 + 30$$ Multiply.
$$P = 70$$ Add.

So 70 feet of molding is required. To find the total cost, we multiply the 70 feet by the price per foot, $6.

$$\text{Cost} = (6)(70) = 420$$

Answer It will cost $420 to install crown molding in the room.

Check To check, we can reverse the process. We can find the length of molding that could be purchased with $420 by dividing 420 by 6.

$$\text{Length purchased} = 420 \div 6 = 70$$

So 70 feet of molding could be purchased with $420. Two 20-foot segments and two 15-foot segments equal 70 feet; so our calculations are accurate.

▶ **Do Your Turn 6**

Now let's solve problems involving *composite shapes*. Composite shapes are shapes formed by putting together two or more fundamental shapes such as rectangles or parallelograms. A composite shape also can be a shape within a shape.

▶ **Your Turn 6**

A fence is to be built around a yard that is 60 feet long by 80 feet wide. A contractor will charge $35 per foot to install the fence. How much will it cost to install the fence?

Answer to Your Turn 6
$9800

Example 7 Shown here is a base plan for a new building. What is the total area occupied by the building?

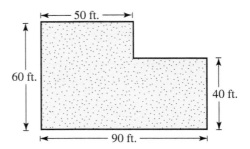

Understand We must find the total area occupied by the building. There are several ways we can determine the area.

We could draw dashed lines to see the area as two smaller rectangles combined as in Figures A and B.

Figure A   **Figure B**

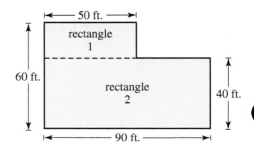

Or, we could imagine a smaller rectangle removed from the corner of a larger rectangle as in Figure C.

Figure C

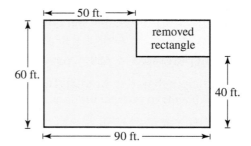

Visualize the area in the way that is easiest for you. We will use Figure A.

Plan Using Figure A, add the areas of rectangle 1 and rectangle 2. To find the area of each rectangle, use the formula $A = lw$ (or $A = bh$).

Execute First, we find the area of rectangle 1, which is 50 feet by 60 feet.

$$\text{Area of rectangle 1} = (50)(60) = 3000 \text{ ft.}^2$$

Now we need the area of rectangle 2, but its length was not shown on the original figure. Using the other dimensions we find the length to be 40 feet, which means rectangle 2 is actually a 40-foot by 40-foot square.

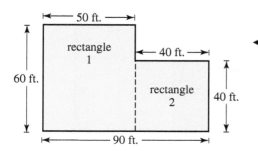

Note Because the bottom is 90 feet across, the top must also be 90 feet across. So the length of rectangle 2 must be $90 - 50 = 40$ feet.

Now we can find its area.

$$\text{Area of rectangle } 2 = (40)(40) = 1600 \text{ ft.}^2$$

The total area occupied by the building is the sum of the two areas.

$$\text{Total area} = 3000 + 1600 = 4600 \text{ ft.}^2$$

Answer The total area is 4600 ft.^{2} ◀ **Note** Make sure you include the units.

Check We will solve the problem again using Figure C from the previous page.

Figure C

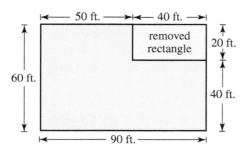

Because the smaller rectangle is removed from the larger one, we subtract its area from the larger rectangle's area to find the area occupied by the building.

$$\text{Area of building} = \text{area of larger rectangle} - \text{area of removed rectangle}$$

The large rectangle is 90 feet by 60 feet.

$$\text{Area of large rectangle} = (90)(60) = 5400 \text{ ft.}^2$$
$$\text{Area of removed rectangle} = (40)(20) = 800 \text{ ft.}^2$$
$$\text{Area of the building} = 5400 - 800 = 4600 \text{ ft.}^2$$

Because we get the same answer, we can feel secure that our calculation for the area occupied by the building is correct.

▶ **Do Your Turn 7**

Example 8   A builder is required to sod the yard of a new home. The lot is a rectangle 75 feet wide by 125 feet long. The house occupies 2500 square feet of the lot. Sod must be purchased in pallets of 500 square feet that cost $85 each. How much will it cost to sod the yard?

Understand We are to find the total cost to sod a yard. Let's underline key words and important information.

A builder is required to sod the yard of a new home. The lot is a <u>rectangle 75 feet wide by 125 feet long</u>. The house <u>occupies 2500 square feet of the lot</u>. Sod must be purchased in <u>pallets of 500 square feet</u> that cost <u>$85 each</u>. How much will it cost to sod the yard?

Because the problem involves rectangles, let's draw a picture.

▶ **Your Turn 7**

Solve.

a. The picture shows two bedrooms and a connecting hallway that are to be carpeted. How many square feet of carpet will be needed?

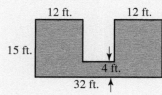

b. The picture shows the driveway and walk of a new house. The builder needs to know how much concrete to plan for. What is the total area of the driveway and walk?

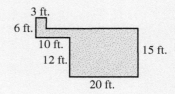

Answers to Your Turn 7
a. 392 ft.^{2} **b.** 339 ft.^{2}

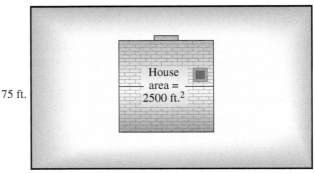

75 ft.

125 ft.

To develop a plan, we ask ourselves questions starting with the question in the problem.

What do we need to know to find the cost of the sod?

Answer: If each pallet costs $85, we need to know how many pallets will be used.

How can we figure the number of pallets required?

Answer: If each pallet covers 500 ft.^{2}, we need to know how many square feet of the lot will be covered with sod.

How can we figure the number of square feet to be covered?

Answer: Because the whole lot, except for the house, is to be covered, we need to subtract the area occupied by the house from the total area of the lot.

We were given the area occupied by the house but not the area of the lot. How can we find the area of the lot?

Answer: Because the lot is a rectangle and the area of a rectangle is found using the formula $A = lw$, we need to multiply the length and width of the lot.

Plan Now we have our starting point. We'll find the area of the lot, then subtract the area occupied by the house, which will give us the area to be covered with sod. From there, we can calculate how many pallets will be needed. Once we have the number of pallets, we can determine the cost.

Execute Calculate the area of the lot:

$$A = lw$$
$$A = (125)(75)$$
$$A = 9375 \text{ ft.}^2$$

Now we can calculate how much area will be covered with sod by subtracting the area occupied by the house.

$$\text{Area to sod} = 9375 - 2500 = 6875 \text{ ft.}^2$$

Next, we need to determine how many pallets will be needed to cover 6875 square feet. Because each pallet covers 500 square feet, we need to divide 500 into 6875 to see how many sets of 500 are in 6875.

$$\text{Number of pallets} = 6875 \div 500 = 13.75$$

This means that 13 whole pallets are needed with 375 square feet uncovered by sod. Because partial pallets cannot be purchased, the builder must buy one more pallet to cover that 375 square feet even though not all of the sod will be used from the extra pallet. So a total of 14 pallets must be purchased.

Because each of the 14 pallets costs $85, we multiply to find the total cost.

$$\text{Total cost} = (14)(85) = \$1190$$

Answer The total cost to sod the yard is $1190.

AUDITORY
Learning Strategy

This type of question-and-answer verbal approach is an excellent strategy for auditory learners.

Check Let's do a quick estimate of the area. We'll round the width to 80 feet and the length to 130 feet and calculate the area.

$$\text{Estimate} = (130)(80) = 10{,}400 \text{ ft.}^2$$

The house occupies 2500 square feet; so let's round that to 3000 square feet. We can now estimate the amount of land to be covered in sod by subtracting the area of the house from our estimate for the whole lot.

$$\text{Approximate area of coverage} = 10{,}400 - 3000 = 7400 \text{ ft.}^2$$

If each pallet covers 500 square feet, about 15 pallets will be needed to cover 7400 square feet.

If we round the cost of each pallet to about $90, the total cost is $90 · 15 pallets = $1350. So we should expect the cost to be approximately $1350. Our actual calculation of $1190 is reasonable given the amount of rounding we did in our estimates.

▶ **Do Your Turn 8**

Example 9   Approximate the volume of the automobile shown.

Understand *Volume* indicates that we are to find the total number of cubic units that completely fill the car. Because we need only an approximation, we can view the automobile as two boxes stacked as shown. The formula for the volume of a box is $V = lwh$.

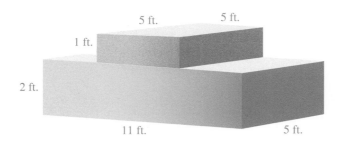

Plan To find the total volume, we add the volumes of the two boxes.

Execute Total volume = volume of bottom box + volume of top box

$$
\begin{aligned}
V &= \quad l \cdot w \cdot h \quad + \quad l \cdot w \cdot h \\
V &= \quad (11)(5)(2) \quad + \quad (5)(5)(1) \\
V &= \quad 110 \qquad\qquad + \quad 25 \\
V &= \quad 135
\end{aligned}
$$

Answer The volume of the car is about 135 ft.^{3}.

Check Let's reverse the process. We start with 135 cubic feet and see if we can work our way back to the original dimensions. The last step in the preceding process was $110 + 25 = 135$; so we can check by subtracting.

$$135 - 25 = 110 \quad \text{This is true; so our addition checks.}$$

▶ **Your Turn 8**

Solve.

a. A house occupies 2000 square feet on a 9000-square-foot lot. The lot is to be landscaped with 1800 square feet of flower beds and the remaining area covered with grass seed. Each bag of grass seed covers 1000 square feet and costs $18. How much will it cost to seed the lot?

b. An 8-foot-long by 4-foot-wide entryway in a new house is to have a tile landing and the rest of the floor covered in hardwood. The tile landing is to be 4 feet long by 2 feet wide. How many square feet of hardwood will be needed? The cost of the tile is $7 per square foot, and the cost of the hardwood is $13 per square foot. What will be the total cost of the floor in the entryway?

Connection In Example 7, to find the area made of two combined rectangles, we added the areas of the rectangles. Similarly, to find the volume of two combined boxes, we add the volumes of the two boxes.

Answers to Your Turn 8
a. $108 (must purchase 6 bags of seed)
b. 24 ft.^{2}; $368

▶ Your Turn 9

Engineers are designing a clean room for a computer chip manufacturer. The room is to be L-shaped (see diagram). In designing the air recycling and filter system, the engineers need to know the volume of the room. Find the volume of the room if the whole room has a 10-foot ceiling.

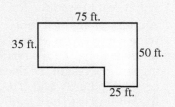

Answer to Your Turn 9
30,000 ft.^{3}

Now let's check 110 cubic feet. Because we multiplied (11)(5)(2), we can check by dividing 110 ÷ 2 ÷ 5 to see if we get 11.

$$110 \div 2 \div 5$$
$$= 55 \div 5$$
$$= 11 \qquad \text{It checks.}$$

Now check 25 cubic feet. Because we multiplied (5)(5)(1) to get 25, we can check by dividing 25 ÷ 1 ÷ 5 to see if we get 5.

$$25 \div 1 \div 5$$
$$= 25 \div 5$$
$$= 5 \qquad \text{It checks.}$$

◀ **Do Your Turn 9**

Examples 8 and 9 suggest the following procedure.

Procedure

To find the area or volume of a composite figure:
1. Identify the shapes or objects that make up the figure.
2. If those shapes or objects combine to form the composite figure, add their areas or volumes. If one shape or object is inside another, subtract the area/volume of the inner shape/object from the area/volume of the outer shape/object.

1.6 Exercises (For Extra Help) MyMathLab®

Objective 1

Prep Exercise 1 Explain how to use a formula.

Prep Exercise 2 What is the formula for finding the perimeter of a rectangle?

For Exercises 1–6, find the perimeter.

1.
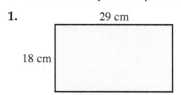
29 cm, 18 cm

2.
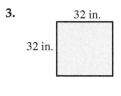
10 ft., 15 ft.

3.
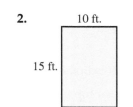
32 in., 32 in.

4.
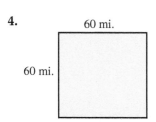
60 mi., 60 mi.

5.

24 km, 8 km

6.
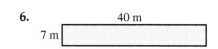
40 m, 7 m

Objective 2

Prep Exercise 3 What is a parallelogram?

Prep Exercise 4 What is the formula for finding the area of a parallelogram?

For Exercises 7–12, find the area.

7.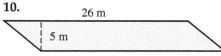
19 m
12 m

8.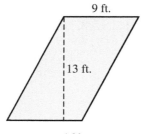
9 ft.
13 ft.

9.
12 in.
12 in.

10.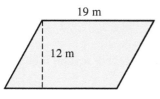
26 m
5 m

11.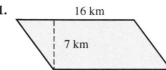
16 km
7 km

12.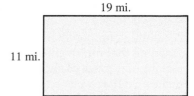
19 mi.
11 mi.

Objective 3

Prep Exercise 5 What is volume?

Prep Exercise 6 What is the formula for finding the volume of a box?

For Exercises 13–18, find the volume.

13.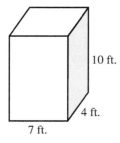
10 ft.
4 ft.
7 ft.

14.
21 m
9 m
6 m

15.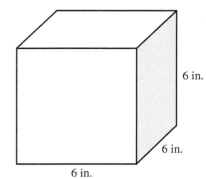
6 in.
6 in.
6 in.

16.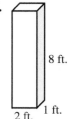
8 ft.
1 ft.
2 ft.

17.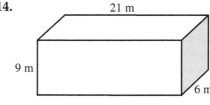
14 km
5 km
3 km

18.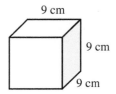
9 cm
9 cm
9 cm

Objectives 4 and 5

Prep Exercise 7 According to the problem-solving outline, what is the first step in solving a problem?

Prep Exercise 8 What is the last step in the problem-solving process?

Prep Exercise 9 How would you calculate the area of a shape that is a combination of two smaller shapes?

Prep Exercise 10 How would you calculate the area of a shape that is formed by cutting away a portion of a larger shape?

For Exercises 19–50, solve.

19. A parallelogram has an area of 144 square meters. If the base is 16 meters, find the height.

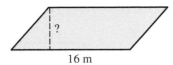

20. A parallelogram has an area of 2408 square centimeters. If the base is 56 centimeters, find the height.

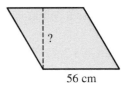

21. A parallelogram has an area of 1922 square inches. If the height is 31 inches, find the base.

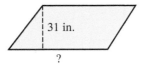

22. A parallelogram has an area of 399 square feet. If the height is 19 feet, find the base.

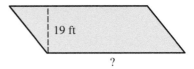

23. A canning company feeds sheets of metal into a machine that molds them into cans. The area of a rectangular piece of sheet metal must be 180 square inches. So that a piece of metal fits into the machine, the width of the sheet of metal must be 9 inches. What must the length be?

24. A search-and-rescue helicopter is assigned to search a 3584-square-mile rectangular region over the Atlantic Ocean. After flying north for 64 miles along the western border of the region, the helicopter turns east (90° turn) to fly along the northern border. How far will the helicopter have to fly on that line to complete the search along the northern border?

25. A cell phone has a volume of 90 cubic centimeters. If the length is 9 centimeters and the width is 5 centimeters, find the height.

26. A digital camera has a volume of 108 cubic centimeters. If the width is 9 centimeters and the height is 6 centimeters, find the length.

27. A box has a volume of 22,320 cubic inches. If the length is 31 inches and the height is 40 inches, find the width.

28. A box has a volume of 30,600 cubic centimeters. If the height is 60 centimeters and the length is 15 centimeters, find the width.

29. An engineer is designing a refrigerator. The desired volume is to be 24 cubic feet. If the length must be 3 feet and width must be 2 feet, what must the height be?

30. A bus is a box with a volume of 2520 cubic feet. If the bus is 35 feet long by 8 feet wide, what is the height of the bus?

31. The Antonov AN 124 is the world's largest commercial cargo aircraft. Its cargo compartment is approximately box-shaped and measures 120 feet long and 14 feet tall. If the approximate volume of the cargo compartment is 35,280 cubic feet, what is the width of the compartment?

32. The volume of the Vehicle Assembly Building where NASA constructs its rockets and shuttles is 5,511,040 cubic meters. If the building is 218 meters long and 160 meters tall, what is its width?

Of Interest

The NASA assembling building is actually a large shell. Because the open space inside is so large, clouds could form in the upper part of the building. NASA had to design a humidity/temperature-control system to keep storms from developing inside the building.

33. Crown molding is to be installed in a 16-foot-wide by 20-foot-long living room. The subcontractor charges $4 per foot to install the molding. What will be the total cost of installing molding for the room?

34. Angela plans to put a wallpaper border just below the ceiling in her bedroom. The room is 12 feet wide by 14 feet long. If the border costs $2 per foot, what will be the total cost?

35. Baseboard molding is to be placed where the walls connect to the floor in the room shown below. An 8-foot section of baseboard molding costs $4. If no partial sections are sold, how much will the total baseboard cost?

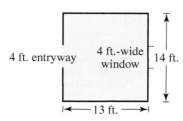

4 ft. entryway   4 ft.-wide window   14 ft.

13 ft.

36. Chair railing is to be placed on the walls around the room mentioned in Exercise 35. The railing will be at a height of 3 feet from the floor. If the railing costs $5 for an 8-foot section and no partial sections are sold, how much will the chair railing cost for the room?

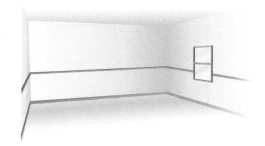

37. Mr. Williams wants to fence a large pasture. The plot plan for the land is shown here. A company charges $5 per foot for the fence and $75 for the 10-foot gate. What will be the total cost?

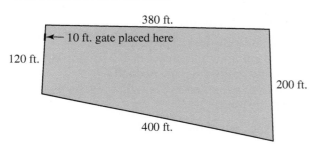

380 ft.
10 ft. gate placed here
120 ft.
200 ft.
400 ft.

38. The Henry family wants to put a fence around their backyard. The plan is shown here. A company charges $8 per foot for the wood picket fence and $50 for each 4-foot-wide gate. What will be the total cost?

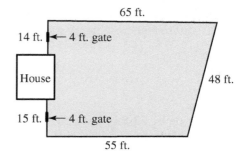

65 ft.
14 ft. 4 ft. gate
House
48 ft.
15 ft. 4 ft. gate
55 ft.

39. Mr. Williams from Exercise 37 decides to investigate the cost of building the fence himself.

  a. If he places wood posts every 8 feet (except where the gate goes), how many posts must he purchase? The posts cost $2 each. What will be the cost for all of the posts?

  b. He will nail four rows of barbed wire along the posts (except where the gate goes). How many feet of barbed wire must he have in all? A roll of barbed wire contains 1320 feet of wire and costs $45. How many rolls will he need? What will be the cost of barbed wire?

  c. Because he'll be nailing four rows of wire to each post, he'll need 4 U-nails for each post. How many nails will he need? If the nails come in boxes of 100, how many boxes will he need? If the boxes cost $3 each, how much will the nails cost?

  d. The gate costs $75. What will be the total cost of materials?

  e. Compare the total cost of materials for Mr. Williams to build the fence himself with the total cost of having the fence company build it. How much does Mr. Williams save if he builds the fence himself? Before committing to build the fence himself, what factors should he consider?

40. Mr. Henry from Exercise 38 decides to investigate the cost of building the fence himself. He decides that 8-foot prefabricated panels are the least expensive option. Each 8-foot panel costs $25. Posts are placed at each end of each section and at each end of each gate. The posts cost $3 each. Rental of an auger to dig the post holes costs $50. The gates cost $45 each. What will be the total cost? (*Hint:* Draw the plan for post placement.)

41. The floor of a house is to be made from 4-foot-wide by 8-foot-long sheets of plywood. The floor plan is shown here. How many sheets of plywood will be needed? If each sheet of wood costs $15, what will be the total cost?

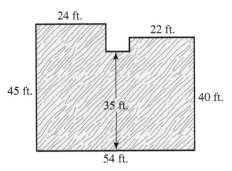

42. The area shown is to be carpeted. How many square feet of carpet will be needed? If the chosen carpet costs $3 for each square foot, how much will it cost to carpet the area?

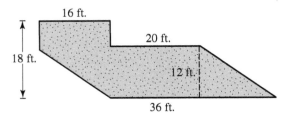

43. The walls of a 24-foot-long by 15-foot-wide room with a 9-foot ceiling are to be painted. The room has a 6-foot-wide by 7-foot-high entryway and two windows each measuring 3 feet high by 2 feet wide. What is the area to be painted? If each gallon of paint covers 400 square feet, how many gallons will be needed?

44. The walls in nine offices in an office building are to be painted. Each office is 10 feet long by 9 feet wide and has an 8-foot ceiling. Each office has a 3-foot-wide by 7-foot-high doorway and a 3-foot-high by 2-foot-wide window.

   a. What is the area to be painted?
   b. If a gallon of paint covers 400 square feet, how many gallons will be needed?
   c. If each gallon costs $14, what will be the total cost of the paint?

45. An 84-foot-wide by 72-foot-long building is on a lot that is 245 feet long by 170 feet wide. The entire area surrounding the building is to be paved for parking. Calculate the area that will be paved.

46. Find the area of the metal (shaded) in the plate shown. The hole in the center is an 8-centimeter by 8-centimeter square.

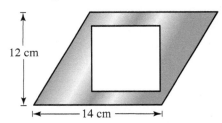

47. A large building has two wings as shown below. What is the total volume of the building?

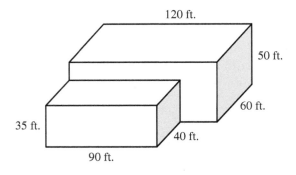

48. Ziggurats are terraced pyramids. Find the volume of the ziggurat shown if the bottom level is 297 feet wide by 297 feet long by 12 feet high, the middle level is 260 feet wide by 260 feet wide by 10 feet high, and the top level is 145 feet wide by 145 feet wide by 9 feet high.

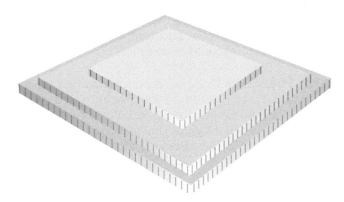

Of Interest

The ancient Babylonians and Assyrians constructed ziggurats as temples. About 2500 years ago, the Babylonian king Nebuchadnezzar built a large ziggurat in the city of Borsippa (75 miles south of modern Baghdad). The massive ziggurat was built using millions of fired mud bricks that were 13 inches by 13 inches by 3 inches each. The ziggurat had seven terraces and was 231 feet tall.

49. The chair shown is constructed from thick foam padding. What volume of padding is needed to form the chair?

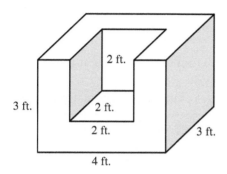

50. Concrete will be poured to form a swimming pool that measures 40 feet long by 30 feet wide by 4 feet deep. If the concrete is 1 foot thick, what volume of concrete is needed to form the pool?

Puzzle Problem An *equilateral triangle* is a triangle with all three sides equal in length. Suppose we construct an equilateral triangle out of three pencils that are of equal length.

We can construct larger equilateral triangles out of the smaller version as shown below.

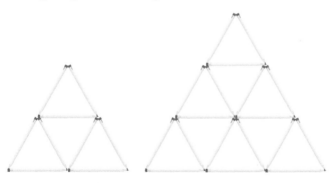

How many pencils would be needed to construct an equilateral triangle like those shown that has 15 pencils along each side?

Review Exercises

[1.2] **1.** Solve and check. $x + 175 = 2104$

[1.3] **2.** Write 2,408,073 in expanded form using powers of 10.

[1.3] **3.** Estimate $452 \cdot 71{,}203$. Then calculate the actual product.

[1.4] **4.** Solve and check. $17 \cdot y = 3451$

[1.5] **5.** Simplify. $5^3 - \left[(9 + 3)^2 - 2^6 \right] + 24 \div 6 \cdot 2$

Chapter 1 Summary and Review Exercises

TACTILE VISUAL AUDITORY **Learning Strategy**

Your study sheet (described in the *To the Student* section of the text) should resemble the summaries at the end of each chapter. To see how well you know the rules and procedures, try duplicating your study sheet from memory. This is sometimes called a *memory dump*.

1.1 Introduction to Numbers, Notation, and Rounding

Definitions/Rules/Procedures	Key Example(s)
A **set** is _____. A **subset** is _____. The **natural numbers** are _____. The **whole numbers** are _____. Complete the following table of place values.	Every natural number is a whole number. Therefore, as the following diagram illustrates, the set of natural numbers is a subset of the set of whole numbers. **Whole Numbers:** 0, 1, 2, 3, . . . **Natural Numbers:** 1, 2, 3, . . .

Place Values

Millions period			Thousands period			Ones period		
Hundred millions				Ten thousands			Hundreds	
								Ones

Definitions/Rules/Procedures	Key Example(s)
To write a number in expanded form, write each _____ followed by its _____ until you reach the last place, which is not followed by a plus sign. **To convert a number from expanded form to standard form,** _____ _____. If the expanded form is missing a place value, insert a 0 for that place value in the standard form.	Write 70,916 in expanded form. **Answer:** 7 ten thousands + 9 hundreds + 1 ten + 6 ones Write 2 millions + 8 ten thousands + 5 thousands + 4 tens + 7 ones in standard form. **Answer:** 2,085,047

[1.1] For Exercises 1 and 2, write the number in expanded form.

1. 5,680,901

2. 42,519

[1.1] For Exercises 3 and 4, write the number in standard form.

3. 9 ten thousands + 8 thousands + 2 hundreds + 7 tens + 4 ones

4. 8 millions + 2 ten thousands + 9 tens + 6 ones

Definitions/Rules/Procedures	Key Example(s)
To write the word name for a whole number, work from left to right through the periods. 1. Write the _____. 2. Write the period name followed by a comma. 3. Repeat steps 1 and 2 for each period except the ones period. For the ones period, write only the word name formed by the digits. Do not follow the ones period with its name. Warning Do not write the word *and* in the word name for a whole number. As we will see in Chapter 6, the word *and* takes the place of a decimal point.	Write the word name for 14,907,156. **Answer:** fourteen million, nine hundred seven thousand, one hundred fifty-six

[1.1] For Exercises 5 and 6, write the word name.

5. 9421

6. 123,405,600

Definitions/Rules/Procedures	Key Example(s)
To graph a whole number on a number line _____ _____.	Graph 2 on a number line. **Answer:**

[1.1] For Exercises 7 and 8, graph the number on a number line.

7. 6

8. 9

Definitions/Rules/Procedures	Key Example(s)
An **equation** is _____ _____. An **inequality** is _____ _____. **To use <, >, or = to make an incomplete statement true:** ▶ If the two numbers are the same, use _____. ▶ If the two numbers are different, choose < or > so that the inequality opens toward the _____ of the two numbers.	Use <, >, or = to write a true statement. **a.** 15 ? 15 **Answer:** 15 = 15 **b.** 20 ? 17 **Answer:** 20 > 17

[1.1] For Exercises 9 and 10, use < or > to write a true statement.

9. 14 ? 19

10. 2930 ? 2899

Definitions/Rules/Procedures	Key Example(s)
To round a number to a given place value: 1. Identify the digit in the given place value. 2. If the digit to the right of the given place value is 5 or greater, _____ _____. If the digit to the right of the given place value is 4 or less, _____ _____. 3. Change all of the digits to the right of the rounded place value to zeros.	Round to the nearest thousand. **a.** 54,259 **Answer:** 54,000 **b.** 127,502 **Answer:** 128,000 **c.** 4801 **Answer:** 5000

[1.1] For Exercises 11–14, use the following graph.

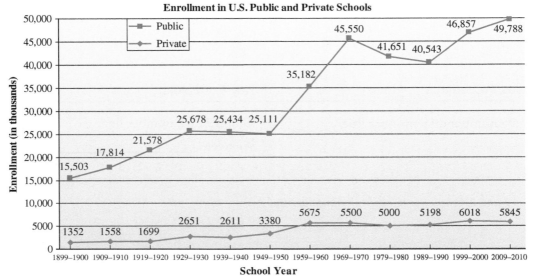

Enrollment in U.S. Public and Private Schools

Source: U.S. Department of Education.

11. What was the private school enrollment for the school year 1949–1950?

12. The United States experienced an increase in public school enrollment from 1899–1900 to 1929–1930, 1949–1950 to 1969–1970, and 1989–1990 to 2009–2010. During which of those three eras was the increase the greatest?

13. Round the public school enrollment for 1969–1970 to the nearest million.

14. Round the private school enrollment for 1939–1940 to the nearest hundred thousand.

1.2 Adding and Subtracting Whole Numbers; Solving Equations

Definitions/Rules/Procedures	Key Example(s)
Addition is the mathematical operation that _____ amounts. **To add whole numbers:** 1. Stack the numbers so that _____ _____.	

Definitions/Rules/Procedures	Key Example(s)
2. Add the digits. If the sum of the digits in any place value (or column) is greater than 9, write the ones digit of that sum in that place-value column and add the remainder of the sum to the digits in the next place-value column. This process is sometimes called regrouping or carrying. **To estimate a sum of whole numbers:** 1. Round _____ _____. 2. Round all of the other numbers to that same place value. 3. Add the rounded numbers.	Estimate $5408 + 913$ by rounding. Then find the actual sum. **Estimate:** $\overset{1}{5400}$ **Actual:** $\overset{1\ 1}{5408}$ $\underline{+\ 900}$ $\underline{+\ 913}$ 6300 6321 The estimate and actual sum agree.

[1.2] For Exercises 15 and 16, estimate by rounding. Then find the actual sum.

15. $45{,}902 + 6819$ **16.** $545 + 9091 + 28 + 30{,}009$

Definitions/Rules/Procedures	Key Example(s)
The **perimeter** of a figure is _____. **To find the perimeter of a figure,** _____ _____.	Find the perimeter. 8 ft. 5 ft. 9 ft. 6 ft. $P = 5 + 8 + 9 + 6$ $P = 28$ ft.

[1.2] For Exercises 17 and 18, find the perimeter of the figure.

17.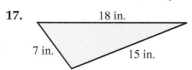

18 in.
7 in.   15 in.

18.

16 ft.
9 ft.   12 ft.
18 ft.

Definitions/Rules/Procedures	Key Example(s)
Subtraction is the mathematical operation that means to _____, to find the _____ between two amounts, or to find a(n) _____. **To subtract whole numbers:** 1. Stack the greater number on top of the smaller number so that the corresponding place values align. 2. Subtract the digits in the bottom number from the digits directly above. If a digit in the top number is less than the digit beneath it, _____.	Estimate $8942 - 765$ by rounding. Then find the actual difference. **Estimate:** 8900 **Actual:** $\overset{13}{\underset{}{8\overset{3}{\cancel{}}\ \overset{12}{\cancel{}}}}\ 894\,2$ $\underline{-\ 800}$ $\underline{-\ 765}$ 8100 8177 The estimate and the actual difference agree.

Definitions/Rules/Procedures	Key Example(s)
To estimate a difference of whole numbers: 1. Round _____ _____. 2. Round all of the other numbers to that same place value. 3. Subtract the rounded numbers.	

[1.2] For Exercises 19 and 20, estimate by rounding. Then find the actual difference.

19. $541,908 - 56,192$

20. $8002 - 295$

Definitions/Rules/Procedures	Key Example(s)
To solve for an unknown addend, write a related _____ equation in which the known _____ is subtracted from the _____.	Solve. $23 + x = 57$ Related Subtraction: $x = 57 - 23$ $x = 34$

[1.2] For Exercises 21 and 22, solve and check.

21. $29 + x = 54$

22. $y + 14 = 203$

Definitions/Rules/Procedures	Key Example(s)
List six key words/phrases that indicate addition. _____ **List six key words/phrases that indicate subtraction.** _____ _____	Juanita earns $1900 per month. She pays $975 for rent, $340 for her car loan, and $225 for utilities each month. How much does she have left after paying those expenses? Solution: total expenses: amount left: 975 340 +225 1540 1900 −1540 360

[1.2] For Exercises 23 and 24, solve.

23. In the morning, Dionne's checking account balance is $349. During her lunch break, she deposits her paycheck of $429 and pays a $72 phone bill. What is her new balance?

24. Using the data in the graph from Exercises 11–14, find the difference between the number of students enrolled in public and private schools during the 1999–2000 school year. Write the answer in standard form.

1.3 Multiplying Whole Numbers; Exponents

Definitions/Rules/Procedures	Key Example(s)
Multiplication is the mathematical operation that _____.	Multiply. **a.** $0 \cdot 15 = 0$ **b.** $271 \cdot 0 = 0$ **c.** $1 \cdot 94 = 94$ **d.** $68 \cdot 1 = 68$

Definitions/Rules/Procedures	Key Example(s)
Multiplicative property of 0: $\quad 0 \cdot n = n \cdot 0 = $ _____ , where n is any number. **Multiplicative property of 1:** $\quad 1 \cdot n = n \cdot 1 = $ _____ , where n is any number. **To multiply two whole numbers,** stack them and then apply _____ . **To estimate a product:** 1. Round _____ _____ . 2. Multiply the rounded factors.	Estimate $583 \cdot 62$ by rounding. Then find the actual product.

Estimate and actual:

$$\text{Estimate:} \quad \begin{array}{r} 600 \\ \times \quad 60 \\ \hline 000 \\ +3600 \\ \hline 36{,}000 \end{array} \qquad \text{Actual:} \quad \begin{array}{r} {\scriptstyle 41} \\ {\scriptstyle 1} \\ 583 \\ \times \quad 62 \\ \hline 1166 \\ +3498 \\ \hline 36{,}146 \end{array}$$

The estimate and the actual product agree.

[1.3] For Exercises 25 and 26, estimate by rounding. Then find the actual product.

25. 72×25

26. 591×307

Definitions/Rules/Procedures	Key Example(s)
Exponential form is a notation having the form b^n, where b is the _____ and n is the _____ . **To evaluate an exponential form with a natural-number exponent,** _____ _____ , then multiply.	Evaluate 2^5. **Solution:** $2^5 = 2 \cdot 2 \cdot 2 \cdot 2 \cdot 2$ $\qquad\qquad = 32$ Write $7 \cdot 7 \cdot 7 \cdot 7$ in exponential form. **Answer:** 7^4

[1.3] For Exercises 27 and 28, evaluate.

27. 2^7

28. 5^3

[1.3] For Exercises 29 and 30, write in exponential form.

29. $10 \cdot 10 \cdot 10 \cdot 10 \cdot 10$

30. $7 \cdot 7 \cdot 7 \cdot 7 \cdot 7 \cdot 7 \cdot 7 \cdot 7$

Definitions/Rules/Procedures	Key Example(s)
A **rectangular array** is _____ _____ . A **square unit** is a(n) _____ square. **Area** is the total number of _____ _____ . A **formula** is an equation that _____ . **To use a formula:** 1. _____ . 2. Solve for the unknown variable. **The formula for the area of a rectangle is** $A = $ _____ . **List six key words/phrases that indicate multiplication.** _____	Hardwood floor is to be installed in a room that measures 15 feet by 24 feet. How many square feet are to be covered? **Solution:** $A = lw$ $A = 15 \text{ ft.} \cdot 24 \text{ ft.}$ $A = 360 \text{ ft.}^2$

[1.3] For Exercises 31 and 32, solve.

31. A computer translates everything into binary code. This means that numbers are written so that each place value contains a 1 or a 0 digit. If a certain computer chip has 6 place values (bits), how many different numbers can be coded?

32. A room measures 16 feet wide by 18 feet long. How many square feet of carpet will be needed to cover the floor of the room?

1.4 Dividing Whole Numbers; Solving Equations

Definitions/Rules/Procedures	Key Example(s)
Division is the mathematical operation that _____ _____.	
A **remainder** is _____ _____.	
Dividing a number by 1: $n \div 1 = \dfrac{n}{1} =$ _____ when n is any number.	Determine the quotient. **a.** $\dfrac{78}{1} = 78$
Dividing any number other than 0 by 0: $n \div 0$, or $\dfrac{n}{0}$, is _____ when $n \neq 0$.	**b.** $\dfrac{25}{0}$ is undefined.
Dividing 0 by any number other than 0: $0 \div n = \dfrac{0}{n} =$ _____ when $n \neq 0$.	**c.** $\dfrac{0}{19} = 0$
If the dividend and divisor are both 0: $0 \div 0$, or $\dfrac{0}{0}$, is _____.	**d.** $\dfrac{0}{0}$ is indeterminate.
Dividing a number (other than 0) by itself: $n \div n = \dfrac{n}{n} =$ _____, when $n \neq 0$.	**e.** $\dfrac{245}{245} = 1$

[1.4] For Exercises 33–36, determine the quotient.

33. $0 \div 13$ **34.** $45/0$ **35.** $24 \div 1$ **36.** $16 \div 16$

Definitions/Rules/Procedures	Key Example(s)
2 is an exact divisor for any _____.	Use the divisibility rules to determine whether 2 is an exact divisor for 24,806. **Answer:** Yes, because 24,806 is even.
3 is an exact divisor if _____.	Use the divisibility rules to determine whether 3 is an exact divisor for 5142. **Answer:** Yes, because $5 + 1 + 4 + 2 = 12$, which is divisible by 3.

Definitions/Rules/Procedures	Key Example(s)
5 is an exact divisor for any number with _____.	Use the divisibility rules to determine whether 5 is an exact divisor for 612. **Answer:** No, because 612 does not have 0 or 5 in the ones place.

[1.4] 37. Use the divisibility rules to determine whether 3 is an exact divisor for 596.

[1.4] 38. Use the divisibility rules to determine whether 5 is an exact divisor for 1780.

Definitions/Rules/Procedures	Key Example(s)
To divide whole numbers, use _____. **To estimate a quotient of whole numbers:** 1. Round _____ _____. 2. Divide the rounded numbers. If the calculation is difficult to perform mentally, try rounding the dividend so that it has two nonzero digits.	Estimate $6409 \div 12$ by rounding. Then find the actual quotient. Estimate: $\begin{array}{r} 600 \\ 10)\overline{6000} \\ -60 \\ \hline 00 \\ -0 \\ \hline 00 \\ -0 \\ \hline 0 \end{array}$ Actual: $\begin{array}{r} 534 \\ 12)\overline{6409} \\ -60 \\ \hline 40 \\ -36 \\ \hline 49 \\ -48 \\ \hline 1 \end{array}$ **Answer:** 534 r1

[1.4] For Exercises 39 and 40, estimate by rounding. Then find the actual quotient.

39. $78{,}413 \div 19$

40. $\dfrac{80{,}156}{38}$

Definitions/Rules/Procedures	Key Example(s)
To solve for an unknown factor, write a related _____ equation in which the _____ is divided by the _____.	Solve. $x \cdot 12 = 180$ Related Division: $x = 180 \div 12$ $\qquad\qquad\qquad x = 15$

[1.4] For Exercises 41 and 42, solve and check.

41. $19 \cdot b = 456$

42. $8 \cdot k = 2448$

Definitions/Rules/Procedures	Key Example(s)
List six key words/phrases that indicate division. _____	A company budgets \$30,000 to distribute evenly among its 48 employees as a bonus. How much does each employee receive? **Solution:** bonus $= 30{,}000 \div 48 = 625$ $$\begin{array}{r} 625 \\ 48\overline{)30{,}000} \\ -28\ 8 \\ \hline 1\ 20 \\ -\ \ 96 \\ \hline 240 \\ -240 \\ \hline 0 \end{array}$$

[1.4] For Exercises 43 and 44, solve.

43. Jana is to administer 120 milliliters of Pitocin in an IV drip over the course of an hour. How many milliliters of Pitocin should the patient receive each minute?

44. Andre has a gross annual salary of \$39,300. What is his gross monthly salary?

1.5 Order of Operations; Mean, Median, and Mode

Definitions/Rules/Procedures	Key Example(s)
Order of Operations Agreement **Perform operations in the following order:** 1. _____, which include parentheses ( ), brackets [ ], braces { }, and fraction bars —— 2. _____ 3. _____ in order as these operations occur from left to right 4. _____ in order as these operations occur from left to right	Simplify. $$26 - [19 - 2(15 - 9)] + 4^2$$ $$= 26 - [19 - 2(6)] + 4^2$$ $$= 26 - [19 - 12] + 4^2$$ $$= 26 - 7 + 4^2$$ $$= 26 - 7 + 16$$ $$= 19 + 16$$ $$= 35$$

[1.5] For Exercises 45–48, simplify.

45. $24 \div 8 \cdot 6 + 5^2$

46. $64 - [(28 - 4) \div 2^2]$

47. $3^3 + \{40 - [25 + 2(14 - 9)]\}$

48. $\dfrac{9^2 - 21}{56 - 2(4 + 1)^2}$

Definitions/Rules/Procedures	Key Example(s)
A **statistic** is a(n) _____ of a set of data. **The mean** of a set of data is the quotient of _____ _____.	

Definitions/Rules/Procedures	Key Example(s)
To find the mean, or average, of a set of data: 1. Find the sum of the values. 2. Divide the sum by _____ . The **median** of a set of data is the _____ number in a(n) _____ list of numbers. **To find the median of a set of data:** 1. List the values in order from least to greatest. 2. If the list has an odd number of values, the median is the _____ . If the list has an even number of values, the median is the _____ . The **mode** of a set of data is _____ . **To find the mode of a set of data**, count the number of repetitions of each score. The score with the _____ repetitions is the mode. **Note:** If no value is repeated, we say that the data has no mode. If there is a tie for the most repetitions, list each of those values as the mode.	A group of eight children have the following weights. Find mean, median, and mode weight. $$55, 84, 60, 64, 74, 69, 60, 62$$ **Solution:** $$\text{Mean} = \frac{55 + 84 + 60 + 64 + 74 + 69 + 60 + 62}{8}$$ $$= \frac{528}{8} \quad \text{Add the weights.}$$ $$= 66 \quad \text{Divide.}$$ Arrange the weights from least to greatest: $$55, 60, 60, 62, 64, 69, 74, 84$$ $$\text{Median} = \frac{62 + 64}{2} = \frac{126}{2} = 63$$ $$\text{Mode} = 60$$

[1.5] 49. The following graph shows the number of new memberships sold per month by a health club. Find the mean, median, and **mode**.

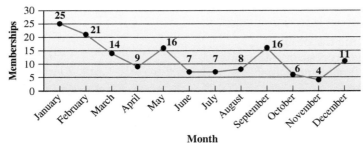

1.6 More with Formulas

Definitions/Rules/Procedures	Key Example(s)
Parallel lines are straight lines lying in the same plane that _____ . A **parallelogram** is a four-sided figure with _____ _____ . A **right angle** is an angle that measures _____ . A **cubic unit** is a(n) _____ cube. **Volume** is the total number of _____ that completely fill an object. **Additional Formulas** Perimeter of a rectangle: $P =$ _____ Area of a parallelogram: $A =$ _____ Volume of a box: $V =$ _____	Find the perimeter of a rectangle with a length of 24 centimeters and a width of 17 centimeters. $$P = 2l + 2w$$ $$P = 2(24) + 2(17)$$ $$P = 48 + 34$$ $$P = 82 \text{ cm}$$

[1.6] 50. Find the area.

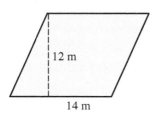

12 m

14 m

[1.6] 51. Find the volume of a box that is 7 feet wide, 6 feet high, and 9 feet long.

Definitions/Rules/Procedures	Key Example(s)
Problem-Solving Outline	Find the area of the shaded region.

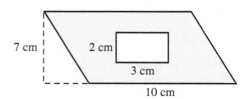

Problem-Solving Outline

1. _____ the problem.

  a. Read the question(s), underline key words, and note what you are to find.

  b. If useful, draw a picture, make a list or table, simulate the situation, or search for a related example problem.

2. _____ your solution strategy. Search for a formula or translate key words to an equation.

3. _____ the plan. Solve the formula or equation.

4. _____ the question. Include appropriate units.

5. _____ the results.

  a. Make sure the answer is reasonable for the situation.

  b. Check calculations by solving the problem a different way, reversing the process, or estimating and verifying that your estimate and actual answer are reasonably close.

To find the area or volume of a composite figure:

1. Identify the shapes or objects that make up the figure.

2. If those shapes or objects combine to form the composite figure, _____ their areas or volumes.

If one shape or object is inside another, _____ the area/volume of the inner shape/object from the area/volume of the outer shape/object.

Find the area of the shaded region.

7 cm   2 cm   3 cm   10 cm

Understand We must find the area of the shaded region, which is a parallelogram with a rectangle removed.

Plan Subtract the area of the rectangle from the area of the parallelogram.

Execute

$$\text{Area} = \text{area of the parallelogram} \\ -\text{area of the rectangle}$$
$$A = (10)(7) - (3)(2)$$
$$A = 70 - 6$$
$$A = 64$$

Answer The area of the shaded region is 64 cm^2.

Check Verify the calculations. The check will be left to the reader.

[1.7] 52. The picture shows the plans for a department store and a parking lot. What is the area of the parking lot?

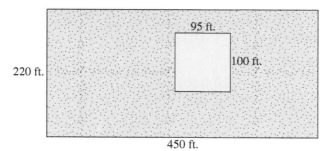

95 ft.

100 ft.

220 ft.

450 ft.

Chapter 1 Practice Test

For Extra Help  **CHAPTER Test Prep** VIDEOS

Step-by-step test solutions are found on the Chapter Test Prep Videos available via the Video Resources on DVD, in **MyMathLab®**, and on **You Tube** ™ (search "CarsonPrealgebra" and click on "Channels").

TACTILE VISUAL AUDITORY   Learning Strategy

Treat a practice test as a dress rehearsal. Have ready the same materials you will use during the actual test. Set a timer for the same amount of time as you will have for the actual test. When you finish, calculate your score and resolve any mistakes you made. Continue taking the practice test until you feel confident with your performance.

[1.1] **1.** What digit is in the ten thousands place in 1,258,987?

1. _____

[1.3] **2.** Write 48,210,907 in expanded form using powers of 10.

2. _____

[1.1] **3.** Use $<$ or $>$ to write a true statement.

$$19,304 \ \ ? \ \ 19,204$$

3. _____

[1.1] **4.** Write the word name for 67,194,210.

4. _____

[1.2] **5.** Estimate the sum by rounding to the nearest hundred thousand.

$$2,346,502 + 481,901$$

5. _____

[1.2] **6.** Estimate $63,209 + 4981$ by rounding. Then find the actual sum.

6. _____

[1.2] **7.** Estimate $480,091 - 54,382$ by rounding. Then find the actual difference.

7. _____

[1.2] **8.** Solve and check. $27 + x = 35$

8. _____

[1.3] **9.** Estimate 357×6910 by rounding. Then find the actual product.

9. _____

[1.3] **10.** Evaluate. 4^3

10. _____

[1.4] **11.** Estimate $45,019 \div 18$ by rounding. Then find the actual quotient.

11. _____

[1.4] **12.** Solve and check. $7 \cdot y = 126$

12. _____

For Exercises 13–17, evaluate.

[1.5] **13.** $9 + 32 \div 4 \cdot 2$

13. _____

[1.5] **14.** $14 + 2^5 - 8 \cdot 3$

14. _____

[1.5] **15.** $20 - 6(3 + 2) \div 15$

15. _____

[1.5] **16.** $\dfrac{(14 + 26)}{8} - [17 - (4 + 9)]$

16. _____

[1.5] **17.** $\dfrac{5^2 + 47}{(4 + 2)^2}$

17. _____

18. _____

[1.1]   **18.** Use the following graph.

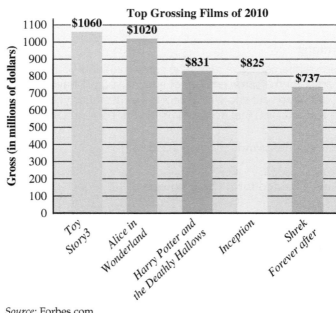

Top Grossing Films of 2010

a. How much did *Toy Story 3* gross? Write the answer in standard form.

b. Round the gross for *Inception* to the nearest ten million.

Source: Forbes.com

19. _____

[1.2]   **19.** Using the graph above, how much more did *Alice in Wonderland* gross than *Harry Potter and the Deathly Hallows*? Write the answer in standard form.

20. _____

[1.4]   **20.** A college admits 2487 new students. All new students must enroll in a college orientation course. If the maximum number of students allowed in a single class is 30, how many classes should be offered to accommodate all of the new students?

21. _____

[1.5]   **21.** A student has six test scores: 85, 98, 85, 100, 97, and 93.
   a. Find the mean.   **b.** Find the median.   **c.** Find the mode.

22. _____

[1.6]   **22.** Find the area of the parallelogram.

9 ft.

18 ft.

23. _____

[1.6]   **23.** A rectangular playground measures 65 feet wide by 80 feet long.
   a. What is the perimeter of the playground?
   b. If fence material costs $6 per foot, how much would a fence cost for the playground?

24. _____

[1.6]   **24.** A walk-in refrigerator at a restaurant is a box measuring 8 feet long, 6 feet wide, and 7 feet high. What is the volume?

25. _____

[1.7]   **25.** A wall 16 feet wide and 9 feet high is to be painted. This wall has an open doorway that is 3 feet wide by 7 feet high. How many square feet must be painted?

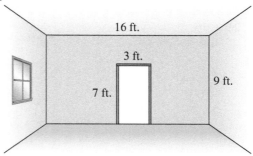

16 ft.

3 ft.

9 ft.

7 ft.

Chapter Overview

To be successful in algebra, we must learn the rules for adding, subtracting, multiplying, and dividing positive and negative numbers. In this chapter, we learn about:

▶ Positive and negative numbers called integers.

▶ Absolute value and additive inverse.

▶ Arithmetic with integers.

▶ Square roots and order of operations with integers.

2.1 Introduction to Integers

Objectives

1 Identify integers.
2 Interpret number lines and graphs with integers.
3 Use < or > to write a true statement.
4 Find the absolute value.

Of Interest

The first known mention of negative numbers is found in the work of an Indian mathematician, Brahmagupta (about A.D. 628). He called them "negative and affirmative quantities." There have been many different ways of indicating negative numbers. The Chinese would use colors: black for negative numbers and red for positive numbers. In sixteenth-century Europe, it was popular to use the letter *m*, for minus, to indicate negative numbers, as in m:7 to indicate −7.

Warm-up

[1.1] 1. Graph 3 on a number line.
[1.1] 2. Use < or > to write a true statement.

          54 ? 35

Objective 1 Identify integers.

In this chapter, we will learn about positive and negative numbers.

A positive number is written with a plus sign or no sign, as in +5 or 5. We read +5 (or 5) as "positive five," "plus five," or "five." All of the natural numbers (1, 2, 3, . . .) are positive numbers. A negative number is written with a minus sign, as in −6, which is read as "negative six" or "minus six." The number 0 has no sign. Zero is neither positive nor negative.

The whole numbers $\{0, 1, 2, 3, \ldots\}$ along with all of the negative counting numbers $\{\ldots, -3, -2, -1\}$ form a set of numbers called the **integers** $\{\ldots, -3, -2, -1, 0, 1, 2, 3, \ldots\}$.

Definition **Integers:** The integers are . . . , −3, −2, −1, 0, 1, 2, 3, . . .

The following diagram shows how the set of integers contains the set of whole numbers and the set of natural numbers.

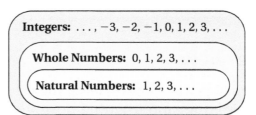

Integers: . . . , −3, −2, −1, 0, 1, 2, 3, . . .

Whole Numbers: 0, 1, 2, 3, . . .

Natural Numbers: 1, 2, 3, . . .

Positive and negative numbers can be used to describe such things as money, temperature, elevation, and direction. The table to the right lists words that indicate a nega-

Situation	Negative number	Positive number
Money	Debt/liability	Credit/asset
Temperature	Below zero	Above zero
Elevation	Below sea level	Above sea level
Direction	Reverse/go down	Forward/go up

tive number or a positive number in each of the listed situations. Relating integers to familiar situations will help us understand arithmetic with signed numbers.

Learning Strategy

One way I study is using highlighters with Post-Its attached. I can color-coordinate my class notes and the textbook by putting a Post-It tab on my notes and on my textbook to connect the two thoughts, especially if a professor mentions specific pages and sections from the textbook. If you can connect what you've read to what you've talked about using a quick and easy reference tool, you won't have to flip pages and search for the right material.

                                                                    —Carlin H.

Answers to Warm-up

1.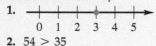

2. 54 > 35

Example 1 Express each amount as a positive or negative integer.

a. Angela owes $480 on her credit card.

Answer: −480

Explanation: Money owed is debt, which is represented by a negative amount.

b. The temperature is 29°C outside.

Answer: 29

Explanation: Because 29°C is above zero, it is represented by a positive amount.

◄ **Note** When writing a positive number, we do not have to write +29. We can write just 29.

c. A research submarine is at a depth of 245 feet.

Answer: −245

Explanation: Because the submarine is below sea level, its position is represented by a negative amount.

▶ **Do Your Turn 1**

Objective 2 Interpret number lines and graphs with integers.

Integers also appear in graphs.

Example 2 Use the following graph.

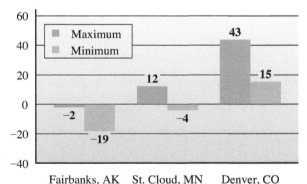

Maximum and Minimum Normal Temperatures in January (degrees Fahrenheit)

a. What is the maximum normal January temperature in St. Cloud, Minnesota?

Answer: 12°F

b. What is the minimum normal January temperature in Fairbanks, Alaska?

Answer: −19°F

▶ **Do Your Turn 2**

In Section 1.1, we learned how to graph whole numbers on number lines. We can now include negative numbers on our number lines. Because the set of integers continues indefinitely in the positive and negative directions, we will place an arrow at each end of the number line.

▶ **Your Turn 1**

Express each amount as a positive or negative integer.

a. Ian has $3250 in a savings account.

b. According to the U.S. National Debt Clock, on 6/15/2010, the national debt was $13,088,213,695,570.

c. The melting point of gold is 1063°C.

d. A submarine is at a depth of 250 feet.

▶ **Your Turn 2**

Use the graph in Example 2.

a. What is the maximum normal January temperature in Fairbanks, Alaska?

b. What is the minimum normal January temperature in Denver, Colorado?

Answers to Your Turn 1
a. 3250   b. −13,088,213,695,570
c. 1063   d. −250

Answers to Your Turn 2
a. −2°F   b. 15°F

Procedure

To graph a number on a number line, draw a dot on the mark for that number.

▶ **Your Turn 3**

Graph the number on a number line.

a. −2   b. 0   c. 5

Example 3   Graph −1 on a number line.

Solution: Draw a dot on the mark for −1.

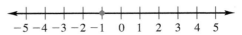

◀ **Do Your Turn 3**

Objective 3 Use < or > to write a true statement.

We can use number lines to compare numbers. Because numbers increase from left to right on a number line, when we compare two integers, the integer that is farther to the right on a number line is the greater integer.

▶ **Your Turn 4**

Use < or > to write a true statement.

a. 15 ? −26

b. −19 ? −17

c. −14 ? −2

d. 0 ? −9

  Learning Strategy

If you are a visual or tactile learner, imagine a building with floors above ground and below ground level, with 0 representing ground level. A positive 4 represents the fourth floor above ground whereas −2 represents the second floor below ground level. The fourth floor above ground is higher than the second floor below ground; so 4 is greater than −2.

Answers to Your Turn 3

a.

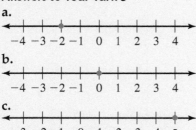

b.

c.

Answers to Your Turn 4

a. 15 > −26   b. −19 < −17
c. −14 < −2   d. 0 > −9

Example 4   Use < or > to write a true statement.

a. 4 ? −2

Answer: 4 > −2

Explanation: 4 is farther to the right on a number line than −2 is; so 4 is the greater number.

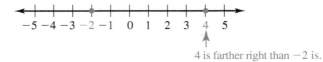

4 is farther right than −2 is.

b. 0 ? −6

Answer: 0 > −6

Explanation: 0 is farther to the right on a number line than −6 is; so 0 is the greater number.

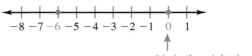

0 is farther right than −6 is.

c. −12 ? −7

Answer: −12 < −7

Explanation: −7 is farther to the right on a number line than −12 is; so −7 is the greater number.

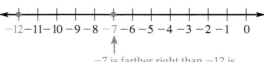

−7 is farther right than −12 is.

◀ **Do Your Turn 4**

Example 4 suggests the following rules for comparing signed numbers:

Rules

Any positive number is greater than any negative number.

0 is greater than any negative number.

When two negative numbers are compared, the negative number closer to 0 is the greater number.

Objective 4 Find the absolute value.

We also can use number lines to understand the **absolute value** of a number.

Definition **Absolute value:** A given number's distance from zero on a number line.

| **Example 5** | Find the absolute value of each number.

a. 5

Answer: 5

Explanation: Because 5 is five steps from 0 on a number line, its absolute value is 5.

b. −6

Answer: 6

Explanation: Because −6 is six steps from 0 on a number line, its absolute value is 6. Notice that it doesn't matter which direction we move; we just count the steps. As an analogy, imagine a size 6 shoe. It doesn't matter which way you point the shoe or where it is located, it's still a size 6 shoe.

c. 0

Answer: 0

Explanation: Because there is no distance from 0 to itself, its absolute value is 0.

▶ **Do Your Turn 5**

Example 5 suggests the following rules:

Rules

The absolute value of a positive number is positive.

The absolute value of a negative number is positive.

The absolute value of zero is zero.

The symbols for absolute value are two vertical lines. For example, the absolute value of 5 is written $|5|$.

▶ **Your Turn 5**

Find the absolute value of each number.

a. 14   b. −27   c. −1347

TACTILE

Learning Strategy

If you are a tactile learner, think of absolute value as physically walking from 0 to the given number and counting the steps you take to get to that number.

Answers to Your Turn 5
a. 14   b. 27   c. 1347

▶ **Your Turn 6**

Simplify.
a. $|-19|$ b. $|90|$ c. $|0|$

Answers to Your Turn 6
a. 19 b. 90 c. 0

Example 6 Simplify.

a. $|-8|$ ◀ **Note** In words, $|-8|$ translates to "the absolute value of negative eight."

Answer: 8

Explanation: Because -8 is 8 steps away from 0, its absolute value is 8.

b. $|27|$

Answer: 27

Explanation: Because 27 is 27 steps away from 0, its absolute value is 27.

◀ **Do Your Turn 6**

2.1 Exercises For Extra Help MyMathLab®

Objective 1

Prep Exercise 1 Write a list indicating the numbers in the set of integers.

For Exercises 1–12, express each amount as a positive or negative integer.

1. Sonja receives a paycheck of $450.

2. Adam's business made a profit of $215,000 last year.

3. Holly owes her friend $35.

4. A financial planner analyzes the Smiths' financial situation and finds that they have a total debt of $75,243.

5. Mount Everest is the tallest mountain in the world. Its peak is 29,035 feet above sea level.

6. A plane's cruising altitude is 35,000 feet above sea level.

Of Interest

Mount Everest is in the central Himalayan range on the border of Tibet and Nepal. Its summit is the highest peak in the world and was first reached in 1953 by Edmund Hillary of New Zealand and Tenzing Norgay, a Nepalese guide.

Source: Microsoft Bookshelf '98.

7. The *Titanic* is lying on the bottom of the North Atlantic Ocean at a depth of approximately 13,200 feet.

8. The tunnel connecting Britain to France is approximately 220 meters below the surface of the English Channel.

Of Interest

The English Channel Tunnel is approximately 31 miles long, 23 miles of which are under water. The rubble from excavating the tunnel increased the size of Britain by 90 acres.

9. Water boils at 100°C.

10. Carbon dioxide (CO_2) freezes at 78°C below 0.

11. The countdown to launch time of a space shuttle is stopped with 20 minutes to go.

12. A college football team gains 40 yards in one play.

Objective 2

For Exercises 13–16, use the following graph.

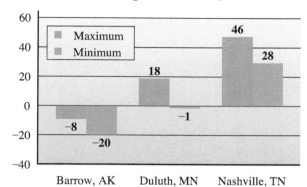

**Maximum and Minimum Normal
Temperatures in January
(degrees Fahrenheit)**

13. What is the maximum normal January temperature in Barrow, Alaska?

14. What is the maximum normal January temperature in Nashville, Tennessee?

15. What is the minimum normal January temperature in Barrow, Alaska?

16. What is the minimum normal January temperature in Duluth, Minnesota?

For Exercises 17–24, use the following graph. In the graph, 0 represents the average critical reading, writing, and math scores, of all students taking the SAT.

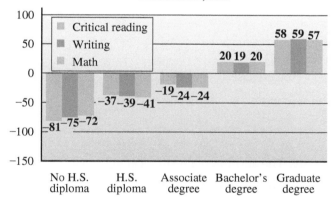

**Amount SAT Scores Differed from Average
According to Parents' Highest Level
of Education, 2009**

Of Interest

The average SAT scores of all students during 2008–2009 were as follows:

Critical reading: 501
Writing: 493
Math: 515

17. What does the graph indicate about the average critical reading score of students whose parents have no high school diploma?

18. What does the graph indicate about the average critical reading score of students whose parents have an associate degree as their highest level of education?

19. What does the graph indicate about the average math score of students whose parents have a high school diploma as their highest level of education?

20. What does the graph indicate about the average math score of students whose parents have an associate degree as their highest level of education?

21. What does the graph indicate about the average writing score of students whose parents have a bachelor's degree as their highest level of education?

22. What does the graph indicate about the average writing score of students whose parents have a graduate degree as their highest level of education?

23. What does the graph indicate about the average math score of students whose parents have a bachelor's degree as their highest level of education?

24. What does the graph indicate about the average math score of students whose parents have a graduate degree as their highest level of education?

Prep Exercise 2   Draw a number line showing the integers from -5 to 5.

For Exercises 25–34, graph each integer on a number line.

25. 8

26. 2

27. 7

28. 5

29. -9

30. -2

31. -6

32. -8

33. 0

34. 9

Objective 3

For Exercises 35–46, use $<$ or $>$ to write a true statement.

35. -20 ? 17

36. 26 ? -18

37. 20 ? 16

38. 12 ? 19

39. -30 ? -16

40. -12 ? -19

41. -9 ? -12

42. -2 ? -7

43. 0 ? -16

44. 0 ? 5

45. -19 ? -17

46. -14 ? -11

Objective 4

Prep Exercise 3 The absolute value of a positive number is _____.

Prep Exercise 4 The absolute value of a negative number is _____.

Prep Exercise 5 The absolute value of zero is _____.

For Exercises 47–56, complete the sentence.

47. The absolute value of 26 is _____.

48. The absolute value of 19 is _____.

49. The absolute value of 21 is _____.

50. The absolute value of 9 is _____.

51. The absolute value of -18 is _____.

52. The absolute value of -4 is _____.

53. The absolute value of -10 is _____.

54. The absolute value of -16 is _____.

55. The absolute value of 0 is _____.

56. The absolute value of -7 is _____.

For Exercises 57–68, simplify.

57. $|14|$

58. $|242|$

59. $|-18|$

60. $|-9|$

61. $|-2004|$

62. $|12|$

63. $|8|$

64. $|-35|$

65. $|0|$

66. $|-200|$

67. $|-47|$

68. $|97|$

Review Exercises

For Exercises 1– 4, perform the indicated operation.

[1.2] 1. $84{,}759 + 9506$

[1.2] 2. $900{,}406 - 35{,}918$

[1.3] 3. 457×609

[1.4] 4. $85{,}314 \div 42$

[1.2] 5. Amelia begins the month with $452 in savings. Look at the list of her deposits and withdrawals for the month. What is her balance at the end of the month?

Deposits	Withdrawals
$45	$220
$98	$ 25
$88	$ 10
$54	

[1.5] 6. Simplify. $8^2 - 30 \div 5 \cdot 3$

2.2 Adding Integers

Objectives

1 Add integers with like signs.
2 Add integers with different signs.
3 Add integers.
4 Find the additive inverse.
5 Solve applications involving addition of integers.

Warm-up

[2.1] 1. Use < or > to write a true statement.
      a. 20 ? −35 b. −15 ? −9

[2.1] 2. Simplify.
      a. $|7|$ b. $|-13|$

Objective 1 Add integers with like signs.

Let's examine some realistic situations to discover how we add signed numbers. First, we will consider situations in which we add two numbers that have the same sign.

Example 1

a. Suppose we have $200 in our checking account and we deposit $40. What would the balance be?

Solution: Because the $40 deposit increases the $200 balance, we add the values to end up with a balance of $240. We can write the following addition statement:

$$200 + 40 = 240$$

We can also use a number line. Think of the first addend, 200, as the starting position. The absolute value of the second addend indicates the distance to move from the starting position, and its sign indicates the direction. Because 40 is positive, we move to the right 40 steps. The final position, 240, is the answer.

+ 40 means move 40 steps to the right

200   240
Starting position   Final position

b. Suppose we have a debt of $200 on a credit card account and make a $40 purchase using the same credit card. What is the balance after the purchase?

Solution: A purchase using a credit card increases the debt balance. Therefore, the $40 purchase adds to the existing $200 debt balance for a total debt of $240. Each debt is written as a negative number; so we can write the following addition statement:

$$-200 + (-40) = -240$$

Note In math sentences, parentheses are placed around negative numbers that *follow* operation symbols. Although parentheses are not needed, they clarify the meaning of the plus and minus signs.

Using a number line, $-200 + (-40)$ means to start at -200 and move 40 steps in the negative direction, which is to the left.

+ (−40) means move 40 steps to the left

−240   −200
Final position   Starting position

Answers to Warm-up
1. a. 20 > −35 b. −15 < −9
2. a. 7 b. 13

Notice that in Example 1, we added the absolute values because we were either increasing assets or increasing debt. On a number line, we were starting positive and moving more positive or starting negative and moving more negative.

Conclusion: When we add two numbers that have the same sign, we add their absolute values and keep the same sign.

▶ **Do Your Turn 1**

Objective 2 Add integers with different signs.

Now consider situations in which we add two numbers that have different signs.

Example 2

a. Suppose we have a debt of $200 on a credit card account and we make a payment of $80. What is the balance?

Solution: The $80 payment is a positive amount added to the account, which decreases the $200 debt. Because the debt exceeds the payment, we still owe $120. We can write the following addition statement:

$$-200 + 80 = -120$$

Using a number line, we start at -200 and move 80 steps to the right, which is not enough to move past 0; so the result is still negative.

Also notice that because of the commutative law of addition, we can write the statement as follows:

$$80 + (-200) = -120$$

It doesn't matter which way the addends are arranged; a $200 debt with $80 in assets results in a debt of $120.

b. Suppose we have a debt of $200 on the credit card account and we make a payment of $240. What is the balance?

Solution: The $240 payment is a positive amount added to the account, which decreases the $200 debt. Because the payment exceeds the debt by $40, we have a $40 credit. We can write the following addition statement:

$$-200 + 240 = 40$$

Using a number line, we start at -200 and move 240 steps to the right. Moving 200 steps to the right puts us at 0; so 240 steps moves 40 steps past 0; making the result 40.

From Example 2, we can draw the following conclusion:

Conclusion: When we add two numbers that have different signs, we subtract their absolute values and keep the sign of the number with the greater absolute value.

▶ **Do Your Turn 2**

▶ **Your Turn 1**

Add.

a. $19 + 74$

b. $-42 + (-39)$

c. $-15 + (-18)$

d. $20 + 13$

▶ **Your Turn 2**

Add.

a. $-18 + 6$

b. $-20 + 28$

c. $30 + (-26)$

d. $14 + (-24)$

e. $15 + (-15)$

TACTILE VISUAL AUDITORY

Learning Strategy

If you are a visual learner, use number lines to solidify your understanding of the rules for adding integers. If you are an auditory learner, try making up a song or rhyme out of the rules. If you are a tactile learner, move your finger or a pencil along a number line to the right for positive numbers and to the left for negative numbers.

Answers to Your Turn 1
a. 93 b. −81 c. −33 d. 33

Answers to Your Turn 2
a. −12 b. 8 c. 4 d. −10 e. 0

Objective 3   Add integers.

We can summarize what we learned in Examples 1 and 2 with the following two rules for adding integers. It is helpful to relate each addition situation to reconciling debts and credits.

> ### Procedure
>
> To add two integers:
> - If they have the same sign, add their absolute values and keep the same sign. (*Think:* Two credits or two debts)
> - If they have different signs, subtract their absolute values and keep the sign of the number with the greater absolute value. (*Think:* Payment toward a debt)

▶ **Your Turn 3**

Explain each situation in terms of debts and credits; then add.

  a. $-40 + (-12)$

b. $28 + 15$

  c. $16 + (-10)$

d. $-26 + 14$

  e. $-22 + 22$

Example 3   Explain each situation in terms of debts and credits; then add.

a. $-40 + (-21)$

Solution: We are to add two debts, which increases the debt; so we add, and the result is still a debt. Using the rules, because the numbers have the same sign, we add the absolute values and keep the negative sign.

$$-40 + (-21) = -61$$

b. $28 + (-16)$

Solution: We have a \$28 payment toward a \$16 debt. We subtract the amount, and because the payment exceeds the debt, the result is a credit. Using the rules, because the numbers have different signs, we subtract their absolute values, and because the positive integer, 28, has the greater absolute value, the result is positive.

$$28 + (-16) = 12$$

c. $10 + (-24)$

Solution: We have a \$10 payment toward a \$24 debt. We subtract the amounts and because the debt exceeds the payments, the result is a debt. Using the rules, because the numbers have different signs, we subtract their absolute values. And because the negative integer, -24, has the greater absolute value, the result is negative.

$$10 + (-24) = -14$$

◀ **Do Your Turn 3**

Answers to Your Turn 3

a. Two debts; so add and the result is a debt/negative; -52.

b. Two credits; so add and the result is a credit/positive; 43.

c. A credit with a debt; so subtract and because the credit, 16, is more than the debt, the result is a credit/positive; 6.

d. A debt with a credit; so subtract and because the debt, -26, is more than the credit, the result is a debt/negative; -12.

e. Paying \$22 on a \$22 debt means that the balance is 0.

Now let's consider how to add more than two integers. There are two popular approaches: (1) We can add from left to right, or (2) we can add all of the positive and negative numbers separately, then reconcile the two results.

Example 4   Add. $-14 + 26 + (-18) + 15 + (-32)$

Solution: First, let's add from left to right.

$$
\begin{aligned}
-14 + 26 &+ (-18) + 15 + (-32) \\
&= 12 + (-18) + 15 + (-32) \quad &&\text{Add} -14 \text{ and } 26 \text{ to get } 12. \\
&= -6 + 15 + (-32) \quad &&\text{Add } 12 \text{ and } -18 \text{ to get } -6. \\
&= 9 + (-32) \quad &&\text{Add } -6 \text{ and } 15 \text{ to get } 9. \\
&= -23 \quad &&\text{Add } 9 \text{ and } -32 \text{ to get } -23.
\end{aligned}
$$

Now let's add all of the positives and negatives separately, then reconcile the result. The commutative and associative properties of addition make this method possible because they allow us to rearrange and regroup addition any way we choose.

$$-14 + 26 + (-18) + 15 + (-32)$$

$$= \underbrace{26 + 15} + \underbrace{(-14) + (-18) + (-32)}$$

Rearrange the addends so that the positives and negatives are grouped separately.

$$= \quad 41 \quad + \quad (-64)$$

Add the positives and negatives separately.

$$= -23$$

Reconcile the positive and negative sums.

▲

Note In terms of credits and debts, we found the total credit to be $41 and the total debt to be $64. Because the debt outweights the credit, the result is a debt of $23.

▶ **Do Your Turn 4**

Objective 4 Find the additive inverse.

In Your Turn 3e, you found that $-22 + 22 = 0$. Because their sum is 0, we say that -22 and 22 are **additive inverses**.

Definition Additive inverses: Two numbers whose sum is zero.

Examples of additive inverses: 15 and -15 because $15 + (-15) = 0$
 -9 and 9   because $-9 + 9 = 0$
 0 and 0   because $0 + 0 = 0$

Notice that 0 is its own additive inverse and that for numbers other than 0, the additive inverses have the same absolute value but *opposite* signs.

> ### Rules
> The sum of two additive inverses is zero.
> The additive inverse of 0 is 0.
> The additive inverse of a nonzero number has the same absolute value but opposite sign.

Example 5 Find the additive inverse.

a. 17

Answer: -17 17 and -17 are additive inverses because $17 + (-17) = 0$. Also, they have the same absolute value but opposite signs.

b. -8

Answer: 8 -8 and 8 are additive inverses because $-8 + 8 = 0$. Also, they have the same absolute value but opposite signs.

▶ **Do Your Turn 5**

Objective 5 Solve applications involving addition of integers.

The key to solving addition problems that involve integers is to simulate the situation.

▶ **Your Turn 4**

Add.
a. $-18 + (-14) + 30 + (-2) + 15$
b. $-38 + 64 + 40 + (-100)$
c. $28 + (-14) + 5 + (-12) + (-10) + 3$

▶ **Your Turn 5**

Find the additive inverse.
a. 6   b. -12

Answers to Your Turn 4
a. 11   b. -34 c. 0

Answers to Your Turn 5
a. -6 b. 12

▶ **Your Turn 6**

The following table lists the assets and debts for the Cromwell family. Calculate the family's net worth.

Assets
Savings = $3528
Checking = $1242
Furniture = $21,358
Stocks = $8749

Debts
Credit card balance = $3718
Mortgage = $55,926
Automobile 1 = $4857
Automobile 2 = $3310

Example 6 Jane's credit card statement showed a closing balance of −$240. Since that time, she made the following transactions. What is her balance now?

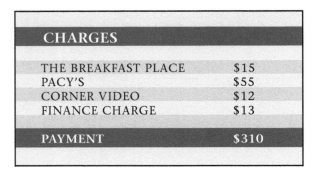

CHARGES	
THE BREAKFAST PLACE	$15
PACY'S	$55
CORNER VIDEO	$12
FINANCE CHARGE	$13
PAYMENT	$310

Solution: The balance is the sum of the closing balance, all charges, and all payments.

$$\text{balance} = -240 + (-15) + (-55) + (-12) + (-13) + 310$$
$$= -335 + 310$$ ◀ **Note** Adding all of the negative numbers computes the total debt, $335.
$$= -25$$

Answer: Jane's current balance is a debt of $25.

◀ **Do Your Turn 6**

▶ **Your Turn 7**

At sunset one cold winter evening in North Dakota, the temperature was −5°F. By midnight, the temperature had dropped another 15°F. What was the temperature at midnight?

Example 7 A research submarine is at a depth of 458 feet. The submarine's sonar detects that the ocean floor is another 175 feet down. What will be the depth if the submarine descends to the ocean floor? The submarine is designed to withstand the pressure from the surrounding water up to a depth of 600 feet. Is it safe to allow the submarine to descend to the ocean floor in this case?

Understand We must determine whether it is safe for the submarine to proceed from its current depth to the ocean floor.

Plan We must first calculate the depth of the ocean floor. We can then assess whether it is beyond the submarine's depth limit of 600 feet. Because −458 feet is the submarine's starting position and it is to proceed down another 175 feet, we can add to get the depth of the ocean floor.

Execute ocean floor depth = −458 + (−175) = −633

Answer The depth of the ocean floor is −633 feet, which is beyond the submarine's 600-foot depth limit. Therefore, it is not safe to allow the submarine to descend to the ocean floor.

Check We can use a vertical number line.

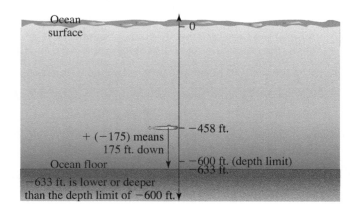

Answer to Your Turn 6
−$32,934

Answer to Your Turn 7
−20°F

◀ **Do Your Turn 7**

Example 8 We can use integers to analyze the forces acting on an object. The resultant force is the sum of the forces acting on the object. A steel beam weighs 250 pounds and has two wires attached to hold it up. Each wire has an upward force measuring 140 pounds. (See the following diagram.) Find the resultant force on the beam. What does this force tell you?

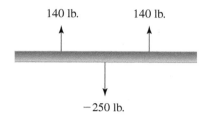

Understand Because weight is a force due to gravity pulling downward on an object, weight has a negative value. The wires pull upward against gravity; so those upward forces have positive values.

Plan To find the resultant force, we compute the sum of the forces.

Execute resultant force $= -250 + 140 + 140$
$\quad\quad\quad\quad\quad\;\; = -250 + 280$ **Add the positive numbers.**
$\quad\quad\quad\quad\quad\;\; = 30$ **Add -250 and 280.**

Answer The resultant force is 30 lb. Because the resultant force is positive, the beam is moving upward.

Check Compute $-250 + 140 + 140$ from left to right.

$\quad\quad -250 + 140 + 140$
$\quad\quad = -110 + 140$ **Add -250 and 140.**
$\quad\quad = 30$ **Add -110 and 140.**

▶ **Do Your Turn 8**

▶ **Your Turn 8**

A 500-pound concrete slab is suspended by four ropes tied to the corners. Each rope has an upward force of 110 pounds. What is the resultant force? What does the resultant force tell you?

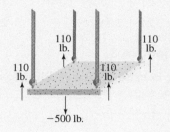

Of Interest

In engineering and physics, the preceding diagram is called *a free-body diagram*. Because forces have a value as well as a direction of action, we refer to a force as a *vector*. We use vectors in situations when both quantity and direction affect the situation. Force, velocity, and acceleration are all vector quantities because they have both value and direction.

Answer to Your Turn 8
-60 lb. Because the resultant force is negative, the slab is moving downward.

2.2 Exercises $\quad$ For Extra Help $\quad$ MyMathLab®

Objective 1

Prep Exercise 1 Explain $-24 + (-6)$ in terms of debts or credits.

For Exercises 1–10, add.
Key thought: Adding integers with the same sign is similar to increasing assets or increasing debts.

1. $14 + 9$ **2.** $74 + 13$ **3.** $-15 + (-8)$ **4.** $-14 + (-16)$

5. $-9 + (-6)$ **6.** $-31 + (-9)$ **7.** $12 + 18 + 16$ **8.** $4 + 15 + 41$

9. $-38 + (-17) + (-21)$ **10.** $-60 + (-19) + (-32)$

Objective 2

Prep Exercise 2 Explain $-20 + 12$ in terms of debts or credits.

For Exercises 11–20, add.
Key thought: Adding integers with different signs is similar to reconciling assets and debts.

11. $28 + (-18)$ **12.** $17 + (-5)$ **13.** $-34 + 20$ **14.** $-25 + 16$

15. $-21 + 35$ **16.** $-35 + 47$ **17.** $35 + (-53)$ **18.** $16 + (-29)$

19. $-24 + 80$ **20.** $75 + (-43)$

Objective 3

Prep Exercise 3 When adding two numbers that have the same sign, we _____ their absolute values and _____ .

Prep Exercise 4 When adding two numbers that have different signs, we _____ their absolute values and _____ .

For Exercises 21–42, add.

21. $48 + 90$ **22.** $35 + 16$ **23.** $-45 + (-27)$ **24.** $-68 + (-42)$

25. $84 + (-23)$ **26.** $68 + (-15)$ **27.** $-15 + 42$ **28.** $-54 + 65$

29. $-81 + 60$ **30.** $-45 + 21$ **31.** $37 + (-58)$ **32.** $62 + (-92)$

33. $-45 + 45$ **34.** $64 + (-64)$

35. $48 + (-18) + 16 + (-12)$ **36.** $69 + (-20) + (-15) + 11$

37. $-36 + (-17) + 94 + (-9)$ **38.** $-42 + 25 + (-61) + (-3)$

39. $25 + (-17) + (-33) + 19 + 6$ **40.** $40 + 16 + (-51) + 3 + (-8)$

41. $-75 + (-14) + 38 + 9 + (-17)$ **42.** $-60 + 18 + (-40) + (-12) + 93$

Objective 4

Prep Exercise 5 The additive inverse of a nonzero number has the same _____ but _____ sign.

Prep Exercise 6 The additive inverse of 0 is _____ .

For Exercises 43–46, find the additive inverse.

43. 18   **44.** 42   **45.** -29 **46.** -37

Objective 5

For Exercise 47–58, solve.

47. Jason's checking account shows a balance of $24. Unfortunately, he forgot about a check for $40 and it clears. The bank then charges $17 for insufficient funds. What is his new balance?

48. Charlene has a current balance of −$1243 on her credit card. During the month, she makes the transactions shown below. What will her balance be at the end of the month?

Charges	
Truman's	$58
Dave's Diner	$13
Fuel 'n Go	$15
Finance charge	$18
Payment	$150

49. The following table lists the assets and debts for the Smith family. Calculate their net worth.

Assets	Debts
Savings = $1498	Credit card balance = $1841
Checking = $2148	Mortgage = $74,614
Furniture = $18,901	Automobile = $5488
Jewelry = $3845	

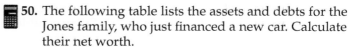

50. The following table lists the assets and debts for the Jones family, who just financed a new car. Calculate their net worth.

Assets	Debts
Savings = $214	Student loans = $15,988
Checking = $1242	Automobile 1 = $4857
Furniture = $21,358	Automobile 2 = $23,410

51. Lea's current credit card balance is −$125. During the month, she makes the following purchases with her card. If she is charged $9 in interest and makes a $325 payment, what will her new balance be at the end of the month?

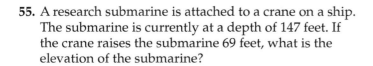

PURCHASES	
JACK'S CAFÉ	$12
MINUTE DRY-CLEAN	$8
TIRES ARE US	$120
PET COMFORT	$22

52. Greg's January closing balance on his credit card was −$1320. In February, he made one payment of $450 and another payment of $700. His finance charges were $19. Greg made purchases at Tommy's for $32 and Petromart for $27. What is his February closing balance?

53. A scientist working during the summer at the South Pole station notes that the temperature is about −21°F. During the winter, she can expect the temperature to be as much as 57°F colder. What temperature might she expect during the winter?

Of Interest

Because of the tilt of the earth, the South Pole experiences daylight for six months (summer) and then night for six months (winter). The sun rises in mid-September and sets in mid-March.

54. The Inuit people in the arctic regions of North America sometimes build igloos of snow blocks. During winter months, the temperatures can be as low as −58°F. A small heater can raise the internal temperature of a snow-block igloo by 108°F. What would the temperature be inside the igloo?

55. A research submarine is attached to a crane on a ship. The submarine is currently at a depth of 147 feet. If the crane raises the submarine 69 feet, what is the elevation of the submarine?

56. A mining team is working 147 feet below ground. If they dig straight down another 28 feet, what is their new position?

57. An elevator weighs 745 pounds. When the elevator is on the first floor of a 12-story building, the steel cable connected to the top of the elevator adds another 300 pounds. Three people enter the elevator. One person weighs 145 pounds; the second, 185 pounds; and the third, 168 pounds. The motor exerts an upward force of 1800 pounds. What is the resultant force?

58. A concrete block is suspended by two cables. The block weighs 500 pounds. Each cable is exerting 250 pounds of upward force. What is the resultant force? What does this mean?

For Exercises 59 and 60, use the following graph, which shows the amount the SAT scores are above or below the national average score according to parental education level in 2009. (Source: The College Board)

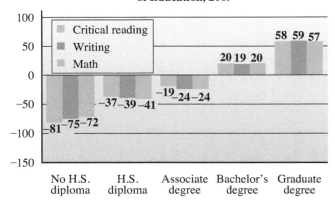

Amount SAT Scores Differed from Average According to Parents' Highest Level of Education, 2009

59. The national average SAT critical reading score in 2009 was 501.

  a. Complete the following table. In each block of the row labeled "Expression," write a numeric expression for the sum of the national average critical reading score and each amount the critical reading score differed from average from the graph. In each block of the row labeled "Sum," calculate the sum, which will be the average critical reading score according to parental educational level.

	No H.S. diploma	H.S. diploma	Associate degree	Bachelor's degree	Graduate degree
Expression					
Sum					

  b. What does the graph suggest about performance on the critical reading portion of the SAT based on parental educational level?

60. The national average SAT math score in 2009 was 515.

  a. Complete the following table. In each block of the row labeled "Expression," write a numeric expression for the sum of the national average math score and each amount the math score differed from average from the graph. In each block of the row labeled "Sum," calculate the sum, which will be the average math score according to parental educational level.

	No H.S. diploma	H.S. diploma	Associate degree	Bachelor's degree	Graduate degree
Expression					
Sum					

  b. What does the graph suggest about performance on the math portion of the SAT based on parental educational level?

Puzzle Problem A snail is at the bottom of a well that is 30 feet deep. The snail climbs up 3 feet per hour, then stops to rest for an hour. During each rest period, the snail slides back down 2 feet. If the snail began climbing at 8 A.M., at what time will the snail get out of the well?

Review Exercises

[1.3] 1. Write 42,561,009 in expanded form using powers of 10.

[1.1] 2. Write the word name for 2,407,006.

[1.2] 3. Subtract. $60,041 - 4596$

[1.2] 4. Solve and check. $15 + n = 28$

[1.5] 5. Simplify. $7 \cdot 5 - 2(14 - 9)$

[1.6] 6. Becky plans to place a fence around her property. A plot plan of the property is shown below. She plans to have two 10-foot gates. The fence company charges $7 per foot for the fence and $80 for each gate. What will be the total cost?

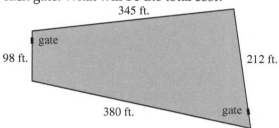

345 ft.

gate

98 ft.

212 ft.

380 ft.

gate

2.3 Subtracting Integers and Solving Equations

Objectives

1 Write subtraction statements as equivalent addition statements.
2 Solve equations containing an unknown addend.
3 Solve applications involving subtraction of integers.

Warm-up

[2.2] 1. Add.
   a. $-16 + (-5)$ b. $-20 + 8$
   c. $14 + (-9)$ d. $10 + (-10)$

[2.2] 2. The additive inverse of -3 is _____.

Objective 1 Write subtraction statements as equivalent addition statements.

In Section 2.2, we learned that when we add two numbers with different signs, we actually subtract their values. This suggests that addition and subtraction are closely related. In fact, any subtraction statement can be written as an equivalent addition statement. This can be quite useful because some subtraction problems involving signed numbers are more challenging to analyze than their equivalent addition statements. Let's consider some examples to determine how to write subtraction as addition.

Example 1

a. Suppose we have $30 in a checking account and we write a check for $42. What is the balance?

Solution: Because the check amount of $42 is a deduction, we subtract it from the initial balance of $30. Because the deduction is $12 more than the amount in the account, the balance is $-\$12$. We can write the following subtraction statement:

$$30 - 42 = -12$$

In Section 2.2, we would have solved this problem using the following addition:

$$30 + (-42) = -12$$

This suggests that $30 - 42$ and $30 + (-42)$ are equivalent. Notice that two things changed going from the subtraction to the addition: (1) the operation symbol changed from $-$ to $+$, and (2) the subtrahend, 42, changed to its additive inverse, -42.

The operation changes from $-$ to $+$. $30 - 42$ The subtrahend changes to its additive inverse.

$$= 30 + (-42)$$
$$= -12$$

b. Suppose we have a balance of $-\$12$ and write a check for $15. What is the balance?

Solution: As in part a, we subtract the check amount from the initial balance. In this case, however, the initial balance is negative, which is a debt; so the deduction of $15 increases the debt, making the balance $-\$27$. We can write the following subtraction statement:

$$-12 - 15 = -27$$

Note that this subtraction is equivalent to the following addition:

$$-12 + (-15) = -27$$

Also, as in part a, the same two things changed going from the subtraction to the addition: (1) the operation symbol changed from − to +, and (2) the subtrahend changed to its additive inverse (15 changed to −15).

$$-12 \ - \ 15$$

The operation changes from − to +.   The subtrahend changes to its additive inverse.

$$= -12 + (-15)$$
$$= -27$$

Connection Subtracting a positive number is equivalent to adding a debt.

Example 1 suggests the following procedure.

Procedure

To write a subtraction statement as an equivalent addition statement:
1. Change the operation symbol from − to +.
2. Change the subtrahend (second number) to its additive inverse.

▶ **Do Your Turn 1**

What if we subtract a negative number? We will see that our procedure still applies.

Example 2

a. Suppose we have a balance of $28 in a checking account. The bank discovers that we were charged a $5 service fee by mistake. The bank indicates that the fee will be corrected. What will the balance be after the correction?

Solution: The service charge is a debt; so it is written as −5. To correct the mistake, the bank must subtract that debt. Subtracting the debt means that $5 is deposited back into the account; so the resulting balance is $33. As a subtraction statement, we would write the following:

$$28 - (-5) = 33$$

The following equivalent addition statement is easier to think through.

$$28 + 5 = 33$$

Two things changed in writing the subtraction as an addition: (1) the operation symbol changed from − to +, and (2) the subtrahend changed to its additive inverse (−5 changed to 5).

$$28 \ - \ (-5)$$

The operation changes from − to +.   The subtrahend changes to its additive inverse.

$$= 28 + \ 5$$
$$= 33$$

b. Suppose we owe a friend $50, and out of kindness, the friend decides to cancel (or remove) $30 of that debt. What is the new balance?

Solution: The $50 debt is written as −50. The $30 is also debt; so it is written as −30. Because the friend is canceling, or removing, that debt, we subtract. If $30 of the $50 debt is removed, we now owe the friend $20; so the balance is −$20. We can write the following subtraction statement:

$$-50 - (-30) = -20$$

▶ **Your Turn 1**

Write as an equivalent addition; then evaluate.

  a. $30 - 48$ **b.** $24 - 82$

  c. $-16 - 25$ **d.** $-33 - 44$

Answers to Your Turn 1
a. $30 + (-48) = -18$
b. $24 + (-82) = -58$
c. $-16 + (-25) = -41$
d. $-33 + (-44) = -77$

Write as an equivalent addition; then evaluate.

a. $42 - (-21)$

b. $-48 - (-10)$

c. $-36 - (-40)$

The following equivalent addition statement is easier to think through.

$$-50 + 30 = -20$$

Again, two things changed: (1) the operation symbol changed from $-$ to $+$, and (2) the subtrahend changed to its additive inverse (-30 changed to 30).

The operation changes from $-$ to $+$. ——— $-50 - (-30)$ ——— The subtrahend changes to its additive inverse.

$$= -50 + 30$$

$$= -20$$

Connection Subtracting a negative number is equivalent to making a deposit.

◀ **Do Your Turn 2**

Objective 2 Solve equations containing an unknown addend.

In Section 1.2, we learned how to solve equations containing an unknown addend. Let's recall the procedure.

Procedure

To solve for an unknown addend, write a related subtraction equation in which the known addend is subtracted from the sum.

Solve and check.

a. $18 + y = 12$

b. $c + 24 = 9$

c. $-15 + d = -23$

d. $k + (-21) = -6$

e. $9 + t = -17$

Example 3   Solve and check.

a. $25 + x = 10$

Solution: $x = 10 - 25$ Write a related subtraction equation in which the known addend, 25, is subtracted from the sum, 10.

$x = 10 + (-25)$ Write the subtraction as an equivalent addition.

$x = -15$ Add. ◀ **Note** Because 10 and -25 have different signs, we subtract their absolute values. The result is negative because -25 has the greater absolute value.

Check: $25 + (-15) \overset{?}{=} 10$ In $25 + x = 10$, replace x with -15.

$10 = 10$ Verify that the equation is true: $25 + (-15) = 10$ is true.

Explanation: Think about the original addition statement. If we start with 25 and end with 10, a deduction has taken place. The only way this happens with addition is when a negative number is added.

b. $-14 + x = -20$

Solution: $x = -20 - (-14)$ Write a related subtraction equation in which the known addend, -14, is subtracted from the sum, -20.

$x = -20 + 14$ Write the subtraction as an equivalent addition.

$x = -6$ Add.

Check: $-14 + (-6) \overset{?}{=} -20$ In $-14 + x = -20$, replace x with -6.

$-20 = -20$ Verify that the equation is true: $-14 + (-6) = -20$ is true.

Explanation: Think through the original addition statement. Starting at -14, to end up at -20, we must continue six more steps in the negative direction. Therefore, the second addend must be -6.

◀ **Do Your Turn 3**

Answers to Your Turn 2
a. $42 + 21 = 63$
b. $-48 + 10 = -38$
c. $-36 + 40 = 4$

Answers to Your Turn 3
a. -6 b. -15 c. -8
d. 15 e. -26

Objective 3 Solve applications involving subtraction of integers.

In Chapter 1, we learned that subtraction problems can involve taking one amount away from another, finding the difference between amounts, or finding an unknown addend. First, let's consider a situation involving taking away. Calculating a business's **net** involves subtracting its total **costs** from its total **revenue.** If the net is positive, we say that it is a **profit;** if the net is negative, we say that it is a **loss.**

Definitions **Net:** Money remaining after subtracting costs from revenue (money made minus money spent).

Cost: Money spent on production, operation, labor, and debts.

Revenue: Income (money made).

Profit: A positive net (when revenue is greater than cost).

Loss: A negative net (when revenue is less than cost).

The definition of net suggests the following formula.

$$\text{Net} = \text{revenue} - \text{cost}$$
$$N = R - C$$

Example 4 The financial report for a business indicates that the total revenue for 2010 was $2,453,530 and the total costs were $2,560,000. What was the net? Did the business experience a profit or loss?

Solution: We must find the net given the revenue and cost. The formula for net is $N = R - C$.

$$N = R - C$$
$$N = 2,453,530 - 2,560,000 \qquad \text{In } N = R - C, \text{ replace } R \text{ with } 2,453,530 \text{ and } C \text{ with } 2,560,000.$$
$$N = -106,470 \qquad \text{Subtract.}$$

Answer: The net for 2010 was $-\$106,470$, which means that the business had a loss of $106,470.

▶ **Do Your Turn 4**

Now let's consider difference problems. In these problems, we must find the amount between two given numbers by subtracting the smaller number from the larger number.

Example 5 On the evening news, a meteorologist says that the current temperature is 25°F. She then says that the evening low could get down to −6°F. How much of a change is this from the current temperature?

Solution: Because we are asked to find the amount of change, we must calculate a difference. Subtract the smaller number, −6, from the larger number, 25.

$$\text{temperature difference} = 25 - (-6)$$
$$= 25 + 6 \qquad \text{Write the subtraction as an equivalent addition statement.}$$
$$= 31$$

Answer: From 25°F to −6°F is a change of 31°F.

▶ **Do Your Turn 5**

▶ **Your Turn 4**

The revenue for one month for a small business was $45,382, and the total costs were $42,295. What was the net? Was it a profit or loss?

▶ **Your Turn 5**

Solve.

a. The surface temperature on the moon at lunar noon can reach 120°C. During the lunar night, the temperature can drop to −190°C. How much of a change in temperature occurs from day to night on the moon?

b. A submarine goes from a depth of −78 feet to a depth of −34 feet. What is the difference in depth?

Answer to Your Turn 4
$3087; profit

Answers to Your Turn 5
a. 310°C b. 44 ft.

VISUAL

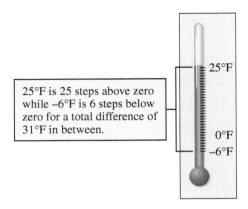

Learning Strategy

If you are a visual learner, draw a thermometer, which is just a vertical number line, to see that the difference in Example 5 is correct.

25°F is 25 steps above zero while –6°F is 6 steps below zero for a total difference of 31°F in between.

In a problem with an unknown addend, we are given a starting amount and must find how much more is needed to reach a new amount.

Example 6   A submarine is at −126 feet. How much must the submarine climb to reach −40 feet?

Understand The word *climb* indicates that we must add an amount to −126 feet to reach a value of −40 feet, but the amount added is unknown. A picture is helpful.

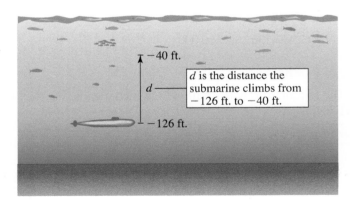

d is the distance the submarine climbs from −126 ft. to −40 ft.

Plan Write a missing addend equation; then solve.

Execute $-126 + d = -40$

$d = -40 - (-126)$ To solve, subtract the known addend, −126, from the sum, −40.

$d = -40 + 126$ Write the subtraction as an equivalent addition.

$d = 86$

Answer The submarine must climb 86 feet to reach a depth of −40 feet.

Check Verify that if the submarine climbs 86 feet from −126 feet, it will be at −40 feet.

$-126 + 86 \overset{?}{=} -40$ In the original equation, replace d with 86.

$-40 = -40$ True; so 86 is the solution.

◄ **Do Your Turn 6**

► **Your Turn 6**

Solve.

a. Barbara currently has a balance of −$58 in her checking account. To avoid any further charges, she must have a minimum balance of $25. How much should she deposit to have the minimum balance?

b. A mining team is on an elevator platform at a depth of −148 feet. They are to ascend to a depth of −60 feet to dig a side tunnel from the main shaft. How much does the elevator need to rise to get them to the appropriate depth?

Connection In the diagram, we can see that the unknown addend is the same as the difference between −40 and −126. This explains why the unknown addend statement, $-126 + d = -40$, can be written as the difference statement, $d = -40 - (-126)$.

Answers to Your Turn 6
a. $83   **b.** 88 ft.

2.3 Exercises For Extra Help MyMathLab®

Objective 1

Prep Exercise 1 Explain how to write subtraction as an equivalent addition.

Prep Exercise 2 In $2 - 8$, which number is the subtrahend? What is its additive inverse?

Prep Exercise 3 Fill in the blanks to complete the equivalent addition. $5 - 13 = 5$ _____.

Prep Exercise 4 Fill in the blanks to complete the equivalent addition. $-9 - 14 = -9$ _____.

For Exercises 1–6, write as an equivalent addition; then evaluate.

1. $18 - 25$ **2.** $30 - 54$ **3.** $-15 - 18$

4. $-21 - 6$ **5.** $0 - 9$ **6.** $0 - 16$

Prep Exercise 5 In $3 - (-7)$, which number is the subtrahend? What is its additive inverse?

Prep Exercise 6 Fill in the blanks to complete the equivalent addition. $12 - (-4) = 12$ _____.

Prep Exercise 7 Fill in the blanks to complete the equivalent addition. $-10 - (-6) = -10$ _____.

For Exercises 7–12, write as an equivalent addition; then evaluate.

7. $20 - (-8)$ **8.** $33 - (-12)$ **9.** $-14 - (-18)$

10. $-20 - (-30)$ **11.** $-15 - (-8)$ **12.** $-26 - (-14)$

For Exercises 13–26, write as an equivalent addition; then evaluate.

13. $-21 - 19$ **14.** $-13 - 28$ **15.** $-4 - (-19)$

16. $-19 - (-24)$ **17.** $31 - 44$ **18.** $9 - 15$

19. $35 - (-10)$ **20.** $26 - (-6)$ **21.** $-28 - (-16)$

22. $-15 - (-12)$ **23.** $0 - 18$ **24.** $0 - 6$

25. $0 - (-5)$ **26.** $0 - (-8)$

Objective 2

Prep Exercise 8 For the following equation, complete the related subtraction equation used to solve the equation.

$x + 10 = -9$

$\qquad x =$ _____

For Exercises 27–38, solve and check.

27. $18 + t = 12$ **28.** $21 + c = 16$ **29.** $d + (-6) = 9$

30. $a + (-14) = 15$ **31.** $-28 + u = -15$ **32.** $-22 + p = -19$

33. $-14 + m = 0$ **34.** $16 + n = 0$ **35.** $h + (-35) = -40$

36. $r + (-29) = -38$ **37.** $-13 + k = -25$ **38.** $-36 + x = -53$

Objective 3

For Exercises 39–56, solve.

39. Belinda has a balance of $126 in her bank account. A check written against her account for $245 arrives at the bank.

    a. What is her balance?

    b. Because Belinda has insufficient funds to cover the check amount, the bank assesses a service charge of $20. What is her balance after the service charge?

40. Brian has a credit of $86 on his account with an electronics store. He uses his credit account to purchase some equipment at a total cost of $585. What is his new balance after the purchase?

41. In 2009, Verizon Communications earned $97,354 million in revenue and had $90,926 million in total expenses. What was the net? Was it a profit or loss? (*Source: Fortune 500*, 2009.)

42. As of 2010, *Avatar* was the top-grossing movie of all time, grossing $2,729,673,452 in total box office sales in the world. If the budget for making the film was $387,000,000, what was the net? Was it a profit or loss? (*Source: Box Office Mojo*, June 2010.)

43. Florence spent a total of $26,458 for her car, including the costs of financing and maintenance. Five years later she sold the car for $4500. What was her net? Was it a profit or loss?

44. Gary bought a new computer in 2006 for $2675. In 2010, he sold it for $250. What was the net? Was it a profit or loss?

For Exercises 45 and 46, use the following graph, which shows the total income and disbursements in millions of dollars from the Social Security trust fund for the indicated years. (Source: Social Security Administration.)

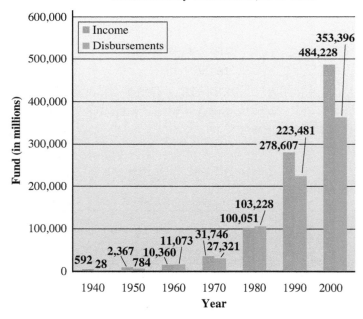

Social Security Trust Funds, 1940–2000

45. Which year had the greatest net loss? What was the net that year?

46. Which year had the greatest net profit? What was the net that year?

47. The temperature at sunset was reported to be 19°F. By midnight, it is reported to be −27°F. What is the amount of the decrease?

48. Liquid nitrogen has a temperature of −208°C. An orange is placed in the liquid nitrogen, which causes the liquid to boil at −196°C. What is the amount of the increase in temperature?

49. Molecular motion is minimized at −273°C. Oxygen goes from liquid to solid at −218°C. After oxygen turns from liquid to solid, how much colder does it need to get for its molecular motion to be minimized?

50. A container of liquid hydrogen fluoride is at a temperature of −90°C. The boiling point for hydrogen fluoride is 19°C. How much must the temperature rise for the hydrogen fluoride to boil?

51. Derrick has a balance of −$37 in his bank account. To avoid further charges, he must have a balance of $30. What is the minimum he can deposit to avoid further charges?

52. Danielle owns a business that is in financial trouble. Her accountant informs her that her net worth is −$5267. Her net worth should be at least $2500 by the end of the month; otherwise, she will be forced to close her business. What is the minimum profit she must clear to remain in business?

53. The lowest level of the Willis Tower in Chicago is at an elevation of −43 feet. The skydeck is at an elevation of 1353 feet. What is the distance between the lowest level and the skydeck?

54. As a result of initial expenses, a new business has a net worth of −$45,000. The goal for the first year is to make enough profit to have a net worth of $15,800. How much must the business make that first year to achieve its goal?

For Exercises 55 and 56, use the following graph, which shows the maximum and minimum normal temperatures in January for several cities in the United States.

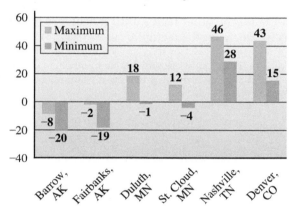

Source: The World Almanac and Book of Facts 2010.

55. a. Complete the following table. In the row labeled "Expression," write a numeric expression for the difference between the maximum and minimum temperature in January for each city. In each block of the row labeled "Difference," calculate the difference between the maximum and minimum temperature in January for each city.

	Barrow, AK	Duluth, MN	Nashville, TN
Expression			
Difference			

b. Which of the cities in part a had the greatest difference in maximum and minimum temperatures?

56. a. Complete the following table. In the row labeled "Expression," write a numeric expression for the difference between the maximum and minimum temperature in January for each city. In each block of the row labeled "Difference," calculate the difference between the maximum and minimum temperature in January for each city.

	Fairbanks, AK	St. Cloud, MN	Denver, CO
Expression			
Difference			

b. Which of the cities in the graph had the greatest difference in maximum and minimum temperatures?

Review Exercises

[1.3] 1. Multiply. $(145)(209)$

[1.4] 3. Divide. $11,597 \div 23$

[1.5] 5. $9 \cdot 7 + 3[4^3 - (9 + 5 \cdot 4)]$

[1.3] 2. Evaluate. $(3)^5$

[1.4] 4. Solve $3 \cdot y = 24$

[1.6] 6. Jacqueline owns 20 square miles of land that borders a national park. The government wants to purchase an 8-mile-long by 2-mile-wide rectangular strip of her land. How much land will she have left if she sells?

2.4 Multiplying and Dividing Integers; Exponents; Square Roots; Solving Equations

Objectives

1 Multiply integers.

2 Evaluate numbers in exponential form.

3 Divide integers.

4 Solve equations containing an unknown factor.

5 Evaluate square roots.

6 Solve applications involving multiplication or division of integers.

Warm-up

[2.3] 1. Write as an equivalent addition; then evaluate.

   a. $12 - 17$ b. $-13 - (-21)$

[1.3, 1.4] 2. Multiply or divide.

   a. 43×26 b. $2889 \div 14$

Objective 1 Multiply integers.

To discover the rules for multiplying and dividing signed numbers, we will look for a pattern in the following list of products. We'll keep the first factor the same and gradually decrease the other factor until it becomes negative. What happens to the product as we decrease the factor?

$2 \cdot 4 = 8$ ◄ From our work with whole numbers in Chapter 1, we know that when we multiply two positive numbers, the product is positive.

$2 \cdot 3 = 6$ ◄ Each time we decrease the second factor by 1, we lose one of the repeatedly added 2s so that the product decreases by 2.

$2 \cdot 2 = 4$

$2 \cdot 1 = 2$

$2 \cdot 0 = 0$ ◄ When we decrease the second factor from 1 to 0, the product must continue to decrease by 2. This is why $2 \cdot 0$ must equal 0, which affirms the rule that any number multiplied by 0 is 0.

$2 \cdot (-1) = -2$ ◄ As we continue, the pattern must continue. Notice that when we decrease the second factor from 0 to -1, we must continue to decrease the product by 2. This explains why multiplying $2 \cdot (-1)$ must equal -2, and so on.

$2 \cdot (-2) = -4$

From this pattern, we can conclude that multiplying a positive number by a negative number gives a negative product. The commutative property of multiplication tells us that rearranging the factors does not affect the product; so $2 \cdot (-2) = (-2) \cdot 2 = -4$. Therefore, multiplying a negative number times a positive number also gives a negative product.

Rule

When multiplying two numbers that have different signs, the product is negative.

► Your Turn 1

Multiply.

a. $9 \cdot (-7)$ b. $-8 \cdot 10$

c. $-4(5)$ d. $6(-3)$

Answers to Your Turn 1

a. -63 b. -80

c. -20 d. -18

Answers to Warm-up

1. a. $12 + (-17) = -5$

   b. $-13 + 21 = 8$

2. a. 1118   b. 206 r5

Example 1 Multiply.

a. $8 \cdot (-6)$

Answer: -48 ◄ **Note** The product is negative because we multiplied two numbers that have different signs.

b. $-12 \cdot 5$

Answer: -60

◄ Do Your Turn 1

Now let's consider multiplying two numbers with the same sign. We have already seen that the product of two positive numbers is positive. To discover the rule for multiplying two negative numbers, we will look for a pattern in another list of products. We begin with a negative number times a positive number and gradually decrease the positive factor until it becomes negative.

$(-2) \cdot 4 = -8$ ◄ We have already established that the product of a negative number and a positive number is negative.

$(-2) \cdot 3 = -6$ ◄ As we decrease the positive factor from 4 to 3, the product actually increases from −8 to −6, which is an increase of 2.

$(-2) \cdot 2 = -4$

$(-2) \cdot 1 = -2$

$(-2) \cdot 0 = 0$

$(-2) \cdot (-1) = 2$ ◄ To continue the same pattern, when we decrease the second factor from 0 to −1, we must continue to increase the product by 2. This means that the product must become positive.

$(-2) \cdot (-2) = 4$

We already knew that when we multiplied two positive numbers, the product would be positive. Now we've discovered that when we multiply two negative numbers, the product is positive as well.

Rule

When multiplying two numbers that have the same sign, the product is positive.

Example 2 Multiply.

a. $-6 \cdot (-8)$

Answer: 48   ◄ **Note** The product is positive because we multiplied two numbers that have the same sign.

b. $(-9)(-7)$

Answer: 63

► Do Your Turn 2

► **Your Turn 2**

Multiply.

a. $-5 \cdot (-8)$

b. $(-4)(-9)$

c. $-10(-13)$

We can now develop rules for multiplying more than two signed numbers. To discover those rules, let's look at some examples where we use the order of operations agreement, which tells us to multiply from left to right.

Example 3 Multiply.

a. $-2(-3)(-4)$

Solution: Multiply from left to right.

$$-2(-3)(-4)$$
$$= 6(-4) \qquad \text{Multiply } -2 \text{ by } -3 \text{ to get positive 6.}$$
$$= -24 \qquad \text{Multiply 6 by } -4 \text{ to get } -24.$$

Answer: -24 ◄ **Note** The product of three negative factors is negative.

Answers to Your Turn 2
a. 40   b. 36   c. 130

▶ Your Turn 3

Multiply.
a. $(-4)(-6)(-2)$
b. $(-1)(-1)(-2)(-5)$
c. $-5(-7)(-2)(-1)$
d. $-3(-5)(2)$
e. $5(-4)(3)(-1)(-1)$

b. $-1(-1)(-1)(-5)$

Solution: Multiply from left to right.

$$-1(-1)(-1)(-5)$$
$$= 1(-1)(-5) \qquad \text{Multiply} -1 \text{ by} -1 \text{ to get positive 1.}$$
$$= -1(-5) \qquad \text{Multiply 1 by} -1 \text{ to get} -1.$$
$$= 5 \qquad \text{Multiply} -1 \text{ by} -5 \text{ to get positive 5.}$$

Answer: 5   ◀ **Note** The product of four negative factors is positive.

In Example 3, we see that the number of negative factors determines the sign of the product.

Rules

When multiplying signed numbers, count the negative factors:
- If there are an *even* number of negative factors, the product is *positive*.
- If there are an *odd* number of negative factors, the product is *negative*.

◀ **Do Your Turn 3**

Look back at Example 3b. Multiplication by -1 can be written without the 1 so that we see only the negative sign.

$$-1(-1)(-1)(-5) = -(-(-(-5))) = 5$$

Note When we do not write a 1 factor, we say that it is "understood."

The result can be found by counting the total number of minus signs and applying the rules we just learned. Because there are an even number of minus signs (four) indicating an even number of negative factors, the result is positive.

▶ Your Turn 4

Simplify.
a. $-(-(-(-23)))$
b. $-(-(-(-(-42))))$

Example 4   Simplify.

a. $-(-8)$

Solution: $-(-8) = 8$ There are an even number of minus signs (two); so the product is positive. Also, $-(-8) = -1(-8) = 8$.

Connection The fact that $-(-8) = 8$ is consistent with what we learned in writing subtractions as equivalent additions. The subtraction $5-(-8)$ is written as $5 + 8$.

b. $-(-(-10))$

Solution: $-(-(-10)) = -10$ There are an odd number of minus signs (three); so the product is negative. Also, $-(-(-10)) = -1(-1)(-10) = -10$.

◀ **Do Your Turn 4**

Answers to Your Turn 3
a. -48 b. 10   c. 70
d. 30   e. -60

Answers to Your Turn 4
a. 23   b. -42

Objective 2 Evaluate numbers in exponential form.

Because exponents mean repeated multiplication, we can extend what we've learned to evaluating exponential forms with negative bases.

Example 5 Evaluate.

a. $(-5)^2$

Solution: $(-5)^2 = (-5)(-5) = 25$

Explanation: The exponent 2 means that we multiply the base -5 by itself. Because we are multiplying two negative numbers, the product is positive.

b. $(-2)^3$

Solution: $(-2)^3 = (-2)(-2)(-2) = -8$

Explanation: The exponent 3 means that we multiply the base -2 as a factor three times. Because we have an odd number of negative factors, the result is negative.

Example 5 suggests the following rules:

> ### Rules
> When evaluating an exponential form that has a negative base:
> * If the exponent is even, the product is positive.
> * If the exponent is odd, the product is negative.

▶ **Do Your Turn 5**

A negative base must be in parentheses, as in $(-2)^4$. Without the parentheses, as in -2^4, the expression is evaluated differently. The minus sign in -2^4 is an understood -1 multiplying 2^4.

$$-2^4 = -1 \cdot 2^4 = -1 \cdot 2 \cdot 2 \cdot 2 \cdot 2 = -16$$

Example 6 Evaluate. -3^4

Solution: $-3^4 = -1 \cdot 3 \cdot 3 \cdot 3 \cdot 3 = -81$

▶ **Do Your Turn 6**

Objective 3 Divide integers.

To understand the rules for determining the sign of a quotient, look at the following multiplications and their related divisions. The third column shows the conclusions that can be drawn.

Multiplication	Related division	Conclusion
$5 \cdot 3 = 15$	$15 \div 3 = 5$	When dividing two numbers that have the same sign, the result is positive.
$6 \cdot (-7) = -42$	$-42 \div (-7) = 6$	
$-2 \cdot 3 = -6$	$-6 \div 3 = -2$	When dividing two numbers that have different signs, the result is negative.
$-4 \cdot (-5) = 20$	$20 \div (-5) = -4$	

Notice that the rules for determining the sign of a quotient follow the same pattern as the rules for determining the sign of a product; so we can combine them as follows:

> ### Rules
> When multiplying or dividing two numbers:
> * If they have the same sign, the result is positive.
> * If they have different signs, the result is negative.

▶ **Your Turn 5**

Evaluate.

a. $(-2)^5$ b. $(-2)^6$

▶ **Your Turn 6**

Evaluate.

a. -7^2 b. -4^3

Answers to Your Turn 5
a. -32 b. 64

Answers to Your Turn 6
a. -49 b. -64

▶ **Your Turn 7**

Divide.

a. $36 \div (-4)$

b. $-40 \div 8$

c. $\dfrac{-28}{-7}$

d. $\dfrac{-54}{9}$

e. $\dfrac{-72}{-12}$

f. $\dfrac{-18}{0}$

Example 7 Divide.

a. $-45 \div 9$

Answer: -5 ◀ **Note** Because -45 and 9 have different signs, their quotient is negative.

b. $\dfrac{-18}{-6}$

Answer: 3   ◀ **Note** Because -18 and -6 have the same sign, their quotient is positive.

c. $-15 \div 0$

Answer: undefined

> **Connection** In Section 1.4, we learned that if the divisor is 0 and the dividend is any number other than 0, the quotient is undefined.

◀ **Do Your Turn 7**

Objective 4 Solve equations containing an unknown factor.

In Section 1.4, we learned how to solve equations involving an unknown factor. Let's recall the procedure.

> **Procedure**
>
> To solve for an unknown factor, write a related division equation in which the product is divided by the known factor.

Example 8 Solve and check.

a. $9x = -63$

Solution: $x = \dfrac{-63}{9}$ Write a related division equation in which the product, -63, is divided by the known factor, 9.

$x = -7$ Divide.   ◀ **Note** Because -63 and 9 have different signs, their quotient is negative.

Check: $9(-7) \stackrel{?}{=} -63$ In $9x = -63$, replace x with -7.

$-63 = -63$ Verify that the equation is true: $9(-7) = -63$ is true.

b. $-30y = -60$

Solution: $y = \dfrac{-60}{-30}$ Write a related division equation in which the product, -60, is divided by the known factor, -30.

$y = 2$ Divide.   ◀ **Note** Because -60 and -30 have the same sign, their quotient is positive.

Check: $-30(2) \stackrel{?}{=} -60$ In $-30y = -60$, replace y with 2.

$-60 = -60$ Verify that the equation is true: $-30(2) = -60$ is true.

Answers to Your Turn 7
a. -9 b. -5 c. 4 d. -6
e. 6 f. undefined

c. $0n = 19$

Solution: $n = \dfrac{19}{0}$ Write a related division equation.

Because $\dfrac{19}{0}$ is undefined, $0n = 19$ has no solution.

▶ Do Your Turn 8

▶ **Your Turn 8**

Solve and check.
 a. $7n = -42$
 b. $-12k = -36$
 c. $-8h = 96$
 d. $0m = -14$

Objective 5 Evaluate square roots.

We have learned that raising a number to the second power means to multiply the number by itself. For example, $5^2 = 5 \cdot 5 = 25$. We also have learned that 5^2 can be read as "five squared" and the answer, 25, is the "square of five." If we were to reverse this operation, we would say that "the **square root** of 25 is 5."

Definition Square root: The square root of a given number is a number whose square is the given number.

Notice that $(-5)^2 = (-5)(-5) = 25$ as well; so 25 has two square roots, 5 and -5. In fact, every positive integer has two square roots—a positive square root and a negative square root. Both of these square roots have the same absolute value.

Suppose we were asked for the square roots of a negative integer such as -25. That would mean to find a number whose square is -25. But the product of two numbers with the same sign is always positive. Therefore, a negative integer has no square roots in the number sets we will study in this course. All of the numbers we will study in this course are in the set of *real* numbers (whole numbers, integers, fractions, and decimals). So we will say that a negative number has no real square roots.

> **Rules**
>
> Every positive integer has two square roots, a positive square root and a negative square root with the same absolute value as the positive square root.
>
> A negative number has no real square roots.

It is helpful to memorize some perfect squares and their square roots. The following table lists the first 20 perfect squares.

Root	Perfect Square	Root	Perfect Square
0	$0^2 = 0$	10	$10^2 = 100$
1	$1^2 = 1$	11	$11^2 = 121$
2	$2^2 = 4$	12	$12^2 = 144$
3	$3^2 = 9$	13	$13^2 = 169$
4	$4^2 = 16$	14	$14^2 = 196$
5	$5^2 = 25$	15	$15^2 = 225$
6	$6^2 = 36$	16	$16^2 = 256$
7	$7^2 = 49$	17	$17^2 = 289$
8	$8^2 = 64$	18	$18^2 = 324$
9	$9^2 = 81$	19	$19^2 = 361$

Connection Notice that the ones places in the perfect squares are identical in the two columns.

$0^2 = 0$ $10^2 = 100$
$1^2 = 1$ $11^2 = 121$
$2^2 = 4$ $12^2 = 144$
$3^2 = 9$ $13^2 = 169$

They are identical because the ones places in the square roots are identical.

Answers to Your Turn 8
a. -6 **b.** 3 **c.** -12
d. no solution

▶ **Your Turn 9**

Find all square roots of each number.

a. 100   **b.** 144   **c.** −36

Example 9   Find all square roots of each number.

a. 81

Answers: 9 and −9   ◀ **Note** We can express both the positive and negative root more concisely as ±9.

Explanation: There are two square roots because $9^2 = 81$ and $(-9)^2 = 81$.

b. −16

Answer: No real roots exist.

Explanation: Squaring any nonzero number, whether positive or negative, always produces a positive product; so no real roots exist.

◀ **Do Your Turn 9**

The *radical sign* $\sqrt{}$ indicates the positive square root, or *principal square root*. The number inside the radical is the *radicand*. For example, $\sqrt{49} = 7$ because 7 is the positive square root of 49. To indicate the negative square root of a number using the radical sign, we place a minus sign in front of the radical, as in $-\sqrt{49} = -7$. We can write the following rules:

Connection Think of $-\sqrt{49}$ as the product of −1 and the positive square root of 49.
$$-\sqrt{49} = -1 \cdot \sqrt{49} = -1 \cdot 7 = -7$$

Rules

$\sqrt{n}$ indicates the positive square root (principal square root) of the radicand, n.
$-\sqrt{n}$ indicates the negative square root of the radicand, n.

▶ **Your Turn 10**

Simplify.

a. $\sqrt{121}$ **b.** $-\sqrt{225}$
c. $\sqrt{-81}$

Example 10   Simplify.

a. $\sqrt{144}$

Answer: 12

Explanation: The radical sign indicates the positive square root.

b. $-\sqrt{9}$

Answer: −3

Explanation: $-\sqrt{9}$ indicates the negative square root of 9. Also, $-\sqrt{9} = -1 \cdot \sqrt{9} = -1 \cdot 3 = -3$.

c. $\sqrt{-36}$

Answer: $\sqrt{-36}$ is not a real number

Explanation: There is no positive number whose square is −36.

◀ **Do Your Turn 10**

Answers to Your Turn 9
a. ±10   **b.** ±12
c. no real roots exist

Answers to Your Turn 10
a. 11   **b.** −15
c. not a real number

Objective 6 Solve applications involving multiplication or division of integers.

| Example 11 | Solve.

a. Jean was traveling for three months. Each month while she was away, $19 in finance charges was posted to her account. What were her total finance charges upon her return?

Solution: A $19 finance charge is a debt; so we write it as $-$19$. Because Jean was charged the same amount each month for three months, we are repeatedly adding the same amount. Repeated addition means that we can multiply. $-$19$ by 3.

$$\text{Total finance charge} = -\$19 \cdot 3 = -\$57$$

b. Rick's total debt is $12,600. To pay off the debt, he makes equal monthly payments for five years. What is his monthly debt payment?

Solution: Because the $12,600 is debt, we write it as $-$12,600$. There are 12 months in a year; so making one payment each month for five years means that he makes $5(12) = 60$ payments over the five years. Splitting the debt into equal payments means that we divide $-12,600$ by 60.

$$\text{Monthly payment} = -\$12,600 \div 60 = -\$210$$

▶ Do Your Turn 11

2.4 Exercises (For Extra Help) MyMathLab®

Objective 1

Prep Exercise 1 When multiplying two numbers that have the same sign, the result is _____.

Prep Exercise 2 When multiplying two numbers that have different signs, the result is _____.

For Exercises 1–26, multiply.

1. $-2 \cdot 16$
2. $-4 \cdot 11$
3. $4 \cdot (-2)$
4. $6 \cdot (-4)$
5. $(-2) \cdot (-32)$
6. $(-4) \cdot (-17)$
7. $-21 \cdot (-8)$
8. $-9 \cdot (-18)$
9. $0 \cdot (-9)$
10. $0 \cdot (-5)$
11. $-13 \cdot 0$
12. $-4 \cdot 0$
13. $15(-1)$
14. $12(-2)$
15. $-19(-20)$
16. $-14(-31)$
17. $(-1)(-5)(-7)$
18. $(-1)(-3)(-6)$
19. $(-1)(-1)(-6)(-9)$
20. $(-1)(-1)(-8)(-7)$
21. $(-1)(-1)(-1)(-1)(-78)$
22. $(-1)(-1)(-1)(-1)(-94)$
23. $-(-13)$
24. $-(-19)$
25. $-(-(-36))$
26. $-(-(-44))$

Objective 2

Prep Exercise 3 If the base of an exponential form is a negative number and the exponent is even, the product is _____.

Prep Exercise 4 If the base of an exponential form is a negative number and the exponent is odd, the product is _____.

For Exercises 27–34, evaluate.

27. $(-3)^2$ **28.** $(-2)^2$ **29.** $(-4)^3$ **30.** $(-5)^3$

31. $(-3)^4$ **32.** $(-2)^4$ **33.** $(-4)^5$ **34.** $(-3)^5$

Prep Exercise 5 Explain how to evaluate -8^2.

For Exercises 35–42, evaluate.

35. -8^2 **36.** -11^2 **37.** -10^6 **38.** -10^4

39. -1^2 **40.** -1^3 **41.** -3^3 **42.** -2^3

Objective 3

Prep Exercise 6 When dividing two num bers that have the same sign, the result is _____.

Prep Exercise 7 When dividing two numbers that have different signs, the result is _____.

For Exercises 43–58, divide.

43. $36 \div (-3)$ **44.** $40 \div (-10)$ **45.** $-81 \div 27$ **46.** $-48 \div 8$

47. $-32 \div (-4)$ **48.** $-100 \div (-20)$ **49.** $0 \div (-2)$ **50.** $0 \div (-12)$

51. $\dfrac{65}{-13}$ **52.** $\dfrac{-96}{8}$ **53.** $\dfrac{-124}{-4}$ **54.** $\dfrac{-91}{-7}$

55. $-\dfrac{28}{4}$ **56.** $-\dfrac{42}{7}$ **57.** $\dfrac{-41}{0}$ **58.** $\dfrac{-27}{0}$

Objective 4

Prep Exercise 8 For the following equation, complete the related division equation used to solve the equation.

$3x = -15$

$x = $ _____

For Exercises 59–76, solve and check.

59. $4x = 12$ **60.** $7p = 28$ **61.** $9x = -18$

62. $5x = -125$ **63.** $-12t = 48$ **64.** $-14g = 42$

65. $-1c = 17$ **66.** $-1c = 12$ **67.** $-6m = -54$

68. $-8n = -32$ **69.** $-18m = 0$ **70.** $-25b = 0$

71. $0v = -14$ **72.** $0b = 28$ **73.** $-2(-5)d = -50$

74. $3(-8)f = 72$ **75.** $-1(-1)(-7)g = -63$ **76.** $(-9)(-1)(2)w = -90$

Objective 5

Prep Exercise 9 Every positive integer has two square roots, a _____ square root and a _____ square root.

Prep Exercise 10 A negative number has _____ real square roots.

For Exercises 77–82, find all square roots of each number.

77. 16   **78.** 49   **79.** 36

80. 25   **81.** -81 **82.** -121

Prep Exercise 11 $\sqrt{n}$ indicates the _____ square root (principal square root) of the radicand, n.

Prep Exercise 12 $-\sqrt{n}$ indicates the _____ square root of the radicand, n.

For Exercises 83–94, simplify.

83. $\sqrt{81}$

84. $\sqrt{36}$

85. $\sqrt{64}$

86. $\sqrt{100}$

87. $\sqrt{49}$

88. $\sqrt{25}$

89. $\sqrt{-169}$

90. $\sqrt{-144}$

91. $-\sqrt{121}$

92. $-\sqrt{256}$

93. $\sqrt{0}$

94. $\sqrt{1}$

Objective 6

For Exercises 95–100, solve.

95. Rohini had a balance of − $214 on her credit card in June. By November, her balance had tripled. What was her balance in November?

96. Michael has five credit accounts. Three of the accounts have a balance of −$100 each. Each of the other two accounts have a balance of −$258. What is Michael's total debt?

97. During one difficult year, the Morrisons had insufficient funds in their checking account on seven occasions, and each time they were assessed a charge appearing as −$17 on their statement. What were the total charges for insufficient funds that year?

98. An oil drilling crew estimated the depth of an oil pocket to be −450 feet. After drilling to three times that depth, they finally found oil. At what depth did they find oil?

99. Alicia borrowed $1656 from her friend and promised to pay him back with equal monthly payments for three years. How much should each payment be?

100. Liam borrowed $25,200 from his parents to pay for college. After graduation, he will pay off the loan by making equal monthly payments for six years. How much will each payment be?

Review Exercises

[1.2] 1. A wallpaper border is to be placed just below the ceiling around a room that is 14 feet wide by 16 feet long. How much border is needed?

[1.5] *For Exercises 2–4, Simplify.*

2. $19 - 6 \cdot 3 + 2^3$

3. $7 \cdot 3^2 - 2(5 + 3)$

4. $12 \cdot 4 - 2[5^2 - (6 + 4)]$

[1.6] 5. Sherry wants to paint a wall in her house. The wall is 15 feet long and 10 feet high and has a window that is 4 feet long by 3 feet wide. How many square feet must she be prepared to paint?

[1.6] 6. What is the volume of the object shown?

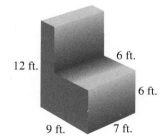

12 ft.   6 ft.   6 ft.   9 ft.   7 ft.

2.5   Order of Operations

Objective

1 Simplify numerical expressions by following the order of operations agreement.

Warm-up

[2.4] 1. Multiply or divide.

    a. $-13 \cdot 7$ **b.** $-72 \div (-8)$

[2.4] 2. Simplify.

    a. $(-3)^3$ **b.** $\sqrt{64}$

Objective 1 Simplify numerical expressions by following the order of operations agreement.

We need to modify the order of operations agreement we learned in Section 1.5 to include absolute value and radicals, which are grouping symbols.

> ### Order of Operations Agreement
> Perform operations in the following order:
> 1. Grouping Symbols: parentheses ( ), brackets [ ], braces { }, absolute value | |, radicals $\sqrt{}$, and fraction bars —
> 2. Exponents or roots in order as these operations occur from left to right
> 3. Multiplication or division in order as these operations occur from left to right
> 4. Addition or subtraction in order as these operations occur from left to right

Many people use the acronym $\overrightarrow{\text{GER}}$ $\overrightarrow{\text{MD}}$ $\overrightarrow{\text{AS}}$ to help remember the order of operations. The arrows over ER, MD, and AS remind us to perform exponents or roots, multiplication or division, and addition or subtraction from left to right.

 Learning Strategy

If you are an auditory learner, try creating a sentence in which the first letter of each word matches the first letter of each operation in the order of operations (g, e, r, m, d, a, s).

Give each runner more doughnuts and soda.

Be careful because a sentence such as the one above seems to indicate that exponents come before roots, multiplication before division, and addition before subtraction, which is not necessarily true. To avoid that confusion, you might try creating a sentence like the one below with hyphenated words to help you remember that exponents/roots, multiplication/division, and addition/subtraction are three stages in which you work from left to right.

Get ex-reality multi-dimensional actor-singers.

▶ **Your Turn 1**

Simplify.

  a. $-24 + (-9) \cdot 3$

  b. $32 - 16 \div (-4)$

  c. $-28 \div 4 + (-15)$

Answers to Your Turn 1
a. -51 b. 36   c. -22

Answers to Warm-up
1. a. -91 b. 9
2. a. -27 b. 8

| **Example 1** | Simplify. $-8 + 6 \cdot (-2)$ |

Solution: $-8 + 6 \cdot (-2)$

        $= -8 + (-12)$ Multiply. $6(-2) = -12$

        $= -20$ Add $-8 + (-12) = -20$

Note Write each new step underneath the previous step instead of outward to the right.

◀ **Do Your Turn 1**

Example 2 Simplify. $-16 - 20 \div (-4) \cdot 6$

Solution: $-16 - 20 \div (-4) \cdot 6$

$= -16 - (-5) \cdot 6$ Divide. $20 \div (-4) = -5$ ◄

$= -16 - (-30)$ Multiply. $-5 \cdot 6 = -30$

$= -16 + 30$ Write the subtraction as an equivalent addition.

$= 14$ Add. $-16 + 30 = 14$

Note We divide before multiplying here because we multiply or divide from left to right in the order they occur.

► **Do Your Turn 2**

► **Your Turn 2**

Simplify.

a. $18 - 48 \div (-6) \cdot (-3)$

b. $20 + (-3) \cdot 4 - (-6)$

c. $-7 \cdot 5 - (-30) \div 5$

d. $12 \cdot (-4) \div (-6) + 9 - 27$

Example 3 Simplify. $(12 - 4) \div (-2) + 1$

Solution: $(12 - 4) \div (-2) + 1$

$= 8 \div (-2) + 1$ Subtract in the parentheses. $12 - 4 = 8$

$= -4 + 1$ Divide. $8 \div (-2) = -4$

$= -3$ Add. $-4 + 1 = -3$

► **Do Your Turn 3**

► **Your Turn 3**

Simplify.

a. $4 - (3 - 5) \cdot 7$

b. $12 \div [4 - (-2)] + (-4)$

c. $-4 + 5(7 - 4)$

Example 4 Simplify. $-|-24 - 5(-2)|$ ◄ **Note** The minus sign to the left of the absolute value symbol is an understood -1 factor. As a reminder, we will write it in the solution steps.

Solution: $-1 \cdot |-24 - 5(-2)|$

$= -1 \cdot |-24 - (-10)|$ Multiply in the absolute value symbols. $5(-2) = -10$

$= -1 \cdot |-24 + 10|$ Write the subtraction as an equivalent addition.

$= -1 \cdot |-14|$ Add. $-24 + 10 = -14$

$= -1 \cdot 14$ Find the absolute value of -14, which is 14.

$= -14$ Multiply. $-1 \cdot 14 = -14$.

► **Do Your Turn 4**

► **Your Turn 4**

Simplify.

a. $|-4 - (6)(-5)|$

b. $-|25 + (-34)|$

c. $-2|48 \div (-6)|$

► **Your Turn 5**

Simplify.

a. $(-4)^2 + (2 - 7) - (-9)$

b. $5^2 \div (-1 + 6) + 1$

c. $2 - [9 - (-2)] + (-1)^3$

Example 5 Simplify. $(-1 + 6) - 2^3 + (-2)$

Solution: $(-1 + 6) - 2^3 + (-2)$

$= 5 - 2^3 + (-2)$ Add in the parentheses. $-1 + 6 = 5$

$= 5 - 8 + (-2)$ Evaluate the exponential form. $2^3 = 8$

$= -3 + (-2)$ Subtract. $5 - 8 = -3$

$= -5$ Add. $-3 + (-2) = -5$

► **Do Your Turn 5**

► **Your Turn 6**

Simplify.

a. $14 + (-2)^3 - 8 \cdot (-3)$

b. $(-5) \cdot 6 \div (-3) - \sqrt{49}$

c. $3^3 - 4 \cdot (-8) + \sqrt{81} - |24 \div (-8)|$

Answers to Your Turn 2
a. -6 b. 14 c. -29 d. -10

Answers to Your Turn 3
a. 18 b. -2 c. 11

Answers to Your Turn 4
a. 26 b. -9 c. -16

Answers to Your Turn 5
a. 20 b. 6 c. -10

Answers to Your Turn 6
a. 30 b. 3 c. 65

Example 6 Simplify. $8 - (-3)^2 + (-15) \div \sqrt{25}$

Solution: $8 - (-3)^2 + (-15) \div \sqrt{25}$

$= 8 - 9 + (-15) \div 5$ Evaluate the exponential form and square root. $(-3)^2 = 9$ and $\sqrt{25} = 5$

$= 8 - 9 + (-3)$ Divide. $-15 \div 5 = -3$

$= -1 + (-3)$ Subtract. $8 - 9 = -1$

$= -4$ Add. $-1 + (-3) = -4$

► **Do Your Turn 6**

▶ **Your Turn 7**

Simplify.

a. $[-2 - (-5)] + [1 + (-3)] \cdot \sqrt{4}$

b. $\sqrt{16} \div 4 - 2(4 - 7)$

c. $12 \div \sqrt{9} + 3(-2 + 4)$

Example 7 Simplify. $\sqrt{36} \div 3 + 2(-1 - 4)$

Solution: $\sqrt{36} \div 3 + 2(-1 - 4)$

$= \sqrt{36} \div 3 + 2[-1 + (-4)]$ In the brackets, write subtraction as equivalent addition.

$= \sqrt{36} \div 3 + 2(-5)$ Add in the parentheses.
$-1 + (-4) = -5$

$= 6 \div 3 + 2(-5)$ Evaluate the square root. $\sqrt{36} = 6$

$= 2 + (-10)$ Multiply and divide from left to right.
$6 \div 3 = 2$ and $2(-5) = -10$

$= -8$ Add. $2 + (-10) = -8$

◀ Do Your Turn 7

▶ **Your Turn 8**

Simplify.

a. $4(9 - 15) + 16 \div [2 + (-10)]$

b. $-20 + 6(3 - 8) - [8 + (-5)]^2$

c. $|19 - 29| \div 2 + [14 - 2(-6)] \div (-13)$

Example 8 Simplify. $2(8 - 12) + (-2)^3 - [20 + (-2)] \div (-6)$

Solution: $2(8 - 12) + (-2)^3 - [20 + (-2)] \div (-6)$

$= 2(-4) + (-2)^3 - 18 \div (-6)$ Calculate in the parentheses and brackets. $8 - 12 = -4$ and $20 + (-2) = 18$

$= 2(-4) + (-8) - 18 \div (-6)$ Evaluate the exponential form. $(-2)^3 = -8$

$= -8 + (-8) - (-3)$ Multiply and divide working from left to right. $2(-4) = -8$ and $18 \div (-6) = -3$

$= -16 - (-3)$ Add. $-8 + (-8) = -16$

$= -16 + 3$ Write the subtraction as an equivalent addition.

$= -13$ Add. $-16 + 3 = -13$

◀ Do Your Turn 8

Example 9 Simplify.

a. $\{15 - 4[2 + (-5)]\} - 40 \div \sqrt{9 + 16}$

Solution: $\{15 - 4[2 + (-5)]\} - 40 \div \sqrt{9 + 16}$

$= \{15 - 4[-3]\} - 40 \div \sqrt{9 + 16}$ Add in the brackets (the innermost grouping symbols). $2 + (-5) = -3$

$= \{15 - [-12]\} - 40 \div \sqrt{9 + 16}$ Multiply in the braces. $4 \cdot [-3] = -12$

$= \{15 + 12\} - 40 \div \sqrt{9 + 16}$ In the braces, write subtraction as addition.

$= 27 - 40 \div \sqrt{9 + 16}$ Add in the braces. $15 + 12 = 27$

$= 27 - 40 \div \sqrt{25}$ Add in the radical. $9 + 16 = 25$

$= 27 - 40 \div 5$ Evaluate the square root. $\sqrt{25} = 5$

$= 27 - 8$ Divide. $40 \div 5 = 8$

$= 19$ Subtract. $27 - 8 = 19$

▲

Note To find the square root of a sum or difference, we must add or subtract first, then find the square root of the sum or difference.

Answers to Your Turn 7
a. -1 b. 7   c. 10

Answers to Your Turn 8
a. -26 b. -59 c. 3

b. $5[(9 - 15) \div (-2)] - \sqrt{16 \cdot 9}$

Solution: $5[(9 - 15) \div (-2)] - \sqrt{16 \cdot 9}$

$= 5[(-6) \div (-2)] - \sqrt{16 \cdot 9}$ Subtract in the parentheses. $9 - 15 = -6$

$= 5[3] - \sqrt{16 \cdot 9}$ Divide in the brackets. $(-6) \div (-2) = 3$

$= 5[3] - \sqrt{144}$ Multiply in the radical. $16 \cdot 9 = 144$

$= 5[3] - 12$ Evaluate the square root. $\sqrt{144} = 12$

$= 15 - 12$ Multiply. $5[3] = 15$

$= 3$ Subtract. $15 - 12 = 3$

Note We could have performed some of the calculations in the same step. For example, we could have computed $5[3]$ and $16 \cdot 9$ in the same step. However, combining steps increases the likelihood of making careless mistakes.

▶ **Do Your Turn 9**

Example 10 Simplify. $\dfrac{8^2 - (-16)}{8 - 4(4 - 6)^3}$

Solution: Simplify the numerator and denominator separately, then divide.

$\dfrac{8^2 - (-16)}{8 - 4(4 - 6)^3}$

$= \dfrac{64 - (-16)}{8 - 4(-2)^3}$ Evaluate the exponential form in the numerator. $8^2 = 64$
   Subtract in the denominator. $4 - 6 = -2$

$= \dfrac{64 + 16}{8 - 4(-8)}$ Write an equivalent addition in the numerator. $64 - (-16) = 64 + 16$
   Evaluate the exponential expression in the denominator. $(-2)^3 = -8$

$= \dfrac{80}{8 - (-32)}$ Add in the numerator. $64 + 16 = 80$
   Multiply in the denominator. $4(-8) = -32$

$= \dfrac{80}{8 + 32}$ In the denominator, write subtraction as equivalent addition. $8 - (-32) = 8 + 32$

$= \dfrac{80}{40}$ Add in the denominator. $8 + 32 = 40$

$= 2$ Divide. $80 \div 40 = 2$

▶ **Do Your Turn 10**

▶ **Your Turn 9**

Simplify.

a. $\{-8 + 3[24 \div (4 + (-10))]\} + (-3)\sqrt{25 \cdot 4}$

b. $(-5)^3 - 4[(-6 + 14) \div (-7 + 6)] + \sqrt{81 \div 9}$

c. $\{[16 - 3(5 + 3)] + [26 - (3 - 9)^2]\} - 4\sqrt{25 - 9}$

d. $\{|-16(-3) - 40| + 7\} \div 3[-6 + \sqrt{121}]^2$

▶ **Your Turn 10**

Simplify.

a. $\dfrac{(2 - 14) + 4^3}{(-3)(-4) - 5^2}$

b. $\dfrac{2 \cdot (-4) - 7^2}{9 \cdot 5 - 3(3 - 7)^2}$

Answers to Your Turn 9
a. -50 b. -90
c. -34 d. 125

Answers to Your Turn 10
a. -4 b. 19

2.5 Exercises For Extra Help MyMathLab®

Objective 1

Prep Exercise 1 List the four stages of the order of operations.

Prep Exercise 2 What is the first step in simplifying the expression $12 + 5 \cdot 2$? Explain.

Prep Exercise 3 What is the first step in simplifying the expression $2(3 - 8) + 9$? Explain.

Prep Exercise 4 What is the first step in simplifying the expression $12 \div (-3) \cdot 4 - 20$? Explain.

For Exercises 1–74, simplify.

1. $3 - 10 \div 2$

2. $-8 \div 2 + 4$

3. $9 + 4 \cdot (-6)$

4. $15 - 5 \cdot 4$

5. $3 - 2 \cdot 4 + 11$

6. $2 + 16 \div 4 - 9$

7. $9 + 20 \div (-4) \cdot 3$

8. $-30 \div 2 \cdot (-3) + 21$

9. $3 - (-2)^2$

10. $(-3)^2 - 8$

11. $-5 + (-4)^2 - 1$

12. $5 - (-5)^2 + 13$

13. $9^2 - 6 \cdot (-4) \div 8$

14. $-14 - (-3)^2 \cdot 4 + 5$

15. $3 + (4 - 6^2)$

16. $-2 - (5 - 2^2)$

17. $3 + (4 - 6)^2$

18. $-2 - (5 - 2)^2$

19. $\sqrt{25 - 9}$

20. $\sqrt{16 + 9}$

21. $\sqrt{25} - \sqrt{9}$

22. $\sqrt{16} + \sqrt{9}$

23. $(-2)^4$

24. -2^4

25. -2^5

26. $(-2)^5$

27. $-(-3)^2$

28. $-(-2)^3$

29. $[-(-3)]^2$

30. $[-(-2)]^3$

31. $9 \div 3 - (-2)^2$

32. $4^3 - 3(-2)$

33. $3 - (-2)(-3)^2$

34. $2 - (-5)(-4)^2$

35. $|18 - 5(-4)|$

36. $|21 \div (-3) + 12|$

37. $-|-43 + 6 \cdot 4|$

38. $-|-16 - (-3)(-7)|$

39. $-5|26 \div 13 - 7 \cdot 2|$

40. $-4|3^2 + (-5)(6)|$

41. $(-3)^2 - 2[3 - 5(1 + 4)]$

42. $5[4 + 3(-2 + 1)] - (-4)^2$

43. $28 \div (-7) + \sqrt{49} + (-3)^3$

44. $25 - 6^2 + (-52) \div \sqrt{169}$

45. $12 + 8\sqrt{81} \div (-6)$

46. $-5\sqrt{36} + 40 \div (-5) \cdot 2 - 14$

47. $4^2 - 7 \cdot 5 + \sqrt{121} - 21 \div (-7)$

48. $(-2)^3 + 3\sqrt{25} - 18 \div 3$

49. $15 + (-3)^3 + (-5)|14 + (-2)9|$

50. $-28 \div 4 + |-3(-2)(-5)| - 2^5$

51. $(-15 + 12) + 5(-6) - [9 - (-12)] \div 3$

52. $39 \div 3 + (24 - 30) - 5^2 + [-21 - (-13)]$

53. $(-2) \cdot (5 - 8)^2 \div 6 + (4 - 2)$

54. $(5 - 7)^2 \div (-2) - (-3) \div 3$

55. $-|38 - 14 \cdot 2| + 44 \div [5 - (-6)] - 2^4$

56. $-3\sqrt{49} + |16 \div (-8)| - (28 - 40) \div (-3)$

57. $\{19 - 2[4 + (-9)]\} - 18 \div \sqrt{25 - 16}$

58. $\{12[-2 - (-5)] - 40\} \div \sqrt{25 - 9}$

59. $[(-14 + 2) + 5] \div 0 + 9^4 \cdot 7$

60. $[19(-2) - (-18)] \div \{15 - 5[2 - (-1)]\}$

61. $\{(-3)^2 + 4](8 - 20) \div (-2)]\} + (-5)\sqrt{25 \cdot 4}$

62. $4\sqrt{16 \cdot 9} - \{(-4)^3 + 2[18 \div (-2) + (4 - (-2))]\}$

63. $\{|-12(-5)-38| + 2\} \div 3[-9 + \sqrt{49}]^3$

64. $-3\{14 - 2|20 - 7(4)|\} + [(3)(-9) - (-21)]^2$

65. $\{2[14 - 11] + \sqrt{(4)(-9)}\} - [12 - 6 \cdot 9]$

66. $5^3 - [19 + 4\sqrt{16 - 25}] + |41 - 50|^2$

67. $\dfrac{6(-4) + 16}{(4 + 7) - 9}$

68. $\dfrac{34 - 5(2)}{7 + (9 - 4)}$

69. $\dfrac{3(-12) + 1}{3^2 - 2}$

70. $\dfrac{20 + 12(-3)}{3^2 - 1}$

71. $\dfrac{(5 - 19) + 3^3}{(-2)(-6) - 5^2}$

72. $\dfrac{2^4 + 3(7 + 19)}{8^2 - (2 \cdot 8 + 1)}$

73. $\dfrac{3 \cdot (-6) - 6^2}{-2(2 - 5)^2}$

74. $\dfrac{(-2)^3 + 5(7 - 15)}{4 \cdot 11 - 2(5 - 9)^2}$

75. Explain the difference between $(-2)^4$ and -2^4.

76. Explain the difference between $-(-3)^2$ and $[-(3)]^2$.

77. Why do both $(-2)^5$ and -2^5 simplify to -32?

78. Why is the answer to Exercise 60 undefined?

For Exercises 79–86, explain the mistake in each problem and work it correctly.

79. $28 - 5(24 - 30)$
$= 28 - 5(-6)$
$= 23(-6)$
$= -138$

80. $24 \div 2 \cdot 3 + 5$
$= 24 \div 6 + 5$
$= 4 + 5$
$= 9$

81. $4 - (9 - 4)^2$
$= 4 - (9 - 16)$
$= 4 - (-7)$
$= 4 + 7$
$= 11$

82. $19 - (-2^5)$
$= 19 - 32$
$= -13$

83. $34 - [3 \cdot 5 - (14 + 8)]$
$= 34 - 3 \cdot 5 - 22$
$= 34 - 15 - 22$
$= 19 - 22$
$= -3$

84. $\dfrac{3^2 - (25 - 4^2)}{2(-7)}$
$= \dfrac{3^2 - (25 - 16)}{-14}$
$= \dfrac{3^2 - 9}{-14}$
$= \dfrac{9 - 9}{-14}$
$= \dfrac{0}{-14}$, which is undefined

85. $\sqrt{169 - 25}$
$= 13 - 5$
$= 8$

86. $-2[16(3 - 5) + 7]$
$= -2[16(-2) + 7]$
$= -2[32 + 7]$
$= -2[39]$
$= -78$

For Excercises 87 and 88, use the following table, which contains the high and low temperatures in degrees Fahrenheit each day from March 14 to March 20 in Nome, Alaska.

	3/14	3/15	3/16	3/17	3/18	3/19	3/20
High	−6	0	19	25	26	29	26
Low	−30	−25	−11	−1	13	17	9

87. Find the mean, median, and mode high temperature.

88. Find the mean, median, and mode low temperature.

Review Exercises

[2.2] 1. Felicia has a balance of $185 in her checking account. If she writes checks for $45, $68, and $95, what is her new balance?

[2.3] 3. A submarine at −98 feet ascends to −25 feet. How far did the submarine ascend?

[2.3] 2. Tina has a television that she bought for $245. She sells it at a yard sale for $32. What was the net? Was it a profit or loss?

[2.4] 4. Jeff's credit card account has a balance of $0. If he uses the card to purchase three dining chairs that cost $235 each, what is the new balance?

2.6   Additional Applications and Problem Solving

Warm-up *Find the sum, difference, product, or quotient.*
[2.4] 1. $(-9)(5)$ **[2.2]** 2. $-15 + 7$
[2.3] 3. $14 - (-10)$ **[2.4]** 4. $-48 \div (-8)$

Objectives

1 Solve problems involving net.

2 Solve problems involving voltage.

3 Solve problems involving average rate.

Warm-up *Find the sum, difference, product, or quotient.*
[2.4] 1. $(-9)(5)$ **[2.2]** 2. $-15 + 7$
[2.3] 3. $14 - (-10)$ **[2.4]** 4. $-48 \div (-8)$

Objective 1 Solve problems involving net.

Recall that the net profit or loss is calculated using the formula $N = R - C$, where N represents the net amount, R represents revenue, and C represents cost.

Example 1   Cassif puts $2000 down on a car, then makes monthly payments of $275 for five years (60 months). After six years, he sells the car for $5200. If he spent a total of $1500 in routine maintenance and minor repairs, what is his net? Is it a profit or loss?

Understand We must find Cassif's net and determine whether it is a profit or loss. The formula for net is $N = R - C$, so we need total revenue and total cost. Cassif's total revenue is the amount he sold the car for, which was $5200. His total cost is the sum of all of the money he spent on the car.

Plan Calculate total cost; then subtract the cost from the total revenue to find the net.

Execute cost = amount down + total of all payments + maintenance cost

$$\text{(60 payments of \$275)}$$

cost =	2000	+	60(275)	+	1500	
cost =	2000	+	16,500	+	1500	
cost =	$20,000					

Now use $N = R - C$ to find the net.

$N = R - C$

$N = 5200 - 20,000$ **Replace *R* with 5200 and *C* with 20,000.**

$N = -14,800$

Answer Cassif's net is $-\$14,800$, which is a loss of $14,800.

Check Reverse the process. The net added to cost should produce the revenue.

$$-14,800 + 20,000 \overset{?}{=} 5200$$
$$5200 = 5200 \qquad \text{\textbf{The net amount checks.}}$$

Subtracting the maintenance costs and total of all payments from the total cost should produce the down payment.

$$20,000 - 1500 - 16,500 \overset{?}{=} 2000$$
$$2000 = 2000 \qquad \text{\textbf{The total cost checks.}}$$

Dividing the total of all payments by 60 should produce the amount of each payment.

$$16,500 \div 60 \overset{?}{=} 275$$
$$275 = 275 \qquad \text{\textbf{The total payments check.}}$$

▶ **Do Your Turn 1**

▶ **Your Turn 1**

Shawna put $1200 down when she bought her car. She made 48 payments of $350 and spent $2500 in maintenance and repairs. Four years after paying off the car, she sold it for $4400. What was her net? Was it a profit or loss?

Note We developed the problem-solving process in Section 1.6. The steps in the process are Understand, Plan, Execute, Answer, and Check.

Answer to Your Turn 1
$-\$16,100$; loss

Answers to Warm-up
1. -45 **2.** -8 **3.** 24 **4.** 6

Objective 2 Solve problems involving voltage.

Georg Simon Ohm was a German scientist who is credited with the discovery of the relationship among voltage, current, and resistance. His famous formula is $V = ir$. The V stands for voltage, which is the electrical pressure in a circuit measured in volts (V). The i stands for current, a measure of electricity moving through a wire, which is measured in amperes, or amps (A). The r stands for resistance to the flow of electricity through a wire, which is measured in ohms (named after Mr. Ohm in honor of his discovery). The symbol for ohms is the Greek letter Ω (omega).

Voltage and current can be negative numbers. A voltage source such as a battery has $+$ and $-$ connections called terminals. A voltage meter measures voltage and has two probes, one for $+$ and one for $-$ terminals/connections. When we connect each probe to the appropriate terminal on the battery, we measure a positive voltage. When we reverse the probes, we measure a negative voltage. Voltage is positive or negative depending on the orientation of the measuring device.

Current is a measure of the flow of electricity. A current meter is essentially the same as a voltage meter. If we connect the probes and the meter indicates negative current, the current is flowing in the opposite direction of the orientation of the probes from the meter.

Note You do not necessarily need to understand the concepts a formula describes to use it (although it makes the formula more interesting). As long as you can match the numbers in a problem with the variables in the formula, you can perform the calculations indicated by the formula.

| **Example 2** | Solve. |

a. Find the voltage in a circuit that has a resistance of 20 ohms and a current of -9 amps.

Solution:

$V = ir$

$V = (-9)(20)$ In $V = ir$, replace i with -9 and r with 20.

$V = -180$ Multiply.

Answer: The voltage is -180 V.

Connection The formulas $V = ir$ and $A = lw$ are mathematically the same because they involve the product of two factors. Many different phenomena have the same mathematical relationship.

b. The voltage in a circuit is measured to be -120 volts across a 40-ohm resistor. Find the current.

Solution:

$$V = ir$$
$$-120 = i(40)$$ In $V = ir$, replace V with -120 and r with 40.
$$-120 \div 40 = i$$ Write a related division equation in which the product, -120, is divided by the known factor, 40.
$$-3 = i$$ Divide.

Answer: The current is -3 A.

◀ **Do Your Turn 2**

▶ **Your Turn 2**

Solve.

a. The current in a circuit is measured to be -12 amps through a 20-ohm resistor. Find the voltage.

b. The voltage is measured to be -9 volts across a 3-ohm resistor. Find the current.

Objective 3 Solve problems involving average rate.

Another situation that has the same mathematical relationship as $V = ir$ and $A = lw$ is the formula that describes the relationship among distance, rate, and time.

If we drive a car 60 miles per hour for two hours, how far do we travel?

60 miles per hour means that we travel 60 miles for each hour we drive; so in two hours, we'll go 120 miles. Notice that to find distance, we multiply the rate by the time; so we can write the following formula:

$$\text{Distance} = \text{rate} \cdot \text{time}$$
$$d = rt$$

Answers to Your Turn 2
a. -240 V   b. -3 A

Example 3   A commuter train travels at an average rate of 30 miles per hour for two hours. How far does the train travel?

Solution: $d = rt$

$\quad\quad\quad d = (30)(2)$ In $d = rt$, replace r with 30 and t with 2.

$\quad\quad\quad d = 60$ Multiply.

Answer: The train travels 60 mi.

▶ **Do Your Turn 3**

Consider the fact that in reality, we cannot drive a car at an exact speed for a lengthy period of time. Our rate will vary slightly as a result of hills, wind resistance, and even our inability to maintain consistant foot pressure on the accelerator. Because rates can vary, we usually consider **average rate.**

Definition   **Average rate:** A measure of the rate at which an object travels a total distance in a total amount of time.

Example 4   A bus leaves at 9 A.M. and travels 40 miles, then stops for an hour. It travels another 110 miles and arrives at its final destination at 1 P.M. Find the average rate.

Understand   To find average rate, we must consider the total distance of the trip and the total travel time. The trip was broken into two parts. We were given the distances for both parts. We were also given a departure time and an arrival time, but there was a one-hour stop that we must take into account. The formula that relates distance, rate, and time is $d = rt$.

Plan   Find the total distance of the trip and total time of the trip; then use the formula $d = rt$ to solve for r.

Execute   The total distance is $40 + 110 = 150$ miles.

The total time from 9 A.M. to 1 P.M. is four hours. However, the bus stopped for an hour; so the actual travel time was three hours.

We now have the total distance of 150 miles and total travel time of three hours; so we can use $d = rt$.

$\quad\quad\quad\quad d = rt$

$\quad\quad\quad 150 = r(3)$ In $d = rt$, replace d with 150 and t with 3.

$\quad\quad 150 \div 3 = r$ Solve for the unknown factor by dividing.

$\quad\quad\quad\quad 50 = r$

Answer   The average rate was 50 mph.

Check   Use $d = rt$ to verify that if the bus traveled an average of 50 miles per hour for three hours, it would travel 150 miles.

$\quad\quad\quad d = (50)(3)$ In $d = rt$, replace r with 50 and t with 3.

$\quad\quad\quad d = 150$ Multiply. It checks.

▶ **Do Your Turn 4**

▶ **Your Turn 3**

A research submarine is lowered at an average speed of 8 feet per second. What will be the submarine's depth after 45 seconds?

▶ **Your Turn 4**

On a vacation trip, Candice leaves at 10 A.M. and travels 85 miles, then takes a 30-minute break. She travels another 120 miles and stops for another 30-minute break. Finally, she travels 70 miles and arrives at her destination at 4 P.M. What was her average rate?

Answer to Your Turn 3
−360 ft.

Answer to Your Turn 4
55 mph

2.6 Exercises

For Extra Help   MyMathLab®

Objective 1

Prep Exercise 1 In the formula $N = R - C$, what does each variable represent?

Prep Exercise 2 What do positive and negative net amounts indicate?

For Exercises 1–20, solve.

1. In 1990, Karen bought a collection of five paintings by a relatively unknown artist. Three of the paintings cost $250 each, and the other two cost $400 each. In 2010, she put the paintings up for auction and they sold for $1500. What was Karen's net? Was it a profit or loss?

2. In 1998, Mario purchased a collection of seven pieces of antique furniture. Four of the pieces cost $1200 each, and the other three cost $1500 each. In 2010, he sold the entire collection at an auction for $8600. What was Mario's net? Was it a profit or loss?

3. Darwin put $800 down when he bought a Buick Century. He made 60 payments of $288 and spent $950 in maintenance and repairs. Three years after paying off the car, he sold it for $4000. What was his net? Was it a profit or loss?

4. In 1955, Malvin put $800 down on a new Chevrolet Bel Air. He made 24 payments of $18. In 2000, he spent $2500 to restore the car to as-new condition. Over the years, he paid $3000 for maintenance. In 2010, he sold the car for $34,000. What was his net? Was it a profit or loss?

5. Lynn takes out a loan to buy a fixer-upper house. She then spends $14,500 in repairs and improvements. She sells the house for $185,000. If the payoff amount for the loan she took out to buy the house is $151,728, what is her net? Is it a profit or loss?

6. Scott takes out a loan to buy a fixer-upper house. He spends $15,475 in repairs and improvements. He sells the house for $194,000 but pays the new owner's $3750 closing costs. If the payoff for the loan that Scott took out to buy the house is $162,324, what is his net? Is it a profit or loss?

7. Greg bought 40 shares of stock at a price of $54 per share. One day he sold 25 shares for $56 per share. Two days later he sold the rest of his shares at a price of $50 per share. What was his net? Was it a profit or loss?

8. Maria bought 50 shares of stock at a price of $36 per share. One day she sold 30 shares for $42 per share. The next day she sold the rest of her shares at a price of $30 per share. What was her net? Was it a profit or loss?

Objective 2

Prep Exercise 3 In the formula $V = ir$, what does each variable represent?

9. An electrical circuit has a resistance of 7 ohms and a current of -9 amps. Find the voltage.

10. An electrical circuit has a resistance of 100 ohms and a current of -12 amps. Find the voltage.

11. The voltage in an electrical circuit measures -220 volts. If the resistance is 5 ohms, find the current.

12. The voltage in an electrical circuit measures -240 volts. If the resistance is 30 ohms, find the current.

13. A technician is told that the resistance in an electrical circuit is 40 ohms. The voltage measures -120 volts, and the current measures -4 amps. Was the technician told the correct resistance? Explain.

14. It is suspected that an incorrect resistor was put into a circuit. The correct resistor should be 12 ohms. A voltage measurement is taken and found to be -60 volts. The current is measured to be -10 amps. Is the resistor correct? Explain.

Objective 3

Pre Exercise 4 What is an average rate?

15. The elevator in the Empire State Building travels from the lobby to the 80th floor, a distance of about 968 feet, in about 44 seconds. What is the average rate of the elevator?

16. A research submarine is lowered at an average rate of 7 feet per second. What will be the submarine's depth after 29 seconds?

Of Interest

Completed in November 1930, the Empire State Building became the tallest human-made structure at a height of 381 meters with 102 floors. The current tallest structure is the Burj Khalifa, in Dubai, completed in 2009. It is 828 meters (2716.5 feet) tall and has 160 stories.

17. The space shuttle travels at a rate of about 17,060 miles per hour while in orbit. How far does the shuttle travel in three hours?

18. A ship travels at an average rate of 20 miles per hour. How far will the ship travel in four hours?

19. On a vacation trip, Devin left at 11 A.M. and traveled 105 miles, then took a 30-minute break. He traveled another 140 miles and stopped for another 30-minute break. Finally, he traveled 80 miles and arrived at his destination at 5 P.M. What was his average speed?

20. On a trip, Corrine left at 7 A.M. and traveled 152 miles, then took a 15-minute break. She traveled another 145 miles and stopped for 45 minutes to eat lunch. Finally, she traveled 135 miles and arrived at her destination at 2 P.M. What was her average speed?

Puzzle Problem A boat is in a harbor at low tide. Over the side hangs a ladder with its bottom step 6 inches below the surface of the water. The ladder steps are 12 inches apart. If the tide rises at a rate of 8 inches per hour, how many steps will be under water after five hours?

Review Exercises

[2.1] 1. Graph $-(-3)$ on a number line.

[2.1] 2. Simplify. $|15|$

For Exercises 3 and 4, solve and check.

[2.3] 3. $k + 76 = -34$

[2.4] 4. $-14m = 56$

[2.5] *For Exercises 5 and 6, simplify.*

    5. $(-2)^5 + 40 \div (-4)(-2)$

    6. $-7 \cdot 6 - 3(9 - \sqrt{64 + 36})$

Chapter 2 Summary and Review Exercises

Learning Strategy

When studying rules, if you are a visual learner, read the rules over and over. If you are an auditory learner, record yourself *saying* the rules (or the rhymes or songs you've made out of the rules). Then listen to your recording over and over. If you are a tactile learner, write the rules repeatedly.

2.1 Introduction to Integers

Definitions/Rules/Procedures	Key Example(s)
The **integers** are _____	Express each amount as a positive or negative integer. **a.** Juan has $2500 in his checking account. **Answer:** 2500 **b.** A shipwreck is found at a depth of 75 feet. **Answer:** −75

[2.1] For Exercises 1 and 2, express each amount as an integer.

1. Mark has a debt of $13,000.

2. Water boils at 212°F.

Definitions/Rules/Procedures	Key Example(s)
To graph a number on a number line, _____ _____ .	Graph −3 on a number line. $\xleftarrow{\quad}\overset{\displaystyle\bullet}{\underset{-4\ -3\ -2\ -1\ \ \ 0\ \ \ 1\ \ \ 2\ \ \ 3\ \ \ 4}{\mid\ \mid\ \mid\ \mid\ \mid\ \mid\ \mid\ \mid\ \mid}}\xrightarrow{\quad}$

[2.1] For Exercises 3 and 4, graph each integer on a number line.

3. −8

4. 5

Definitions/Rules/Procedures	Key Example(s)
Any positive number is _____ any negative number. 0 is _____ any negative number. When two negative numbers are compared, the negative number closer to _____ is the greater number.	Use < or > to write a true statement. a. 13 ? −16 **Answer:** 13 > −16 b. −12 ? 0 **Answer:** −12 < 0 c. −19 ? −7 **Answer:** −19 < −7

[2.1] For Exercises 5 and 6, use < or > to write a true statement.

5. −15 ? 0

6. −26 ? −35

Definitions/Rules/Procedures	Key Example(s)
The **absolute value** of a number is _____.	The absolute value of 7 is 7.
The absolute value of a positive number is _____.	The absolute value of -13 is 13.
The absolute value of a negative number is _____.	Simplify.
The absolute value of zero is _____.	**a.** $\|15\| = 15$ **b.** $\|-9\| = 9$ **c.** $\|0\| = 0$

7. The absolute value of 41 is _____.

8. Simplify. $\|-16\|$

2.2 Adding Integers

Definitions/Rules/Procedures	Key Example(s)
To add two integers: If they have the same sign, _____ _____. (*Think*: Two credits or two debts) If they have different signs, _____ _____. (*Think*: Payment toward a debt)	Add. **a.** $5 + 9 = 14$ **b.** $-5 + (-9) = -14$ **c.** $-5 + 9 = 4$ **d.** $5 + (-9) = -4$

[2.2] For Exercises 9–12, add.

9. $-12 + (-14)$

10. $-16 + 25$

11. $18 + (-30)$

12. $-14 + 30 + (-25) + 16$

Definitions/Rules/Procedures	Key Example(s)
Additive inverses are two numbers whose sum is _____. The additive inverse of 0 is _____. The additive inverse of a nonzero number has the same _____ but _____ sign.	Find the additive inverse of the given integer. **a.** 25 **Solution:** -25 **b.** -16 **Solution:** 16

[2.2] For Exercises 13 and 14, find the additive inverse.

13. -27

14. 32

2.3 Subtracting Integers and Solving Equations

Definitions/Rules/Procedures	Key Example(s)
To write a subtraction statement as an equivalent addition statement: 1. Change the operation symbol from $-$ to $+$. 2. Change the _____.	Write as an equivalent addition; then evaluate. **a.** $5 - 9 = 5 + (-9) = -4$ **b.** $-5 - 9 = -5 + (-9) = -14$ **c.** $5 - (-9) = 5 + 9 = 14$ **d.** $-5 - (-9) = -5 + 9 = 4$

[2.3] For Exercises 15–18, write as an equivalent addition; then evaluate.

15. $24 - 31$

16. $-20 - 17$

17. $-13 - (-19)$

18. $-27 - (-22)$

Definitions/Rules/Procedures	Key Example(s)
To solve for an unknown addend, write a related subtraction equation in which the _____ is subtracted from the _____.	Solve. $$x + 30 = 18$$ $$x = 18 - 30$$ $$x = -12$$

[2.3] For Exercises 19 and 20, solve.

19. $14 + x = -27$

20. $n + (-12) = -7$

2.4 Multiplying and Dividing Integers; Exponents; Square Roots; Solving Equations

Definitions/Rules/Procedures	Key Example(s)
When multiplying or dividing two numbers: If they have the same sign, the result is _____. If they have different signs, the result is _____. **When multiplying signed numbers,** count the negative factors. If there are an *even* number of negative factors, the product is _____. If there are an *odd* number of negative factors, the product is _____.	Multiply or divide. **a.** $4 \cdot 6 = 24$ **b.** $(-4)(-6) = 24$ **c.** $4(-6) = -24$ **d.** $(-4)(6) = -24$ **e.** $18 \div 9 = 2$ **f.** $-18 \div (-9) = 2$ **g.** $18 \div (-9) = -2$ **h.** $-18 \div 9 = -2$ **i.** $3(-2)(-5) = 30$ **j.** $(-1)(-1)(-12) = -12$ **k.** $-(-(-(-34))) = 34$

[2.4] For Exercises 21–26, multiply or divide.

21. $-5(-12)$

22. $6 \cdot (-9)$

23. $-48 \div 8$

24. $\dfrac{-63}{-9}$

25. $(-1)(-3)(-6)(-2)$

26. $-(-(-(-(-16))))$

Definitions/Rules/Procedures	Key Example(s)
When evaluating an exponential form that has a negative base: If the exponent is even, the product is _____. If the exponent is odd, the product is _____.	Evaluate. **a.** $(-2)^4 = (-2)(-2)(-2)(-2) = 16$ **b.** $(-2)^5 = (-2)(-2)(-2)(-2)(-2) = -32$ **c.** $-2^4 = -1 \cdot 2^4 = -1 \cdot 2 \cdot 2 \cdot 2 \cdot 2 = -16$

[2.4] For Exercises 27–30, evaluate.

27. $(-3)^4$

28. $(-10)^3$

29. -4^4

30. -3^3

Definitions/Rules/Procedures	Key Example(s)
To solve for an unknown factor, write a related division equation in which the _____ is divided by the _____.	Solve $6x = -24$ $\qquad x = -24 \div 6$ $\qquad x = -4$

[2.4] For Exercises 31 and 32, solve.

31. $7k = -63$

32. $(-10)h = -80$

Definitions/Rules/Procedures	Key Example(s)
The **square root** of a given number is a number whose _____ is the given number. Every positive integer has two square roots, a _____ square root and a _____ square root with the same absolute value as the positive square root. A negative number has _____ real square roots. $\sqrt{n}$ indicates the _____ square root (principal square root) of the radicand, n. $-\sqrt{n}$ indicates the _____ square root of the radicand, n.	Find all square roots of 64. **Answer:** 8 and -8 (or ± 8) Evaluate. **a.** $\sqrt{81} = 9$ **b.** $-\sqrt{36} = -6$ **c.** $\sqrt{-25}$ is not a real number

33. Find all square roots of 64.

[2.4] For Exercises 34–36, simplify.

34. $\sqrt{100}$

35. $-\sqrt{121}$

36. $\sqrt{-36}$

2.5 Order of Operations

Definitions/Rules/Procedures	Key Example(s)
Perform operations in the following order: 1. _____: parentheses ( ), brackets [ ], braces { }, absolute value | |, radicals $\sqrt{}$, and fraction bars ——. 2. _____ in order as they occur from left to right 3. _____ in order as they occur from left to right 4. _____ in order as they occur from left to right Note: Use G $\overrightarrow{ER}$ $\overrightarrow{MD}$ $\overrightarrow{AS}$ to remember the order.	Simplify. $\{(-4)^2 + 5[(9 - 15) \div (-2)]\} + (-3)\sqrt{16 \cdot 9}$ $= \{(-4)^2 + 5[(-6) \div (-2)]\} + (-3)\sqrt{144}$ $= \{16 + 5[3]\} + (-3) \cdot 12$ $= \{16 + 15\} + (-36)$ $= 31 + (-36)$ $= -5$

[2.5] For Exercises 37–42, simplify.

37. $-16 - 28 \div (-7)$

38. $-|26 - 3(-2)|$

39. $13 + 4(6 - 15) + 2^3$

40. $-3\sqrt{49} + 4(2 - 6)^2$

41. $[-12 + 4(15 - 10)] + \sqrt{100 - 36}$

42. $\dfrac{4 + (5 + 3)^2}{3(-13) + 5}$

Formulas

[2.3] Net: $N =$ _____

[2.6] Voltage: $V =$ _____

[2.6] An **average rate** is a measure of the rate at which an object travels a(n) _____ distance in a(n) _____ amount of time.

[2.6] Distance: $d =$ _____

For Exercises 43–52, solve.

43. William's initial credit card balance was $-\$1100$. He has made the following transactions. What is his new balance?

CHARGES	
MUSIC SHOP	$23
AUTOMART	$14
BURGER HEAVEN	$9
FINANCE CHARGE	$15
PAYMENT	$900

44. A concrete block is suspended by two cables. The block weighs 600 pounds. Each cable is exerting 300 pounds of upward force. What is the resultant force? What does this mean?

45. The financial report for a business indicates that the total revenue for 2010 was $1,648,200 and the total costs were $928,600. What was the net? Did the business have a profit or loss?

46. The temperature at sunset was reported to be 12°F. By midnight, the temperature was reported to be −19°F. What is the amount of the decrease?

47. Arturo has a balance of −$45 in his bank account. To avoid further charges, he must have a balance of $25. What is the minimum he must deposit to avoid further charges?

48. Branford has a debt of $4272. He agrees to make equal monthly payments for two years to repay the debt. How much is each payment?

49. A circuit has a resistance of 12 ohms and a current of −3 amps. Calculate the voltage.

50. The voltage in a circuit measures −110 volts. If the resistance is 10 ohms, find the current.

51. Jacquelyn drives at an average rate of 65 miles per hour for three hours. How far does she travel?

52. Steve begins a trip at 7 A.M. and drives 150 miles. After a 30-minute break, he drives another 110 miles. After another 30-minute break, he drives 40 miles. If he arrives at 1 P.M., what was his average rate of speed?

Chapter 2 Practice Test

For Extra Help  | **Test Prep**
VIDEOS
Step-by-step test solutions are found on the Chapter Test Prep Videos available via the Video Resources on DVD, in **MyMathLab®**, and on **YouTube** (search "CarsonPrealgebra" and click on "Channels").

TACTILE VISUAL AUDITORY

Learning Strategy

When you believe you are ready to take a practice test, instead of working out the problems on your first pass through the test, write the definition, rule, or procedure that applies to each problem. This approach will help you determine how prepared you are. Once you are certain of how to solve each problem, work through the test repeatedly until you can quickly and effortlessly solve every problem with no mistakes.

[2.1]   **1.** Graph -5 and 2 on a number line.

1. _____

For Exercises 2–12, simplify.

[2.1]   **2.** $|26|$

2. _____

[2.2]   **3.** $17 + (-29)$

3. _____

[2.2]   **4.** $-31 + (-14)$

4. _____

[2.3]   **5.** $20 - 34$

5. _____

[2.3]   **6.** $-16 - 19$

6. _____

[2.3]   **7.** $-30 - (-14)$

7. _____

[2.4]   **8.** $9(-12)$

8. _____

[2.4]   **9.** $-30 \div 6$

9. _____

[2.4]   **10.** $\dfrac{-48}{-12}$

10. _____

[2.4]   **11.** $(-2)(-6)(-7)$

11. _____

[2.4]   **12.** $-(-18)$

12. _____

[2.3]   **13.** Solve and check. $-19 + k = 25$

13. _____

[2.4]   **14.** Solve and check. $6n = -54$

14. _____

[2.4]   **15.** Evaluate. $(-4)^3$

15. _____

[2.4]   **16.** Evaluate. -2^2

16. _____

[2.4]   **17.** Find all square roots of 81.

17. _____

For Exercises 18–25, simplify.

18. _____

[2.4]   **18.** $\sqrt{144}$

19. _____

[2.4]   **19.** $\sqrt{-25}$

20. _____

[2.5]   **20.** $28 - 6 \cdot 5$

21. _____

[2.5]   **21.** $19 - 4(7 + 2) + 2^4$

22. _____

[2.5]   **22.** $|28 \div (2 - 6)| + 2(4 - 5)$

23. _____

[2.5]   **23.** $[18 \div 2 + (4 - 6)] - \sqrt{49}$

24. _____

[2.5]   **24.** $\dfrac{5^2 + 3(-12)}{36 \div 2 - 18}$

25. _____

[2.5]   **25.** $\dfrac{(-7)^2 + 11}{(-2)(5) + \sqrt{64}}$

26. _____

[2.3]   **26.** Allison's initial credit card balance was $-\$487$. She has made the following transactions. What is her new balance?

CHARGES	
CLOTHING BOUTIQUE	$76
FINANCE CHARGES	$14
PAYMENT	$125

27. _____

[2.4]   **27.** During a very difficult month, Natasha overdrafted her account five times. Each overdraft charge appears as $-\$14$ on her statement. What were the total overdraft charges that month?

28. _____

[2.6]   **28.** Howard put $8000 down when he bought a new car. He made 48 payments of $476 and spent $347 on maintenance and repair. Three years later he sold the car for $15,500. What is his net? Is it a profit or loss?

29. _____

[2.6]   **29.** The current in a circuit measures -8 amps. If the resistance is 12 ohms, find the voltage.

30. _____

[2.6]   **30.** Lashanda drives at an average rate of 62 miles per hour. How far will she travel in two hours?

Chapters 1–2 Cumulative Review Exercises

For Exercises 1–4, answer true or false.

[1.3] **1.** $3 \cdot 4 + 3 \cdot 5 = 3(4 + 5)$

[2.1] **2.** -6 is a whole number.

[2.2] **3.** The sum of a negative number and a positive number is always negative.

[2.3] **4.** $-9 - 12 = -9 + (-12)$

For Exercises 5–8, complete the rule.

[1.2] **5.** $2 + 3 = 3 + 2$ illustrates the _____ property of _____.

[1.3] **6.** $9 \cdot (4 \cdot 3) = (9 \cdot 4) \cdot 3$ illustrates the _____ property of _____.

[2.4] **7.** When multiplying or dividing two numbers that have different signs, the result is _____.

[2.5] **8.** Perform operations in the following order:
1. _____
2. _____ from left to right
3. _____ from left to right
4. _____ from left to right

[1.1] **9.** Write the word name for 409,254,006.

[1.2] **10.** Estimate $49,902 + 6519$ by rounding.

[2.1] **11.** Graph -4 on a number line.

[2.1] **12.** Use $<$ or $>$ to write a true statement.

$$-930 \ \boxed{?} \ -932$$

For Exercises 13–22, simplify.

[2.1] **13.** $|16|$

[2.2] **14.** $287 + 48 + (-160) + (-82)$

[2.3] **15.** $-64 - (-14)$

[2.4] **16.** $-1(12)(-6)(-2)$

[2.4] **17.** $(-2)^6$

[1.4] **18.** $\dfrac{1208}{6}$

[2.4] **19.** $-105 \div 7$

[2.4] **20.** $\sqrt{121}$

[2.5] **21.** $-15 - 2[36 \div (3 + 15)]$

[2.5] **22.** $-4(2)^3 + [18 - 12(2)] - \sqrt{25 \cdot 4}$

For Exercises 23 and 24, solve and check.

[1.2] **23.** $29 + x = 54$

[2.4] **24.** $-18c = 126$

$\begin{bmatrix} 1.2 \\ 1.6 \end{bmatrix}$ **25.** Find the perimeter and area.

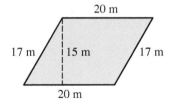

[1.6] **26.** Find the volume.

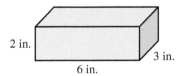

For Exercises 27–30, solve.

[1.5] **27.** The test scores for the students in a chemistry class are shown in the following table. Find the mean, median, and mode score.

Student	Score	Student	Score
Brown, D.	85	Lane, K.	80
Driscol, R.	95	Manning, E.	76
Fisher, B.	75	Parker, A.	72
Green, T.	50	Sims, P.	95
Hoover, H.	68	Timmons, W.	100
Jennings, P.	90	Walker, H.	74

[1.6] **28.** A landscaper needs to know the area of a yard. A plot plan showing the position of the house on the lot is shown. Find the area of the yard.

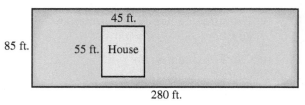

[2.3] **29.** Use the following graph, which shows the warmest and coldest average temperatures for the North and South poles.

Average High and Low Temperatures

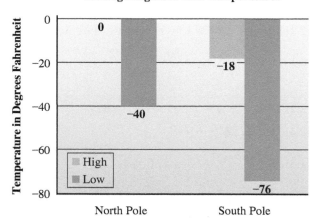

a. How much higher are the average lows at the North Pole than at the South Pole?
b. What is the difference in the average high and low temperatures at the South Pole?

[2.6] **30.** Sonya put $1500 down on a new car. She made 60 payments of $376 and spent $1450 on maintenance and repairs. Three years after paying off the car, she sold it for $8500. What was her net? Was it a profit or loss?

CHAPTER

3

Expressions and Polynomials

Chapter Overview

In Chapters 1 and 2, we set a foundation with numbers, variables, and symbols of arithmetic. In Chapter 3, we will put all of those pieces together to build expressions. Everything we learn in this chapter boils down to one of two actions that we can perform with expressions:

► Evaluate expressions.
► Rewrite expressions.

3.1 Translating and Evaluating Expressions

Objectives

1 Differentiate between an expression and an equation.
2 Translate word phrases to expressions.
3 Evaluate expressions.

Connection Think of equations as complete sentences and expressions as phrases or incomplete sentences. In grammar, we learn that a complete sentence must have a subject and a verb. The verb in an equation is the equal sign. Because expressions do not have an equal sign, they do not have a verb and therefore are incomplete.

▶ **Your Turn 1**

Determine whether each of the following is an expression or an equation and explain why.
a. $E = mc^2$
b. $19 - 2(3 + 8)$
c. $\dfrac{\sqrt{b^2 - 4ac}}{2a}$
d. $y = mx + b$

Answers to Your Turn 1
a. Equation; it contains an equal sign.
b. Expression; it does not contain an equal sign.
c. Expression; it does not contain an equal sign.
d. Equation; it contains an equal sign.

Answers to Warm-up
1. -6 **2.** 76 ft.

Warm-up

[2.5] 1. Simplify. $2\sqrt{81} - 8[6^2 \div (3 + 9)]$
[1.6] 2. Use the formula $P = 2l + 2w$ to find the perimeter of a rectangular room that measures 21 feet by 17 feet.

Objective 1 Differentiate between an expression and an equation.

In Chapters 1 and 2, we learned about the basic building blocks of algebra: constants, variables, and arithmetic symbols. Those building blocks form **expressions** and **equations**.

Definitions **Expression:** A constant; a variable; or any combination of constants, variables, and arithmetic symbols that describes a calculation.

Equation: A mathematical relationship that contains an equal sign.

For example, P and $2l + 2w$ are expressions because they have no equal sign, whereas the formula $P = 2l + 2w$ is an equation because it has an equal sign.

Example 1   Determine whether each of the following is an expression or an equation and explain why.

a. $16 \cdot 5 = 80$

Solution: $16 \cdot 5 = 80$ is an equation because it contains an equal sign. More specifically, it is a *numeric* equation because it does not contain variables.

b. $d = rt$

Solution: $d = rt$ is an equation because it contains an equal sign. More specifically, it is an *algebraic* equation because it contains variables.

c. ir

Solution: ir is an expression because it does not contain an equal sign. More specifically, it is an *algebraic* expression because it contains variables.

◀ **Do Your Turn 1**

Objective 2 Translate word phrases to expressions.

A key strategy in solving problems is to translate words in the problem to math symbols. The first step is to identify the unknown amount. If a variable is not already given, look for words such as *some number* and *a number* and select a variable to represent that unknown number. The following table shows some basic phrases and their translations.

Translating Basic Phrases

Addition	Translation
the sum of x and three	$x + 3$
h plus k	$h + k$
t added to seven	$7 + t$
three more than some number	$n + 3$
y increased by two	$y + 2$

Subtraction	Translation
the difference of three and x	$3 - x$
h minus k	$h - k$
seven subtracted from t	$t - 7$
three less than a number	$n - 3$
two decreased by y	$2 - y$

Note Addition is commutative; so the order of the addends does not matter.

the sum of x and three can be written as $x + 3$ or $3 + x$.

Note Subtraction is not commutative; so the order of the translation matters. We must translate each key phrase exactly as shown above. Notice that the amount before *subtracted from* and *less than* is the subtrahend; so it appears after the minus sign.

Multiplication	Translation
the product of three and x	$3x$
h times k	hk
some number multiplied by five	$5n$
twice a number	$2n$
triple some number	$3n$

Division	Translation
the quotient of x and three	$x \div 3$ or $\dfrac{x}{3}$
h divided by k	$h \div k$ or $\dfrac{h}{k}$
h divided into k	$k \div h$ or $\dfrac{k}{h}$

Note Multiplication is commutative; so the order of the factors does not matter.

h times k can be written as *hk* or *kh*.

Note Division is not commutative; so the order of the translation matters. Notice that the amount before *divided into* is the divisor; so it appears after the division sign.

Exponents	Translation
c squared	c^2
the square of b	b^2
k cubed	k^3
the cube of b	b^3
n to the fourth power	n^4
y raised to the fifth power	y^5

Roots	Translation
the square root of x	$\sqrt{x}$

In translating the key words *sum, difference, product,* and *quotient,* the word *and* separates the parts and can therefore be translated to the operation symbol indicated by the key word *sum, difference, product,* or *quotient.*

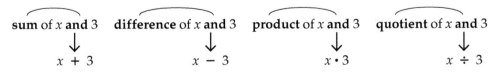

Translating Combinations of Basic Phrases

Now let's consider combinations of the basic phrases.

Translate to a variable expression.

a. twelve more than eight times n

b. the product of five and m subtracted from nine

c. sixteen less than the quotient of some number and two

d. the square root of m divided by the cube of n

Example 2   Translate to a variable expression.

a. seven added to the product of four and x

Translation:

Note In a product of a number and variable, the number is written first and the multiplication symbol can be eliminated. Instead of $4 \cdot x$, we can write $4x$, which is simpler.

Answer: $7 + 4x$ or $4x + 7$

b. ten subtracted from the quotient of some number and three

Translation:

Note The subtrahend is the amount to the left of *subtracted from* or *less than*.

Answer: $n \div 3 - 10$ or $\dfrac{n}{3} - 10$

c. twelve less than the square of some number

Translation:

Answer: $n^2 - 12$

d. nine divided into the square root of some number

Translation: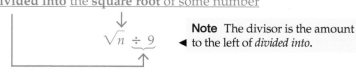

Note The divisor is the amount to the left of *divided into*.

Answer: $\sqrt{n} \div 9$ or $\dfrac{\sqrt{n}}{9}$

◀ **Do Your Turn 2**

Translating Phrases Involving Parentheses

Some word phrases indicate an order of operations that require parentheses in the translation. This happens when the phrase indicates that a sum or difference is to be calculated before a higher-order operation such as multiplication, division, exponent, or square root is performed.

Answers to Your Turn 2

a. $12 + 8n$ or $8n + 12$

b. $9 - 5m$

c. $n \div 2 - 16$ or $\dfrac{n}{2} - 16$

d. $\sqrt{m} \div n^3$ or $\dfrac{\sqrt{m}}{n^3}$

Example 3 Translate the words to a variable expression.

a. four times the sum of y and five.

Translation: four **times** the **sum** of y **and** five

$$4 \cdot \qquad (y + 5)$$

Note A sum is the result of addition; so this phrase says to add 5 to an unknown number before multiplying that result by 4. To make the addition occur before multiplication, we place the addition in parentheses.

Answer: $4(y + 5)$

b. the product of negative six and the difference of some number and two

Translation: the **product** of negative six **and** the **difference** of some number **and** two

$$-6 \cdot \qquad\qquad (n - 2)$$

Answer: $-6(n - 2)$

c. the sum of x squared and three divided by the difference of y and four

Translation: the **sum** of x squared **and** three **divided by** the **difference** of y **and** four

$$(x^2 + 3) \qquad \div \qquad (y - 4)$$

Answer: $(x^2 + 3) \div (y - 4)$ or $\dfrac{x^2 + 3}{y - 4}$

Note With the fraction line, we calculate the top and bottom separately before dividing; so parentheses are not needed.

▶ **Do Your Turn 3**

Objective 3 Evaluate expressions.

We can perform two actions with an expression:

1. Evaluate.
2. Rewrite.

In this section, we will focus on evaluating expressions. The rest of Chapter 3 will deal with rewriting expressions.

Procedure

To evaluate an expression:
1. Replace the variables with the corresponding given values.
2. Calculate using the order of operations agreement.

Example 4 Evaluate.

a. $x^2 + 5x - 7$; $x = -3$

Solution: $x^2 + \quad 5x - 7$

$(-3)^2 + 5(-3) - 7$ **Replace x with -3 using parentheses.**

$= 9 + 5(-3) - 7$ **Simplify the exponential form.** $(-3)^2 = 9$

$= 9 + (-15) - 7$ **Multiply.** $5(-3) = -15$

$= -6 - 7$ **Add.** $9 + (-15) = -6$

$= -13$ **Subtract.** $-6 - 7 = -13$

▶ **Your Turn 3**

a. three multiplied by the sum of x and y

b. negative four times the difference of some number and nine

c. the product of m and the square root of n divided by the difference of h and k

Warning When replacing a variable with a number, use parentheses to avoid changing the intended calculation. For example, when evaluating x^2 when $x = -3$, write $(-3)^2$, not -3^2. Remember, $(-3)^2 = (-3)(-3) = 9$, which is the intended calculation, whereas -3^2 is not the same because it means $-1 \cdot 3^2 = -1 \cdot 3 \cdot 3 = -9$.

Answers to Your Turn 3
a. $3(x + y)$
b. $-4(n - 9)$
c. $m\sqrt{n} \div (h - k)$ or $\dfrac{m\sqrt{n}}{h - k}$

▶ **Your Turn 4**

Evaluate.

a. $x^2 - 4x + 10; x = -2$

b. $-x + y(x - y); x = -6, y = 2$

c. $-2n^2 - |6np|; n = -3, p = 2$

d. $-4a + \sqrt{2a - b}; a = 9, b = 2$

e. $\dfrac{12 - x}{5 - x}; x = -2$

f. $\dfrac{x + y}{2x - y}; x = 3, y = 6$

b. $-x + 5(x + y); x = -3, y = 1$

Solution: $-(-3) + 5[(-3) + (1)]$ Replace x with -3 and y with 1, using parentheses.

$= -(-3) + 5[-2]$ Add in the brackets. $-3 + 1 = -2$

$= 3 + (-10)$ Simplify. $-(-3) = 3$ and $5[-2] = -10$

$= -7$ Add. $3 + (-10) = -7$

c. $|r^2 - p| + 4r; r = -2, p = 6$

Solution: $|(-2)^2 - (6)| + 4(-2)$ Replace r with -2 and p with 6, using parentheses.

$= |4 - (6)| + 4(-2)$ Simplify the exponential form in the absolute value. $(-2)^2 = 4$

$= |-2| + 4(-2)$ Subtract in the absolute value. $4 - 6 = -2$

$= 2 + 4(-2)$ Evaluate the absolute value. $|-2| = 2$

$= 2 + (-8)$ Multiply. $4(-2) = -8$

$= -6$ Add. $2 + (-8) = -6$

d. $7\sqrt{m + n} + mn; m = -6, n = 10$

Solution: $7\sqrt{(-6) + (10)} + (-6)(10)$ Replace m with -6 and n with 10, using parentheses.

$= 7\sqrt{4} + (-6)(10)$ Add in the radical. ◀ **Note** When a radical contains a sum or difference, we treat the sum or difference as if parentheses were around them.

$= 7 \cdot 2 + (-6)(10)$ Find the principal square root.

$= 14 + (-60)$ Multiply.

$= -46$ Add.

e. $\dfrac{3h^2}{k + 5}; h = 2, k = -5$

Solution: $\dfrac{3(2)^2}{(-5) + 5}$ Replace h with 2 and k with -5, using parentheses.

$= \dfrac{3(4)}{0}$ Simplify the numerator and denominator separately.

$= \dfrac{12}{0}$, which is undefined

Explanation: When we replace k with -5, the divisor simplifies to 0. Because the divisor is 0 and the dividend is not 0, the expression is undefined.

Answers to Your Turn 4

a. 22   b. -10 c. -54
d. -32 e. 2   f. undefined

◀ **Do Your Turn 4**

3.1 Exercises   For Extra Help   MyMathLab®

Objective 1

Prep Exercise 1 What is the difference between an equation and an expression?

For Exercises 1–12, indicate whether each of the following is an expression or an equation.

1. $19 + 3 \cdot 2 = 25$

2. $|14 - 29| = 15$

3. $13 + 5\sqrt{169}$

4. $\dfrac{19 + 9}{4 - 2^3}$

5. $25 - x^2$

6. $(x + 3)^2$

7. $9 + 2(x + 3) = 14 - x$

8. $-6x + 5 = 3x - 13$

9. $x + y = 9$

10. $x^2 + 5xy - 2 = 7$

11. $mx + b$

12. $\sqrt{1 - v^2/c^2}$

Objective 2

Prep Exercise 2 List five key words/phrases that indicate addition.

Prep Exercise 3 List five key words/phrases that indicate subtraction.

Prep Exercise 4 List three key words that indicate multiplication.

Prep Exercise 5 List three key words that indicate division.

Prep Exercise 6 What key words for subtraction and division translate in reverse order of what we read?

Prep Exercise 7 The phrase *seven times the sum of some number and eight* translates to $7(n + 8)$. Why are parentheses needed around the addition?

For Excercises 13–36, translate to a variable expression.

13. six plus y

14. n added to five

15. eight minus some number

16. ten decreased by some number

17. fifteen subtracted from a number

18. twenty less than a number

19. negative seven times some number

20. some number multiplied by negative eleven

21. negative twelve divided by k

22. m divided into negative forty

23. negative five plus the product of seven and some number

24. negative eight added to the product of ten and some number

25. fifteen less than the quotient of twenty and some number

26. twelve subtracted from the quotient of some number and two

27. the product of x and the square of y

28. the product of m and the square root of n

29. the quotient of the square root of h and k

30. the quotient of u cubed and t

31. negative six times the sum of some number and three

32. negative five multiplied by the sum of eight and some number

33. eight multiplied by the difference of one and a number

34. ten times the difference of a number and two

35. the sum of x cubed and five divided into the square root of y

36. the difference of six and m squared divided by the product of u and t

Objective 3

For Exercises 37–60, evaluate.

Prep Exercise 8 Explain how to evaluate an expression.

37. $4a - 9; a = 4$

38. $3b + 12; b = 2$

39. $3x + 8; x = -5$

40. $5y - 7; y = -2$

41. $-2m + n; m = 4, n = 7$

42. $3z - 2w; z = 5, w = 2$

43. $2x - 4y; x = 3, y = -2$

44. $3p - 2q; p = 4, q = -3$

45. $3y - 5(y + 2); y = 4$

46. $-2x + 3(x - 4); x = -2$

47. $a^2 + 9a - 2; a = -2$

48. $c^2 - 3c + 7; c = -3$

49. $3t^2 - 4u + 1; t = -2, u = 4$

50. $2r^2 + 4s - 4; r = -3, s = 5$

51. $b^2 - 4ac; b = -4, a = -3, c = 2$

52. $b^2 - 4ac; b = 3, a = -5, c = -1$

53. $|x^2 + y|; x = 2, y = -10$

54. $|r - p^3|; r = -7, p = -2$

55. $\sqrt{m} + \sqrt{n}; m = 144, n = 25$

56. $\sqrt{x} - \sqrt{y}; x = 25, y = 16$

57. $\sqrt{x^2 + y^2}; x = -3, y = 4$

58. $\sqrt{mn} - 5n^2; m = 27, n = 3$

59. $\dfrac{3x + 5}{7 - 2y}; x = 10, y = 1$

60. $\dfrac{2p - 6}{3 + 4r}; p = -8, r = 2$

For Exercises 61–66, complete each table by evaluating the expression using the given values for the variable.

61.

x	$3x - 1$
-4	
-2	
0	
2	
4	

62.

y	$-2y + 3$
-4	
-2	
0	
2	
4	

63.

m	$m^2 - 2m + 4$
-1	
0	
1	
2	
3	

64.

n	$n^2 + 2n - 1$
-3	
-2	
-1	
0	
1	

65.

y	$|y + 2|$
-4	
-3	
-2	
-1	
0	

66.

t	$|2t - 4|$
-2	
0	
2	
4	
6	

For Exercises 67–70, translate and evaluate.

67. six less than the product of negative five and a number
   a. Translate to a variable expression.
   b. Evaluate the expression when the variable is -4.

68. negative fourteen plus the quotient of a number and two
   a. Translate to a variable expression.
   b. Evaluate the expression when the variable is -10.

69. negative three times the difference of eight and some number
   a. Translate to a variable expression.
   b. Evaluate the expression when the variable is -2.

70. twelve minus the sum of one and some number
   a. Translate to a variable expression.
   b. Evaluate the expression when the variable is -5.

  Learning Strategy

If you find that you are unsure of how to work a particular problem, look for examples in the text and your notes relating to that problem. If you still cannot figure out the problem, mark it and move on. When you've completed all that you can on your own, seek help to complete the exercises.

Review Exercises

[1.1] 1. What digit is in the thousands place in 472,603?

[1.1] 2. Write 279 in expanded form.

[1.3] 3. Write $7 \cdot 7 \cdot 7 \cdot 7$ in exponential form.

[1.3] 4. Evaluate 2^6.

[1.3] 5. Write $3 \cdot 10^5$ in standard form.

[2.3] 6. Write $3 - 7 - 9$ as addition.

3.2 Introduction to Polynomials; Combining Like Terms

Warm-up

[3.1] 1. Is $5x^2 + 7x + 2$ an expression or an equation?

[3.1] 2. Evaluate $2x^2 + 7x + 9$ when $x = 10$.

Objective 1 Identify monomials.

As we saw in Section 3.1, there are many types of expressions. We will now focus on expressions known as *polynomials*. Polynomials are algebraic expressions that are similar to whole numbers written in expanded notation.

For example, in expanded form, we see that 279 is like the expression $2x^2 + 7x + 9$.

Expanded form, written with base 10: $279 = 2 \cdot 10^2 + 7 \cdot 10 + 9$

Polynomial form, written with base x: $2x^2 \quad + \quad 7x \quad + 9$

The expression $2x^2 + 7x + 9$ is a *polynomial*. To define *polynomial,* we must first define **monomial**. Think about the prefixes involved. *Poly* means many and *mono* means one.

Definition Monomial: An expression that is a constant or a product of a constant and variables that are raised to whole number powers.

Example 1 Determine whether the expression is a monomial. Explain.

a. $5x$

Explanation: $5x$ is a monomial because it is a product of a constant, 5, and a variable, x, that has a whole number power, 1.

From Example 1a, we can make the following conclusion:

Conclusion: Any number or variable with no apparent exponent has an understood exponent of 1.

$$5x = 5^1 x^1$$

We write $5x$ instead of $5^1 x^1$ because it is simpler; that is, it has fewer symbols.

b. $-4xy^3$

Explanation: From what we learned in Example 1a, $-4xy^3 = (-4)^1 x^1 y^3$; so it is a monomial because it is a product of a constant, -4, and variables, x and y, that have whole number powers, 1 and 3.

c. x^6

Explanation: x^6 is a monomial because it is a product of a constant, 1, and a variable, x, that has a whole number power, 6.

From Example 1c, we can make the following conclusion:

Conclusion: If a numerical factor is not apparent in a monomial, it is understood to be 1.

$$x^6 = 1x^6$$

We write x^6 instead of $1x^6$ because it is simpler.

d. 8

Explanation: 8 is a monomial because it is a constant.

Objectives

1 Identify monomials.

2 Identify the coefficient and degree of a monomial.

3 Identify polynomials and their terms.

4 Identify the degree of a polynomial.

5 Write polynomials in descending order of degree.

6 Simplify polynomials in one variable by combining like terms.

Learning Strategy

In class, pay extra attention to the examples that the professor gives you and write down the problem, answer, and explanation. Then, when you go back to study for the exam, take all of the examples from class and create a sample test for yourself. After taking the sample test, you can go back to see the correct answers and the explanations for the problems that your professor thinks are most important.

—Kim D.

Answers to Warm-up
1. expression
2. 279

▶ **Your Turn 1**

Determine whether the expression is a monomial. Explain.

a. $4x^7$

b. $-9mn$

c. -6

d. $\dfrac{3x^2}{y}$

e. $9a + 5$

▶ **Your Turn 2**

Identify the coefficient and degree of each monomial.

a. m^6

b. $-7x^2y^3$

c. $14ab$

d. -10

Answers to Your Turn 1

a. $4x^7$ is a monomial because it is a product of a constant, 4, and a variable, x, with a whole number power, 7.

b. $-9mn$ is a monomial because it is a product of a constant, -9, and variables, m and n, that both have a whole number power, 1.

c. -6 is a monomial because it is a constant.

d. $\dfrac{3x^2}{y}$ is not a monomial because it has a variable, y, as a divisor.

e. $9a + 5$ is not a monomial because addition is involved.

Answers to Your Turn 2

a. coefficient: 1, degree: 6

b. coefficient: -7, degree: 5

c. coefficient: 14, degree: 2

d. coefficient: -10, degree: 0

e. $\dfrac{5}{xy}$

Explanation: $\dfrac{5}{xy}$ is not a monomial because it is not a product of a constant and variables raised to whole number powers. Rather, it is a quotient of a constant and variables. The variables make up the divisor.

f. $3x^2 + 5x - 4$

Explanation: $3x^2 + 5x - 4$ is not a monomial because it is not a product of a constant and variables raised to whole number powers. Rather, addition and subtraction are involved.

From Examples 1e and 1f, we can make the following conclusion:

Conclusion: Monomials do not have variables in a divisor, nor do they contain addition or subtraction.

◀ **Do Your Turn 1**

Objective 2 Identify the coefficient and degree of a monomial.

Two important parts of a monomial are its **coefficient** and its **degree.**

Definitions Coefficient: The numerical factor in a monomial.

Degree of a monomial: The sum of the exponents of all variables in the monomial.

Example 2 Identify the coefficient and degree of each monomial.

a. $-7x^2$

Answer: Coefficient: -7

Degree: 2

Explanation: The coefficient is -7 because it is the numerical factor. The degree is 2 because it is the exponent for the variable x.

b. $-xy^3$

Answer: Coefficient: -1

Degree: $1 + 3 = 4$

Explanation: From the conclusions we made in Examples 1b and 1c, we can express $-xy^3$ as $-1x^1y^3$. In this alternative form, we can see that the numerical factor is -1, which is the coefficient. We can also see that the variables' exponents are 1 and 3. Because degree is the sum of the variables' exponents, we add 1 and 3, which equals 4.

c. 17

Answer: Coefficient: 17

Degree: 0

Explanation: The monomial 17 is equivalent to $17x^0$. They are equivalent because $x^0 = 1$, which means $17x^0 = 17(1) = 17$. In the alternative form, it is apparent that 17 is the coefficient because it is the numerical factor and 0 is the degree because it is the variable's exponent.

◀ **Do Your Turn 2**

Objective 3 Identify polynomials and their terms.

Now we can define a **polynomial.**

Definition **Polynomial:** A monomial or a sum of monomials.

All of the following expressions are polynomials.

$$2x^3$$

$$5x + 7$$

$$9x^3 - 4x^2 + 8x - 6$$

$$2x^2 + 5xy + 8y$$

In $9x^3 - 4x^2 + 8x - 6$, it would seem that the subtraction signs contradict the word *sum* in the definition. However, in Chapter 2, we learned that subtraction can be written as addition; so $9x^3 - 4x^2 + 8x - 6$ can be expressed as a sum.

$$9x^3 - 4x^2 + 8x - 6 = 9x^3 + (-4x^2) + 8x + (-6)$$

Mathematicians prefer to write expressions with the fewest symbols possible. Because $9x^3 - 4x^2 + 8x - 6$ has fewer symbols than $9x^3 + (-4x^2) + 8x + (-6)$ has, it is the preferred form. An expression written with the fewest symbols possible is said to be in **simplest form.**

Definition **Simplest form:** An expression written with the fewest symbols possible.

The polynomial $9x^3 - 4x^2 + 8x - 6$ has only one variable, x; so we say that it is a **polynomial in one variable.** The polynomial $2x^2 + 5xy + 8y$ has two variables, x and y; so we call it a **multivariable polynomial.**

Definitions **Polynomial in one variable:** A polynomial in which every variable term has the same variable.

Multivariable polynomial: A polynomial with more than one variable.

$p^2 - 5p + 2$ is a polynomial in one variable.

$q^2 + 8qt$ is a multivariable polynomial.

In this text, we will focus on polynomials in one variable.

Now let's identify the monomials, or *terms,* of a polynomial. If a polynomial contains subtraction, you may find it helpful to first write the polynomial as a sum.

Example 3 Identify the terms in the polynomial and their coefficients.

a. $9x^3 - 4x^2 + 8x - 6$

Solution: $9x^3 - 4x^2 + 8x - 6 = 9x^3 + (-4x^2) + 8x + (-6)$ Write the polynomial as a sum.

First term: $9x^3$ Coefficient: 9
Second term: $-4x^2$ Coefficient: -4
Third term: $8x$ Coefficient: 8
Fourth term: -6 Coefficient: -6

▶ Your Turn 3

Identify the terms in the polynomial and their coefficients.

a. $12x^2 - 8x + 9$
b. $4m^3 + 5m^2 - 10m - 11$

b. $4y^4 - 8y^3 - y + 2$

Solution: $4y^4 - 8y^3 - y + 2 = 4y^4 + (-8y^3) + (-y) + 2$ Write the polynomial as a sum.

First term: $4y^4$ Coefficient: 4
Second term: $-8y^3$ Coefficient: -8
Third term: $-y$ Coefficient: -1
Fourth term: 2   Coefficient: 2

Note You can identify the sign of each term's coefficient without writing the polynomial as a sum by looking at the operation signs. A plus sign not only means add, but also means that the coefficient of the term to its right is positive. A minus sign not only means subtract, but also means that the coefficient of the term to its right is negative.

◀ Do Your Turn 3

Some polynomials have special names that are based on the number of terms in the polynomial.

Monomial: A **single**-term polynomial
Binomial: A polynomial that has exactly **two** terms
Trinomial: A polynomial that has exactly **three** terms

Note The prefix of each special name corresponds to the number of terms in the polynomial.

Polynomials that have more than three terms have no special name.

▶ Your Turn 4

Indicate whether the polynomial is a monomial, binomial, or trinomial or has no special name.

a. $5x + 9$
b. $4x^3 - 2x^2 + 5x - 6$
c. $8x^3$
d. $x^2 + 2x - 9$

Example 4   Indicate whether the polynomial is a monomial, binomial, or trinomial or has no special name.

a. $5x^3$

Answer: $5x^3$ is a monomial because it is a single term.

b. $-3x^2 + 5x$

Answer: $-3x^2 + 5x$ is a binomial because it has two terms.

c. $4m^2 + 6m - 9$

Answer: $4m^2 + 6m - 9$ is a trinomial because it has three terms.

d. $x^3 + 4x^2 - 9x + 2$

Answer: $x^3 + 4x^2 - 9x + 2$ has no special name because it has more than three terms.

◀ Do Your Turn 4

Objective 4 Identify the degree of a polynomial.

Earlier in this section, we learned that the degree of a monomial is the sum of the variables' exponents. For a polynomial with more than one term, the degree is the greatest degree of any of its terms.

Definition Degree of a polynomial: The greatest degree of any of the terms in the polynomial.

Answers to Your Turn 3
a. $12x^2$, coefficient 12; $-8x$, coefficient -8; 9, coefficient 9
b. $4m^3$, coefficient 4; $5m^2$, coefficient 5; $-10m$, coefficient -10; -11, coefficient -11

Answers to Your Turn 4
a. binomial
b. no special name
c. monomial
d. trinomial

Example 5 Identify the degree of each polynomial.

a. $5x^3 + 9x^6 - 10x^2 + 8x - 2$

Answer: Degree: 6

Explanation: Look at the degree of each term.

$$5x^3 \quad + \quad 9x^6 \quad - \quad 10x^2 \quad + \quad 8x \quad - \quad 2$$

degree is 3, degree is 6, degree is 2, degree is 1, degree is 0

The greatest degree of all of the terms is 6.

b. $12a^7 - a^5 - 3a^9 + 4a - 16$

Answer: Degree: 9

Explanation: The greatest degree of all of the terms is 9.

▶ **Do Your Turn 5**

Objective 5 Write polynomials in descending order of degree.

The commutative property of addition tells us that rearranging addends does not affect a sum. Because a polynomial is a sum, rearranging its terms does not change it. For example, $6x^2 + 9 + 5x$ and $6x^2 + 5x + 9$ are equivalent.

To make comparing polynomials easier, mathematicians prefer writing them in descending order of degree. This means that we want the term with the greatest degree written first, then the term with the next greatest degree, and so on. So $6x^2 + 5x + 9$ is in descending order of degree because the terms go from degree 2 to degree 1 to degree 0.

Procedure

To write a polynomial in descending order of degree, write the term with the greatest degree first, then the term with the next greatest degree, and so on.

Example 6 Write each polynomial in descending order of degree.

a. $5x^3 + 4x^2 - 3x^5 - 7 + 9x$

Solution: $5x^3 + 4x^2 - 3x^5 - 7 + 9x$

Answer: $-3x^5 + 5x^3 + 4x^2 + 9x - 7$ Move $-3x^5$ to the front. Swap -7 and $9x$.

Note $5x^3$ is understood to have a positive coefficient; so when it is no longer the first term, a plus sign is placed to its left.

b. $-16b^3 + 5b^6 - 13b^4 - 12b + 10b^2 + 19$

Solution: $-16b^3 + 5b^6 - 13b^4 - 12b + 10b^2 + 19$

Answer: $5b^6 - 13b^4 - 16b^3 + 10b^2 - 12b + 19$ Move $-16b^3$ to follow $-13b^4$. Swap $-12b$ and $10b^2$.

Note Because $5b^6$ has a positive coefficient, when it is written as the first term, the plus sign is no longer needed.

▶ **Do Your Turn 6**

▶ **Your Turn 5**

Identify the degree of each polynomial.

a. $-2m^5 + 13m^7 - 8m^3 + 4m + 17$

b. $3x^{12} + 5x^6 + 8x^2 - 10x - 7$

▶ **Your Turn 6**

Write each polynomial in descending order of degree.

a. $9m^4 + 7m^9 - 3m^2 + 15 - 6m^7 + 2m$

b. $-14y^2 - 3y + 17 + 9y^3 - 13y^5$

c. $24 + 9x^2 - 15x + 8x^5 - 7x^3 - 4x^4$

Answers to Your Turn 5
a. 7 b. 12

Answers to Your Turn 6
a. $7m^9 - 6m^7 + 9m^4 - 3m^2 + 2m + 15$
b. $-13y^5 + 9y^3 - 14y^2 - 3y + 17$
c. $8x^5 - 4x^4 - 7x^3 + 9x^2 - 15x + 24$

Objective 6 Simplify polynomials in one variable by combining like terms.

Remember that monomials are also referred to as terms. We can simplify polynomials that contain **like terms.**

Definition Like terms: Constant terms or variable terms that have the same variables raised to the same exponents.

Examples of like terms		Examples of unlike terms	
8 and 4		$7x$ and $7y$	(different variables)
$3x$ and $2x$	◄ **Note** The coefficients	$10x^2$ and $3x$	(different exponents)
$5y^2$ and $-9y^2$	of like terms can be different.	$-5x^2y$ and $12xy^2$	(different exponents)
xy and $6xy$		$6a$ and $9A$	(different variables)

Now suppose we have a polynomial that contains like terms, such as $3x + 2x$. Because multiplication is repeated addition, $3x$ means that three x's are added together and $2x$ means that two x's are added together. Rewriting $3x + 2x$ as a sum of all of those x's gives the following:

$$3x \quad + \quad 2x$$
$$= \underbrace{x + x + x}_{} + \underbrace{x + x}_{}$$
$$= \qquad 5x$$

Note We have a total of five x's repeatedly added; so we can represent that sum as $5x$. ◄

Notice that the coefficient of the result, 5, is the sum of the coefficients of the like terms, $3 + 2 = 5$. Because $5x$ has fewer symbols than $3x + 2x$ has, we say that we have *simplified* $3x + 2x$. In fact, $5x$ is in *simplest form* because it cannot be written with fewer symbols. Our example suggests the following procedure:

> **Procedure**
>
> To combine like terms, add the coefficients and keep the variables and their exponents the same.

Connection We can affirm that $3x + 2x = 5x$ by evaluating $3x + 2x$ and $5x$ using the same value of x. For example, let's evaluate $3x + 2x$ when $x = 4$.

$$3(4) + 2(4) = 12 + 8$$
$$= 20$$

Evaluating $5x$ when $x = 4$ also gives a result of 20.

$$5(4) = 20$$

Equivalent expressions always yield the same result when we evaluate them using the same variable values. This also shows why simplifying an expression is useful: the simplified expression is much easier to evaluate.

▶ **Your Turn 7**

Combine like terms.

a. $6x + 9x$
b. $-m^2 - 12m^2$
c. $4y^2 - 20y^2$
d. $12x^3 - 12x^3$

Example 7 Combine like terms.

a. $4x + 9x$

Solution: $4x + 9x = 13x$

Add the coefficients, 4 and 9, and keep the variable, x, and its exponent, 1, the same. Remember, an exponent of 1 is not written.

Note When adding two numbers that have the same sign, as in Examples 7a and 7b, we add their absolute values and keep the same sign. ◄

b. $-2y^3 - 5y^3$

Solution: $-2y^3 - 5y^3 = -7y^3$

Add the coefficients, -2 and -5, and keep the variable, y, and its exponent, 3, the same.

c. $-12n^2 + 7n^2$

Solution: $-12n^2 + 7n^2 = -5n^2$

Add the coefficients, -12 and 7, and keep the variable, n, and its exponent, 2, the same.

Note When adding two numbers that have different signs, as in Examples 7c and 7d, we subtract their absolute values and keep the sign of the number with the greater absolute value. ◄

Answers to Your Turn 7

a. $15x$
b. $-13m^2$
c. $-16y^2$
d. 0

d. $-4m + 4m$

Solution: $-4m + 4m = 0m = 0$ -4 and 4 are additive inverses; so their sum is 0 and 0 times m is 0.

Connection Adding measurements that have the same unit is like combining like terms. Suppose we have to calculate the perimeter of the rectangle shown.

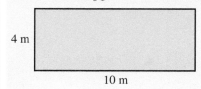

4 m

10 m

$P = 4 \text{ m} + 10 \text{ m} + 4 \text{ m} + 10 \text{ m}$

$P = 28 \text{ m}$

This perimeter problem is similar to a problem in which we must combine like terms in a polynomial where all of the variables are m.

$$4m + 10m + 4m + 10m = 28m$$

Notice that if the variables were different, such as $4cm + 10m + 4cm + 10m$, we would not be able to combine the terms with different variables. In a similar way, we cannot combine measurements with different units. If the width were 4 cm and the length 10 m, the perimeter would no longer be 28 m because cm and m represent different distance measurements.

$$4 \text{ cm} + 10 \text{ m} + 4 \text{ cm} + 10 \text{ m} \neq 28$$

▶ **Do Your Turn 7**

Sometimes there are several like terms in the same polynomial.

Example 8   Simplify and write the resulting polynomial in descending order of degree.

$$9x^2 + 5 + 7x + 2x^2 - 8 - 7x$$

Solution:

$= 9x^2 + 2x^2 + 7x - 7x + 5 - 8$ Collect like terms as needed.

$= \quad 11x^2 \quad + \quad 0 \quad - \quad 3$ Notice that $7x$ and $-7x$ have a sum of 0.

$= 11x^2 - 3$ We can drop 0 and bring the terms together.

Alternative Solution: Some people prefer to combine like terms without first collecting them, as shown next.

$9x^2 + 5 + 7x + 2x^2 - 8 - 7x$

$= 11x^2 + \quad 0 \quad - 3$

$= 11x^2 - 3$

THINK $9x^2 + 2x^2 = 11x^2$

$7x + (-7x) = 0$

$5 + (-8) = -3$

▶ **Do Your Turn 8**

▶ **Your Turn 8**

Simplify and write the resulting polynomial in descending order of degree.

a. $7x^4 - 4x^3 - 13x^4 + 12 + 4x^3 - 15$

b. $12k^4 - 19k + k^2 + 19k - 6 - 5k^2 + 2k^4$

VISUAL

Learning Strategy

When combining like terms, mark through the terms that you combine so that you know what you've combined and what you have left. Also, combine the terms in order of degree, combining the highest-degree terms first, and so on. This will put the resulting polynomial in descending order of degree.

Answers to Your Turn 8
a. $-6x^4 - 3$
b. $14k^4 - 4k^2 - 6$

3.2 Exercises For Extra Help MyMathLab®

Objective 1

Prep Exercise 1 What is a monomial? Give three examples.

For Exercises 1–12, determine whether the expression is a monomial. Explain.

1. $5x^3$ **2.** $7y^2$ **3.** $8m + 5$ **4.** $3t^2 + 5t - 6$

5. $-x$ **6.** y **7.** -7 **8.** 19

9. $\dfrac{4x^2}{5y}$ **10.** $\dfrac{-mn}{k}$ **11.** $-9x^2y$ **12.** $5m^3n^7p$

Objective 2

Prep Exercise 2 What is the coefficient of a monomial?   **Prep Exercise 3** What is the degree of a monomial?

For Exercises 13–24, identify the coefficient and degree of each monomial.

13. $3x^8$ **14.** $5t^2$ **15.** $-9x$ **16.** $-m$

17. 8 **18.** -7 **19.** xy^2 **20.** $-a^2b^3c$

21. $-2t^3u^4$ **22.** $15mnp$ **23.** -1 **24.** k

Objective 3

Prep Exercise 4 What is a polynomial? Give an example.

For Exercises 25–30, identify the terms in the polynomial and their coefficients.

25. $5x^2 + 8x - 7$ **26.** $4y^3 - 2y - 9$ **27.** $6t - 1$

28. $9n^4 - 8n^2 + 7n - 1$ **29.** $-6x^3 + x^2 - 9x + 4$ **30.** $-9x^2 - x + 5$

Prep Exercise 5 A binomial has _____ terms, and a trinomial has _____ terms.

For Exercises 31–42, indicate whether the expression is a monomial, binomial, or trinomial or has no special name.

31. $5x + 2$

32. $9y^2 + 5y - 6$

33. $18x^2$

34. 2

35. $6x^3 - 14x^2 + 9x - 8$

36. $2a - 6b$

37. $16t^2 - 8t - 5$

38. $a^5 - 6a^3 + 7a^2 - a + 2$

39. -5

40. $b^3 + 7b^2 - 12b$

41. $y^3 - 8$

42. $m^4 - 5m^3 + m^2 + 6m - 9$

Objective 4

Prep Exercise 6 What is the degree of a polynomial?

For Exercises 43–48, identify the degree of each polynomial.

43. $6x^3 - 9x^2 + 8x + 2$

44. $8y^4 + 2y - 5y^6 + y^3 + 4$

45. $-m^3 - 8m^5 + 3m^2 - 2m^6$

46. $5 + 6a + 3a^3 + 4a^2 - a^5$

47. $19t^9 + 4t^3 + 25t^{12} - 11t^5$

48. $-b^2 - 16b^8 + 5b^3 + 2b^4 - 22$

Objective 5

Prep Exercise 7 How do you write a polynomial in descending order of degree?

For Exercises 49–54, write each polynomial in descending order of degree.

49. $14t^6 + 9t^3 + 5t^2 - 8t^4 - 1$

50. $6x + 19 - 4x^3 + 3x^2$

51. $9 - 18y^3 + 12y - 10y^2 + y^5$

52. $21m - 6m^7 + 9m^2 + 13 - 8m^5$

53. $7a^3 + 9a^5 - 6 + 2a^2 - a$

54. $2n^4 - 5n^3 - n + 2 + n^6 - 7n^8$

Objective 6

Prep Exercise 8 What are like terms? Give an example of a pair of like terms.

For Exercises 55–66, determine whether the monomials are like terms.

55. $8x$ and $-5x$

56. $3y$ and y

57. $-5m$ and $-7n$

58. $2a$ and $2b$

59. $4n^2$ and $7n^2$

60. $8h^3$ and h^3

61. $-9K$ and $-10K$

62. T and $-T$

63. $5n$ and $9N$

64. $6xY$ and $5xy$

65. 14 and -6

66. 1 and -5

Prep Exercise 9 What does it mean for an expression to be in simplest form?

Prep Exercise 10 How do you combine like terms?

For Exercises 67–82, simplify by combining like terms.

67. $3x + 9x$

68. $5a + 2a$

69. $3y + y$

70. $b + 7b$

71. $12m - 7m$

72. $10x - 2x$

73. $9t - 12t$

74. $7k - 18k$

75. $7x^3 - 2x^3$

76. $17c^4 - c^4$

77. $-10t^2 + 9t^2$

78. $-14u^3 + 13u^3$

79. $-4m^5 - 4m^5$

80. $-j^3 - j^3$

81. $2b^2 - 2b^2$

82. $-9x^2 + 9x^2$

For Exercises 83–94, simplify and write the resulting polynomial in descending order of degree.

83. $9x^2 + 5x + 2x^2 - 3x + 4$

84. $7y^2 + 4y - 3y^2 - y$

85. $7a - 8a^3 - a + 4 - 5a^3$

86. $3b - 9b^4 + 4b^4 - 7b$

87. $-9c^2 + c + 4c^2 - 5c$

88. $4m^7 - 9m^5 - 12m^5 + 7m^7$

89. $-3n^2 - 2n + 7 - 4n^2$

90. $p^3 + 9p^2 - 17 + 3p^2$

91. $12m^2 + 5m^3 - 11m^2 + 3 + 2m^3 - 7$

92. $4y + 15 - y^2 - 16 + 5y + 6y^2$

93. $-7x + 5x^4 - 2 + 3x^4 + 7x + 10 - 3x^2$

94. $-10b^3 + 7b^4 + 1 + b^3 - 2b^2 - 12 + b^4$

Review Exercises

For Exercises 1–4, simplify.

[2.2] 1. $-7 + 5$

[2.3] 2. $-9 - 7$

[2.3] 3. $8 - 15$

[2.3] 4. $-15 + 8 - 21 + 2$

[2.3] 5. Is $3 - 5$ the same as $5 - 3$? Explain.

[2.3] 6. Is $-5 + 3$ the same as $3 + (-5)$? Explain.

[2.4] 7. Is $(-3)(5)$ the same as $(5)(-3)$? Explain.

[1.6] 8. Find the perimeter of the shape.

4 m

10 m

3.3   Adding and Subtracting Polynomials

Warm-up *Combine like terms and write the polynomial in descending order of degree.*

[3.2] 1. $3x^2 + 5x + 2 + 2x^2 + 3x + 1$

[3.2] 2. $8x^2 - 9x + 4 - 6x^2 - x + 7$

Objective 1 Add polynomials in one variable.

We can add and subtract polynomials in the same way we add and subtract numbers. However, instead of adding digits in like place values, we add like terms. Consider the following comparison:

Numeric addition:

$352 + 231 = 583$

We can stack like place values to find the sum:

$$
\begin{array}{r}
352 \\
+\ 231 \\
\hline
583
\end{array}
$$

Polynomial addition:

$(3x^2 + 5x + 2) + (2x^2 + 3x + 1) = 5x^2 + 8x + 3$

We can stack like terms to find the sum:

$$
\begin{array}{r}
3x^2 + 5x + 2 \\
+\ 2x^2 + 3x + 1 \\
\hline
5x^2 + 8x + 3
\end{array}
$$

Actually, we do not have to stack polynomials to add them as long as we combine all of the like terms.

Procedure

To add polynomials, combine like terms.

Example 1 Add and write the resulting polynomial in descending order of degree.

a. $(5x^3 + 9x^2 + 2x + 3) + (3x^3 + 5x + 1)$

Solution: $(5x^3 + 9x^2 + 2x + 3) + (3x^3 + 5x + 1)$

$= 8x^3 + 9x^2 + 7x + 4$ Combine like terms.

Explanation: We combined like terms in order of degree.

First, we combined the degree-three terms. $5x^3 + 3x^3 = 8x^3$

The second polynomial had no degree-two term; so we brought down the $9x^2$.

Next, we combined the degree-one terms. $2x + 5x = 7x$

Finally, we combined the degree-zero terms. $3 + 1 = 4$

Connection As we saw in Section 3.2, a polynomial such as $3x^2 + 5x + 2$ can be transformed to the number 352 by replacing x with 10.

$$3x^2 + 5x + 2$$
$$3(10)^2 + 5(10) + 2$$
$$= 300 + 50 + 2$$
$$= 352$$

Connection If we stack the polynomials in Example 1, we must leave a blank space under the $9x^2$ because there is no degree-two term in the second polynomial. It's like adding $5923 + 3051$.

$$
\begin{array}{r}
5x^3 + 9x^2 + 2x + 3 \\
+\ 3x^3 \qquad + 5x + 1 \\
\hline
8x^3 + 9x^2 + 7x + 4
\end{array}
\qquad
\begin{array}{r}
5923 \\
+\ 3051 \\
\hline
8974
\end{array}
$$

As the polynomials get more complex, these blanks become more common, which is why stacking is not the best method for all cases of adding polynomials.

Answers to Warm-up
1. $5x^2 + 8x + 3$
2. $2x^2 - 10x + 11$

▶ **Your Turn 1**

Add and write the resulting polynomial in descending order of degree.

a. $(2x + 8) + (6x + 1)$

b. $(6t^3 + 4t + 2) + (2t^4 + t + 1)$

c. $(9x^2 - 2x + 5) + (3x^2 - 8x - 4)$

d. $(9m^3 + 14m^2 + 4m - 8) + (2m^3 + 3m^2 - 4m - 1)$

b. $(4x^5 - 9x + 6) + (2x^5 - 3x - 2)$

Solution: $(4x^5 - 9x + 6) + (2x^5 - 3x - 2)$

$$= 6x^5 - 12x + 4 \quad \text{Combine like terms.}$$

Explanation: First, we combined the degree-five terms. $4x^5 + 2x^5 = 6x^5$

Then, we combined degree-one terms. $-9x + (-3x) = -12x$

Finally, we combined degree-zero terms. $6 + (-2) = 4$

Note We determine the signs in the answer based on the outcome of combining. Remember, the sign to the left of the term is the sign of the term.

If the outcome is positive, as with 4, we place a plus sign to the left of the term. Although $6x^5$ is positive, it is the first term in the answer; so no plus sign is needed to its left.

If the outcome is negative, as with $-12x$, we place a minus sign to the left of the term.

◀ **Do Your Turn 1**

▶ **Your Turn 2**

Write an expression in simplest form for the perimeter of the shape shown.

$x + 1$ $x + 1$

$3x - 2$

Example 2   Write an expression in simplest form for the perimeter of the rectangle shown.

Solution: To find the perimeter, we need to find the total distance around the shape. Instead of numerical measurements for the length and width, we have expressions. In this case, the length is the expression $3x - 1$ and the width is the expression $2x + 5$. We can only write an expression for the perimeter; we cannot get a numerical answer until we have a value for the variable.

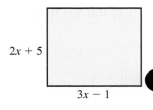

$2x + 5$

$3x - 1$

$$
\begin{aligned}
\text{Perimeter} &= \text{length} + \text{width} + \text{length} + \text{width} \\
&= (3x - 1) + (2x + 5) + (3x - 1) + (2x + 5) \\
&= \underbrace{3x + 2x + 3x + 2x}_{} \ \underbrace{- 1 + 5 - 1 + 5}_{} \quad \text{Collect like terms.} \\
&= \qquad\quad 10x \qquad\qquad\quad + 8 \qquad\quad \text{Combine like terms.}
\end{aligned}
$$

Answer: The expression for the perimeter is $10x + 8$.

Warning $10x + 8$ is in simplest form. It might be tempting to try to write it as $18x$, but $10x$ and 8 are not like terms and cannot be combined.

◀ **Do Your Turn 2**

Objective 2 Subtract polynomials in one variable.

Subtracting polynomials is tricky because they have signs involved, like the integers. Recall that when we subtracted integers, we found it simpler to write the subtraction statements as equivalent addition statements. We can apply this same principle to subtracting polynomials.

Let's compare polynomial subtraction to numeric subtraction.

Numeric subtraction:	**Polynomial subtraction:**
$896 - 254$	$(8x^2 + 9x + 6) - (2x^2 + 5x + 4)$

Like addition, we can align the like place values or like terms in a stacked form.

$$
\begin{array}{r}
896 \\
- \ 254 \\
\hline
642
\end{array}
\qquad\qquad
\begin{array}{r}
8x^2 + 9x + 6 \\
- \ 2x^2 + 5x + 4 \\
\hline
6x^2 + 4x + 2
\end{array}
$$

Answers to Your Turn 1

a. $8x + 9$

b. $2t^4 + 6t^3 + 5t + 3$

c. $12x^2 - 10x + 1$

d. $11m^3 + 17m^2 - 9$

Answer to Your Turn 2

$5x$

In the numeric subtraction, we subtracted digits in the like place value. In the polynomial subtraction, we subtracted like terms. Note that subtracting 4 from 6 is the same as adding 6 and -4, subtracting $5x$ from $9x$ is the same as adding $9x$ and $-5x$, and subtracting $2x^2$ from $8x^2$ is the same as adding $8x^2$ and $-2x^2$. Therefore, we would write the following equivalent addition:

$$
\begin{array}{r}
8x^2 + 9x + 6 \\
+ \quad -2x^2 - 5x - 4 \\
\hline
6x^2 + 4x + 2
\end{array}
$$

or in sentence form:

$$(8x^2 + 9x + 6) + (-2x^2 - 5x - 4) = 6x^2 + 4x + 2$$

Conclusion: We can write polynomial subtraction as equivalent polynomial addition by changing the signs of each term in the subtrahend (second) polynomial.

Change the operation from $-$ to $+$.

$$(8x^2 + 9x + 6) - (\ 2x^2 + 5x + 4)$$

Change the sign of each term in the subtrahend.

$$= (8x^2 + 9x + 6) + (-2x^2 - 5x - 4)$$

Our exploration suggests the following procedure:

Procedure

To subtract polynomials:

1. Write the subtraction expression as an equivalent addition expression.
   a. Change the operation symbol from a $-$ to a $+$.
   b. Change the subtrahend (second polynomial) to its additive inverse by changing the sign of each of its terms.
2. Combine like terms.

Example 3   Subtract and write the resulting polynomial in descending order of degree.

a. $(7x^2 + 8x + 5) - (3x^2 + 7x + 1)$

Solution: Write an equivalent addition expression; then combine like terms.

Change the operation from $-$ to $+$.

$$(7x^2 + 8x + 5) - (\ 3x^2 + 7x + 1)$$

Change the sign of each term in the subtrahend.

$$= (7x^2 + 8x + 5) + (-3x^2 - 7x - 1)$$
$$= 4x^2 + x + 4 \qquad \text{Combine like terms.}$$

Explanation: After writing the equivalent addition, we combined like terms in order of degree.

First, we combined the degree-two terms.

$$7x^2 + (-3x^2) = 4x^2$$

Second, we combined the degree-one terms.

$$8x + (-7x) = x$$

Third, we combined the degree-zero terms.

$$5 + (-1) = 4$$

Connection We use the distributive property to change all of the signs in a polynomial. If we disregard the first polynomial, we have the following:

$$-(2x^2 + 5x + 4)$$
$$= -1(2x^2 + 5x + 4)$$

The distributive property tells us that we can distribute the -1 (or minus sign) to each term inside the parentheses.

$$= -1 \cdot 2x^2 - 1 \cdot 5x - 1 \cdot 4$$
$$= -2x^2 - 5x - 4$$

Connection The polynomial subtraction in part a is equivalent to the following numeric subtraction:

$$
\begin{array}{r}
785 \\
- \ 371 \\
\hline
414
\end{array}
$$

Notice that the numeric result, 414, corresponds to the polynomial result, $4x^2 + x + 4$.

Although we could use the stacking method for subtracting polynomials, it is not the best method because we sometimes have missing terms or terms that are not like terms.

▶ **Your Turn 3**

Subtract and write the resulting polynomial in descending order of degree.

a. $(7x + 3) - (3x + 1)$

b. $(8t^2 + 9t + 3) - (7t^2 + t + 2)$

c. $(9x^2 - 7x + 5) - (2x^2 - 3x + 8)$

d. $(17x^4 + 9x^3 - 10x^2 - 12) - (13x^4 + 9x^3 - 3x^2 + 2)$

Answers to Your Turn 3

a. $4x + 2$

b. $t^2 + 8t + 1$

c. $7x^2 - 4x - 3$

d. $4x^4 - 7x^2 - 14$

b. $(12x^3 + 10x - 9) - (4x^3 - 5x + 3)$

Solution: Write an equivalent addition expression; then combine like terms.

$$(12x^3 + 10x - 9) - (\ \ 4x^3 - 5x + 3\ )$$

Change the operation from − to +.

Change the sign of each term in the subtrahend.

$$= (12x^3 + 10x - 9) + (-4x^3 + 5x - 3)$$

$$= 8x^3 + 15x - 12 \qquad \text{Combine like terms.}$$

Explanation: After writing the equivalent addition, we combined like terms in order of degree.

First, we combined the degree-three terms. $12x^3 + (-4x^3) = 8x^3$

Then, we combined the degree-one terms. $10x + 5x = 15x$

Finally, we combined the degree-zero terms. $-9 + (-3) = -12$

◀ **Do Your Turn 3**

3.3 Exercises For Extra Help MyMathLab®

Objective 1

Prep Exercise 1 Explain how to add two polynomials.

For Exercises 1–18, add and write the resulting polynomial in descending order of degree.

1. $(2x + 5) + (3x + 1)$

2. $(5x + 2) + (3x + 7)$

3. $(7y - 4) + (2y - 1)$

4. $(9m - 3) + (4m - 6)$

5. $(10x + 7) + (3x - 9)$

6. $(7p + 4) + (9p - 7)$

7. $(4t - 11) + (t + 13)$

8. $(3g - 4) + (g + 2)$

9. $(2x^2 + x + 3) + (5x + 7)$

10. $(7x^2 + 2x + 1) + (6x + 5)$

11. $(9n^2 - 14n + 7) + (6n^2 + 2n - 4)$

12. $(8t^2 + 12t - 10) + (2t^2 - 8t - 2)$

13. $(4p^2 - 7p - 9) + (3p^2 + 4p + 5)$

14. $(9a^2 + 3a - 5) + (5a^2 - 7a - 6)$

15. $(-3b^3 + 2b^2 - 9b) + (-5b^3 - 3b^2 + 11)$

16. $(9c^3 - 4c^2 + 5) + (-7c^2 - 10c + 7)$

17. $(3a^3 - a^2 + 10a - 4) + (6a^2 - 9a + 2)$

18. $(9x^3 + 4x^2 + x - 11) + (3x^2 - 8x - 5)$

For Exercises 19–22, write an expression in simplest form for the perimeter of each shape.

19.
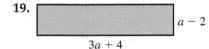
$a - 2$
$3a + 4$

20.
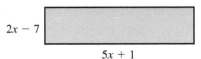
$2x - 7$
$5x + 1$

21.
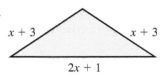
$x + 3$ $x + 3$
$2x + 1$

22.
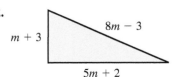
$m + 3$ $8m - 3$
$5m + 2$

For Exercises 23–26, explain the mistake and work the problem correctly.

23. $(9y^3 + 4y^2 - 6y) + (y^3 - 2y^2)$
$= 10y^3 - 2y^2 - 6y$

24. $(10a^4 + 5a^3 - 11) + (5a^4 + 3a^2 - 4)$
$= 15a^4 + 3a^2 - 15$

25. $(8x^6 - 9x^4 + 15x^3 - 1) + (x^5 - 8x^3 - 3)$
$= 9x^6 - 9x^4 + 7x^3 - 4$

26. $(n^3 + 2n^2 - 19) + (n^3 - 2n + 4)$
$= 2n^3 - 15$

Objective 2

Prep Exercise 2 Explain how to write a subtraction of two polynomials as an equivalent addition.

For Exercises 27–42, subtract and write the resulting polynomial in descending order of degree.

27. $(5x + 9) - (3x + 2)$

28. $(7m + 4) - (2m + 1)$

29. $(8t + 3) - (5t + 3)$

30. $(6a + 7) - (6a + 5)$

31. $(10n + 1) - (2n + 7)$

32. $(9x + 8) - (3x + 10)$

33. $(5x - 6) - (2x + 1)$

34. $(2m + 5) - (7m - 3)$

35. $(6x^2 + 8x - 9) - (9x + 2)$

36. $(8n^2 - 7n - 5) - (2n - 3)$

37. $(8a^2 - 10a + 2) - (-4a^2 + a + 8)$

38. $(y^2 + 3y - 9) - (-6y^2 + 8y - 1)$

39. $(7x^2 + 5x - 8) - (7x^2 + 9x + 2)$

40. $(2t^2 - t + 7) - (3t^2 - t - 6)$

41. $(-8m^3 + 9m^2 - 17m - 2) - (-8m^3 + 5m^2 + m - 6)$

42. $(5u^3 - 4u^2 + u - 10) - (-2u^3 + 4u^2 + u + 7)$

For Exercises 43–46, explain the mistake and work the problem correctly.

43. $(5x^2 - 6x) - (3x^2 - 4x)$
$= (5x^2 - 6x) + (-3x^2 - 4x)$
$= 2x^2 - 10x$

44. $(10y^2 - 5y) - (y^2 + 3y)$
$= (10y^2 - 5y) + (-y^2 - 3y)$
$= 9y^2 + 8y$

45. $(-9t^2 + 4t + 11) - (5t^3 + 9t^2 + 6)$
$= (-9t^2 + 4t + 11) + (-5t^3 - 9t^2 - 6)$
$= -5t^3 + 4t + 5$

46. $(x^4 - 6x^2 + 3x) - (x^4 + 2x^3 - 4x)$
$= (x^4 - 6x^2 + 3x) + (-x^4 - 2x^3 + 4x)$
$= -2x^4 - 2x^3 - 6x^2 + 7x$

For Exercises 47–50, complete the table. The first row will contain variable expressions only. To find the sum, add the two given expressions. To find the difference, subtract the second expression from the first. After finding the variable expressions for the sum and difference, evaluate each variable expression using the given values.

47.

	Expression 1	Expression 2	Sum	Difference
x	$5x + 4$	$3x - 7$		
4				
-2				

48.

	Expression 1	Expression 2	Sum	Difference
t	$2t - 1$	$6t - 3$		
3				
-5				

49.

	Expression 1	Expression 2	Sum	Difference
n	$n^2 + 3n - 1$	$4n^2 - n - 7$		
2				
-3				

50.

	Expression 1	Expression 2	Sum	Difference
y	$2y^2 - 5y - 3$	$3y^2 - 5y + 2$		
5				
-1				

Puzzle Problem A trog is a mythical creature that reproduces in a peculiar way. An adult trog will produce one newborn every hour. A newborn trog spends its first hour growing to an adult. If we place one newborn trog in a cage, how many trogs will exist after six hours? After seven hours? Explain the pattern.

 Learning Strategy

After completing an assignment without using your notes or study sheet, quiz yourself by doing additional exercises like those assigned. Monitor how quickly you recall the solution process for each problem. If you have trouble recalling the process, you need to spend more time studying the procedures.

Review Exercises

For Exercises 1–4, evaluate.

[2.4] 1. $2 \cdot (-341)$

[2.4] 2. $(-3)^4$

[2.4] 3. $(-2)^5$

[1.3] 4. $2^3 \cdot 2^4$

[1.6] 5. A rectangular swimming pool is 25 meters long by 15 meters wide. What is the area of the pool?

[1.3] 6. A lunch menu has a choice of five main dishes and four side dishes. How many different meals can a customer order?

3.4   Exponent Rules; Multiplying Polynomials

Warm-up

[2.4] **1.** Multiply.

    a. $-7 \cdot 9$ **b.** $(-8)(-6)$

[2.5] **2.** Simplify using the order of operations agreement. $2^3 \cdot 2^4$

Objective 1 Multiply monomials.

To multiply monomials, we need to understand how to multiply exponential forms that have the same base, as in $2^3 \cdot 2^4$. One way to simplify $2^3 \cdot 2^4$ is to use the order of operations agreement.

$$2^3 \cdot 2^4 = 8 \cdot 16 \quad \text{Evaluate exponential forms.}$$
$$= 128 \quad \text{Multiply.}$$

However, there is an alternative. We can express the result in exponential form by expanding each of 2^3 and 2^4 into its factored form.

$$\overset{2^3 \qquad \quad 2^4}{= \overbrace{2 \cdot 2 \cdot 2} \cdot \overbrace{2 \cdot 2 \cdot 2 \cdot 2}}$$

Because a total of seven 2s are multiplied, we
◄ can express the product as 2^7.

$$= \qquad 2^7$$

Note that $2^7 = 128$. Notice in the alternative method that the resulting exponent is the *sum* of the original exponents.

$$2^3 \cdot 2^4 = 2^{3+4} = 2^7$$

Our example suggests the following rule:

> **Rule**
>
> To multiply exponential forms that have the same base, we can add the exponents and keep the same base.
>
> **In math language:** $a^m \cdot a^n = a^{m+n}$

Example 1 Multiply. $(x^2)(x^3)$

Solution: $(x^2)(x^3) = x^{2+3} = x^5$ Add the exponents and keep the same base.

Check: We can expand the exponential forms into their factored forms.

$$(x^2)(x^3)$$

x^2 means two x's. x^3 means three x's.

$$= \overbrace{(x \cdot x)} \; \overbrace{(x \cdot x \cdot x)}$$

A total of five x's are multiplied
together; so we can express the
product as x^5.

$$= x^5$$

► **Do Your Turn 1**

Objectives

1 Multiply monomials.

2 Simplify monomials raised to a power.

3 Multiply a polynomial by a monomial.

4 Multiply polynomials.

► **Your Turn 1**

Multiply.
a. $x^3 \cdot x^4$
b. $(n^6)(n^2)$
c. $m^5(m)$

Answers to Your Turn 1
a. x^7 **b.** n^8 **c.** m^6

Answers to Warm-up
1. a. -63 b. 48
2. 128

Let's apply this rule to monomials that have coefficients, such as $(2x^4)(3x^6)$. We can use the commutative property of multiplication to separate the coefficients and variables. We can then multiply the coefficients and use our rule for exponents.

$$(2x^4)(3x^6) = 2 \cdot 3 \cdot x^4 \cdot x^6$$
$$= 6x^{4+6} \qquad \text{Multiply coefficients and add exponents of the like bases, } x.$$
$$= 6x^{10}$$

We can confirm our result by expanding the original expression so that the exponential forms are in factored form.

$$(2x^4)(3x^6) = 2 \cdot x \cdot x \cdot x \cdot x \cdot 3 \cdot x \cdot x \cdot x \cdot x \cdot x \cdot x$$
$$= 2 \cdot 3 \cdot x \cdot x \cdot x \cdot x \cdot x \cdot x \cdot x \cdot x \cdot x \cdot x$$
$$= 6x^{10}$$

◀ **Note** There are 10 factors of x.

Connection Multiplying monomials such as $(2x^4)(3x^6)$ is similar to multiplying numbers in expanded form. If we replace the x's with the number 10, we have expanded form.

$$(2x^4)(3x^6)$$
$$(2 \cdot 10^4)(3 \cdot 10^6)$$

Multiplying numbers in expanded form must work the same way as multiplying monomials. We can multiply the $2 \cdot 3$ to get 6 and add the exponents of the like bases.

$$2 \cdot 3 \cdot 10^{4+6} = 6 \cdot 10^{10}$$

To confirm our result, we can write the expanded forms in standard form and then multiply.

$$(2 \cdot 10^4)(3 \cdot 10^6) = (20,000)(3,000,000)$$
$$= 60,000,000,000$$
$$= 6 \cdot 10^{10} \qquad \text{The product has a 6 followed by ten 0s.}$$

Our example suggests the following procedure:

Procedure

To multiply monomials:

1. Multiply the coefficients.
2. Add the exponents of the like variables.

▶ **Your Turn 2**

Multiply.
a. $4x^5 \cdot 2x^8$
b. $(4 \cdot 10^5)(2 \cdot 10^8)$
c. $12m^2(m^8)$
d. $-2y^2(4y)(9y^4)$
e. $2b^3(-4b^2)(3b^5)$

Answers to Your Turn 2
a. $8x^{13}$ b. $8 \cdot 10^{13}$
c. $12m^{10}$ d. $-72y^7$
e. $-24b^{10}$

Example 2 Multiply. $-3x^5(4x^2)(5x)$

Solution: $-3x^5(4x^2)(5x) = -3 \cdot 4 \cdot 5x^{5+2+1}$ Multiply coefficients and add exponents of the like bases, x.

$$= -60x^8$$

Check: We can expand the original expression so that the exponential forms are in factored form.

$$-3x^5(4x^2)(5x) = -3 \cdot x \cdot x \cdot x \cdot x \cdot x \cdot 4 \cdot x \cdot x \cdot 5 \cdot x$$

$$= -3 \cdot 4 \cdot 5 \cdot x \cdot x \cdot x \cdot x \cdot x \cdot x \cdot x \cdot x$$

Use the commutative property of multiplication.

$$= -60x^8$$

Multiply the numbers and count the factors of *x*, which we write as an exponent, 8.

▶ **Do Your Turn 2**

Connection Multiplying distance measurements in area and volume is like multiplying monomials.

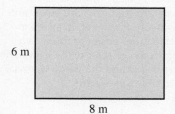

The area of the rectangle shown is

$$A = bh$$
$$A = (8\,\text{m})(6\,\text{m})$$
$$A = 48\,\text{m}^2$$

The area calculation is similar to multiplying two monomials with *m* as the variable.

Multiply: $(8m)(6m) = 48m^2$

The volume of the box shown is

$$V = lwh$$
$$V = (5\,\text{m})(3\,\text{m})(4\,\text{m})$$
$$V = 60\,\text{m}^3$$

The volume calculation is similar to multiplying three monomials with *m* as the variable.

Multiply: $(5m)(3m)(4m) = 60m^3$

Objective 2 Simplify monomials raised to a power.

To simplify a monomial raised to a power, we need to understand how to simplify an exponential form raised to a power, as in $(2^3)^2$. To simplify $(2^3)^2$, we could use the order of operations agreement and evaluate the exponential form within the parentheses first.

$$(2^3)^2 = (2 \cdot 2 \cdot 2)^2$$
$$= 8^2$$
$$= 64$$

However, there is an alternative. The outer exponent, 2, indicates to multiply 2^3 by itself.

$$(2^3)^2 = 2^3 \cdot 2^3$$
$$= 2^{3+3}$$
$$= 2^6$$

◀ **Note** Because this is a multiplication of exponential forms that have the same base, we can add the exponents.

Note that $2^6 = 64$ and that the exponent, 6, in the result is the *product* of the original exponents, 3 and 2.

$$(2^3)^2 = 2^{3 \cdot 2} = 2^6$$

Our example suggests the following rule:

> ### Rule
> To simplify an exponential form raised to a power, we can multiply the exponents and keep the same base.
>
> **In math language:** $(a^m)^n = a^{mn}$

How might we simplify a monomial raised to a power, as in $(2x^3)^4$? The exponent, 4, means that $2x^3$ is a factor 4 times.

$$(2x^3)^4 = 2x^3 \cdot 2x^3 \cdot 2x^3 \cdot 2x^3$$
$$= 2 \cdot 2 \cdot 2 \cdot 2 \cdot x^{3+3+3+3}$$
$$= 16x^{12}$$

Note $(2x^3)^4$ is read "two x to the third power, raised to the fourth power." Both 3 and 4 are exponents and are referred to as 3rd *power* and 4th *power*, respectively. However, in an effort to make the explanations easier to follow, we will use the word *power* to refer to the exponent outside the parentheses and the word *exponent* to refer to the variable's exponent.

Notice that the final coefficient is the result of evaluating 2^4 and that the final exponent is the product of the power and the variable's initial exponent; so we could write

$$(2x^3)^4 = 2^4 x^{3 \cdot 4} = 16x^{12}$$

Our example suggests the following procedure:

> ### Procedure
> To simplify a monomial raised to a power:
> 1. Evaluate the coefficient raised to that power.
> 2. Multiply each variable's exponent by the power.

► **Your Turn 3**

Simplify.
a. $(3x^2)^4$
b. $(5a^6)^3$
c. $(-2m^4)^5$
d. $(-2t^3)^6$

Example 3   Simplify. $(-4x^2)^3$

Solution: $(-4x^2)^3 = (-4)^3 x^{2 \cdot 3}$
$\qquad\qquad\quad = -64x^6$

Evaluate the coefficient, -4, raised to the power, 3, and multiply the variable's exponent, 2, by the power, 3.

◄ **Do Your Turn 3**

Objective 3 Multiply a polynomial by a monomial.

To multiply numbers such as $2 \cdot 34$, we stack and use the distributive property to multiply each digit in the 34 by 2.

$$\begin{array}{r} 34 \\ \times\ 2 \\ \hline 68 \end{array}$$

Connection In Chapter 1, we learned that the distributive property is

$$a(b + c) = ab + ac.$$

Now let's multiply $2 \cdot 34$ without stacking. We can write 34 as $30 + 4$ so that the expression looks more like the distributive property.

$$2(34) = 2(30 + 4)$$

Using the distributive property to distribute the 2 to the 30 and the 4, we have

$$2(30 + 4) = 2 \cdot 30 + 2 \cdot 4$$
$$= 60 + 8 = 68$$

Answers to Your Turn 3
a. $81x^8$ b. $125a^{18}$
c. $-32m^{20}$ d. $64t^{18}$

Now let's look at a product involving a polynomial such as $2(3x + 4)$. We use the distributive property just as we did for $2(30 + 4)$. Compare the numeric and algebraic versions.

Numeric: $2(30 + 4) = 2 \cdot 30 + 2 \cdot 4$ Algebraic: $2(3x + 4) = 2 \cdot 3x + 2 \cdot 4$
$\qquad\qquad\qquad = 60 + 8$ $\qquad\qquad\qquad\qquad = 6x + 8$
$\qquad\qquad\qquad = 68$

Notice that if we let $x = 10$ in the algebraic version, we get the numeric version.

Procedure

To multiply a polynomial by a monomial, use the distributive property to multiply each term in the polynomial by the monomial.

Example 4 Multiply.

a. $2x(3x^2 + 4x + 1)$

Solution: We multiply each term in the polynomial by $2x$.

$$2x \qquad (3x^2 \quad + \quad 4x \quad + \quad 1)$$
$$2x \cdot 3x^2$$
$$2x \cdot 4x$$
$$2x \cdot 1$$

$= 2x \cdot 3x^2 + 2x \cdot 4x + 2x \cdot 1$ **Distribute.**

$= 6x^3 + 8x^2 + 2x$ **Multiply.** ◄

Note $2x = 2x^1$ and $4x = 4x^1$; so $2x \cdot 3x^2 = 2x^1 \cdot 3x^2 = 6x^{1+2} = 6x^3$ and $2x \cdot 4x = 2x^1 \cdot 4x^1 = 8x^{1+1} = 8x^2$.

b. $-3x^2(4x^2 + 5x - 6)$

Solution: Multiply each term in the polynomial by $-3x^2$. Watch the signs because we are multiplying by a monomial with a negative coefficient.

$$-3x^2 \qquad (4x^2 \quad + \quad 5x \quad - \quad 6)$$
$$-3x^2 \cdot 4x^2$$
$$-3x^2 \cdot 5x$$
$$-3x^2 \cdot (-6)$$

$= -3x^2 \cdot 4x^2 - 3x^2 \cdot 5x - 3x^2 \cdot (-6)$

$= -12x^4 - 15x^3 + 18x^2$

Note Remember, the sign to the left of a term indicates addition or subtraction *and* the sign of the term. For example, $+5x$ means that we add $5x$ and that $5x$ is positive. Similarly, -6 means that we subtract 6 and that the term is -6.

Note When we multiply a polynomial by a monomial with a negative coefficient, the signs of the terms in the resulting polynomial will be the opposite of the signs of the terms in the original polynomial.

$$-3x^2 \,(4x^2 \ + \ 5x \ - \ 6)$$
$$= -12x^4 \ - \ 15x^3 \ + \ 18x^2$$

▶ **Do Your Turn 4**

▶ **Your Turn 4**

Multiply.
a. $2x(4x^2 + 3)$
b. $5a^2(6a^2 - 7a + 2)$
c. $-4n^3(8n - 9)$
d. $-7m^2(4m^2 - m + 3)$

Connection Multiplying $2x(3x^2 + 4x + 1)$ is essentially the same as this numeric multiplication:

$20(300 + 40 + 1)$
$\quad = 20 \cdot 300 +$
$\qquad 20 \cdot 40 + 20 \cdot 1$
$\quad = 6000 + 800 + 20$

If you think of x^3 as thousands place, x^2 as hundreds, and x as tens, the algebraic result is the same.

$2x(3x^2 + 4x + 1)$
$\quad = 6x^3 + 8x^2 + 2x$

Answers to Your Turn 4
a. $8x^3 + 6x$
b. $30a^4 - 35a^3 + 10a^2$
c. $-32n^4 + 36n^3$
d. $-28m^4 + 7m^3 - 21m^2$

Objective 4 Multiply polynomials.

Multiplying two binomials is like multiplying a pair of two-digit numbers. For example, $(x + 2)(x + 3)$ is like $(12)(13)$. To see how to multiply polynomials, let's consider the procedure for multiplying $(12)(13)$.

$$
\begin{array}{r}
12 \\
\times\ 13 \\
\hline
36 \\
+\ 12 \\
\hline
156
\end{array}
$$

Note We think to ourselves "3 times 2 is 6, then 3 times 1 is 3," which creates 36. We then move to the 1 in the tens place of the 13 and do the same thing. Because this 1 digit is in the tens place, it really means 10; so when we multiply this 10 times 12, it makes 120. We usually omit writing the 0 in the ones place and write 12 in the next two places.

Remember, we are using the distributive property. We multiply each digit in one number by each digit in the other number and shift underneath as we move to each new place.

The same process applies to the binomials. We can stack them as we did the previous numbers. However, we'll stack only to see how the method works. You'll find that as the polynomials get more complex, the stacking method becomes too tedious.

$$
\begin{array}{r}
x + 2 \\
x + 3 \\
\hline
3x + 6 \\
+\ x^2 + 2x \\
\hline
x^2 + 5x + 6
\end{array}
$$

Note We think "3 times 2 is 6, then 3 times x is $3x$." Now move to the x and think "x times 2 is $2x$, then x times x is x^2." Notice how we shifted so that the $2x$ and $3x$ line up. This is because they are like terms. It is the same as lining up the tens column when we work with whole numbers. Note that the numeric result, 156, is basically the same as the algebraic result, $x^2 + 5x + 6$.

Our examples suggest the following procedure:

Procedure

To multiply two polynomials:

1. Multiply each term in the second polynomial by each term in the first polynomial.
2. Combine like terms.

▶ Your Turn 5

Multiply.

a. $(x + 4)(x + 2)$
b. $(t + 3)(4t - 1)$
c. $(2y - 5)(3y + 1)$
d. $(n - 6)(7n - 3)$

Answers to Your Turn 5

a. $x^2 + 6x + 8$
b. $4t^2 + 11t - 3$
c. $6y^2 - 13y - 5$
d. $7n^2 - 45n + 18$

Example 5   Multiply.

a. $(x + 5)(x + 1)$

Solution: Multiply each term in $x + 1$ by each term in $x + 5$.

$$
(x + 5)\ (x + 1)
$$

$$
\begin{aligned}
&= x \cdot x + x \cdot 1 + 5 \cdot x + 5 \cdot 1 && \text{Distribute } x \text{ to } x \text{ and 1, then 5 to } x \text{ and 1.} \\
&= x^2 + x + 5x + 5 && \text{Multiply.} \\
&= x^2 + 6x + 5 && \text{Combine like terms. } x + 5x = 6x
\end{aligned}
$$

Explanation: First, we multiplied $x \cdot x$ to get x^2.

Then we multiplied $x \cdot 1$ to get x.

Next, we multiplied $5 \cdot x$ to get $5x$.

Finally, we multiplied $5 \cdot 1$ to get 5.

To finish, we combined like terms x and $5x$ to get $6x$.

b. $(x + 4)(x - 3)$

Solution: Multiply each term in $x - 3$ by each term in $x + 4$.

$$
\begin{array}{c}
\overbrace{}^{x\,\cdot\,(-3)} \\[-2pt]
\overbrace{}^{x\,\cdot\,x} \\[-2pt]
(x + 4) \quad (x - 3) \\[-2pt]
\underbrace{}_{4\,\cdot\,x} \\[-2pt]
\underbrace{}_{4\,\cdot\,(-3)}
\end{array}
$$

$\begin{aligned}
&= x \cdot x + x \cdot (-3) + 4 \cdot x + 4 \cdot (-3) && \text{Distribute } x \text{ to } x \text{ and } -3, \text{ then } 4 \text{ to } x \text{ and } -3. \\
&= x^2 \quad \underbrace{-3x \qquad + 4x}_{} \quad - 12 && \text{Multiply.} \\
&= x^2 \quad + \qquad x \qquad\qquad -12 && \text{Combine like terms. } -3x + 4x = x
\end{aligned}$

Explanation: First, we multiplied $x \cdot x$ to get x^2.

Then we multiplied $x \cdot (-3)$ to get $-3x$.

Next, we multiplied $4 \cdot x$ to get $4x$.

Finally, we multiplied $4 \cdot (-3)$ to get -12.

To finish, we combined the like terms $-3x$ and $4x$ to get x.

▶ **Do Your Turn 5**

Learning Strategy

Depending on your learning style, you may use different techniques to remember the procedure for multiplying two polynomials.

Tactile learners might imagine a party where two groups of people (two polynomials) meet each other. Every member of one group must shake hands with (multiply) each member of the second group.

Visual learners might want to draw arrows or lines connecting the terms that are multiplied, as we did in the two examples.

Auditory learners might want to use the word *FOIL*, which stands for **F**irst **O**uter **I**nner **L**ast. We can use Example 5a to demonstrate.

First terms: $x \cdot x = x^2$

$(\overset{\frown}{x + 5})(x + 1)$

Outer terms: $x \cdot 1 = x$

$(\overset{\frown}{x + 5)(x + 1})$

Inner terms: $5 \cdot x = 5x$

$(x + \underset{\smile}{5)(x} + 1)$

Last terms: $5 \cdot 1 = 5$

$(x + \underset{\smile}{5)(x + 1})$

Warning FOIL makes sense only when multiplying two binomials. When multiplying polynomials with more than two terms, follow the original procedure of multiplying *each term* in the second polynomial by *each term* in the first polynomial.

Now let's multiply special binomials called **conjugates.**

Definition Conjugates: Binomials that differ only in the sign of the second term.

Examples of conjugates: $x + 9$ and $x - 9$ ◀ **Note** In conjugates, the first
                              $-5y - 7$ and $-5y + 7$ terms are the same and the second
                                                          terms are additive inverses.

▶ **Your Turn 6**

State the conjugate of each binomial.
  a. $x + 7$
  b. $2a - 5$
  c. $-9t + 1$
  d. $-6h - 5$

Example 6   State the conjugate of each binomial.

a. $x - 2$ b. $3x + 5$ c. $-4n + 3$

Answer: $x + 2$ **Answer:** $3x - 5$ **Answer:** $-4n - 3$

◀ **Do Your Turn 6**

When we multiply conjugates, a special pattern emerges. Consider $(3x + 5)(3x - 5)$.

$(3x + 5)(3x - 5) = 3x \cdot 3x + 3x \cdot (-5) + 5 \cdot 3x + 5 \cdot (-5)$ Distribute.
(Use FOIL.)

$\qquad\qquad\qquad = 9x^2 - 15x + 15x - 25$ Multiply.

$\qquad\qquad\qquad = 9x^2 + 0 - 25$ Combine like terms.
$\qquad\qquad\qquad\qquad\qquad\qquad\qquad -15x + 15x = 0$

$\qquad\qquad\qquad = 9x^2 \;-\; 25$ Simplify.

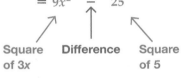

Square   Difference   Square   ◀
of 3x   of 5

Note When we multiply conjugates, the like terms are additive inverses; so the product is a difference of squares.

Our example suggests the following rule:

Rule

The product of conjugates is a difference of two squares.

In math language: $(a + b)(a - b) = a^2 - b^2$

▶ **Your Turn 7**

Multiply.
  a. $(x + 9)(x - 9)$
  b. $(4n - 1)(4n + 1)$
  c. $(-2y + 5)(-2y - 5)$

Example 7   Multiply. $(2x + 7)(2x - 7)$

Solution: We are to multiply conjugates; so we can use the rule $(a + b)(a - b) = a^2 - b^2$.

$\qquad\qquad (a + b)(a - b) = a^2 - b^2$

$(2x + 7)(2x - 7) = (2x)^2 - (7)^2$ Use $(a + b)(a - b) = a^2 - b^2$, where a is 2x and b is 7.

$\qquad\qquad\qquad\quad = 4x^2 - 49$ Square 2x and 7.

◀ **Do Your Turn 7**

Answers to Your Turn 6
a. $x - 7$
b. $2a + 5$
c. $-9t - 1$
d. $-6h + 5$

Answers to Your Turn 7
a. $x^2 - 81$
b. $16n^2 - 1$
c. $4y^2 - 25$

Example 8   Multiply. $(x + 4)(2x^2 + 5x - 3)$

Solution: We multiply each term in $2x^2 + 5x - 3$ by each term in $x + 4$.

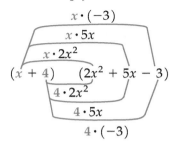

Warning FOIL does *not* make sense here because there are too many terms. Remember, FOIL describes how to multiply only two binomials.

$$= x \cdot 2x^2 + x \cdot 5x + x \cdot (-3) + 4 \cdot 2x^2 + 4 \cdot 5x + 4 \cdot (-3)$$
$$= 2x^3 \quad + 5x^2 \quad - 3x \quad + 8x^2 \quad + 20x \quad - 12$$
$$= 2x^3 + 13x^2 + 17x - 12$$

Explanation: We first multiplied each term in the trinomial by x. We then multiplied each term in the trinomial by 4. Finally, we combined like terms. Notice that we had two pairs of like terms. The $5x^2$ and $8x^2$ combined to equal $13x^2$, and $-3x$ and $20x$ combined to equal $17x$.

▶ **Do Your Turn 8**

Connection One of the ways to quickly check whether you have the correct number of terms in your resulting unsimplified polynomial is to use the knowledge you developed in Chapter 1. Recall that to get the total number of possible combinations in a problem with multiple slots to fill, you can multiply the number of items that can fill the first slot by the number of items that can fill the next slot, and so on.

Let's look at the resulting unsimplified polynomial from Example 8.

Slot 1 contains one of the two terms from $x + 4$. $\quad x \cdot 2x^2 + x \cdot 5x + x \cdot (-3) + 4 \cdot 2x^2 + 4 \cdot 5x + 4 \cdot (-3) \quad$ Slot 2 contains one of the three terms from $2x^2 + 5x - 3$.

Each term has two slots, one for a term from the binomial $(x + 4)$ and the other for a term from the trinomial $(2x^2 + 5x - 3)$. The binomial has two possible terms that can fill the first slot; the trinomial has three. To find the total number of possible combinations, we multiply $2 \cdot 3$ and get 6, the number of terms in the resulting unsimplified polynomial.

If we multiply a trinomial by a trinomial, we should have $3 \cdot 3 = 9$ terms in the resulting polynomial before simplifying.

Example 9   Multiply. $3x^2(2x - 3)(x + 4)$

Solution: As with multiplying three numbers, we can multiply three expressions from left to right.

$$3x^2(2x - 3)(x + 4)$$
$$= [3x^2 \cdot 2x - 3x^2 \cdot 3](x + 4) \qquad \text{Distribute } 3x^2 \text{ to } 2x \text{ and } -3.$$
$$= [6x^3 - 9x^2](x + 4) \qquad \text{Multiply.}$$
$$= 6x^3 \cdot x + 6x^3 \cdot 4 - 9x^2 \cdot x - 9x^2 \cdot 4 \qquad \text{Distribute. (Use FOIL.)}$$
$$= 6x^4 + 24x^3 - 9x^3 - 36x^2 \qquad \text{Multiply.}$$
$$= 6x^4 + 15x^3 - 36x^2 \qquad \text{Combine like terms.}$$

▶ **Your Turn 8**

Multiply.
  a. $(x + 3)(4x^2 - 5x + 1)$
  b. $(2n - 1)(3n^2 - 7n - 5)$

Answers to Your Turn 8
a. $4x^3 + 7x^2 - 14x + 3$
b. $6n^3 - 17n^2 - 3n + 5$

▶ **Your Turn 9**

Multiply.

a. $2x(3x - 2)(x + 1)$
b. $-a^2(a + 3)(2a - 1)$
c. $t^3(t - 1)(t + 1)$
d. $-2p^4(2p - 3)(3p + 1)$

Alternatively, because of the associative property of multiplication, we can multiply the two binomials first, then multiply the result by the monomial.

$$3x^2(2x - 3)(x + 4)$$
$$= 3x^2[2x \cdot x + 2x \cdot 4 - 3 \cdot x - 3 \cdot 4] \quad \text{Distribute. (Use FOIL.)}$$
$$= 3x^2[2x^2 + 8x - 3x - 12] \quad \text{Multiply.}$$
$$= 3x^2[2x^2 + 5x - 12] \quad \text{Combine like terms.}$$
$$= 3x^2 \cdot 2x^2 + 3x^2 \cdot 5x + 3x^2 \cdot (-12) \quad \begin{array}{l}\text{Distribute } 3x^2 \text{ to } 2x^2, \\ 5x, \text{ and } -12.\end{array}$$
$$= 6x^4 + 15x^3 - 36x^2 \quad \text{Multiply.}$$

◀ **Do Your Turn 9**

▶ **Your Turn 10**

Write an expression for the area of the rectangle shown.

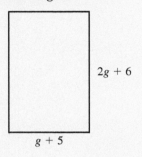

$2g + 6$

$g + 5$

Example 10   Write an expression for the area of the rectangle shown.

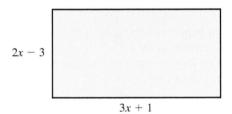

$2x - 3$

$3x + 1$

Solution: To find the area of a rectangle, we multiply length times width. In this case, the length and width are expressions; so we can write only an expression for the area. We cannot get a numerical answer until we are given a value for x.

$$\begin{aligned}\text{Area} &= \text{length} \cdot \text{width} \\ &= (3x + 1)(2x - 3) \\ &= 3x \cdot 2x + 3x \cdot (-3) + 1 \cdot 2x + 1 \cdot (-3) \quad \text{Distribute. (Use FOIL.)} \\ &= 6x^2 - 9x + 2x - 3 \quad \text{Multiply.} \\ &= 6x^2 - 7x - 3 \quad \text{Combine like terms.}\end{aligned}$$

Answer: The expression for the area of the rectangle is $6x^2 - 7x - 3$.

◀ **Do Your Turn 10**

▶ **Your Turn 11**

Write an expression for the volume of the box shown.

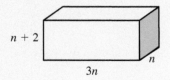

$n + 2$

$3n$

n

Example 11   Write an expression for the volume of the box shown.

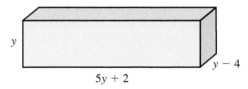

y

$y - 4$

$5y + 2$

Solution: To find the volume of a box, we multiply the length, width, and height. As in Example 10, these dimensions are expressions; so we can write only an expression for the volume.

Answers to Your Turn 9
a. $6x^3 + 2x^2 - 4x$
b. $-2a^4 - 5a^3 + 3a^2$
c. $t^5 - t^3$
d. $-12p^6 + 14p^5 + 6p^4$

Answer to Your Turn 10
$2g^2 + 16g + 30$

Answer to Your Turn 11
$3n^3 + 6n^2$

Volume $=$ length $\cdot$ width $\cdot$ height

$\quad = (5y + 2)(y - 4)(y)$

$\quad = (5y + 2)(y \cdot y - 4 \cdot y)$ **Distribute y to y and -4.**

$\quad = (5y + 2)(y^2 - 4y)$ **Multiply.**

$\quad = 5y \cdot y^2 + 5y \cdot (-4y) + 2 \cdot y^2 + 2 \cdot (-4y)$ **Distribute. (Use FOIL.)**

$\quad = 5y^3 - 20y^2 + 2y^2 - 8y$ **Multiply.**

$\quad = 5y^3 - 18y^2 - 8y$ **Combine like terms.**

Answer: The expression for the volume of the box is $5y^3 - 18y^2 - 8y$.

▶ **Do Your Turn 11**

Note When multiplying three expressions we use the associative property, multiplying two of them first and then multiplying the result by the third.

3.4 Exercises For Extra Help MyMathLab®

Objective 1

Prep Exercise 1 To multiply exponential forms that have the same base, we can _____ the exponents and keep the same base.

Prep Exercise 2 Explain how to multiply monomials.

For Exercises 1–18, multiply.

1. $x^3 \cdot x^4$

2. $y^3 \cdot y^5$

3. $t \cdot t^9$

4. $m^4 \cdot m$

5. $7a^2 \cdot 3a$

6. $9b^3 \cdot 4b$

7. $2n^3 \cdot 5n^4$

8. $4x^5 \cdot 3x^3$

9. $-8u^4 \cdot 3u^2$

10. $9x^2(-4x)$

11. $-y^5(-8y^2)$

12. $-3n^4(-7n^6)$

13. $5x(9x^2)(2x^4)$

14. $2a^2(5a)(3a)$

15. $3m(-3m^2)(4m)$

16. $u^3(7u^4)(-8u^3)$

17. $3y^4(-5y^3)(-4y^2)$

18. $-6k^5(2k)(-4k^2)$

For Exercises 19 and 20, explain the mistake and work the problem correctly.

19. $9x^3 \cdot 5x^4$

$\quad = 45x^{12}$

20. $-3m^2 \cdot 10m^5$

$\quad = -30m^{10}$

Objective 2

Prep Exercise 3 To simplify an exponential form raised to a power, we can _____ the exponents and keep the same base.

Prep Exercise 4 Explain how to simplify a monomial raised to a power.

For Exercises 21–34, simplify.

21. $(2x^3)^2$

22. $(3a^4)^3$

23. $(-7m^5)^2$

24. $(-4n^2)^2$

25. $(-2y^6)^5$

26. $(-3t^7)^3$

27. $(-4x^5)^4$

28. $(-5y^3)^4$

29. $(5y^6)^2$

30. $(6a^4)^3$

31. $(-5v^6)^3$

32. $(-2j^6)^5$

33. $(3x)(2x^3)^4$

34. $7a^3 \cdot (6a)^2$

For Exercises 35 and 36, explain the mistakes and work the problem correctly.

35. $(3x^4)^4$
$\quad = 3x^8$

36. $(-2b^4)^3$
$\quad = -6b^7$

Objective 3

Prep Exercise 5 Explain how to multiply a polynomial by a monomial.

For Exercises 37–56, multiply.

37. $4(2x + 3)$ **38.** $9(3x + 1)$ **39.** $5(3x - 7)$ **40.** $8(2t - 3)$

41. $-3(5t + 2)$ **42.** $-6(4y + 5)$ **43.** $-4(7a - 9)$ **44.** $-2(5m - 6)$

45. $6u(3u + 4)$ **46.** $3x(7x - 1)$ **47.** $-2a(3a + 1)$ **48.** $-5y(2y - 4)$

49. $-2x^3(5x - 8)$ **50.** $7y^2(4y - 3)$ **51.** $8x(2x^2 + 3x - 4)$ **52.** $2n(7n^2 - 5n + 3)$

53. $-x^2(5x^2 - 6x + 9)$ **54.** $-t^3(3t^2 + t - 8)$ **55.** $-2p^2(3p^2 + 4p - 5)$ **56.** $-3a^2(4a^2 - a + 3)$

Objective 4

Prep Exercise 6 Explain how to multiply two polynomials.

For Exercises 57–70, multiply.

57. $(x + 4)(x + 2)$ **58.** $(a + 3)(a + 1)$ **59.** $(m - 3)(m + 5)$

60. $(t + 7)(t - 2)$ **61.** $(y - 8)(y + 1)$ **62.** $(u + 4)(u - 9)$

63. $(3x + 2)(4x - 5)$ **64.** $(5t - 4)(3t + 1)$ **65.** $(4x - 1)(3x - 5)$

66. $(2a - 7)(a - 6)$ **67.** $(3t - 5)(4t + 1)$ **68.** $(h + 2)(4h - 3)$

69. $(a - 7)(2a - 5)$ **70.** $(2x - 5)(3x - 1)$

Prep Exercise 7 What are conjugates? Give an example of a pair of conjugates.

For Exercises 71–78, state the conjugate of each binomial.

71. $x - 7$ **72.** $y + 2$ **73.** $2x + 5$ **74.** $3t - 8$

75. $-b + 2$ **76.** $-u + 4$ **77.** $-6x - 9$ **78.** $-8y - 1$

Prep Exercise 8 The product of two binomials that are conjugates is a(n) _____ of squares.

For Exercises 79–84, multiply.

79. $(x + 3)(x - 3)$ **80.** $(a - 4)(a + 4)$ **81.** $(5t + 6)(5t - 6)$

82. $(2m + 7)(2m - 7)$ **83.** $(-6x - 1)(-6x + 1)$ **84.** $(-h + 9)(-h - 9)$

For Exercises 85 and 86, explain the mistakes and work the problem correctly.

85. $(2x - 5)(3x + 1)$
$\quad = 6x + 2x - 15x + 5$
$\quad = -7x + 5$

86. $(a - 6)(a - 6)$
$\quad = a^2 + 36$

For Exercises 87–96, multiply.

87. $(x + 5)(2x^2 - 3x + 1)$ **88.** $(a + 2)(5a^2 + 3a - 4)$

89. $(2y - 3)(4y^2 + y - 6)$ **90.** $(3m - 7)(m^2 - 5m - 2)$

91. $2a(a - 1)(a + 2)$ **92.** $3c(c + 4)(c - 2)$

93. $-x^2(2x + 3)(x + 1)$ **94.** $-p^3(p + 2)(4p + 1)$

95. $-3q^2(3q - 1)(3q + 1)$ **96.** $-2t^3(5t + 2)(5t - 2)$

For Exercises 97 and 98, write an expression for the area of each rectangle.

97.

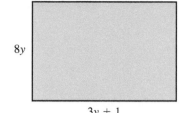

$8y$

$3y + 1$

98.

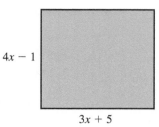

$4x - 1$

$3x + 5$

For Exercises 99 and 100, write an expression for the volume of each box.

99.

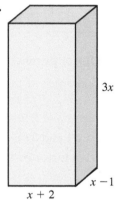

$3x$

$x - 1$

$x + 2$

100.

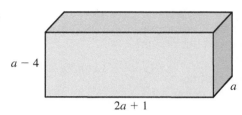

$a - 4$

a

$2a + 1$

For Exercises 101–104, complete the table. To find the product in the first row, multiply the two variable expressions. After finding the variable expressions for the product, evaluate each variable expression using the given values.

101.

	Expression 1	Expression 2	Product
x	$5x$	$-3x$	
2			
-3			

102.

	Expression 1	Expression 2	Product
t	$4t$	$t + 3$	
3			
-2			

103.

	Expression 1	Expression 2	Product
n	$n + 3$	$n - 7$	
1			
-3			

104.

	Expression 1	Expression 2	Product
y	$2y + 3$	$3y - 1$	
10			
-1			

Puzzle Problem If there are about 10^{11} stars in a galaxy and about 10^{11} galaxies in the visible universe, about how many stars are in the visible universe?

Review Exercises

[1.4] **1.** Divide. $846 \div 2$

[2.4] **2.** Find the missing factor. $-6 \cdot (?) = 24$

[1.6] **3.** A rectangular room has an area of 132 square feet. If one wall is 12 feet long, find the length of the other wall.

[1.6] **4.** The packing crate shown has a volume of 210 cubic feet. Find the missing length.

[3.3] **5.** Combine like terms. $3x - 7x$

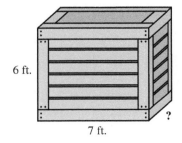

6 ft.

7 ft.

?

3.5 Prime Numbers and GCF

Objectives

1 Determine whether a number is prime, composite, or neither.

2 Find the prime factorization of a given number.

3 Find all factors of a given number.

4 Find the greatest common factor of a given set of numbers by listing.

5 Find the greatest common factor of a given set of numbers using prime factorization.

6 Find the greatest common factor of a set of monomials.

Warm-up

[1.3] **1.** Multiply. $2 \cdot 2 \cdot 3 \cdot 5 \cdot 7$

[1.4] **2.** Divide. $91 \div 7$

Objective 1 Determine whether a number is prime, composite, or neither.

In this section, we learn about special numbers called **prime numbers.**

Definition **Prime number:** A natural number that has exactly two different factors: 1 and the number itself.

Let's look at the first few natural numbers to develop a partial list of primes.

Is 1 prime? No, because a prime number must have exactly two *different* factors and 1's factors are the same—1 and 1.

Is 2 prime? Yes, because 2 is a natural number whose only factors are 1 and 2 (itself).

Is 3 prime? Yes, because 3 is a natural number whose only factors are 1 and 3 (itself).

Is 4 prime? No, because 4 has factors other than 1 and 4 (2 is also a factor of 4).

Continuing in this way gives the following list of prime numbers:

$$2, 3, 5, 7, 11, 13, 17, 19, 23, 29, 31, 37, 41, 43, \ldots$$

Notice that 2 is the only even prime number and that the list of prime numbers continues without end. Numbers greater than 1 that are not prime, such as 4, 6, and 9, are called **composite numbers.**

Definition **Composite number:** A natural number that has factors other than 1 and itself.

Here is a list of composites:

$$4, 6, 8, 9, 10, 12, 14, 15, 16, 18, 20, 21, 22, 24, 25, \ldots$$

Notice that 1 is neither prime nor composite because its only factors are 1 and itself (1), the same factors. Also notice that every composite number is divisible by at least one prime number. This suggests a method for determining whether a given number is prime or composite. We divide the given number by each prime number sequentially until we find a prime that goes evenly into the number, or determine that no prime divides it evenly. It may seem rather tedious, but the divisibility rules for 2, 3, and 5 that we learned in Section 1.4 help. Also, as we shall see, you don't have to go as far along the list of primes as you might think.

To demonstrate, let's consider the number 91, which we will see is composite. The first three prime numbers are 2, 3, and 5. We can determine whether 91 is divisible by 2, 3, or 5 using the divisibility rules from Section 1.4.

91 is not divisible by 2 because 91 is not an even number.

91 is not divisible by 3 because the sum of its digits is $9 + 1 = 10$, and 10 is not divisible by 3.

91 is not divisible by 5 because the digit in the ones place is not 0 or 5.

Answers to Warm-up

1. 420 **2.** 13

Now test 7. Because 7 does not have a simple divisibility rule, we use long division.

$$\begin{array}{r} 13 \\ 7\overline{)91} \\ \underline{-7} \\ 21 \\ \underline{-21} \\ 0 \end{array}$$

Answer: Because 91 is evenly divisible by 7, it is a composite number.

Now consider 127, which we will see is prime. First, we can determine whether 127 is divisible by 2, 3, or 5 using the divisibility rules from Section 1.4.

> 127 is not divisible by 2 because 127 is not an even number.
>
> 127 is not divisible by 3 because the sum of its digits is $1 + 2 + 7 = 10$ and 10 is not divisible by 3.
>
> 127 is not divisible by 5 because the digit in the ones place is not 0 or 5.

Now test 7 using long division.

$$\begin{array}{r} 18 \\ 7\overline{)127} \\ \underline{-7} \\ 57 \\ \underline{-56} \\ 1 \end{array}$$

Because 7 didn't go into 127 evenly, we try the next prime, 11.

$$\begin{array}{r} 11 \\ 11\overline{)127} \\ \underline{-11} \\ 17 \\ \underline{-11} \\ 6 \end{array}$$

It seems like this process of division could go on forever. However, we can stop at dividing by 11 because the quotient is equal to the divisor. Continuing to a larger divisor would only produce a smaller quotient than 11. We already tried the primes less than 11 and found no prime less than 11 that divides 127 evenly. It is therefore pointless to continue.

Answer: Because we found no prime number that divides 127 evenly, it must be a prime number.

From what we've learned in testing 91 and 127, we can write the following procedure for determining whether a given number is prime or composite:

Procedure

To determine whether a given number is prime or composite, divide it by each prime number sequentially in ascending order and consider the results.

- If the given number is divisible by the prime, stop. The given number is a composite number.
- If the given number is not divisible by the prime, consider the quotient.
  - ▶ If the quotient is greater than the current prime divisor, repeat the process with the next prime in the list of prime numbers.
  - ▶ If the quotient is equal to or less than the current prime divisor, stop. The given number is a prime number.

Determine whether the number is prime, composite, or neither.
a. 39 **b.** 119 **c.** 1 **d.** 107

Example 1   Determine whether 157 is prime or composite.

Solution: Divide by the list of primes.

Is 157 divisible by 2? No, because 157 is odd.
Is 157 divisible by 3? No, because $1 + 5 + 7 = 13$ and this sum is not divisible by 3.
Is 157 divisible by 5? No, because 157 does not end in 0 or 5.
Is 157 divisible by 7?

$$
\begin{array}{r}
22 \\
7{\overline{)157}} \\
-14 \\
\hline
17 \\
-14 \\
\hline
3
\end{array}
$$

◀ No, 157 is not divisible by 7. Notice that the quotient, 22, is greater than the divisor, 7. This means that we need to repeat the process with the next prime, which is 11.

Is 157 divisible by 11?

$$
\begin{array}{r}
14 \\
11{\overline{)157}} \\
-11 \\
\hline
47 \\
-44 \\
\hline
3
\end{array}
$$

◀ No, 157 is not divisible by 11. Because the quotient, 14, is greater than the divisor, 11, we must go to the next prime, which is 13.

Is 157 divisible by 13?

$$
\begin{array}{r}
12 \\
13{\overline{)157}} \\
-13 \\
\hline
27 \\
-26 \\
\hline
1
\end{array}
$$

◀ No, 157 is not divisible by 13. Notice that the quotient, 12, is less than the divisor, 13. This means that we can stop testing and conclude that 157 is a prime number.

Answer: 157 is a prime number.

◀ **Do Your Turn 1**

Of Interest

Eratosthenes (about 276–296 B.C.) was a Greek mathematician who lived in the ancient city of Alexandria, in what is now northern Egypt. One of his contributions was a method for finding prime numbers. He referred to the method as a *sieve* because it sifted out the composite numbers leaving only prime numbers. The idea of the method is to write odd natural numbers beginning with 3, and mark through those that are divisible by each number in the list. For instance, start with 3 and mark out all odd numbers divisible by 3. Then write 5 and mark out all odd numbers divisible by 5, and so on.

$$3, 5, 7, \cancel{9}, 11, 13, \cancel{15}, 17, 19, \cancel{21}, 23, \cancel{25}, \cancel{27}, 29, 31, \cancel{33}, \cancel{35}, 37, \cancel{39}, 41, 43, \cancel{45}, \ldots$$

Source: D. E. Smith; *History of Mathematics*, 1953

Objective 2   Find the prime factorization of a given number.

We have learned that every composite number is divisible by at least one prime number. It turns out that we can write any composite number as a product of only prime factors. We call this expression a **prime factorization.**

Definition   **Prime factorization:** A product written with prime factors only.

Answers to Your Turn 1
a. composite   **b.** composite
c. neither   **d.** prime

For example, the prime factorization of 20 is $2 \cdot 2 \cdot 5$, which we can write in exponential form as $2^2 \cdot 5$.

There are several popular methods for finding a number's prime factorization. The factor tree method is one of the most flexible. We will use 20 to demonstrate. First, draw two branches below the given number.

At the end of these branches, we place two factors whose product is the given number. For 20, we could use 2 and 10 or 4 and 5. It is helpful to circle the prime factors as they appear.

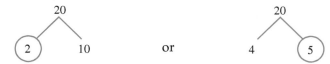

We then keep breaking down all of the composite factors until we have only prime factors.

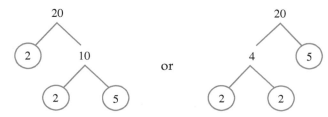

All of the circled prime numbers in the factor tree form the prime factorization; so $20 = 2 \cdot 2 \cdot 5 = 2^2 \cdot 5$. Notice that it didn't matter whether we used 2 and 10 or 4 and 5 to begin the factor tree. Either way, we ended up with two 2s and one 5 circled.

Procedure

To find the prime factorization of a composite number using a factor tree:
1. Draw two branches below the number.
2. Find two factors whose product is the given number and place them at the end of the two branches.
3. Repeat steps 1 and 2 for every composite factor.
4. Write a multiplication expression containing all of the prime factors.

Example 2   Find the prime factorization. Write the answer in exponential form.

a. 84

Solution: Use a factor tree.

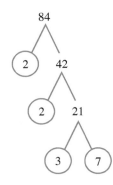

◄ Because 84 is even, we know it is divisible by 2; so let's start with 2 and 42.

◄ Because 42 is even, let's use 2 again. It breaks down to 2 and 21.

◄ Finally, break down 21 to 3 and 7. Because there are no more composite numbers to break down, we're done.

Answer: $2^2 \cdot 3 \cdot 7$

▶ **Your Turn 2**

Find the prime factorization. Write your answer in exponential form.
a. 64 b. 105 c. 560 d. 910

Answers to Your Turn 2
a. 2^6 b. $3 \cdot 5 \cdot 7$
c. $2^4 \cdot 5 \cdot 7$ d. $2 \cdot 5 \cdot 7 \cdot 13$

We could have started the factor tree for 84 many different ways and ended up with the same result. For example, we could have started with 4 and 21, 7 and 12, or 3 and 28.

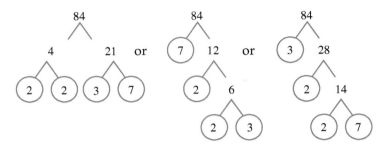

No matter how we start, we end up with the same result each time: The prime factorization of 84 is $2^2 \cdot 3 \cdot 7$. In fact, no other number has that prime factorization because the product of $2^2 \cdot 3 \cdot 7$ is always 84.

Conclusion: No two numbers have the same prime factorization. In math language, we say that prime factorizations are unique.

b. 3500

Solution: Use a factor tree. Because 3500 is divisible by 10, let's start by dividing out the tens.

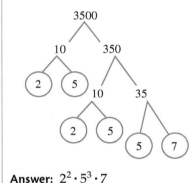

Answer: $2^2 \cdot 5^3 \cdot 7$

Note We can check the prime factorization by multiplying the prime factors together. ◄ The product should be the given number.

◄ **Do Your Turn 2**

Objective 3 Find all factors of a given number.

In mathematics, we often need to consider the factors, or divisors, of a number. Recall that factors of a given number are numbers that divide the given number with no remainder. Let's consider how to find all possible factors of a given number.

► Your Turn 3

List all factors of each number.
  a. 36   **b.** 60   **c.** 84

Answers to Your Turn 3
a. 1, 2, 3, 4, 6, 9, 12, 18, 36
b. 1, 2, 3, 4, 5, 6, 10, 12, 15, 20, 30, 60
c. 1, 2, 3, 4, 6, 7, 12, 14, 21, 28, 42, 84

Example 3   List all factors of 24.

Solution: To list all of the factors, we'll divide by the natural numbers 1, 2, 3, and so on, until we have them all. We can start with 1 and 24, then go to 2 and 12, and so on.

$$1 \cdot 24$$
$$2 \cdot 12$$
$$3 \cdot 8$$
$$4 \cdot 6$$

Explanation: Once we establish 2 and 12, all remaining factors must be between 2 and 12. When we establish 3 and 8, all remaining factors must be between 3 and 8. When we establish 4 and 6, any remaining factors must be between 4 and 6. The only natural number between 4 and 6 is 5. Because 24 is not divisible by 5, we have found all factors.

Answer: 1, 2, 3, 4, 6, 8, 12, 24

▶ **Do Your Turn 3**

Objective 4 Find the greatest common factor of a given set of numbers by listing.

Suppose a fence company has a job to enclose three sides of a yard. The left side is 32 feet long, the center is 40 feet, and the right side is 24 feet. The company is to use prefabricated sections. What is the longest section it can use so that no partial sections are needed?

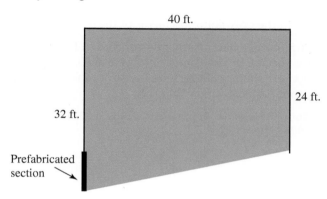

Notice that several sizes would work. The company could use 2-foot sections, 4-foot sections, and 8-foot sections because 32, 40, and 24 are divisible by those numbers. But the longest possible section would be the 8-foot section. The number 8 is the greatest number that divides 32, 40, and 24 with no remainder and is therefore called the **greatest common factor.**

Definition **Greatest common factor:** The greatest number that divides all given numbers with no remainder.

Note The greatest common factor is sometimes called the greatest common divisor.

We abbreviate greatest common factor as GCF. One way to find the GCF for a set of numbers is by listing factors.

Procedure

To find the greatest common factor by listing:
1. List all factors for each given number.
2. Search the lists for the greatest factor common to all lists.

Example 4 Find the GCF of 24 and 60.

Solution: We listed all factors of 24 in Example 3, and you listed all factors of 60 in *Your Turn 3b.*

Factors of 24: 1, 2, 3, 4, 6, 8, 12, 24

Factors of 60: 1, 2, 3, 4, 5, 6, 10, 12, 15, 20, 30, 60

Notice that 12 is the greatest factor common to both lists.

Answer: GCF = 12

Learning Strategy

If you write the factor pairs in column form, as illustrated in Example 3, to avoid excluding a factor in your final list, imagine traveling down the left column, then up the right column, writing the factors in that order.

1 · 24
2 · 12
3 · 8
4 · 6

▶ **Your Turn 4**

Find the GCF by listing.
a. 36 and 54
b. 32 and 45
c. 28, 32, and 40

Answers to Your Turn 4
a. 18   b. 1   c. 4

We can check the greatest common factor by dividing it into the given numbers.

These quotients share no common factors except the number 1.

$$
\begin{array}{r}
2 \\
12\overline{)24} \\
-24 \\
\hline
0
\end{array}
\qquad\qquad
\begin{array}{r}
5 \\
12\overline{)60} \\
-60 \\
\hline
0
\end{array}
$$

Notice that 12 divides 24 and 60 with no remainder; so it is a common factor. We can tell that 12 is the *greatest* common factor because 1 is the only common factor of the quotients.

Conclusion: If you divide each given number by the GCF, each remainder will be 0 and the only common factor of the quotients will be 1.

Suppose we had mistakenly concluded that 6 was the GCF of 24 and 60. When we check 6 by dividing it into 24 and 60, we see that it is a common factor because the remainders are 0, but not the GCF because the quotients have a common factor other than 1.

Both of these quotients are divisible by 2. This means that 6 is not the greatest common factor.

$$
\begin{array}{r}
4 \\
6\overline{)24} \\
-24 \\
\hline
0
\end{array}
\qquad\qquad
\begin{array}{r}
10 \\
6\overline{)60} \\
-60 \\
\hline
0
\end{array}
$$

This technique for checking the GCF also offers a very easy way to fix a mistake. Multiplying the incorrect answer by the greatest common factor of the quotients in the check produces the correct GCF. For example, multiplying our incorrect answer, 6, by the greatest common factor of the quotients, which is 2, produces the GCF: $6 \cdot 2 = 12$.

◀ **Do Your Turn 4**

Objective 5   Find the greatest common factor of a given set of numbers using prime factorization.

For smaller numbers, listing is a good method, but for larger numbers, there is a more efficient method to find the GCF using prime factorization. Consider the prime factorizations for the numbers from Example 4.

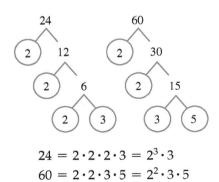

$$24 = 2 \cdot 2 \cdot 2 \cdot 3 = 2^3 \cdot 3$$
$$60 = 2 \cdot 2 \cdot 3 \cdot 5 = 2^2 \cdot 3 \cdot 5$$

Connection You may wonder why we learn several methods for solving the same type of problem. Think of methods as if they were tools. A good analogy is a saw. For small jobs where you need to cut through only a few pieces of wood, a hand saw is a great tool and simple to use. However, for a big job involving a great deal of cutting or large pieces of wood, a power saw, although it takes more effort to set up, is more efficient to use. The listing method for GCF, like the hand saw, is a good method for simpler problems involving small numbers. The prime factorization method is like the power saw. It's a good method for the bigger problems involving large numbers or many numbers.

We already know the answer is 12 from our listing method in Example 4. Let's look at the prime factorization for 12 to see if we can discover a rule for finding the primes to get the GCF.

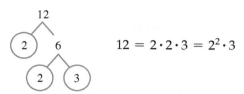

$$12 = 2 \cdot 2 \cdot 3 = 2^2 \cdot 3$$

Notice that 12 contains two factors of 2 and one factor of 3, which are the factors common to the factorizations of 24 and 60. This suggests that we only use primes that are common to all factorizations involved. The factor 5 is not common to both 24's and 60's factorizations; so it is not included in the GCF.

$$24 = 2 \cdot 2 \cdot 2 \cdot 3 = 2^3 \cdot 3$$
$$60 = 2 \cdot 2 \cdot 3 \cdot 5 = 2^2 \cdot 3 \cdot 5$$
$$\text{GCF} = 2 \cdot 2 \cdot 3 \qquad = 2^2 \cdot 3 = 12$$

We could also compare exponents for a particular common prime factor to determine how many of that factor should be included in the GCF. For example, 2^2 is included instead of 2^3 because 2^2 has the lesser exponent. This suggests the following:

Procedure

To find the greatest common factor using prime factorization:
1. Write the prime factorization of each number in exponential form.
2. Create a factorization for the GCF that has only those prime factors common to all factorizations, each raised to the least of its exponents.
3. Multiply.

> **Note** If there are no common prime factors, the GCF is 1.

Example 5 Find the GCF of 3024 and 2520 using prime factorization.

Solution: First, we write the prime factorization of 3024 and 2520 in exponential form.

$$3024 = 2^4 \cdot 3^3 \cdot 7$$
$$2520 = 2^3 \cdot 3^2 \cdot 5 \cdot 7$$

The common prime factors are 2, 3, and 7. Now we compare the exponents of each of these common factors. For 2, the least exponent is 3. For 3, the least exponent is 2. For 7, the least exponent is 1. The GCF is the product of 2^3, 3^2, and 7.

$$\text{GCF} = 2^3 \cdot 3^2 \cdot 7$$
$$= 8 \cdot 9 \cdot 7$$
$$= 504$$

Let's check by dividing 504 into 3024 and 2520. We should find that 504 divides both with no remainder and that 1 is the only factor common to both quotients.

The only factor common to both quotients is 1.

$$
\begin{array}{r}
6 \\
504\overline{)3024} \\
-3024 \\
\hline
0
\end{array}
\qquad\qquad
\begin{array}{r}
5 \\
504\overline{)2520} \\
-2520 \\
\hline
0
\end{array}
$$

← No remainder →

504 checks. It is the GCF.

▶ **Do Your Turn 5**

▶ **Your Turn 5**

Find the GCF using prime factorization.
a. 84 and 48
b. 28 and 140
c. 60 and 77
d. 42, 63, and 105

Answers to Your Turn 5
a. 12 **b.** 28 **c.** 1 **d.** 21

▶ **Your Turn 6**

A rectangular kitchen floor measures 18 feet long by 16 feet wide. What is the largest-size square tile that can be used to cover the floor without cutting or overlapping the tiles?

Example 6 The floor of a 42-foot by 30-foot room is to be covered with colored squares like a checkerboard. All of the squares must be of equal size with whole number dimensions and may not be cut or overlapped to fit inside the room. Find the dimensions of the largest possible square that can be used.

Solution: Because we want the largest possible square, we must find the GCF of 42 and 30.

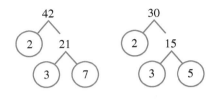

$$42 = 2 \cdot 3 \cdot 7$$
$$30 = 2 \cdot 3 \cdot 5$$
$$\text{GCF} = 2 \cdot 3 = 6$$

Answer: The largest square that can be used is 6 ft. by 6 ft.

◀ **Do Your Turn 6**

Objective 6 Find the greatest common factor of a set of monomials.

Now let's find the greatest common factor of a set of monomials. We use the same prime factorization method that we have learned and treat variables as prime factors.

▶ **Your Turn 7**

Find the GCF of the monomials.
 a. $20x$ and $32x^3$
 b. $36m^4$ and $12m^2$
 c. $16x^3$ and 30
 d. $24h^5$, $16h^4$, and $48h^3$

Example 7 Find the GCF of the monomials.

a. $18x^4$ and $12\,x^3$

Solution: Find the prime factorization of each monomial. Treat the variables as prime factors.

$18x^4 = 2 \cdot 3^2 \cdot x^4$ Write the prime factorization of $18x^4$ in exponential form.
$12x^3 = 2^2 \cdot 3 \cdot x^3$ Write the prime factorization of $12x^3$ in exponential form.

The common prime factors are 2, 3, and x. The least exponent for 2 is 1; so we have a single factor of 2 in the GCF. The least exponent for 3 is also 1. The least exponent for x is 3.

$$\text{GCF} = 2 \cdot 3 \cdot x^3 = 6x^3$$

Connection In Section 3.6, we will learn how to factor the greatest common monomial factor out of the terms in a given polynomial.

b. $24n^2$ and $40n^5$

Solution: $24n^2 = 2^3 \cdot 3 \cdot n^2$ Write the prime factorization of $24n^2$ in exponential form.
$40n^5 = 2^3 \cdot 5 \cdot n^5$ Write the prime factorization of $40n^5$ in exponential form.

The common prime factors are 2 and n. The least exponent for 2 is 3, and the least exponent for n is 2.

$$\text{GCF} = 2^3 \cdot n^2 = 8n^2$$

c. $14t^4$ and 45

Solution: $14t^4 = 2 \cdot 7 \cdot t^4$ Write the prime factorization of $14t^4$ in exponential form.
$45 = 3^2 \cdot 5$ Write the prime factorization of 45 in exponential form.

There are no common prime factors; so the GCF is 1.

Answer to Your Turn 6
2 ft. by 2 ft.

Answers to Your Turn 7
 a. $4x$ **b.** $12m^2$ **c.** 2 **d.** $8h^3$

d. $18y^4$, $27y^3$, and $54y^2$

Solution: $18y^4 = 2 \cdot 3^2 \cdot y^4$ Write the prime factorization of $18y^4$ in exponential form.

$27y^3 = 3^3 \cdot y^3$ Write the prime factorization of $27y^3$ in exponential form.

$54y^2 = 2 \cdot 3^3 \cdot y^2$ Write the prime factorization of $54y^2$ in exponential form.

The common prime factors are 3 and y. The least exponent for 3 is 2, and the least exponent for y is 2. Although 2 appears in two of the factorizations, it is not in all three; so it is not included in the GCF.

$$\text{GCF} = 3^2 \cdot y^2 = 9y^2$$

▶ **Do Your Turn 7**

3.5 Exercises   For Extra Help   MyMathLab®

Objective 1

Prep Exercise 1 What is a prime number? Give three examples.

Prep Exercise 2 What is a composite number? Give three examples.

For Exercises 1–18, determine whether the number is prime, composite, or neither.

1. 49	**2.** 93	**3.** 89	**4.** 71
5. 1	**6.** 0	**7.** 151	**8.** 179
9. 153	**10.** 187	**11.** 247	**12.** 253
13. 311	**14.** 389	**15.** 409	**16.** 421

17. Are all prime numbers odd? Explain.

18. Are all odd numbers prime? Explain.

Objective 2

Prep Exercise 3 Give three pairs of factors that can be used to start the following factor tree.

For Exercises 19–34, find the prime factorization. Write the answer in exponential form.

19. 80	**20.** 72	**21.** 156	**22.** 120
23. 268	**24.** 180	**25.** 200	**26.** 324
27. 343	**28.** 220	**29.** 975	**30.** 462
31. 378	**32.** 875	**33.** 952	**34.** 1404

Objective 3

For Exercises 35–40, list all factors of each number.

35. 90	**36.** 40
37. 81	**38.** 80
39. 120	**40.** 54

Objective 4

Prep Exercise 4 What is the greatest common factor of a set of numbers?

For Exercises 41–46, find the GCF by listing.

41. 24 and 60

42. 40 and 32

43. 81 and 65

44. 48 and 77

45. 72 and 120

46. 80 and 100

Objective 5

Prep Exercise 5 The prime factorization of the GCF of a set of numbers contains only those prime factors common to all factorizations, each raised to the _____ of its exponents.

For Exercises 47–58, find the GCF using prime factorization.

47. 140 and 196

48. 240 and 150

49. 130 and 78

50. 324 and 270

51. 336 and 504

52. 270 and 675

53. 99 and 140

54. 252 and 143

55. 60, 120, and 140

56. 120, 300, and 420

57. 64, 160, and 224

58. 315, 441, and 945

For Exercises 59–62, solve.

59. A painter sections a 60-inch by 70-inch canvas into a grid of squares to be painted individually. What is the largest square possible so that all are of equal size and no overlapping or cutting of any region occurs?

60. The ceiling of a 45-foot-long by 36-foot-wide lobby of a new hotel is to be sectioned into a grid of square regions by wooden beams. What is the largest square region possible so that all regions are of equal size and no overlapping or cutting of any region takes place?

61. A landscaper is planning a sprinkler system that will require precut PVC pipe to be connected and placed in the ground. The trenches he will dig are to be 40 feet, 32 feet, and 24 feet in length. What is the longest length of PVC pipe he can use that will fit the length of each trench exactly and not require any further cutting? How many of the precut pipes will he use in each trench?

62. A professor has three classes. The first class has 30 people, the second class has 42 people, and the third class has 24 people. She wants to break the classes into project groups so that all groups are the same size no matter which class they are in. What is the largest-size group she can make? How many groups will be in each class?

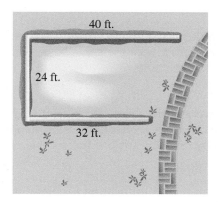

Objective 6

For Exercises 63–76, find the GCF of the monomials.

63. $20x$ and 30

64. 18 and $9t$

65. $8x^2$ and $14x^5$

66. $12a^5$ and $15a$

67. $24h^3$ and $35h^4$

68. $55m^5$ and $36m^4$

69. $27n^4$ and 49

70. 15 and $42r^3$

71. $56x^9$ and $72x^7$

72. $48t^5$ and $60t^8$

73. $18n^2$, $24n^3$, and $30n^4$

74. $20y^3$, $30y^4$, and $40y^6$

75. $42x^4$, $35x^6$, and $28x^3$

76. $36m^5$, $48m^4$, and $60m^6$

Review Exercises

[1.3] 1. Write 5,784,209 in expanded form using powers of 10.

[1.4] 2. Divide. $34{,}328 \div 17$

[1.3] 3. Evaluate. 2^7

[1.3] 4. A college has 265 employees. The college wants to assign an ID to each employee. The proposed ID would contain one of the digits 0–9 and one of the letters A–Z. Is the proposed ID system sufficient?

[2.4] 5. A company must produce square plastic panels that have an area of 36 square feet. What must be the dimensions of the square panels?

3.6 Exponent Rules; Introduction to Factoring

Objectives

1 Divide monomials.

2 Divide a polynomial by a monomial.

3 Find an unknown factor.

4 Factor the GCF out of a polynomial.

Connection We could simplify $2^7 \div 2^4$ using the order of operations.

$$2^7 \div 2^4 = 128 \div 16 = 8$$

The prime factorization of 8 is 2^3.

$$2^7 \div 2^4 = 128 \div 16 = 8$$
$$= 2^3$$

Note Remember that division can be written in fraction form.

$$2^7 \div 2^4 = \frac{2^7}{2^4}$$

► **Your Turn 1**

Divide.

a. $3^9 \div 3^2$

b. $\dfrac{a^{12}}{a^3}$

c. $y^6 \div y^5$

Answers to Your Turn 1
a. 3^7 b. a^9 c. y

Answers to Warm-up
1. $12x$ **2.** 5

Warm-up

[3.5] 1. Find the GCF of $48x^3$ and $60x$.

[1.4, 2.4] 2. Solve for the unknown factor. $40 = 8 \cdot (?)$

Objective 1 Divide monomials.

In Section 3.4, we learned how to multiply exponential forms. We concluded that when we multiply exponential forms that have the same base, we can add the exponents and keep the same base. For example,

$$2^3 \cdot 2^4 = 2^{3+4} = 2^7$$

Now let's consider how to divide exponential forms. Because multiplication and division are inverse operations, we can write a related division statement for the preceding multiplication statement by dividing the product by one of the factors.

$$2^3 \cdot 2^4 = 2^7 \qquad \text{implies} \qquad 2^7 \div 2^4 = 2^3$$

Because the exponents were added in the multiplication problem, it follows that they should be subtracted in the division problem.

$$2^7 \div 2^4 = 2^{7-4} = 2^3$$

Note that we subtracted the divisor's exponent from the dividend's exponent. This order is important because subtraction is not commutative.

> **Rule**
>
> To divide exponential forms that have the same base, we can subtract the divisor's exponents from the dividend's exponent and keep the same base.
>
> **In math language:** $a^m \div a^n = \dfrac{a^m}{a^n} = a^{m-n}$ when $a \neq 0$
>
> **Note** $a \neq 0$ because if a were replaced with 0, we would have $\dfrac{0^m}{0^n}$. In Section 1.4, we learned that when both the dividend and divisor are 0, the quotient is indeterminate.

Example 1 Divide. $x^6 \div x^2$

Solution: $x^6 \div x^2 = x^{6-2} = x^4$ Subtract the divisor's exponent from the dividend's exponent and keep the same base.

Check: Multiplying the answer, x^4, by the divisor, x^2, should equal the dividend, x^6.

$$x^4 \cdot x^2 = x^{4+2} = x^6$$

◄ **Do Your Turn 1**

Now let's look at exponential forms that have the same base and the same exponent, as in $\dfrac{2^3}{2^3}$. Using the rule that we developed, we subtract the divisor's exponent from the dividend's exponent.

$$\frac{2^3}{2^3} = 2^{3-3} = 2^0$$

What does it mean to have 0 as an exponent? To answer that question, we could evaluate $\frac{2^3}{2^3}$ using the order of operations. The result must be equal to 2^0.

$$\frac{2^3}{2^3} = \frac{8}{8} = 1$$

Because both 2^0 and 1 are correct results of the same problem, they must be equal.

$$2^0 = 1$$

Conclusions: In the division of exponential forms that have the same base and the same exponent, the result is 1. The result of raising any nonzero number to the exponent 0 is 1.

Note that the situation 0^0 would arise from a division situation where two exponential forms have 0 as the base and both have the same exponent.

$$\frac{0^n}{0^n} = 0^{n-n} = 0^0$$

In Section 1.4, we learned that $\frac{0}{0}$ is indeterminate; therefore, 0^0 is indeterminate.

Conclusion: 0^0 is indeterminate.

> ### Rules
>
> The result of raising any nonzero number to the exponent 0 is 1.
>
> **In math language:** $a^0 = 1$ when $a \neq 0$
>
> 0^0 is indeterminate.

Example 2 Simplify.

a. $\dfrac{n^3}{n^3}$ (Assume that $n \neq 0$.)

Solution: $\dfrac{n^3}{n^3} = n^{3-3} = n^0 = 1$ Subtract the divisor's exponent from the dividend's exponent. Because $n \neq 0$, the result is 1.

b. 9^0

Solution: $9^0 = 1$ The result of raising a nonzero number to the exponent 0 is 1.

▶ **Do Your Turn 2**

In Example 2a, we had to assume that $n \neq 0$ to avoid the potential for an indeterminate quotient. In all future division problems involving variables, assume that the divisor's variables do not equal 0.

Now let's divide monomials that have coefficients, such as $\dfrac{20m^2}{4m}$. When we multiplied monomials, we multiplied the coefficients and added the exponents of the like bases. Because multiplication and division are inverse operations, it follows that to divide monomials, we divide the coefficients and subtract the exponents of the like bases.

$$\frac{20m^2}{4m} = \frac{20}{4}m^{2-1} = 5m$$

Connection In Section 1.3, we looked at a pattern in the powers of 10 to see that $10^0 = 1$. In $10^3 = 1000$, $10^2 = 100$, and $10^1 = 10$, we see that the exponent corresponds to the number of 0 digits after the 1 in the product. Therefore, 10^0 has no 0 digits after the 1 in its product; so $10^0 = 1$.

▶ **Your Turn 2**

Simplify.

a. $\dfrac{x^6}{x^6}$

b. 5^0

c. $(-71)^0$

Answers to Your Turn 2
a. 1   **b.** 1   **c.** 1

Note that we can check by multiplying the quotient by the divisor. The result should be the dividend.

$$5m \cdot 4m = 5 \cdot 4m^{1+1} = 20m^2$$

Procedure

To divide monomials that have the same variable:

1. Divide the coefficients.
2. Subtract the exponent of the variable in the divisor from the exponent of the variable in the dividend.

   Note If the variables have the same exponent, they divide out.

▶ Your Turn 3

Divide.

a. $10u^7 \div 2u^3$

b. $\dfrac{28x^3}{-4x}$

Connection We can check Example 3a by multiplying $-2a^6$ by $-4a$ to see if we get $8a^7$.

$$-2a^6 \cdot (-4a)$$
$$= (-2)(-4)a^{6+1}$$
$$= 8a^7$$

Try checking Example 3b.

Example 3   Divide.

a. $8a^7 \div (-4a)$

Solution: $8a^7 \div (-4a) = [8 \div (-4)]a^{7-1}$ **Divide the coefficients and subtract the exponents of the like bases.**

$$= -2a^6$$ **Simplify.**

b. $\dfrac{16m^2}{8m^2}$

Solution: $\dfrac{16m^2}{8m^2} = \dfrac{16}{8}m^{2-2}$ **Divide the coefficients and subtract the exponents of the like bases.**

$$= 2m^0$$ **Simplify.**

$$= 2(1)$$ **Replace m^0 with 1.**   ◀ **Note** Because the quotient of m^2 and m^2 is 1, no m appears in the result. Consequently, we say that m^2 "divides out."

$$= 2$$ **Multiply.**

◀ Do Your Turn 3

Objective 2 Divide a polynomial by a monomial.

Recall that every multiplication operation can be checked using division. Consider the following multiplication:

$$(3x + 4) \cdot 2 = 3x \cdot 2 + 4 \cdot 2$$ ◀ **Note** We used the distributive property to multiply each term in $3x + 4$ by 2.
$$= 6x + 8$$

To check our result, we would divide $6x + 8$ by 2 and the result *must* be $3x + 4$.

$$(6x + 8) \div 2 = 6x \div 2 + 8 \div 2$$ ◀ **Note** To get $3x + 4$, we must divide each term in $6x + 8$ by 2.
$$= 3x + 4$$

We can also write the division using a fraction bar.

$$\frac{6x + 8}{2} = \frac{6x}{2} + \frac{8}{2}$$
$$= 3x + 4$$

Procedure

To divide a polynomial by a monomial, divide each term in the polynomial by the monomial.

Answers to Your Turn 3
a. $5u^4$ b. $-7x^2$

Example 4   Divide.

a. $(14x^2 + 16x) \div (2x)$

Solution: $(14x^2 + 16x) \div (2x)$

$= 14x^2 \div (2x) + 16x \div (2x)$ Divide each term in the polynomial, $14x^2 + 16x$, by the monomial, $2x$.

$= (14 \div 2)x^{2-1} + (16 \div 2)x^{1-1}$ Divide the coefficients and subtract the exponents of the like bases.

$= 7x + 8$

b. $\dfrac{15x^3 - 10x^2}{5x^2}$

Solution: $\dfrac{15x^3 - 10x^2}{5x^2} = \dfrac{15x^3}{5x^2} - \dfrac{10x^2}{5x^2}$ Divide each term in $15x^3 - 10x^2$ by $5x^2$.

$= \dfrac{15}{5}x^{3-2} - \dfrac{10}{5}x^{2-2}$ Divide the coefficients and subtract the exponents of the like bases.

$= 3x - 2$

c. $(18x^4 - 12x^3 + 6x^2) \div (-6x)$

Solution: $(18x^4 - 12x^3 + 6x^2) \div (-6x)$

$= 18x^4 \div (-6x) - 12x^3 \div (-6x) +$ Divide each term in
$6x^2 \div (-6x)$ $18x^4 - 12x^3 + 6x^2$ by $-6x$.

$= [18 \div (-6)]x^{4-1} + [-12 \div (-6)]x^{3-1} +$
$[6 \div (-6)]x^{2-1}$

$= -3x^3 + 2x^2 - x$

▲

Note The coefficient of $-12x^3$ is -12; so when we divide that term by $-6x$, we get positive $2x^2$, which is written as $+2x^2$ in the resulting polynomial.

d. $\dfrac{24x^9 - 20x^7 + 4x^6}{4x^6}$

Solution:

$\dfrac{24x^9 - 20x^7 + 4x^6}{4x^6} = \dfrac{24x^9}{4x^6} - \dfrac{20x^7}{4x^6} + \dfrac{4x^6}{4x^6}$ Divide each term in $24x^9 - 20x^7 + 4x^6$ by $4x^6$.

$= \dfrac{24}{4}x^{9-6} - \dfrac{20}{4}x^{7-6} + \dfrac{4}{4}x^{6-6}$

$= 6x^3 - 5x + 1$

▶ **Do Your Turn 4**

Warning Remember that although multiplication is commutative, division is not.

$2 \cdot (3x + 4)$ is equivalent to $(3x + 4) \cdot 2$.

$(6x + 8) \div 2$ is not equivalent to $2 \div (6x + 8)$.

You will learn how to divide polynomials by polynomials in other courses.

Objective 3 Find an unknown factor.

Now let's use what we've learned about dividing monomials to find unknown factors. Remember that we find an unknown factor by dividing the product by the known factor.

▶ **Your Turn 4**

Divide.

a. $(9x + 21) \div 3$

b. $\dfrac{52y^4 - 39y^2}{13y^2}$

c. $(18x^4 - 12x^3 + 6x^2) \div (-3x)$

d. $\dfrac{-28b^{13} + 49b^8 - 63b^5}{-7b^5}$

Connection We can check Example 4a by multiplying the quotient, $7x + 8$, by the divisor, $2x$. The result should be the dividend, $14x^2 + 16x$.

$(7x + 8) \cdot 2x$
$= 7x \cdot 2x + 8 \cdot 2x$
$= 14x^2 + 16x$

Try checking parts b–d.

Note Dividing by a monomial with a negative coefficient, as in Example 4c, causes the signs in the resulting polynomial to be opposite the signs in the dividend polynomial.

Answers to Your Turn 4
a. $3x + 7$
b. $4y^2 - 3$
c. $-6x^3 + 4x^2 - 2x$
d. $4b^8 - 7b^3 + 9$

▶ **Your Turn 5**

Find the unknown factor.

a. $7m^2 \cdot (?) = 21m^6$

b. $-8y \cdot (?) = 40y^2$

Example 5   Find the unknown factor. $6x^3 \cdot (?) = 24x^5$

Solution: We solve for a missing factor by writing a related division sentence.

$$(?) = \frac{24x^5}{6x^3}$$ Divide the product, $24x^5$, by the known factor, $6x^3$.

$$(?) = \frac{24}{6}x^{5-3}$$ Divide the coefficients and subtract the exponents of the like bases.

$$(?) = 4x^2$$ Simplify.

Check: Verify that multiplying $6x^3$ by $4x^2$ gives $24x^5$.

$$6x^3 \cdot 4x^2 = 6 \cdot 4x^{3+2} = 24x^5$$ Multiply the coefficients and add the exponents. It checks.

◀ **Do Your Turn 5**

Now let's find unknown factors that are multiple-term polynomials.

▶ **Your Turn 6**

Find the unknown factor.

a. $8x - 10 = 2 \cdot (?)$

b. $12h^2 + 18h = (?) \cdot 6h$

c. $9y^5 - 7y^4 + 2y^3 = (?)y^3$

d. $24n^5 - 30n^3 - 12n^2$
$= 6n^2(?)$

Example 6   Find the unknown factor.

a. $3 \cdot (?) = 6x + 15$

Solution: To find an unknown factor, we divide the product by the known factor.

$$(?) = \frac{6x + 15}{3}$$ Divide the product, $6x + 15$, by the known factor, 3.

$$(?) = \frac{6x}{3} + \frac{15}{3}$$ Divide each term in the polynomial by the monomial.

$$(?) = 2x + 5$$ Simplify.

Check: Verify that multiplying $2x + 5$ by 3 gives $6x + 15$.

$$3 \cdot (2x + 5) = 3 \cdot 2x + 3 \cdot 5$$ Use the distributive property.
$$= 6x + 15$$ Simplify. It checks.

b. $16x^4 - 12x^3 + 8x^2 = 4x^2 \cdot (?)$

Solution: $\dfrac{16x^4 - 12x^3 + 8x^2}{4x^2} = (?)$ Divide the product by the known factor.

$$\frac{16x^4}{4x^2} - \frac{12x^3}{4x^2} + \frac{8x^2}{4x^2} = (?)$$ Divide each term in the polynomial by the monomial.

$$4x^2 - 3x + 2 = (?)$$ Simplify.

Check: Verify that multiplying $4x^2 - 3x + 2$ by $4x^2$ gives $16x^4 - 12x^3 + 8x^2$.

$$4x^2 \cdot (4x^2 - 3x + 2) = 4x^2 \cdot 4x^2 + 4x^2 \cdot (-3x) + 4x^2 \cdot 2$$ Use the distributive property.

$$= 16x^4 - 12x^3 + 8x^2$$ Simplify. It checks.

◀ **Do Your Turn 6**

Answers to Your Turn 5

a. $3m^4$ b. $-5y$

Answers to Your Turn 6

a. $4x - 5$

b. $2h + 3$

c. $9y^2 - 7y + 2$

d. $4n^3 - 5n - 2$

Objective 4 Factor the GCF out of a polynomial.

All of the work we've done to this point has been to develop the concept of factoring. When we rewrite a number or an expression as a product of factors, we say that we are writing the number or expression in **factored form.**

Definition   Factored form: A number or an expression written as a product of factors.

For example, $3(2x + 5)$ is factored form for the polynomial $6x + 15$. We can check that it is its factored form by multiplying.

$$3(2x + 5) = 3 \cdot 2x + 3 \cdot 5 \quad \text{Use the distributive property.}$$
$$= 6x + 15$$

When we factor, we begin with the product, $6x + 15$, and try to find the factored form, $3(2x + 5)$. First, notice that the monomial factor, 3, is the GCF of the terms in $6x + 15$. Our first step in factoring will be to find the GCF of the terms in the polynomial. That GCF will be one of two factors. Suppose we didn't already know that the other factor in the example is $2x + 5$. We could write the following statement with an unknown factor:

$$6x + 15 = 3 \cdot (?)$$

Notice that the preceding statement is the same as in Example 6a. Look back at how we found the unknown factor. We said $(?) = \dfrac{6x + 15}{3}$; so in our factored form, we can replace the unknown factor with $\dfrac{6x + 15}{3}$. Finding that quotient reveals the unknown factor to be $2x + 5$.

$$6x + 15 = 3 \cdot \left(\frac{6x + 15}{3} \right)$$

Replace the unknown factor with $\dfrac{6x + 15}{3}$.

◄ **Note** The unknown factor is the quotient of the given polynomial and the GCF.

$$= 3 \cdot \left(\frac{6x}{3} + \frac{15}{3} \right) \quad \text{Divide each term by 3.}$$

$$= 3(2x + 5) \quad \text{Divide to complete the factored form.}$$

Our example suggests the following procedure:

Procedure

To factor a monomial GCF out of a polynomial:
1. Find the GCF of the terms in the polynomial.
2. Rewrite the polynomial as a product of the GCF and the quotient of the polynomial and the GCF.

$$\text{Polynomial} = \text{GCF}\left(\frac{\text{polynomial}}{\text{GCF}} \right)$$

Connection Keep in mind that when we factor, we are simply writing the original expression in a different form, called *factored form*. The factored form expression is equal to the original expression. If we were given a value for the variable(s), it wouldn't matter which version of the expression we used to evaluate that value. Both the factored form and the original polynomial form will give the same result.

For example, let's evaluate $6x + 15$ and $3(2x + 5)$ when $x = 4$ (any number would do).

$$
\begin{array}{ll}
\qquad 6x + 15 & \qquad 3(2x + 5) \\
\qquad 6 \cdot 4 + 15 & \qquad 3(2 \cdot 4 + 5) \\
= 24 + 15 & \qquad = 3(8 + 5) \\
& \qquad = 3(13) \\
= 39 \longleftarrow \text{Same result} \longrightarrow & = 39
\end{array}
$$

▶ **Your Turn 7**

Factor.
a. $12y - 9$
b. $10x^3 + 15x$
c. $20a^4 + 30a^3 - 40a^2$

Connection In Section 3.5, we learned that the GCF contains common prime factors each raised to its least exponent. In the case of $18n^2$ and $45n$, the common primes are 3 and n. The least exponent for 3 is 2 and for n is 1; so the GCF $= 3^2 \cdot n = 9n$.

Example 7   Factor.

a. $12y - 8$

Solution: First, we find the GCF of $12y$ and -8, which is 4.

$$12y - 8 = 4\left(\frac{12y - 8}{4}\right)$$ Rewrite the polynomial as a product of the GCF and the quotient of the polynomial and the GCF.

$$= 4\left(\frac{12y}{4} - \frac{8}{4}\right)$$ Divide each term in the polynomial by the GCF.

$$= 4(3y - 2)$$ Simplify in the parentheses to complete the factored form.

Check: Verify that the factored form multiplies to equal the original polynomial.

$$4(3y - 2) = 4 \cdot 3y - 4 \cdot 2$$ Use the distributive property.
$$= 12y - 8$$ It checks.

b. $18n^2 + 45n$

Solution: The GCF of $18n^2$ and $45n$ is $9n$.

$$18n^2 + 45n = 9n\left(\frac{18n^2 + 45n}{9n}\right)$$ Rewrite the polynomial as a product of the GCF and the quotient of the polynomial and the GCF.

$$= 9n\left(\frac{18n^2}{9n} + \frac{45n}{9n}\right)$$ Divide each term in the polynomial by the GCF.

$$= 9n(2n + 5)$$ Simplify in the parentheses to complete the factored form.

We can check by multiplying the factored form to verify that the product is the original polynomial. The check will be left to the reader.

c. $24x^6 + 18x^4 - 30x^3$

Solution: The GCF of $24x^6$, $18x^4$, and $-30x^3$ is $6x^3$.

$$24x^6 + 18x^4 - 30x^3 = 6x^3\left(\frac{24x^6 + 18x^4 - 30x^3}{6x^3}\right)$$ Rewrite the polynomial as a product of the GCF and the quotient of the polynomial and the GCF.

$$= 6x^3\left(\frac{24x^6}{6x^3} + \frac{18x^4}{6x^3} - \frac{30x^3}{6x^3}\right)$$ Divide each term in the polynomial by the GCF.

$$= 6x^3(4x^3 + 3x - 5)$$ Simplify in the parentheses to complete the factored form.

The check will be left to the reader.

◀ **Do Your Turn 7**

Answers to Your Turn 7
a. $3(4y - 3)$
b. $5x(2x^2 + 3)$
c. $10a^2(2a^2 + 3a - 4)$

3.6 Exercises For Extra Help MyMathLab®

Objective 1

Prep Exercise 1 To divide exponential forms that have the same base, we can _____ the exponents and keep the same base.

Prep Exercise 2 Explain how to divide monomials that have the same variable.

For Exercises 1–16, divide.

1. $x^9 \div x^2$

2. $a^{11} \div a^5$

3. $\dfrac{m^7}{m}$

4. $\dfrac{y^5}{y}$

5. $u^4 \div u^4$

6. $k^6 \div k^6$

7. $12a^7 \div (3a^4)$

8. $18w^5 \div (9w^2)$

9. $-40n^9 \div (-2n)$

10. $-33h^{12} \div (-11h)$

11. $12x^5 \div (-4x)$

12. $38m^7 \div (-2m^4)$

13. $\dfrac{-36b^7}{9b^5}$

14. $\dfrac{54x^5}{-27x^4}$

15. $14b^4 \div (-b^4)$

16. $-25h^7 \div (h^7)$

Objective 2

Prep Exercise 3 Explain how to divide a polynomial by a monomial.

For Exercises 17–32, divide.

17. $(9a + 6) \div 3$

18. $(4x + 10) \div 2$

19. $\dfrac{14c - 8}{-2}$

20. $\dfrac{9t - 18}{-9}$

21. $(12x^2 + 8x) \div (4x)$

22. $(20y^2 + 10y) \div (5y)$

23. $\dfrac{-63d^2 + 49d}{7d}$

24. $\dfrac{-15R^2 + 25R}{5R}$

25. $(16a^5 + 24a^3) \div (4a^2)$

26. $(21b^6 - 36b^4) \div (3b^3)$

27. $\dfrac{30a^4 - 24a^3}{6a^3}$

28. $\dfrac{55p^7 - 66p^5}{11p^5}$

29. $\dfrac{16x^4 - 8x^3 + 4x^2}{4x}$

30. $\dfrac{20y^5 - 15y^3 - 25y}{5y}$

31. $\dfrac{32c^9 + 16c^5 - 40c^3}{-4c^2}$

32. $\dfrac{24n^8 - 12n^5 + 30n^3}{-6n^2}$

Objective 3

Prep Exercise 4 Explain how to find an unknown factor.

For Exercises 33–48, find the unknown factor.

33. $5a \cdot (?) = 10a^3$

34. $7m^2 \cdot (?) = 21m^5$

35. $(?) \cdot (-6x^3) = 42x^7$

36. $(?) \cdot 10y^4 = -40y^9$

37. $-36u^5 = (?) \cdot (-9u^4)$

38. $-48x^4 = -6x \cdot (?)$

39. $8x + 12 = 4 \cdot (?)$

40. $12y - 6 = 6 \cdot (?)$

41. $5a^2 - a = a (?)$

42. $10m^4 + m^3 = (?) \, m^3$

43. $18x^3 - 30x^4 = 6x^3 (?)$

44. $44n^3 + 33n^6 = 11n^3 (?)$

45. $6t^5 - 9t^4 + 12t^2 = 3t^2 (?)$

46. $25k^7 - 30k^5 - 20k^2 = 5k^2 (?)$

47. $45n^4 - 36n^3 - 18n^2 = 9n^2(?)$

48. $28y^5 - 56y^3 + 35y^2 = 7y^2(?)$

For Exercises 49–52, find the unknown side length.

49. $A = 63m^2$

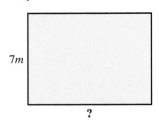

7m

?

50. $A = 24x^5$

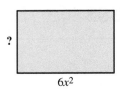

?

$6x^2$

51. $V = 60t^5$

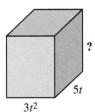

?

5t

$3t^2$

52. $V = 120a^7$

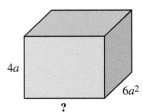

4a

$6a^2$

?

Objective 4

Prep Exercise 5   Explain how to factor a monomial out of the terms of a polynomial.

For Exercises 53–70, factor.

53. $8x - 4$

54. $14y + 7$

55. $10a + 20$

56. $8m - 20$

57. $2n^2 + 6n$

58. $5x^3 - 10x^2$

59. $7x^3 - 3x^2$

60. $12t^2 + 5t$

61. $20r^5 - 24r^3$

62. $32b^8 + 24b^4$

63. $12x^3 + 20x^2 + 32x$

64. $28y^4 - 42y^3 + 35y^2$

65. $9a^7 - 12a^5 + 18a^3$

66. $20x^5 - 15x^3 - 25x$

67. $14m^8 + 28m^6 + 7m^5$

68. $15a^2b^5 + 25a^2b^3 - 30a^2b^2$

69. $10x^9 - 20x^5 - 40x^3$

70. $45u^9 - 18u^6 + 27u^3$

Review Exercises

[1.2] **1.** Find the perimeter of the shape.

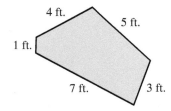

4 ft.

5 ft.

1 ft.

7 ft.   3 ft.

[1.6] **2.** Find the area of the parallelogram.

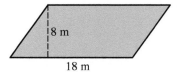

8 m

18 m

[1.6] **3.** Find the volume of a box with a length of 9 inches, a width of 3 inches, and a height of 4 inches.

[3.4] **5.** Multiply. $(x + 4)(x - 6)$

[3.3] **4.** Subtract. $(6x^3 - 9x^2 + x - 12) - (4x^3 + x + 5)$

3.7 Additional Applications and Problem Solving

Warm-up

[3.6] 1. Factor. $2l + 2w$

[3.1] 2. Evaluate $-16x^2 + y$
    when $x = 3$ and $y = 200$.

**Objective 1 Solve polynomial problems involving perimeter,
                area, and volume.**

We can use mathematics to describe situations. If we can write an expression
that describes the situation, we can experiment with numbers without having to do real experiments. To illustrate this process, let's write expressions
for perimeter, area, and volume, then evaluate those expressions using given
numbers.

Objectives

1 Solve polynomial problems
   involving perimeter, area,
   and volume.

2 Solve surface area
   problems.

3 Solve problems involving
   a falling object.

4 Solve net-profit problems.

Example 1 Write an expression in simplest form for the
perimeter of the triangle. Find the perimeter if x is 7 feet.

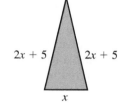

Understand We are to write an expression for the perimeter of a triangle given expressions for its side lengths. We
must then find the perimeter if x is 7 feet.

Plan Because perimeter is the sum of the side lengths, we
will add the expressions for the side lengths. Then we will
evaluate the perimeter expression with $x = 7$.

Execute Perimeter $= x + 2x + 5 + 2x + 5$

$\qquad\qquad\qquad = 5x + 10$ Combine like terms.

Now evaluate $5x + 10$ when x is 7.

$$5(7) + 10 \qquad \text{Replace } x \text{ with 7.}$$
$$= 35 + 10 \qquad \text{Multiply.}$$
$$= 45 \qquad \text{Add.}$$

Answer The perimeter expression is $5x + 10$; when x is 7 ft., the perimeter is
45 ft.

Connection Because $x + 2x + 5 + 2x + 5$ and $5x + 10$ are equivalent,
they produce the same result when $x = 7$, but the simplified expression is
easier to evaluate.

Evaluating the unsimplified expression $x + 2x + 5 + 2x + 5$	Evaluating the simplified expression $5x + 10$
$7 + 2(7) + 5 + 2(7) + 5$	$5(7) + 10$
$= 7 + 14 + 5 + 14 + 5$	$= 35 + 10$
$= 45$	$= 45$

Check Evaluate the expressions for the side lengths with $x = 7$ to get numeric
values for each side length. Then verify that the perimeter is 45 feet by adding
those side lengths.

Answers to Warm-up
1. $2(l + w)$ **2.** 56

▶ **Your Turn 1**

Solve.

a. Write an expression in simplest form for the perimeter of the shape. Find the perimeter if d is 9 meters.

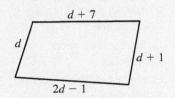

b. Write an expression in simplest form for the area of the shape. Find the area if w is 4 inches.

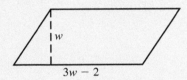

c. Write an expression in simplest form for the volume of the box. Find the volume if h is 2 centimeters.

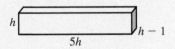

Replace x with 7 in each expression.

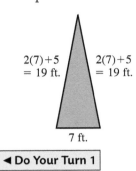

Now find the perimeter.

$P = 7 + 19 + 19$ Add the side lengths.

$P = 45$ feet   The result is the same.

◀ **Do Your Turn 1**

Objective 2 Solve surface area problems.

In Section 1.6, we developed the formula for the volume of a box. Now let's develop a formula for the **surface area** of a box.

Definition Surface area: The total number of square units that completely cover the outer shell of an object.

Think of surface area as finding the area of the skin or outer shell of an object. To develop a formula for the surface area of a box, we will cut and unfold a box as shown below so that we can see all of its surfaces.

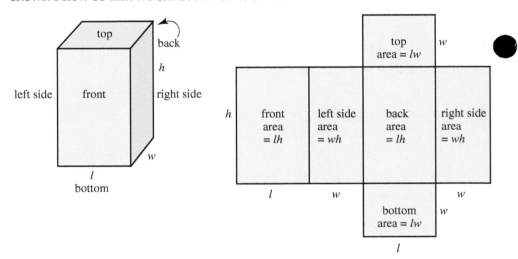

Notice that a box has six rectangular surfaces: top, bottom, front, back, left side, and right side. To find the surface area, we add the areas of the six rectangular surfaces.

Surface area = top + bottom + front + back + left side + right side
of a box   area   area   area   area   area   area

$SA = lw + lw + lh + lh + wh + wh$

We can simplify the polynomial that describes the surface area of a box by combining like terms.

$$SA = 2lw + 2lh + 2wh$$

We could also factor out a common factor, 2.

$$SA = 2(lw + lh + wh)$$

Connection Remember, like terms have the same variables raised to the same exponents. The two lw terms are like, as are the two lh terms and the two wh terms.

Answers to Your Turn 1
a. $5d + 7$; 52 m
b. $3w^2 - 2w$; 40 in.^{2}
c. $5h^3 - 5h^2$; 20 cm^3

Example 2 A company produces metal boxes that are 2 feet long by 3 feet wide by 1 foot high. How many square feet of metal are needed to produce one box?

Solution: $SA = 2(lw + lh + wh)$

$SA = 2[(2)(3) + (2)(1) + (3)(1)]$ In $SA = 2(lw + lh + wh)$, replace *l* with 2, *w* with 3, and *h* with 1.

$SA = 2[6 + 2 + 3]$ Multiply in the brackets.

$SA = 2[11]$ Add in the brackets.

$SA = 22$ Multiply.

Answer: Each box will require 22 ft.^{2} of metal.

▶ **Do Your Turn 2**

Objective 3 Solve problems involving a falling object.

The polynomial $-16t^2 + h_0$ describes the height in feet of a falling object at any time during the fall. The variable h_0 represents the initial height of the object in feet from the ground. The variable *t* represents the number of seconds after the object is dropped from that initial height.

If we choose *h* to represent the height of the object after falling for *t* seconds, we can write the following formula:

$$h = -16t^2 + h_0$$

Galileo Galilei discovered that all objects fall at the same rate. Therefore, our formula for the height of an object after falling for *t* seconds is true for all objects regardless of their weight. Note, however, that the formula does not take into account air resistance. If we drop a feather and a lead weight at the same time, air resistance will keep the feather from reaching the ground at the same time as the lead weight. However, if we perform the same experiment in a vacuum, the feather and lead weight will reach the ground at the same time.

Of Interest

Galileo Galilei (1564–1642) was an Italian mathematician and astronomer. He contributed to the study of pendulums, falling objects, projectile motion, and astronomy. Using the newly invented telescope, he discovered moons around Jupiter, sun spots, and the phases of Venus. His observations affirmed the recently published Copernican view that the planets orbit the sun. This view was in opposition to the prevailing view supported by the Catholic Church that the sun, planets, and stars orbit a stationary Earth. Because Galileo publicly supported the Copernican system, he was accused of heresy and in 1633 was put on trial. Offered leniency if he would recant his support of the Copernican system, he did so and spent the remainder of his life under house arrest.

Example 3 A skydiver jumps from a plane at an altitude of 10,000 feet and deploys her parachute after 20 seconds of free fall. What was her altitude upon deploying the parachute?

Solution: $h = -16t^2 + h_0$

$h = -16(20)^2 + 10,000$ In $h = -16t^2 + h_0$, replace *t* with 20 and h_0 with 10,000.

$h = -16(400) + 10,000$ Square 20.

$h = -6400 + 10,000$ Multiply. $-16(400) = -6400$

$h = 3600$ Add.

Answer: The skydiver deployed her parachute at an altitude of 3600 ft.

▶ **Do Your Turn 3**

▶ **Your Turn 2**

A carpenter builds wooden chests that are 2 feet wide by 3 feet long by 2 feet high. He finishes the outside of the chests with stain. What is the surface area of the chest? What total area must he plan to cover if he intends to put three coats of stain on the box?

▶ **Your Turn 3**

A marble is dropped from a 100-foot tower. How far is the marble from the ground after 2 seconds?

Answers to Your Turn 2
32 ft.^{2}; 96 ft.^{2}

Answer to Your Turn 3
36 ft.

Objective 4 Solve net-profit problems.

In Section 2.3, we developed the formula for net profit or loss given revenue and cost.

$$N = R - C$$

Let's consider the situation where we are given expressions for revenue and cost. We can write an expression for net by subtracting the cost expression from the revenue expression.

Example 4 The expression $7b + 9d + 1265$ describes the revenue for a toy manufacturer, where b represents the number of toy bears sold and d represents the number of toy dogs sold. The expression $8b + 3d + 742$ describes the cost of producing the toy bears and toy dogs.

a. Write an expression in simplest form for the net.

Solution: $N = R - C$

$N = (7b + 9d + 1265) - (8b + 3d + 742)$ In $N = R - C$, replace R with $7b + 9d + 1265$ and C with $8b + 3d + 742$.

▲

Note Use parentheses when replacing a variable with a value or an expression.

$N = (7b + 9d + 1265) + (-8b - 3d - 742)$ Write as an equivalent addition.

$N = -b + 6d + 523$ Combine like terms.

Answer: The expression that describes net is $-b + 6d + 523$.

We can check by adding the expression for net with the expression for cost. We should get the expression for revenue.

$$R = N + C$$
$$R = (-b + 6d + 523) + (8b + 3d + 742)$$
$$R = 7b + 9d + 1265$$

b. In one month, the company sold 2345 bears and 3687 dogs. Using the expression for net from part a, find the net profit or loss.

Solution: $-b + 6d + 523$

$-(2345) + 6(3687) + 523$ In $-b + 6d + 523$, replace b with 2345 and d with 3687.

$= -2345 + 22{,}122 + 523$ Evaluate the additive inverse of 2345 and multiply 6(3687).

$= 20{,}300$ Add from left to right.

Answer: The business made a profit of $20,300.

Connection Think about the terms in the expressions for revenue, cost, and net. In the revenue expression, $7b + 9d + 1265$, the $7b$ term represents the revenue from selling toy bears and $9d$ represents the revenue from selling toy dogs. The coefficient, 7, means that each toy bear sells for $7 and the 9 means that each toy dog sells for $9. The constant, 1265, is an income that does not depend on the number of toy bears or dogs sold. It could be interest or fees of some kind.

In the cost expression, $8b + 3d + 742$, the $8b$ term represents the cost of producing the toy bears and $3d$ represents the cost of producing the toy dogs. The coefficient, 8, means that each bear costs $8 to produce and 3 means that each dog costs $3 to produce. The constant, 742, is a cost that does not depend on the number of toy bears or dogs produced. It could be utility and operation costs for the plant.

▶ **Your Turn 4**

A business sells two types of small motors. The revenue from the sales of motor A and motor B is described by $85A + 105B + 215$. The total cost of producing the two motors is described by $45A + 78B + 345$.

a. Write an expression in simplest form for the net.

b. In one month, the business sells 124 of motor A and 119 of motor B. Using the expression for net from part a, find the net profit or loss.

Answers to Your Turn 4
a. $40A + 27B - 130$
b. $8043 profit

When we developed the expression for net by subtracting the cost expression from the revenue expression, we combined like terms. In combining like terms, we subtracted the coefficients. Subtracting $8b$ from $7b$ equals $-b$. Because the $8 cost of producing each bear is more than the $7 revenue from the sale of each bear, the resulting -1 coefficient means that the company has a net loss of $1 for each bear sold. Subtracting $3d$ from $9d$ equals $6d$. Because the $3 cost is less than the $9 revenue, the company has a net profit of $6 for each toy dog sold. Subtracting 742 from 1265 equals 523. Because the constant cost of $742 is less than the constant revenue of $1265, the $523 is a profit.

Putting it all together:

$$(7b + 9d + 1265) - (8b + 3d + 742) = -b \quad + \quad 6d \quad + \quad 523$$

          ↑   ↑   ↑

        Net loss of $1   **Net profit**   **Net profit of**
        for each toy   **of $6 for**   **$523 from**
        bear sold   **each toy**   **constant**
                           dog sold   **revenue and**
                                       costs

▶ **Do Your Turn 4**

3.7 Exercises For Extra Help MyMathLab®

Objective 1

Prep Exercise 1 Explain how to find the perimeter of a shape.

For Exercises 1–6, write an expression in simplest form for the perimeter of each shape.

1.
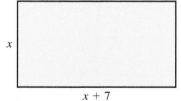

2.
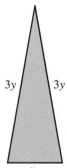

3.
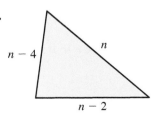

4.
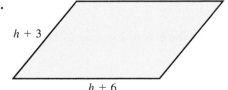

5.

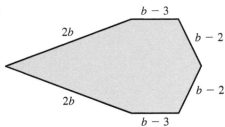

6.

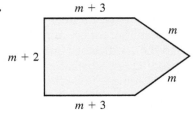

7. Find the perimeter of the rectangle in Exercise 1 if x is

    a. 19 inches
    b. 27 centimeters

8. Find the perimeter of the triangle in Exercise 2 if y is

    a. 15 miles
    b. 22 meters

9. Find the perimeter of the triangle in Exercise 3 if n is

    a. 12 feet
    b. 8 yards

10. Find the perimeter of the parallelogram in Exercise 4 if h is

    a. 6 kilometers
    b. 15 centimeters

11. Find the perimeter of the shape in Exercise 5 if b is

    a. 10 inches
    b. 20 feet

12. Calculate the perimeter of the shape in Exercise 6 if m is

    a. 14 miles
    b. 32 meters

Prep Exercise 2 Explain how to find the area of a parallelogram.

For Exercises 13 and 14, write an expression in simplest form for the area of each shape.

13.

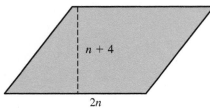

14.

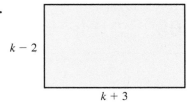

15. Find the area for the parallelogram in Exercise 13 if n is

    a. 6 kilometers
    b. 12 feet

16. Find the area of the rectangle in Exercise 14 if k is

    a. 8 inches
    b. 13 centimeters

Prep Exercise 3 Explain how to find the volume of a box.

For Exercises 17 and 18, write an expression in simplest form for the volume of each box.

17.

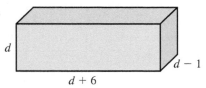

18.

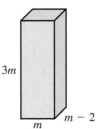

19. Find the volume of the box in Exercise 17 if d is

    a. 3 inches
    b. 5 feet

20. Find the volume of the box in Exercise 18 if m is

    a. 7 miles
    b. 14 inches

Objective 2

Prep Exercise 4 What is the formula for the surface area of a box?

For Exercises 21–32, solve.

21. A company produces cardboard boxes. Each box is 2 feet long by 2 feet wide by 3 feet high. What is the area of the cardboard required for each box?

22. A cube that is 5 centimeters along each side is to be covered with paper. What is the total area that will be covered by the paper?

23. Therese designs and builds metal sculptures. One particular sculpture is to be a large cube that will rest on one of its corners. She will cover the cube with panels that are 1 square foot in size. If the cube is to be 16 feet on each side, how many panels will she need to cover the cube?

24. The main body of the GOES-10 satellite is a box that measures 6 feet wide by 7 feet long by 8 feet high. The box is covered with a skin of a thin metal alloy. What is the area covered by the metal alloy?

Of Interest

GOES stands for Geostationary Operational Enviromental Satellite. *Geostationary* means that these satellites orbit the earth along the equator at the same speed as the earth's rotation so that they remain over the same position at all times. They are used to monitor the weather.

Objective 3

Prep Exercise 5 What is the formula $h = -16t^2 + h_0$ used for, and what does each variable represent?

25. The leaning tower of Pisa is 180 feet tall. When Galileo dropped a cannonball from the top of the leaning tower of Pisa, what was the height of the cannonball after 2 seconds? After 3 seconds?

26. Victoria Falls is a waterfall on the Zambezi river in south central Africa that is 355 feet high. If a coin was dropped from the top of the falls, how high would the coin be after 3 seconds? After 4 seconds?

27. A skydiver jumps from a plane at 12,000 feet and deploys her parachute after 22 seconds of free fall. What is her altitude at the time she deploys her parachute?

28. Two skydivers jump from a plane at an altitude of 16,000 feet. One skydiver deploys his parachute after 27 seconds; the other, 2 seconds later. What were their respective altitudes at the time each deployed his parachute?

Of Interest

Construction on the tower at Pisa began in 1173 and continued for more than 200 years due to continuous structural problems and soft, unstable soil. Because of the soft soil, one side of the tower began sinking before the first three stories were completed, creating a noticeable lean. The lean is estimated to be increasing at a rate of 1 millimeter per year.

Objective 4

Prep Exercise 6 Given polynomials for revenue and cost, how is net profit or loss determined?

29. A company produces two different tables, a round top and a square top. If r represents the number of round-top tables and s represents the number of square-top tables, then $145r + 215s + 100$ describes the revenue from the sales of the two types of table. The polynomial $110r + 140s + 345$ describes the cost of producing the two types of table.

    a. Write an expression in simplest form for the net.

    b. In one month, the company sells 120 round-top tables and 106 square-top tables. Find the net profit or loss.

30. Calvin installs tile. The expression $9lw + 45$ describes the revenue in dollars he receives for tiling a rectangular room, where l represents the length of the room and w represents the width. The expression $3lw + t$ describes his cost in dollars for tiling a rectangular room, where t represents the number of tiles required.

  a. Write an expression in simplest form for the net.
  b. Calvin gets a job to tile a 12-foot-wide by 14-foot-long room. He uses 672 tiles. Find his net profit or loss.

31. Candice makes reed baskets in three sizes: small, medium, and large. She sells the small baskets for $5, medium for $9, and large for $15. Each small basket costs her $2 to make; each medium, $4; and each large, $7.

  a. Write a polynomial that describes the revenue she receives from the sale of all three baskets.
  b. Write a polynomial that describes her cost to produce all three baskets.
  c. Write an expression in simplest form for her net.
  d. In one day, she sells 6 small, 9 medium, and 3 large baskets. Find her net profit or loss.

32. Devon paints portraits in three sizes: small, medium, and large. He receives $50 for a small, $75 for a medium, and $100 for a large portrait. It costs him $22 for each small, $33 for each medium, and $47 for each large portrait.

  a. Write a polynomial that describes the revenue he receives from the sale of all three sizes of portraits.
  b. Write a polynomial that describes his cost for all three sizes.

  c. Write an expression in simplest form for his net.
  d. In one month, he sells 4 small, 6 medium, and 2 large portraits. Find his net profit or loss.

Puzzle Problem How many cubic feet of dirt are in a hole 4 feet wide, 6 feet long, and 3 feet deep?

Review Exercises

[3.1] 1. Evaluate $\sqrt{2a + 7} - b^3$ when $a = 21$ and $b = -2$.

[3.2] 2. What is the coefficient of $-x^3$?

[3.2] 3. What is the degree of 5?

[3.2] 4. Combine like terms.
$$8x^3 - 9x^2 + 11 - 12x^2 + 2x - 9x^3 + 3 - 2x$$

[3.4] 5. Multiply. $-4(3a + 7)$

[3.6] 6. Evaluate. 6^0

Chapter 3 Summary and Review Exercises

Learning Strategy

Create a "master" study guide for each exam and incorporate your text-book, notebook, homework, and anything else used for class. Include all of the concepts you need to know and make it detailed enough so that you won't have to go back to your books or notes. Share study guides with your friends so that you can find things that you've missed or didn't understand. Before each exam, create a list of concepts you need to know. Meet with friends to create sample questions for each concept and test each other until you complete the list.

—Kim D.

Complete each incomplete definition, rule, or procedure, study the key examples, and then work the related exercises.

3.1 Translating and Evaluating Expressions

Definitions/Rules/Procedures	Key Example(s)
An **expression** is a constant; a variable; or any combination of constants, variables, and arithmetic symbols that _____. An **equation** is a mathematical relationship that _____. To translate a word phrase to an expression, identify the unknowns, constants, and _____; then write the corresponding symbolic form. **To evaluate an expression,** 1. _____ with the corresponding given values. 2. _____ using the order of operations agreement.	Example of an expression: $3x^2 + 7x + 5$ Example of an equation: $9x + 7 = 4x - 8$ Translate to a variable expression. **a.** nine subtracted from the product of four and some number **Answer:** $4n - 9$ **b.** the sum of x and two divided by the difference of y squared and three **Answer:** $(x + 2) \div (y^2 - 3)$ or $\dfrac{x + 2}{y^2 - 3}$ Evaluate $-2x + 8$ when $x = 3$. **Solution:** $-2(3) + 8$ Replace x with 3. $\quad\quad\quad = -6 + 8$ Multiply. $\quad\quad\quad = 2$ Add.

[3.1] For Exercises 1 and 2, translate to a variable expression.

1. eight less than negative two times m

2. the product of five and the difference of some number and twelve

[3.1] For Exercises 3 and 4, evaluate the expression using the given values.

3. $5n^2 - 9n + 2; n = -2$

4. $\dfrac{-3ab + 4}{a - b^2}; a = 2, b = -4$

3.2 Introduction to Polynomials; Combining Like Terms

Definitions/Rules/Procedures	Key Example(s)
A **monomial** is an expression that is a(n) _____ _____ that are raised to whole number powers. A **coefficient** is the _____ in a monomial. The **degree of a monomial** is the _____ _____ in the monomial.	$5x^2$ is a monomial. Its coefficient is 5, and its degree is 2. $-mn^3$ is a monomial. Its coefficient is -1, and its degree is 4. $9x + 2$ is not a monomial because it contains a sum. $\dfrac{7h}{k}$ is not a monomial because it contains division by a variable.

[3.2] **5.** Is $9x^3$ a monomial? Explain.

[3.2] **6.** Is $4x - 5$ a monomial? Explain.

[3.2] For Exercises 7–10, identify the coefficient and degree of each monomial.

7. $18x$

8. y^3

9. -9

10. $-3m^5n$

Definitions/Rules/Procedures	Key Example(s)
A **polynomial** is a(n) _____ or a(n) _____. A **polynomial in one variable** is a polynomial in which _____. A **multivariable polynomial** is a polynomial with _____. The **degree of a polynomial** is the _____ of any of the terms in the polynomial. **To write a polynomial in descending order of degree**, write the term with the _____ degree first, then the term with the _____ degree, and so on.	$9x^2 - 11x - 16$ is a polynomial in one variable. More specifically, it is a trinomial because it has three terms. $7xy + 13$ is a multivariable polynomial. More specifically, it is a binomial because it has two terms. $20x + 9x^2 - 16 + 5x^3$ is a polynomial in one variable. It has no special name because it has more than three terms. Determine the degree of $20x + 9x^2 - 16 + 5x^3$ and write it in descending order of degree. Degree: 3 Descending order: $5x^3 + 9x^2 + 20x - 16$

[3.2] For Exercises 11–14, indicate whether the expression is a monomial, binomial, or trinomial, or has no special name.

11. $9y - 5$

12. $4x^3 + 9x^2 - x + 7$

13. $-3a$

14. $x^2 + 5x - 4$

[3.2] **15.** What is the degree of $7a^4 - 9a + 13 + 5a^6 - a^2$?

[3.2] **16.** Write $7a^4 - 9a + 13 + 5a^6 - a^2$ in descending order of degree.

Definitions/Rules/Procedures	Key Example(s)
An expression is in **simplest form** if it is written with the _____ symbols possible. **Like terms** are _____ terms or variable terms that have _____. **To combine like terms**, add the _____ and keep the _____ and their _____ the same.	Simplify and write the resulting polynomial in descending order of degree. $\quad 7x + 15x^3 - 12 + 4x^2 - 5x^3 - 6 - 8x$ $= 15x^3 - 5x^3 + 4x^2 + 7x - 8x - 12 - 6$ Collect like terms. $= 10x^3 + 4x^2 - x - 18$ Combine like terms.

[3.2] 17. Explain why $4y^3$ and $9y^3$ are like terms.   **[3.2] 18.** Explain why $7a^2$ and $7a$ are not like terms.

[3.2] For Exercises 19 and 20, simplify and write the resulting polynomial in descending order of degree.

19. $3a^2 + 2a - 4a^2 - 1$ **20.** $3x^2 - 5x + x^7 + 13 + 5x - 6x^7$

3.3 Adding and Subtracting Polynomials

Definitions/Rules/Procedures	Key Example(s)
To add polynomials, _____ .	Add. $$(3x^2 - 5x + 1) + (7x^2 - 4x - 6)$$ $$= 3x^2 + 7x^2 - 5x - 4x + 1 - 6$$ $$= 10x^2 - 9x - 5$$

[3.3] For Exercises 21 and 22, add and write the resulting polynomial in descending order of degree.

21. $(x - 4) + (2x + 9)$ **22.** $(y^4 + 2y^3 - 8y + 5) + (3y^4 - 2y - 9)$

Definitions/Rules/Procedures	Key Example(s)
To subtract polynomials: Write the subtraction expression as an equivalent addition expression. a. Change the operation symbol from a $-$ to a $+$. b. Change the subtrahend (second polynomial) to its additive inverse by _____ .	Subtract. $$(5a^3 - 9a + 2) - (a^3 - 6a - 7)$$ $$= (5a^3 - 9a + 2) + (-a^3 + 6a + 7)$$ $$= 5a^3 - a^3 - 9a + 6a + 2 + 7$$ $$= 4a^3 - 3a + 9$$

[3.3] For Exercises 23 and 24, subtract and write the resulting polynomial in descending order of degree.

23. $(a^2 - a) - (-2a^2 + 4a)$ **24.** $(19h^3 - 4h^2 + 2h - 1) - (6h^3 + h^2 + 7h + 2)$

3.4 Exponent Rules; Multiplying Polynomials

Definitions/Rules/Procedures	Key Example(s)
To multiply exponential forms that have the same base, we can _____ the exponents and _____ . In math language: $a^m \cdot a^n =$ _____ . **To multiply monomials,** 1. Multiply the _____ . 2. _____ of the like variables.	Multiply. **a.** $t^5 \cdot t^4 = t^{5+4} = t^9$ **b.** $-6a^4 \cdot 5a^3 = -6 \cdot 5a^{4+3}$ $= -30a^7$

[3.4] For Exercises 25–28, multiply.

25. $m^3 \cdot m^4$ **26.** $2x \cdot (-5x^4)$ **27.** $-4y \cdot 3y^4$ **28.** $-6t^3 \cdot (-t^5)$

Definitions/Rules/Procedures	Key Example(s)
To simplify an exponential form raised to a power, we can _____ the exponents and _____ . In math language: $(a^m)^n =$ _____ . **To simplify a monomial raised to a power:** 1. Evaluate the _____ raised to that power. 2. Multiply each _____ by the power.	Simplify. **a.** $(x^2)^4 = x^{2 \cdot 4} = x^8$ **b.** $(-4k^5)^3 = (-4)^3 k^{5 \cdot 3}$ $= -64k^{15}$

[3.4] For Exercises 29–32, simplify.

29. $(5x^4)^3$ **30.** $(-2y^3)^2$ **31.** $(4t^3)^3$ **32.** $(-2a^4)^3$

Definitions/Rules/Procedures	Key Example(s)

To multiply a polynomial by a monomial, use the distributive property to multiply _____ _____ .
 	
To multiply two polynomials:
 	
1. Multiply _____ in the second polynomial by _____ in the first polynomial.
 	
2. _____ like terms.
 	
Conjugates are _____ that differ only in the _____ .
 	
The product of conjugates is a(n) _____ .
 	
In math language: $(a + b)(a - b) =$ _____ .
	Multiply.
	
	a. $2m(m^4 - 5m^2 + 3)$
	     $= 2m \cdot m^4 + 2m \cdot (-5m^2) + 2m \cdot 3$
	     $= 2m^5 - 10m^3 + 6m$
	
	b. $(2x + 3)(5x - 1)$
	     $= 2x \cdot 5x + 2x \cdot (-1) + 3 \cdot 5x + 3(-1)$
	     $= 10x^2 - 2x + 15x - 3$
	     $= 10x^2 + 13x - 3$
	
	c. $(3x + 5)(3x - 5) = (3x)^2 - (5)^2$
	          $= 9x^2 - 25$

[3.4] For Exercises 33–40, multiply.

33. $2(3x - 4)$ **34.** $-4y(4y - 5)$

35. $3n(5n^2 - n + 7)$ **36.** $(a + 5)(a - 7)$

37. $(2x - 3)(x + 1)$ **38.** $(2y - 1)(5y - 8)$

39. $(3t + 4)(3t - 4)$ **40.** $(u + 2)(u^2 - 5u + 3)$

[3.4] For Exercises 41 and 42, state the conjugate of each binomial.

41. $3y - 4$ **42.** $-7x + 2$

3.5 Prime Numbers and GCF

Definitions/Rules/Procedures	Key Example(s)
A **prime number** is a(n) _____ number that has two different factors: _____ and _____ . A **composite number** is a(n) _____ number that has _____ .	Determine whether 37 is prime or composite. Divide by the list of primes. Is 37 divisible by 2? No, because 37 is odd. Is 37 divisible by 3? No, because $3 + 7 = 10$ and this sum is not divisible by 3. Is 37 divisible by 5? No, because 37 does not end with 5 or 0.

Definitions/Rules/Procedures (*continued*)	Key Example(s)

To determine whether a given number is prime or composite, divide it by each prime number sequentially in ascending order and consider the results.
 	
• If the given number is divisible by the prime number, stop. The given number is a(n) _____ number.
• If the given number is not divisible by the prime number, consider the quotient.
  ▶ If the quotient is greater than the current prime divisor, repeat the process with the next prime in the list of prime numbers.
  ▶ If the quotient is equal to or less than the current prime divisor, stop. The given number is a(n) _____ number.
	Is 37 divisible by 7?
	
	$$\begin{array}{r} 5 \\ 7\overline{)37} \\ -35 \\ \hline 2 \end{array}$$
	
	37 is not divisible by 7. The quotient is smaller than the divisor. Stop test! 37 is prime.

[3.5] *For Exercises 43 and 44, determine whether the number is prime, composite, or neither.*

43. 119

44. 97

Definitions/Rules/Procedures	Key Example(s)
A **prime factorization** is a product written with _____ factors only. **To find the prime factorization of a composite number using a factor tree:** 1. Draw two branches below the number. 2. Find two _____ and place them at the end of the two branches. 3. Repeat steps 1 and 2 for every composite factor. 4. Write a multiplication expression containing all of the prime factors.	Find the prime factorization. 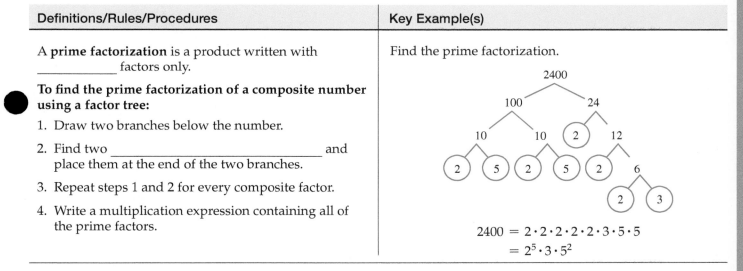 $$2400 = 2 \cdot 2 \cdot 2 \cdot 2 \cdot 2 \cdot 3 \cdot 5 \cdot 5$$ $$= 2^5 \cdot 3 \cdot 5^2$$

[3.5] *For Exercises 45 and 46, find the prime factorization. Write the answer in exponential form.*

45. 360

46. 4200

Definitions/Rules/Procedures	Key Example(s)
The **greatest common factor** of a set of numbers is the _____ number that _____ all given numbers with no remainder. **To find the greatest common factor by listing:** 1. List _____ factors for each given number. 2. Search the lists for the _____ factor common to all lists.	Find the GCF of 32 and 40 by listing. Factors of 32: 1, 2, 4, 8, 16, 32 Factors of 40: 1, 2, 4, 5, 8, 10, 20, 40 GCF = 8

(continued)

Definitions/Rules/Procedures (*continued*)	Key Example(s)
To find the greatest common factor using prime factorization: 1. Write the prime factorization of each number in exponential form. 2. Create a factorization for the GCF that has only those prime factors _____, each raised to the _____ of its exponents. 3. Multiply. Note: If there are no common prime factors, the GCF is _____ .	Find the GCF. **a.** 84 and 120 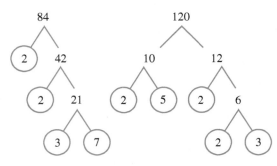 $$84 = 2^2 \cdot 3 \cdot 7$$ $$120 = 2^3 \cdot 3 \cdot 5$$ $$\text{GCF} = 2^2 \cdot 3 = 2 \cdot 2 \cdot 3 = 12$$ **b.** $36m^3$ and $54m^2$ 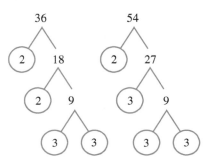 $$36m^3 = 2^2 \cdot 3^2 \cdot m^3$$ $$54m^2 = 2 \cdot 3^3 \cdot m^2$$ $$\text{GCF} = 2 \cdot 3^2 \cdot m^2 = 18m^2$$

[3.5] For Exercises 47–50, find the GCF.

47. 140 and 196

48. 45 and 28

49. $48x^5$ and $36x^6$

50. $18a^2$, $30a^4$, and $24a^3$

3.6 Exponent Rules; Introduction to Factoring

Definitions/Rules/Procedures	Key Example(s)
To divide exponential forms that have the same base, we can _____ their exponents and _____ . In math language: $a^m \div a^n = \dfrac{a^m}{a^n} = $ _____ when $a \neq 0$. **The result of raising any nonzero number to the exponent 0 is** _____ .	Divide. **a.** $t^9 \div t^2 = t^{9-2} = t^7$ **b.** $\dfrac{y^6}{y^2} = y^{6-2} = y^4$ Evaluate. **a.** $245^0 = 1$ **b.** $(-5)^0 = 1$

Definitions/Rules/Procedures (*continued*)	Key Example(s)
In math language: $a^0 = $ _____ when $a \neq 0$. **To divide monomials** that have the same variable: 1. Divide the _____. 2. _____ of the variable in the divisor from the exponent of the variable in the dividend. **To divide a polynomial by a monomial,** divide _____ in the polynomial by the monomial.	Divide. **a.** $-18m^3 \div 2m = -18 \div 2 \cdot m^{3-1}$ $\qquad\qquad\qquad = -9m^2$ **b.** $\dfrac{30x^4}{5x^4} = \dfrac{30}{5}x^{4-4}$ $\qquad = 6x^0$ $\qquad = 6(1)$ $\qquad = 6$ Divide. **a.** $(20x^3 + 28x^2) \div (4x^2)$ $\qquad = 20x^3 \div (4x^2) + 28x^2 \div (4x^2)$ $\qquad = (20 \div 4)x^{3-2} + (28 \div 4)x^{2-2}$ $\qquad = 5x + 7$ **b.** $\dfrac{14y^4 - 28y^3 + 35y^2}{7y^2}$ $\qquad = \dfrac{14y^4}{7y^2} - \dfrac{28y^3}{7y^2} + \dfrac{35y^2}{7y^2}$ $\qquad = \dfrac{14}{7}y^{4-2} - \dfrac{28}{7}y^{3-2} + \dfrac{35}{7}y^{2-2}$ $\qquad = 2y^2 - 4y + 5$

[3.6] For Exercises 51–54, divide.

51. $r^8 \div r^2$

52. $\dfrac{20x^5}{-5x^2}$

53. $(14t^2 + 18t) \div 2t$

54. $\dfrac{-30x^6 + 24x^4 - 12x^2}{-6x^2}$

Definitions/Rules/Procedures	Key Example(s)
To find an unknown factor, divide the _____ by the _____.	Find the unknown factor. $30x^3 - 45x^2 + 10x = 5x\,(\,?\,)$ $\dfrac{30x^3 - 45x^2 + 10x}{5x} = (\,?\,)$ Divide the product by the known factor. $\dfrac{30x^3}{5x} - \dfrac{45x^2}{5x} + \dfrac{10x}{5x} = (\,?\,)$ Divide each term in the polynomial by the monomial. $6x^2 - 9x + 2 = (\,?\,)$ Simplify.

[3.6] For Exercises 55–58, find the unknown factor.

55. $6m^3 \cdot (?) = -54m^8$

56. $12x^6 + 9x^4 = 3x^4 \cdot (?)$

57. $40y^6 - 30y^4 - 20y^2 = 10y^2 (?)$

58. Find an expression for the unknown length in the rectangle shown if its area is $42b^6$.

$3b^5$

?

Definitions/Rules/Procedures	Key Example(s)
A number or an expression is in **factored form** if it is written as a(n) _____. **To factor a monomial GCF out of a polynomial:** 1. Find the GCF of the terms in the polynomial. 2. Rewrite the polynomial as a product of the GCF and the _____ of the polynomial and the GCF. Polynomial = _____	Factor. **a.** $24x - 15 = 3\left(\dfrac{24x - 15}{3}\right)$ $\qquad\qquad = 3\left(\dfrac{24x}{3} - \dfrac{15}{3}\right)$ $\qquad\qquad = 3(8x - 5)$ **b.** $9u^7 + 27u^5 - 18u^3$ $\qquad = 9u^3\left(\dfrac{9u^7 + 27u^5 - 18u^3}{9u^3}\right)$ $\qquad = 9u^3\left(\dfrac{9u^7}{9u^3} + \dfrac{27u^5}{9u^3} - \dfrac{18u^3}{9u^3}\right)$ $\qquad = 9u^3(u^4 + 3u^2 - 2)$

[3.6] For Exercises 59–62, factor.

59. $30x + 12$

60. $12y^2 + 20y$

61. $9n^4 - 15n^3$

62. $18x^3 + 24x^2 - 36x$

3.7 Additional Applications and Problem Solving

Definitions/Rules/Procedures	Key Example(s)
Surface area is the total number of square units that _____ of an object. **Formulas** Area of a parallelogram: $A =$ _____ Volume of a box: $V =$ _____ Surface area of a box: $SA =$ _____ Height of a falling object: $h =$ _____ Net profit or loss: $N =$ _____	A 20-inch by 15-inch by 3-inch box is to be gift wrapped with paper. How much paper is required? **Solution:** $SA = 2lw + 2lh + 2wh$ $\quad = 2(20)(15) + 2(20)(3) + 2(15)(3)$ Replace *l* with 20, *w* with 15, and *h* with 3. $\quad = 600 + 120 + 90$ Multiply. $\quad = 810 \text{ in.}^2$ Add.

[3.7] 63. a. Write an expression in simplest form for the perimeter of the rectangle shown.

    b. Find the perimeter if x is 9 meters.

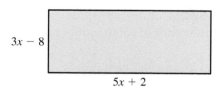

[3.7] 64. a. Write an expression in simplest form for the area of the parallelogram shown.

    b. Find the area of the parallelogram if m is 8 inches.

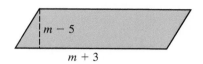

[3.7] 65. a. Write an expression in simplest form for the volume of the box shown.

    b. Find the volume of the box if n is 4 feet.

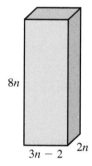

[3.7] 66. A company produces cardboard boxes that are 4 feet long by 3 feet wide by 2 feet high. How many square feet of cardboard are needed to produce one box?

[3.7] 67. A coin is dropped from 150 feet. What is the height of the coin after 2 seconds?

[3.7] 68. A company produces two different tennis racquets, a normal size and an oversize version. If n represents the number of normal-sized racquets and b represents the number of oversize racquets, the polynomial $65n + 85b + 345$ describes the revenue from the sales of the two racquets. The polynomial $22n + 42b + 450$ describes the cost of producing the two racquets.

    a. Write an expression in simplest form for net.

    b. In one week, the company sells 487 normal-size racquets and 246 oversize racquets. Find the net profit or loss.

Chapter 3 Practice Test

For Extra Help

Step-by-step test solutions are found on the Chapter Test Prep Videos available via the Video Resources on DVD, in MyMathLab®, and on You Tube™ (search "CarsonPrealgebra" and click on "Channels").

1._____

2._____

3._____

4._____

5._____

6._____

7._____

8._____

9._____

10._____

11._____

12._____

13._____

14._____

15._____

16._____

17._____

[3.1] **1.** Translate to a variable expression.
Negative seven times the difference of some number and five.

[3.1] **2.** Evaluate $2n^3 - 3p^2$ when $n = -2$ and $p = -3$.

[3.2] **3.** Identify the coefficient and degree of $-y^4$.

[3.2] **4.** Indicate whether $4x^3 - 9$ is a monomial, binomial, or trinomial or has no special name.

[3.2] **5.** What is the degree of $x^2 - 9x + 5x^4 - x^3$?

[3.2] **6.** Simplify and write the resulting polynomial in descending order of degree. $10a^4 + 3a^2 - 5a + 2a^2 + 5a - 11a^4$

[3.3] **7.** Add and write the resulting polynomial in descending order of degree. $(2y^3 - 8y + 5) + (3y^3 - 2y - 9)$

[3.3] **8.** Subtract and write the resulting polynomial in descending order of degree. $(12y^5 + 5y^3 - 7y - 3) - (6y^5 + y^2 - 7y + 1)$

[3.4] **9.** Multiply. $(-4u^5) \cdot (-5u^2)$

[3.4] **10.** Simplify. $(-2a^3)^5$

[3.4] For Exercises 11–13, multiply.

11. $-2t(t^3 - 3t - 7)$

12. $(x - 6)(x + 6)$

13. $-2a^2(a + 1)(2a - 1)$

[3.4] **14.** What is the conjugate of $6x - 5$?

[3.5] **15.** Determine whether 91 is prime, composite, or neither.

[3.5] **16.** Find the prime factorization of 340.

[3.5] **17.** Find the GCF of $60h^5$ and $48h^7$.

[3.6] For Exercises 18 and 19, divide.

18. $m^6 \div m^4$

18. _____

19. $\dfrac{16x^5 - 18x^3 + 22x^2}{-2x}$

19. _____

20. Find the unknown factor. $-8x^4 \cdot (?) = -40x^7$

20. _____

[3.6] For Exercises 21 and 22, factor.

21. $12x - 20$

21. _____

22. $10y^4 - 18y^3 + 14y^2$

22. _____

[3.7] **23. a.** Write an expression in simplest form for the area of the parallelogram.
   b. Calculate the area if n is 7 centimeters.

23. a. _____

   b. _____

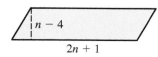

$n - 4$

$2n + 1$

[3.7] **24.** An artist creates a metal box that is 10 inches long by 8 inches wide by 4 inches high. How many square inches of metal are needed to make the box?

24. _____

[3.7] **25.** A company makes two different styles of frames for eyeglasses. One style allows for round lenses; the other, for rectangular lenses. If a represents the number of round-style frames and b represents the number of rectangular-style frames, then $145a + 176b$ describes the revenue from the sale of the frames. The polynomial $61a + 85b$ describes the cost of producing the frames.
   a. Write an expression in simplest form for the net.
   b. In one month, the company sold 128 round frames and 115 rectangular frames. Find the net profit or loss.

25. a. _____

   b. _____

Chapters 1–3 Cumulative Review Exercises

For Exercises 1–4, answer true or false.

[1.1] **1.** 0 is a natural number.

[2.1] **2.** The absolute value of 8 is -8.

[2.1] **3.** The additive inverse of 7 is 7.

[3.2] **4.** $15y^2$ is a monomial.

For Exercises 5–8, complete the rule.

[2.2] **5.** The sum of two negative numbers is a(n) _____ number.

[2.4] **6.** The product of a negative number and a positive number is a(n) _____ number.

[3.2] **7.** To combine like terms, add the _____ and keep the _____ and their _____ the same.

[3.4] **8.** $a^m \cdot a^n =$ _____

[1.3] **9.** Estimate $682 \cdot 246$ by rounding to the nearest hundred.

[2.1] **10.** Graph -9 on a number line.

[3.2] **11.** What is the coefficient of $-x^2$?

[3.2] **12.** What is the degree of $7t^2 - 9t + 4t^5 + 12$?

For Exercises 13–16, simplify.

[2.2] **13.** $-20 + (-8)$

[2.4] **14.** $(-5)^3$

[2.5] **15.** $\sqrt{16 \cdot 25}$

[2.5] **16.** $2[9 + (3 - 16)] - 35 \div 5$

[3.1] **17.** Evaluate $\sqrt{x - y}$ when $x = 100$ and $y = 36$.

[3.3] **18.** Subtract. $(4y^3 - 6y^2 + 9y - 8) - (3y^2 - 12y - 2)$

[3.4] **19.** Multiply. $(b - 8)(2b + 3)$

[3.5] **20.** Find the prime factorization of 360.

[3.6] **21.** Divide. $\dfrac{18n^5}{3n^2}$

[3.6] **22.** Factor. $12m^4 - 18m^3 + 24m^2$

For Exercises 23 and 24, solve and check.

[1.2] **23.** $x + 19 = 25$

[2.4] **24.** $-16x = -128$

[1.6] **25.** Find the area of the parallelogram.

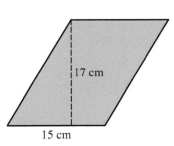

17 cm

15 cm

[3.7] **26.** Write an expression in simplest form for the perimeter of the rectangle.

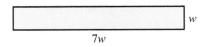

w

$7w$

For Exercises 27–30, solve.

[1.5] **27.** The graph to the right shows the monthly cost of natural gas over one year for the Jones family. Find the mean, median, and mode monthly cost.

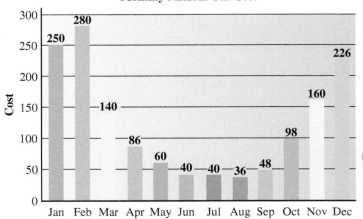

Monthly Natural Gas Cost

[2.2] **28.** Carla has a balance of −$353 in a credit card account. If she makes a payment of $150, then makes two purchases at $38 each, what is her new balance?

[3.7] **30.** A company produces two different car alarm systems, basic and advanced. If b represents the number of basic alarms and a represents the number of advanced alarms, $225b + 345a + 200$ describes the revenue from sales of the two types of alarm systems. The polynomial $112b + 187a + 545$ describes the cost of producing the two alarm systems.
   a. Write an expression in simplest form for the net.
   b. In one month, the company sells 88 basic and 64 advanced alarm systems. Find the net profit or loss.

[3.7] **29.** A 10-foot-wide by 12-foot-long by 7-foot-high box is to be made out of plastic. What is the surface area of the box?

4

Equations

Chapter Overview

In Chapter 3, we learned about expressions, which are like phrases, or incomplete sentences. We learned to evaluate or re-write an expression. In Chapter 4, we will combine expressions with the equal sign to form equations, which are like complete sentences. We will learn to *solve* equations.

4.1 Equations and Their Solutions

Warm-up *[3.1] Evaluate.*
1. $2(a + 6); a = -5$
2. $6a - 4(3a + 7); a = -5$

Objective 1 Differentiate between an expression and an equation.

In Section 3.1, we considered the difference between an expression and an equation. We learned that an equation contains an equal sign, whereas an expression does not.

Expressions: $3x + 5$ Equations: $3x + 5 = 9$
$7x^2 - 8x + 1$ $7x^2 - 8x + 1 = 5x - 2$

Think of expressions as phrases and equations as complete sentences. The expression $3x + 5$ is read "three x plus five," which is not a complete sentence. The equation $3x + 5 = 9$ is read "three x plus five is equal to nine" or simply "three x plus five is nine," which are complete sentences. Notice that the equal sign translates to the verb *is*.

$$3x + 5 = 9$$
$$\updownarrow$$
"Three x plus five is nine."

Example 1   Determine whether each of the following is an expression or an equation.

a. $7x^2 + 9x + 5$

Answer: $7x^2 + 9x + 5$ is an expression because it does not contain an equal sign.

b. $9x - 10 = 3x + 1$

Answer: $9x - 10 = 3x + 1$ is an equation because it contains an equal sign.

▶ **Do Your Turn 1**

Objective 2 Check to see if a number is a solution for an equation.

In Chapter 3, we learned that we can perform two actions with expressions:

1. Evaluate (replace each variable with a given number and perform the arithmetic).
2. Rewrite (simplify, distribute, or factor).

With an equation, our goal is to **solve** it, which means to find its **solution** or solutions.

Definitions Solve: To find the solution or solutions for an equation.

Solution: A number that makes an equation true when it replaces the variable in the equation.

For example, 5 is a solution for the equation $x + 3 = 8$ because the equation is true when 5 replaces x.

$$x + 3 = 8$$
$$\downarrow$$
$$5 + 3 = 8 \quad \text{True}$$

Showing that $x + 3 = 8$ is true when x is replaced by 5 serves as a *check* for the solution.

Objectives

1 Differentiate between an expression and an equation.

2 Check to see whether a number is a solution for an equation.

Learning Strategies

Before class, read the material to be covered. It is much easier to understand your teacher having read the new material rather than hearing it all for the first time in class.
—Kim D.

During lectures, I think it is helpful to have your book open and read along with what your teacher is saying. This can make both the book and your teacher easier to understand.
—Hannah J.

▶ **Your Turn 1**

Determine whether each of the following is an expression or an equation.
  a. $5x - 8$
  b. $7y + 9 = 15$
  c. $P = 2l + 2w$
  d. $4^2 + 3(9 - 1) + \sqrt{25}$

Answers to Your Turn 1
a. expression   b. equation
c. equation   d. expression

Answers to Warm-up
1. 2 2. 2

Of Interest

The modern form of the equal sign is attributed to English mathematician Robert Recorde in his book *Wetstone of Witte*, published in 1557. He wrote,

I will sette as I doe often in woorke vse, a paire of paralleles, or Gemowe [twin] lines of one lengthe, thus: =, bicause noe .2. thynges, can be moare equalle.

Recorde's symbol was not immediately popular. The popular symbol of the time was ∝, which was used into the eighteenth century.

Source: D. E. Smith, *History of Mathematics*, 1953

Procedure

To check to see if a number is a solution for an equation:
1. Replace the variable(s) with the number.
2. Simplify both sides of the equation as needed. If the resulting equation is true, the number is a solution.

► **Your Turn 2**

Check to see if the given number is a solution for the given equation.
a. $12 = 9b - 5$; check $b = 2$
b. $m^2 + 5 = 21$; check $m = -4$

Connection In previous chapters, we placed results below the expression from which they came.

We wrote: $2(3) - 5$
$$= 6 - 5$$
$$= 1$$

By bringing results down, we prepared for this checking technique, which compares the result of the left-hand expression with the result of the right-hand expression.

Example 2 Check to see if the given number is a solution for the given equation.

a. $2x - 5 = 1$; check $x = 3$

Solution: $2x - 5 = 1$
$$2(3) - 5 \overset{?}{=} 1 \qquad \text{Replace } x \text{ with 3.}$$
$$6 - 5 \overset{?}{=} 1 \qquad \text{Simplify.}$$
$$1 = 1 \qquad \text{True; so 3 is a solution.}$$

Note The $\overset{?}{=}$ symbol means that we are asking whether the left side of the equation equals the right side.

b. $13 = y^2 + 7$; check $y = -2$

Solution: $13 = y^2 + 7$
$$13 \overset{?}{=} (-2)^2 + 7 \qquad \text{Replace } y \text{ with } -2.$$
$$13 \overset{?}{=} 4 + 7 \qquad \text{Simplify.}$$
$$13 = 11 \qquad \text{False; so } -2 \text{ is not a solution.}$$

◄ **Do Your Turn 2**

Now let's consider equations that have variable expressions on both sides of the equation, as in $-3n + 8 = 2n - 17$. With these equations, a number is a solution if both expressions produce the same numeric result.

Example 3 Check to see if the given number is a solution for the given equation.

a. $-3n + 8 = 2n - 17$; check $n = 4$

Solution: $-3n + 8 = 2n - 17$
$$-3(4) + 8 \overset{?}{=} 2(4) - 17 \qquad \text{Replace } n \text{ with 4.}$$
$$-12 + 8 \overset{?}{=} 8 - 17 \qquad \text{Simplify.}$$
$$-4 = -9 \qquad \text{False; so 4 is not a solution.}$$

Note We simplify each side of the equation separately until we can determine whether it is true or false.

b. $3x + 6 = 2(x - 1)$; check $x = -8$

Solution: $3x + 6 = 2(x - 1)$
$$3(-8) + 6 \overset{?}{=} 2((-8) - 1) \qquad \text{Replace } x \text{ with } -8.$$
$$-24 + 6 \overset{?}{=} 2(-9) \qquad \text{Simplify.}$$
$$-18 = -18 \qquad \text{True; so } -8 \text{ is a solution.}$$

Answers to Your Turn 2
a. 2 is not a solution.
b. -4 is a solution.

c. $2(a + 6) = 6a - 4(3a + 7)$; check $a = -5$

Solution: $2(a + 6) = 6a - 4(3a + 7)$

$2(-5 + 6) \overset{?}{=} 6(-5) - 4(3(-5) + 7)$ Replace a with -5.

$2(1) \overset{?}{=} 6(-5) - 4(-15 + 7)$ Simplify.

$2 \overset{?}{=} 6(-5) - 4(-8)$

$2 \overset{?}{=} -30 - (-32)$

$2 \overset{?}{=} -30 + 32$

$2 = 2$ True; so -5 is a solution.

Note We keep bringing down 2 until we resolve the other side. ▶

▶ **Do Your Turn 3**

▶ **Your Turn 3**

Check to see if the given number is a solution for the given equation.

a. $5m - 29 = -8m - 3$; check $m = 2$

b. $5x - 6 = 4(x + 2)$; check $x = -2$

c. $5x - 2(x + 3) = 7(x + 2)$; check $x = 4$

VISUAL

🔺 Learning Strategy

Instead of writing $\overset{?}{=}$ to separate the left and right sides, you may prefer to draw a straight line down from the equal sign in the original equation.

$$3x + 6 = 2(x - 1)$$

$$\begin{array}{c|c} 3(-8) + 6 & 2((-8) - 1) \\ -24 + 6 & 2(-9) \\ -18 & -18 \end{array}$$ True; so -8 is a solution.

Answers to Your Turn 3
a. 2 is a solution.
b. -2 is not a solution.
c. 4 is not a solution.

4.1 Exercises   For Extra Help   MyMathLab®

Objective 1

Prep Exercise 1 What is the difference between an expression and an equation?

For Exercises 1–6, determine whether each of the following is an expression or an equation.

1. $9x + 7 = 5$

2. $7x - 8$

3. $2n + 8m^3$

4. $10x^2 = 12x + 1$

5. $14 + 2^5 - 2(3 + 7)$

6. $2y = 16$

Objective 2

Prep Exercise 2 What is a solution to an equation?

Prep Exercise 3 What does it mean to solve an equation?

Prep Exercise 4 Explain how to check to see if a number is a solution for an equation.

For Exercises 7–26, check to see if the given number is a solution for the given equation.

7. $x + 9 = 14$; check $x = -5$

8. $c + 7 = -2$; check $c = -9$

9. $3t = -9$; check $t = -9$

10. $2p = -18$; check $p = -8$

11. $3y + 1 = 13$; check $y = 3$

12. $5d - 1 = 19$; check $d = 4$

13. $6x - 7 = 41$; check $x = 8$

14. $-5a + 19 = 9$; check $a = 2$

15. $3m + 10 = -2m$; check $m = 5$

16. $18 - 7n = 2n$; check $n = -2$

17. $x^2 - 15 = 2x$; check $x = -3$

18. $x^2 - 15 = 2x$; check $x = 5$

19. $b^2 = 5b + 6$; check $b = -3$

20. $b^2 = 5b + 6$; check $b = -1$

21. $-5y + 20 = 4y + 20$; check $y = 0$

22. $10n - 14 = -n + 19$; check $n = 1$

23. $-2(x - 3) - x = 4(x + 5) - 21$; check $x = 3$

24. $3(x - 5) - 1 = 11 - 2(x + 1)$; check $x = 5$

25. $a^3 - 5a + 6 = a^2 + 3(a - 1)$; check $a = -2$

26. $2t(t + 1) = t^3 - 5t$; check $t = -3$

Puzzle Problem A farmer has a fox, a goat, and a large head of cabbage to transport across a river. His boat is so small that it can hold only the farmer and the fox, goat, or cabbage at a time. If the farmer leaves the fox with the goat, the fox will eat the goat. If he leaves the goat with the cabbage, the goat will eat the cabbage. Somehow he transports them all across safely in the boat. Explain the process.

Review Exercises

For Exercises 1 and 2, simplify.

[3.2] **1.** $2x - 19 + 3x + 7$

[3.2] **2.** $7y + 4 - 12 - 9y$

For Exercises 3 and 4, multiply.

[3.4] **3.** $5(m + 3)$

[3.4] **4.** $-1(3m - 4)$

[1.2] **5.** Solve and check. $x + 8 = 15$

4.2 The Addition Principle of Equality

Objectives

1 Determine whether a given equation is linear.

2 Solve linear equations in one variable using the addition principle of equality.

3 Solve application problems.

Warm-up
[4.1] 1. Check to see if -2 is a solution for $7x - 9 = 6x - 11$.
[3.2, 3.4] 2. Simplify. $3(y - 5) - 2y$

Objective 1 Determine whether a given equation is linear.

There are many types of equations. We will focus on **linear equations.**

Definition Linear equation: An equation in which each variable term is a monomial of degree 1.

This means that linear equations contain constants and variable terms that have a single variable raised to an exponent of 1. All other equations are called *nonlinear equations.*

Example 1   Determine whether the given equation is linear or nonlinear.

a. $5x + 9 = 24$

Answer: $5x + 9 = 24$ is a linear equation because the variable term, $5x$, is a monomial of degree 1.

b. $2x - 9x^2 = 15x + 7$

Answer: $2x - 9x^2 = 15x + 7$ is nonlinear because it has a variable term, $-9x^2$, with a degree other than 1. The degree of $-9x^2$ is 2.

▶ **Do Your Turn 1**

▶ **Your Turn 1**

Determine whether the given equation is linear or nonlinear.
a. $2x - 8 = 14$
b. $7y = 21$
c. $x^3 = 5x + 2$
d. $y = x^2 - 5$
e. $y = 2x + 1$

If an equation has the same variable throughout, we say that it is an equation in one variable.

$$2x + 9 = 5x - 12 \text{ is a linear equation in one variable, } x.$$
$$y^2 - 2y + 5 = 8 \text{ is a nonlinear equation in one variable, } y.$$

If an equation has two different variables, we say that it is an equation in two variables. Such an equation is linear as long as the degree of every variable term is 1.

$$2x - 3y = 6 \text{ is a linear equation in two variables, } x \text{ and } y.$$
$$y = x^3 \text{ is a nonlinear equation in two variables.}$$

Objective 2 Solve linear equations in one variable using the addition principle of equality.

In Chapters 1 and 2, we solved equations such as $x + 7 = 9$ by writing a related subtraction equation. Now we will build on what we learned and develop a method, called the **balance method,** for solving more complex equations. To understand this method, imagine an equation as a balanced scale with the equal sign acting as the pivot point.

◀ **Note** The solution to $x + 7 = 9$ is 2 because $2 + 7 = 9$.

Answers to Your Turn 1
a. linear
b. linear
c. nonlinear
d. nonlinear
e. linear

Answers to Warm-up
1. -2 is a solution. 2. $y - 15$

Like a scale, adding an amount to one side unbalances the equation. For example, adding 4 to the left side of the equation tips it out of balance.

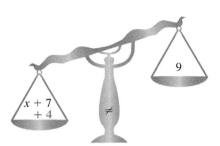

Note The equation is now unbalanced because its solution is no longer 2. Simplifying the left side gives $x + 11 = 9$, and 2 is no longer the solution because $2 + 11 \neq 9$.

However, if the same amount is added to *both* sides, like a scale, the equation stays balanced.

Note The equation is now balanced because its solution is still 2. Simplifying both sides gives $x + 11 = 13$, and 2 is the solution because $2 + 11 = 13$.

We say that the equation is balanced because adding 4 to both sides of the equation did not change the solution. This rule is called the **addition principle of equality**.

> ## The Addition Principle of Equality
> Adding the same amount to both sides of an equation does not change the equation's solution(s).

Connection The balance method is an extension of the related equations method we used in Chapters 1 and 2. Related equations method:

$$x + 7 = 9$$
$$x = 9 - 7$$
$$x = 2$$

With both methods, we subtract 7 from 9 to get the solution 2.

Note Adding an additive inverse is equivalent to subtraction; so we can also say that we subtract the term we want to eliminate from both sides.

Now let's use the principle more strategically to solve equations. When solving an equation in one variable, the goal is to manipulate the equation so that the variable is isolated on one side of the equation and the solution appears on the other side. To isolate x in $x + 7 = 9$, we need to eliminate 7; so let's add -7 to both sides.

$$
\begin{array}{rl}
x + 7 = & 9 \\
\underline{-7 \quad} & \underline{-7} \\
x + 0 = & 2 \\
x = & 2
\end{array}
$$

Note Adding -7 to both sides eliminates 7 from the left side, isolating the variable x. On the right side, we see the solution, 2.

Notice that $+7$ and -7 are additive inverses, which is the key to using the addition principle of equality.

> ## Procedure
> To use the addition principle of equality to eliminate a term from one side of an equation, add the additive inverse of that term to both sides of the equation.

Example 2 Solve and check.

a. $x - 8 = -15$

Solution: To isolate x, eliminate -8 by adding 8 to both sides.

> **Note** Because the sum of additive inverses is always 0 and the sum of the variable and 0 is always the variable, many people skip this step and go straight to the solution step. ▶

$$x - 8 = -15$$
$$\underline{+\ 8 \qquad +\ 8} \qquad \text{Add 8 to both sides.}$$
$$x + 0 = \quad -7$$
$$x = -7 \qquad \text{The sum of } x \text{ and 0 is } x.$$

Check: $x - 8 = -15$
$$-7 - 8 \overset{?}{=} -15 \qquad \text{In the original equation, replace } x \text{ with } -7.$$
$$-15 = -15 \qquad \text{True; so } -7 \text{ is the solution.}$$

b. $14 = m + 5$

Solution: To isolate m, eliminate 5 by adding -5 to both sides (or subtracting 5 from both sides).

$$14 = m + 5$$
$$\underline{-5 \qquad\quad -5} \qquad \text{Add } -5 \text{ to both sides. (Subtract 5 from both sides.)}$$
$$9 = m + 0$$
$$9 = m \qquad \text{The sum of } m \text{ and 0 is } m.$$

Check: $14 = m + 5$
$$14 \overset{?}{=} 9 + 5 \qquad \text{In the original equation, replace } m \text{ with 9.}$$
$$14 = 14 \qquad \text{True; so 9 is the solution.}$$

▶ **Do Your Turn 2**

Some equations contain expressions that can be simplified. If like terms appear on the same side of an equation, we combine the like terms before isolating the variable.

Example 3 Solve and check. $5n - 14 - 4n = -12 + 7$

Solution: On the left side of the equation, we have the like terms $5n$ and $-4n$. On the right side, we have the like terms -12 and 7. We will combine those like terms first, then isolate n.

$$5n - 14 - 4n = -12 + 7$$

Combine $5n$ and $-4n$ to get n. $n - 14 = -5$ Combine -12 and 7 to get -5.
$$\underline{+\ 14 \quad +\ 14} \qquad \text{Add 14 to both sides to isolate } n.$$
$$n + 0 = \quad 9$$
$$n = 9$$

Check: $5n - 14 - 4n = -12 + 7$
$$5(9) - 14 - 4(9) \overset{?}{=} -12 + 7 \qquad \text{In the original equation, replace } n \text{ with 9.}$$
$$45 - 14 - 36 \overset{?}{=} -5 \qquad \text{Simplify each side.}$$
$$31 - 36 \overset{?}{=} -5$$
$$-5 = -5 \qquad \text{True; so 9 is the solution.}$$

▶ **Do Your Turn 3**

▶ **Your Turn 2**

Solve and check.
a. $16 = x - 5$
b. $n + 13 = 20$
c. $-13 = y - 7$
d. $21 = -9 + k$

Learning Strategy

If you are a visual learner, try writing the amount added to or subtracted from both sides of an equation in a different color than the rest of the equation. These steps are shown in blue in the text to illustrate how this looks.

▶ **Your Turn 3**

Solve and check.
a. $4x + 5 - 3x = 12 - 13$
b. $25 - 18 = 6c + 2 - 5c$

Note It may be tempting to check a solution using an equation that occurs in one of the solution steps instead of the original equation. For example, in Example 3, it would be easier to check 9 using $n - 14 = -5$ instead of using ◀ $5n - 14 - 4n = -12 + 7$. However, remember that we are solving the original equation; so we always check using the original equation.

Answers to Your Turn 2
a. $x = 21$ b. $n = 7$
c. $y = -6$ d. $k = 30$

Answers to Your Turn 3
a. $x = -6$ b. $c = 5$

If an equation contains parentheses, we can use the distributive property to eliminate them. After using the distributive property, we usually have to simplify further by combining like terms that are on the same side of the equation.

► Your Turn 4

Solve and check.

a. $14 - 17 = 4(u - 3) - 3u$

b. $7h - 2(3h + 4) = 15 - 12$

Example 4   Solve and check. $3(y - 5) - 2y = -20 + 2$

Solution: Use the distributive property, combine like terms, and then isolate y.

$$3(y - 5) - 2y = -20 + 2$$

Distribute to clear parentheses. $$3y - 15 - 2y = -18$$ Combine -20 and 2 to get -18.

Combine $3y$ and $-2y$ to get y.

$$y - 15 = -18$$
$$\underline{+15 \qquad +15}$$ Add 15 to both sides to isolate y.
$$y + 0 = -3$$
$$y = -3$$

Check:
$$3(y - 5) - 2y = -20 + 2$$
$$3((-3) - 5) - 2(-3) \stackrel{?}{=} -20 + 2$$ In the original equation, replace y with -3.
$$3(-8) - 2(-3) \stackrel{?}{=} -18$$ Simplify each side.
$$-24 - (-6) \stackrel{?}{=} -18$$
$$-24 + 6 \stackrel{?}{=} -18$$
$$-18 = -18$$ True; so -3 is the solution.

◄ Do Your Turn 4

Objective 3 Solve application problems.

Let's put what we've learned about using the addition principle of equality into the context of solving problems involving an unknown addend. You may recall that the way to recognize a missing addend problem is to look for key words such as *How much more is needed?*

► Your Turn 5

Translate to an equation; then solve.

a. A patient must receive 350 cubic centimeters of a medication in three injections. He has received two injections of 110 cubic centimeters each. How much medication should the patient receive in the third injection?

b. Daryl has a balance of $-\$568$ on a credit card. How much must he pay to bring his balance to $-\$480$?

Example 5   Laura wants to buy a car stereo that costs $275. She currently has $142. How much more does she need?

Understand We are given the total required and the amount she currently has, and we must find how much she needs. Adding the amount needed to her current amount, $142, should equal the total required, $275. The amount needed is an unknown addend.

Plan We will translate to an equation with an unknown addend, where x represents the unknown addend, and then solve.

Execute Current amount + needed amount = 275

$$142 \qquad + \qquad x \qquad = 275$$
$$142 + x = 275$$
$$\underline{-142 \qquad -142}$$ Subtract 142 from both sides.
$$0 + x = 133$$
$$x = 133$$

Answers to Your Turn 4
a. $u = 9$ b. $h = 11$

Answers to Your Turn 5
a. $110 + 110 + x = 350$; 130 cc
b. $-568 + x = -480$; $88

Answer Laura needs $133 more to buy the stereo.

Check Does $142 plus the additional $133 equal $275?

$$142 + 133 \overset{?}{=} 275$$
$$275 = 275 \quad \text{It checks.}$$

▶ **Do Your Turn 5**

Of Interest

The word *algebra* first appeared in a book by Arab mathematician Mohammed ibn Musa al-Khowarizmi (about A.D. 825), entitled *al-jabr w'al-muqabalah*. The word *al-jabr* translates to *the reunification*. *Al-jabr* went through many spelling changes and finally emerged as our modern word *algebra*. When the Moors brought the word to Spain, it came to mean the reunification of broken bones, an *algebrista* being one who sets broken bones.

Source: D. E. Smith, *History of Mathematics*, 1953

4.2 Exercises For Extra Help MyMathLab®

Objective 1

Prep Exercise 1 Explain why $6x + 2 = 5x + 12$ is a linear equation.

Prep Exercise 2 Explain why $x^2 + 4 = 13$ is a nonlinear equation.

For Exercises 1–12, determine whether the given equation is linear or nonlinear.

1. $9x - 7 = 4x + 3$
2. $7y = 14$
3. $t^2 - 5 = 20$
4. $2n - 9 = n^3 + 6n - 1$
5. $5u - 17 = u^2 + 1$
6. $14n - 9n = 5n + 3$
7. $2x + y = 5$
8. $y = x^2 - 5$
9. $t = u^3 + 2$
10. $x = -6$
11. $y = 3x + 2$
12. $(y - 5) = 3(x - 1)$

Objective 2

Prep Exercise 3 What does the addition principle of equality say?

Prep Exercise 4 Explain how to use the addition principle of equality to solve $x + 9 = 16$.

For Exercises 13–32, solve and check.

13. $n + 14 = 20$
14. $25 = m + 9$
15. $10 = c + 17$
16. $d + 28 = 16$
17. $x + 10 = -7$
18. $-6 = y + 19$
19. $h - 14 = 2$
20. $13 = k - 6$
21. $-7 = x - 15$
22. $-5 = t - 11$
23. $10 = 5x - 2 - 4x$
24. $6y - 8 - 5y = 4$
25. $9y - 8y + 11 = 12 - 2$
26. $3r - 2r + 13 = 16 - 7$
27. $6 - 19 = 4y - 11 - 3y$
28. $2x - 3 - x = 9 - 17$
29. $7(m - 2) - 6m = -13 + 5$
30. $-19 + 12 = 5(n - 4) - 4n$
31. $-4 - 17 = 16y - 3(5y + 6)$
32. $13x - 4(3x + 2) = -6 - 13$

Objective 3

For Exercises 33–46, translate to an equation and solve.

33. Yolanda is to close on her new house in three weeks. The amount she must have at closing is $4768. She currently has $3295. How much more does she need?

34. Nikki has a balance of −$457 on her credit card. What payment should she make to get the balance to −$325?

35. A patient is to receive 450 cubic centimeters of a medication in three injections. The first injection is to be 200 cubic centimeters; the second injection, 180 cubic centimeters. How much must the third injection be?

36. An entry in a chemist's notebook is smudged. The entry is the initial temperature measurement of a chemical in an experiment. In the experiment, a substance was introduced to the chemical, decreasing its temperature by 19°C. The final temperature is listed as 165°C. What was the initial temperature?

37. Use the following graph, which shows the national average tuition and fees in public two-year colleges for each school year from 2004/2005 to 2008/2009.

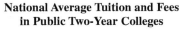
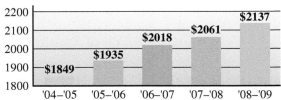

National Average Tuition and Fees in Public Two-Year Colleges

Source: U.S. Department of Education.

a. Complete the following table. In the row labeled "Equation," write an equation that can be used to find the amount of increase in tuition for the years indicated. Use the variable x to represent the increase in tuition. Solve the equation and write the amount the tuition increased in the row labeled "Tuition increase."

	'04/'05 to '05/'06	'05/'06 to '06/'07	'06/'07 to '07/'08	'07/'08 to '08/'09
Equation				
Tuition increase				

b. Which school year had the greatest increase in average tuition from the previous school year?

38. The following graph shows the low and high temperatures for each day from March 1 through March 4 in Fairbanks, Alaska.

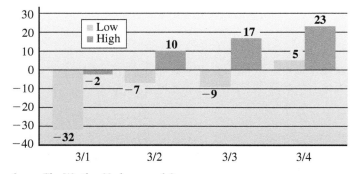

Source: The Weather Underground, Inc.

a. Complete the following table. In the row labeled "Equation," for each day, write an equation that can be used to find the temperature increase from the low to the high temperature. Use the variable t to represent the temperature increase for each day. Solve the equation and write the temperature increase in the row labeled "Temperature increase."

	March 1	March 2	March 3	March 4
Equation				
Temperature increase				

b. Which day had the greatest increase in temperature?

39. Jerry sells medical equipment. The company quota is set at $10,500 each month. The spreadsheet below shows Jerry's sales as of the end of the second week of July. How much more does Jerry need to sell to make the July quota? Do you think Jerry will meet the quota? Why or why not?

Date	Item No.	Quantity	Price	Total Sale
7/3	45079	2	$800	$1600
7/8	47002	1	$4500	$4500
7/13	39077	3	$645	$1935

40. Marc is a server at a restaurant. His rent of $675 is due at the end of the week. Currently, he has $487 in the bank. At the end of his shift, he finds that he made $85. How much more does he need in order to pay his rent? If he has two more shifts before rent is due, do you think he will make enough to pay the rent? Explain.

41. Connie is considering a new apartment. The kitchen and dining area are shown below. What is the length of the dining area? Is this area large enough to accommodate a 4-foot by 4-foot square table and four chairs?

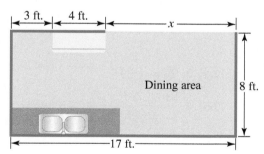

42. At an archaeological dig, the archaeologists suspect that there may be a secret passage or small chamber between two chambers of a tomb. The archaeologists have mapped and documented distances. What is the distance between the two chambers?

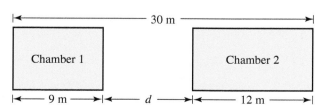

43. The perimeter of the triangle shown is 84 centimeters. Find the length of the missing side.

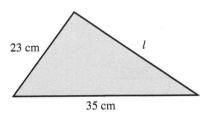

23 cm l

35 cm

44. The perimeter of the shape shown below is 255 inches. Find the length of the side labeled n.

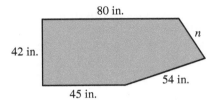

80 in.

42 in. n

45 in.   54 in.

45. For a person to take more than the standard deduction on taxes, his or her itemized deductions must exceed the standard deduction. Because the Colemans' filing status is married filing jointly, their standard deduction is $11,400 in 2010. If the Colemans already have the following itemized deductions, how much are they lacking to reach the level of the standard deduction?

    Medical and dental expenses: $2420
    Taxes paid: $2447
    Charitable donations: $1200

46. The Smith family will be filing their tax return jointly. They have the following itemized deductions. How much are they lacking to reach the level of the standard deduction, which is $11,400?

    Mortgage interest: $5600
    Medical and dental expenses: $1650
    Taxes paid: $1538
    Charitable donations: $450

Review Exercises

For Exercises 1–4, evaluate.

[2.4] **1.** $5(-8)$

[2.4] **2.** $24 \div (-3)$

[2.5] **3.** $12 + 3(-5 - 2)$

[2.5] **4.** $\dfrac{16 + 2(3 - 5)}{-2(5) + 4}$

[1.4] **5.** Solve and check. $5x = 15$

4.3 The Multiplication Principle of Equality

Warm-up *Solve and check.*

[4.2] **1.** $4(2n - 3) - 7n = -4$

[2.4] **2.** $6x = -30$

Objective 1 Solve equations using the multiplication principle of equality.

In Section 4.2, we learned that the addition principle of equality states that adding the same amount to or subtracting the same amount from both sides of an equation does not change the equation's solution. Likewise, we can multiply or divide both sides of an equation by the same number without affecting the equation's solution. This rule is called the **multiplication principle of equality.**

The Multiplication Principle of Equality

Multiplying (or dividing) both sides of an equation by the same amount does not change the equation's solution(s).

While the addition principle of equality is used to eliminate terms from one side of an equation, the multiplication principle of equality is used to eliminate a variable's coefficient. For example, to solve $2n = 10$, we need to eliminate the coefficient 2 to isolate the variable n. Dividing both sides of the equation by 2 eliminates the 2 but leaves the variable, n.

$$\frac{2n}{2} = \frac{10}{2} \qquad \text{Divide both sides by 2.}$$

$$1n = 5 \qquad \text{Divide.}$$

$$n = 5 \qquad \text{The product of 1 and } n \text{ is } n.$$

We can summarize how to use the multiplication principle of equality to eliminate a coefficient with the following procedure:

Procedure

To use the multiplication principle of equality to eliminate a coefficient, divide both sides of the equation by the coefficient.

Example 1   Solve and check.

a. $7x = 21$

Solution: To isolate x, eliminate the coefficient 7 by dividing both sides by 7.

$$\frac{7x}{7} = \frac{21}{7} \qquad \text{Divide both sides by 7.}$$

$$1x = 3 \qquad \text{Divide.}$$

$$x = 3 \qquad \text{The product of 1 and } x \text{ is } x.$$

Check: $\quad 7x = 21$

$\qquad 7(3) \stackrel{?}{=} 21 \qquad$ In the original equation, replace x with 3.

$\qquad\quad 21 = 21 \qquad$ True; so 3 is the solution.

Objectives

1 Solve equations using the multiplication principle of equality.

2 Solve equations using both the addition and multiplication principles of equality.

3 Solve application problems.

Connection In Chapter 5, we will see that dividing by 2 is equivalent to multiplying by $\frac{1}{2}$. Because division can be written as multiplication, it is simpler to call the principle the *multiplication* principle of equality.

Connection The balance method is an extension of the related equations method we used in Chapters 1 and 2. Related equations method:

$$7x = 21$$

$$x = 21 \div 7$$

$$x = 3$$

With both methods, we divide 21 by 7 to get the solution 3.

Answers to Warm-up
1. $n = 8$ 2. $x = -5$

▶ **Your Turn 1**

Solve and check.

a. $5y = 45$

b. $6n = -24$

c. $-2x = 16$

d. $-a = -6$

---

b. $-6m = 30$

Solution:

Note The quotient of a nonzero number and itself is always 1, and the product of that 1 and the variable is always the variable; so many people skip this step and go straight to the solution step.

▶

$$\frac{-6m}{-6} = \frac{30}{-6}$$ Divide both sides by -6.

$$1m = -5$$

$$m = -5$$

Check: $-6m = 30$

$$-6(-5) \stackrel{?}{=} 30 \quad \text{In the original equation, replace } m \text{ with } -5.$$

$$30 = 30 \quad \text{True; so } -5 \text{ is the solution.}$$

c. $-y = 8$

Solution: As we learned in Chapter 3, the minus sign in front of the variable is understood to mean that the coefficient is -1.

$$\frac{-1y}{-1} = \frac{8}{-1} \quad \text{Divide both sides by } -1.$$

$$1y = -8 \quad \text{Divide.}$$

$$y = -8 \quad \text{The product of 1 and } y \text{ is } y.$$

Check: $-1y = 8$

$$-1(-8) \stackrel{?}{=} 8 \quad \text{In the original equation, replace } y \text{ with the solution, } -8.$$

$$8 = 8 \quad \text{True; so } -8 \text{ is the solution.}$$

◀ **Do Your Turn 1**

Objective 2 Solve equations using both the addition and multiplication principles of equality.

Now let's solve equations that require both the addition principle and the multiplication principle of equality. First, we consider equations of the form $ax + b = c$. Our method is to use the addition principle of equality to isolate the variable term. Then we use the multiplication principle of equality to eliminate the variable's coefficient, which isolates the variable.

▶ **Your Turn 2**

Solve and check.

a. $4x - 1 = -21$

b. $-7b + 1 = 22$

c. $12 = 17 + 5x$

d. $-4 = 32 - 9y$

Example 2   Solve and check.

a. $-2x - 9 = 7$

Solution:

Connection After adding 9 to both sides, we have $-2x = 16$, which is the equation in Your Turn 1c.

▶

$$-2x - 9 = 7$$

$$\underline{+9 \quad +9} \quad \text{Add 9 to both sides to}$$

$$-2x + 0 = 16 \quad \text{isolate the } -2x \text{ term.}$$

$$-2x = 16$$

$$\frac{-2x}{-2} = \frac{16}{-2} \quad \text{Divide both sides by } -2 \text{ to}$$

$$\text{eliminate the } -2 \text{ coefficient.}$$

$$1x = -8$$

$$x = -8$$

Answers to Your Turn 1

a. $y = 9$ b. $n = -4$

c. $x = -8$ d. $a = 6$

Answers to Your Turn 2

a. $x = -5$ b. $b = -3$

c. $x = -1$ d. $y = 4$

Check: $-2x - 9 = 7$

$-2(-8) - 9 \overset{?}{=} 7$ In the original equation, replace x with -8.

$16 - 9 \overset{?}{=} 7$ Simplify.

$7 = 7$ True; so -8 is the solution.

b. $12 = 26 - 7x$

Solution: $12 = \quad 26 - 7x$

$\dfrac{-26 \qquad -26}{-14 = \quad 0 - 7x}$ Subtract 26 from both sides to isolate the $-7x$ term.

$\dfrac{-14}{-7} = \dfrac{-7x}{-7}$ Divide both sides by -7 to eliminate the -7 coefficient.

$2 = 1x$

$2 = x$

Check: $12 = 26 - 7x$

$12 \overset{?}{=} 26 - 7(2)$ In the original equation, replace x with 2.

$12 \overset{?}{=} 26 - 14$ Simplify.

$12 = 12$ True; so 2 is the solution.

▶ **Do Your Turn 2**

Suppose variable terms appear on both sides of the equation, as in $7x - 1 = 3x - 21$. We use the addition principle of equality to get the variable terms together on the same side of the equation so that it is in the form $ax + b = c$. We choose one of the variable terms (it does not matter which one) and then use the addition principle of equality to eliminate that term. Although it does not matter which term we choose, eliminating the variable term with the lesser coefficient has its advantages. The rest of the steps are just like Example 2.

Example 3 Solve and check. $7x - 1 = 3x - 21$

Solution: Because like variable terms appear on opposite sides of the equation, we use the addition principle of equality to clear one of them.

Note Choosing to eliminate $3x$ instead of $7x$ avoids getting a negative coefficient after the like terms are combined.

▶ $7x - 1 = 3x - 21$

$\dfrac{-3x \qquad\quad -3x}{4x - 1 = \quad 0 - 21}$ Subtract $3x$ from both sides.

$4x - 1 = -21$

$\dfrac{+1 \qquad +1}{4x + 0 = -20}$ Add 1 to both sides to isolate the $4x$ term.

$\dfrac{4x}{4} = \dfrac{-20}{4}$ Divide both sides by 4 to eliminate the 4 coefficient.

$1x = -5$

$x = -5$

Check: $7x - 1 = 3x - 21$

$7(-5) - 1 \overset{?}{=} 3(-5) - 21$ In the original equation, replace x with -5.

$-35 - 1 \overset{?}{=} -15 - 21$ Simplify.

$-36 = -36$ True; so -5 is the solution.

▶ **Your Turn 3**

Solve and check.

a. $9b + 2 = 5b + 26$

b. $n + 12 = 6 - 2n$

Connection After clearing the $3x$ term, we have the same equation as in Your Turn 2a.

Answers to Your Turn 3
a. $b = 6$ b. $n = -2$

Alternative Solution: Subtracting $7x$ from both sides instead of $3x$ as the first step does not change the solution. Afterwards, we add 21 to both sides and then divide both sides by -4 instead of 4 in the last step.

$$7x - 1 = 3x - 21$$

$$\underline{-7x \qquad\qquad -7x}$$ Subtract $7x$ from both sides.

$$0 - 1 = -4x - 21$$

$$-1 = -4x - 21$$

$$\underline{+21 \qquad\qquad +21}$$ Add 21 to both sides.

$$20 = -4x + 0$$

$$\frac{20}{-4} = \frac{-4x}{-4}$$ Divide both sides by -4.

$$-5 = 1x$$

$$-5 = x$$

◄ **Note** Choosing to eliminate $7x$ instead of $3x$ gives a negative coefficient after the like terms are combined.

◄ **Do Your Turn 3**

As we learned in Section 4.2, sometimes the expressions in an equation can be simplified before we use the addition principle of equality. First, let's consider equations that have like terms on the same side of the equation.

► **Your Turn 4**

Solve and check.

a. $5t + 13 - t = 3 + 6t - 12$

b. $15 - 2y + 6 = 12y + 30 - 5y$

Example 4   Solve and check. $9n + 6 - 2n = 16 + 10n - 4$

Solution: Because like terms are already on the same sides of the equation, we combine those like terms first.

$$9n + 6 - 2n = 16 + 10n - 4$$

$$7n + 6 = 12 + 10n$$ Combine like terms.

$$\underline{-7n \qquad\qquad -7n}$$ Subtract $7n$ from both sides.

$$0 + 6 = 12 + 3n$$

$$\underline{-12 \quad -12}$$ Subtract 12 from both sides to isolate $3n$.

$$-6 = 0 + 3n$$

$$\frac{-6}{3} = \frac{3n}{3}$$ Divide both sides by 3.

$$-2 = 1n$$

$$-2 = n$$

Check:

$$9n + 6 - 2n = 16 + 10n - 4$$

$$9(-2) + 6 - 2(-2) \overset{?}{=} 16 + 10(-2) - 4$$ In the original equation, replace n with -2.

$$-18 + 6 - (-4) \overset{?}{=} 16 + (-20) - 4$$

$$-12 - (-4) \overset{?}{=} -4 - 4$$

$$-12 + 4 \overset{?}{=} -8$$

$$-8 = -8$$ True; so -2 is the solution.

◄ **Do Your Turn 4**

Finally, we consider equations that contain parentheses.

Answers to Your Turn 4

a. $t = 11$ b. $y = -1$

Example 5   Solve and check. $15 - 2(y - 3) = 6(2y + 5) - 5y$

Solution: We simplify each side first by using the distributive property to eliminate the parentheses and combining like terms.

$$15 - 2(y - 3) = 6(2y + 5) - 5y$$

$15 - 2y + 6 = 12y + 30 - 5y$	Distribute -2 and 6.
$21 - 2y = 7y + 30$	Combine like terms.
$\underline{+2y \quad +2y}$	Add $2y$ to both sides.
$21 + 0 = 9y + 30$	
$\underline{-30 \qquad\qquad -30}$	Subtract 30 from both sides to isolate $9y$.
$-9 \quad = 9y + 0$	
$\dfrac{-9}{9} = \dfrac{9y}{9}$	Divide both sides by 9.
$-1 = 1y$	
$-1 = y$	

Connection After distributing, we have the same equation as in Your Turn 4b.

▶ **Your Turn 5**

Solve and check.

a. $3(3t - 5) + 13 = 4(t - 3)$

b. $9 - 5(x + 4) =$
    $7x - (6x + 5)$

Check:

$$15 - 2(y - 3) = 6(2y + 5) - 5y$$

$15 - 2((-1) - 3) \stackrel{?}{=} 6(2(-1) + 5) - 5(-1)$	In the original equation, replace y with -1.
$15 - 2(-4) \stackrel{?}{=} 6(-2 + 5) - 5(-1)$	Simplify.
$15 - (-8) \stackrel{?}{=} 6(3) - 5(-1)$	
$15 + 8 \stackrel{?}{=} 18 - (-5)$	
$23 \stackrel{?}{=} 18 + 5$	
$23 = 23$	True; so -1 is the solution.

▶ **Do Your Turn 5**

We can summarize everything we have learned about solving linear equations in one variable with the following procedure:

Procedure

To solve a linear equation in one variable:

1. Simplify each side of the equation.
   a. Distribute to eliminate parentheses.
   b. Combine like terms.
2. Use the addition principle of equality so that all variable terms are on one side of the equation and all constants are on the other side. Then combine like terms.

> **Note** Eliminate the variable term with the lesser coefficient to avoid negative coefficients.

3. Use the multiplication principle of equality to eliminate the coefficient of the variable.

Objective 3 Solve application problems.

In Section 4.2, we revisited many of the application problems that we solved earlier in the text and used the addition principle of equality to solve those problems.

Answers to Your Turn 5
a. $t = -2$ b. $x = -1$

Now we will solve application problems using the multiplication principle of equality. The formulas we will use involve multiplying variables.

$$\text{Area of a parallelogram: } A = bh$$
$$\text{Volume of a box: } V = lwh$$
$$\text{Distance: } d = rt$$
$$\text{Voltage: } V = ir$$

Notice that in all of these formulas, we multiply numbers to achieve the result. In fact, the area of a parallelogram, distance, and voltage are all mathematically identical because they are found by multiplying two numbers together. The only difference with volume is that we multiply three numbers.

► **Your Turn 6**

A building code in a certain region requires that upstairs windows have a minimum area of 864 square inches. Doris wants an upstairs window that is 18 inches wide. What must the length be?

Example 6 Juanita is driving at a rate of 59 feet per second (about 40 miles per hour). A sign is 177 feet ahead. How much time until she passes the sign?

Solution: $d = rt$

$177 = 59t$ In $d = rt$, replace d with 177 and r with 59, then solve for t.

$\dfrac{177}{59} = \dfrac{59t}{59}$ Divide both sides by 59 to isolate t.

$3 = 1t$

$3 = t$

Answer: Juanita passes the sign after 3 sec. (The time is in units of seconds because the rate was in terms of feet per *second*.)

◄ **Do Your Turn 6**

► **Your Turn 7**

Solve.

a. The length of a rectangle is 207 centimeters. Find the width of the rectangle if the perimeter is 640 centimeters.

b. The formula $C = 8f + 50$ describes the total cost of a fence with a gate, where C is the total cost and f is the number of feet to be enclosed. If Mario and Rosa have a budget of $1810, how many feet of fence can they afford?

▲

Note In the formula $C = 8f + 50$, the coefficient 8 represents the price per foot of fence. It costs $8 for each foot of fence. The 50 represents the cost of the gate that was mentioned. Therefore, the total cost is $8 times the number of feet of fence plus the $50 cost of the gate.

Answers to Your Turn 6
48 in.

Answers to Your Turn 7
a. 113 cm   **b.** 220 ft.

Some problems involve using both the addition and multiplication principles of equality. If a formula has multiplication *and* addition or subtraction, then we will likely have to use both principles to solve for the unknown amount. Some formulas where this would apply are listed below.

$$\text{Perimeter of a rectangle: } P = 2l + 2w$$
$$\text{Surface area of a box: } SA = 2lw + 2lh + 2wh$$

Example 7 The total material allotted for the construction of a metal box is 3950 square inches. The length is to be 25 inches, and the width is to be 40 inches. Find the height.

Solution:

$3950 = 2(25)(40) + 2(25)h + 2(40)h$ In $SA = 2lw + 2lh + 2wh$, replace SA with 3950, l with 25, and w with 40; then solve for h.

$3950 = 2000 + 50h + 80h$ Simplify.

$3950 = 2000 + 130h$ Combine like terms.

$\underline{-2000 \quad -2000}$ Subtract 2000 from both sides.

$1950 = \quad 0 + 130h$

$\dfrac{1950}{130} = \dfrac{130h}{130}$ Divide both sides by 130.

$15 = 1h$

$15 = h$

Answer: The height must be 15 in.

◄ **Do Your Turn 7**

4.3 Exercises For Extra Help MyMathLab®

Objective 1

Prep Exercise 1 What does the multiplication principle of equality say?

Prep Exercise 2 Explain how to use the multiplication principle of equality to solve $-3x = 18$.

For Exercises 1–10, solve using the multiplication principle and check.

1. $3x = 21$
2. $5t = 20$
3. $4a = -36$
4. $8m = -48$
5. $-7b = 77$
6. $-9x = 108$
7. $-14a = -154$
8. $-11c = -132$
9. $-12y = -72$
10. $-15n = 45$

Objective 2

Prep Exercise 3 What is the first step in solving $3x - 7 = 20$?

For Exercises 11–16, solve and check.

11. $2x + 9 = 23$
12. $5k - 4 = 31$
13. $29 = -6b - 13$
14. $45 = 15 - 10h$
15. $9u - 17 = -53$
16. $-7t + 11 = -73$

Prep Exercise 4 In solving $5x - 10 = 3x + 4$, why might adding $-3x$ to both sides be a better choice than adding $-5x$ to both sides of the equation?

For Exercises 17–24, solve and check.

17. $7x - 16 = 5x + 6$
18. $3b + 13 = 7b + 17$
19. $9k + 19 = -3k - 17$
20. $-4y + 15 = -9y - 20$
21. $-2h + 17 - 8h = 15 - 7h - 10$
22. $5 - 8x - 24 = 7x + 11 - 5x$
23. $13 - 9h - 15 = 7h + 22 - 8h$
24. $-4m - 7 + 13m = 12 + 2m - 5$

Prep Exercise 5 What is the first step in solving $4y + 2(y - 4) = 13 + 3(y - 5)$?

Prep Exercise 6 In $10x - (3x + 2) = 14x - 6 + x$, what number is distributed to eliminate the parentheses?

For Exercises 25–30, solve and check.

25. $3(t - 4) = 5(t + 1) - 1$
26. $-2(x + 7) - 9 = 5x + 12$
27. $9 - 2(3x + 5) = 4x + 3(x - 9)$
28. $11n - 4(2n - 3) = 18 + 5(n + 2)$
29. $15 + 3(3u - 1) = 16 - (2u - 7)$
30. $8y - (3y + 7) = 14 - 4(y - 6)$

For Exercises 31–34, the check for each equation indicates that a mistake was made. Find and correct the mistake.

31.
$$5x - 11 = 7x - 9$$
$$11 = 2x - 9$$
$$20 = 2x$$
$$\frac{20}{2} = \frac{2x}{2}$$
$$10 = x$$

Check:
$$5x - 11 = 7x - 9$$
$$5(10) - 11 \stackrel{?}{=} 7(10) - 9$$
$$50 - 11 \stackrel{?}{=} 70 - 9$$
$$39 \neq 61$$

32. $6y + 11 = -3y + 38$ Check: $6y + 11 = -3y + 38$
$\quad\quad \underline{-11\quad\quad\quad -11}$ $6(3) + 11 \overset{?}{=} -3(3) + 38$
$\quad 6y + 0 = -3y + 27$ $18 + 11 \overset{?}{=} 9 + 38$
$\quad\quad 6y = -3y + 27$ $29 \neq 47$
$\quad\quad 9y = \quad 0 + 27$
$\quad\quad \dfrac{9y}{9} = \dfrac{27}{9}$
$\quad\quad 1y = 3$
$\quad\quad y = 3$

33. $6 - 2(x + 5) = 3x - (x - 8)$ Check: $6 - 2(x + 5) = 3x - (x - 8)$
$\quad 6 - 2x - 10 = 3x - x - 8$ $6 - 2(1 + 5) \overset{?}{=} 3(1) - (1 - 8)$
$\quad\quad -4 - 2x = 2x - 8$ $6 - 2(6) \overset{?}{=} 3 - (-7)$
$\quad\quad \underline{+ 2x \quad + 2x}$ $6 - 12 \overset{?}{=} 3 + 7$
$\quad -4 + 0 = 4x - 8$ $-6 \neq 10$
$\quad\quad -4 = 4x - 8$
$\quad\quad 4 = 4x + 0$
$\quad\quad \dfrac{4}{4} = \dfrac{4x}{4}$
$\quad\quad 1 = 1x$
$\quad\quad 1 = x$

34. $2x + 3(x - 3) = 10x - (3x - 11)$ Check: $2x + 3(x - 3) = 10x - (3x - 11)$
$\quad 2x + 3x - 9 = 10x - 3x + 11$ $2(10) + 3(10 - 3) \overset{?}{=} 10(10) - (3(10) - 11)$
$\quad\quad 5x - 9 = 7x + 11$ $20 + 3(7) \overset{?}{=} 100 - (30 - 11)$
$\quad\quad -2x - 9 = 0 + 11$ $20 + 21 \overset{?}{=} 100 - 19$
$\quad\quad -2x - 9 = 11$ $41 \neq 81$
$\quad\quad \underline{+ 9 \quad + 9}$
$\quad -2x + 0 = 20$
$\quad\quad \dfrac{-2x}{2} = \dfrac{20}{2}$
$\quad\quad x = 10$

Objective 3

For Exercises 35–52, solve for the unknown amount.

35. A standard sheet of blueprint paper has an area of 864 square inches. If its length is 36 inches, what is its width? (Use $A = lw$.)

36. The Star Spangled Banner flown over Fort McHenry during the Battle of Baltimore in 1814 for which the song that became the U.S. national anthem was written has an area of 1260 square feet. If the width of the flag is 30 feet, what is the length? (Use $A = lw$.)

Of Interest

Major George Armistead was the commander of Fort McHenry during the battle for Baltimore in 1814. Despite heavy bombardment by British ships in the harbor, he refused to remove the flag. Francis Scott Key, an American being held aboard one of the British ships, was so moved by seeing the flag still flying in the morning that he wrote the poem that became the U.S. national anthem. The flag became an heirloom of the Armistead family and was displayed rarely until it was donated to the Smithsonian by Eben Appleton, George Armistead's grandson, in 1907.

37. King Tut's burial chamber is a box with a volume of 1683 cubic feet. The height of the chamber is 9 feet, and the width is 11 feet. Find the length of the chamber. (Use $V = lwh$.)

38. A fish tank is to have a volume of 4320 cubic inches. The length is to be 20 inches; the width, 12 inches. Find the height. (Use $V = lwh$.)

39. Margaret is driving along a highway at 30 feet per second. Her exit is 120 feet in front of her. How much time until she reaches the exit? (Use $d = rt$.)

40. The Jones family is planning a 455-mile trip. If they travel at an average speed of 65 miles per hour, what will their travel time be? (Use $d = rt$.)

41. The voltage in a circuit is −220 volts. If the current is −5 amps, find the resistance. (Use $V = ir$.)

42. The voltage in a circuit is −110 volts. If the resistance is 55 ohms, find the current. (Use $V = ir$.)

43. The U.S.S. *Arizona* memorial is a rectangle with a perimeter of 1218 feet. If the width of the memorial is 88 feet, find the length. (Use $P = 2l + 2w$.)

44. The reflecting pool in Washington, D.C., is a rectangle with a perimeter of 4392 feet. If the length of the pool is 2029 feet, find the width. (Use $P = 2l + 2w$.)

Of Interest

Completed in 1962, the U.S.S. *Arizona* memorial was built upon the central part of the sunken remains of the U.S.S. *Arizona* in Pearl Harbor, Hawaii. It was built to honor all those who died in the attack on Pearl Harbor on December 7, 1941. The attack prompted the United States to enter World War II.

45. The surface area of a large cardboard box is to be 11,232 square inches. The length is to be 36 inches, and the width is to be 28 inches. Find the height. (Use $SA = 2lw + 2lh + 2wh$.)

46. An aluminum box with a surface area of 392 square inches is used as a chassis for all of the electronics in an amplifier. If the length of the box is 18 inches and the height is 2 inches, what is the width? (Use $SA = 2lw + 2lh + 2wh$.)

47. The formula $C = 28h + 40$ describes the total cost for a plumber to visit a home, where C is the total cost and h is the number of hours on site. How long can the plumber work for Li if Li has $152?

48. The formula $C = 6h + 32$ describes the total cost of renting a tiller, where C is the total cost and h is the number of hours rented. How many hours can the tiller be rented for $50?

49. The formula $v = v_i + at$ describes the final velocity (speed) of an object after being accelerated, where v_i is the initial velocity, a is the acceleration, t is the number of seconds the object is accelerated, and v is the final velocity. Suppose the initial velocity of a car is 30 miles per hour and the car accelerates at a rate of 2 miles per hour per second. How long will it take the car to reach a speed of 40 miles per hour?

50. A sports car is said to be able to go from 0 to 60 miles per hour in 4 seconds. What acceleration is required? (Use $v = v_i + at$.)

51. The expression $54b + 1215$ describes the cost of materials for a certain computer chip, where b is the number of chips produced. The expression $25b + 4200$ describes the labor costs. How many chips can be produced with a total cost budget of $54,000?

52. Two different circuits operate in a television. The equation $V = 7i + 8 - 4i$ describes the output voltage of the first circuit. The equation $V = 3(3i - 1) - 13$ describes the output voltage of the second circuit. The variable i represents the current. The design calls for both circuits to have the same output voltage. What must the current in amperes be?

Puzzle Problem

The letters below correspond to digits. Find the digit that corresponds to each letter.

```
  F O R T Y
    T E N
+   T E N
---------
  S I X T Y
```

Review Exercises

[1.1] **1.** Write the word name for 6,784,209.

[3.6] **2.** Evaluate 7^0.

[3.1] **3.** Evaluate $x^2 - 9x + 7$ when $x = -3$.

[3.5] **4.** Find the prime factorization of 240.

[3.6] **5.** Factor. $24x^5 - 30x^4 + 18x^3$

4.4 Translating Word Sentences to Equations

Objective

1 Translate sentences to equations using key words; then solve.

Objective

1 Translate sentences to equations using key words; then solve.

Warm-up *[3.1] Translate to a variable expression.*
1. twelve less than a number
2. seven more than the product of three and r

Objective 1 Translate sentences to equations using key words; then solve.

In Section 3.1, we learned to translate word phrases to expressions. Once translated, the expression could be evaluated if we were given numbers for the variables. Now we will translate complete sentences to equations. After translating, we will solve the equation. Because we have learned to solve only equations containing addition, subtraction, and multiplication, we will translate sentences with only those key words. The following table reviews some basic phrases and their translations.

Note Remember, *some number* or *a number* means *an unknown number*; so we use a variable. We can select any variable we like.

Addition	Translation
the sum of x and three	$x + 3$
h plus k	$h + k$
t added to seven	$7 + t$
three more than some number	$n + 3$
y increased by two	$y + 2$

Subtraction	Translation
the difference of three and x	$3 - x$
h minus k	$h - k$
seven subtracted from t	$t - 7$
three less than a number	$n - 3$
two decreased by y	$2 - y$

Note Addition is commutative; so the order of the addends does not matter.

the sum of x and three can be writen as $x + 3$ or $3 + x$.

Note Subtraction is not commutative; so the order of the translation matters. We must translate each key phrase exactly as shown above. Notice that the amount before *subtracted from* and *less than* is the subtrahend; so it appears after the minus sign.

Multiplication	Translation
the product of three and x	$3x$
four times k	$4k$
some number multiplied by five	$5n$

Multiplication	Translation
twice a number	$2n$
triple some number	$3n$

Note Multiplication is commutative; so the order of the factors does not matter.

four times k can be $4k$ or $k \cdot 4$. However, remember that coefficients are written to the left of a variable; so $4k$ is more appropriate.

In addition to the key words for addition, subtraction, and multiplication from Section 3.1, we also need to know the key words that indicate an equal sign.

Key Words for an Equal Sign

is equal to	produces
is the same as	yields
is	results in

Answers to Warm-up
1. $n - 12$
2. $7 + 3r$

Procedure

To translate a word sentence to an equation, identify the unknown(s), constants, and key words, then write the corresponding symbolic form.

Example 1 Translate to an equation; then solve.

a. The sum of seventeen and a number is equal to fifteen.

Solution: Translate: <u>The sum of seventeen and a number</u> <u>is equal to</u> <u>fifteen</u>.

$$17 + n \qquad = \qquad 15$$

Solve: $17 + n = 15$

$\underline{-17= -17}$ Subtract 17 from both sides to isolate n.

$0 + n = -2$

$n = -2$

Check:

$17 + (-2) \overset{?}{=} 15$ **Verify that the sum of 17 and -2 is equal to 15.**

$15 = 15$ **True; so -2 is correct.**

◀ **Note** If you have translated correctly, the check relates to the original problem. In this case, the check answers this question: Is the sum of seventeen and negative two equal to fifteen? That is true; so -2 is the solution not only to the equation, but also to the original problem.

b. Twelve less than some number is ten.

Solution: Translate: <u>Twelve less than some number</u> <u>is</u> <u>ten</u>.

$$n - 12 \qquad = 10$$

Solve: $n - 12 = 10$

$\underline{+12 +12}$ **Add 12 to both sides to isolate n.**

$n + 0 = 22$

$n = 22$

◀ **Note** The key words *less than* require careful translation. In the sentence, the word *twelve* comes before the words *less than* and the words *some number* come after. This order is reversed in the translation.

Check: $22 - 12 \overset{?}{=} 10$ **Verify that 12 less than 22 is 10.**

$10 = 10$ **True; so 22 is correct.**

c. The product of seven and y is negative thirty-five.

Solution: Translate: <u>The product of seven and y</u> <u>is</u> negative thirty-five.

$$7 \ \cdot \ y = -35$$

Solve: $\dfrac{7y}{7} = \dfrac{-35}{7}$ **Divide both sides by 7 to isolate y.**

$1y = -5$

$y = -5$

Check: $7 \cdot (-5) \overset{?}{=} -35$ **Verify that the product of 7 and -5 is -35.**

$-35 = -35$ **True; so -5 is correct.**

▶ **Do Your Turn 1**

▶ **Your Turn 1**

Translate to an equation; then solve.

a. Fifteen more than a number is forty.

b. The difference of some number and twenty-one is negative six.

c. Twelve times some number is negative seventy-two.

Answers to Your Turn 1
a. $n + 15 = 40$; $n = 25$
b. $n - 21 = -6$; $n = 15$
c. $12n = -72$; $n = -6$

Now let's translate sentences that involve more than one operation.

Translate to an equation; then solve.

a. Eight less than the product of three and x is equal to thirteen.

b. Negative five times n plus eighteen is twenty-eight.

Example 2   Seven more than the product of three and r is equal to nineteen. Translate to an equation; then solve.

Solution: Translate: <u>Seven more than</u> the product of three and r <u>is equal to</u> nineteen.

$$7 + \qquad\qquad 3 \cdot r \qquad = \qquad 19$$

$$\text{Solve:} \quad 7 + 3r = 19$$

$$\underline{-7 \qquad\quad -7} \qquad \text{Subtract 7 from both sides to isolate } 3r.$$

$$0 + 3r = 12$$

$$\frac{3r}{3} = \frac{12}{3} \qquad \text{Divide both sides by 3 to isolate } r.$$

$$1r = 4$$

$$r = 4$$

Check: $7 + 3 \cdot 4 \overset{?}{=} 19$ Verify that 7 more than the product of 3 and 4 is equal to 19.

$7 + 12 \overset{?}{=} 19$ Multiply.

$19 = 19$ True; so 4 is correct.

◀ **Do Your Turn 2**

Finally, let's translate sentences that require parentheses.

Translate to an equation; then solve.

a. Negative six times the sum of y and three is equal to forty subtracted from five times y.

b. The difference of eight times b and nine is the same as triple the sum of b and two.

Example 3   Twice the difference of five and n is the same as seventeen subtracted from seven times n. Translate to an equation; then solve.

Solution: Translate:

<u>Twice</u> <u>the difference of five and n</u> <u>is the same as</u> <u>seventeen subtracted from</u> <u>seven times n.</u>

$$2 \cdot \qquad (5 - n) \qquad = \qquad 7n \quad - 17$$

$$\text{Solve:} \quad 2(5 - n) = 7n - 17$$

$$10 - 2n = 7n - 17 \qquad \text{Distribute 2.}$$

$$\underline{+2n \qquad\quad +2n} \qquad \text{Add } 2n \text{ to both sides.}$$

$$10 + 0 = 9n - 17$$

$$10 = 9n - 17$$

$$\underline{+17 \qquad\quad +17} \qquad \text{Add 17 to both sides to isolate } 9n.$$

$$27 = 9n + 0$$

$$\frac{27}{9} = \frac{9n}{9} \qquad \text{Divide both sides by 9 to isolate } n.$$

$$3 = 1n$$

$$3 = n$$

Check: $2(5 - 3) \overset{?}{=} 7(3) - 17$ Verify that twice the difference of 5 and 3 is the same as 17 subtracted from 7 times 3.

$2(2) \overset{?}{=} 21 - 17$

$4 = 4$ True; so 3 is correct.

◀ **Do Your Turn 3**

Answers to Your Turn 2

a. $3x - 8 = 13; x = 7$

b. $-5n + 18 = 28; n = -2$

Answers to Your Turn 3

a. $-6(y + 3) = 5y - 40; y = 2$

b. $8b - 9 = 3(b + 2); b = 3$

4.4 Exercises For Extra Help MyMathLab®

Objective 1

Prep Exercise 1 List five key words/phrases that indicate addition.

Prep Exercise 2 List five key words/phrases that indicate subtraction.

Prep Exercise 3 List three key words/phrases that indicate multiplication.

Prep Exercise 4 List three key words/phrases that indicate an equal sign.

For Exercises 1–24, translate to an equation and solve.

1. Five more than some number is equal to negative seven.

2. The difference of some number and nine is equal to four.

3. Six less than a number is fifteen.

4. The sum of a number and twelve is negative twenty-seven.

5. Some number increased by seventeen is negative eight.

6. Some number decreased by twenty-four is negative seven.

7. The product of negative three and some number is twenty-one.

8. Negative eight times some number is equal to forty.

9. A number multiplied by nine is negative thirty-six.

10. Negative seven times a number is equal to negative forty-two.

11. Four more than the product of five and x yields fourteen.

12. Eighteen subtracted from the product of negative seven and y is equal to three.

13. The difference of negative six times m and sixteen is fourteen.

14. The sum of eight times b and twenty is negative four.

15. Thirty-nine minus five times x is equal to the product of eight and x.

16. Forty less than the product of three and y is equal to seven times y.

17. The sum of seventeen and four times t is the same as the difference of six times t and nine.

18. The difference of ten times n and seven is equal to twenty-five less than the product of four and n.

19. Two times the difference of b and eight is equal to five plus nine times b.

20. Nine more than the product of eight and m is the same as three times the sum of m and thirteen.

21. Six times x plus five times the difference of x and seven is equal to nineteen minus the sum of x and six.

22. Two times r subtracted from seven times the sum of r and one is equal to three times the difference of r and five.

23. Fourteen less than negative eight times the difference of y and 3 is the same as the difference of y and 5 subtracted from the product of negative two and y.

24. The sum of n and three subtracted from twelve times n is the same as negative eleven plus the product of 2 and the difference of r and five.

For Exercises 25–30, explain the mistake in the translation. Then write the correct translation.

25. Seven less than some number is fifteen.
Translation: $7 - n = 15$

26. Nineteen subtracted from some number is eleven.
Translation: $19 - n = 11$

27. Two times the sum of x and thirteen is equal to negative ten.
Translation: $2x + 13 = -10$

28. Twelve subtracted from five times x is the same as three times the difference of x and four.
Translation: $5x - 12 = 3x - 4$

29. Sixteen decreased by the product of six and n is the same as twice the difference of n and four.
Translation: $6n - 16 = 2(n - 4)$

30. Negative three times the sum of y and five is equal to twice y minus the difference of y and one.
Translation: $-3(y + 5) = 2y - (1 - y)$

Review Exercises

[3.6] 1. Find the GCF of $30x^3y^5$ and $24xy^7z$.

[3.6] 2. Factor the polynomial. $18b^5 - 27b^3 + 54b^2$

[3.7] 3. Write an expression in simplest form for the perimeter of the rectangle.

w

$3w$

[3.1] 4. Evaluate the perimeter expression from Review Exercise 3 when w is 6 feet.

[1.6] 5. Jasmine sells two sizes of holly shrub, large and small. One day she sold 7 large holly shrubs at $30 each and some small holly shrubs at $20 each. If she sold a total of 16 holly shrubs, how many of the small shrubs did she sell? How much money did she make from the sale of the holly shrubs altogether?

4.5   Applications and Problem Solving

Warm-up *[4.4] Translate to an equation; then solve.*
1. Ten more than twice some number is thirty-four.
2. The sum of four times x and seven is the same as eight less than x.

Objective 1 Solve problems involving two unknown amounts.

Let's extend what we learned about translating sentences to problems that have two or more unknown amounts. In general, if there are two unknowns, there will be two relationships. We will use the following procedure as a guide.

Procedure

To solve problems with two or more unknowns:
1. Determine which unknown will be represented by a variable.

   Note Let the unknown that is acted on be represented by the variable.

2. Use one of the relationships to describe the other unknown(s) in terms of the variable.
3. Use the other relationship to write an equation.
4. Solve the equation.

Example 1 One number is six more than another number. The sum of the numbers is forty. Find the two numbers.

Understand We are to find two unknown numbers given two relationships.

Plan Select a variable for one of the unknowns, translate to an equation, and solve the equation.

Execute Use the first relationship to determine which unknown will be represented by the variable.

Relationship 1: One number is six more than another number.

$$\text{one number} = 6 + \text{another number}$$
$$= 6 + n$$

Note Because "another number" is the unknown that is acted on (6 is added to it), we let it be represented by n.

Now we can use the second relationship to write an equation.

Relationship 2: The sum of the numbers is forty.

$$\text{one number} + \text{another number} = 40$$

$$6 + n \quad + \quad\quad n \quad\quad = 40$$

$6 + 2n = 40$	Combine like terms.
$\underline{-6 \qquad\qquad -6}$	Subtract 6 from both sides.
$0 + 2n = 34$	
$\dfrac{2n}{2} = \dfrac{34}{2}$	Divide both sides by 2.
$1n = 17$	
$n = 17$	

▶ **Your Turn 1**

Translate to an equation; then solve.

a. One number is five more than another number. The sum of the numbers is forty-seven. Find the numbers.

b. One number is twice another. The sum of the numbers is thirty. Find the numbers.

Answers to Your Turn 1
a. $5 + n + n = 47$; 21 and 26
b. $2n + n = 30$; 10 and 20

Answers to Warm-up
1. $10 + 2n = 34$; $n = 12$
2. $4x + 7 = x - 8$; $x = -5$

Answer The second number, "another number," is 17. To find the first number, we use relationship 1.

$$\text{one number} = 6 + n$$
$$= 6 + 17 \quad \text{Replace } n \text{ with 17.}$$
$$= 23$$

Check Verify that 17 and 23 satisfy both relationships: 23 is 6 more than 17, and the sum of 23 and 17 is 40.

◀ **Do Your Turn 1**

Sometimes the relationships needed to solve a problem are not stated in an obvious manner. This often occurs in problems in which the definition of a geometry term provides the needed relationship. For example, *perimeter* indicates to add the lengths of the sides of the shape.

▶ **Your Turn 2**

Translate to an equation; then solve.

A TV is to have a rectangular screen with a length that is 5 inches less than twice the width and a perimeter of 44 inches. What are the dimensions of the screen?

Example 2   A carpenter is asked to make a rectangular frame out of an 8-foot strip of wood. The length of the frame must be three times the width. What must the dimensions be?

Understand Draw a picture.

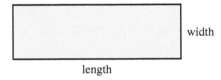

length

width

List the relationships:

Relationship 1: The length is three times the width.

Relationship 2: The total frame material is 8 feet. (This means that the perimeter must be 8 feet.)

There are two unknown amounts, the length and the width. Because we know how to solve linear equations in only one variable, we must choose the length or width to be the variable and write our equation in terms of that variable.

Plan Translate to an equation using the key words; then solve the equation.

Execute Remember that there are two amounts missing, length and width, and we'll have to select one of those amounts to be our variable.

Relationship 1: Length must be three times the width.

$$\text{length} = 3 \cdot \text{width}$$
$$= 3 \cdot w$$
$$= 3w$$

Note Because width is multiplied by 3, we will let width be represented by w ◀

Now we use the second relationship to write an equation to solve.

Relationship 2: The perimeter must be 8 feet.

Note Recall that *perimeter* indicates to add the lengths of all of the sides.

▶
$$\text{length} + \text{width} + \text{length} + \text{width} = 8$$
$$3w + w + 3w + w = 8$$
$$8w = 8 \quad \text{Combine like terms.}$$
$$\frac{8w}{8} = \frac{8}{8} \quad \text{Divide both sides by 8 to isolate } w.$$
$$1w = 1$$
$$w = 1$$

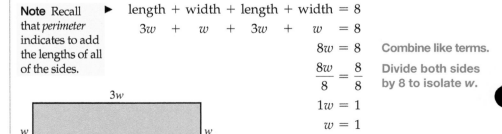

Answer to Your Turn 2
$2w - 5 + w + 2w - 5 + w = 44$;
9 in. wide by 13 in. long

Answer The width, w, is 1 ft. To determine the length, we use relationship 1.

$$\text{length} = 3w$$
$$= 3(1) \quad \text{Replace } w \text{ with 1.}$$
$$= 3$$

The frame must have a length of 3 ft. and a width of 1 ft.

Check Verify that the length is three times the width and that the perimeter is 8 feet. A length of 3 feet is three times a width of 1 foot. The perimeter is $3 + 1 + 3 + 1 = 8$ feet.

▶ **Do Your Turn 2**

Let's consider problems involving **isosceles** and **equilateral** triangles.

Definitions Isosceles triangle: A triangle with two sides of equal length.

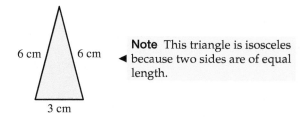

6 cm   6 cm

3 cm

Note This triangle is isosceles ◀ because two sides are of equal length.

Equilateral triangle: A triangle with all three sides of equal length.

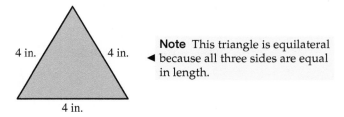

4 in.   4 in.

4 in.

Note This triangle is equilateral ◀ because all three sides are equal in length.

Example 3 Suppose the length of each equal-length side of an isosceles triangle is twice the length of the base. If the perimeter is 75 centimeters, what are the lengths of the base and the sides of equal length?

Understand Isosceles triangles have two sides of equal length.

Note These two sides are of equal length. We are told that the length of each is twice the length of the base.

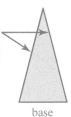

base

We are also told that the perimeter is 75 centimeters.

Plan Translate the relationships, write an equation, and solve.

▶ **Your Turn 3**

Translate to an equation; then solve.

Suppose each of the equal-length sides of an isosceles triangle is 3 inches more than the base. The perimeter is 30 inches. What are the lengths of the base and the sides of equal length?

Answer to Your Turn 3
$b + 3 + b + 3 + b = 30;$
8 in., 11 in., and 11 in.

Execute

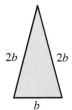

Relationship 1: The length of each equal-length side is twice the length of the base.

$$\text{each equal-length side} = 2 \cdot \textbf{length of the base}$$
$$= 2b$$

▲

Note Because the length of the base is acted on (multiplied by 2), we will use the variable b to represent its length.

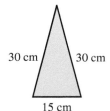

Now we use relationship 2 to write an equation we can solve.

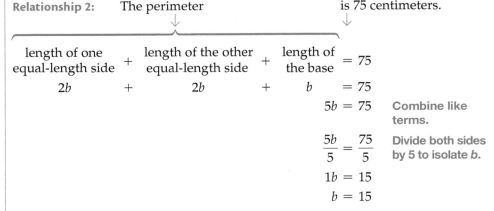

Relationship 2: The perimeter is 75 centimeters.

length of one equal-length side	+	length of the other equal-length side	+	length of the base	= 75	
$2b$	+	$2b$	+	b	= 75	

$$5b = 75 \quad \text{Combine like terms.}$$
$$\frac{5b}{5} = \frac{75}{5} \quad \text{Divide both sides by 5 to isolate } b.$$
$$1b = 15$$
$$b = 15$$

Answer The length of the base is 15 cm. To determine the length of each equal-length side, we use relationship 1.

$$\text{each equal-length side} = 2b$$
$$= 2(15) \quad \textbf{Replace } b \textbf{ with 15.}$$
$$= 30$$

The length of each equal-length side is 30 cm.

Check Because 30 centimeters is twice 15 centimeters, the equal-length sides are in fact twice the length of the base, which verifies relationship 1. The perimeter is $30 + 30 + 15 = 75$ centimeters, which verifies relationship 2.

◀ **Do Your Turn 3**

Let's consider problems involving angles. Angles are formed when two lines or line segments intersect or touch. Line segments AB and BC form an angle in the following figure. The measure of the angle is 40°.

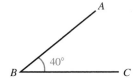

Note The symbol $\angle$ means *angle*. To name this angle, we can write all three letters with B in ◀ the center (for example, $\angle ABC$ or $\angle CBA$). Or we can simply write $\angle B$. To indicate the measure of the angle, we write $\angle B = 40°$.

Problems involving angle measurements may indicate that two unknown angles are **congruent, complementary,** or **supplementary.**

Definition Congruent angles: Angles that have the same measurement.

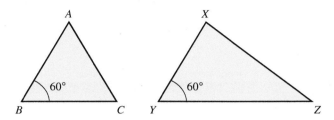

In the two triangles, $\angle B$ is congruent to $\angle Y$ because they have the same measurement of 60°. The symbol for congruent is $\cong$.

$$\angle B \cong \angle Y$$

Definition Complementary angles: Two angles whose measurements sum to 90°.

In the following figure, $\angle ABD$ and $\angle DBC$ are complementary angles because $32° + 58° = 90°$.

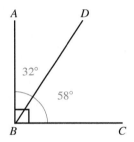

Definition Supplementary angles: Two angles whose measurements sum to 180°.

A straight line forms an angle that measures 180°. Any line that intersects a straight line cuts that 180° angle into two angles that are supplementary. For example, $\angle ABD$ and $\angle DBC$ are supplementary angles because $20° + 160° = 180°$.

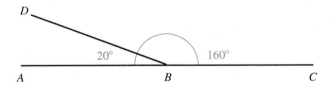

Let's look at a problem containing one of these terms.

Example 4   In designing a roof frame, the architect wants the angled beam to meet the horizontal truss so that the outer angle measurement is 30° more than twice the inner angle measurement. What are the angle measurements?

Understand We must find the inner and outer angle measurements. The sketch for this situation is shown. (Notice that it is similar to the preceding figure.)

▶ **Your Turn 4**

Translate to an equation; then solve.

a. Two angles are to be constructed from metal beams so that when they are joined, they form a 90° angle. One of the angles is to be 6° less than three times the other angle. What are the angle measurements?

b. Two electric doors are designed to swing open simultaneously when activated. The design calls for the doors to open so that the angle made between each door and the wall is the same. Because the doors are slightly different in weight, the expression $15i - 1$ describes the angle for one of the doors and $14i + 5$ describes the angle for the other door, where i represents the current supplied to the controlling motors. What current must be supplied so that the door angles are the same?

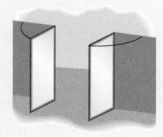

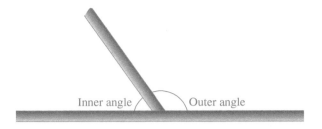

Inner angle   Outer angle

Relationship 1: The outer angle measurement is 30° more than twice the inner angle measurement.

Relationship 2: Looking at the picture, we see that the inner and outer angles are supplementary, which means that the sum of the measurements is 180°.

Plan Translate the relationships to an equation; then solve.

Execute

Relationship 1: The outer angle measurement is 30° more than twice the inner angle measurement.

outer angle $= 30 + 2 \cdot$ inner angle
$\qquad\qquad = 30 + 2a$

Note Because the inner angle measurement is multiplied by 2, we let the variable represent the inner angle measurement. Let's use the letter a.

Now we use relationship 2 to write an equation we can solve.

Relationship 2: inner angle $+$ outer angle $= 180$

$$a \quad + \quad 30 + 2a \;=\; 180$$
$$3a + 30 = 180 \qquad \text{Combine like terms.}$$
$$\underline{\;-30 \qquad -30\;} \qquad \begin{array}{l}\text{Subtract 30 from both sides}\\ \text{to isolate } 3a.\end{array}$$
$$3a + 0 = 150$$
$$\frac{3a}{3} = \frac{150}{3} \qquad \begin{array}{l}\text{Divide both sides by 3 to}\\ \text{isolate } a.\end{array}$$
$$1a = 50$$
$$a = 50$$

Answer Because a represents the inner angle measurement, we can say that the inner angle measures 50°. To find the outer angle measurement, we use relationship 1.

$$\text{outer angle} = 30 + 2a$$
$$= 30 + 2(50) \qquad \text{Replace } a \text{ with 50.}$$
$$= 30 + 100$$
$$= 130$$

The inner angle measurement is 50°, and the outer angle measurement is 130°.

Check Verify the relationships. The outer angle measuring 130° is 30° more than twice the inner angle measuring 50°. The sum of the inner and outer angle measurements is $50° + 130° = 180°$, which means that they are in fact supplementary.

◀ **Do Your Turn 4**

Answers to Your Turn 4
a. $3a - 6 + a = 90$; 24° and 66°
b. $15i - 1 = 14i + 5$; 6 A

Objective 2 Use a table in solving problems with two unknown amounts.

We have seen how drawing a picture, making a list, and using key words have helped us understand a problem. Let's now consider problems in which a table is helpful in organizing information. These problems will describe two items. We will be given the value of each item and asked to find the number of items that combine to yield a total amount. Our tables will have four columns, as shown below.

Categories	Value	Number	Amount

In the first column, categories, we describe each item. In the second column, value, we list the given value of each item. In the third column, number, we describe the number of each item. Each entry in the amount column is found using the following relationship:

$$\text{Value} \cdot \text{number} = \text{amount}$$

For the number column, you will be given either a relationship about the number of items or a total number of items to split. Let's consider an example to see how it all fits together.

Example 5 Salvador is an artist. A company produces prints of two of his paintings. The first print sells for $45; the second print, for $75. He is told that during the first day of sales, the company sold six more of the $75 prints than the $45 prints, with a total income of $1410. How many of each print were sold?

Understand We know that the total income is $1410. From this we can say the following:

$$\begin{matrix} \text{income from} \\ \text{first print} \end{matrix} + \begin{matrix} \text{income from} \\ \text{second print} \end{matrix} = 1410$$

How can we describe the income from each print? If we knew the number of prints sold, we could multiply that number times the selling price to get the income from that particular print.

$$\text{Selling price} \cdot \text{number of prints} = \text{income from the print}$$

Because the problem describes two items (two prints), gives their value (selling price), and asks us to find the number of each print, we will use a four-column table to organize the information. The amount column contains the income from each print.

Categories	Selling price	Number of prints	Income
First print	45	n	$45n$
Second print	75	$n + 6$	$75(n + 6)$

We were given these.

We selected n to represent the number of the first print, then translated *six more of the $75 print than the $45 print* to $n + 6$.

We multiplied straight across because:

$$\left(\begin{matrix} \text{selling} \\ \text{price} \end{matrix}\right) \cdot \left(\begin{matrix} \text{number} \\ \text{of prints} \end{matrix}\right) = \left(\begin{matrix} \text{income from} \\ \text{each print} \end{matrix}\right)$$

The expressions in the last column in the table describe the income from the sale of each print.

▶ **Your Turn 5**

Complete a four-column table, write an equation, and solve.

a. A marketing manager wants to research the sale of two different sizes of perfume. The small bottle sells for $35; the large bottle, for $50. The report indicates that the number of large bottles sold was 24 less than the number of small bottles sold and that the total sales were $4070. How many bottles of each size were sold?

b. A bottling company produces 12-ounce and 16-ounce bottles. In one day, the company produced three times as many 12-ounce bottles as 16-ounce bottles. If the company produced a total of 7280 ounces of the beverage, how many 12-ounce and 16-ounce bottles were filled?

Answers to Your Turn 5
a. $35n + 50(n - 24) = 4070$; 62 small bottles, 38 large bottles
b. $12(3x) + 16x = 7280$; 420 12-oz. bottles, 140 16-oz. bottles

Plan Translate the information in the table to an equation; then solve.

Execute Now we can use our initial relationship.

$$\begin{array}{l}\text{income from} \\ \text{first print}\end{array} + \begin{array}{l}\text{income from} \\ \text{second print}\end{array} = 1410$$

$$45n \;+\; 75(n+6) = 1410$$

$$45n + 75n + 450 = 1410 \qquad \text{Distribute.}$$

$$120n + 450 = 1410 \qquad \text{Combine like terms.}$$

$$120n + 450 = 1410$$

$$\underline{\quad -450 \quad -450} \qquad \text{Subtract 450 from both sides.}$$

$$120n + \quad 0 = \quad 960$$

$$\frac{120n}{120} = \frac{960}{120} \qquad \text{Divide both sides by 120 to isolate } n.$$

$$1n = 8$$

$$n = 8$$

Answer Use the relationships from the table about the number of each print. Because n represents the number of \$45 prints, 8 of the \$45 prints were sold. Because $n + 6$ represents the number of \$75 prints, $8 + 6 = 14$ of the \$75 prints were sold.

Check The number of \$75 prints sold, 14, is six more than the number of \$45 prints sold, 6. Also, 8 of the \$45 prints and 14 of the \$75 prints make a total income of \$1410.

$$\text{Total income} = 8(45) + 14(75)$$

$$= 360 + 1050$$

$$= 1410 \qquad \text{It checks.}$$

◀ **Do Your Turn 5**

▶ **Your Turn 6**

Complete a four-column table, write an equation, and solve.

a. A farmer has a total of 17 pigs and chickens. The combined number of legs is 58. How many pigs and how many chickens are there?

b. A tire factory produces two different-size tires. Tire A costs \$37 to produce while tire B costs \$42. At the end of the day, management gets the following report:

tires produced = 355
total cost = \$13,860

How many of each tire were produced?

Answers to Your Turn 6
a. $4n + 2(17 - n) = 58$;
   12 pigs, 5 chickens
b. $37(355 - B) + 42B = 13,860$;
   210 of tire A, 145 of tire B

Example 6 Jasmine sells two sizes of holly shrubs. The larger size sells for \$30; the smaller size, for \$20. She knows that she sold 16 shrubs for a total of \$390, but she forgot how many of each size. How many of each size did she sell?

Understand We are given the price or value of two different-sized shrubs. We are also given a total amount of income from the sale of the two different shrubs and a total number of shrubs to split up. We can say:

income from small + income from large = total income

Because the problem describes two items (large and small shrubs), gives their value (selling price), and asks us to find the number of each size shrub, we will use a four-column table to organize the information. The tricky part is filling in the number column.

Because Jasmine sold a total of 16 shrubs, we can write the following equation:

number of small + number of large = 16

If we knew one of the numbers, we could find the other by subtracting it from 16. We can write a related subtraction equation.

number of small = 16 − number of large

◀ **Note** We could also say:

$$\begin{array}{l}\text{number} \\ \text{of} \\ \text{large}\end{array} = 16 - \begin{array}{l}\text{number} \\ \text{of} \\ \text{small}\end{array}$$

Either way works.

Choosing L to represent the number of large shrubs, we can complete the table.

Categories	Selling price	Number	Income
Small	20	$16 - L$	$20(16 - L)$
Large	30	L	$30L$

We were given these.

In general, if you are given a total number of items, select one of the categories to be the variable. The other will be:

total number − variable

Because selling price · number = income, we multiply straight across the columns to get each expression in the income column.

Plan Translate to an equation; then solve.

Execute We can now use our initial relationship.

$$\text{income from small} + \text{income from large} = \text{total income}$$

$$20(16 - L) \quad + \quad 30L \quad = 390$$

$$320 - 20L + 30L = 390 \qquad \text{Distribute.}$$

$$320 + 10L = 390 \qquad \text{Combine like terms.}$$

$$320 + 10L = 390$$

$$\underline{-320 \qquad\qquad -320} \qquad \text{Subtract 320 from both}$$
$$0 + 10L = 70 \qquad\quad \text{sides to isolate } 10L.$$

$$\frac{10L}{10} = \frac{70}{10} \qquad \text{Divide both sides by}$$
$$\qquad\qquad\qquad 10 \text{ to isolate } L.$$

$$1L = 7$$

$$L = 7$$

Answer Number of large shrubs sold: $L = 7$

Number of small shrubs sold: $16 - L = 16 - 7 = 9$

Check Verify that Jasmine sold 16 shrubs for $390. Because she sold 7 large-size shrubs and 9 small-size shrubs, she did indeed sell 16 shrubs. Now verify that 7 large-size shrubs at $30 each and 9 small-size shrubs at $20 each is a total income of $390.

$$\text{total income} = 7(30) + 9(20)$$

$$= 210 + 180$$

$$= 390 \qquad\qquad \text{It checks.}$$

▶ **Do Your Turn 6**

Tips for Table Problems

You could be given a relationship about the number of items in each category, as in Example 5, or a total number of items to split up, as in Example 6. If you are given a relationship, use key words to translate to expressions. If you are given a total number of items to split up, choose one of the categories to be the variable; the other will be total number − variable.

4.5 Exercises
For Extra Help   MyMathLab®

Objective 1

Prep Exercise 1 Suppose a problem gives the following information: The length of a rectangle is five more than the width. Which is easier, representing the length or the width with a variable? What expression describes the other unknown?

For Exercises 1–22, translate to an equation and solve.

1. One number is four more than a second number. The sum of the numbers is thirty-four. Find the numbers.

2. One number is eight less than another number. The sum of the numbers is seventy-two. Find the numbers.

3. One number is three times another number. The sum of the numbers is thirty-two. Find the numbers.

4. One number is four times another number. The sum of the numbers is thirty-five. Find the numbers.

5. An architect believes that the optimal design for a new building is a rectangular shape where the length is four times the width. Budget restrictions force the building perimeter to be 300 feet. What will be the dimensions of the building?

6. Miriam hangs 84 feet of wallpaper border around a rectangular room. The width of the room is 4 feet less than the length. What are the dimensions?

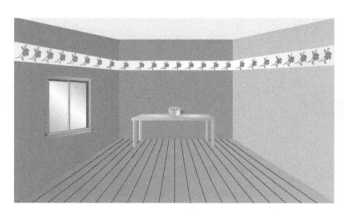

7. After completing the wood trim on her kitchen cabinets, Candice has 48 inches of trim left. She decides to make a picture frame out of the wood. She wants to make the frame so that the length is 4 inches less than three times the width. What must the dimensions be?

8. The Johnsons can afford 220 feet of fence material. They want to fence in a rectangular area with a length that is 20 feet more than twice the width. What will the dimensions be?

Prep Exercise 2 What is an equilateral triangle?

Prep Exercise 3 What is an isosceles triangle?

9. A frame section for the roof of a building is an isosceles triangle. The length of each equal-length side is 11 meters less than twice the length of the base. The perimeter is 18 meters. What are the lengths of the base and the sides of equal length?

10. A section of a public park is in the shape of an isosceles triangle. The length of each equal-length side is three times the length of the base. If the perimeter is 210 feet, find the lengths of the base and the sides of equal length.

11. A corporate logo has an equilateral triangle sitting atop a rectangle. The length of each side of the equilateral triangle is equal to the length of the rectangle. The width of the rectangle is 10 feet less than three times the length. The perimeter of the triangle is equal to the perimeter of the rectangle. What are the dimensions of each shape?

12. A refrigerator magnet is to have the shape of an isosceles triangle sitting atop a rectangle. The length of the base of the triangle is equal to the width of the rectangle. The other two sides of the triangle are 2 centimeters more than the base. The length of the rectangle is twice its width. If the perimeter of the magnet is 39 centimeters, what are all of the dimensions?

Prep Exercise 4 What are complementary angles?

Prep Exercise 5 What are supplementary angles?

Prep Exercise 6 Suppose a problem gives the following information: One angle is 20° less than a second angle. Which is easier, representing the first angle measurement or the second angle measurement with a variable? What expression describes the other unknown?

13. A security laser is placed 3 feet above the floor in the corner of a rectangular room. It is aimed so that its beam contacts a detector 3 feet above the floor in the opposite corner. The beam forms two angles in the corner as shown. The measure of the angle on one side of the beam is four times the measure of the angle on the other side. Find the angle measurements. (Assume that the angle formed by the walls is 90°.)

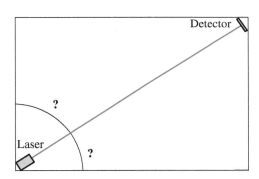

14. A security camera is placed in the corner of a room in a museum. The camera is set so that the measure of the angle between it and one of the walls is 16° less than the measure of the angle made with the other wall. Find the angle measurements. (Assume that the angle formed by the walls is 90°.)

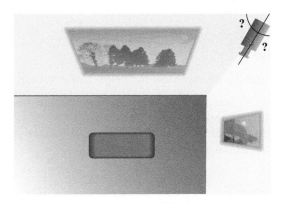

15. In an experiment, a laser contacts a flat detector so that one angle between the laser beam and the detector measures 15° more than four times the measure of the other angle. What are the two angle measurements?

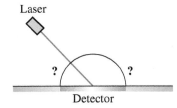

16. A suspension bridge has a steel beam that connects to the flat surface of the bridge, forming two angles. The measure of the smaller of the two angles is 38° less than the measure of the other angle. What are the two angle measurements?

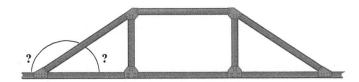

17. A laser beam strikes a flat mirror, forming an angle with the mirror. The beam is deflected at the same angle. The angle between the entering and exiting beams measures 10° more than three times the measure of the angle formed between the beam and the mirror. What are the three angle measurements?

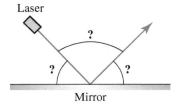

18. A billiard ball strikes the bumper at an angle. Because the ball has side spin, it is deflected at an angle that measures 6° less than the initial angle. The measure of the angle formed between the initial and deflected lines of travel is 10° more than twice the measure of the initial angle. What are the three angle measurements?

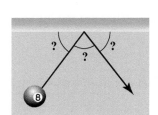

19. The sum of the angle measurements in any triangle is 180°. One of the angles measures 43°; the measure of the second is two more than twice the measure of the third. Find the two unknown angle measurements.

20. The sum of the angle measurements in any triangle is 180°. Suppose one of the angle measurements of a triangle measures 5° more than three times the measure of another angle. The third angle measures 115°. Find the two unknown angle measurements.

21. A decorative shelf is to be in the shape of a triangle, with the second angle measuring 10° more than the first and the third angle measuring 7° less than the first. What are the three angle measurements?

22. A mounting apparatus for a telescope is to be in the shape of a triangle. The measure of the second angle in the triangle is to be twice the measure of the first angle. The third angle is to measure 12° less than the first. What are the three angle measurements?

Objective 2

For Exercises 23–26, complete each table. Do not solve.

23. At a craft fair, Ellyn sold ornaments that she had created. She sold small ornaments for $3 and large ornaments for $7. In one day, she sold 8 more small ornaments than large ornaments.

Categories	Selling price	Number of ornaments sold	Income
Small ornament			
Large ornament			

24. A movie theater sells two sizes of popcorn: large boxes and small boxes. A large box sells for $8 and a small box, for $6. In one day, the theater sold 50 more small boxes of popcorn than large boxes.

Categories	Selling price	Number of boxes sold	Income
Small box			
Large box			

25. Ellyn from Exercise 23 also sold charms for necklaces. The large-sized charms sold for $6; the small-sized charms sold for $4. In one day, she sold 36 charms.

Categories	Selling price	Number of charms sold	Income
Small charm			
Large charm			

26. The movie theater in Exercise 24 offers a discount for children under 12. The regular price of a ticket is $10, and the children's ticket is $6. For one movie, the theater sold 45 tickets.

Categories	Selling price	Number of tickets sold	Income
Regular ticket			
Children's ticket			

For Exercises 27–34, complete a four-column table, write an equation, and solve.

27. Monique sells makeup products. One of the products is a lotion that comes in two different-size bottles. The smaller-size bottle sells for $8; the larger-size bottle, for $12. She remembers that she sold 5 more smaller bottles than larger bottles that day but doesn't remember exactly how many of each. She also remembers that the total sales of the lotion was $260. How many of each size did she sell?

Categories	Value	Number	Amount

28. Byron is a wedding photographer. He develops two different-size prints for a newlywed couple. He sells the large prints for $12 each and the small prints for $7 each. He develops twice as many small prints as large prints at a total cost of $156. How many of each size print did he sell?

Categories	Value	Number	Amount

29. Ian has some $5 bills and $10 bills in his wallet. If he has 19 bills worth a total of $125, how many of each bill is in his wallet?

Categories	Value	Number	Amount

30. A paper mill produces two different sizes of paper. It costs $5 to produce a box of the smaller size and $7 to produce a box of the larger size. Management gets a report that 455 boxes were produced at a total cost of $2561. How many boxes of each size were produced?

Categories	Value	Number	Amount

31. Nina has two different types of foreign bills to exchange for American currency. She is told that the more valuable bill is worth three times that of the other bill. She has 12 of the more valuable bills and 18 of the other bills and receives a total of $108. What is the value of each foreign bill?

32. Cal has two different stocks. He notices in the newspaper that one of his stocks is $3 more valuable than the other. He has 24 shares of the more valuable stock and 22 shares of the other stock. His total assets in the stocks are $348. How much is each stock worth?

33. A pet store sells cockatiels at two different prices. Hand-fed cockatiels are more expensive than parent-raised cockatiels. If a customer buys one of each, the combined price is $205. In one month, the store sold 6 hand-fed cockatiels and 9 parent-raised cockatiels for a total of $1470. What is the price of each?

34. Andre sells furniture. Two different chairs go with a dining set, a chair with arms or a chair without arms. One armchair and one chair without arms sell together for $350. One day he sold 5 armchairs and 17 chairs without arms for a total of $3250. How much does each chair sell for?

Puzzle Problem   At 3:00:05 P.M., what are the approximate angles made between the second hand and the minute and hour hands?

Review Exercises

[4.1] 1. Is -3 a solution for $4x - 8 = 5x - 6$?

[4.2] 2. Is $x^2 + 5 = 9$ a linear equation? Explain.

For Exercises 3–5, solve and check.

$\begin{bmatrix} 2.3 \\ 4.2 \end{bmatrix}$ **3.** $m + 19 = 12$

$\begin{bmatrix} 2.4 \\ 4.3 \end{bmatrix}$ **4.** $-15n = 75$

[4.3] 5. $9x - 7(x + 2) = -3x + 1$

Chapter 4 Summary and Review Exercises

Learning Strategy

If I have a reliable friend in the class who is equally as prepared as I am, we study together by quizzing each other as if we are instructors explaining concepts to new students. We try to have these study sessions two nights before a test so that we have time to study more if we find our knowledge lacking in a particular area.

—Alison R.

4.1 Equations and Their Solutions

Definitions/Rules/Procedures	Key Example(s)
To **solve** an equation means to find the _____ or _____ for the equation. A **solution** is a number that makes an equation_____ when it replaces the variable in the equation. **To check to see if a number is a solution for an equation:** 1. _____ the variable(s) with the number. 2. Simplify both sides of the equation as needed. If the resulting equation is _____, the number is a solution.	Check to see if -2 is a solution for $x - 8x = 14$. $$x - 8x = 14$$ $(-2)-8(-2) \overset{?}{=} 14$ Replace x with -2. $(-2)-(-16) \overset{?}{=} 14$ $-2 + 16 \overset{?}{=} 14$ $14 = 14$ The resulting equation is true; so -2 is a solution. Check to see if 4 is a solution for $2y - 9 = 6y - 23$. $$2y - 9 = 6y - 23$$ $2(4) - 9 \overset{?}{=} 6(4) - 23$ Replace y with 4. $8 - 9 \overset{?}{=} 24 - 23$ $-1 \neq 1$ The resulting equation is false; so 4 is not a solution.

[4.1] For Exercises 1–4, check to see if the given number is a solution for the given equation.

1. $-7n + 12 = 5$; check $n = 1$

2. $14 - 4y = 3y$; check $y = -2$

3. $9r - 17 = -r + 23$; check $r = 5$

4. $5(x - 2) - 3 = 10 - 4(x - 1)$; check $x = 3$

4.2 The Addition Principle of Equality

Definitions/Rules/Procedures	Key Example(s)
A **linear equation** is an equation in which each variable term is a monomial of degree _____. The **addition principle of equality** says that adding _____ of an equation does not change the equation's solution(s). **To use the addition principle of equality to eliminate a term from one side of an equation**, add the _____ of that term to both sides of the equation.	$2x + 3 = 15x - 8$ is linear because each variable term is a monomial of degree 1. $5x^2 + 2 = 9$ is not linear because the degree of $5x^2$ is 2. Solve and check. $t + 15 = 7$ $t + 15 = 7$ $\underline{-15 \quad -15}$ Add -15 to (or subtract 15 from) both sides. $t + 0 = -8$ $t = -8$ **Check:** $t + 15 = 7$ $(-8) + 15 \overset{?}{=} 7$ $7 = 7$ True; so -8 is the solution.

[4.2] For Exercises 5 and 6, solve and check.

5. $n + 19 = 27$

6. $y - 6 = -8$

[4.2] For Exercises 7 and 8, translate to an equation and solve.

7. The temperature at 5 A.M. was reported to be $-15°F$. By 11 A.M., the temperature had risen to $28°F$. How much did the temperature rise?

8. Kari has a balance of $-\$547$ on a credit card. How much should she pay on the account to get a balance of $-\$350$?

4.3 The Multiplication Principle of Equality

Definitions/Rules/Procedures	Key Example(s)
The **multiplication principle of equality** says that multiplying (or dividing) both sides of an equation by _____ does not change the equation's solution(s).	Solve and check.

To use the multiplication principle of equality to eliminate a coefficient, _____ both sides of the equation by the coefficient.

To solve a linear equation in one variable:

1. Simplify each side of the equation.

   a. _____ to eliminate parentheses.

   b. _____ like terms.

2. Use the addition principle of equality so that all _____ are on one side of the equation and all constants are on the other side. Then combine like terms.

> **Note** Eliminate the variable term with the lesser coefficient to avoid negative coefficients.

3. Use the multiplication principle of equality to eliminate the _____ of the variable.

a. $-7x = 42$

$$\frac{-7x}{-7} = \frac{42}{-7} \quad \text{Divide both sides by } -7.$$

$$1x = -6$$

$$x = -6$$

Check: $\quad -7x = 42$

$$-7(-6) \overset{?}{=} 42$$

$$42 = 42 \quad \text{True; so } -6 \text{ is the solution.}$$

b. $12 - (2x + 7) = 13x - 5(x - 7)$

$$12 - 2x - 7 = 13x - 5x + 35 \quad \text{Distribute.}$$

$$5 - 2x = 8x + 35 \quad \text{Combine like terms.}$$

$$\underline{+2x \quad +2x} \quad \text{Add 2x to both sides.}$$

$$5 + 0 = 10x + 35$$

$$5 = 10x + 35$$

$$\underline{-35 \qquad\qquad -35} \quad \text{Subtract 35 from both sides.}$$

$$-30 = 10x + 0$$

$$\frac{-30}{10} = \frac{10x}{10} \quad \text{Divide both sides by 10.}$$

$$-3 = 1x$$

$$-3 = x$$

Check:

$$12 - (2x + 7) = 13x - 5(x - 7)$$

$$12 - (2(-3) + 7) \overset{?}{=} 13(-3) - 5(-3 - 7)$$

$$12 - (-6 + 7) \overset{?}{=} 13(-3) - 5(-10)$$

$$12 - 1 \overset{?}{=} -39 - (-50)$$

$$11 \overset{?}{=} -39 + 50$$

$$11 = 11 \quad \text{True; so } -3 \text{ is the solution.}$$

[4.3] For Exercises 9–18, solve and check.

9. $-4k = 36$

10. $-8m = -24$

11. $2x + 11 = -3$

12. $-3h - 8 = 19$

13. $9t - 14 = 3t + 4$

14. $20 - 5y = 34 + 2y$

15. $10m - 17 + m = 2 + 12m - 20$

16. $-16v - 18 + 7v = 24 - v + 6$

17. $4x - 3(x + 5) = 16 - 7(x + 1)$

18. $9y - (2y + 3) = 4(y - 5) + 2$

[4.3] For Exercises 19 and 20, solve for the unknown amount.

19. A swimming pool is to be designed to have a volume of 5000 cubic feet. The length of the pool is to be 50 feet, and the width is to be 25 feet. Find the depth (height). (Use $V = lwh$.)

20. A box company shapes flat pieces of cardboard into boxes. The flat pieces have a total surface area of 136 square feet. Assuming that the length is 3 feet and the height is 4 feet, find the width. (Use $SA = 2lw + 2lh + 2wh$.)

4.4 Translating Word Sentences to Equations

Definitions/Rules/Procedures	Key Example(s)
To translate a word sentence to an equation, identify the _____, constants, and key words, then write the corresponding symbolic form.	Translate to an equation; then solve. Negative six added to four times some number is equal to twenty-two. $$\text{Translation: } -6 + 4n = 22$$ Solve: $\underline{+6}$ $\underline{+6}$ Add 6 to both sides. $$0 + 4n = 28$$ $$\frac{4n}{4} = \frac{28}{4} \quad \text{Divide both sides by 4.}$$ $$1n = 7$$ $$n = 7$$

[4.4] For Exercises 21–24, translate to an equation and solve.

21. Fifteen minus seven times a number is twenty-two.

22. The difference of five times x and four is the same as the product of three and x.

23. Twice the sum of n and twelve is equal to eight less than negative six times n.

24. Twelve minus three times the difference of x and seven is the same as three less than the product of six and x.

4.5 Applications and Problem Solving

Definitions/Rules/Procedures	Key Example(s)
To solve problems with two or more unknowns: 1. Determine which unknown will be represented by a variable. **Note** Let the unknown that is _____ be represented by the variable. 2. Use one of the relationships to describe the other unknown(s) in terms of the variable. 3. Use the other relationship to write an equation. 4. Solve the equation.	The length of a rectangular yard is 15 feet more than the width. Assuming that the perimeter of the yard is 150 feet, find the length and the width. **Solution:** Relationship 1: The length is 15 feet more than the width. Translation: length $= 15 + w$ Note that w represents the width.

(continued)

Definitions/Rules/Procedures (*continued*)	Key Example(s)
An **isosceles triangle** is a triangle with _____ sides of equal length.	Relationship 2: The perimeter is 150 feet.
An **equilateral triangle** is a triangle with _____ sides of equal length.	Translation:

An **isosceles triangle** is a triangle with _____ sides of equal length.

An **equilateral triangle** is a triangle with _____ sides of equal length.

Congruent angles have _____ measurement.

Complementary angles are two angles whose measurements sum to _____.

Supplementary angles are two angles whose measurements sum to _____.

Relationship 2: The perimeter is 150 feet.

Translation:

$$\text{length} + \text{width} + \text{length} + \text{width} = 150$$
$$15 + w + w + 15 + w + w = 150$$

$30 + 4w = 150$	Combine like terms.
$\underline{-30 \qquad\qquad -30}$	Subtract 30 from both sides.
$0 + 4w = 120$	
$\dfrac{4w}{4} = \dfrac{120}{4}$	Divide both sides by 4.
$w = 30$	

Answer: The width is 30 ft.
  The length is $15 + 30 = 45$ ft.

[4.5] For Exercises 25–32, translate to an equation and solve.

25. One number is twelve more than another number. The sum of the numbers is forty-two. Find the numbers.

26. One number is five times another number. The sum of the numbers is thirty-six. Find the numbers.

27. A field is developed so that the length is 2 meters less than three times the width. The perimeter is 188 meters. Find the dimensions of the field.

28. A window over the entrance to a library is an isosceles triangle. The length of each equal-length side is 38 inches longer than the base. If the perimeter is 256 inches, what are the lengths of the base and the sides of equal length?

29. A security laser is placed in the corner of a room near the floor and aimed at a detector across the room. The angle made with the wall on one side of the beam measures 16° more than the angle made with the wall on the other side of the beam. What are the angle measurements?

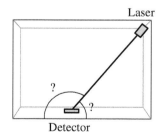

30. The support to a basketball goal is angled for safety. The measure of the angle on the side facing away from the goal is 15° less than twice the measure of the angle made on the goal side. Find the angle measurements.

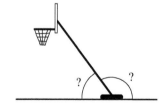

31. Karen has nine more $10 bills than $20 bills. If she has a total of $330, how many of each bill does she have?

32. A store sells two different-size bags of mulch. The large bag costs $6, and the small bag costs $4. In one day, the store sold 27 bags of mulch for a total of $146. How many of each size were sold?

Chapter 4 Practice Test

For Extra Help | CHAPTER **Test Prep** VIDEOS

Step-by-step test solutions are found on the Chapter Test Prep Videos available via the Video Resources on DVD, in **MyMathLab**®, and on You Tube ▪ (search "CarsonPrealgebra" and click on "Channels").

1._____

2._____

3._____

4._____

5._____

6._____

7._____

8._____

9._____

10._____

11._____

12._____

13._____

14._____

15._____

[4.1]   **1.** Is $8y - 17 = 7$ an expression or equation? Why?

[4.1]   **2.** Check to see if -4 is a solution for $-9 = 3x + 5$.

[4.1]   **3.** Check to see if 2 is a solution for $3x - 11 = 2(x - 2) - 5$.

[4.2]   **4.** Is $9x + 7 = x^2 - 4$ linear or nonlinear? Why?

$\begin{bmatrix} 4.2 \\ 4.3 \end{bmatrix}$ *For Exercises 5–12, solve and check.*

5. $n - 15 = -7$

6. $-9m = 54$

7. $-6y + 3 = -21$

8. $9k + 5 = 17 - 3k$

9. $-13x + 26 = 11 - 8x$

10. $4t - 13 + t = 11 + 6t - 3$

11. $7(u - 2) + 12 = 4(u - 2)$

12. $6 + 3(k + 4) = 5k + (k - 9)$

For Exercises 13–20, translate to an equation and solve.

[4.2]   **13.** Daryl owes $458 in taxes. He currently has $375. How much more does he need?

[4.3]   **14.** A farmer needs a fenced pasture with an area of 96,800 square yards. The width must be 242 yards. Find the length. (Use $A = lw$.)

[4.4]   **15.** Nine subtracted from four times a number is equal to twenty-three. Find the number.

[4.4]   **16.** Three times the difference of x and five is the same as nine less than four times x. Find x.

16._____

[4.5]   **17.** One number is three times another number. The sum of the numbers is forty-four. Find the numbers.

17._____

[4.5]   **18.** An entrance to a new restaurant is an isosceles triangle. The length of each equal-length side is 9 inches more than twice the length of the base. The perimeter is 258 inches. Find the lengths of the base and the sides of equal length.

18._____

[4.5]   **19.** A microphone is placed on a stand on the floor of a stage so that the angle made by the microphone and the floor on one side measures 30° less than the angle made on the other side of the microphone. Find the two angle measurements.

19._____

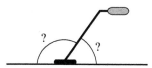

[4.5]   **20.** A music store sells two models of the same guitar. Model A costs $450, and model B costs $675. In one week, the store sold 12 of those guitars for a total of $6300. How many of each model did the store sell? (Use a four-column table.)

20._____

Chapters 1–4 Cumulative Review Exercises

For Exercises 1–4, answer true or false.

[1.2] **1.** $(5 + 4) + 2 = 5 + (4 + 2)$

[2.1] **2.** $-65 \le -68$

[3.5] **3.** 91 is a prime number.

[4.2] **4.** $4y - 9 = 3(y + 5)$ is a linear equation.

For Exercises 5 and 6, complete the rule.

[2.4] **5.** The quotient of two negative numbers is a(n) _____ number.

[3.4] **6.** To multiply exponential forms that have the same base, we can _____ the exponents and keep the same base.

[1.1] **7.** Graph 7 on a number line.

[1.4] **8.** Estimate $5826 \div 224$ by rounding so that there is only one nonzero digit in each number.

[3.2] **9.** What is the coefficient and degree of $-8x^3$?

[3.2] **10.** What is the degree of $4x^2 - 8x + 9x^5 - 6$?

[2.5] For Exercises 11–14, simplify.

11. $3 + (-4)^3$

12. $|2 - 3(7)|$

13. $2(-9) - 20 \div (5)$

14. $7^2 + [8 - 5(6)][6 + 2(7)]$

[3.1] **15.** Evaluate $2x - 5\sqrt{x + y}$ when $x = 9$ and $y = 16$.

[3.3] **16.** Simplify and write the resulting polynomial in descending order of degree.
$7t^3 - 10t + 14t^2 - 7 + t^3 - 15t$

[3.3] **17.** Subtract. $(10x^3 - 7x^2 - 15) - (4x^3 + x - 8)$

[3.5] **18.** Multiply. $(3x - 5)(3x + 5)$

[3.5] **19.** Find the prime factorization of 420.

[3.6] **20.** Factor. $20n^5 + 15mn^3 - 10n^2$

$\begin{bmatrix} 4.2 \\ 4.3 \end{bmatrix}$ *For Exercises 21–24, solve and check.*

21. $-7t = 42$

22. $30 = 2b - 14$

23. $9y - 13 = 4y + 7$

24. $6(n + 2) = 3n - 9$

For Exercises 25–28, solve.

[1.6] **25.** Find the area of the shape.

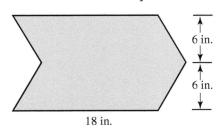

6 in.

6 in.

18 in.

[2.2] **26.** The following table lists the assets and debts for the Goodman family. Calculate their net worth.

Assets	Debts
Savings = $985	Credit card balance = $2345
Checking = $1862	Mortgage = $75,189
Furniture = $12,006	Automobile = $4500

[3.7] **27.** Write an expression in simplest form for the area.

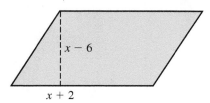

For Exercises 29 and 30, translate to an equation and solve.

[4.4] **29.** Fifteen more than the product of six and n is equal to three.

[4.3] **28.** A wooden jewelry box has a lid that is 8 inches wide by 9 inches long. The surface area of the box is 314 square inches. What is the height of the box? (Use $SA = 2lw + 2lh + 2wh$.)

[4.4] **30.** The length of a rectangle is 4 feet more than the width. If the perimeter is 44 feet, find the length and width.

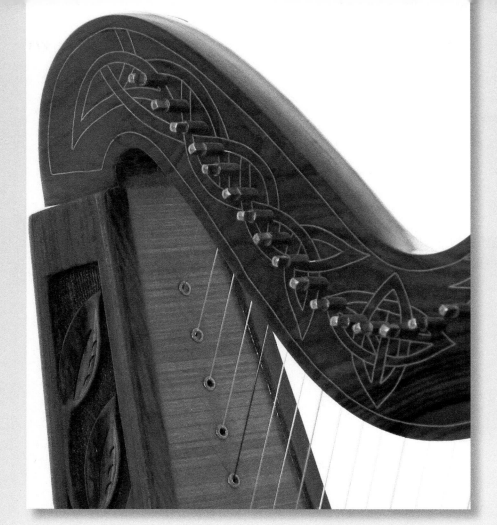

Fractions and Rational Expressions

Chapter Overview

We have developed the number set of integers (Chapters 1 and 2), learned about variable expressions (Chapter 3), and learned how to solve linear equations (Chapter 4). Now in Chapter 5, we will add to our knowledge of numbers by studying fractions. We will also revisit all that we learned about expressions and equations, but use fractions in those expressions and equations.

5.1 Introduction to Fractions

Warm-up

[1.4] 1. Divide. $385 \div 9$

[2.1] 2. Simplify. $|-6|$

Objective 1 Name the fraction represented by a shaded region.

We have seen division written in **fraction** form such as $\frac{8}{2}$, where 8 is the dividend and 2 is the divisor.

When talking about $\frac{8}{2}$ as a fraction, we call 8 the **numerator** and 2 the **denominator**.

Definitions **Fraction:** A quotient of two numbers or expressions a and b having the form $\frac{a}{b}$, where $b \neq 0$.

Numerator: The number written in the top position of a fraction.

Denominator: The number written in the bottom position of a fraction.

$$\frac{8}{2} \quad \begin{array}{l} \leftarrow \text{numerator} \\ \leftarrow \text{denominator} \end{array}$$

Notice that because $\frac{8}{2}$ is division, it can be simplified to 4. Also, in $\frac{8}{2}$, the numerator value is *greater than* the denominator value; so it is called an **improper fraction.** We will learn more about improper fractions later. First, we focus on **proper fractions** such as $\frac{3}{5}$ in which the numerator value is *less than* the denominator value.

Definitions **Improper fraction:** A fraction in which the absolute value of the numerator is greater than or equal to the absolute value of the denominator.

Proper fraction: A fraction in which the absolute value of the numerator is less than the absolute value of the denominator.

A proper fraction describes a part of a whole. For example, the figure to the right represents 1 acre of land that has been divided into five lots of equal size.

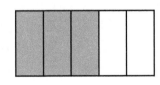

The three shaded lots have been sold; so the fraction $\frac{3}{5}$ describes the portion of the acre that has been sold.

The two unshaded lots represent the fraction of the acre that has not been sold, which is $\frac{2}{5}$.

Now let's identify some fractions.

| **Example 1** | Name the fraction represented by the shaded region. |

a.

Answer: $\frac{7}{15}$

Objectives

1. Name the fraction represented by a shaded region.
2. Write improper fractions as mixed numbers.
3. Write mixed numbers as improper fractions.
4. Graph fractions or mixed numbers on a number line.
5. Write equivalent fractions.
6. Use <, >, or = to write a true statement.

Answers to Warm-up
1. 42 r7   2. 6

▶ **Your Turn 1**

Name the fraction represented by the shaded region.

a.

b.

▶ **Your Turn 2**

Write the fraction for each situation.

a. According to a pamphlet, 1 in 5 teenagers smokes cigarettes. What fraction of teenagers smoke?

b. A football quarterback throws 28 passes during a game and 19 of those passes are completed. What fraction of the quarterback's passes were completed? What fraction were not completed?

c. Margaret spends 8 hours each day at her office. Her company allows 1 of those hours to be a lunch break. What fraction of her time at the office is spent at lunch? What fraction is spent working?

▶ **Your Turn 3**

Simplify.

a. $-\dfrac{10}{2}$ b. $\dfrac{2}{2}$ c. $\dfrac{8}{1}$ d. $\dfrac{1}{0}$

Answers to Your Turn 1

a. $\dfrac{11}{12}$ b. $\dfrac{1}{4}$

Answers to Your Turn 2

a. $\dfrac{1}{5}$ b. $\dfrac{19}{28}, \dfrac{9}{28}$ c. $\dfrac{1}{8}, \dfrac{7}{8}$

Answers to Your Turn 3

a. -5 b. 1   c. 8   d. undefined

Explanation: There are 15 equal-size divisions in the whole region, and 7 are shaded.

b.

Answer: $\dfrac{3}{8}$

Explanation: There are 8 equal-size divisions in the whole region, and 3 are shaded.

◀ **Do Your Turn 1**

Example 2   In a group of 35 people at a conference, 17 are wearing glasses. What fraction of the people at the conference are wearing glasses? What fraction are not wearing glasses?

Answer: The fraction of the people at the conference that are wearing glasses is $\frac{17}{35}$. The fraction of the people at the conference that are not wearing glasses is $\frac{18}{35}$.

Explanation: Because a person can either wear or not wear glasses, there are only two possible categories. This means that if 17 out of the 35 people are wearing glasses, everyone else is not. We can subtract 17 from 35 to get the number of people that are not wearing glasses. There are $35 - 17 = 18$ people who are not wearing glasses out of 35 total people; therefore, the fraction is $\frac{18}{35}$.

◀ **Do Your Turn 2**

Objective 2 Write improper fractions as mixed numbers.

Now let's think about what improper fractions mean. We know that $\dfrac{8}{2}$ means division; so $\dfrac{8}{2} = 4$. This suggests that we can use division to rewrite improper fractions.

Example 3   Simplify.

a. $\dfrac{6}{6}$

Solution: $\dfrac{6}{6} = 1$

Note Because fractions also mean division, a negative fraction's negative sign can be placed by the fraction bar, in the numerator, or in the denominator.

$$-\dfrac{5}{1} = \dfrac{-5}{1} = \dfrac{5}{-1} = -5$$

b. $-\dfrac{5}{1}$

Solution: $-\dfrac{5}{1} = -5$

c. $\dfrac{9}{0}$

Solution: undefined

Note Undefined expressions are to be avoided. That is why, in defining a fraction as $\dfrac{a}{b}$, we say that $b \neq 0$.

◀ **Do Your Turn 3**

Now what if the numerator is not evenly divisible by the denominator, as with $\frac{9}{4}$?

Think about money. A quarter is one-fourth of a dollar. If we have 9 quarters and every 4 quarters equals 1 dollar, we have 2 dollars plus 1 quarter left over.

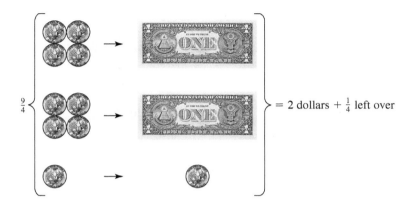

We can write the following result and streamline by eliminating the plus sign.

$$\frac{9}{4} = 2 + \frac{1}{4} = 2\frac{1}{4}$$ **Note** $2\frac{1}{4}$ is read as "two and one-fourth" and means 2 wholes plus $\frac{1}{4}$ of another whole.

This result, $2\frac{1}{4}$, is called a **mixed number.**

Definition   Mixed number: An integer combined with a fraction.

Notice that we grouped the 9 quarters into 2 groups of 4 quarters with 1 quarter left over, which is division.

$$\text{denominator} \rightarrow \quad \begin{array}{r} 2 \quad \leftarrow \text{quotient} \\ 4\overline{)9} \quad \leftarrow \text{numerator} \\ \underline{-8} \\ 1 \quad \leftarrow \text{remainder} \end{array}$$

So long division can be used to write an improper fraction as a mixed number.

Procedure

To write an improper fraction as a mixed number:

1. Divide the denominator into the numerator.
2. Write the result in the following form:

$$quotient \; \frac{remainder}{original \; denominator}$$

> ▶ **Your Turn 4**

Write each improper fraction as a mixed number.

a. $\dfrac{13}{3}$ b. $\dfrac{32}{5}$ c. $-\dfrac{37}{4}$ d. $\dfrac{-79}{10}$

Example 4 Write the improper fraction as a mixed number.

a. $\dfrac{20}{3}$

Solution: Divide 20 by 3.

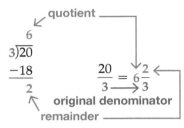

Connection $\frac{20}{3}$ is the same as $20 \div 3$. Mixed numbers are another way to write quotients that have remainders. Instead of 6 r2, we can write $6\frac{2}{3}$.

b. $-\dfrac{41}{7}$

Solution: Divide 41 by 7 and write the negative sign with the integer in the mixed number.

$$
\begin{array}{r}
5 \\
7)\overline{41} \\
-35 \\
\hline
6
\end{array}
\qquad
-\dfrac{41}{7} = -5\dfrac{6}{7}
$$

◀ **Do Your Turn 4**

As we learned in the Connection box in Example 4a, we can write the result of a division as a mixed number. Remember, the divisor is the denominator.

> ▶ **Your Turn 5**

Divide and write the quotient as a mixed number.

a. $49 \div 2$

b. $139 \div 13$

c. $-920 \div 9$

d. $-3249 \div (-16)$

Example 5 Divide $-2035 \div (-19)$ and write the quotient as a mixed number.

Solution: Because the quotient of two negative numbers is positive, we can divide as we would with two positive numbers.

$$
\begin{array}{r}
107 \\
19)\overline{2035} \\
-19 \\
\hline
13 \\
-0 \\
\hline
135 \\
-133 \\
\hline
2
\end{array}
\qquad
107\dfrac{2}{19}
$$

◀ **Do Your Turn 5**

Answers to Your Turn 4

a. $4\dfrac{1}{3}$ b. $6\dfrac{2}{5}$ c. $-9\dfrac{1}{4}$ d. $-7\dfrac{9}{10}$

Answers to Your Turn 5

a. $24\dfrac{1}{2}$ b. $10\dfrac{9}{13}$ c. $-102\dfrac{2}{9}$

d. $203\dfrac{1}{16}$

Objective 3 Write mixed numbers as improper fractions.

In Example 4a, we wrote $\frac{20}{3}$ as $6\frac{2}{3}$. How can we verify that $6\frac{2}{3}$ is correct? Because we divided 20 by 3 to get $6\frac{2}{3}$, we can check the division by multiplying.

$$\text{quotient} \cdot \text{divisor} + \text{remainder} = \text{dividend}$$

$$
\begin{array}{ccccccc}
6 & \cdot & 3 & + & 2 & \overset{?}{=} & 20 \\
& & 18 & + & 2 & \overset{?}{=} & 20 \\
& & & & 20 & = & 20
\end{array}
$$

Yes, it checks.

In mixed-number format, it looks like this:

1. Multiply. $3 \cdot 6 = 18$ **2. Add the product to the numerator. $18 + 2 = 20$**

$$6\frac{2}{3} = \frac{3 \cdot 6 + 2}{3} = \frac{20}{3}$$

3. Keep the same denominator.

Procedure

To write a mixed number as an improper fraction:

1. Multiply the absolute value of the integer by the denominator.
2. Add the product from step 1 to the numerator. The result is the numerator of the improper fraction.
3. Keep the same denominator and sign.

Example 6   Write the mixed number as an improper fraction.

a. $5\dfrac{7}{8}$

Solution: $5\dfrac{7}{8} = \dfrac{8 \cdot 5 + 7}{8} = \dfrac{40 + 7}{8} = \dfrac{47}{8}$ ◀ **Note** We have shown the steps. However, many people perform the multiplication and addition mentally.

b. $-9\dfrac{3}{7}$

Solution: $-9\dfrac{3}{7} = -\dfrac{7 \cdot 9 + 3}{7} = -\dfrac{63 + 3}{7} = -\dfrac{66}{7}$

Warning Do not make the mistake of thinking that we multiply 7 times -9 to get -63, then add 3 to get -60. Think of the negative sign as applying to the entire mixed number.

$$-9\frac{3}{7} = -\left(9\frac{3}{7}\right) = -\left(\frac{66}{7}\right) = -\frac{66}{7}$$

▶ **Do Your Turn 6**

Every number we have learned about so far is a **rational number.**

Definition Rational number: A number that can be expressed as a ratio of integers.

In math language: A number that can be expressed in the form $\dfrac{a}{b}$, where a and b are integers and $b \neq 0$.

Examples: $\dfrac{3}{5}$ is a rational number because 3 and 5 are integers.

$4\dfrac{2}{3}$ is a rational number because it can be written as $\dfrac{14}{3}$ and 14 and 3 are integers.

-7 is a rational number because it can be written as $\dfrac{-7}{1}$ and -7 and 1 are integers.

▶ **Your Turn 6**

Write the mixed number as an improper fraction.

a. $10\dfrac{5}{6}$ b. $-11\dfrac{2}{9}$ c. 19

Answers to Your Turn 6

a. $\dfrac{65}{6}$ b. $-\dfrac{101}{9}$ c. $\dfrac{19}{1}$

As our examples suggest, every integer, fraction, and mixed number is a rational number and is therefore contained in the *set of rational numbers.* It might seem that every number is a rational number. However, when we learn about circles, we will discover a number that is not rational.

Objective 4 Graph fractions or mixed numbers on a number line.

To graph fractions or mixed numbers on a number line, we can use the following procedure:

Procedure

Graphing Fractions or Mixed Numbers on a Number Line

Proper fraction: Draw evenly spaced tick marks between 0 and 1 (if positive) or 0 and −1 (if negative) to divide that segment into the number of equal-sized divisions equal to the denominator. Then draw a dot on the tick mark indicated by the numerator.

Improper fraction: Write it as a mixed number and follow the process for a mixed number.

Mixed number: Draw evenly spaced tick marks between the integer and the next greater integer (if positive) or next lesser integer (if negative) to divide that segment into the number of equal-sized divisions equal to the denominator. Then draw a dot on the tick mark indicated by the numerator.

Hint: The number of tick marks needed is always one less than the denominator.

Connection The fractional marks on a number line are like the marks on a ruler. A typical ruler is marked so that halves, quarters, eighths, and sixteenths of an inch are shown.

▶ **Your Turn 7**

Graph each fraction on a number line.

a. $\dfrac{3}{4}$ b. $2\dfrac{5}{6}$ c. $-\dfrac{3}{8}$ d. $-1\dfrac{4}{5}$

Example 7   Graph the fraction on a number line.

a. $\dfrac{1}{4}$

Solution:

The denominator is 4; so draw 3 evenly spaced marks to divide the segment between 0 and 1 into 4 equal spaces. The numerator is 1; so draw a dot on the first mark to the right of 0.

b. $-\dfrac{5}{8}$

Solution:

The denominator is 8; so draw 7 evenly spaced marks to divide the segment between 0 and −1 into 8 equal spaces. The numerator is 5; so draw a dot on the fifth mark to the left of 0.

c. $-3\dfrac{5}{8}$

Solution:

The denominator is 8; so draw 7 evenly spaced marks to divide the segment between −3 and −4 into 8 equal spaces. The numerator is 5; so draw a dot on the fifth mark to the left of −3.

◀ **Do Your Turn 7**

Answers to Your Turn 7

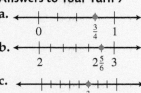

Objective 5 Write equivalent fractions.

Look at the following rulers aligned vertically. They show how the same position can be named using **equivalent fractions.**

Definition Equivalent fractions: Fractions that name the same number.

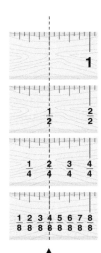

Note This dashed line shows fractions equivalent to 1.

$$1 = \frac{2}{2} = \frac{4}{4} = \frac{8}{8}$$

Note This dashed line shows fractions equivalent to $\frac{1}{2}$.

$$\frac{1}{2} = \frac{2}{4} = \frac{4}{8}$$

Equivalent fractions are produced by multiplying or dividing the numerator and denominator of a fraction by the same number. For example, if we multiply the numerator and denominator of $\frac{1}{2}$ by 2, we produce the equivalent fraction $\frac{2}{4}$.

$$\frac{1}{2} = \frac{1 \cdot 2}{2 \cdot 2} = \frac{2}{4}$$

Similarly, dividing the numerator and denominator of $\frac{2}{4}$ by 2 gives the equivalent fraction $\frac{1}{2}$.

$$\frac{2}{4} = \frac{2 \div 2}{4 \div 2} = \frac{1}{2}$$

Procedure

To write an equivalent fraction, multiply or divide both the numerator and denominator by the same nonzero number.

Example 8 Fill in the blank so that the fractions are equivalent.

a. $\frac{3}{8} = \frac{?}{16}$ **Note** We are given both denominators. Multiplying 8 by 2 produces ◀ 16; so multiplying 3 by 2 will produce the unknown numerator.

Solution: $\frac{3}{8} = \frac{3 \cdot 2}{8 \cdot 2} = \frac{6}{16}$ **Multiply the numerator and denominator of $\frac{3}{8}$ by 2.**

b. $\frac{20}{24} = \frac{5}{?}$ ◀ **Note** We are given both numerators. Dividing 20 by 4 gives 5; so dividing 24 by 4 will give the unknown denominator.

Solution: $\frac{20}{24} = \frac{20 \div 4}{24 \div 4} = \frac{5}{6}$ **Divide the numerator and denominator of $\frac{20}{24}$ by 4.**

▶ **Do Your Turn 8**

▶ **Your Turn 8**

Fill in the blank so that the fractions are equivalent.

a. $\frac{5}{9} = \frac{?}{36}$ b. $-\frac{3}{4} = \frac{?}{12}$

c. $\frac{28}{35} = \frac{4}{?}$ d. $\frac{-24}{42} = \frac{-4}{?}$

Answers to Your Turn 8
a. 20   b. −9   c. 5   d. 7

Objective 6 Use $<$, $>$, or $=$ to write a true statement.

We can easily compare fractions that have the same denominator by looking at their numerators. For example, $\frac{2}{5}$ is less than $\frac{3}{5}$ because 2 is less than 3. In symbols, we write the following:

$$\frac{2}{5} < \frac{3}{5}$$

But what if the fractions have different denominators as in $\frac{1}{2}$ and $\frac{1}{3}$? We could draw pictures to compare.

From the pictures, we see that $\frac{1}{2}$ is greater than $\frac{1}{3}$. In symbols, we write the following:

$$\frac{1}{2} > \frac{1}{3}$$

$\frac{1}{2} = $

$\frac{1}{3} = $

Alternatively, we could write fractions equivalent to $\frac{1}{2}$ and $\frac{1}{3}$ with a common denominator, then compare their numerators. Because we multiply to write equivalent fractions, the common denominator must have both 2 and 3 as factors. That is, the common denominator must be a **multiple** of the denominators.

Definition Multiple: A number that is evenly divisible by a given number.

Multiples of 2 are 2, 4, 6, 8, 10, . . .

Multiples of 3 are 3, 6, 9, 12, 15, . . .

Note We will discuss common multiples in more detail later. For now, we will find a common denominator by multiplying the denominators.

Notice that a common multiple for 2 and 3 is 6, which we can find by multiplying 2 times 3. To write $\frac{1}{2}$ with a denominator of 6, we multiply its numerator and denominator by 3. To write $\frac{1}{3}$ with a denominator of 6, we multiply its numerator and denominator by 2.

$$\frac{1}{2} = \frac{1 \cdot 3}{2 \cdot 3} = \frac{3}{6} \quad \text{and} \quad \frac{1}{3} = \frac{1 \cdot 2}{3 \cdot 2} = \frac{2}{6}$$

Comparing the numerators of the equivalent fractions, we see that 3 is greater than 2; so $\frac{3}{6} > \frac{2}{6}$, which means that $\frac{1}{2} > \frac{1}{3}$.

Procedure

To compare two fractions:

1. Write equivalent fractions that have a common denominator.
2. Compare the numerators of the equivalent fractions.

Example 9 Use $<$, $>$, or $=$ to write a true statement.

a. $\dfrac{5}{8}$? $\dfrac{4}{7}$

Solution: Write equivalent fractions that have a common denominator; then compare numerators. A common multiple for 8 and 7 is 56; so the common denominator is 56.

$$\frac{5}{8} = \frac{5 \cdot 7}{8 \cdot 7} = \frac{35}{56} \qquad \text{Write } \tfrac{5}{8} \text{ with a denominator of 56 by multiplying its numerator and denominator by 7.}$$

$$\frac{4}{7} = \frac{4 \cdot 8}{7 \cdot 8} = \frac{32}{56} \qquad \text{Write } \tfrac{4}{7} \text{ with a denominator of 56 by multiplying its numerator and denominator by 8.}$$

Comparing numerators, we see that 35 is greater than 32; so $\frac{5}{8}$ is greater than $\frac{4}{7}$.

Answer: $\dfrac{5}{8} > \dfrac{4}{7}$

Connection When comparing fractions, we compare only the rewritten numerators and don't really need to know the common denominators. Each rewritten numerator is the product of the original numerator and the other fraction's denominator. Because that multiplication forms an X pattern, those rewritten numerators are often called *cross products*. Consider Example 9a again.

$$\frac{35}{56} \xleftarrow{} \frac{5\cdot 7}{8\cdot 7} \xleftarrow{} \frac{5}{8}\times\frac{4}{7} \xrightarrow{} \frac{4\cdot 8}{7\cdot 8} \xrightarrow{} \frac{32}{56}$$

Because 35 is the greater cross product and is the rewritten numerator of $\frac{5}{8}$, $\frac{5}{8}$ is the greater fraction.

b. $\dfrac{-7}{8}$? $\dfrac{-9}{11}$

Solution: Write equivalent fractions that have a common denominator. Multiplying the denominators gives us a common denominator of 88.

$$\frac{-7}{8} = \frac{-7\cdot 11}{8\cdot 11} = \frac{-77}{88} \qquad \text{Write } -\frac{7}{8} \text{ with a denominator of 88.}$$

$$\frac{-9}{11} = \frac{-9\cdot 8}{11\cdot 8} = \frac{-72}{88} \qquad \text{Write } -\frac{9}{11} \text{ with a denominator of 88.}$$

Now compare the numerators. Because -72 is closer to 0 than to -77, -72 is the greater of the two numerators. Therefore, $\frac{-72}{88}$, or $\frac{-9}{11}$, is the greater fraction.

Answer: $\dfrac{-7}{8} < \dfrac{-9}{11}$

Using the cross products method described in the connection box gives the same result.

$$\frac{-77}{88} = \frac{-7\cdot 11}{8\cdot 11} \xleftarrow{} \frac{-7}{8} ? \frac{-9}{11} \xrightarrow{} \frac{-9\cdot 8}{11\cdot 8} = \frac{-72}{88}$$

Comparing cross products, -72 is greater than -77; so $\frac{-9}{11}$ is the greater fraction.

c. $4\dfrac{5}{6}$? $4\dfrac{7}{9}$

Solution: Because the integers are equal, the fractions will determine which mixed number is greater. Let's use the cross products method to compare the fractions.

$$\frac{45}{54} = \frac{5\cdot 9}{6\cdot 9} \xleftarrow{} \frac{5}{6} ? \frac{7}{9} \xrightarrow{} \frac{7\cdot 6}{9\cdot 6} = \frac{42}{54}$$

Because $\frac{45}{54} > \frac{42}{54}$, we can say that $\frac{5}{6} > \frac{7}{9}$ and therefore that $4\frac{5}{6} > 4\frac{7}{9}$.

Answer: $4\dfrac{5}{6} > 4\dfrac{7}{9}$

▶ **Do Your Turn 9**

▶ **Your Turn 9**

Use $<$, $>$, or $=$ to write a true statement.

a. $\dfrac{3}{7}$? $\dfrac{4}{9}$ b. $2\dfrac{5}{11}$? $2\dfrac{4}{9}$

c. $-\dfrac{2}{7}$? $-\dfrac{12}{42}$

d. $-3\dfrac{9}{10}$? $-3\dfrac{8}{9}$

◀ **Note** Remember that when you are comparing two negative numbers, the greater number has the smaller absolute value.

Answers to Your Turn 9

a. $\dfrac{3}{7} < \dfrac{4}{9}$ b. $2\dfrac{5}{11} > 2\dfrac{4}{9}$

c. $-\dfrac{2}{7} = -\dfrac{12}{42}$ d. $-3\dfrac{9}{10} < -3\dfrac{8}{9}$

5.1 Exercises   For Extra Help   MyMathLab®

Learning Strategy

Doing homework every night is a necessity. Math is a subject that builds on itself and you don't want to get behind. I learn better by doing, so working through problems, even repeatedly, has proven to be an effective study technique for me. I think repetition is really helpful when it comes to learning material.

—Morgan L.

Objective 1

Prep Exercise 1 Identify the numerator and denominator of $\frac{3}{5}$.

Prep Exercise 2 Explain the difference between an improper fraction and a proper fraction.

Prep Exercise 3 When using a fraction to describe the portion of a figure that has been shaded, if the denominator is the number of equal-sized divisions in a figure, what is the numerator?

For Exercises 1–6, name the fraction represented by the shaded region.

1.

2.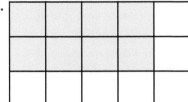

3.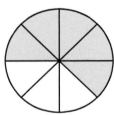

4.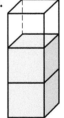

5.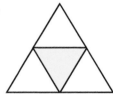

6.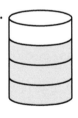

For Exercises 7–12, write the fraction for each situation.

7. **a.** In an experiment, a subject is blindfolded and her nose pinned so that she cannot smell. She could accurately identify 5 out of 16 foods by taste. What fraction did she accurately identify?
   b. Still blindfolded but without her nose pinned, she could then identify 15 out of 16 foods. What fraction did she accurately identify?
   c. What can we conclude from the experiment?

   d. Are there any flaws in the experiment?

8. **a.** In a memory experiment, a subject is shown 20 words in sequence. Each word is shown for 3 seconds. The subject is asked to write down all of the words he can remember in any order. The subject was able to remember 11 words. What fraction of the 20 words was he able to remember?
   b. The experiment is repeated with new words and the subject is asked to write the words in the order in which they appeared. The subject was able to write 5 words in order. What fraction did he get correct?
   c. What can you conclude from the experiment?

   d. Are there any flaws in this experiment?

9. A certain fish lays 800 eggs out of which only 17 survive to adulthood. What fraction describes the portion that survives to adulthood? What fraction describes the portion that does not survive to adulthood?

10. At a certain university, of the 179 entering students that declared premed as their major, only 15 completed the premed program. What fraction completed the program? What fraction did not? What can you conclude about this program?

11. Of the 258 students who have ever taken math with Mrs. Jones, 249 passed. What fraction passed? What fraction did not pass? What can you conclude about taking math with Mrs. Jones as the instructor?

12. An ad claims that a certain medicine is safe according to the FDA. You read an article that claims the medicine caused cancer in laboratory rats. As you read, you see that 9 out of 50 rats got cancer. What fraction of the rats got cancer? What can you conclude about the claims of the FDA versus the claims of the ad?

For Exercises 13–16, use the following graph, which shows the performance of 10th-grade students on a basic skills test at a particular high school.

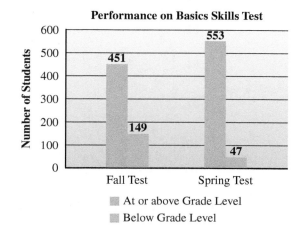

Performance on Basics Skills Test

13. What fraction of the students scored at or above grade level on the fall test?

14. What fraction of the students scored at or above grade level on the spring test?

15. What fraction of the students scored below grade level on the fall test?

16. What fraction of the students scored below grade level on the spring test?

Objective 2

Prep Exercise 4 Explain how to write an improper fraction as a mixed number.

For Exercises 17–24, simplify.

17. $\dfrac{23}{1}$

18. $\dfrac{-45}{1}$

19. $\dfrac{-6}{-6}$

20. $\dfrac{19}{19}$

21. $\dfrac{0}{16}$

22. $\dfrac{0}{-18}$

23. $\dfrac{-2}{0}$

24. $\dfrac{12}{0}$

For Exercises 25–32, write the improper fraction as a mixed number.

25. $\dfrac{30}{7}$

26. $\dfrac{52}{9}$

27. $\dfrac{85}{4}$

28. $\dfrac{19}{2}$

29. $-\dfrac{64}{5}$

30. $-\dfrac{97}{6}$

31. $\dfrac{103}{-8}$

32. $\dfrac{-111}{20}$

For Exercises 33–40, divide and write the quotient as a mixed number.

33. $59 \div 6$

34. $73 \div 10$

35. $140 \div 11$

36. $163 \div 15$

37. $-839 \div 8$

38. $-865 \div 7$

39. $-5629 \div (-14)$

40. $-8217 \div (-16)$

Objective 3

Prep Exercise 5 Explain how to write a mixed number as an improper fraction.

For Exercises 41–48, write the mixed number as an improper fraction.

41. $5\frac{1}{6}$ **42.** $4\frac{3}{8}$ **43.** 11 **44.** $13\frac{1}{2}$

45. $-9\frac{7}{8}$ **46.** -24 **47.** $-1\frac{9}{20}$ **48.** $-15\frac{3}{4}$

Objective 4

Prep Exercise 6 When graphing a positive proper fraction on a number line, how many evenly spaced tick marks do you draw between 0 and 1?

For Exercises 49–60, graph the fraction on a number line.

49. $\frac{1}{3}$ **50.** $\frac{3}{5}$ **51.** $-\frac{5}{8}$

52. $-\frac{1}{2}$ **53.** $1\frac{1}{6}$ **54.** $4\frac{2}{3}$

55. $-1\frac{3}{16}$ **56.** $-5\frac{7}{10}$ **57.** $\frac{9}{5}$

58. $\frac{10}{3}$ **59.** $-\frac{9}{2}$ **60.** $-\frac{17}{6}$

Objective 5

Prep Exercise 7 To write an equivalent fraction, _____ or _____ both the numerator and denominator by the same nonzero number.

For Exercises 61–68, fill in the blank so that the fractions are equivalent.

61. $\frac{5}{9} = \frac{?}{27}$ **62.** $\frac{3}{5} = \frac{?}{35}$ **63.** $\frac{21}{36} = \frac{?}{12}$ **64.** $\frac{12}{42} = \frac{?}{7}$

65. $-\frac{9}{15} = -\frac{18}{?}$ **66.** $-\frac{28}{36} = -\frac{7}{?}$ **67.** $\frac{-6}{16} = \frac{?}{80}$ **68.** $-\frac{24}{60} = \frac{2}{?}$

Objective 6

Prep Exercise 8 To compare two fractions: 1. Write equivalent fractions that have a common denominator.
2. Compare the _____ of the equivalent fractions.

For Exercises 69–80, use $<$, $>$, or $=$ to make a true statement.

69. $\frac{4}{9}$? $\frac{2}{5}$ **70.** $\frac{8}{11}$? $\frac{5}{7}$ **71.** $\frac{9}{16}$? $\frac{12}{18}$ **72.** $\frac{15}{20}$? $\frac{7}{9}$

73. $2\frac{5}{6}$? $2\frac{10}{12}$ **74.** $4\frac{9}{24}$? $4\frac{3}{8}$ **75.** $-\frac{4}{15}$? $-\frac{6}{17}$ **76.** $-\frac{7}{8}$? $-\frac{9}{10}$

77. $-\frac{9}{12}$? $-\frac{15}{20}$ **78.** $-\frac{6}{15}$? $-\frac{8}{20}$ **79.** $-5\frac{4}{7}$? $-5\frac{5}{9}$ **80.** $-6\frac{2}{9}$? $-6\frac{5}{23}$

Review Exercises

[2.4] **1.** Divide. $-48 \div (-12)$ [3.5] **2.** Find the prime factorization of 840.

[3.5] **3.** Find the GCF of 24 and 60.   [3.6] **4.** Find the GCF of $40x^5$ and $56x^2$.

[3.6] **5.** Factor. $40x^5 - 56x^2$

5.2 Simplifying Fractions and Rational Expressions

Warm-up

[3.5] 1. Find the prime factorization of 60.

[5.1] 2. Write $-\dfrac{44}{7}$ as a mixed number.

Objective 1 Simplify fractions to lowest terms.

In Section 5.1, we learned how to write equivalent fractions. In this section, we will learn to write equivalent fractions that are in **lowest terms.**

Definition **Lowest terms:** A fraction is in lowest terms when the greatest common factor of its numerator and denominator is 1.

For example, the fraction $\frac{2}{3}$ is in lowest terms because the greatest common factor of 2 and 3 is 1. The fraction $\frac{8}{12}$ is not in lowest terms because the greatest common factor of 8 and 12 is 4 (not 1).

Recall from Section 5.1 that we can write an equivalent fraction by dividing both its numerator and denominator by any common factor. The process of writing an equivalent fraction in lowest terms is often called *simplifying* or *reducing* the fraction. A fraction written in lowest terms is also said to be in *simplest form*. So for $\frac{8}{12}$, dividing by 4, which is the greatest common factor, gives the equivalent fraction in lowest terms.

$$\frac{8}{12} = \frac{8 \div 4}{12 \div 4} = \frac{2}{3} \quad \leftarrow \text{This is the equivalent fraction in lowest terms.}$$

Notice that we could have also divided by the common factor 2 twice to get lowest terms.

$$\frac{8}{12} = \frac{8 \div 2}{12 \div 2} = \frac{4}{6} = \frac{4 \div 2}{6 \div 2} = \frac{2}{3} \quad \blacktriangleleft \textbf{Note} \text{ Dividing by 2 twice is the same as dividing by 4, the GCF.}$$

Whether in stages or all at once, when simplifying a fraction to lowest terms, we are dividing out the greatest common factor of the numerator and denominator.

Procedure

To simplify a fraction to lowest terms, divide the numerator and denominator by their greatest common factor.

Example 1 Simplify $\dfrac{18}{24}$ to lowest terms.

Solution: $\dfrac{18}{24} = \dfrac{18 \div 6}{24 \div 6} = \dfrac{3}{4}$ Divide both 18 and 24 by their GCF, 6.   ◀   **Note** $\frac{3}{4}$ is in lowest terms because the greatest common factor for 3 and 4 is 1.

▶ **Do Your Turn 1**

We can also simplify fractions using prime factorization. In Section 3.6, we learned that the GCF contains all of the primes common to the numbers' prime factorizations; so dividing out all of the common prime factors is a way of dividing out

Objectives

1 Simplify fractions to lowest terms.

2 Simplify improper fractions or fractions within mixed numbers.

3 Simplify rational expressions.

Connection In Section 3.5, we learned that the greatest common factor is the greatest number that divides all given numbers evenly.

▶ **Your Turn 1**

Simplify to lowest terms.

a. $\dfrac{30}{40}$ b. $-\dfrac{24}{54}$

Answers to Your Turn 1

a. $\dfrac{3}{4}$ b. $-\dfrac{4}{9}$

Answers to Warm-up

1. $2 \cdot 2 \cdot 3 \cdot 5$ or $2^2 \cdot 3 \cdot 5$

2. $-6\dfrac{2}{7}$

the GCF. Let's simplify $\frac{18}{24}$ again, but this time we will replace 18 and 24 with their prime factorizations and then divide out all of the common prime factors.

$$\frac{18}{24} = \frac{\overset{1}{2} \cdot 3 \cdot \overset{1}{3}}{2 \cdot 2 \cdot 2 \cdot \underset{1}{3}}$$
Dividing out the common factors, 2 and 3, divides out the GCF, 6.

$$= \frac{1 \cdot 3 \cdot 1}{1 \cdot 2 \cdot 2 \cdot 1} = \frac{3}{4}$$
Multiplying the remaining factors gives the equivalent fraction in lowest terms.

Procedure

To simplify a fraction to lowest terms using primes:

1. Replace the numerator and denominator with their prime factorizations.
2. Divide out all of the common prime factors.
3. Multiply the remaining factors.

This method is especially useful for simplifying fractions with large numerators or denominators.

► **Your Turn 2**

Simplify to lowest terms.

a. $\dfrac{48}{84}$

b. $-\dfrac{140}{196}$

Example 2 Simplify $-\dfrac{72}{90}$ to lowest terms.

Solution: Because 72 and 90 are large numbers, we will use the prime factorization method. We need the prime factorization of 72 and 90.

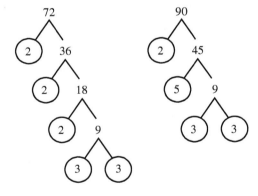

Now replace 72 and 90 with their prime factorizations, divide out all of the common primes, and multiply the remaining factors.

$$-\frac{72}{90} = -\frac{\overset{1}{2} \cdot 2 \cdot 2 \cdot \overset{1}{3} \cdot \overset{1}{3}}{2 \cdot 3 \cdot 3 \cdot 5} = -\frac{4}{5}$$

◄ **Do Your Turn 2**

Example 3 According to a survey of 182 people, 112 responded that they had seen a particular movie. Write, in lowest terms, the fraction of people in the survey who had not seen the movie. Interpret the results.

Solution: Because 112 people saw the movie, we first subtract 112 from 182 to find the number of people who had not seen the movie.

Number of people who did not see the movie = 182 − 112 = 70. If 70 out of 182 people did not see the movie, the fraction of people who did not see the movie is $\frac{70}{182}$.

Answers to Your Turn 2

a. $\dfrac{4}{7}$ b. $-\dfrac{5}{7}$

Now we need to simplify to lowest terms.

$$\frac{70}{182} = \frac{\overset{1}{2} \cdot 5 \cdot \overset{1}{7}}{\underset{1}{2} \cdot \underset{1}{7} \cdot 13} = \frac{5}{13}$$

Answer: $\frac{5}{13}$ of the people surveyed had not seen the movie. This means that 5 out of any 13 people in the survey did *not* see the movie.

Of Interest

If the survey was given to people of different ethnic groups, genders, and ages, we should be able to ask any 13 people of those surveyed if they had seen the movie and we would expect 5 people to say no. If a survey is not accurately conducted, we say that the survey's results are *skewed*.

▶ **Do Your Turn 3**

Objective 2 Simplify improper fractions or fractions within mixed numbers.

What if an improper fraction is not in lowest terms or a mixed number contains a fraction that can be simplified?

Example 4 Write $\frac{26}{12}$ as a mixed number.

Solution:

Method 1: $\frac{26}{12} = \frac{\overset{1}{2} \cdot 13}{\underset{1}{2} \cdot 2 \cdot 3} = \frac{13}{6}$ Simplify to lowest terms.

$\qquad\qquad = 2\frac{1}{6}$ Write the mixed number. Its fraction is in lowest terms.

Method 2: $\frac{26}{12} = 2\frac{2}{12}$ Write as a mixed number.

$\qquad\qquad = 2\frac{\overset{1}{2}}{\underset{1}{2} \cdot 2 \cdot 3} = 2\frac{1}{6}$ Simplify the mixed number's fraction.

▶ **Do Your Turn 4**

Objective 3 Simplify rational expressions.

Let's apply what we've learned about simplifying to a special class of expressions called **rational expressions.**

Definition **Rational expression:** A fraction in which the numerator and denominator are polynomials.

Some examples of rational expressions are as follows:

$$\frac{4x^3}{8x} \qquad \frac{5x}{7x + 1} \qquad \frac{2x^2 - 3x + 4}{8x - 5}$$

Connection A rational number is a number that can be expressed as a ratio of integers, whereas a rational expression is a ratio of polynomials.

We will consider only rational expressions that contain monomials, as in $\frac{4x^3}{8x}$. We simplify such rational expressions the same way we simplify fractions.

▶ **Your Turn 3**

Write each requested fraction in lowest terms.

a. In a survey of 300 people, 252 responded that they believe there is life elsewhere in the universe. What fraction of people in the survey believe that life exists elsewhere in the universe?

b. According to the *CIA World Fact Book*, in 2003, about 60,000,000 out of 300,000,000 people living in the United States were between 0 and 14 years of age. What fraction of people living in the United States were between 0 and 14 years of age?

▶ **Your Turn 4**

Write each fraction as a mixed number and simplify.

a. $\frac{28}{6}$ b. $\frac{80}{15}$

c. $-\frac{54}{24}$ d. $\frac{-180}{24}$

Answers to Your Turn 3

a. $\frac{21}{25}$ b. $\frac{1}{5}$

Answers to Your Turn 4

a. $4\frac{2}{3}$ b. $5\frac{1}{3}$ c. $-2\frac{1}{4}$ d. $-7\frac{1}{2}$

Connection Dividing out the common factor of x relates to the rule for dividing exponential forms with the same base. We can subtract the exponents and keep the same base.

$$\frac{x^3}{x} = x^{3-1} = x^2$$

▶ **Your Turn 5**

Simplify to lowest terms.

a. $\dfrac{9x^4}{24x^3}$

b. $\dfrac{10a^5}{20a}$

c. $-\dfrac{4x^3y}{30x}$

d. $\dfrac{-15m^4}{20m^2n}$

Example 5   Simplify $\dfrac{4x^3}{8x}$ to lowest terms.

Solution:

$$\frac{4x^3}{8x} = \frac{2 \cdot 2 \cdot x \cdot x \cdot x}{2 \cdot 2 \cdot 2 \cdot x}$$ Write the numerator and denominator in factored form.

$$= \frac{\overset{1}{2} \cdot \overset{1}{2} \cdot x \cdot x \cdot \overset{1}{x}}{\underset{1}{2} \cdot \underset{1}{2} \cdot 2 \cdot \underset{1}{x}}$$ Divide out two 2s and one x.

$$= \frac{x^2}{2}$$ Multiply the remaining factors.

◀ **Do Your Turn 5**

Connection We can use what we have learned about simplifying fractions to better understand the rule we learned in Section 3.6 about dividing exponential forms with the same base. The rule read as follows: When dividing exponential forms that have the same base, we can subtract the divisor's exponent from the dividend's exponent and keep the same base.

$$\frac{2^3}{2} = 2^3 \div 2 = 2^{3-1} = 2^2$$

If we write all of the factors and divide out the common 2, we have

$$\frac{2^3}{2} = \frac{\overset{1}{2} \cdot 2 \cdot 2}{\underset{1}{2}} = 2^2.$$

Conclusion: Subtracting exponents corresponds to dividing out common factors.

We can also get a better understanding of what happens when the exponents are the same.

$$\frac{3^2}{3^2} = 3^2 \div 3^2 = 3^{2-2} = 3^0$$

If we write all of the factors and divide out the 3s, we have $\dfrac{3^2}{3^2} = \dfrac{\overset{1}{3} \cdot \overset{1}{3}}{\underset{1}{3} \cdot \underset{1}{3}} = 1.$

Conclusion: Because $\dfrac{3^2}{3^2}$ equals both 3^0 and 1, we can conclude that $3^0 = 1$.

What if the variables in the denominator have greater exponents than the like variables in the numerator?

▶ **Your Turn 6**

Simplify to lowest terms.

a. $\dfrac{14x^4y}{18x^6}$ b. $\dfrac{20m^2}{40m^5}$

c. $-\dfrac{13a^6b}{26a^4b^4}$ d. $\dfrac{-10hk^2}{5k^7}$

Answers to Your Turn 5

a. $\dfrac{3x}{8}$ b. $\dfrac{a^4}{2}$

c. $-\dfrac{2x^2y}{15}$ d. $\dfrac{-3m^2}{4n}$

Answers to Your Turn 6

a. $\dfrac{7y}{9x^2}$ b. $\dfrac{1}{2m^3}$

c. $-\dfrac{a^2}{2b^3}$ d. $\dfrac{-2h}{k^5}$

Example 6   Simplify $\dfrac{-16p^3r^2s}{28p^5r^2}$ to lowest terms.

Solution:

$$\frac{-16p^3r^2s}{28p^5r^2} = \frac{-2 \cdot 2 \cdot 2 \cdot 2 \cdot p \cdot p \cdot p \cdot r \cdot r \cdot s}{2 \cdot 2 \cdot 7 \cdot p \cdot p \cdot p \cdot p \cdot p \cdot r \cdot r}$$ Write the numerator and denominator in factored form.

$$= \frac{-2 \cdot \overset{1}{2} \cdot \overset{1}{2} \cdot 2 \cdot \overset{1}{p} \cdot \overset{1}{p} \cdot \overset{1}{p} \cdot \overset{1}{r} \cdot \overset{1}{r} \cdot s}{2 \cdot \underset{1}{2} \cdot \underset{1}{7} \cdot \underset{1}{p} \cdot \underset{1}{p} \cdot \underset{1}{p} \cdot p \cdot p \cdot \underset{1}{r} \cdot \underset{1}{r}}$$ Divide out two 2s, three p's, and two r's.

$$= \frac{-4s}{7p^2}$$ Multiply the remaining factors.

▲

Note p^2 is in the denominator because there were two more factors of p in the denominator than in the numerator.

◀ **Do Your Turn 6**

5.2 Exercises | For Extra Help | MyMathLab®

Objective 1

Prep Exercise 1 A fraction is in lowest terms when the
_____ of its numerator and
denominator is _____.

Prep Exercise 2 Explain how to simplify a fraction
to lowest terms.

For Exercises 1–20, simplify to lowest terms.

1. $\dfrac{25}{30}$

2. $\dfrac{12}{28}$

3. $\dfrac{-14}{35}$

4. $\dfrac{-16}{24}$

5. $\dfrac{26}{52}$

6. $\dfrac{15}{45}$

7. $-\dfrac{24}{40}$

8. $-\dfrac{20}{45}$

9. $\dfrac{66}{88}$

10. $\dfrac{32}{80}$

11. $\dfrac{-57}{76}$

12. $\dfrac{-51}{85}$

13. $\dfrac{120}{140}$

14. $\dfrac{196}{210}$

15. $\dfrac{182}{234}$

16. $\dfrac{221}{357}$

17. $\dfrac{-121}{187}$

18. $-\dfrac{270}{900}$

19. $\dfrac{-360}{480}$

20. $\dfrac{-210}{294}$

For Exercises 21–24, write each requested fraction in lowest terms.

21. 35 seconds is what fraction of a minute?

22. 25 minutes is what fraction of an hour?

23. 9 hours is what fraction of a day?

24. If we consider a year to be 365 days, what fraction of a year is 45 days?

Of Interest

A year is the time it takes the earth to make one orbit around the sun, which is actually $365\frac{1}{4}$ days.

25. In a survey, 248 women were asked if they used a certain cosmetic product. 96 responded that they had used the product.
    a. What fraction of the women surveyed said that they used the product?
    b. Interpret the results.

    c. What, if any, flaws do you see in the survey?

26. A marketing firm conducted an experiment to determine which of two commercials was more effective. It randomly selected 400 people and paid 200 of them to watch one of the commercials and the other 200 to watch the other commercial. It then asked those viewers whether they would buy the product. 180 said yes in the first group, while 148 said yes in the second group.
    a. What fraction of the first group said yes?
    b. What fraction of the second group said yes?
    c. Which commercial should be used?

    d. What, if any, flaws do you see in the experiment?

27. A bank manager discovers that out of 500 transactions completed by a particular teller, 48 mistakes were made. In what fraction of the transactions did the teller make a mistake? In what fraction of the transactions did the teller make no mistake?

28. A company produces 540 computer chips on a particular day. Quality control inspectors discover 21 defective chips. What lowest-terms fraction of the chips produced is defective? What lowest-terms fraction is not defective?

29. Use the following graph, which shows a sales representative's travel expenses for one year.

Travel Expenses

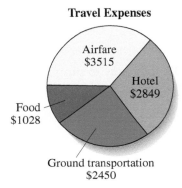

Airfare $3515

Hotel $2849

Food $1028

Ground transportation $2450

a. Find the representative's total travel expenses.
b. What fraction of the travel expenses went toward food?
c. What fraction of the travel expenses went toward ground transportation?
d. What fraction of the travel expenses went toward hotel?
e. What fraction of the travel expenses went toward airfare?

30. Use the following graph, which shows the amount the Gardner family spent on transportation in one year.

Gardner Family Transportation Expenses

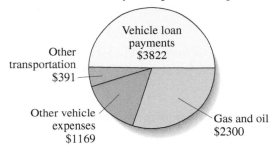

Vehicle loan payments $3822

Other transportation $391

Other vehicle expenses $1169

Gas and oil $2300

a. How much money did the Gardner family spend on transportation?
b. What fraction of the money spent on transportation went toward loan payments?
c. What fraction of the money spent on transportation was spent on gas and oil?
d. What fraction of the money spent on transportation was spent on other vehicle expenses?
e. What fraction of the money spent on transportation was spent on other transportation?

Objective 2

For Exercises 31–38, write as a mixed number and simplify.

31. $\dfrac{30}{8}$

32. $\dfrac{69}{9}$

33. $\dfrac{-50}{15}$

34. $\dfrac{-84}{16}$

35. $\dfrac{116}{28}$

36. $\dfrac{168}{105}$

37. $\dfrac{-186}{36}$

38. $\dfrac{486}{-63}$

Objective 3

For Exercises 39–50, simplify to lowest terms.

39. $\dfrac{10x}{32}$

40. $\dfrac{15y^2}{27}$

41. $\dfrac{-x^3}{xy}$

42. $\dfrac{-a^4b}{a^3b^5}$

43. $-\dfrac{6m^4n}{15m^7}$

44. $-\dfrac{30h^2k}{18h^4k}$

45. $\dfrac{9t^2u}{36t^5u^2}$

46. $\dfrac{7x^3y}{35xy^2}$

47. $\dfrac{21a^6bc}{35a^5b^2}$

48. $\dfrac{10m^7n}{30m^9n^4p}$

49. $\dfrac{-14a^4b^5c^2}{28a^{10}bc^7}$

50. $\dfrac{8x^2yz^3}{-32xy^4z^3}$

Review Exercises

[2.4] 1. Multiply. $165 \cdot (-91)$

[2.4] 2. Multiply. $(-12)(-6)$

[3.4] 3. Multiply. $(5x^2)(7x^3)$

[3.4] 4. Simplify. $(-2x^3)^5$

[1.6] 5. Find the area of the figure.

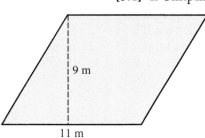

9 m

11 m

5.3 Multiplying Fractions, Mixed Numbers, and Rational Expressions

Warm-up

[2.4] 1. Multiply. $-7 \cdot 9$

[5.2] 2. Simplify $-\dfrac{36}{48}$ to lowest terms.

Objective 1 Multiply fractions.

Let's see what we can discover about how to multiply fractions. Suppose a recipe calls for $\frac{1}{4}$ of a cup of oil but we want to make only half of the recipe. We must find half of $\frac{1}{4}$. Consider the following picture of $\frac{1}{4}$.

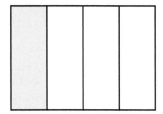

If we want half of $\frac{1}{4}$, we need to cut the single fourth into two pieces, then shade one of those two pieces.

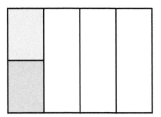

To name the fraction represented by half of the fourth, we need equal-size divisions. Extending the line across makes equal-size divisions.

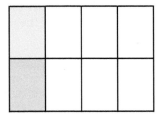

Notice that the darker shaded region is half of the fourth that we originally shaded and is $\frac{1}{8}$ of the whole picture; so we can write the following statement:

$$\text{Half of } \frac{1}{4} \text{ is } \frac{1}{8}.$$

We can translate this statement to an equation. The word *of*, when preceded by a fraction, indicates multiplication, and *is* indicates an equal sign.

$$\text{Half of } \frac{1}{4} \text{ is } \frac{1}{8}.$$

$$\frac{1}{2} \cdot \frac{1}{4} = \frac{1}{8}$$

Objectives

1 Multiply fractions.

2 Multiply and simplify fractions.

3 Multiply mixed numbers.

4 Multiply rational expressions.

5 Simplify fractions raised to a power.

6 Solve applications involving multiplying fractions.

7 Find the area of a triangle.

8 Find the radius and diameter of a circle.

9 Find the circumference of a circle.

Answers to Warm-up
1. -63

2. $-\dfrac{3}{4}$

Our example suggests that to multiply fractions, we multiply numerator by numerator and denominator by denominator.

▶ **Your Turn 1**

Multiply.

a. $\dfrac{3}{4} \cdot \dfrac{7}{8}$

b. $\dfrac{1}{5} \cdot \left(-\dfrac{4}{9}\right)$

c. $-\dfrac{5}{6} \cdot \left(-\dfrac{1}{6}\right)$

d. $\dfrac{1}{-4} \cdot \left(\dfrac{-1}{5}\right)$

▲

Note Remember, $\dfrac{1}{-4}$ is equivalent to $-\dfrac{1}{4}$ and $\dfrac{-1}{5}$ is equivalent to $-\dfrac{1}{5}$.

Example 1 Multiply.

a. $\dfrac{3}{4} \cdot \dfrac{5}{8}$

Solution: $\dfrac{3}{4} \cdot \dfrac{5}{8} = \dfrac{3 \cdot 5}{4 \cdot 8} = \dfrac{15}{32}$ Multiply numerator by numerator and denominator by denominator. Notice that $\frac{15}{32}$ is in lowest terms.

b. $-\dfrac{2}{3} \cdot \dfrac{1}{5}$

Solution: $-\dfrac{2}{3} \cdot \dfrac{1}{5} = -\dfrac{2 \cdot 1}{3 \cdot 5} = -\dfrac{2}{15}$ Multiply numerator by numerator and denominator by denominator. The product is negative because the two fractions have different signs. Notice that $-\frac{2}{15}$ is in lowest terms.

c. $-\dfrac{5}{7} \cdot \left(-\dfrac{9}{11}\right)$

Solution: $-\dfrac{5}{7} \cdot \left(-\dfrac{9}{11}\right) = \dfrac{5 \cdot 9}{7 \cdot 11} = \dfrac{45}{77}$ Multiply numerator by numerator and denominator by denominator. The product is positive because the two fractions have the same sign. Notice that $\frac{45}{77}$ is in lowest terms.

◀ **Do Your Turn 1**

Objective 2 Multiply and simplify fractions.

As a rule, fraction answers are expected to be in lowest terms. Let's multiply $\dfrac{5}{6} \cdot \dfrac{3}{4}$ and then simplify the product.

$$\dfrac{5}{6} \cdot \dfrac{3}{4} = \dfrac{15}{24} \qquad \text{Multiply.}$$

$$= \dfrac{\overset{1}{3} \cdot 5}{2 \cdot 2 \cdot 2 \cdot \underset{1}{3}} = \dfrac{5}{8} \qquad \text{Simplify } \tfrac{15}{24} \text{ to lowest terms by dividing out the common factor, 3.}$$

Alternatively, because the original numerators are factors of the resulting numerator and the original denominators are factors of the resulting denominator, we can divide out common factors *before* multiplying.

$$\dfrac{5}{6} \cdot \dfrac{3}{4} = \dfrac{5}{2 \cdot \underset{1}{3}} \cdot \dfrac{\overset{1}{3}}{2 \cdot 2} \qquad \text{Replace each numerator and denominator with its prime factorization, and divide out the common factor, 3.}$$

$$= \dfrac{5 \cdot 1}{2 \cdot 1 \cdot 2 \cdot 2} = \dfrac{5}{8} \qquad \text{Multiply the remaining factors to get the same product as above.}$$

Or some students prefer to divide out the common factor(s) as follows:

$$\dfrac{5}{\underset{2}{6}} \cdot \dfrac{\overset{1}{3}}{4} = \dfrac{5}{8} \qquad◀ \begin{array}{l}\textbf{Note} \text{ For now, we will write the numerators and} \\ \text{denominators in prime factored form so that the} \\ \text{factors are easier to see. In Sections 5.7 and 5.8,} \\ \text{we'll leave out the prime factored form.}\end{array}$$

The following procedure summarizes our methods for multiplying fractions.

Answers to Your Turn 1

a. $\dfrac{21}{32}$ **b.** $-\dfrac{4}{45}$ **c.** $\dfrac{5}{36}$ **d.** $\dfrac{1}{20}$

Procedure

To multiply fractions:

Method 1:

1. Multiply the numerators and multiply the denominators.
2. Simplify the product.

Method 2:

1. Divide out factors of any numerator with identical factors of any denominator.
2. Multiply the remaining numerator factors and multiply the remaining denominator factors.

Example 2 Multiply $\frac{6}{8} \cdot \frac{10}{15}$. Write the product in lowest terms.

Solution:

Method 1: $\frac{6}{8} \cdot \frac{10}{15} = \frac{60}{120}$ Multiply numerators and multiply denominators.

$$= \frac{\overset{1}{\cancel{2}} \cdot \overset{1}{\cancel{2}} \cdot \overset{1}{\cancel{3}} \cdot \overset{1}{\cancel{5}}}{\underset{1}{\cancel{2}} \cdot \underset{1}{\cancel{2}} \cdot 2 \cdot \underset{1}{\cancel{3}} \cdot \underset{1}{\cancel{5}}}$$ Divide out the common factors.

$$= \frac{1}{2}$$ Multiply the remaining factors.

Method 2: $\frac{6}{8} \cdot \frac{10}{15} = \frac{2 \cdot 3}{2 \cdot 2 \cdot 2} \cdot \frac{2 \cdot 5}{3 \cdot 5}$ Replace each numerator and denominator with its prime factorization.

$$= \frac{\overset{1}{\cancel{2}} \cdot \overset{1}{\cancel{3}}}{\underset{1}{\cancel{2}} \cdot 2 \cdot \underset{1}{\cancel{2}}} \cdot \frac{\overset{1}{\cancel{2}} \cdot \overset{1}{\cancel{5}}}{\underset{1}{\cancel{3}} \cdot \underset{1}{\cancel{5}}}$$ Divide out the common factors.

$$= \frac{1}{2}$$ Multiply the remaining factors.

Connection No matter which method we use, we divide out the same common factors—in this case, two 2s, a 3, and a 5.

► **Do Your Turn 2**

► **Your Turn 2**

Multiply. Write the product in lowest terms.

a. $\frac{5}{9} \cdot \frac{3}{10}$

b. $\frac{6}{8} \cdot \frac{12}{30}$

c. $-\frac{10}{15} \cdot \frac{6}{16}$

d. $\frac{-18}{20} \cdot \frac{30}{-32}$

Objective 3 Multiply mixed numbers.

Now that we know how to multiply fractions, we can multiply mixed numbers by first writing them as improper fractions.

Procedure

To multiply mixed numbers:

1. Write the mixed numbers as improper fractions.
2. Divide out factors of any numerator with identical factors of any denominator.
3. Multiply the remaining numerator factors and multiply the remaining denominator factors.
4. Simplify. If the product is an improper fraction, write it as a mixed number in simplest form.

When multiplying mixed numbers, it is helpful to estimate the product to get an idea of what to expect. As with any estimate, we first round the numbers so that the calculation is easy to perform mentally.

Answers to Your Turn 2

a. $\frac{1}{6}$ b. $\frac{3}{10}$ c. $-\frac{1}{4}$ d. $\frac{27}{32}$

▶ **Your Turn 3**

Estimate and then find the actual product expressed as a mixed number in simplest form.

a. $3\dfrac{1}{3} \cdot 1\dfrac{4}{5}$

b. $5\dfrac{2}{3} \cdot 2\dfrac{1}{7}$

c. $\dfrac{2}{5} \cdot 12$

d. $20 \cdot 3\dfrac{3}{4}$

Example 3 Estimate and then find the actual product expressed as a mixed number in simplest form.

a. $3\dfrac{1}{6} \cdot 10\dfrac{1}{2}$

Estimate: Round the numbers so that the calculation is easy to perform mentally. We will round these mixed numbers to the nearest integer.

$3\frac{1}{6}$ rounds down to 3 because it is closer to 3 than to 4.

$10\frac{1}{2}$ rounds up to 11 because it is halfway between 10 and 11 and we agree to round up.

 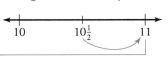

$3 \cdot 11 = 33$ The actual product should be around 33.

Actual: $3\dfrac{1}{6} \cdot 10\dfrac{1}{2} = \dfrac{19}{6} \cdot \dfrac{21}{2}$ Write as improper fractions.

$= \dfrac{19}{2 \cdot \overset{1}{\cancel{3}}} \cdot \dfrac{\overset{1}{\cancel{3}} \cdot 7}{2}$ Divide out a common factor, 3.

$= \dfrac{133}{4}$ Multiply.

$= 33\dfrac{1}{4}$ Write as a mixed number.

The actual product agrees with our estimate.

b. $20 \cdot \dfrac{3}{8}$

Estimate: The nearest integer to $\frac{3}{8}$ is 0. But $20 \cdot 0 = 0$, which is not a helpful estimate. We get a more useful estimate by rounding $\frac{3}{8}$ to a more "comfortable" fraction near its value, such as $\frac{1}{2}$ or $\frac{1}{4}$. We will use $\frac{1}{2}$.

$20 \cdot \dfrac{3}{8}$

For 20, no rounding is necessary.   We round $\dfrac{3}{8}$ to $\dfrac{1}{2}$.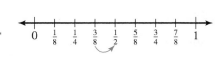

$20 \cdot \dfrac{1}{2} = 10$

The actual product should be around 10.

Actual: $20 \cdot \dfrac{3}{8} = \dfrac{20}{1} \cdot \dfrac{3}{8}$ Write as improper fractions.

$= \dfrac{\overset{1}{\cancel{2}} \cdot \overset{1}{\cancel{2}} \cdot 5}{1} \cdot \dfrac{3}{\underset{1}{\cancel{2}} \cdot \underset{1}{\cancel{2}} \cdot 2}$ Divide out the common factors, which are two 2s.

$= \dfrac{15}{2}$ Multiply.

$= 7\dfrac{1}{2}$ Write the improper fraction as a mixed number.

The actual product agrees with our estimate.

◀ **Do Your Turn 3**

Answers to Your Turn 3

a. 6; 6   b. 12; $12\dfrac{1}{7}$

c. 6; $4\dfrac{4}{5}$ d. 80; 75

Example 4   Multiply. Write the product as a mixed number in simplest form.

a. $-3\frac{1}{5}\cdot 2\frac{2}{9}$

Solution: $-3\frac{1}{5}\cdot 2\frac{2}{9} = -\frac{16}{5}\cdot\frac{20}{9}$ Write as improper fractions.

$$= -\frac{16}{\underset{1}{\underset{5}{5}}}\cdot\frac{2\cdot 2\cdot\overset{1}{5}}{9}$$ Divide out the common factor, 5.

$$= -\frac{64}{9}$$ Multiply. The product is negative because the two mixed numbers have different signs.

$$= -7\frac{1}{9}$$ Write the improper fraction as a mixed number.

b. $2\frac{3}{8}\cdot(-12)$

Solution: $2\frac{3}{8}\cdot(-12) = \frac{19}{8}\cdot\left(-\frac{12}{1}\right)$ Write as improper fractions.

$$= \frac{19}{\underset{1}{2}\cdot\underset{1}{2}\cdot 2}\cdot\left(-\frac{\overset{1}{2}\cdot\overset{1}{2}\cdot 3}{1}\right)$$ Divide out the common factors, which are two 2s.

$$= -\frac{57}{2}$$ Multiply. The product is negative because the two numbers have different signs.

$$= -28\frac{1}{2}$$ Write the improper fraction as a mixed number.

▶ **Do Your Turn 4**

Objective 4 Multiply rational expressions.

We multiply rational expressions the same way we multiply fractions. Although we can use either of the methods we have learned, it is usually easier to divide out the common factors first and then multiply (method 2).

Example 5   Multiply. Write the product in lowest terms. $-\frac{3a}{5b}\cdot\left(-\frac{10ab^2}{9}\right)$

Solution:

$$-\frac{3a}{5b}\cdot\left(-\frac{10ab^2}{9}\right) = -\frac{3\cdot a}{5\cdot b}\cdot\left(-\frac{2\cdot 5\cdot a\cdot b\cdot b}{3\cdot 3}\right)$$ Write the prime factorization of each numerator and denominator.

$$= -\frac{\overset{1}{3}\cdot a}{\underset{1}{5}\cdot\underset{1}{b}}\cdot\left(-\frac{2\cdot\overset{1}{5}\cdot a\cdot\overset{1}{b}\cdot b}{3\cdot 3}\right)$$ Divide out a common 3, a common 5, and a common b factor.

$$= \frac{2a^2b}{3}$$ Multiply the remaining numerator and denominator factors. The product is positive because the two expressions have the same sign.

▶ **Do Your Turn 5**

Objective 5 Simplify fractions raised to a power.

Remember that an exponent indicates the number of times to use the base as a factor.

▶ **Your Turn 4**

Multiply. Write the product as a mixed number in simplest form.

a. $-4\frac{1}{8}\cdot 5\frac{1}{2}$

b. $\left(-2\frac{3}{5}\right)\left(-7\frac{1}{4}\right)$

c. $\left(2\frac{4}{7}\right)\left(-11\frac{2}{5}\right)$

d. $\left(\frac{-4}{9}\right)(-21)$

▶ **Your Turn 5**

Multiply. Write the product in lowest terms.

a. $\frac{x}{6}\cdot\frac{9}{x^2}$

b. $\frac{10a^3}{3}\cdot\frac{12}{a}$

c. $\frac{18tu}{5}\cdot\left(-\frac{15}{27u^4}\right)$

d. $\frac{-6}{13hk^2}\cdot\left(-\frac{26h}{9k}\right)$

Answers to Your Turn 4
a. $-22\frac{11}{16}$ b. $18\frac{17}{20}$
c. $-29\frac{11}{35}$ d. $9\frac{1}{3}$

Answers to Your Turn 5
a. $\frac{3}{2x}$ b. $40a^2$ c. $-\frac{2t}{u^3}$ d. $\frac{4}{3k^3}$

▶ **Your Turn 6**

Simplify.

a. $\left(\dfrac{2}{5}\right)^3$

b. $\left(\dfrac{1}{2}\right)^6$

c. $\left(-\dfrac{3}{4}\right)^3$

d. $\left(\dfrac{-5}{6}\right)^2$

Example 6 Simplify. $\left(\dfrac{2}{3}\right)^4$

Solution: $\left(\dfrac{2}{3}\right)^4 = \dfrac{2}{3}\cdot\dfrac{2}{3}\cdot\dfrac{2}{3}\cdot\dfrac{2}{3} = \dfrac{16}{81}$ Write the base $\frac{2}{3}$ as a factor four times; then multiply.

In Example 6, we multiplied four 2s to get 16 in the numerator and four 3s to get 81 in the denominator; so another way we can think of $\left(\dfrac{2}{3}\right)^4$ is as $\dfrac{2^4}{3^4}$, which suggests the following procedure:

> ## Procedure
> To evaluate a fraction raised to a power, evaluate both the numerator and denominator raised to that power.
>
> **In math language:** $\left(\dfrac{a}{b}\right)^n = \dfrac{a^n}{b^n}$ when $b \neq 0$

Warning

$\dfrac{2^4}{3}$ and $\left(\dfrac{2}{3}\right)^4$ are not equivalent.

$\dfrac{2^4}{3} = \dfrac{2\cdot 2\cdot 2\cdot 2}{3} = \dfrac{16}{3}$, whereas $\left(\dfrac{2}{3}\right)^4 = \dfrac{2\cdot 2\cdot 2\cdot 2}{3\cdot 3\cdot 3\cdot 3} = \dfrac{16}{81}$.

◀ **Do Your Turn 6**

▶ **Your Turn 7**

Simplify.

a. $\left(\dfrac{7x^3}{9}\right)^2$

b. $\left(\dfrac{3x^2}{4z}\right)^3$

c. $\left(-\dfrac{2a^4}{3c^3}\right)^4$

d. $\left(\dfrac{-m^3}{2p}\right)^5$

Example 7 Simplify. $\left(\dfrac{-2x^2}{5y}\right)^3$

Solution:

$\left(\dfrac{-2x^2}{5y}\right)^3 = \dfrac{(-2x^2)^3}{(5y)^3}$ Write both the numerator and denominator raised to the third power.

$= \dfrac{(-2x^2)\cdot(-2x^2)\cdot(-2x^2)}{5y\cdot 5y\cdot 5y}$ Write $-2x^2$ as a factor three times and $5y$ as a factor three times.

$= \dfrac{-8x^6}{125y^3}$ Multiply coefficients and add exponents for the like bases.

Alternatively, we could simplify this expression using our procedure for simplifying a monomial raised to a power.

$\left(\dfrac{-2x^2}{5y}\right)^3 = \dfrac{(-2x^2)^3}{(5y)^3}$ Write both the numerator and denominator raised to the 3rd power.

$= \dfrac{(-2)^3 x^{2\cdot 3}}{5^3 y^{1\cdot 3}}$ Write the coefficient to the 3rd power and multiply each variable's exponent by the power.

$= \dfrac{-8x^6}{125y^3}$ Simplify.

Connection In Section 3.5, we learned how to multiply monomials and how to simplify a monomial raised to a power.

◀ **Do Your Turn 7**

Answers to Your Turn 6

a. $\dfrac{8}{125}$ b. $\dfrac{1}{64}$ c. $-\dfrac{27}{64}$ d. $\dfrac{25}{36}$

Answers to Your Turn 7

a. $\dfrac{49x^6}{81}$ b. $\dfrac{27x^6}{64z^3}$

c. $\dfrac{16a^{16}}{81c^{12}}$ d. $\dfrac{-m^{15}}{32p^5}$

Objective 6 Solve applications involving multiplying fractions.

One of the difficulties in translating application problems that involve fractions is identifying the arithmetic operation. To identify multiplication, look for language indicating that we are to find a fraction *of* an amount. Whenever the word *of* is directly preceded by a fraction and followed by an amount, we multiply the amount by the fraction. Consider the following sentence:

$$\frac{3}{4} \text{ of the 28 people in the class earned an A.}$$

Because *of* is preceded by the fraction $\frac{3}{4}$ and followed by the amount 28, to calculate the number of people in the class who earned an A, we multiply 28 by $\frac{3}{4}$.

The word *of* does not always indicate multiplication. Sometimes *of* indicates a fraction. Consider the following sentence.

21 out of 28 people in the class   and   21 of 28 people in the class earned
earned an A.   an A.

In *21 of 28,* the word *of* indicates the fraction $\frac{21}{28}$. Whenever the word *of* is preceded by a whole number, it indicates a fraction.

Conclusion: The word *of* indicates multiplication when preceded by a fraction and indicates a fraction when preceded by a whole number.

The two uses of the word *of* are related.

$\frac{3}{4}$ of the 28 people in the class earned an A.   21 of 28 people in the class earned an A.

$\frac{3}{4} \cdot 28 = \begin{array}{l}\text{the number of people in the} \\ \text{class that earned an A}\end{array}$ $\frac{21}{28} = \begin{array}{l}\text{the fraction of the people in} \\ \text{the class that earned an A}\end{array}$

$\dfrac{3}{\underset{1}{\cancel{4}}} \cdot \dfrac{\overset{7}{\cancel{28}}}{1} = 21$ $\dfrac{21}{28} = \dfrac{3}{4}$

Example 8   An advertisement claims that 4 out of 5 dentists choose Crest toothpaste. If you were to visit a dental conference that has 345 dentists in attendance, how many would you expect to choose Crest toothpaste based on the ad's claim?

Understand 4 out of 5 is the fraction $\frac{4}{5}$. If the claim of the ad is accurate, we can make the following statement:

$$\frac{4}{5} \text{ of the 345 dentists at the conference should choose Crest.}$$

Plan We can translate our sentence to an equation and solve.

Execute $\frac{4}{5}$ of the 345 dentists at the conference should choose Crest.

$\underset{\downarrow}{\dfrac{4}{5}} \cdot 345 = n$ ◀ **Note** The variable n represents the number of dentists out of the 345 dentists who should choose Crest.

$\dfrac{4}{5} \cdot \dfrac{345}{1} = n$

$\dfrac{4}{\underset{1}{\cancel{5}}} \cdot \dfrac{3 \cdot \overset{1}{\cancel{5}} \cdot 23}{1} = n$

$276 = n$

Answer If we were to survey the dentists at the conference, we should find that out of the 345 dentists, 276 would choose Crest.

▶ **Your Turn 8**

Solve.

a. A doctor says that 8 out of 10 patients receiving a particular treatment have no side effects. If a hospital gives 600 patients this treatment, how many can be expected to have no side effects? How many can be expected to have side effects?

b. A population study finds that there are $2\frac{1}{3}$ children for each household in a given area. If there are 6252 households in the area, how many children are there?

Answers to Your Turn 8
a. 480; 120   b. 14,588

Check 276 out of 345 dentists should be the same ratio as 4 out of 5. We can verify by simplifying $\frac{276}{345}$ to $\frac{4}{5}$.

◀ Do Your Turn 8

▶ **Your Turn 9**

Solve.

a. A study estimates that $\frac{2}{3}$ of all Americans own a car and that $\frac{1}{4}$ of these cars are blue. What fraction of the population drives a blue car?

b. $\frac{5}{8}$ of the people in a survey said that they watch more than 2 hours of television each day. Of these people, $\frac{7}{10}$ were female. What fraction of all those surveyed were females who watch more than 2 hours of television each day?

Example 9 $\frac{3}{4}$ of the students taking a particular history course passed the course. Of these, $\frac{1}{3}$ earned an A. What fraction of all students taking the course earned an A?

Understand The students who earned an A are a fraction of the group that passed. We can make the following statement:

$$\frac{1}{3} \text{ of the } \frac{3}{4} \text{ that passed got an A.}$$

Plan We can translate the sentence to an equation and then solve.

Execute Let a represent the fraction of students who earned an A.

$$\frac{1}{3} \text{ of the } \frac{3}{4} \text{ that passed got an A.}$$

$$\frac{1}{3} \cdot \frac{3}{4} = a$$

Note The variable a represents the fraction of the students in the course who received an A.

$$\frac{1}{\overset{1}{3}} \cdot \frac{\overset{1}{3}}{4} = a$$

$$\frac{1}{4} = a$$

Answer $\frac{1}{4}$ of all students in the class not only passed but also earned an A.

Check We can use a picture to check.

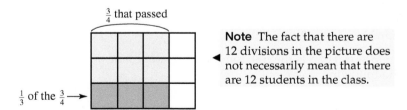

$\frac{3}{4}$ that passed

$\frac{1}{3}$ of the $\frac{3}{4}$ →

Note The fact that there are 12 divisions in the picture does not necessarily mean that there are 12 students in the class.

The darker shaded region represents $\frac{3}{12}$ of the whole picture, which simplifies to $\frac{1}{4}$.

◀ Do Your Turn 9

Objective 7 Find the area of a triangle.

In Section 1.6, we learned that the area of a parallelogram is $A = bh$. We can use this fact to develop the formula for the area of a triangle. By joining any two identical triangles as shown, we can create a parallelogram.

Answers to Your Turn 9

a. $\frac{1}{6}$ b. $\frac{7}{16}$

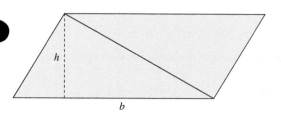

Note Because the triangles are identical, their angle measurements are identical. Rotating one triangle and joining it to the other along their corresponding side forms a four-sided figure with opposing angles that have the same measurement with opposing sides that are parallel. Therefore, the figure is a parallelogram.

 Learning Strategy

If you are a tactile learner, try creating the parallelogram just described. To do this, draw a triangle on a sheet of paper. Place a second sheet of paper behind the first. While holding the second paper tightly to the first, cut out the triangle you've drawn so that an identical triangle is cut out of the second sheet of paper. By rotating one of the two identical triangles and joining them along the corresponding side, you will create a parallelogram similar to the one shown.

Because the triangles are identical, one of them occupies half of the area of the parallelogram.

Area of a triangle = half of the area of a parallelogram with the same base and height

$$A = \frac{1}{2} \cdot bh$$

$$= \frac{1}{2} bh$$

Example 10   A recycling company produces triangular stickers that are $5\frac{1}{2}$ inches along the base and $4\frac{3}{4}$ inches high. Find the area.

Solution: Use the formula $A = \frac{1}{2} bh$.

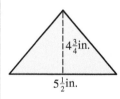

$A = \frac{1}{2}\left(5\frac{1}{2}\right)\left(4\frac{3}{4}\right)$ Replace b with $5\frac{1}{2}$ and h with $4\frac{3}{4}$.

$A = \frac{1}{2}\left(\frac{11}{2}\right)\left(\frac{19}{4}\right)$ Write the mixed numbers as improper fractions.

$A = \frac{209}{16}$ Multiply.

$A = 13\frac{1}{16}$ Write the result as a mixed number.

Answer: The area is $13\frac{1}{16}$ in.^{2}

▶ **Do Your Turn 10**

Objective 8 Find the radius and diameter of a circle.

Now let's learn about **circles.** The distance from the center of a circle to any point on the circle is the same for all points on the circle. This distance is called the **radius** (plural, *radii*). The distance across a circle along a straight line that passes through the center is the **diameter.**

▶ **Your Turn 10**

Find the area of each triangle.

a.

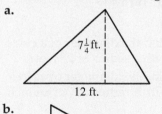

b.

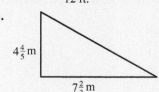

Answers to Your Turn 10

a. $43\frac{1}{2}$ ft.^{2} **b.** $18\frac{2}{5}$ m^2

Definitions Circle: A collection of points that are equally distant from a central point, called the *center*.

Radius: The distance from the center to any point on the circle.

Diameter: The distance across a circle along a straight line through the center.

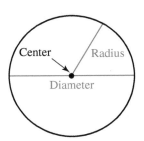

We can state the following relationships for diameter and radius:

The diameter is twice the radius.

The radius is half the diameter.

We can translate these relationships to equations.

The diameter is twice the radius.

$$d = 2 \cdot r$$

The radius is half the diameter.

$$r = \frac{1}{2} \cdot d$$

► **Your Turn 11**

Solve.

a. Find the diameter of a circle with a radius of $2\frac{3}{5}$ feet.

b. Find the diameter of a circle with a radius of $\frac{7}{8}$ of an inch.

Example 11   Find the diameter of the circle shown.

Solution: The radius is $3\frac{5}{8}$ centimeters. The diameter is twice the radius; so we can use $d = 2r$.

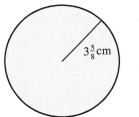

$3\frac{5}{8}$ cm

$d = 2 \cdot 3\frac{5}{8}$ In $d = 2r$, replace r with $3\frac{5}{8}$.

$d = \dfrac{\overset{1}{2}}{1} \cdot \dfrac{29}{\underset{4}{8}}$ Write the mixed number as an improper fraction. Then divide out the common factor, 2.

$d = \dfrac{29}{4}$ Multiply.

$d = 7\dfrac{1}{4}$ Write the result as a mixed number.

Answer: The diameter is $7\frac{1}{4}$ cm.

◄ **Do Your Turn 11**

Answers to Your Turn 11

a. $5\dfrac{1}{5}$ ft.   **b.** $1\dfrac{3}{4}$ in.

Example 12   Find the radius of the circle shown.

Solution: The diameter is 17 feet. The radius is half the diameter; so we can use $r = \frac{1}{2}d$.

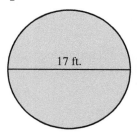

17 ft.

$r = \dfrac{1}{2} \cdot 17$ In $r = \frac{1}{2}d$, replace d with 17.

$r = \dfrac{1}{2} \cdot \dfrac{17}{1}$ Write 17 as an improper fraction.

$r = \dfrac{17}{2}$ Multiply.

$r = 8\dfrac{1}{2}$ Write the result as a mixed number.

Answer: The radius is $8\frac{1}{2}$ ft.

▶ Do Your Turn 12

Objective 9 Find the circumference of a circle.

Now that we have learned how diameter and radius are related, we can learn about **circumference**.

Definition Circumference: The distance around a circle.

When ancient mathematicians began experimenting with circles, they found an interesting property. When they measured the circumferences and diameters of different-sized circles and considered the ratio of circumference to diameter, they found that the fractions were nearly identical no matter how large or small the circles. The fractions were so close that mathematicians concluded that the ratio must be the same for all circles and the slight differences could be accounted for by the mathematicians' inability to accurately measure the circumference and diameter. It turns out that the value of this fraction cannot be expressed exactly, only closely approximated. We call such a number an **irrational number.**

Definition Irrational number: A number that cannot be expressed exactly as a fraction.

In math language: A number that cannot be expressed in the form $\frac{a}{b}$, where a and b are integers and $b \neq 0$.

Because the ratio of the circumference to the diameter is an irrational number, mathematicians decided to use a symbol to represent the value. They chose the Greek letter π (pronounced pī, as in *pie*).

Definition π: An irrational number that is the ratio of the circumference of a circle to its diameter.

In math language: $\pi = \dfrac{C}{d}$, where C represents the circumference and d represents the diameter of any circle.

A circle with a diameter of 7 inches will have a circumference of about 22 inches. This means that the value of π is approximately $\frac{22}{7}$. Symbolically, we write $\pi \approx \frac{22}{7}$. The symbol $\approx$ is read "is approximately equal to."

▶ Your Turn 12

Solve.

a. Find the radius of a circle with a diameter of 25 feet.

b. Find the radius of a circle with a diameter of $9\frac{3}{8}$ meters.

Answers to Your Turn 12

a. $12\dfrac{1}{2}$ ft.   **b.** $4\dfrac{11}{16}$ m

Remember, $\frac{22}{7}$ is only an approximation for the value of π. If we were to try to measure the circumference of a circle with a diameter of 7 inches, we would find the distance to be close to but not exactly 22 inches. As we wrapped the measuring tape around the circle, we would find that the circumference mark would never line up with an exact mark on the measuring tape no matter how precisely the tape could measure.

We can now use this approximate value of π to calculate circumference, given the diameter or radius of a circle. We can rearrange the relationship and say

$$C = \pi d$$

Or because $d = 2r$, we can say

$$C = 2\pi r$$

▶ **Your Turn 13**

Solve.

a. Find the circumference of a circle with a radius of $1\frac{1}{6}$ feet.

b. Find the circumference of a circle with a diameter of $5\frac{5}{6}$ meters.

Answers to Your Turn 13

a. $7\frac{1}{3}$ ft. b. $18\frac{1}{3}$ m

Example 13 Find the circumference of the circle shown.

Solution: We can use the formula $C = \pi d$ or $C = 2\pi r$. Because we are given the radius, let's use $C = 2\pi r$.

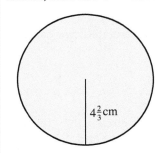

$4\frac{2}{3}$ cm

$C \approx 2 \cdot \frac{22}{7} \cdot 4\frac{2}{3}$ In $C = 2\pi r$, replace π with $\frac{22}{7}$ and r with $4\frac{2}{3}$.

$C \approx \frac{2}{1} \cdot \frac{\overset{2}{22}}{\underset{1}{7}} \cdot \frac{14}{3}$ Write as improper fractions and divide out a common factor of 7.

$C \approx \frac{88}{3}$ Multiply.

$C \approx 29\frac{1}{3}$ Write as a mixed number.

Answer: The circumference is about $29\frac{1}{3}$ cm.

Note When we replace π with $\frac{22}{7}$, the calculation becomes an approximation; so we use the $\approx$ symbol instead of an equal sign.

◀ **Do Your Turn 13**

5.3 Exercises For Extra Help MyMathLab®

Objective 1

Prep Exercise 1 Explain how to multiply two fractions.

For Exercises 1–8, multiply.

1. $\frac{2}{5} \cdot \frac{4}{7}$

2. $\frac{5}{8} \cdot \frac{1}{4}$

3. $-\frac{1}{6} \cdot \frac{1}{9}$

4. $-\frac{1}{4} \cdot \frac{1}{10}$

5. $-\frac{3}{4} \cdot \left(-\frac{5}{7}\right)$

6. $-\frac{9}{10} \cdot \left(-\frac{3}{10}\right)$

7. $\frac{7}{100} \cdot \frac{7}{10}$

8. $\frac{6}{11} \cdot \frac{10}{19}$

Objective 2

Prep Exercise 2 To simplify before multiplying fractions, divide out any _____ factor with any _____ factor.

For Exercises 9–24, multiply. Write the product in lowest terms.

9. $\dfrac{2}{3} \cdot \dfrac{5}{6}$

10. $\dfrac{3}{8} \cdot \dfrac{2}{7}$

11. $\dfrac{6}{7} \cdot \dfrac{14}{15}$

12. $\dfrac{5}{9} \cdot \dfrac{12}{25}$

13. $-\dfrac{8}{12} \cdot \dfrac{10}{20}$

14. $-\dfrac{14}{18} \cdot \dfrac{9}{21}$

15. $\dfrac{24}{32} \cdot \dfrac{26}{30}$

16. $\dfrac{15}{33} \cdot \dfrac{22}{27}$

17. $\dfrac{16}{38} \cdot \left(\dfrac{-57}{80}\right)$

18. $\dfrac{36}{42} \cdot \left(-\dfrac{28}{45}\right)$

19. $\dfrac{19}{20} \cdot \dfrac{16}{38}$

20. $\dfrac{25}{34} \cdot \dfrac{17}{30}$

21. $-\dfrac{18}{40} \cdot \left(-\dfrac{28}{30}\right)$

22. $-\dfrac{22}{35} \cdot \left(-\dfrac{15}{33}\right)$

23. $\dfrac{36}{-40} \cdot \left(\dfrac{-30}{54}\right)$

24. $\dfrac{-60}{81} \cdot \dfrac{45}{-72}$

Objective 3

Prep Exercise 3 Explain how to multiply two mixed numbers.

For Exercises 25–36, estimate and find the actual product expressed as a mixed number in simplest form.

25. $3\dfrac{1}{5} \cdot 1\dfrac{3}{4}$

26. $4\dfrac{1}{6} \cdot 3\dfrac{2}{5}$

27. $\dfrac{5}{6}(28)$

28. $12 \cdot \dfrac{7}{8}$

29. $4\dfrac{5}{8} \cdot 16$

30. $27 \cdot 2\dfrac{4}{9}$

31. $-6\dfrac{1}{8} \cdot \dfrac{4}{7}$

32. $2\dfrac{7}{10} \cdot \left(-8\dfrac{1}{3}\right)$

33. $-5\dfrac{2}{3} \cdot \left(-7\dfrac{1}{5}\right)$

34. $-3\dfrac{9}{16} \cdot \left(-6\dfrac{1}{8}\right)$

35. $\dfrac{-3}{10}(45)$

36. $-18 \cdot \dfrac{7}{24}$

Objective 4

For Exercises 37–48, multiply. Write the product in lowest terms.

37. $\dfrac{x^2}{5} \cdot \dfrac{2}{3}$

38. $\dfrac{3y}{7} \cdot \dfrac{2}{5}$

39. $\dfrac{4x}{9} \cdot \dfrac{3x}{8}$

40. $\dfrac{2}{15a} \cdot \dfrac{5}{6a^2}$

41. $\dfrac{xy}{10} \cdot \dfrac{4y^3}{9}$

42. $\dfrac{3m^2}{14} \cdot \dfrac{7mn}{12}$

43. $-\dfrac{5hk^3}{9} \cdot \dfrac{3}{4h}$

44. $\dfrac{9u}{20t^2} \cdot \left(-\dfrac{4tu^2}{15}\right)$

45. $-\dfrac{10x^4y}{11z} \cdot \left(-\dfrac{22z}{14x^2}\right)$

46. $\dfrac{-a^4b^3}{5c^3} \cdot \dfrac{10c^5}{9b}$

47. $\dfrac{9m^3}{25n^4} \cdot \left(\dfrac{-15n}{18m^2}\right)$

48. $-\dfrac{2t^6}{28t} \cdot \left(\dfrac{-14u^3}{18uv}\right)$

Objective 5

For Exercises 49–60, simplify.

49. $\left(\dfrac{5}{6}\right)^2$

50. $\left(\dfrac{4}{9}\right)^2$

51. $\left(-\dfrac{3}{4}\right)^3$

52. $\left(-\dfrac{1}{5}\right)^3$

53. $\left(\dfrac{-1}{2}\right)^6$

54. $\left(\dfrac{-2}{3}\right)^4$

55. $\left(\dfrac{x}{2}\right)^3$

56. $\left(\dfrac{3}{y}\right)^4$

57. $\left(\dfrac{2x^2}{3}\right)^3$

58. $\left(\dfrac{t^4}{4y}\right)^3$

59. $\left(-\dfrac{m^3n}{3p^2}\right)^4$

60. $\left(\dfrac{-2xy^4}{z^3}\right)^5$

Objective 6

For Exercises 61 and 62, use the following graph, which shows the portion of home energy spending that goes toward each category.

Home Energy Spending

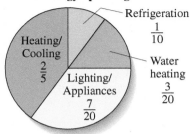

61. Suppose a family spent $1640 per year on energy for their home.

    a. How much was spent heating and cooling their home?
    b. How much was spent for lighting and appliances?
    c. How much was spent heating water?
    d. How much was spent on refrigeration?

62. Suppose a family spent $1860 per year on energy for their home.

    a. How much was spent heating and cooling their home?
    b. How much was spent for lighting and appliances?
    c. How much was spent heating water?
    d. How much was spent on refrigeration?

For Exercises 63–70, solve. Write all answers in lowest terms.

63. A shipping company estimates that $\frac{1}{8}$ of the produce shipped will go bad during shipment. If a single truck carries about 12,480 pieces of fruit, how many can be expected to go bad during shipment?

64. A single share of a certain computer company's stock is listed at $6\frac{3}{8}$ per share. What is the total value of 500 shares of the stock?

65. $\frac{5}{6}$ of a company's employees live within a 15-mile radius of the company. Of these employees, $\frac{3}{4}$ live within a 10-mile radius. What fraction of all employees live within a 10-mile radius?

66. $\frac{3}{4}$ of a lot is to be landscaped. $\frac{2}{3}$ of the landscaped area is to be covered in sod. What fraction of the lot is to be covered in sod?

67. On Tanya's phone bill, 24 out of the 30 long-distance calls were in-state. $\frac{5}{6}$ of the in-state calls were to her parents. What fraction of all of her long-distance calls were to her parents? How many calls were to her parents?

68. 16 out of the 24 people enrolled in a CPR course plan to go into a health-related field. $\frac{5}{8}$ of the people who intend to pursue a health field plan to go into nursing. What fraction of all of the people enrolled plan to become nurses? How many people is this?

69. On a standard-size guitar, the length of the strings between the bridge and the nut is $25\frac{1}{2}$ inches. Guitar makers must place the 12th fret at exactly half the length of the string. How far from the bridge or nut should a guitar maker measure to place the 12th fret?

70. A recipe calls for $1\frac{1}{3}$ cups of sugar. How much sugar should be used in making half of the recipe?

Of Interest

Placing one's finger at the 12th fret on any string produces a pitch that is one octave higher than the pitch of the string. To create an octave on any stringed instrument, the length of the string must be halved.

For Exercises 71 and 72, use d = rt.

71. A cyclist averages 16 miles per hour for $2\frac{2}{3}$ hours. What distance did the cyclist travel?

72. If a family drove $4\frac{3}{4}$ hours at an average rate of 65 miles per hour, how far did they travel?

Objective 7

Prep Exercise 4 What is the formula for the area of a triangle, and what does each variable represent?

For Exercises 73 and 74, find the area of each triangle.

73.

$8\frac{1}{2}$ ft.

12 ft.

74.

$6\frac{5}{8}$ in.

$9\frac{1}{4}$ in.

75. The siding in the triangular portion of a dormer needs to be replaced. What is the area covered by the siding?

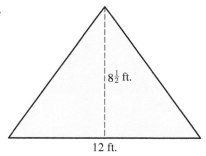

$3\frac{1}{3}$ ft.

$5\frac{1}{2}$ ft.

76. A company sells a designer triangular dinner plate. Find the area of the plate.

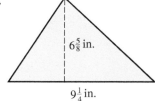

$9\frac{1}{2}$ in.

11 in.

Objectives 8 and 9

Prep Exercise 5 What are the radius and diameter of a circle?

Prep Exercise 6 How are radius and diameter related?

Prep Exercise 7 What is the circumference of a circle?

Prep Exercise 8 What is an irrational number?

Prep Exercise 9 What is π?

Prep Exercise 10 What are the two formulas for calculating the circumference of a circle?

For Exercises 77–80, solve. Use $\frac{22}{7}$ for π.

77. A giant sequoia tree has a diameter of 9 meters.

   a. What is its radius?
   b. What is the circumference?

78. Big Ben is the name of the clock in the tower of London's Westminster Palace. The face of the clock is a circle with a diameter of 23 feet.

   a. What is the radius?
   b. What is the circumference?

79. The Fermi National Accelerator Laboratory is a circular tunnel that is used to accelerate elementary particles. The radius of the tunnel is $\frac{273}{440}$ miles.

   a. What is the diameter?
   b. What is the circumference?

80. The radius of the bottom of the Liberty Bell is $22\frac{87}{88}$ inches.

   a. What is the diameter?
   b. What is the circumference?

Puzzle Problem The Milky Way galaxy in which our solar system exists is a disk shape with spiral arms. The diameter of the galaxy is approximately 100,000 light-years. If our solar system is about $\frac{2}{3}$ of the distance from the center to the edge of the galaxy, how far is our solar system from the center?

Review Exercises

[2.4] 1. Divide. $1836 \div (-9)$

[2.4] 2. Evaluate. $\sqrt{256}$

[3.6] 3. Factor. $9x^2 - 12x$

[2.4] 4. Solve and check. $-6x = 30$

[1.4] 5. A 12-foot board is to be cut into 6 pieces. How long will each piece be?

5.4 Dividing Fractions, Mixed Numbers, and Rational Expressions

Warm-up *[5.3] Multiply. Write the product in lowest terms.*

1. $-\dfrac{9}{10}\cdot\dfrac{8}{21}$

2. $\left(-5\dfrac{1}{3}\right)\left(-4\dfrac{1}{2}\right)$

Objectives

1 Divide fractions.
2 Divide mixed numbers.
3 Divide rational expressions.
4 Find the square root of a fraction.
5 Solve equations involving fractions.
6 Solve applications involving division of fractions.

Objective 1   Divide fractions.

Suppose we want change for a $20 bill in $5 bills. We can determine the number of $5 bills we receive using division. Because $20 \div 5 = 4$, we would receive four $5 bills.

Now suppose we want the change in quarters. To determine the number of quarters in $20, we can use a division statement similar to the one we used for finding the number of $5 bills in $20. Remember, a quarter is $\frac{1}{4}$ of a dollar; so we can write the following division statement:

$$\text{number of quarters in \$20} = 20 \div \frac{1}{4}$$

Alternatively, because there are 4 quarters in each dollar, we can find the number of quarters in $20 by multiplying 20 by 4.

$$\text{number of quarters in \$20} = 20 \cdot 4 = 80 \text{ quarters}$$

This means that $20 \div \frac{1}{4}$ and $20 \cdot 4$ are equivalent.

$$\text{number of quarters in \$20} = 20 \div \frac{1}{4} = 20 \cdot 4 = 80 \text{ quarters}$$

In our example, $\frac{1}{4}$ and 4 are **multiplicative inverses**, or **reciprocals.**

Definition   Reciprocals: Two numbers whose product is 1.

Notice that the product of $\frac{1}{4}$ and 4 is 1.

$$\frac{1}{4}\cdot 4 = \frac{1}{\overset{1}{\cancel{4}}}\cdot\frac{\overset{1}{\cancel{4}}}{1} = 1$$

Example 1	Find the reciprocal.

a. $\dfrac{3}{4}$

Note To find the reciprocal, *invert* the numerator and denominator. In other words, write the numerator in the denominator and the denominator in the numerator.

Answer: $\dfrac{4}{3}$

Explanation: The product of $\frac{3}{4}$ and $\frac{4}{3}$ is 1.

$$\frac{\overset{1}{\cancel{3}}}{\underset{1}{\cancel{4}}}\cdot\frac{\overset{1}{\cancel{4}}}{\underset{1}{\cancel{3}}} = 1$$

b. $\dfrac{1}{5}$

Answer: 5

Explanation: The product of $\frac{1}{5}$ and 5 is 1.

$$\frac{1}{5}\cdot 5 = \frac{1}{\underset{1}{\cancel{5}}}\cdot\frac{\overset{1}{\cancel{5}}}{1} = 1$$

Connection Remember that we use additive inverses to write subtraction statements as equivalent addition statements. Similarly, we use *multiplicative inverses* to write division statements as equivalent multiplication statements. Also, additive inverses *undo* one another in that their sum is 0. Similarly, multiplicative inverses *undo* one another in that their product is 1.

Answers to Warm-up
1. $-\dfrac{12}{35}$ 2. 24

▶ **Your Turn 1**

Find the reciprocal.

a. $\dfrac{3}{8}$ b. $\dfrac{1}{4}$ c. -2 d. $-\dfrac{5}{11}$

c. 6

Answer: $\dfrac{1}{6}$

Explanation: The product of 6 and $\frac{1}{6}$ is 1. $6 \cdot \dfrac{1}{6} = \dfrac{\overset{1}{\cancel{6}}}{1} \cdot \dfrac{1}{\underset{1}{\cancel{6}}} = 1$

d. $-\dfrac{5}{7}$

Answer: $-\dfrac{7}{5}$ ◀ **Note** The sign stays the same. The reciprocal of a number has the same sign as the number.

Explanation: The product of $-\frac{5}{7}$ and $-\frac{7}{5}$ is 1. $-\dfrac{\overset{1}{\cancel{5}}}{7} \left(-\dfrac{\overset{1}{7}}{\underset{1}{\cancel{5}}} \right) = 1$

e. 0

Answer: 0 has no reciprocal.

Explanation: The product of 0 and any number is 0; hence, the product can never equal 1.

◀ **Do Your Turn 1**

We use reciprocals in our procedure for dividing fractions.

> ## Procedure
>
> To divide fractions:
>
> 1. Write an equivalent multiplication by changing the operation symbol from division to multiplication and changing the divisor to its reciprocal.
> 2. Multiply. Use either method described in the procedure for multiplying fractions. (Section 5.3)
> 3. Simplify. If the result is an improper fraction, write it as a mixed number in simplest form.

Example 2   Divide. Write the quotient in lowest terms.

a. $\dfrac{5}{8} \div \dfrac{3}{4}$

Solution: $\dfrac{5}{8} \div \dfrac{3}{4} = \dfrac{5}{8} \cdot \dfrac{4}{3}$ Write an equivalent multiplication statement.

$= \dfrac{5}{\underset{1}{2} \cdot \underset{1}{2} \cdot 2} \cdot \dfrac{\overset{1}{2} \cdot \overset{1}{2}}{3}$ Divide out the common factors.

$= \dfrac{5}{6}$ Multiply.

b. $\dfrac{-4}{9} \div (-12)$

Solution: $\dfrac{-4}{9} \div (-12) = \dfrac{-4}{9} \cdot \left(-\dfrac{1}{12} \right)$ Write an equivalent multiplication statement.

$= \dfrac{-\overset{1}{2} \cdot \overset{1}{2}}{9} \cdot \left(-\dfrac{1}{\underset{1}{2} \cdot \underset{1}{2} \cdot 3} \right)$ Divide out the common factors.

$= \dfrac{1}{27}$ Multiply.

Connection The quotient of two numbers that have the same sign is positive.

Answers to Your Turn 1

a. $\dfrac{8}{3}$ b. 4   c. $-\dfrac{1}{2}$ d. $-\dfrac{11}{5}$

Warning It may be tempting to divide out common factors in the division statement, but we can divide out common factors *only* when multiplying.

Recall that we can write division using a fraction line. For example, we can write $18 \div 3 = 6$ as $\frac{18}{3} = 6$. If we express the division of fractions using fraction form, we call the expression a **complex fraction.**

Definition **Complex fraction:** A fractional expression with a fraction in the numerator and/or denominator.

For example, $\dfrac{\frac{5}{8}}{\frac{3}{4}}$ is a complex fraction equivalent to $\dfrac{5}{8} \div \dfrac{3}{4}$.

Because most people prefer the regular division form, if we are given a complex fraction, we rewrite it in that form.

▶ **Do Your Turn 2**

Objective 2 Divide mixed numbers.

We learned in Section 5.3 that when we multiply mixed numbers, we first write them as improper fractions. The same process applies to division.

Procedure

To divide mixed numbers:
1. Write the mixed numbers as improper fractions.
2. Write an equivalent multiplication by changing the operation symbol from division to multiplication and changing the divisor to its reciprocal.
3. Multiply. Use either method described in the procedure for multiplying fractions. (Section 5.3)
4. Simplify. If the result is an improper fraction, write it as a mixed number in simplest form.

Example 3 Estimate and then find the actual quotient expressed as a mixed number in simplest form.

a. $8\dfrac{2}{5} \div 2\dfrac{1}{4}$

Estimate: Round each mixed number to the nearest integer; then divide.

$8\frac{2}{5}$ rounds down to 8 because it is closer to 8 than to 9.　　　$2\frac{1}{4}$ rounds down to 2 because it is closer to 2 than to 3.

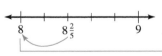　　　　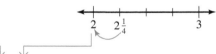

$8 \div 2 = 4$　　The actual quotient should be around 4.

Actual: $8\dfrac{2}{5} \div 2\dfrac{1}{4} = \dfrac{42}{5} \div \dfrac{9}{4}$　　Write the mixed numbers as improper fractions.

$= \dfrac{42}{5} \cdot \dfrac{4}{9}$　　Write an equivalent multiplication statement.

$= \dfrac{2 \cdot 3 \cdot 7}{5} \cdot \dfrac{4}{\underset{1}{3 \cdot 3}}$　　Divide out the common factor 3.

$$= \frac{56}{15}$$ Multiply.

$$= 3\frac{11}{15}$$ Write the improper fraction as a mixed number.

The actual quotient agrees with our estimate.

b. $\dfrac{-5\dfrac{3}{7}}{3\dfrac{1}{3}}$

Estimate: Round each mixed number to the nearest integer; then divide.

$-5\frac{3}{7}$ rounds to -5 because it is closer to -5 than to -6. $3\frac{1}{3}$ rounds to 3 because it is closer to 3 than to 4.

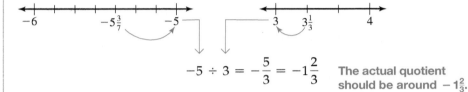

$$-5 \div 3 = -\frac{5}{3} = -1\frac{2}{3}$$ The actual quotient should be around $-1\frac{2}{3}$.

Actual: $-5\dfrac{3}{7} \div 3\dfrac{1}{3} = -\dfrac{38}{7} \div \dfrac{10}{3}$ Write the mixed numbers as improper fractions.

$$= -\frac{38}{7} \cdot \frac{3}{10}$$ Write an equivalent multiplication statement.

$$= -\frac{\overset{1}{2} \cdot 19}{7} \cdot \frac{3}{\underset{1}{2} \cdot 5}$$ Divide out the common factor, 2.

$$= -\frac{57}{35}$$ Multiply.

$$= -1\frac{22}{35}$$ Write the improper fraction as a mixed number.

The actual quotient agrees with our estimate.

c. $4\dfrac{3}{5} \div \dfrac{3}{10}$

Estimate: The nearest integer to $\dfrac{3}{10}$ is 0. But dividing by 0 is undefined; so we round $\dfrac{3}{10}$ to a more comfortable fraction near its value, such as $\dfrac{1}{3}$.

$4\frac{3}{5}$ rounds to 5 because it is closer to 5 than to 4.   We round $\frac{3}{10}$ to $\frac{1}{3}$.

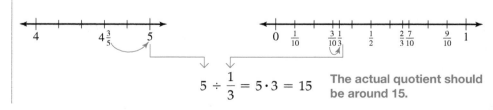

$$5 \div \frac{1}{3} = 5 \cdot 3 = 15$$ The actual quotient should be around 15.

Actual: $4\dfrac{3}{5} \div \dfrac{3}{10} = \dfrac{23}{5} \div \dfrac{3}{10}$ Write the mixed number as an improper fractions.

$$= \dfrac{23}{5} \cdot \dfrac{10}{3}$$ Write an equivalent multiplication statement.

$$= \dfrac{23}{\underset{1}{5}} \cdot \dfrac{2 \cdot \overset{1}{5}}{3}$$ Divide out the common factor, 5.

$$= \dfrac{46}{3}$$ Multiply.

$$= 15\dfrac{1}{3}$$ Write the improper fraction as a mixed number.

The actual quotient agrees with our estimate.

▶ **Do Your Turn 3**

Objective 3 Divide rational expressions.

To divide rational expressions, we follow the same procedure as with numeric fractions.

Example 4   Divide. Write the quotient in lowest terms. $-\dfrac{6a^3b}{35c^4} \div \dfrac{15a}{14c^2}$

Solution: $-\dfrac{6a^3b}{35c^4} \div \dfrac{15a}{14c^2} = -\dfrac{6a^3b}{35c^4} \cdot \dfrac{14c^2}{15a}$ Write an equivalent multiplication statement.

$$= -\dfrac{2 \cdot 3 \cdot \overset{1}{a} \cdot a \cdot a \cdot b}{5 \cdot 7 \cdot \underset{1}{c} \cdot \underset{1}{c} \cdot c \cdot c} \cdot \dfrac{2 \cdot 7 \cdot \overset{1}{c} \cdot \overset{1}{c}}{\underset{1}{3} \cdot 5 \cdot \underset{1}{a}}$$ Write in factored form and divide out common factors.

$$= -\dfrac{4a^2b}{25c^2}$$ Multiply the remaining factors.

▶ **Do Your Turn 4**

Objective 4 Find the square root of a fraction.

Remember that a square root of a given number is a number whose square is the given number. For example,

$$\sqrt{\dfrac{9}{16}} = \dfrac{3}{4} \text{ because } \left(\dfrac{3}{4}\right)^2 = \dfrac{3}{4} \cdot \dfrac{3}{4} = \dfrac{3^2}{4^2} = \dfrac{9}{16}.$$

Our example suggests that we can find the square root of a fraction by finding the square root of the numerator and denominator separately.

$$\sqrt{\dfrac{9}{16}} = \dfrac{\sqrt{9}}{\sqrt{16}} = \dfrac{3}{4}$$

If a numerator or denominator is not a perfect square, we try simplifying first to see if the simplified number is a perfect square. Consider $\sqrt{\dfrac{6}{24}}$.

$$\sqrt{\dfrac{6}{24}} = \sqrt{\dfrac{1}{4}} = \dfrac{\sqrt{1}}{\sqrt{4}} = \dfrac{1}{2}$$

▶ **Your Turn 3**

Estimate and then find the actual quotient expressed as a mixed number in simplest form.

a. $5\dfrac{1}{4} \div 2\dfrac{1}{3}$ b. $\dfrac{7\dfrac{2}{7}}{7\dfrac{1}{2}}$

c. $-13\dfrac{1}{5} \div 5\dfrac{1}{10}$

d. $-4\dfrac{1}{2} \div \left(-11\dfrac{2}{5}\right)$

▶ **Your Turn 4**

Divide. Write the quotient in lowest terms.

a. $\dfrac{8}{13m} \div \dfrac{12m}{5}$

b. $\dfrac{10x^3}{14} \div \dfrac{15x}{9}$

c. $-\dfrac{11n^3}{12p^4} \div \dfrac{33n^3}{18p}$

d. $-\dfrac{21c}{30a^3b} \div \left(-\dfrac{14bc^3}{20a^4}\right)$

Answers to Your Turn 3

a. $2\dfrac{1}{2}; 2\dfrac{1}{4}$ b. $\dfrac{7}{8}; \dfrac{34}{35}$

c. $-2\dfrac{3}{5}; -2\dfrac{10}{17}$ d. $\dfrac{5}{11}; \dfrac{15}{38}$

Answers to Your Turn 4

a. $\dfrac{10}{39m^2}$ b. $\dfrac{3x^2}{7}$

c. $\dfrac{-1}{2p^3}$ d. $\dfrac{a}{b^2c^2}$

Our examples suggest the following procedure:

> **Procedure**
>
> To find the square root of a fraction, try the following:
> Find the square root of the numerator and denominator separately; then simplify.
> *or*
> Simplify the fraction; then find the square root of the quotient.

Note Some expressions such as ▶
$\sqrt{\frac{5}{7}}$ cannot be simplified using
either technique. You will learn
how to simplify those expressions
in other courses.

▶ **Your Turn 5**

Simplify.

a. $\sqrt{\dfrac{4}{9}}$ b. $\sqrt{\dfrac{49}{100}}$

c. $\sqrt{\dfrac{125}{5}}$ d. $\sqrt{\dfrac{392}{8}}$

Example 5   Simplify.

a. $\sqrt{\dfrac{25}{36}}$

Solution: $\sqrt{\dfrac{25}{36}} = \dfrac{\sqrt{25}}{\sqrt{36}} = \dfrac{5}{6}$ Find the square root of the numerator and denominator separately.

b. $\sqrt{\dfrac{45}{5}}$

Solution: $\sqrt{\dfrac{45}{5}} = \sqrt{9} = 3$ 45 and 5 are not perfect squares, but $\frac{45}{5}$ simplifies to 9, which is a perfect square.

◀ **Do Your Turn 5**

Objective 5 Solve equations involving fractions.

In Chapter 4, we solved equations such as $2x = 10$ by dividing both sides of the equation by 2 to eliminate the coefficient. We called that action the multiplication principle of equality, which seemed odd because we were dividing. Now we can make sense of it. Dividing by a number is equivalent to multiplying by its reciprocal; so we could solve $2x = 10$ by multiplying both sides by $\frac{1}{2}$.

Dividing by 2	**Multiplying by $\dfrac{1}{2}$**
$\dfrac{\overset{1}{\cancel{2}}x}{\underset{1}{\cancel{2}}} = \dfrac{10}{2}$ Divide both sides by 2.	$\dfrac{1}{\underset{1}{\cancel{2}}} \cdot \overset{1}{\cancel{2}}x = \dfrac{1}{\underset{1}{\cancel{2}}} \cdot \overset{5}{\cancel{10}}$ Multiply both sides by $\frac{1}{2}$.
$x = 5$	$x = 5$

We choose whichever approach seems easiest. For integer coefficients, dividing both sides by the coefficient is preferred. However, for a fraction coefficient, multiplying both sides by its reciprocal is preferred.

Example 6   Solve and check. $\dfrac{3}{4}x = 2\dfrac{5}{8}$

Solution: First, write the mixed number as an improper fraction, so that the equation is $\frac{3}{4}x = \frac{21}{8}$. Then eliminate the coefficient $\frac{3}{4}$ by multiplying both sides by its reciprocal, $\frac{4}{3}$.

$$\frac{4}{3} \cdot \frac{3}{4}x = \frac{21}{8} \cdot \frac{4}{3}$$ Multiply both sides by $\frac{4}{3}$.

$$\frac{\overset{1}{\cancel{4}}}{\underset{1}{\cancel{3}}} \cdot \frac{\overset{1}{\cancel{3}}}{\underset{1}{\cancel{4}}}x = \frac{\overset{1}{\cancel{3}} \cdot 7}{2 \cdot 2 \cdot \cancel{2}} \cdot \frac{\overset{1}{\cancel{2} \cdot \cancel{2}}}{\cancel{3}}$$ Divide out the common factors.

$$1x = \frac{7}{2}$$ Multiply.

$$x = \frac{7}{2}, \text{ or } 3\frac{1}{2}$$

Answers to Your Turn 5

a. $\dfrac{2}{3}$ b. $\dfrac{7}{10}$ c. 5   d. 7

Check: $\frac{3}{4}x = 2\frac{5}{8}$

$\frac{3}{4} \cdot \frac{7}{2} \stackrel{?}{=} 2\frac{5}{8}$ Replace x in $\frac{3}{4}x = 2\frac{5}{8}$ with $\frac{7}{2}$ (or $3\frac{1}{2}$) and verify that the equation is true.

$\frac{21}{8} \stackrel{?}{=} 2\frac{5}{8}$

$2\frac{5}{8} = 2\frac{5}{8}$ True; so $\frac{7}{2}$ is the solution.

▶ **Do Your Turn 6**

Example 7   Solve and check. $\frac{m}{5} = -3\frac{1}{10}$

Solution: Because $\frac{m}{5}$ is the same as $\frac{1m}{5}$, and $\frac{1}{5}m$, the coefficient is $\frac{1}{5}$. To isolate m, we multiply both sides of the equation by the reciprocal of $\frac{1}{5}$, which is $\frac{5}{1}$.

$\frac{m}{5} = -\frac{31}{10}$ Write $-3\frac{1}{10}$ as an improper fraction, $-\frac{31}{10}$.

$\frac{5}{1} \cdot \frac{m}{5} = -\frac{31}{10} \cdot \frac{5}{1}$ Multiply both sides by $\frac{5}{1}$.

$\frac{\overset{1}{5}}{1} \cdot \frac{m}{\underset{1}{5}} = -\frac{31}{2 \cdot \overset{}{5}} \cdot \frac{\overset{1}{5}}{1}$ Divide out the common factors.

$1m = -\frac{31}{2}$ Multiply.

$m = -\frac{31}{2}$, or $-15\frac{1}{2}$

Check: $\frac{m}{5} = -3\frac{1}{10}$

$\frac{-\frac{31}{2}}{5} \stackrel{?}{=} -3\frac{1}{10}$ Replace m in $\frac{m}{5} = -3\frac{1}{10}$ with $-\frac{31}{2}$ (or $-15\frac{1}{2}$) and verify that the equation is true.

$-\frac{31}{2} \div 5 \stackrel{?}{=} -3\frac{1}{10}$

$-\frac{31}{2} \cdot \frac{1}{5} \stackrel{?}{=} -3\frac{1}{10}$

$-\frac{31}{10} \stackrel{?}{=} -3\frac{1}{10}$

$-3\frac{1}{10} = -3\frac{1}{10}$ True; so $-\frac{31}{2}$ is the solution.

▶ **Do Your Turn 7**

▶ **Your Turn 6**

Solve and check.

a. $\frac{3}{4}a = \frac{2}{3}$ b. $\frac{5}{6}y = \frac{3}{10}$

c. $\frac{4}{9}k = -\frac{8}{15}$ d. $\frac{-4}{15} = -\frac{1}{6}n$

▶ **Your Turn 7**

Solve and check.

a. $\frac{x}{10} = 6$ b. $\frac{y}{7} = \frac{5}{14}$

c. $\frac{-3}{4} = \frac{n}{8}$ d. $\frac{h}{-9} = -2\frac{1}{3}$

Note We get a complex fraction ◀ when we replace m with $-\frac{31}{2}$. Remember, the complex fraction means that we divide.

Answers to Your Turn 6

a. $a = \frac{8}{9}$ b. $y = \frac{9}{25}$

c. $k = -1\frac{1}{5}$ d. $n = 1\frac{3}{5}$

Answers to Your Turn 7

a. $x = 60$ b. $y = 2\frac{1}{2}$

c. $n = -6$ d. $h = 21$

Objective 6 Solve applications involving division of fractions.

Let's return to the problem about changing a $20 bill into quarters. At the beginning of this section, we translated the problem to division: $20 \div \frac{1}{4}$. You should note that $20 is the amount that we split up and that the size of each division is a quarter, $\frac{1}{4}$. We can also translate to an equation with an unknown factor.

$$\text{Value of the currency} \cdot \text{number of coins/bills} = \text{total money}$$

$$\frac{1}{4} \qquad \cdot \qquad n \qquad = \qquad 20$$

$$\frac{1}{4}n = 20$$

$$\frac{\overset{1}{\cancel{4}}}{1} \cdot \frac{1}{\underset{1}{\cancel{4}}}n = \frac{20}{1} \cdot \frac{4}{1} \qquad \text{Multiply both sides by 4.}$$

$$n = 80$$

Note We end up multiplying 20 by 4 just as we did earlier in this section.

Conclusion: Whenever we are given a total amount and either the size of the parts or the number of parts, we can write an equation with an unknown factor using the following formula:

$$\text{Size of each part} \cdot \text{number of parts} = \text{whole amount}$$

Of course, if you recognize the situation for what it is, go directly to the related division equation.

$$\text{Size of each part} = \text{whole amount} \div \text{number of parts}$$
or
$$\text{Number of parts} = \text{whole amount} \div \text{size of each part}$$

Example 8 Solve.

a. A board is $35\frac{3}{4}$ inches long. The board is to be cut into pieces that are each $6\frac{1}{2}$ inches long. How many pieces can be cut?

Solution: We will write an equation with an unknown factor, then solve.

$$\text{size of each part} \cdot \text{number of parts} = \text{whole amount}$$
$$6\frac{1}{2} \qquad \cdot \qquad n \qquad = \qquad 35\frac{3}{4}$$

Translate to an equation where n represents the number of parts.

$$\frac{13}{2}n = \frac{143}{4} \qquad \text{Write the mixed numbers as improper fractions.}$$

$$\frac{2}{13} \cdot \frac{13}{2}n = \frac{143}{4} \cdot \frac{2}{13} \qquad \text{Multiply both sides by } \frac{2}{13}.$$

$$\frac{\overset{1}{\cancel{2}}}{\underset{1}{\cancel{13}}} \cdot \frac{\overset{1}{\cancel{13}}}{\underset{1}{\cancel{2}}}n = \frac{11 \cdot \overset{1}{\cancel{13}}}{\underset{1}{\cancel{2}} \cdot 2} \cdot \frac{\overset{1}{\cancel{2}}}{\underset{1}{\cancel{13}}} \qquad \text{Divide out the common factors.}$$

$$1n = \frac{11}{2} \qquad \text{Multiply.}$$

$$n = 5\frac{1}{2} \qquad \text{Write as a mixed number.}$$

Answer: The board can be cut into $5\frac{1}{2}$ pieces that are each $6\frac{1}{2}$ inches long.

▲

Note This answer means that we could cut 5 whole pieces that are each $6\frac{1}{2}$ inches long and would be left with $\frac{1}{2}$ of another $6\frac{1}{2}$-inch piece.

Connection We could check Example 8a or 8b by multiplying. For example, in 8a, multiplying $6\frac{1}{2}$ by the solution $5\frac{1}{2}$ should produce the length of the whole board, $35\frac{3}{4}$.

$$6\frac{1}{2} \cdot 5\frac{1}{2} \overset{?}{=} 35\frac{3}{4}$$

$$\frac{13}{2} \cdot \frac{11}{2} \overset{?}{=} 35\frac{3}{4}$$

$$\frac{143}{4} \overset{?}{=} 35\frac{3}{4}$$

$$35\frac{3}{4} = 35\frac{3}{4} \qquad \text{It checks.}$$

b. The circumference of a circle is $20\frac{3}{7}$ inches. Find the radius.

Solution:

$$C = 2\pi r$$

$$20\frac{3}{7} \approx 2\left(\frac{22}{7}\right)r$$

In the formula $C = 2\pi r$, replace C with $20\frac{3}{7}$ and π with $\frac{22}{7}$, then solve for r.

$$\frac{143}{7} \approx \frac{2}{1}\left(\frac{22}{7}\right)r$$

Write improper fractions.

$$\frac{143}{7} \approx \frac{44}{7}r$$

Multiply.

$$\frac{7}{44} \cdot \frac{143}{7} \approx \frac{7}{44} \cdot \frac{44}{7}r$$

Multiply both sides by $\frac{7}{44}$.

$$\frac{\overset{1}{7}}{2 \cdot 2 \cdot \underset{1}{11}} \cdot \frac{\overset{1}{111} \cdot 13}{\underset{1}{7}} \approx \frac{\overset{1}{7}}{44} \cdot \frac{\overset{1}{44}}{\underset{1}{7}}r$$

Divide out the common factors.

$$\frac{13}{4} \approx 1r$$

Multiply.

$$\frac{13}{4} \quad \text{or} \quad 3\frac{1}{4} \approx r$$

Answer: The radius of the circle is approximately $3\frac{1}{4}$ inches.

▶ **Do Your Turn 8**

▶ **Your Turn 8**

Solve.

a. How many $4\frac{1}{2}$-ounce servings are in $30\frac{1}{4}$ ounces of cereal?

b. A circle has a circumference of $25\frac{41}{56}$ inches. What is the diameter?

Answers to Your Turn 8

a. $6\frac{13}{18}$ servings   **b.** $8\frac{3}{16}$ in.

5.4 Exercises   For Extra Help   MyMathLab®

Objective 1

Prep Exercise 1 What are reciprocals? Give an example of a pair of reciprocals.

For Exercises 1–8, find the reciprocal.

1. $\frac{2}{3}$

2. $\frac{3}{8}$

3. $\frac{1}{6}$

4. $-\frac{1}{9}$

5. -15

6. 7

7. $-\frac{5}{4}$

8. $-\frac{9}{7}$

Prep Exercise 2 Explain how to divide fractions.

For Exercises 9–20, divide. Write the quotient in lowest terms.

9. $\frac{1}{2} \div \frac{1}{6}$

10. $\frac{5}{8} \div \frac{2}{3}$

11. $\frac{3}{10} \div \frac{5}{6}$

12. $\frac{9}{16} \div \frac{3}{4}$

13. $\frac{14}{15} \div \left(-\frac{7}{12}\right)$

14. $-\frac{10}{13} \div \frac{5}{12}$

15. $\frac{-7}{12} \div (-14)$

16. $-6 \div \left(-\frac{3}{4}\right)$

17. $\dfrac{\frac{2}{5}}{\frac{8}{15}}$

18. $\dfrac{\frac{9}{10}}{\frac{21}{10}}$

19. $\dfrac{12}{-\frac{2}{3}}$

20. $\dfrac{-\frac{3}{5}}{18}$

Objective 2

Prep Exercise 3 Explain how to divide mixed numbers.

For Exercises 21–28, estimate and find the actual quotient expressed as a mixed number in simplest form.

21. $5\frac{1}{6} \div \frac{4}{9}$

22. $7\frac{1}{3} \div \frac{3}{8}$

23. $9\frac{1}{2} \div 3\frac{3}{4}$

24. $2\frac{7}{12} \div 2\frac{1}{6}$

25. $-2\frac{9}{16} \div 10\frac{1}{4}$

26. $4\frac{2}{9} \div \left(-8\frac{1}{3}\right)$

27. $\dfrac{-12\frac{1}{2}}{-6\frac{3}{4}}$

28. $\dfrac{-10\frac{2}{3}}{-2\frac{5}{6}}$

Objective 3

For Exercises 29–34, divide. Write the quotient in lowest terms.

29. $\dfrac{7}{15x} \div \dfrac{1}{6x^3}$

30. $\dfrac{ab}{10} \div \dfrac{4b^3}{25}$

31. $\dfrac{9m^5}{20n} \div \dfrac{12m^2n}{25}$

32. $\dfrac{t^2u^4}{18v^3} \div \left(\dfrac{-10t^5}{12v}\right)$

33. $\dfrac{14x^6}{28y^4z} \div \left(\dfrac{-8x^2y}{18z}\right)$

34. $-\dfrac{13h^6}{20hk} \div \left(\dfrac{-65h^3k}{25k^2}\right)$

Objective 4

Prep Exercise 4 What are two approaches for finding the square root of a fraction?

For Exercises 35–42, simplify.

35. $\sqrt{\dfrac{64}{81}}$

36. $\sqrt{\dfrac{49}{100}}$

37. $\sqrt{\dfrac{121}{36}}$

38. $\sqrt{\dfrac{169}{100}}$

39. $\sqrt{\dfrac{180}{5}}$

40. $\sqrt{\dfrac{960}{15}}$

41. $\sqrt{\dfrac{1053}{13}}$

42. $\sqrt{\dfrac{3400}{34}}$

Objective 5

Prep Exercise 5 Explain how to use the multiplication principle of equality to solve an equation such as $\frac{4}{5}x = \frac{7}{9}$.

For Exercises 43–50, solve and check.

43. $\dfrac{3}{4}x = 12$

44. $\dfrac{2}{5}a = 10$

45. $\dfrac{5}{8} = \dfrac{-3}{16}y$

46. $\dfrac{7}{10} = \dfrac{-1}{5}x$

47. $\dfrac{n}{12} = -\dfrac{3}{20}$

48. $-\dfrac{1}{16} = \dfrac{k}{14}$

49. $\dfrac{-5m}{6} = \dfrac{-5}{21}$

50. $-\dfrac{9}{40} = -\dfrac{3u}{10}$

Objective 6

For Exercises 51–62, solve.

51. A child is to receive $1\frac{1}{3}$ teaspoons of amoxicillin (an antibiotic). How many doses can be given if the bottle contains 40 teaspoons? If the instructions are to give the dosage twice a day for 14 days, is the amount of amoxicillin in the bottle correct? Explain.

52. A cereal box contains $9\frac{1}{4}$ cups of cereal. The label indicates that a serving size is $\frac{3}{4}$ of a cup. How many servings are in a box of the cereal? The label indicates that there are about 12 servings per container. Is this accurate?

53. A baker had only enough ingredients to make $10\frac{1}{2}$ cups of batter for muffins. If he is to make a dozen muffins of equal size, how much of the batter should he place in each cup of the muffin pan?

54. Ellyn makes curtains. She wants to make 4 curtains of equal length from 78 inches of material. How long should each piece be?

 55. A rectangular field has an area of $2805\frac{1}{2}$ square feet and a width of $30\frac{1}{2}$ feet. What is the length?

56. A photograph has an area of $18\frac{1}{3}$ square inches and a length of $5\frac{1}{2}$ inches. What is its width?

57. The desired area for a triangular garden is $14\frac{1}{2}$ square feet. If the base is to be $6\frac{1}{2}$ feet, what must the height be?

58. A triangular company logo is to be $1\frac{3}{4}$ square inches. If it is to be $2\frac{1}{3}$ inches high, what must its base be?

59. The largest radio telescope in the world is the Arecibo dish located in Puerto Rico. It has a circumference of $3111\frac{2}{5}$ feet. What is the diameter?

60. A circular track has a circumference of $\frac{1}{4}$ mile. What is the radius?

61. Using the odometer in his car, Ron finds that the distance from his home to his office building is $18\frac{3}{10}$ miles. If it takes him $\frac{2}{5}$ hour to drive from home to his office building, what is his average rate?

62. How much time does it take an F-16 fighter plane traveling at Mach 2 to travel 20 miles? Mach 2 is about $\frac{2}{5}$ mile per second, which is twice the speed of sound.

Review Exercises

[3.2] 1. Combine like terms.
$$5t^3 - 8t + 9t^2 + 4t^3 - 16 - 5t$$

[3.4] 2. Simplify. $2^2 \cdot 3 \cdot 5 \cdot x^3 \cdot y$

[3.5] 3. Find the prime factorization of 378.

[3.6] 4. Factor. $32x^2 + 16x$

[5.1] 5. Use $<, >,$ or $=$ to write a true statement.
$$\frac{5}{6} \quad ? \quad \frac{10}{12}$$

5.5 Least Common Multiple

Objectives

1 Find the least common multiple (LCM) by listing.

2 Find the LCM using prime factorization.

3 Find the LCM of a set of monomials.

4 Write fractions as equivalent fractions with the least common denominator (LCD).

5 Write rational expressions as equivalent expressions with the LCD.

Connection In Section 5.6, we will learn that to add or subtract fractions, we must make sure they have a common denominator. If we are to add or subtract fractions that do not have a common denominator, we will rewrite them so that they do. We will use the LCM of the denominators as their common denominator.

Warm-up

[3.5] **1.** Find the GCF of 36 and 60 using prime factorization.

[5.4] **2.** Divide. $2\dfrac{1}{10} \div \left(-\dfrac{3}{4}\right)$

Objective 1 Find the least common multiple (LCM) by listing.

In Section 5.1, we learned to compare fractions by writing equivalent fractions with a common multiple denominator. The common multiple was found by multiplying the denominators. Although it was a quick method, the common multiples were often large numbers. When adding and subtracting fractions, we prefer to use the **least common multiple** as a common denominator to keep the numbers involved as small as possible.

Definition Least common multiple (LCM): The smallest natural number that is divisible by all of the given numbers.

For example, look at the multiples of 4 and 6.

Multiples of 4: 4, 8, 12, 16, 20, 24, 28, 32, 36, . . .
Multiples of 6: 6, 12, 18, 24, 30, 36, 42, 48, . . .

Note The least common multiple is 12 because it is the smallest number that is evenly divisible by both 4 and 6.

Our example suggests that to find the LCM of a set of numbers, we list the multiples of each number. However, we could speed up the process by listing only multiples of the greatest given number 6, until we find one that is evenly divisible by the other given number, 4. In this case, the LCM of 4 and 6 is 12.

Procedure

To find the LCM by listing, list multiples of the greatest given number until you find a multiple that is divisible by all of the other given numbers.

Example 1 Find the LCM of 36 and 120 by listing.

Solution: We list the multiples of the greater number, 120, until we find a number that is also a multiple of 36. Multiples of 120 are 120, 240, 360, and so on.

120	240	360
▲	▲	▲
120 is not evenly divisible by 36 ($120 \div 36 = 3\,r12$); so go to the next multiple of 120.	240 is not evenly divisible by 36 ($240 \div 36 = 6\,r24$); so go to the next multiple of 120.	360 is evenly divisible by 36 ($360 \div 36 = 10$); so it is the LCM.

Answer: $\text{LCM}(36, 120) = 360$

We can check the LCM by dividing it by the given numbers and considering the quotients' common factors.

$$\begin{array}{r} 10 \\ 36\overline{)360} \\ -36 \\ \hline 00 \\ -0 \\ \hline 0 \end{array}$$

◄ **Note** We know that 360 is the LCM because 1 is the only common factor for the quotients 10 and 3. ►

$$\begin{array}{r} 3 \\ 120\overline{)360} \\ -360 \\ \hline 0 \end{array}$$

Answers to Warm-up

1. 12

2. $-2\dfrac{4}{5}$

If any factor other than 1 divides all of the quotients evenly, the result is not the least common multiple. For example, suppose we mistakenly get 720 as our answer. Look at the check.

$$\begin{array}{r} 20 \\ 36\overline{)720} \\ -72 \\ \hline 00 \\ -0 \\ \hline 0 \end{array}$$

◄ **Note** The quotients, 20 and 6, are both divisible by 2. This means that we overshot the LCM by a factor of 2. ►

$$\begin{array}{r} 6 \\ 120\overline{)720} \\ -720 \\ \hline 0 \end{array}$$

To correct the mistake, we can divide our incorrect answer by the factor common to the quotients. In this case, that factor is 2; so dividing 720 by 2 gives the correct answer, 360.

► **Do Your Turn 1**

Objective 2 Find the LCM using prime factorization.

We can also find the LCM of a set of numbers using the prime factorizations of those numbers. To see how this works, let's look at the numbers from Example 1: 36 and 120 and their LCM, 360.

$$36 = 2 \cdot 2 \cdot 3 \cdot 3$$
$$120 = 2 \cdot 2 \cdot 2 \cdot 3 \cdot 5$$
$$\text{LCM}(36, 120) = 360 = 2 \cdot 2 \cdot 2 \cdot 3 \cdot 3 \cdot 5$$

Because 36 and 120 are factors of 360, their prime factorizations are in the prime factorization of 360.

$$360 = 2 \cdot \overbrace{2 \cdot 2 \cdot 3 \cdot 3}^{36} \cdot 5 \qquad\qquad 360 = \overbrace{2 \cdot 2 \cdot 2 \cdot 3}^{120} \cdot 3 \cdot 5$$

Therefore, we can build the prime factorization of the LCM out of the prime factorizations of the given numbers. We use each prime factor the most it occurs in the factorizations. Because 120 has the most 2s and 5s, those three 2s and one 5 go into the LCM's prime factorization. Because 36 has the most 3s, those two 3s go into the LCM's prime factorization. Multiplying the three 2s, two 3s, and one 5 gives the LCM, 360.

$$36 = 2 \cdot 2 \cdot 3 \cdot 3 \qquad\qquad 120 = 2 \cdot 2 \cdot 2 \cdot 3 \cdot 5$$

$$\text{LCM} = 2 \cdot 2 \cdot 2 \cdot 3 \cdot 3 \cdot 5 = 360$$

◄ **Note** The two 2s in 36 are not used because 120 has more 2s. Likewise, the one 3 in 120 is not used because 36 has more 3s.

It is helpful to write the factorizations in exponential form because we can easily compare the exponents and use the greatest of each prime's exponents.

$$36 = 2^2 \cdot 3^2 \qquad\qquad 120 = 2^3 \cdot 3 \cdot 5$$

$$\text{LCM} = 2^3 \cdot 3^2 \cdot 5 = 360$$

► **Your Turn 1**

Find the LCM by listing.
a. 18 and 24
b. 21 and 35
c. 3, 4, and 10
d. 4, 6, and 9

Answers to Your Turn 1
a. 72 **b.** 105 **c.** 60 **d.** 36

Warning Don't confuse LCM with GCF. When we use primes to find a GCF, we include only those primes common to all factorizations. For an LCM, we include each prime factor.

Procedure

To find the LCM of a set of numbers using prime factorization:

1. Find the prime factorization of each given number.
2. Write a factorization that contains each prime factor the greatest number of times it occurs in the factorizations. (If using exponential forms, use each prime raised to its greatest exponent in the factorizations.)
3. Multiply to get the LCM.

▶ **Your Turn 2**

Find the LCM using prime factorization.

a. 36 and 80
b. 30 and 75
c. 20, 35, and 40
d. 26, 40, and 65

Example 2 Find the LCM of 24, 90, and 70 using prime factorization.

Solution: We find the prime factorization of 24, 90, and 70, then write a factorization that contains each prime factor raised to the greatest exponent that occurs in the factorizations.

$$24 = 2^3 \cdot 3$$
$$90 = 2 \cdot 3^2 \cdot 5$$
$$70 = 2 \cdot 5 \cdot 7$$
$$\text{LCM}(24, 90, 70) = 2^3 \cdot 3^2 \cdot 5 \cdot 7$$
$$= 8 \cdot 9 \cdot 5 \cdot 7$$
$$= 2520$$

2^3 has the greatest exponent for 2, 3^2 has the greatest exponent for 3, 5 has the greatest exponent for 5, and 7 has the greatest exponent for 7.

◀ **Do Your Turn 2**

The listing method is simple, direct, and useful when the given numbers are small. The prime factorization method is more complicated, but useful, when the given numbers are large. Of course, we can use either method to find the LCM, but the idea is to select the best method for the situation.

Objective 3 Find the LCM of a set of monomials.

To find the LCM of a group of monomials, we follow the same procedure and treat the variables as prime factors.

▶ **Your Turn 3**

Find the LCM.

a. $6a^3$ and $12a$
b. $9m^2n$ and $12m^4$
c. $10hk^5$ and $8h^2k$

Example 3 Find the LCM of $18x^3y$ and $24xz^2$.

Solution: We use the prime factorization method.

$$18x^3y = 2 \cdot 3^2 \cdot x^3 \cdot y$$
$$24xz^2 = 2^3 \cdot 3 \cdot x \cdot z^2$$
$$\text{LCM}(18x^3y, 24xz^2) = 2^3 \cdot 3^2 \cdot x^3 \cdot y \cdot z^2$$
$$= 72x^3yz^2$$

2^3 has the greatest exponent for 2, 3^2 has the greatest exponent for 3, x^3 has the greatest exponent for x, y has the greatest exponent for y, and z^2 has the greatest exponent for z.

◀ **Do Your Turn 3**

Objective 4 Write fractions as equivalent fractions with the least common denominator (LCD).

In Section 5.1, we learned that multiplying both the numerator and denominator of a fraction by the same nonzero number produces an equivalent fraction. We used this rule to rewrite two fractions so that they have a common denominator, which allowed us to easily compare the fractions. Now let's rewrite fractions so that they have the least common multiple of their denominators as their common denominator. When we use the LCM as a common denominator, we say that we are using the **least common denominator (LCD)**.

Answers to Your Turn 2
a. 720 b. 150 c. 280 d. 520

Answers to Your Turn 3
a. $12a^3$ b. $36m^4n$ c. $40h^2k^5$

Definition **Least common denominator (LCD):** The least common multiple of the denominators.

For example, the LCD of $\frac{3}{4}$ and $\frac{5}{6}$ is 12 because 12 is the least common multiple of the denominators, 4 and 6. Now let's rewrite the fractions so that 12 is their denominator. We need to determine what to multiply each fraction by so that we have equivalent fractions with 12 as the denominator.

$$\frac{3}{4} = \frac{3 \cdot ?}{4 \cdot ?} = \frac{}{12} \qquad\qquad \frac{5}{6} = \frac{5 \cdot ?}{6 \cdot ?} = \frac{}{12}$$

Looking at the denominator, we have an unknown factor statement, $4 \cdot ? = 12$. To find the unknown factor, we can divide: $? = 12 \div 4 = 3$. So to write a fraction equivalent to $\frac{3}{4}$ with a denominator of 12, we multiply the numerator and denominator of $\frac{3}{4}$ by 3.

Here we need to find the unknown factor in $6 \cdot ? = 12$. To find that unknown factor, we divide: $? = 12 \div 6 = 2$. So to write a fraction equivalent to $\frac{5}{6}$ with a denominator of 12, we multiply the numerator and denominator of $\frac{5}{6}$ by 2.

$$\frac{3}{4} = \frac{3 \cdot 3}{4 \cdot 3} = \frac{9}{12} \qquad\qquad \frac{5}{6} = \frac{5 \cdot 2}{6 \cdot 2} = \frac{10}{12}$$

Example 4 Write $\frac{7}{12}$ and $\frac{2}{15}$ as equivalent fractions with the least common denominator.

Solution: The LCD is the LCM of the denominators; so we need to find the LCM of 12 and 15. We will use the listing method, listing multiples of 15 until we find a number that is also a multiple of 12.

15	30	45	60
15 is not evenly divisible by 12; so go to the next multiple of 15.	30 is not evenly divisible by 12; so go to the next multiple of 15.	45 is not evenly divisible by 12; so go to the next multiple of 15.	60 is evenly divisible by 12; so 60 is the LCM.

Now we write $\frac{7}{12}$ and $\frac{2}{15}$ as equivalent fractions with 60 as their denominator.

$$\frac{7}{12} = \frac{7 \cdot ?}{12 \cdot ?} = \frac{}{60} \qquad\qquad \frac{2}{15} = \frac{2 \cdot ?}{15 \cdot ?} = \frac{}{60}$$

We must find the factor that multiplies by 12 to equal 60; so we divide 60 by 12.

$$? = 60 \div 2 = 5$$

To write the equivalent fraction, we multiply the numerator and denominator by 5.

We must find the factor that multiplies by 15 to equal 60; so we divide 60 by 15.

$$? = 60 \div 15 = 4$$

To write the equivalent fraction, we multiply the numerator and denominator by 4.

$$\frac{7}{12} = \frac{7 \cdot 5}{12 \cdot 5} = \frac{35}{60} \qquad\qquad \frac{2}{15} = \frac{2 \cdot 4}{15 \cdot 4} = \frac{8}{60}$$

▶ **Do Your Turn 4**

Objective 5 Write rational expressions as equivalent expressions with the LCD.

Rational expressions are rewritten in the same way as numeric fractions.

▶ **Your Turn 4**

Write as equivalent fractions with the LCD.

a. $\frac{2}{3}$ and $\frac{1}{4}$

b. $\frac{5}{8}$ and $\frac{7}{12}$

c. $\frac{2}{5}$ and $\frac{3}{10}$

d. $\frac{9}{20}$ and $\frac{8}{15}$

Answers to Your Turn 4

a. $\frac{8}{12}$ and $\frac{3}{12}$

b. $\frac{15}{24}$ and $\frac{14}{24}$

c. $\frac{4}{10}$ and $\frac{3}{10}$

d. $\frac{27}{60}$ and $\frac{32}{60}$

▶ **Your Turn 5**

Write as equivalent expressions with the LCD.

a. $\dfrac{5}{9}$ and $\dfrac{2}{3a}$

b. $\dfrac{3}{4x}$ and $\dfrac{7}{6x^2}$

c. $-\dfrac{m}{n^4}$ and $\dfrac{3}{mn}$

d. $\dfrac{11}{12t^2}$ and $-\dfrac{7u}{9tv}$

Answers to Your Turn 5

a. $\dfrac{5a}{9a}$ and $\dfrac{6}{9a}$

b. $\dfrac{9x}{12x^2}$ and $\dfrac{14}{12x^2}$

c. $-\dfrac{m^2}{mn^4}$ and $\dfrac{3n^3}{mn^4}$

d. $\dfrac{33v}{36t^2v}$ and $-\dfrac{28tu}{36t^2v}$

Example 5   Write $\frac{3}{8x}$ and $\frac{5}{6x^2}$ as equivalent rational expressions with the LCD.

Solution: The LCD is the LCM of the denominators; so we need to find the LCM of $8x$ and $6x^2$.

The LCM of 8 and 6 is 24. For the variables, x^2 has the greater exponent; so the LCD is $24x^2$. Now we write $\frac{3}{8x}$ and $\frac{5}{6x^2}$ with $24x^2$ as the denominator.

$$\frac{3}{8x} = \frac{3 \cdot \,?}{8x \cdot \,?} = \frac{}{24x^2} \qquad\qquad \frac{5}{6x^2} = \frac{5 \cdot \,?}{6x^2 \cdot \,?} = \frac{}{24x^2}$$

▲
We must find the factor that multiplies by $8x$ to equal $24x^2$; so we divide $24x^2$ by $8x$.

$$? = 24x^2 \div 8x = 3x$$

To write the equivalent rational expressions, we multiply the numerator and denominator by $3x$.

▼
$$\frac{3}{8x} = \frac{3 \cdot 3x}{8x \cdot 3x} = \frac{9x}{24x^2}$$

▲
We must find the factor that multiplies by $6x^2$ to equal $24x^2$; so we divide $24x^2$ by $6x^2$.

$$? = 24x^2 \div 6x^2 = 4$$

To write the equivalent rational expressions, we multiply the numerator and denominator by 4.

▼
$$\frac{5}{6x^2} = \frac{5 \cdot 4}{6x^2 \cdot 4} = \frac{20}{24x^2}$$

◀ **Do Your Turn 5**

5.5 Exercises   For Extra Help   MyMathLab®

Objective 1

Prep Exercise 1 What is the least common multiple (LCM) of a set of numbers?

Prep Exercise 2 Explain how to find the least common multiple by listing.

For Exercises 1–8, find the LCM by listing.

1. 10 and 6
2. 15 and 9
3. 12 and 36
4. 8 and 32
5. 20 and 30
6. 16 and 20
7. 6, 9, and 15
8. 6, 10, and 15

Objective 2

Prep Exercise 3 Explain how to find the least common multiple using prime factorization.

For Exercises 9–20, find the LCM using prime factorization.

9. 18 and 24
10. 24 and 50
11. 63 and 28
12. 42 and 56
13. 52 and 28
14. 68 and 56
15. 180 and 200
16. 210 and 420
17. 26, 30, and 39
18. 42, 56, and 80
19. 28, 32, and 60
20. 36, 49, and 72

Objective 3

For Exercises 21–28, find the LCM.

21. $12x$ and $8y$
22. $9a$ and $6b$
23. $10y^3$ and $6y$
24. $14t^2$ and $5t$
25. $16mn$ and $8m$
26. $20h^5k^3$ and $15h^2k$
27. $18x^2y$ and $12xy^3$
28. $24a^4b$ and $30ab^2$

Objective 4

Prep Exercise 4 What is the least common denominator (LCD) of two fractions?

Prep Exercise 5 What do we multiply the numerator and denominator of $\frac{5}{6}$ by to produce an equivalent fraction with a denominator of 24?

For Exercises 29–36, write as equivalent fractions with the LCD.

29. $\frac{3}{10}$ and $\frac{5}{6}$

30. $\frac{2}{15}$ and $\frac{4}{9}$

31. $\frac{7}{12}$ and $\frac{11}{36}$

32. $\frac{1}{8}$ and $\frac{15}{32}$

33. $\frac{1}{20}$ and $\frac{17}{30}$

34. $\frac{3}{16}$ and $\frac{9}{20}$

35. $\frac{3}{4}, \frac{1}{6},$ and $\frac{7}{9}$

36. $\frac{2}{5}, \frac{5}{6},$ and $\frac{13}{15}$

Objective 5

Prep Exercise 6 What do we multiply the numerator and denominator of $\frac{2x}{3}$ by to produce an equivalent rational expression with a denominator of $18x$?

For Exercises 37–44, write as equivalent expressions with the LCD.

37. $\frac{7}{12x}$ and $\frac{3}{4}$

38. $\frac{5}{9b}$ and $\frac{2}{3b}$

39. $\frac{9}{16mn}$ and $\frac{3n}{8m}$

40. $\frac{1}{14t^2}$ and $\frac{2}{5}$

41. $\frac{7}{10y^3z}$ and $\frac{5z}{6y}$

42. $\frac{-13}{20h^5k^3}$ and $\frac{4}{15h^2k}$

43. $\frac{z}{18x^2y}$ and $\frac{-5}{12xy^3}$

44. $\frac{5}{24a^4b}$ and $\frac{c}{30ab^2}$

Puzzle Problem Planetary conjunctions occur when two or more planets form a rough line with the sun. Using rough estimates of the orbital periods of the following planets, calculate the time that will elapse between alignments of these planets.

Planet	Orbit Period
Earth	1 year
Mars	2 years
Jupiter	12 years
Saturn	30 years
Uranus	84 years

Of Interest

Because all of the planets do not orbit in the same plane, it is virtually impossible for more than two planets to form a straight line with the sun. However, if we were to view the solar system from above, the planets could form an apparent line (conjunction) with the sun. Theoretically, this type of alignment for all nine planets could occur every 334,689,064 years. When conjunction dates are calculated, we relax the criterion that the planets be in a straight line and choose an acceptable angle of dispersion.

The conjunction that occurred in May 2000 included Mercury, Venus, Earth, the moon, Mars, Jupiter, and Saturn, with an angle of dispersion of about 15 degrees in the sky.

Review Exercises

For Exercises 1 and 2, simplify.

[2.2] 1. $16 + (-28)$

[2.3] 2. $-8 - (-19)$

[3.2] 3. Simplify and write the resulting polynomial in descending order of degree.
$14x^2 - 16x + 25x^2 - 12 + 4x^3 + 9$

For Exercises 4–6, solve and check.

[4.2] 4. $m + 14 = 6$

[4.2] 5. $x - 37 = 16$

[4.3] 6. $-3a = \frac{12}{5}$

5.6 Adding and Subtracting Fractions, Mixed Numbers, and Rational Expressions

Objectives

1 Add and subtract fractions with the same denominator.

2 Add and subtract rational expressions with the same denominator.

3 Add and subtract fractions with different denominators.

4 Add and subtract rational expressions with different denominators.

5 Add and subtract mixed numbers.

6 Add and subtract negative mixed numbers.

7 Solve equations.

8 Solve applications.

Warm-up

[2.3] **1.** Write $-16 - (-9)$ as an equivalent addition, then evaluate.

[5.5] **2.** Write $\dfrac{3}{8}$ and $\dfrac{5}{12}$ as equivalent fractions with their LCD.

Objective 1 Add and subtract fractions with the same denominator.

To see how we add fractions that have the same denominator, consider the fact that when we add two halves, we get a whole.

Notice that if we add the numerators and keep the same denominator, we'll have $\frac{2}{2}$, which simplifies to 1.

$$\frac{1}{2} + \frac{1}{2} = \frac{1+1}{2} = \frac{2}{2} = 1$$

Warning It is tempting to treat addition like multiplication and add numerator plus numerator and denominator plus denominator. But when we follow this approach, we have a half plus a half equals a half.

$$\frac{1}{2} + \frac{1}{2} = \frac{1+1}{2+2} = \frac{2}{4} = \frac{1}{2} \qquad \text{This can't be true.}$$

The only situation where we can add an amount to a number and end up with the same number is $n + 0 = n$.

Our example suggests the following procedure:

> ### Procedure
> To add or subtract fractions with the same denominator:
> 1. Add or subtract the numerators and keep the same denominator.
> 2. Simplify.

Example 1   Add or subtract.

a. $\dfrac{1}{8} + \dfrac{3}{8}$

Solution: $\dfrac{1}{8} + \dfrac{3}{8} = \dfrac{1+3}{8} = \dfrac{4}{8} = \dfrac{1}{2}$ Add the numerators and keep the same denominator; then simplify.

b. $\dfrac{8}{9} - \dfrac{2}{9}$

Solution: $\dfrac{8}{9} - \dfrac{2}{9} = \dfrac{8-2}{9} = \dfrac{6}{9} = \dfrac{2}{3}$ Subtract the numerators and keep the same denominator; then simplify.

◄ **Do Your Turn 1**

▶ **Your Turn 1**

Add or subtract.

a. $\dfrac{5}{7} + \dfrac{1}{7}$

b. $\dfrac{3}{16} + \dfrac{1}{16}$

c. $\dfrac{9}{11} - \dfrac{3}{11}$

d. $\dfrac{7}{8} - \dfrac{5}{8}$

Answers to Your Turn 1

a. $\dfrac{6}{7}$ b. $\dfrac{1}{4}$ c. $\dfrac{6}{11}$ d. $\dfrac{1}{4}$

Answers to Warm-up

1. $-16 + 9 = -7$

2. $\dfrac{9}{24}$ and $\dfrac{10}{24}$

Example 2   Add.

a. $-\dfrac{3}{10} + \left(-\dfrac{1}{10}\right)$

Solution:

$-\dfrac{3}{10} + \left(-\dfrac{1}{10}\right) = \dfrac{-3 + (-1)}{10} = \dfrac{-4}{10} = -\dfrac{2}{5}$

Add the numerators and keep the same denominator; then simplify. These fractions have the same sign; so we add their absolute values and keep the same sign.

Note Remember, to indicate a negative fraction, we can place the minus sign to the left of the numerator, denominator, or fraction bar. In an answer, we will place the sign to the left of the fraction bar.

b. $\dfrac{-7}{12} + \dfrac{5}{12}$

Solution:

$\dfrac{-7}{12} + \dfrac{5}{12} = \dfrac{-7 + 5}{12} = \dfrac{-2}{12} = -\dfrac{1}{6}$

Add the numerators and keep the same denominator; then simplify. These fractions have different signs; so we subtract their absolute values and keep the sign of the number with the greater absolute value.

▶ **Do Your Turn 2**

Example 3   Subtract.

a. $\dfrac{5}{9} - \dfrac{7}{9}$

Solution:

Note To write a subtraction statement as an equivalent addition statement, we change the subtraction sign to an addition sign and change the subtrahend to its additive inverse. ▶

$\dfrac{5}{9} - \dfrac{7}{9} = \dfrac{5 - 7}{9}$

Subtract the numerators and keep the same denominator.

$= \dfrac{5 + (-7)}{9}$

Write an equivalent addition expression.

$= -\dfrac{2}{9}$

Add.

b. $\dfrac{-3}{16} - \left(\dfrac{-5}{16}\right)$

Solution:

$\dfrac{-3}{16} - \left(\dfrac{-5}{16}\right) = \dfrac{-3 - (-5)}{16}$

Subtract the numerators and keep the same denominator.

$= \dfrac{-3 + 5}{16}$

Write an equivalent addition expression.

$= \dfrac{2}{16}$

Add.

$= \dfrac{1}{8}$

Simplify.

▶ **Do Your Turn 3**

▶ **Your Turn 2**

Add.

a. $\dfrac{-1}{7} + \dfrac{-3}{7}$

b. $-\dfrac{5}{12} + \left(-\dfrac{5}{12}\right)$

c. $\dfrac{11}{24} + \left(-\dfrac{5}{24}\right)$

d. $\dfrac{-7}{16} + \dfrac{1}{16}$

▶ **Your Turn 3**

Subtract.

a. $\dfrac{1}{9} - \dfrac{5}{9}$

b. $-\dfrac{3}{15} - \dfrac{2}{15}$

c. $\dfrac{1}{7} - \left(-\dfrac{4}{7}\right)$

d. $-\dfrac{9}{10} - \left(-\dfrac{1}{10}\right)$

Answers to Your Turn 2

a. $-\dfrac{4}{7}$ b. $-\dfrac{5}{6}$ c. $\dfrac{1}{4}$ d. $-\dfrac{3}{8}$

Answers to Your Turn 3

a. $-\dfrac{4}{9}$ b. $-\dfrac{1}{3}$ c. $\dfrac{5}{7}$ d. $-\dfrac{4}{5}$

Objective 2 Add and subtract rational expressions with the same denominator.

To add or subtract rational expressions, we follow the same procedure as for numeric fractions. Keep in mind that we can add or subtract only like terms. Remember, like terms have the same variable(s) raised to the same exponent(s). If the numerators contain unlike terms, the resulting numerator is a polynomial whose terms are those unlike terms.

▶ **Your Turn 4**

Add.

a. $\dfrac{3}{8m} + \dfrac{7}{8m}$

b. $\dfrac{5t}{u} + \dfrac{2}{u}$

c. $\dfrac{x+7}{9} + \dfrac{2-x}{9}$

d. $\dfrac{3t^2+8}{5} + \dfrac{4t^2-t-10}{5}$

Example 4 Add.

a. $\dfrac{3x}{10} + \dfrac{x}{10}$

Solution: $\dfrac{3x}{10} + \dfrac{x}{10} = \dfrac{3x+x}{10} = \dfrac{4x}{10}$ The denominators are the same; so add the numerators and keep the same denominator.   ◀ **Note** Because $3x$ and x are like terms, they can be combined to equal $4x$.

$= \dfrac{2x}{5}$ Simplify.

b. $\dfrac{x}{3} + \dfrac{2}{3}$

Solution: $\dfrac{x}{3} + \dfrac{2}{3} = \dfrac{x+2}{3}$ The numerators, x and 2, are not like terms; so the resulting numerator is the binomial $x + 2$.

c. $\dfrac{x^2-3x+1}{7} + \dfrac{x+4}{7}$

Solution: $\dfrac{x^2-3x+1}{7} + \dfrac{x+4}{7} = \dfrac{(x^2-3x+1)+(x+4)}{7}$

$= \dfrac{x^2-2x+5}{7}$ ◀ **Note** To add polynomials, we combine like terms. Because x^2 has no like terms, we write it as is in the sum. Combining $-3x$ with x gives $-2x$. Combining 1 with 4 gives 5.

◀ **Do Your Turn 4**

Now let's consider subtraction of rational expressions that have the same denominator. We can write subtraction as an equivalent addition. Remember, if the numerators contain unlike terms, the resulting numerator is a polynomial whose terms are the unlike terms.

Example 5 Subtract.

a. $\dfrac{4n}{9} - \left(\dfrac{-5}{9}\right)$

Solution:

$\dfrac{4n}{9} - \left(\dfrac{-5}{9}\right) = \dfrac{4n-(-5)}{9}$ The denominators are the same; so subtract the numerators and keep the same denominator.

$= \dfrac{4n+5}{9}$ Write as an equivalent addition expression.   ◀ **Note** We cannot combine $4n$ and 5 because they are not like terms. Because $4n + 5$ cannot be factored, the expression cannot be simplified.

Answers to Your Turn 4

a. $\dfrac{5}{4m}$ b. $\dfrac{5t+2}{u}$ c. 1

d. $\dfrac{7t^2-t-2}{5}$

b. $\dfrac{8x^2 - 5x + 3}{y} - \dfrac{2x - 7}{y}$

Solution:

$$\dfrac{8x^2 - 5x + 3}{y} - \dfrac{2x - 7}{y} = \dfrac{(8x^2 - 5x + 3) - (2x - 7)}{y}$$

Note To write the equivalent addition, we must find the additive inverse of the subtrahend, $2x - 7$. To write the additive inverse of a polynomial, we change all of the signs in the polynomial.

▶ $= \dfrac{(8x^2 - 5x + 3) + (-2x + 7)}{y}$ Write as an equivalent addition expression.

$= \dfrac{8x^2 - 7x + 10}{y}$ ◀ **Note** $8x^2$ has no like terms. $-5x$ and $-2x$ combine to equal $-7x$, and 3 and 7 combine to equal 10.

▶ **Do Your Turn 5**

Objective 3 Add and subtract fractions with different denominators.

To add or subtract fractions that have different denominators, we first need to find a common denominator. We will apply what we learned in Section 5.5 about rewriting fractions as equivalent fractions with the LCD.

Procedure

To add or subtract fractions with different denominators:

1. Write the fractions as equivalent fractions with a common denominator.

   Note Any common multiple of the denominators will do, but using the LCD makes the numbers more manageable.

2. Add or subtract the numerators and keep the common denominator.
3. Simplify.

Example 6 Add. $\dfrac{7}{12} + \dfrac{2}{15}$

Solution: We must write the fractions with a common denominator, then add the numerators and keep the common denominator. The LCD for 12 and 15 is 60. (See Example 4, Section 5.5.)

Note To write an equivalent fraction with 60 as the denominator, we rewrite $\frac{7}{12}$ by multiplying its numerator and denominator by 5.

▶ $= \dfrac{7(5)}{12(5)} + \dfrac{2(4)}{15(4)}$ ◀ **Note** To write an equivalent fraction with 60 as the denominator, we rewrite $\frac{2}{15}$ by multiplying its numerator and denominator by 4.

$= \dfrac{35}{60} + \dfrac{8}{60}$

$= \dfrac{35 + 8}{60}$

$= \dfrac{43}{60}$

▶ **Do Your Turn 6**

Objective 4 Add and subtract rational expressions with different denominators.

We add and subtract rational expressions using the same procedure as for adding and subtracting numeric fractions.

► **Your Turn 7**

Add or subtract.

a. $\dfrac{3m}{4} + \dfrac{2}{3m}$

b. $\dfrac{7}{3x} - \dfrac{9}{6x^3}$

c. $\dfrac{5}{6a} + \dfrac{3}{4}$

d. $\dfrac{4}{5y^2} - \dfrac{1}{3y}$

Example 7   Add or subtract.

a. $\dfrac{2x}{5} + \dfrac{3}{x}$

Solution: We write the rational expressions with a common denominator, then add the numerators and keep the common denominators. Keep in mind that if the numerators are not like terms, we will have to express the sum as a polynomial.

For 5 and 1, the LCD is 5. For the x's, x^1 has the greater exponent; so the LCD is $5x^1$, or $5x$.

Note To write an equivalent fraction with $5x$ as the denominator, we rewrite $\frac{2x}{5}$ by multiplying its numerator and denominator by x. ►

$$= \dfrac{2x(x)}{5(x)} + \dfrac{3(5)}{x(5)}$$

◄ **Note** To write an equivalent fraction with $5x$ as the denominator, we rewrite $\frac{3}{x}$ by multiplying its numerator and denominator by 5.

$$= \dfrac{2x^2}{5x} + \dfrac{15}{5x}$$

$$= \dfrac{2x^2 + 15}{5x}$$

◄ **Note** Because $2x^2$ and 15 are not like terms, we express their sum as a polynomial. Because we cannot factor $2x^2 + 15$, we cannot simplify the rational expression.

b. $\dfrac{3}{8x} - \dfrac{5}{6x^2}$

Solution: The LCM for 8 and 6 is 24. For the x's, x^2 has the greater exponent; so the LCD is $24x^2$. (See Example 5, Section 5.5.)

Note To write an equivalent fraction with $24x^2$ as the denominator, we rewrite $\frac{3}{8x}$ by multiplying its numerator and denominator by $3x$. ►

$$= \dfrac{3(3x)}{8x(3x)} - \dfrac{5(4)}{6x^2(4)}$$

◄ **Note** To write an equivalent fraction with $24x^2$ as the denominator, we rewrite $\frac{5}{6x^2}$ by multiplying its numerator and denominator by 4.

$$= \dfrac{9x}{24x^2} - \dfrac{20}{24x^2}$$

$$= \dfrac{9x - 20}{24x^2}$$

◄ **Note** Because $9x$ and 20 are not like terms, we express their difference as a polynomial. Because we cannot factor $9x - 20$, we cannot simplify the rational expression.

◄ **Do Your Turn 7**

Objective 5 Add and subtract mixed numbers.

We will consider two methods for adding and subtracting mixed numbers: (1) We can write the mixed numbers as improper fractions, then add or subtract. (2) We can leave them as mixed numbers and add or subtract the integers and fractions separately. First, let's look at adding mixed numbers. Consider $2\frac{1}{5} + 3\frac{2}{5}$.

Method 1: Write them as improper fractions.

$$2\frac{1}{5} + 3\frac{2}{5} = \frac{11}{5} + \frac{17}{5} = \frac{28}{5} = 5\frac{3}{5}$$

Answers to Your Turn 7

a. $\dfrac{9m^2 + 8}{12m}$ b. $\dfrac{14x^2 - 9}{6x^3}$

c. $\dfrac{9a + 10}{12a}$ d. $\dfrac{12 - 5y}{15y^2}$

Method 1: Because mixed numbers are, by definition, the result of addition, we can use the commutative property of addition and add the integers and fractions separately.

$$2\frac{1}{5} \;+\; 3\frac{2}{5}$$

$$= \left(2 + \frac{1}{5}\right) + \left(3 + \frac{2}{5}\right) \qquad \text{Use the definition of mixed numbers.}$$

$$= 2 + 3 + \frac{1}{5} + \frac{2}{5} \qquad \text{Use the commutative property of addition.}$$

$$= 5 + \frac{3}{5} \qquad \text{Add the integers and fractions separately.}$$

$$= 5\frac{3}{5} \qquad \text{Write the sum as a mixed number.}$$

Note We are showing all of these steps only to illustrate why we can add the integers and fractions separately. From here on, we will no longer show all of these steps.

Example 8 Estimate $5\frac{3}{4} + 4\frac{2}{3}$; then find the actual sum expressed as a mixed number in lowest terms.

Estimate: Round the numbers so that the calculation is easy to perform mentally. We will round these mixed numbers to the nearest integer.

$5\frac{3}{4}$ rounds to 6 because it is closer to 6 than to 5.

$4\frac{2}{3}$ rounds to 5 because it is closer to 5 than to 4.

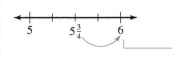

$$6 + 5 = 11 \qquad \text{The actual sum should be around 11.}$$

Actual:

Method 1: $5\frac{3}{4} + 4\frac{2}{3} = \dfrac{23}{4} + \dfrac{14}{3}$ Write as improper fractions. Notice that the LCD is 12.

Note To write $\frac{23}{4}$ with a denominator of 12, we multiply the numerator and denominator by 3.

$$= \frac{23(3)}{4(3)} + \frac{14(4)}{3(4)}$$

$$= \frac{69}{12} + \frac{56}{12}$$

Note To write $\frac{14}{3}$ with a denominator of 12, we multiply the numerator and denominator by 4.

$$= \frac{125}{12} \qquad \text{Add the numerators.}$$

$$= 10\frac{5}{12} \qquad \text{Write the result as a mixed number.}$$

Method 2: $5\frac{3}{4} + 4\frac{2}{3} = 5\dfrac{3(3)}{4(3)} + 4\dfrac{2(4)}{3(4)}$ Write equivalent fractions with the LCD, 12.

$$= 5\frac{9}{12} + 4\frac{8}{12}$$

$$= 9\frac{17}{12} \qquad \text{Add integers and fractions separately.}$$

$$= 10\frac{5}{12} \qquad \begin{array}{l}\text{Simplify the improper fraction}\\\text{within the mixed number.}\end{array}$$

The actual sum agrees with our estimate.

▶ **Do Your Turn 8**

▶ **Your Turn 8**

Estimate; then find the actual sum expressed as a mixed number in lowest terms.

a. $4 + 9\frac{2}{7}$

b. $6\frac{1}{8} + 3\frac{5}{8}$

c. $5\frac{5}{6} + 7\frac{1}{3}$

d. $3\frac{3}{4} + 8\frac{4}{5}$

Note An improper fraction in a mixed number is not considered simplest form. Because $\frac{17}{12} = 1\frac{5}{12}$, we add the 1 to the 9 so that we have $10\frac{5}{12}$. Writing all of the steps, we have $9\frac{17}{12} = 9 + \frac{17}{12} = 9 + 1\frac{5}{12} = 10\frac{5}{12}$.

Answers to Your Turn 8

a. $13\frac{2}{7}$ b. $9\frac{3}{4}$

c. $13\frac{1}{6}$ d. $12\frac{11}{20}$

Now let's look at subtracting mixed numbers. Consider $6\frac{4}{9} - 2\frac{1}{9}$.

Method 1: Write the mixed numbers as improper fractions; then subtract.

$$6\frac{4}{9} - 2\frac{1}{9} = \frac{58}{9} - \frac{19}{9} = \frac{39}{9} = 4\frac{3}{9} = 4\frac{1}{3}$$

Method 2: Subtract the integers and fractions separately. We will show why here, but will leave out these "extra" steps in the future.

$$6\frac{4}{9} \quad - \quad 2\frac{1}{9}$$

Note We are including these extra steps to show why we can subtract the integers and fractions separately.

$$= \left(6 + \frac{4}{9}\right) - \left(2 + \frac{1}{9}\right)$$ Use the definition of mixed numbers.

$$= \left(6 + \frac{4}{9}\right) + \left(-2 - \frac{1}{9}\right)$$ Change the subtraction to addition.

$$= 6 - 2 + \frac{4}{9} - \frac{1}{9}$$ Use the commutative property of addition to separate the integers and fractions.

$$= 4 + \frac{3}{9}$$ Subtract the integers and fractions separately.

$$= 4\frac{1}{3}$$ Write the mixed number and simplify the fraction.

Example 9   Estimate $7\frac{1}{4} - 2\frac{5}{6}$; then find the actual difference expressed as a mixed number in simplest form.

Estimate: Round the numbers so that the calculation is easy to perform mentally. We will round these mixed numbers to the nearest integer.

$7\frac{1}{4}$ rounds to 7 because it is closer to 7 than to 8.

$2\frac{5}{6}$ rounds to 3 because it is closer to 3 than to 2.

$7 - 3 = 4$ The actual difference should be around 4.

Actual:

Method 1:

$$7\frac{1}{4} - 2\frac{5}{6} = \frac{29}{4} - \frac{17}{6}$$ Write as improper fractions.

$$= \frac{29(3)}{4(3)} - \frac{17(2)}{6(2)}$$ Write equivalent fractions with the LCD, 12.

$$= \frac{87}{12} - \frac{34}{12}$$

$$= \frac{53}{12}$$ Subtract numerators and keep the LCD.

$$= 4\frac{5}{12}$$ Write as a mixed number.

Method 1:

$$7\frac{1}{4} - 2\frac{5}{6} = 7\frac{1(3)}{4(3)} - 2\frac{5(2)}{6(2)}$$ Write equivalent fractions with the LCD, 12.

$$= 7\frac{3}{12} - 2\frac{10}{12}$$

We have a problem now. Although we can subtract the 2 from 7, if we subtract $\frac{10}{12}$ from $\frac{3}{12}$, we will get a negative number. We need to rewrite $7\frac{3}{12}$ so that its fraction is greater than $\frac{10}{12}$. The steps look like this.

Rewrite 7 as 6 + 1.   Add $\frac{12}{12}$ and $\frac{3}{12}$.

$$7\frac{3}{12} = 6 + 1 + \frac{3}{12} = 6 + \frac{12}{12} + \frac{3}{12} = 6\frac{15}{12}$$

Rewrite 1 as $\frac{12}{12}$.

Note These steps can be performed mentally by subtracting 1 from the integer, then adding the denominator to the original numerator to get the new numerator.

$$7\frac{3}{12} = 6\frac{12+3}{12} = 6\frac{15}{12}$$

Now we can complete the subtraction.

$$= 6\frac{15}{12} - 2\frac{10}{12}$$ Write $7\frac{3}{12}$ as $6\frac{15}{12}$ by taking 1 from 7 and adding it to $\frac{3}{12}$.

$$= 4\frac{5}{12}$$ Subtract the integers and fractions separately.

The actual difference agrees with our estimate.

▶ **Do Your Turn 9**

We can summarize how to add or subtract mixed numbers with the following procedure:

Procedure
To add or subtract mixed numbers:

Method 1: Write as improper fractions; then add or subtract those improper fractions.

Method 2: Add or subtract the integers and fractions separately.

Objective 6 Add and subtract negative mixed numbers.

We have seen that using the second method to add mixed numbers sometimes leads to a mixed number sum that must be simplified because it contains an improper fraction. Similarly, when we subtract using the second method, we sometimes have to rewrite one of the mixed numbers so that the fraction parts can be subtracted. But using the first method avoids these extra steps. Because these extra steps can be distracting in the adding or subtracting of negative mixed numbers, we will use only the first method.

Example 10 Subtract. $-9\frac{1}{6} - \left(-2\frac{1}{3}\right)$

Solution: First, we will write an equivalent addition. Also, to avoid potential extra steps, we will write the mixed numbers as improper fractions.

$$-9\frac{1}{6} - \left(-2\frac{1}{3}\right) = -9\frac{1}{6} + 2\frac{1}{3}$$ Write an equivalent addition using the additive inverse of the subtrahend.

$$= -\frac{55}{6} + \frac{7}{3}$$ Write the mixed numbers as improper fractions. Note that the LCD is 6.

▶ **Your Turn 9**

Subtract.

a. $9\frac{7}{8} - 4\frac{5}{8}$

b. $7\frac{9}{10} - 1\frac{1}{5}$

c. $10\frac{1}{5} - 3\frac{3}{4}$

d. $8 - 5\frac{1}{4}$

Answers to Your Turn 9

a. $5\frac{1}{4}$ b. $6\frac{7}{10}$ c. $6\frac{9}{20}$ d. $2\frac{3}{4}$

▶ **Your Turn 10**

Add or subtract.

a. $5\dfrac{3}{4} + \left(-5\dfrac{1}{8}\right)$

b. $7\dfrac{1}{2} - 10\dfrac{1}{3}$

c. $-5\dfrac{7}{10} - 2\dfrac{1}{4}$

d. $-8\dfrac{2}{9} - \left(-\dfrac{5}{6}\right)$

Note $-\frac{55}{6}$ already has 6 as its denominator; so we do not need to rewrite it.

$$\blacktriangleright \quad = -\dfrac{55}{6} + \dfrac{7(2)}{3(2)} \quad \blacktriangleleft$$

Note To write $\frac{7}{3}$ as an equivalent fraction with 6 as the denominator, we multiply the numerator and denominator by 2.

$$= -\dfrac{55}{6} + \dfrac{14}{6}$$

$$= -\dfrac{41}{6} \qquad \text{Add.}$$

$$= -6\dfrac{5}{6} \qquad \text{Write the result as a mixed number.}$$

◀ **Do Your Turn 10**

Objective 7 Solve equations.

In Section 5.4, we revisited the multiplication principle of equality and used it to eliminate fraction coefficients. Now let's revisit the addition principle of equality. In Section 4.2, we learned to use the addition principle of equality to eliminate a term from one side of an equation by adding its additive inverse to both sides of the equation.

▶ **Your Turn 11**

Solve and check.

a. $\dfrac{1}{3} + y = \dfrac{3}{4}$

b. $a - \dfrac{1}{5} = \dfrac{5}{8}$

c. $m + 3\dfrac{1}{2} = -6\dfrac{1}{4}$

d. $-4\dfrac{1}{7} = t - \dfrac{2}{5}$

Example 11 Solve and check. $x + \dfrac{3}{8} = \dfrac{5}{6}$

Solution:

$$x + \dfrac{3}{8} = \dfrac{5}{6}$$

Note $\frac{3}{8} - \frac{3}{8} = 0$, which isolates x on the left side.

$$\blacktriangleright \quad x + \dfrac{3}{8} - \dfrac{3}{8} = \dfrac{5}{6} - \dfrac{3}{8} \qquad \text{Subtract } \tfrac{3}{8} \text{ from both sides to isolate } x.$$

$$x + 0 = \dfrac{5(4)}{6(4)} - \dfrac{3(3)}{8(3)} \qquad \text{Write equivalent fractions with the LCD, 24.}$$

$$x = \dfrac{20}{24} - \dfrac{9}{24}$$

$$x = \dfrac{11}{24} \qquad \text{Subtract.}$$

Check:

$$x + \dfrac{3}{8} = \dfrac{5}{6}$$

$$\dfrac{11}{24} + \dfrac{3}{8} \overset{?}{=} \dfrac{5}{6} \qquad \text{In the original equation, replace } x \text{ with } \tfrac{11}{24}.$$

$$\dfrac{11}{24} + \dfrac{3(3)}{8(3)} \overset{?}{=} \dfrac{5}{6} \qquad \text{Write equivalent fractions with the LCD, 24.}$$

$$\dfrac{11}{24} + \dfrac{9}{24} \overset{?}{=} \dfrac{5}{6}$$

$$\dfrac{20}{24} \overset{?}{=} \dfrac{5}{6} \qquad \text{Add the fractions.}$$

$$\dfrac{5}{6} = \dfrac{5}{6} \qquad \text{Simplify. The equation is true; so } \tfrac{11}{24} \text{ is the solution.}$$

◀ **Do Your Turn 11**

Answers to Your Turn 10

a. $\dfrac{5}{8}$ b. $-2\dfrac{5}{6}$

c. $-7\dfrac{19}{20}$ d. $-7\dfrac{7}{18}$

Answers to Your Turn 11

a. $y = \dfrac{5}{12}$ b. $a = \dfrac{33}{40}$

c. $m = -9\dfrac{3}{4}$ d. $t = -3\dfrac{26}{35}$

Objective 8 Solve applications.

What clues can we look for that indicate that we need to add or subtract? With addition, we look for words indicating that we must find a total. For subtraction, we look for clues indicating that we must find a difference. Another clue for addition/subtraction situations is that the fractions represent distinct categories.

Example 12 In a poll, $\frac{1}{4}$ of the respondents said that they agreed with the passing of a certain bill, $\frac{1}{5}$ said that they disagreed, and the rest said that they had no opinion. What fraction of the respondents had no opinion?

Understand There are three distinct ways to respond to the poll: agree, disagree, or have no opinion. Because the three categories together make up the whole group that responded, we can write the following equation:

$$\boxed{\text{respondents who agreed}} \;+\; \boxed{\text{respondents who disagreed}} \;+\; \boxed{\text{respondents who had no opinion}} \;=\; \boxed{\text{all respondents}}$$

Because we are dealing with fractions, *all respondents* means a whole amount, which is the number 1. The number 1 always represents the whole amount.

Plan Use the preceding formula to write an equation; then solve.

Execute Let n represent the number of respondents that had no opinion.

$$\boxed{\text{respondents who agreed}} \;+\; \boxed{\text{respondents who disagreed}} \;+\; \boxed{\text{respondents who had no opinion}} \;=\; \boxed{\text{all respondents}}$$

$$\frac{1}{4} \;+\; \frac{1}{5} \;+\; n \;=\; 1$$

$$\frac{1(5)}{4(5)} + \frac{1(4)}{5(4)} + n = \frac{1(20)}{1(20)} \qquad \text{Write equivalent fractions with the LCD, 20.}$$

$$\frac{5}{20} + \frac{4}{20} + n = \frac{20}{20}$$

$$\frac{9}{20} + n = \frac{20}{20} \qquad \text{Combine fractions on the left side.}$$

$$\frac{9}{20} - \frac{9}{20} + n = \frac{20}{20} - \frac{9}{20} \qquad \text{Subtract } \frac{9}{20} \text{ from both sides to isolate } n.$$

$$n = \frac{11}{20}$$

Answer $\frac{11}{20}$ had no opinion.

Check Add the three fractions together. The whole group should be the sum, which is represented by the number 1.

$$\frac{1}{4} + \frac{1}{5} + n = 1$$

$$\frac{1}{4} + \frac{1}{5} + \frac{11}{20} \stackrel{?}{=} 1 \qquad \text{Replace } n \text{ with } \frac{11}{20}.$$

$$\frac{1(5)}{4(5)} + \frac{1(4)}{5(4)} + \frac{11}{20} \stackrel{?}{=} 1 \qquad \text{Write equivalent fractions with the LCD, 20.}$$

$$\frac{5}{20} + \frac{4}{20} + \frac{11}{20} \stackrel{?}{=} 1$$

$$\frac{20}{20} \stackrel{?}{=} 1 \qquad \text{Add.}$$

$$1 = 1 \qquad \text{Simplify. It checks.}$$

▶ **Do Your Turn 12**

▶ **Your Turn 12**

Solve.

a. A report claims that $\frac{3}{8}$ of all cars are blue, $\frac{1}{4}$ are red, and the rest are other colors. What fraction of all cars are other colors?

b. At the opening of the stock market one day, a company's stock was worth $\$5\frac{3}{8}$ per share. At the close of the stock market that day, the stock was worth $\$6\frac{1}{4}$. How much did the stock increase in value?

Note If you are not sure how to approach a problem containing fractions, try replacing the fractions with whole numbers and read the problem again. Suppose Example 12 read as follows:

In a poll, 25 of the respondents said that they agreed with the passing of a certain bill, 20 said that they disagreed, and the rest said that they had no opinion. How many of the respondents had no opinion?

If we knew the total number of people in the poll, to solve, we could add the number that agreed plus the number that disagreed, then subtract from the total number of people polled. With fractions, we do not need the total number because it is always represented by 1. Therefore, to solve our equation, we added the fraction that agreed plus the fraction that disagreed, then subtracted the sum from all respondents, which is represented by 1.

Answers to Your Turn 12

a. $\dfrac{3}{8}$ b. $\$\dfrac{7}{8}$

5.6 Exercises For Extra Help MyMathLab®

Objectives 1 and 2

Prep Exercise 1 Explain how to add or subtract fractions with the same denominator.

For Exercises 1–18, add or subtract.

1. $\dfrac{2}{7} + \dfrac{3}{7}$

2. $\dfrac{1}{13} + \dfrac{6}{13}$

3. $-\dfrac{4}{9} + \left(-\dfrac{2}{9}\right)$

4. $-\dfrac{2}{15} + \left(-\dfrac{8}{15}\right)$

5. $\dfrac{10}{17} - \dfrac{4}{17}$

6. $\dfrac{9}{11} - \dfrac{1}{11}$

7. $\dfrac{6}{35} - \dfrac{13}{35}$

8. $\dfrac{1}{30} - \dfrac{4}{30}$

9. $\dfrac{9}{x} + \dfrac{3}{x}$

10. $\dfrac{4}{ab} + \dfrac{16}{ab}$

11. $\dfrac{8x^2}{9} - \dfrac{2x^2}{9}$

12. $\dfrac{x}{6} + \dfrac{1}{6}$

13. $\dfrac{4m}{n} - \dfrac{6m}{n}$

14. $\dfrac{4}{5h} - \dfrac{k}{5h}$

15. $\dfrac{3x^2 + 4x}{7y} + \dfrac{x^2 - 7x + 1}{7y}$

16. $\dfrac{9t - 12}{11u^2} + \dfrac{6t + 5}{11u^2}$

17. $\dfrac{7n^2 + 12}{5m} - \dfrac{2n^2 + 3}{5m}$

18. $\dfrac{h^2 - 3h + 5}{k} - \dfrac{h^2 + 4h - 1}{k}$

Objectives 3 and 4

Prep Exercise 2 Explain how to add or subtract fractions with different denominators.

Prep Exercise 3 Suppose you are given the problem $\frac{3}{4} + \frac{5}{6}$ and you know that the LCD is 12. Explain how to write the equivalent fractions so that they have the LCD.

For Exercises 19–34, add or subtract.

19. $\dfrac{3}{10} + \dfrac{5}{6}$

20. $\dfrac{2}{15} - \dfrac{4}{9}$

21. $\dfrac{7}{12} - \dfrac{11}{36}$

22. $\dfrac{1}{8} + \dfrac{15}{32}$

23. $\dfrac{1}{20} + \dfrac{17}{30}$

24. $\dfrac{3}{16} - \dfrac{9}{20}$

25. $\dfrac{3}{4} + \dfrac{1}{6} + \dfrac{7}{9}$

26. $\dfrac{2}{5} + \dfrac{5}{6} + \dfrac{13}{15}$

27. $\dfrac{7x}{12} + \dfrac{3x}{8}$

28. $\dfrac{a}{9} + \dfrac{2a}{3}$

29. $\dfrac{9}{16m} - \dfrac{3}{8m}$

30. $\dfrac{7}{10x} - \dfrac{5}{6x}$

31. $\dfrac{13}{20} + \dfrac{4}{5h}$

32. $\dfrac{7}{4x} + \dfrac{5}{12}$

33. $\dfrac{2}{3n^2} - \dfrac{7}{9n}$

34. $\dfrac{1}{2t} - \dfrac{3}{5t^2}$

Objective 5

Prep Exercise 4 Explain how to add or subtract mixed numbers.

For Exercises 35–42, estimate and find the actual sum expressed as a mixed number in simplest form.

35. $3\dfrac{4}{9} + 7$

36. $4\dfrac{2}{5} + \dfrac{1}{5}$

37. $2\dfrac{1}{4} + 5\dfrac{1}{4}$

38. $6\dfrac{1}{7} + 1\dfrac{5}{7}$

39. $5\dfrac{5}{6} + 1\dfrac{2}{3}$

40. $3\dfrac{4}{5} + 7\dfrac{1}{4}$

41. $6\dfrac{5}{8} + 3\dfrac{7}{12}$

42. $9\dfrac{1}{2} + 10\dfrac{4}{9}$

For Exercises 43–50, estimate and find the actual difference expressed as a mixed number in simplest form.

43. $9\frac{7}{8} - \frac{1}{8}$

44. $4\frac{5}{9} - 3$

45. $11\frac{7}{12} - 2\frac{5}{12}$

46. $8\frac{3}{7} - 2\frac{1}{7}$

47. $5\frac{1}{8} - 1\frac{5}{6}$

48. $6\frac{2}{5} - 2\frac{3}{4}$

49. $8\frac{1}{3} - 7\frac{5}{6}$

50. $10\frac{5}{12} - 6\frac{5}{8}$

Objective 6

For Exercises 51–58, add or subtract.

51. $6\frac{3}{4} + \left(-2\frac{1}{3}\right)$

52. $-5\frac{1}{2} + 2\frac{1}{8}$

53. $-7\frac{5}{6} - 2\frac{2}{3}$

54. $-\frac{4}{5} - 3\frac{1}{2}$

55. $\frac{5}{8} - 4\frac{1}{4}$

56. $6\frac{7}{8} - \left(-3\frac{1}{4}\right)$

57. $-7\frac{3}{4} - \left(-1\frac{1}{8}\right)$

58. $-2\frac{1}{6} - \left(-3\frac{1}{2}\right)$

Objective 7

For Exercises 59–66, solve and check.

59. $x + \frac{3}{5} = \frac{7}{10}$

60. $\frac{1}{4} + y = \frac{5}{8}$

61. $n - \frac{4}{5} = \frac{1}{4}$

62. $\frac{5}{8} = k - \frac{1}{6}$

63. $-\frac{1}{6} = b + \frac{3}{4}$

64. $-\frac{2}{3} = h - \frac{1}{8}$

65. $-3\frac{1}{2} + t = -4\frac{1}{5}$

66. $-2\frac{1}{6} = m - \frac{3}{4}$

Objective 8

For Exercises 67–76, solve.

67. A $\frac{1}{4}$-inch-thick slab of maple is glued to a $\frac{5}{8}$-inch-thick slab of mahogany to form what will become the body of an electric guitar. After the two slabs are glued together, what is the total thickness?

68. Amanda wants to hang a picture frame so that the bottom of the frame is $54\frac{1}{2}$ inches from the floor. The hanger on the back of the picture is $11\frac{5}{8}$ inches from the bottom of the frame. How high should she place the nail?

69. A tabletop is $\frac{3}{4}$ inch thick. During sanding, $\frac{1}{16}$ inch is removed off the top. How thick is the tabletop after sanding?

70. A class prepares an evaporation experiment. They place a ruler vertically in a bucket and then fill the bucket with water to a depth of $10\frac{1}{2}$ inches. After several days, they return to find that the water is $9\frac{7}{8}$ inches deep. How much did the water level decrease?

71. Students taking a certain introductory course are considered successful if they earn a grade of A, B, or C. The following graph shows how students taking the course performed one semester.

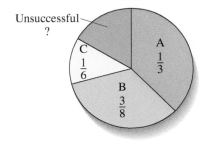

Student Performance

a. What fraction of the students who took the course earned an A or a B?

b. What fraction of the students who took the course were unsuccessful?

72. A poll is taken to assess a governor's approval rating. The following graph shows how people responded.

Governor's Rating Poll

a. What fraction of the respondents said "excellent" or "good"?

b. What fraction said "poor"?

73. A wooden frame is to be $20\frac{3}{4}$ inches long by $13\frac{1}{2}$ inches wide. What total length of wood is needed to make the frame?

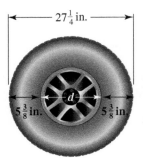

74. A game preserve is in the shape of the triangle shown. What is the total distance around the park?

75. What is the width, w, of the hallway in the following floor plan?

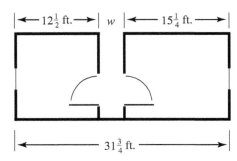

76. What is the diameter, d, of the hubcap on the tire shown?

Review Exercises

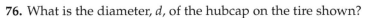

[2.5] 1. Simplify. $3^2 - 5[2 + (18 \div (-2))] + \sqrt{49}$

[3.1] 2. Evaluate $x^2 - 2x + \sqrt{x}$ when $x = 4$.

[3.1] 3. Evaluate $y^2 + 2y + 3$ when $y = -2$.

[3.3] 4. Add. $(4x^2 + 5x - 6) + (6x^2 - 2x + 1)$

[3.3] 5. Subtract. $(8y^3 - 4y + 2) - (9y^3 + y^2 + 7)$

[3.4] 6. Multiply. $(-6b^3)(9b)$

[3.4] 7. Multiply. $(x + 3)(2x - 5)$

[3.6] 8. Divide. $\dfrac{-16x^7}{4x^5}$

[3.7] 9. Write an expression in simplest form for the volume of the box shown.

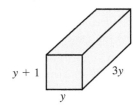

5.7 Order of Operations; Evaluating and Simplifying Expressions

Warm-up

[3.1] 1. Evaluate $4x^2 - 5y$ when $x = -3$ and $y = 8$.

[5.6] 2. Subtract. $7\dfrac{3}{5} - 2\dfrac{3}{4}$

Objective 1 Use the order of operations agreement to simplify expressions containing fractions and mixed numbers.

Objectives

1 Use the order of operations agreement to simplify expressions containing fractions and mixed numbers.

2 Find the mean, median, and mode of a set of values.

3 Evaluate expressions.

4 Find the area of a trapezoid.

5 Find the area of a circle.

6 Simplify polynomials containing fractions.

Example 1   Simplify. $\dfrac{3}{4} + 2\dfrac{4}{5} \cdot \left(-\dfrac{5}{6}\right)$

Solution: Follow the order of operations agreement. We multiply first.

$\dfrac{3}{4} + 2\dfrac{4}{5} \cdot \left(-\dfrac{5}{6}\right)$

$= \dfrac{3}{4} + \dfrac{14}{5} \cdot \left(-\dfrac{5}{6}\right)$ Write $2\dfrac{4}{5}$ as an improper fraction so that we can multiply.

$= \dfrac{3}{4} + \dfrac{\overset{7}{\cancel{14}}}{\cancel{5}} \cdot \left(-\dfrac{\cancel{5}}{\underset{3}{\cancel{6}}}\right)$ Divide out a common factor of 2 in 14 and 6; then divide out the common 5s.

$= \dfrac{3}{4} + \left(-\dfrac{7}{3}\right)$ Multiply the remaining factors.

$= \dfrac{3(3)}{4(3)} + \left(-\dfrac{7(4)}{3(4)}\right)$ Write equivalent fractions with the LCD, 12.

$= \dfrac{9}{12} + \left(-\dfrac{28}{12}\right)$

$= -\dfrac{19}{12}$ Add.

$= -1\dfrac{7}{12}$ Write as a mixed number.

▶ **Do Your Turn 1**

▶ **Your Turn 1**

Simplify.

a. $\dfrac{5}{8} - \dfrac{2}{9} \cdot \dfrac{3}{4}$

b. $3\dfrac{1}{2} + \dfrac{4}{5} \div \dfrac{3}{10}$

c. $\dfrac{1}{6} + \left(-4\dfrac{1}{3}\right) \cdot 2\dfrac{1}{2}$

d. $-5\dfrac{3}{8} - \dfrac{7}{10} \div \left(-\dfrac{7}{30}\right)$

Example 2   Simplify. $\left(\dfrac{1}{2}\right)^3 - 6\left(\dfrac{3}{5} + \dfrac{2}{3}\right)$

Solution: First, we find the sum in the parentheses. The LCD of $\dfrac{3}{5}$ and $\dfrac{2}{3}$ is 15.

$\left(\dfrac{1}{2}\right)^3 - 6\left(\dfrac{3}{5} + \dfrac{2}{3}\right) = \left(\dfrac{1}{2}\right)^3 - 6\left(\dfrac{3(3)}{5(3)} + \dfrac{2(5)}{3(5)}\right)$ Write equivalent fractions with the LCD, 15.

$= \left(\dfrac{1}{2}\right)^3 - 6\left(\dfrac{9}{15} + \dfrac{10}{15}\right)$

$= \left(\dfrac{1}{2}\right)^3 - 6 \cdot \dfrac{19}{15}$ Add.

$= \dfrac{1}{8} - 6 \cdot \dfrac{19}{15}$ Evaluate the expression with the exponent. $\left(\dfrac{1}{2}\right)^3 = \dfrac{1}{2} \cdot \dfrac{1}{2} \cdot \dfrac{1}{2} = \dfrac{1}{8}$

Answers to Your Turn 1

a. $\dfrac{11}{24}$ b. $6\dfrac{1}{6}$ c. $-10\dfrac{2}{3}$

d. $-2\dfrac{3}{8}$

Answers to Warm-up

1. -4 2. $4\dfrac{17}{20}$

▶ **Your Turn 2**

Simplify.

a. $\left(\dfrac{1}{3}\right)^2 + 16\left(\dfrac{1}{4} + \dfrac{3}{8}\right)$

b. $\left(\dfrac{1}{25} - \dfrac{4}{5}\right) - \left(\dfrac{2}{5}\right)^2$

c. $\left(1\dfrac{2}{3}\right)^2 + 2\dfrac{1}{2} \div \dfrac{5}{8}$

d. $\left(\dfrac{1}{2} + \dfrac{1}{3}\right) \div \left(\dfrac{3}{4} - \dfrac{1}{8}\right)$

$= \dfrac{1}{8} - \dfrac{\overset{2}{\cancel{6}}}{1} \cdot \dfrac{19}{\underset{5}{\cancel{15}}}$ Divide out the common factors.

$= \dfrac{1}{8} - \dfrac{38}{5}$ Multiply.

$= \dfrac{1(5)}{8(5)} - \dfrac{38(8)}{5(8)}$ Write equivalent fractions with the LCD, 40.

$= \dfrac{5}{40} - \dfrac{304}{40}$

$= -\dfrac{299}{40}$ Subtract.

$= -7\dfrac{19}{40}$ Write the result as a mixed number.

◀ **Do Your Turn 2**

Objective 2 Find the mean, median, and mode of a set of values.

Let's revisit mean, median, and mode from Chapter 1.

Example 3 An agricultural student decides to study pumpkin growth in her area. The following table contains the circumference of six pumpkins from a particular farm. Find the mean, median, and mode circumference.

$37\dfrac{1}{2}$ in.	$34\dfrac{3}{4}$ in.	$40\dfrac{3}{8}$ in.	$36\dfrac{5}{8}$ in.	$40\dfrac{3}{8}$ in.	$34\dfrac{1}{4}$ in.

Solution:

$\text{mean} = \left(37\dfrac{1}{2} + 34\dfrac{3}{4} + 40\dfrac{3}{8} + 36\dfrac{5}{8} + 40\dfrac{3}{8} + 34\dfrac{1}{4}\right) \div 6$ Divide the sum of the measurements by the number of measurements, 6.

$= \left(37\dfrac{4}{8} + 34\dfrac{6}{8} + 40\dfrac{3}{8} + 36\dfrac{5}{8} + 40\dfrac{3}{8} + 34\dfrac{2}{8}\right) \div 6$ Write each fraction with the LCD, 8.

Note This mixed number will be written as an improper fraction so that we can divide; so there is no need to simplify the improper fraction.

▶ $= 221\dfrac{23}{8} \div 6$ Add the mixed numbers.

$= \dfrac{1791}{8} \cdot \dfrac{1}{6}$ Write the mixed number as an improper fraction and write the division as multiplication.

$= \dfrac{\overset{597}{\cancel{1791}}}{8} \cdot \dfrac{1}{\underset{2}{\cancel{6}}}$ Divide out the common factor, 3.

$= \dfrac{597}{16}, \text{ or } 37\dfrac{5}{16}$ Multiply.

Answer: The mean circumference is $37\dfrac{5}{16}$ in.

Answers to Your Turn 2

a. $10\dfrac{1}{9}$ b. $-\dfrac{23}{25}$

c. $6\dfrac{7}{9}$ d. $1\dfrac{1}{3}$

Median:

$34\dfrac{1}{4}, 34\dfrac{3}{4}, 36\dfrac{5}{8}, 37\dfrac{1}{2}, 40\dfrac{3}{8}, 40\dfrac{3}{8}$ Write the measurements in order from least to greatest.

Because there is an even number of measurements, the median is the mean of the middle two numbers, $36\frac{5}{8}$ and $37\frac{1}{2}$.

$$\text{median} = \left(36\frac{5}{8} + 37\frac{1}{2}\right) \div 2$$

$$= \left(36\frac{5}{8} + 37\frac{4}{8}\right) \div 2 \qquad \text{Write each fraction with the LCD, 8.}$$

$$= 73\frac{9}{8} \div 2 \qquad \text{Add the mixed numbers.}$$

$$= \frac{593}{8} \cdot \frac{1}{2} \qquad \text{Write the mixed number as an improper fraction and write the division as multiplication.}$$

$$= \frac{593}{16}, \text{ or } 37\frac{1}{16} \qquad \text{Multiply.}$$

Answer: The median circumference is $37\frac{1}{16}$ inches.

Mode: The mode is the number that occurs most often, which is $40\frac{3}{8}$ inches.

▶ **Do Your Turn 3**

Objective 3 Evaluate expressions.

Now let's evaluate algebraic expressions when we are given variable values that are fractions. In Section 3.1, we learned that to evaluate an expression, we replace each variable with the corresponding given value and then use the order of operations agreement to simplify the numeric expression.

Example 4   Evaluate $4xy^2 + z$ when $x = \frac{3}{4}$, $y = \frac{2}{3}$, and $z = 3\frac{1}{2}$.

Solution:

$$4\left(\frac{3}{4}\right)\left(\frac{2}{3}\right)^2 + 3\frac{1}{2} \qquad \text{Replace } x \text{ with } \frac{3}{4}, y \text{ with } \frac{2}{3}, \text{ and } z \text{ with } 3\frac{1}{2}.$$

$$= \frac{4}{1}\left(\frac{3}{4}\right)\frac{4}{9} + 3\frac{1}{2} \qquad \text{Evaluate the expression with the exponent and write 4 as } \frac{4}{1}.$$

$$= \frac{\overset{1}{4}}{1}\left(\frac{3}{\underset{1}{4}}\right)\frac{4}{9} + 3\frac{1}{2} \qquad \text{Divide out the common factors.}$$

$$= \frac{4}{3} + 3\frac{1}{2} \qquad \text{Multiply.}$$

Note We could have written $3\frac{1}{2}$ as $\frac{7}{2}$ and then added the improper fractions. Choose the method that is easiest for you.

$$= 1\frac{1}{3} + 3\frac{1}{2} \qquad \text{Write } \frac{4}{3} \text{ as a mixed number in order to add.}$$

$$= 1\frac{1(2)}{3(2)} + 3\frac{1(3)}{2(3)} \qquad \text{Write equivalent fractions with the LCD, 6.}$$

$$= 1\frac{2}{6} + 3\frac{3}{6}$$

$$= 4\frac{5}{6} \qquad \text{Add.}$$

▶ **Do Your Turn 4**

▶ **Your Turn 3**

A tree farmer measures the circumference of a sample of trees after eight years. The following table contains the circumferences of six trees in the farm. Find the mean, median, and mode circumference.

$18\frac{1}{2}$ in.	$16\frac{3}{4}$ in.
$17\frac{3}{4}$ in.	$18\frac{5}{8}$ in.
$19\frac{1}{4}$ in.	$17\frac{1}{4}$ in.

▶ **Your Turn 4**

Evaluate each expression using the given values.

a. mv^2; $m = 3\frac{1}{4}$, $v = 10$

b. $\frac{1}{2}at^2$; $a = -9\frac{4}{5}$, $t = 2$

c. $a^2 - 5b$; $a = \frac{3}{4}$, $b = \frac{7}{20}$

d. $2m(n + 3)$; $m = \frac{5}{8}$, $n = 1\frac{1}{3}$

Answers to Your Turn 3

mean: $18\frac{1}{48}$ in.;

median: $18\frac{1}{8}$ in.; no mode

Answers to Your Turn 4

a. 325   b. $-19\frac{3}{5}$

c. $-1\frac{3}{16}$ d. $5\frac{5}{12}$

Objective 4 Find the area of a trapezoid.

We have learned formulas for finding the area of a parallelogram $(A = bh)$ and a triangle $(A = \frac{1}{2}bh)$. Now let's develop a formula for finding the area of a **trapezoid**.

Definition Trapezoid: A four-sided figure with one pair of parallel sides.

The following figures are examples of trapezoids.

In each figure, notice that the top and bottom sides are parallel but the other sides are not. Because the top and bottom sides are different lengths, we will label them a and b.

To develop the formula for the area of a trapezoid, we use the same process we used for finding the formula for the area of a triangle. First, we make two identical trapezoids. By inverting one of the trapezoids, then joining the corresponding sides, we create a parallelogram, as shown here.

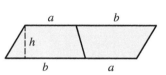

The area of this parallelogram is found by multiplying the base and height. Notice that the base is $a + b$ and the height is h; so the area of the parallelogram is $h(a + b)$. However, we really want the area of one of the trapezoids, which is half the area of the parallelogram.

Conclusion: The formula for the area of a trapezoid is $A = \dfrac{1}{2}h(a + b)$.

▶ **Your Turn 5**

Find the area of the trapezoid shown.

a.

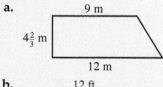

9 m

$4\frac{2}{3}$ m

12 m

b.

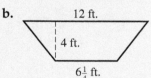

12 ft.

4 ft.

$6\frac{1}{4}$ ft.

Answers to Your Turn 5

a. $49\,\text{m}^2$ **b.** $36\dfrac{1}{2}\text{ft.}^2$

Example 5   Find the area of the trapezoid shown.

6 ft.

$2\frac{1}{2}$ ft.

7 ft.

Solution: Use the formula $A = \dfrac{1}{2}h(a + b)$.

$$A = \frac{1}{2}h(a + b)$$

$$A = \frac{1}{2}\left(2\frac{1}{2}\right)(6 + 7)$$ Replace h with $2\frac{1}{2}$, a with 6, and b with 7.

$$A = \frac{1}{2}\left(2\frac{1}{2}\right)(13)$$ Add 6 and 7 inside the parentheses.

$$A = \frac{1}{2}\left(\frac{5}{2}\right)\left(\frac{13}{1}\right)$$ Write as improper fractions.

$$A = \frac{65}{4}$$ Multiply.

$$A = 16\frac{1}{4}$$

Answer: The area is $16\frac{1}{4}$ ft.^{2}

Note Remember, area is always in square units. Because the distance units are in terms of feet, the area unit is in terms of square feet.

◀ **Do Your Turn 5**

Objective 5 Find the area of a circle.

In Section 5.3, we developed the formula for the circumference of a circle. Now let's develop a formula for the area of a circle. As we did for triangles and trapezoids, to develop the formula for the area of a circle, we will transform a circle into a parallelogram.

To make a circle look like a parallelogram, we cut it up like a pizza. First, we cut the top half into six pizza slices and fan those slices out. Then we cut the bottom half of the circle the same way.

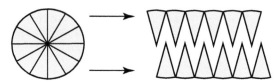

Next, we fit the slices together like teeth. The resulting figure closely resembles a parallelogram. This parallelogram has the same area as the circle.

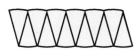 **Note** If we were to cut more slices, the *slight curves* along the top and bottom would be less prominent. In fact, if we were to use extremely slender slices, the curves would be almost imperceptible.

Because the area of a parallelogram is found by multiplying base and height, we need to relate the base and height of the parallelogram back to the circle. If our circle is a pizza, the height goes from the center to the crust; so the height corresponds to the radius of the circle. The base distance corresponds to the crust of the bottom half of the pizza; so the base is half the circumference of the circle.

The height corresponds to the radius.

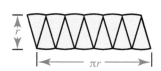

The base corresponds to half the circumference. The full circumference is found by $2\pi r$; so half the circumference is πr.

Because the area of a parallelogram is found by multiplying base and height, we can multiply the base, πr, by the height, r, to write the formula for the area of a circle.

$$A = \text{base} \cdot \text{height}$$
$$A = \pi r \cdot r$$
$$A = \pi r^2$$

Conclusion: The formula for the area of a circle with radius r is $A = \pi r^2$.

Example 6 Find the area of a circle with a diameter of 5 meters.

Solution: We are given the diameter, but $A = \pi r^2$ requires radius; so we must first find the radius.

$$r = \frac{1}{2}d$$

$$r = \frac{1}{2} \cdot 5 \qquad \text{Replace } d \text{ with 5.}$$

$$r = \frac{1}{2} \cdot \frac{5}{1}$$

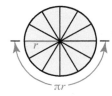 **Note** Because we must square the radius to find area, we will leave the fraction as an improper fraction.

$$r = \frac{5}{2}$$

▶ **Your Turn 6**

Solve.

a. Find the area of a circle with a radius of 6 yards.

b. Find the area of a circle with a diameter of 7 inches.

Now that we know that the radius is $\dfrac{5}{2}$ meters, we can find the area using $A = \pi r^2$.

$$A = \pi r^2$$

Note We approximate π ▶ as $\frac{22}{7}$. When we replace π with $\frac{22}{7}$, the entire calculation becomes an approximation; so we use $\approx$ instead of an equal sign.

$$A \approx \frac{22}{7} \cdot \left(\frac{5}{2}\right)^2 \qquad \text{Replace } \pi \text{ with } \tfrac{22}{7} \text{ and } r \text{ with } \tfrac{5}{2}.$$

$$A \approx \frac{22}{7} \cdot \frac{25}{4} \qquad \text{Square } \tfrac{5}{2} \text{ to get } \tfrac{25}{4}.$$

$$A \approx \frac{\overset{11}{22}}{7} \cdot \frac{25}{\underset{2}{4}} \qquad \begin{array}{l}\text{Divide out the common}\\ \text{factor of 2 in 22 and 4.}\end{array}$$

$$A \approx \frac{275}{14} \qquad \text{Multiply.}$$

$$A \approx 19\frac{9}{14} \qquad \text{Write the result as a mixed number.}$$

Answer: The area is $19\dfrac{9}{14}$ m². ◀ **Note** Area is always in square units. Because the distance unit for the radius is meters, the area unit is square meters.

◀ **Do Your Turn 6**

Objective 6 Simplify polynomials containing fractions.

We learned in Section 3.2 that we can simplify expressions by combining like terms. Like terms have the same variables raised to the same exponents. To combine like terms, add the coefficients and keep the variables and their exponents the same.

▶ **Your Turn 7**

Combine like terms.

a. $\dfrac{5}{8}x + \dfrac{1}{2} - \dfrac{3}{4}x^2 + \dfrac{2}{3} - \dfrac{1}{4}x$

b. $\dfrac{1}{6}m^2 - 3m + \dfrac{2}{3}m^2 + \dfrac{1}{7} - \dfrac{1}{2}m$

| **Example 7** | Combine like terms. $\dfrac{1}{4}a^3 - \dfrac{3}{5}a + \dfrac{1}{6}a^3 + \dfrac{1}{2}a$ |

Solution: Add coefficients for the like terms. Because we cannot combine an a^3 term with an a term, we need to get common denominators only for the fractions we will actually combine.

$$\frac{1}{4}a^3 - \frac{3}{5}a + \frac{1}{6}a^3 + \frac{1}{2}a$$

$$= \frac{1}{4}a^3 + \frac{1}{6}a^3 - \frac{3}{5}a + \frac{1}{2}a \qquad \begin{array}{l}\text{Use the commutative property}\\ \text{to group the like terms together.}\\ \text{Remember, this is optional.}\end{array}$$

$$= \frac{1(3)}{4(3)}a^3 + \frac{1(2)}{6(2)}a^3 - \frac{3(2)}{5(2)}a + \frac{1(5)}{2(5)}a \qquad \begin{array}{l}\text{Write equivalent fractions.}\\ \text{The LCD for the } a^3 \text{ coefficients}\\ \text{is 12. The LCD for the } a \text{ coeffi-}\\ \text{cients is 10.}\end{array}$$

$$= \frac{3}{12}a^3 + \frac{2}{12}a^3 - \frac{6}{10}a + \frac{5}{10}a$$

$$= \frac{5}{12}a^3 - \frac{1}{10}a \qquad \text{Add or subtract coefficients.}$$

Note Because a^3 and a are not like terms, we cannot combine further. The expression is in simplest form.

◀ **Do Your Turn 7**

Answers to Your Turn 6

a. $113\dfrac{1}{7}$ yd.² **b.** $38\dfrac{1}{2}$ in.²

Answers to Your Turn 7

a. $-\dfrac{3}{4}x^2 + \dfrac{3}{8}x + \dfrac{7}{6}$

b. $\dfrac{5}{6}m^2 - \dfrac{7}{2}m + \dfrac{1}{7}$

In Section 3.3, we learned how to add and subtract polynomials. To add polynomials, we combine like terms. To subtract, we write an equivalent addition expression using the additive inverse of the subtrahend, which is the second polynomial.

Example 8   Add or subtract. $\left(\dfrac{3}{4}h - \dfrac{5}{6}\right) - \left(\dfrac{1}{3}h - \dfrac{2}{5}\right)$

Solution: Write an equivalent addition expression by changing the operation to addition and the subtrahend to its additive inverse. To write the additive inverse of a polynomial, we change the sign of each term in the polynomial.

$$\left(\frac{3}{4}h - \frac{5}{6}\right) - \left(\frac{1}{3}h - \frac{2}{5}\right)$$

$$= \left(\frac{3}{4}h - \frac{5}{6}\right) + \left(-\frac{1}{3}h + \frac{2}{5}\right) \qquad \text{Write an equivalent addition expression.}$$

$$= \frac{3}{4}h - \frac{1}{3}h - \frac{5}{6} + \frac{2}{5} \qquad \text{Collect like terms.}$$

$$= \frac{3(3)}{4(3)}h - \frac{1(4)}{3(4)}h - \frac{5(5)}{6(5)} + \frac{2(6)}{5(6)} \qquad \text{Write equivalent fractions with the LCD.}$$

$$= \frac{9}{12}h - \frac{4}{12}h - \frac{25}{30} + \frac{12}{30}$$

$$= \frac{5}{12}h - \frac{13}{30} \qquad \text{Combine the like terms.}$$

▶ **Do Your Turn 8**

We can also multiply monomials and polynomials with fractional coefficients. Consider multiplying monomials first. We learned in Section 3.4 that to multiply monomials, we multiply the coefficients and add the exponents of the like bases.

Example 9   Multiply. $\left(-\dfrac{3}{10}m^4\right)\left(\dfrac{2}{9}m\right)$

Solution: Multiply the coefficients and add the exponents of the like bases.

$$\left(-\frac{3}{10}m^4\right)\left(\frac{2}{9}m\right) = -\frac{\overset{1}{\cancel{3}}}{\underset{5}{\cancel{10}}} \cdot \frac{\overset{1}{\cancel{2}}}{\underset{3}{\cancel{9}}}m^{4+1} \qquad \begin{array}{l}\text{Divide out the common factors in the co-}\\ \text{efficients and add the exponents of the}\\ \text{variables.}\end{array}$$

$$= -\frac{1}{15}m^5 \qquad \begin{array}{l}\text{Multiply the coefficients and simplify the ad-}\\ \text{dition of the exponents.}\end{array}$$

▶ **Do Your Turn 9**

Next, as we learned in Section 3.4, when we multiply a polynomial by a monomial, we use the distributive property and multiply each term inside the polynomial by the monomial.

▶ **Your Turn 8**

Add or subtract.

a. $\left(\dfrac{3}{8}n - \dfrac{2}{3}\right) + \left(\dfrac{1}{4}n - \dfrac{1}{6}\right)$

b. $\left(\dfrac{2}{5}x^2 + \dfrac{5}{16}\right) - \left(\dfrac{1}{10}x^2 - \dfrac{3}{4}\right)$

▶ **Your Turn 9**

Multiply.

a. $\left(\dfrac{1}{4}a\right)\left(\dfrac{2}{7}a^5\right)$

b. $\left(\dfrac{-5}{12}h^4\right)\left(\dfrac{6}{-11}h^3\right)$

Answers to Your Turn 8

a. $\dfrac{5}{8}n - \dfrac{5}{6}$ b. $\dfrac{3}{10}x^2 + \dfrac{17}{16}$

Answers to Your Turn 9

a. $\dfrac{1}{14}a^6$ b. $\dfrac{5}{22}h^7$

▶ **Your Turn 10**

Multiply.

a. $\dfrac{2}{3}\left(\dfrac{3}{7}a^3 - 9a^2\right)$

b. $-\dfrac{5}{12}\left(6m^2 + \dfrac{2}{5}mn\right)$

Example 10 Multiply. $\dfrac{3}{5}\left(\dfrac{2}{3}x - 10\right)$

Solution: We use the distributive property and multiply each term in $\frac{2}{3}x - 10$ by $\frac{3}{5}$.

$$\dfrac{3}{5}\left(\dfrac{2}{3}x - 10\right) = \dfrac{3}{5} \cdot \dfrac{2}{3}x - \dfrac{3}{5} \cdot \dfrac{10}{1} \qquad \text{Distribute } \tfrac{3}{5}.$$

$$= \dfrac{\overset{1}{3}}{5} \cdot \dfrac{2}{\underset{1}{3}}x - \dfrac{3}{\underset{1}{5}} \cdot \dfrac{\overset{2}{10}}{1} \qquad \text{Divide out the common factors.}$$

$$= \dfrac{2}{5}x - 6 \qquad \text{Multiply.}$$

◀ **Do Your Turn 10**

Now let's multiply polynomials. In Section 3.4, we learned that when we multiply polynomials, we multiply each term in the second polynomial by each term in the first polynomial.

▶ **Your Turn 11**

Multiply.

a. $\left(3a - \dfrac{1}{8}\right)\left(2a - \dfrac{1}{4}\right)$

b. $\left(\dfrac{1}{5}m - 2\right)\left(\dfrac{1}{5}m + 2\right)$

Example 11 Multiply. $\left(\dfrac{3}{4}x - 5\right)\left(\dfrac{1}{2}x + 8\right)$

Solution: We multiply each term in the second polynomial by each term in the first polynomial.

$$\left(\dfrac{3}{4}x - 5\right)\left(\dfrac{1}{2}x + 8\right)$$

(with arrows: $\frac{3}{4}x \cdot 8$, $\frac{3}{4}x \cdot \frac{1}{2}x$, $-5 \cdot \frac{1}{2}x$, $-5 \cdot 8$)

$$= \dfrac{3}{4}x \cdot \dfrac{1}{2}x + \dfrac{3}{4}x \cdot 8 - 5 \cdot \dfrac{1}{2}x - 5 \cdot 8 \qquad \text{Multiply each term in the second polynomial by each term in the first polynomial.}$$

$$= \dfrac{3}{4}x \cdot \dfrac{1}{2}x + \dfrac{3}{\underset{1}{4}}x \cdot \dfrac{\overset{2}{8}}{1} - \dfrac{5}{1} \cdot \dfrac{1}{2}x - 5 \cdot 8 \qquad \text{Divide out the common 4 in 4 and 8.}$$

$$= \dfrac{3}{8}x^2 + 6x - \dfrac{5}{2}x - 40 \qquad \text{Multiply.}$$

$$= \dfrac{3}{8}x^2 + \dfrac{6(2)}{1(2)}x - \dfrac{5}{2}x - 40 \qquad \text{Write the coefficients of the like terms with their LCD, 2.}$$

$$= \dfrac{3}{8}x^2 + \dfrac{12}{2}x - \dfrac{5}{2}x - 40$$

$$= \dfrac{3}{8}x^2 + \dfrac{7}{2}x - 40 \qquad \text{Subtract coefficients of the like terms.}$$

◀ **Do Your Turn 11**

Answers to Your Turn 10

a. $\dfrac{2}{7}a^3 - 6a^2$

b. $-\dfrac{5}{2}m^2 - \dfrac{1}{6}mn$

Answers to Your Turn 11

a. $6a^2 - a + \dfrac{1}{32}$

b. $\dfrac{1}{25}m^2 - 4$

5.7 Exercises For Extra Help MyMathLab®

Objective 1

Prep Exercise 1 In $\dfrac{1}{4} + 2 \cdot \dfrac{5}{8}$, which operation is performed first?

For Exercises 1–14, simplify.

1. $\dfrac{1}{4} + 2 \cdot \dfrac{5}{8}$

2. $\dfrac{1}{6} - 5 \cdot \dfrac{7}{10}$

3. $3\dfrac{1}{5} - \dfrac{1}{2} \div \dfrac{5}{6}$

4. $5\dfrac{1}{3} + \dfrac{3}{4} \div \dfrac{5}{12}$

5. $2\dfrac{1}{2} - \left(\dfrac{3}{4}\right)^2$

6. $\left(\dfrac{2}{3}\right)^2 + 2\dfrac{1}{6}$

7. $\dfrac{1}{4} - \dfrac{2}{3}\left(6 - 1\dfrac{1}{2}\right)$

8. $\dfrac{2}{5} + 9\left(\dfrac{1}{3} + \dfrac{1}{6}\right)$

9. $5\dfrac{3}{4} - 2\sqrt{\dfrac{80}{5}}$

10. $7\dfrac{1}{4} + \dfrac{3}{4}\sqrt{\dfrac{4}{9}}$

11. $\left(\dfrac{1}{2}\right)^3 + 2\dfrac{1}{4} - 5\left(\dfrac{3}{10} + \dfrac{1}{5}\right)$

12. $\left(\dfrac{2}{3}\right)^2 \div \dfrac{1}{3}\left(1\dfrac{1}{3} + \dfrac{5}{6}\right)$

13. $5\left(\dfrac{1}{2} - 3\dfrac{4}{5}\right) - 2\left(\dfrac{1}{6} + 4\right)$

14. $1\dfrac{1}{4} - \left(2\dfrac{1}{4}\right)\left(\dfrac{1}{3} + \dfrac{1}{2}\right)^2$

Objective 2

Prep Exercise 2 Explain how to find the mean of a set of numbers.

15. A biologist studying trout in a particular stream catches six specimens. The table lists the length, in inches, and weight, in pounds, of the fish.

  a. Find the mean length and weight.

  b. Find the median length and weight.

  c. Find the mode length and weight.

Length	Weight	Length	Weight
6	$\dfrac{3}{4}$	$9\dfrac{1}{4}$	2
$10\dfrac{1}{4}$	$2\dfrac{1}{4}$	12	3
$8\dfrac{3}{4}$	$1\dfrac{3}{4}$	$9\dfrac{1}{4}$	$1\dfrac{3}{4}$

16. A biologist at a fish farm measures eight specimens at eight months of age. The table lists the length, in inches, and weight, in pounds, of each fish.

  a. Find the mean length and weight.

  b. Find the median length and weight.

  c. Find the mode length and weight.

Length	Weight	Length	Weight
$5\dfrac{5}{8}$	1	5	$\dfrac{3}{4}$
$4\dfrac{5}{8}$	$\dfrac{1}{2}$	$5\dfrac{1}{4}$	$\dfrac{7}{8}$
$6\dfrac{1}{4}$	$1\dfrac{3}{8}$	6	$1\dfrac{3}{8}$
$4\dfrac{1}{2}$	$\dfrac{3}{8}$	$4\dfrac{5}{8}$	$\dfrac{5}{8}$

17. In thoroughbred horse racing, times in the quarter mile are expressed to the nearest fifth of a second. In 1973, Secretariat won the Kentucky Derby with the fastest time ever. The table contains his time for each quarter mile of the race.

  a. What is the total distance of the race?
  b. What was Secretariat's final time?
  c. What was Secretariat's average time in the quarter mile?
  d. What was his median time in the quarter mile?
  e. What was the mode time?

Quarter Mile	Time (in seconds)
First	$25\dfrac{1}{5}$
Second	24
Third	$23\dfrac{4}{5}$
Fourth	$23\dfrac{2}{5}$
Fifth	23

Of Interest

Generally, horses run fastest out of the gate in a race and then slow as they tire. Secretariat's record-breaking win was all the more remarkable because he ran each quarter mile faster than the previous one.

18. Secretariat also won the Preakness Stakes in 1973. The table contains his time for each quarter mile of the race.

   a. What is the total distance of the race?
   b. What was Secretariat's final time?
   c. What was Secretariat's average time in the quarter mile?
   d. What was his median time in the quarter mile?
   e. What was the mode time?

Quarter Mile	Time (in seconds)
First	$24\frac{2}{5}$
Second	$23\frac{4}{5}$
Third	$23\frac{1}{5}$
Fourth	$24\frac{1}{5}$
Fifth	$18\frac{4}{5}$

Objective 3

Prep Exercise 3 What is the first step in evaluating $\frac{1}{2}d$ when $d = 5\frac{3}{4}$?

For Exercises 19–26, evaluate the expression using the given values.

19. $x + vt$; $x = 2\frac{1}{2}, v = 30, t = \frac{1}{4}$

20. $x + vt$; $x = 6\frac{4}{5}, v = 40, t = \frac{1}{6}$

21. mv^2; $m = 1\frac{1}{4}, v = \frac{3}{4}$

22. mv^2; $m = 6\frac{1}{2}, v = -\frac{2}{5}$

23. $\frac{1}{2}at^2$; $a = -9\frac{4}{5}, t = \frac{1}{2}$

24. $\frac{1}{2}at^2$; $a = -32\frac{1}{5}, t = 2\frac{1}{2}$

25. $2xyz$; $x = \frac{1}{6}, y = -1\frac{3}{4}, z = -\frac{8}{7}$

26. $2xyz$; $x = -\frac{1}{5}, y = -2\frac{1}{3}, z = -\frac{10}{7}$

Objective 4

Prep Exercise 4 What is a trapezoid?

Prep Exercise 5 What is the formula for the area of a trapezoid, and what does each variable in it represent?

For Exercises 27–30, find the area.

27.
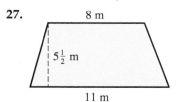
8 m, $5\frac{1}{2}$ m, 11 m

28.
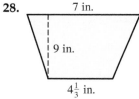
7 in., 9 in., $4\frac{1}{3}$ in.

29.
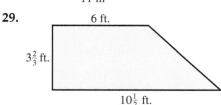
6 ft., $3\frac{2}{3}$ ft., $10\frac{1}{2}$ ft.

30.
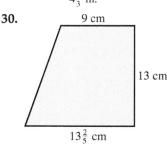
9 cm, 13 cm, $13\frac{2}{5}$ cm

Objective 5

Prep Exercise 6 What is the formula for the area of a circle, and what does each symbol in it represent?

For Exercises 31–34, use $\frac{22}{7}$ for π.

31. Find the area of a circle with a radius of 14 inches.

32. Find the area of a circle with a radius of 3 feet.

33. Find the area of a circle with a diameter of $2\frac{1}{2}$ meters.

34. Find the area of a circle with a diameter of $10\frac{1}{3}$ centimeters.

35. Find the area of the shape shown.

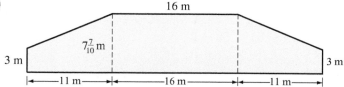

16 m

$7\frac{7}{10}$ m

3 m

3 m

11 m

16 m

11 m

36. Find the area of the shaded region. The diameter of the circle is $4\frac{1}{2}$ inches.

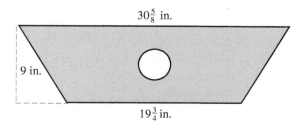

$30\frac{5}{8}$ in.

9 in.

$19\frac{3}{4}$ in.

Objective 6

For Exercises 37 and 38, combine like terms and write the resulting polynomial in descending order of degree.

37. $\dfrac{1}{2}x^2 - \dfrac{3}{4}x - \dfrac{1}{3}x^2 - 6x$

38. $\dfrac{3}{8}n^3 - 3n^3 - \dfrac{1}{4} - 2$

For Exercises 39–42, add or subtract and write the resulting polynomial in descending order of degree.

39. $\left(5y^3 - \dfrac{4}{5}y^2 + y - \dfrac{1}{6}\right) + \left(y^3 + 3y^2 - \dfrac{2}{3}\right)$

40. $\left(a^4 + \dfrac{1}{4}a^2 - a - \dfrac{1}{6}\right) + \left(\dfrac{3}{10}a^3 + a^2 - \dfrac{2}{3}a + 4\right)$

41. $\left(\dfrac{4}{5}t^3 + \dfrac{2}{3}t^2 - \dfrac{1}{6}\right) - \left(\dfrac{7}{10}t^3 + \dfrac{1}{4}t^2 + \dfrac{1}{2}\right)$

42. $\left(2x^4 + \dfrac{3}{5}x^2 - \dfrac{1}{8}x + 1\right) - \left(6x^4 + \dfrac{1}{4}x^2 - \dfrac{1}{2}x - \dfrac{1}{3}\right)$

For Exercises 43–48, multiply.

43. $\left(-\dfrac{1}{6}m\right)\left(\dfrac{3}{5}m^3\right)$

44. $\left(\dfrac{-7}{15}x^4\right)\left(\dfrac{5}{-21}x^2\right)$

45. $\dfrac{5}{8}\left(\dfrac{4}{5}t^2 - \dfrac{2}{3}t\right)$

46. $-\dfrac{1}{15}\left(3h^2 - \dfrac{3}{4}k^2\right)$

47. $\left(x - \dfrac{1}{2}\right)\left(x + \dfrac{1}{4}\right)$

48. $\left(\dfrac{1}{4}u - 1\right)\left(\dfrac{1}{2}u + 2\right)$

Puzzle Problem   The original size of Khufu's pyramid in Egypt was 754 feet along the base with a start height of 481 feet. It was originally covered with blocks cut so that all of the faces were smooth. (Most of these blocks are now gone.) What was the total surface area of the four triangular faces of Khufu's pyramid?

Review Exercises

For Exercises 1 and 2, solve and check.

[4.3] 1. $4x - 9 = 11$

[4.3] 2. $5(n - 3) = 3n + 7$

[4.3] 3. Florence studies a map and finds that her trip will be 240 miles. If she averages 60 miles per hour, how long will it take her to get to her destination? (Use $d = rt$.)

[4.4] 4. Six times the sum of n and 2 is equal to 3 less than n. Translate to an equation and solve.

[4.5] 5. David has $105 all in tens and fives. If he has 15 bills, how many tens and how many fives does he have?

5.8 Solving Equations

Objectives

1 Use the LCD to eliminate fractions from equations.

2 Translate sentences to equations; then solve.

3 Solve applications involving one unknown.

4 Solve applications involving two unknowns.

Warm-up

[4.3] 1. Solve and check. $7x + 2(x - 5) = 5x - 2$

[5.7] 2. Multiply. $12\left(\dfrac{3}{4}x - \dfrac{2}{3}\right)$

Objective 1 Use the LCD to eliminate fractions from equations.

In Sections 5.4 and 5.6, we reviewed solving equations using the multiplication principle of equality and the addition principle of equality. Now let's consider equations that require both principles. We will also develop a way to use the multiplication principle of equality to simplify equations that contain fractions.

Example 1 Solve and check. $\dfrac{3}{4}x - \dfrac{2}{3} = \dfrac{1}{6}$

Solution: First, we use the addition principle to isolate the variable term. Then we use the multiplication principle to clear the remaining coefficient.

$$\frac{3}{4}x - \frac{2}{3} = \frac{1}{6}$$

$$\frac{3}{4}x - \frac{2}{3} + \frac{2}{3} = \frac{1}{6} + \frac{2}{3} \qquad \text{Add } \tfrac{2}{3} \text{ to both sides.}$$

Note $-\tfrac{2}{3}$ and $\tfrac{2}{3}$ are additive inverses; so their sum is 0. ▶

$$\frac{3}{4}x + 0 = \frac{1}{6} + \frac{2(2)}{3(2)} \qquad \text{Write equivalent fractions with the LCD, 6.}$$

$$\frac{3}{4}x = \frac{1}{6} + \frac{4}{6}$$

$$\frac{3}{4}x = \frac{5}{6} \qquad \text{Add the fractions.}$$

$$\overset{1}{\underset{3}{\cancel{\frac{4}{3}}}} \cdot \overset{1}{\underset{1}{\cancel{\frac{3}{4}}}}x = \overset{2}{\underset{3}{\cancel{\frac{5}{6}}}} \cdot \overset{}{\underset{}{\cancel{\frac{4}{3}}}} \qquad \text{Multiply both sides by } \tfrac{4}{3} \text{ to isolate } x. \text{ Divide out the common factors.}$$

Note $\tfrac{3}{4}$ and $\tfrac{4}{3}$ are multiplicative inverses/reciprocals; so their product is 1. ▶

$$1x = \frac{10}{9} \qquad \text{Multiply.}$$

$$x = 1\frac{1}{9} \qquad \text{Write the solution as a mixed number.}$$

Check:

$$\frac{3}{4}x - \frac{2}{3} = \frac{1}{6}$$

$$\frac{3}{4}\left(1\frac{1}{9}\right) - \frac{2}{3} \overset{?}{=} \frac{1}{6} \qquad \text{In the original equation, replace } x \text{ with } 1\tfrac{1}{9}.$$

$$\frac{3}{4} \cdot \frac{10}{9} - \frac{2}{3} \overset{?}{=} \frac{1}{6} \qquad \text{Write } 1\tfrac{1}{9} \text{ as an improper fraction.}$$

$$\underset{2}{\overset{1}{\cancel{\frac{3}{4}}}} \cdot \underset{3}{\overset{5}{\cancel{\frac{10}{9}}}} - \frac{2}{3} \overset{?}{=} \frac{1}{6} \qquad \text{Divide out common factors.}$$

$$\frac{5}{6} - \frac{2}{3} \overset{?}{=} \frac{1}{6} \qquad \text{Multiply.}$$

Answers to Warm-up
1. $x = 2$ **2.** $9x - 8$

$$\frac{5}{6} - \frac{2(2)}{3(2)} \stackrel{?}{=} \frac{1}{6} \qquad \text{Write equivalent fractions with the LCD, 6.}$$

$$\frac{5}{6} - \frac{4}{6} \stackrel{?}{=} \frac{1}{6}$$

$$\frac{1}{6} = \frac{1}{6} \qquad \text{Subtract. True; so } 1\tfrac{1}{9} \text{ is the solution.}$$

Example 1 illustrates the approach to solving equations that we learned in Chapter 4. But working with fractions can be tedious. We can use the multiplication principle of equality to eliminate the fractions, leaving us with integers, which simplifies the process considerably.

Remember, the multiplication principle of equality simply says that multiplying both sides of an equation by the same nonzero amount does not change its solution(s). The principle does not say that we have to wait until the last step of the process to use it. If we multiply both sides by a multiple of all of the denominators, they will divide out. To keep the numbers as small as possible, it is best to use the LCD (least common denominator).

Let's eliminate the fractions in the equation from Example 1 by multiplying both sides of the equation by the LCD, 12.

$$\frac{3}{4}x - \frac{2}{3} = \frac{1}{6}$$

$$12\left(\frac{3}{4}x - \frac{2}{3}\right) = \left(\frac{1}{6}\right)12 \qquad \text{Multiply both sides by 12.}$$

$$\frac{12}{1} \cdot \frac{3}{4}x - \frac{12}{1} \cdot \frac{2}{3} = \frac{1}{6} \cdot \frac{12}{1} \qquad \text{Distribute.}$$

$$\frac{\overset{3}{\cancel{12}}}{1} \cdot \frac{3}{\cancel{4}}x - \frac{\overset{4}{\cancel{12}}}{1} \cdot \frac{2}{\cancel{3}} = \frac{1}{\cancel{6}} \cdot \frac{\overset{2}{\cancel{12}}}{1}$$

Note After dividing out the common factors, we are left with an equation that contains only integers, which is easier for most people to ◄ work with.

$$9x - 8 = 2$$

$$\underline{+\ 8 \quad +\ 8} \qquad \text{Add 8 to both sides to isolate } 9x.$$

$$9x + 0 = 10$$

$$\frac{9x}{9} = \frac{10}{9} \qquad \text{Divide both sides by 9 to isolate } x.$$

$$1x = \frac{10}{9} \qquad \text{Simplify.}$$

Note Our alternative method leads to the same solution. ►

$$x = 1\frac{1}{9} \qquad \text{Write the solution as a mixed number.}$$

Although we used a different method to solve the equation, the check is the same as in Example 1.

Warning When using the LCD to eliminate fractions, it is tempting to check the solution in the rewritten equation (or another later step). In Example 1, it would certainly be easier to check using $9x - 8 = 2$ instead of $\frac{3}{4}x - \frac{2}{3} = \frac{1}{6}$, and as long as we made no mistakes, the solution would check in $9x - 8 = 2$. However, suppose we made a mistake in getting to $9x - 8 = 2$ and then made no mistakes thereafter. Our incorrect answer would check in $9x - 8 = 2$, but it would not check in the original equation. Therefore, always use the original equation to check.

We can modify the procedure for solving equations that we developed in Chapter 4 to include this procedure for eliminating fractions.

▶ **Your Turn 1**

Solve and check.

a. $\frac{5}{8}y - 3 = \frac{1}{2}$

b. $\frac{1}{4} + \frac{1}{8}m = \frac{5}{16}$

c. $-\frac{7}{9} = \frac{2}{3}n + \frac{1}{6}$

d. $\frac{-1}{7} = -2 - \frac{3}{5}k$

Procedure

To solve a linear equation in one variable:

1. Simplify each side of the equation.
   a. Distribute to eliminate parentheses.
   b. Eliminate fractions by multiplying both sides of the equation by the LCD of all of the fractions. (Optional)
   c. Combine like terms.
2. Use the addition principle of equality so that all variable terms are on one side of the equation and all constants are on the other side. Then combine like terms.

> **Note** Eliminate the variable term with the lesser coefficient to avoid negative coefficients.

3. Use the multiplication principle of equality to eliminate the coefficient of the variable.

◀ **Do Your Turn 1**

Example 2 Solve and check. $\frac{2}{3}(x - 3) = \frac{1}{2}x - \frac{2}{5}$

Solution: We will begin by distributing the $\frac{2}{3}$ to eliminate the parentheses. We will then eliminate the fractions by multiplying by their LCD, 30.

$$\frac{2}{3}(x - 3) = \frac{1}{2}x - \frac{2}{5}$$

$$\frac{2}{3}x - \frac{2}{3} \cdot \frac{\overset{1}{\cancel{3}}}{1} = \frac{1}{2}x - \frac{2}{5} \quad \text{Distribute } \frac{2}{3} \text{ to eliminate the parentheses. Then divide out common factors.}$$

$$\frac{2}{3}x - 2 = \frac{1}{2}x - \frac{2}{5} \quad \text{Multiply.}$$

$$30\left(\frac{2}{3}x - 2\right) = \left(\frac{1}{2}x - \frac{2}{5}\right)30 \quad \text{Eliminate the fractions by multiplying both sides by the LCD, 30.}$$

$$\frac{\overset{10}{\cancel{30}}}{1} \cdot \frac{2}{\cancel{3}}x - 30 \cdot 2 = \frac{\overset{15}{\cancel{30}}}{1} \cdot \frac{1}{\cancel{2}}x - \frac{\overset{6}{\cancel{30}}}{1} \cdot \frac{2}{\cancel{5}} \quad \text{Distribute 30 and divide out the common factors.}$$

$$20x - 60 = 15x - 12 \quad \text{Multiply.}$$

$$\underline{-15x \qquad\quad -15x} \quad \text{Subtract 15x from both sides.}$$

$$5x - 60 = \quad 0 - 12$$

$$5x - 60 = -12$$

$$\underline{+60 \qquad +60} \quad \text{Add 60 to both sides to isolate the 5x term.}$$

$$5x - 0 = \quad 48$$

$$\frac{5x}{5} = \frac{48}{5} \quad \text{Divide both sides by 5 to isolate } x.$$

$$1x = \frac{48}{5} \quad \text{Simplify.}$$

$$x = 9\frac{3}{5} \quad \text{Write the solution as a mixed number.}$$

> **Note** We could have multiplied by the LCD as the first step, but this can be tricky with the parentheses; so we chose to clear the parentheses first. We will show how to multiply this equation by the LCD in the first step on the next page.

Answers to Your Turn 1

a. $y = 5\frac{3}{5}$ b. $m = \frac{1}{2}$

c. $n = -1\frac{5}{12}$ d. $k = -3\frac{2}{21}$

As we mentioned in the previous note, we could have eliminated the fractions in the first step before distributing to clear the parentheses. We show this approach below.

Note Because $\frac{2}{3}(x-3)$ is a product, when we multiply it by 30, we multiply the first two factors, 30 and $\frac{2}{3}$, to get 20, then multiply the third factor, $x-3$, by that result.

$$\frac{2}{3}(x-3) = \frac{1}{2}x - \frac{2}{5}$$

$$30 \cdot \frac{2}{3}(x-3) = 30 \cdot \left(\frac{1}{2}x - \frac{2}{5}\right)$$ Multiply both sides by 30 to eliminate the fractions.

$$\frac{\overset{10}{30}}{1} \cdot \frac{2}{\underset{1}{3}}(x-3) = \frac{\overset{15}{30}}{1} \cdot \frac{1}{\underset{1}{2}}x - \frac{\overset{6}{30}}{1} \cdot \frac{2}{\underset{1}{5}}$$ **Note** Because $\frac{1}{2}x - \frac{2}{5}$ is a difference, we distribute the 30 to each term.

$$20(x-3) = 15x - 12$$ Multiply.

$$20x - 60 = 15x - 12$$ Distribute 20.

Note This equation is the same as the one in the sixth step of our previous solution. The remaining steps would match our previous solution.

Check: Replace x in the original equation with $9\frac{3}{5}$ and verify that the equation is true. This check will be left to the reader.

▶ **Do Your Turn 2**

Objective 2 Translate sentences to equations; then solve.

We have learned some new key words associated with fractions. In Section 5.3, we learned that the key word *of* indicates multiplication when preceded by a fraction. Also, the word *reciprocal* indicates to invert the number.

$$\frac{2}{5} \textbf{ of } \text{a number} \quad \text{translates to} \quad \frac{2}{5}n$$

$$\text{The } \textbf{reciprocal} \text{ of } n \quad \text{translates to} \quad \frac{1}{n}$$

The key words we've learned in the past, such as *sum, difference, less than,* and *product,* translate the same regardless of the types of numbers involved.

Example 3 The sum of $2\frac{2}{3}$ and $\frac{3}{5}$ of n is the same as 2 less than $\frac{1}{3}$ of n. Translate to an equation; then solve for n.

Solution: We must translate using the key words and solve for n. The word *sum* indicates addition. *Less than* indicates subtraction in reverse order of the sentence. $\frac{3}{5}$ of n and $\frac{1}{3}$ of n indicate multiplication because *of* is preceded by the fraction in both cases.

Note Remember, when used with the key words *sum, difference, product,* and *quotient,* the word *and* becomes the operation symbol.

Translate: The **sum** of $2\frac{2}{3}$ **and** $\frac{3}{5}$ of n is the same as 2 **less than** $\frac{1}{3}$ of n.

$$2\frac{2}{3} + \frac{3}{5}n \quad = \quad \frac{1}{3}n - 2$$

▶ **Your Turn 3**

Translate to an equation; then solve.

a. $\frac{2}{3}$ of a number is $3\frac{5}{6}$.

b. The difference of x and $\frac{4}{9}$ is equal to $-3\frac{1}{6}$.

c. 3 more than $\frac{3}{4}$ of k is equal to $\frac{1}{8}$.

d. $\frac{1}{2}$ of the difference of n and 4 is equal to $\frac{2}{5}$ of n.

Solve: $15\left(\frac{8}{3} + \frac{3}{5}n\right) = \left(\frac{1}{3}n - 2\right)15$ | Eliminate the fractions by multiplying both sides by the LCD, 15.

$$\frac{\overset{5}{15}}{1} \cdot \frac{8}{\underset{1}{3}} + \frac{\overset{3}{15}}{1} \cdot \frac{3}{\underset{1}{5}}n = \frac{\overset{5}{15}}{1} \cdot \frac{1}{\underset{1}{3}}n - \frac{15}{1} \cdot 2$$ | Distribute 15 and divide out the common factors.

$$40 + 9n = 5n - 30$$ | Multiply.

$$\underline{ -5n \quad\quad -5n}$$ | Subtract 5n from both sides.

$$40 + 4n = 0 - 30$$ | Subtract 40 from both sides.

$$\underline{-40 \ \ -40}$$

$$0 + 4n = -70$$

$$\frac{4n}{4} = \frac{-70}{4}$$ | Divide both sides by 4.

$$1n = -\frac{35}{2}$$ | Simplify.

$$n = -17\frac{1}{2}$$ | Write the solution as a mixed number.

◀ **Do Your Turn 3**

Objective 3 Solve applications involving one unknown.

Example 4   **a.** Find the length a in the trapezoid shown if the area is 9 square feet.

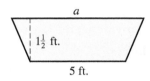

Solution: The given area is for a trapezoid; so we use

$$A = \frac{1}{2}h(a + b).$$

$$9 = \frac{1}{2}\left(1\frac{1}{2}\right)(a + 5)$$ | In $A = \frac{1}{2}h(a + b)$, replace A with 9, h with $1\frac{1}{2}$, and b with 5.

$$9 = \frac{1}{2}\left(\frac{3}{2}\right)(a + 5)$$ | Write the mixed number as an improper fraction.

$$9 = \frac{3}{4}(a + 5)$$ | Multiply $\frac{1}{2}$ and $\frac{3}{2}$.

$$4 \cdot 9 = \frac{\overset{1}{4}}{1} \cdot \frac{3}{\underset{1}{4}}(a + 5)$$ | Eliminate the fractions by multiplying both sides by the LCD, 4.

$$36 = 3(a + 5)$$ | Multiply.

$$36 = 3a + 15$$ | Distribute 3.

$$\underline{-15 \ -15}$$ | Subtract 15 from both sides to isolate the 3a term.

$$21 = 3a + 0$$

$$\frac{21}{3} = \frac{3a}{3}$$ | Divide both sides by 3 to isolate a.

Answers to Your Turn 3

a. $\frac{2}{3}n = 3\frac{5}{6}; n = 5\frac{3}{4}$

b. $x - \frac{4}{9} = -3\frac{1}{6}; x = -2\frac{13}{18}$

c. $3 + \frac{3}{4}k = \frac{1}{8}; k = -3\frac{5}{6}$

d. $\frac{1}{2}(n - 4) = \frac{2}{5}n; n = 20$

$$7 = a$$ | Simplify.

Answer: The length a is 7 feet.

b. Juan is taking a course in which his grade is determined by the average (mean) of four tests. He has the following test scores in the course: 68, 85, and 90. To get a B, he must have a test average (mean) of 80 on the four tests. What is the minimum score he needs on the last test to get a B?

Solution: To calculate a mean, we divide the sum of the scores by the number of scores.

$$\frac{\text{sum of the scores}}{\text{number of the scores}} = \text{mean}$$

Let x represent the unknown fourth test score.

$$\frac{68 + 85 + 90 + x}{4} = 80$$

$$\frac{243 + x}{4} = 80 \qquad \text{Add in the numerator.}$$

$$4 \cdot \frac{243 + x}{4} = 80 \cdot 4 \qquad \begin{array}{l}\text{Eliminate the denominator}\\ \text{by multiplying both sides by 4.}\end{array}$$

$$243 + x = 320$$

$$\underline{-243 \qquad\quad -243} \qquad \text{Subtract 243 from both sides.}$$

$$0 + x = 77$$

$$x = 77$$

Answer: If Juan scores 77 on the last test, the mean of the four tests will be exactly 80 and he will receive a B.

▶ **Do Your Turn 4**

Objective 4 Solve applications involving two unknowns.

In Chapter 4, we developed techniques for solving problems that have two unknowns. In general, these problems have two relationships that we translate to an equation.

Example 5 A small company has two salespeople. In one particular week, Benjamin's total sales are half the amount sold by Florence. If the total amount of sales for the week is $12,000, how much did each person sell?

Understand We are to find the amount that Benjamin sold and the amount that Florence sold. Two sentences describe relationships.

Relationship 1: Benjamin's total sales are half the amount sold by Florence.

Relationship 2: . . . the total amount of sales for the week is $12,000 . . .

Plan and Execute Translate the relationships to an equation using f as the amount sold by Florence. Then solve.

Relationship 1: Benjamin's total sales are <u>half</u> <u>the amount sold by Florence.</u>

$$\text{Benjamin's sales} = \frac{1}{2} \cdot f$$

Relationship 2: . . . the total amount of sales for the week is $12,000 . . .

$$\text{Benjamin's sales} + \text{Florence's sales} = \$12{,}000$$

$$\frac{1}{2}f \qquad + \qquad f \qquad = 12{,}000$$

$$2\left(\frac{1}{2}f + f\right) = (12{,}000)2 \qquad \begin{array}{l}\text{Multiply both sides by}\\ \text{2 to clear the fraction.}\end{array}$$

▶ **Your Turn 4**

a. Find the length a in the trapezoid shown if the area is $4\frac{1}{2}$ square meters.

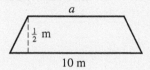

b. Trina's grade in a course is determined by the average (mean) of five tests. Her scores on the first four tests are 80, 92, 88, and 94. To get an A, she must have a final average (mean) of 90 or higher. What is the minimum score she needs on the fifth test to receive an A?

Answers to Your Turn 4
a. $a = 8\,\text{m}$ **b.** 96

▶ **Your Turn 5**

Two boards are joined. One board is $\frac{2}{3}$ the length of the other. The two boards combine to be $7\frac{1}{2}$ feet. What are the lengths of both boards?

$$2 \cdot \frac{1}{2}f + 2 \cdot f = 12{,}000 \cdot 2 \qquad \text{Distribute the 2.}$$

$$f + 2f = 24{,}000 \qquad \text{Multiply.}$$

$$3f = 24{,}000 \qquad \text{Combine like terms.}$$

$$\frac{3f}{3} = \frac{24{,}000}{3} \qquad \begin{array}{l}\text{Divide both sides by 3}\\ \text{to clear the coefficient.}\end{array}$$

$$f = 8000$$

Answer Because f represents Florence's sales, we can say that her sales amounted to \$8000. If Benjamin's sales were $\frac{1}{2}$ of Florence's, his sales totaled only \$4000.

Check Notice that the sum of their sales is in fact \$12,000.

$$\$4000 + \$8000 = \$12{,}000$$

◀ **Do Your Turn 5**

In Chapter 4, we solved problems with two unknowns involving geometry terms such as *perimeter, complementary angles*, and *supplementary angles*. Let's look at these problems again.

Example 6 Figure $ABCD$ is a rectangle. If $\angle DBC$ is $14\frac{1}{3}°$ less than $\angle ABD$, what are the angle measurements?

Understand We must find the angle measurements. It seems that we are given only one relationship.

$$\angle DBC \text{ is } 14\tfrac{1}{3}° \text{ less than } \angle ABD.$$

The other relationship is implied by the shape. Because the shape is a rectangle, each corner angle measures 90°. Therefore, $\angle DBC$ and $\angle ABD$ are complementary, which means that the sum of their measurements is 90°.

Plan and Execute Translate the relationships to an equation using a as the measurement of $\angle ABD$. Then solve.

Relationship 1: $\angle DBC$ is $14\frac{1}{3}°$ less than $\angle ABD$.

$$\angle DBC = a - 14\tfrac{1}{3}$$

Relationship 2: The sum of the angle measurements is 90.

$$\angle ABD + \overbrace{\angle DBC} = 90$$

$$a \;+\; a - 14\tfrac{1}{3} = 90$$

$$2a - 14\tfrac{1}{3} = 90 \qquad \text{Combine like terms.}$$

$$2a - 14\tfrac{1}{3} + 14\tfrac{1}{3} = 90 + 14\tfrac{1}{3} \qquad \text{Add } 14\tfrac{1}{3} \text{ to both sides.}$$

$$2a + 0 = 104\tfrac{1}{3} \qquad \text{Add.}$$

$$\frac{1}{2} \cdot 2a = \frac{313}{3} \cdot \frac{1}{2} \qquad \text{Multiply both sides by } \tfrac{1}{2} \text{ to isolate } a.$$

$$1a = \frac{313}{6} \qquad \text{Simplify.}$$

$$a = 52\tfrac{1}{6} \qquad \text{Write the result as a mixed number.}$$

Note We could have eliminated the fraction by multiplying both sides by 3. We chose not to as a reminder that eliminating fractions from an equation is optional.

Answer to Your Turn 5

$4\frac{1}{2}$ ft., 3 ft.

Answer Because $\angle ABD$ is represented by the variable a, we can say that $\angle ABD = 52\frac{1}{6}°$. Because $\angle DBC = a - 14\frac{1}{3}$, we can find its value by replacing a with $52\frac{1}{6}$ and then subtracting.

$$\angle DBC = 52\frac{1}{6} - 14\frac{1}{3}$$

$$\angle DBC = 52\frac{1}{6} - 14\frac{1(2)}{3(2)} \qquad \text{Write equivalent fractions with the LCD, 6.}$$

$$\angle DBC = 52\frac{1}{6} - 14\frac{2}{6}$$

$$\angle DBC = 51\frac{7}{6} - 14\frac{2}{6} \qquad \text{Rewrite } 52\frac{1}{6} \text{ so that we can subtract.}$$

$$\angle DBC = 37\frac{5}{6} \qquad \text{Subtract.}$$

$\angle ABD$ is $52\frac{1}{6}°$ and $\angle DBC$ is $37\frac{5}{6}°$.

Check Verify that the sum of the angle measurements is 90°.

$$\angle ABD + \angle DBC = 90$$

$$52\frac{1}{6} + 37\frac{5}{6} \overset{?}{=} 90$$

Note $\frac{6}{6}$ is 1, which adds to 89 to get 90. ▶

$$89\frac{6}{6} \overset{?}{=} 90$$

$$90 = 90$$

▶ **Do Your Turn 6**

▶ **Your Turn 6**

a. The length of a rectangular frame is $3\frac{1}{2}$ inches longer than the width. If the perimeter is 42 inches, what are the dimensions of the frame?

b. A piece of steel is welded to a horizontal truss, forming two angles with the truss. The larger of the two angles is $2\frac{1}{2}°$ more than twice the smaller angle. Find the angle measurements.

c. A rectangle has a width that is half the length. The perimeter is $37\frac{1}{2}$ meters. What are the dimensions of the rectangle?

Answers to Your Turn 6

a. $8\frac{3}{4}$ in. by $12\frac{1}{4}$ in.

b. $59\frac{1}{6}°$, $120\frac{5}{6}°$

c. $12\frac{1}{2}$ m by $6\frac{1}{4}$ m

5.8 Exercises   For Extra Help   MyMathLab®

Objective 1

Prep Exercise 1 Explain how to use the multiplication principle of equality to eliminate fractions from an equation.

For Exercises 1–20, use the LCD to simplify the equation, then solve and check.

1. $x + \frac{3}{4} = \frac{1}{2}$

2. $y + \frac{4}{9} = \frac{2}{5}$

3. $p - \frac{1}{6} = \frac{1}{3}$

4. $q - \frac{2}{3} = \frac{3}{7}$

5. $\frac{5}{6} + c = \frac{3}{5}$

6. $\frac{1}{3} + d = \frac{3}{4}$

7. $\frac{3}{4}a = \frac{1}{2}$

8. $\frac{2}{5}b = \frac{1}{10}$

9. $\frac{6}{35}f = \frac{8}{15}$

10. $\frac{14}{15}g = \frac{21}{20}$

11. $3b - \frac{1}{4} = \frac{1}{2}$

12. $\frac{2}{3}y - 8 = \frac{1}{6}$

13. $\frac{3}{4}x + \frac{1}{6} = \frac{1}{2}$

14. $\frac{1}{5}p + \frac{2}{3} = 2$

15. $\frac{4}{5} - \frac{a}{2} = \frac{3}{4} - 1$

16. $\dfrac{1}{3} - \dfrac{b}{4} = \dfrac{1}{2} - 2$

17. $\dfrac{1}{8} + n = \dfrac{5}{6}n - \dfrac{2}{3}$

18. $\dfrac{1}{6} + m = \dfrac{3}{4}m + \dfrac{1}{3}$

19. $\dfrac{1}{2}(x - 6) = \dfrac{1}{4}x - \dfrac{2}{5}$

20. $\dfrac{5}{6}(k - 8) = \dfrac{1}{5}k - \dfrac{1}{3}$

Objective 2

Prep Exercise 2 In the following sentence, the word *of* translates to which operation of arithmetic?

$$\tfrac{4}{5} \text{ of some number is } 2\tfrac{1}{2}.$$

Prep Exercise 3 Explain the mistake in the translation. Then translate correctly, but do not solve.

$$\tfrac{2}{3} \text{ of the sum of some number and 6 is } 8\tfrac{1}{2}.$$

Translation: $\tfrac{2}{3}n + 6 = 8\tfrac{1}{2}$

For Exercises 21–28, translate to an equation and solve.

21. $\tfrac{3}{8}$ of some number is $4\tfrac{5}{6}$.

22. The product of $2\tfrac{1}{3}$ and some number is $-4\tfrac{1}{2}$.

23. $8\tfrac{7}{10}$ less than y is $-2\tfrac{1}{2}$.

24. $6\tfrac{2}{3}$ more than m is $5\tfrac{1}{12}$.

25. $3\tfrac{1}{4}$ more than twice n is $-\tfrac{1}{6}$.

26. $\tfrac{5}{8}$ less than $3\tfrac{4}{5}$ times k is $\tfrac{9}{16}$.

27. $\tfrac{3}{4}$ of the sum of b and 10 is equal to $1\tfrac{1}{6}$ added to b.

28. $\tfrac{2}{5}$ of the difference of 1 and h is the same as $1\tfrac{1}{4}$ times h.

Objective 3

Prep Exercise 4 What are the formulas for the area of a triangle and the area of a trapezoid?

Prep Exercise 5 How do you find the mean of a set of scores?

For Exercises 29–42, solve.

29. Find the length b in the triangle shown if the area is $7\tfrac{41}{50}$ square centimeters.

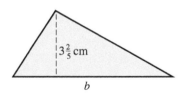

$3\tfrac{2}{5}$ cm

b

30. Find the length a in the trapezoid shown if the area is $122\tfrac{1}{2}$ square inches.

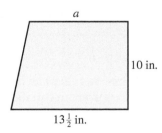

a

10 in.

$13\tfrac{1}{2}$ in.

31. Daniel's grade in a course is determined by the average (mean) of five tests. His scores on the first four tests are 86, 96, 90, and 88. To get an A, he must have a final average of 90 or higher. What is the minimum score he needs on the fifth test to receive an A?

32. Heather's grade in a course is determined by the average (mean) of six tests. Her scores on the first five tests are 83, 78, 82, 70, and 76. To get a B, she must have a final average of 80 or higher. What is the minimum score she needs on the sixth test to receive a B?

33. If Lonnie has a test average (mean) of 94 or higher, she will be exempt from the final exam. She has taken three out of the four tests for the course. Her scores on the three tests are 88, 92, and 90. What is the lowest she can score on the last test to be exempt from the final exam?

34. If Fernando has a test average (mean) of 90 or higher, he will be exempt from the final exam. He has taken four out of the five tests for the course. His scores on those four tests are 85, 88, 94, and 86. What is the lowest he can score on the fifth test to be exempt from the final exam?

Objective 4

Prep Exercise 6 Exercise 35 contains two relationships used to build an equation. Write the sentence that describes each relationship.

Prep Exercise 7 In Exercise 35, what are the two unknown amounts?

Prep Exercise 8 In Exercise 35, which unknown amount should be represented by the variable? Why?

35. Mary is part of a team of social workers trying to raise funds to build an outreach center for a neighborhood. Mary alone raises $2\frac{1}{2}$ times what the rest of the team does. The total amount raised was $31,542. How much did Mary raise? How much did the rest of the team raise?

36. Rosa discovers that her salary is $\frac{5}{6}$ of Rick's salary, yet they have the same job. The difference in their salaries is $4700. What are each of their salaries?

37. Two chemical storage tanks hold the same amount. Tank A is $\frac{3}{4}$ full, while tank B is $\frac{2}{3}$ full. The two tanks currently have a combined total of 1700 gallons of CO_2. How much do the tanks hold?

38. A computer has two equal-size hard drives. Drive C is $\frac{4}{5}$ full, while drive D is only $\frac{1}{4}$ full. If the two drives have a combined total of $6\frac{3}{10}$ gigabytes stored, how much does each drive hold?

39. In a survey, $\frac{3}{4}$ of the people contacted agreed to answer a single yes-or-no question. Of these people, $\frac{1}{5}$ answered yes and the rest answered no. What fraction of all people contacted said no?

40. In a political poll, $\frac{2}{5}$ of the respondents said that they would vote for a certain candidate. Of these respondents, $\frac{1}{3}$ were men. What fraction of all respondents that voted for the candidate were women?

41. Two boards come together as shown. The smaller angle measurement is $\frac{1}{3}$ the measurement of the larger angle. Find the angle measurements.

42. At exactly 3:00:10, what are the angles made between the second hand and the minute and hour hands of a clock?

Puzzle Problem Two people are in a desert together. They have half a gallon (8 cups) of water in a canteen to split between them. They decide to divide the water so that they each get 4 cups. The problem is that one person has a container that holds 3 cups while the other person has a container that holds 5 cups. How can they use their containers and the canteen to get 4 cups each?

Review Exercises

[5.1] **1.** Use <, >, or = to write a true sentence.
$$\frac{6}{17} \ ? \ \frac{7}{19}$$

[5.2] **2.** Simplify to lowest terms. $\dfrac{9x}{24x^2}$

[5.3] **3.** Multiply. $\dfrac{-4}{9} \cdot \dfrac{21}{30}$

[5.4] **4.** Divide. $\dfrac{14a}{15ab^4} \div \dfrac{21}{25ab}$

[5.6] **5.** Add. $\dfrac{7}{8} + 5\dfrac{2}{3}$

Chapter 5 Summary and Review Exercises

5.1 Introduction to Fractions

Definitions/Rules/Procedures	Key Example(s)
A **fraction** is a quotient of two numbers or expressions a and b having the form _____, where $b \neq 0$.	Name the fraction in lowest terms represented by the shaded region.
The **numerator** of a fraction is the number written in the _____ position.	
The **denominator** of a fraction is the number written in the _____ position.	
An **improper fraction** is a fraction in which the absolute value of the numerator is _____ the absolute value of the denominator.	**Answer:** $\dfrac{4}{10} = \dfrac{2}{5}$
A **proper fraction** is a fraction in which the absolute value of the numerator is _____ the absolute value of the denominator.	**Simplify.** **a.** $\dfrac{8}{1} = 8$ **b.** $\dfrac{60}{60} = 1$ **c.** $\dfrac{14}{0}$ is undefined

[5.1] For Exercises 1 and 2, name, in lowest terms, the fraction represented by the shaded region.

1.

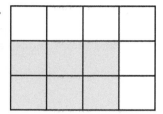

2.

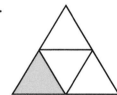

[5.1] For Exercises 3 and 4, simplify.

3. $\dfrac{18}{1}$

4. $\dfrac{-14}{0}$

Definitions/Rules/Procedures	Key Example(s)
A **mixed number** is a(n) _____ combined with a(n) _____.	Write $\frac{19}{8}$ as a mixed number.
A **rational number** is a number that can be expressed in the form $\dfrac{a}{b}$, where a and b are _____ and $b \neq 0$.	$\begin{array}{r} 2 \\ 8\overline{)19} \\ -16 \\ \hline 3 \end{array} \qquad 2\dfrac{3}{8}$
To write an improper fraction as a mixed number: 1. Divide the denominator into the numerator. 2. Write the result in the following form: $\boxed{?}\ \dfrac{?}{\textit{original denominator}}$	Write $5\frac{7}{8}$ as an improper fraction. $5\dfrac{7}{8} = \dfrac{8 \cdot 5 + 7}{8} = \dfrac{40 + 7}{8} = \dfrac{47}{8}$ **Note** Many people do not write these steps, performing the calculations mentally.

Definitions/Rules/Procedures	Key Example(s)
To write a mixed number as an improper fraction: 1. Multiply the absolute value of the integer by the denominator. 2. Add the product from step 1 to the _____. The result is the _____ of the improper fraction. 3. Keep the same denominator and sign.	

[5.1] For Exercises 5 and 6, write each improper fraction as a mixed number.

5. $\dfrac{40}{9}$

6. $-\dfrac{29}{4}$

[5.1] For Exercises 7 and 8, write each mixed number as an improper fraction.

7. $6\dfrac{2}{3}$

8. $-5\dfrac{1}{2}$

Definitions/Rules/Procedures	Key Example(s)
To graph a proper fraction on a number line, draw evenly spaced tick marks between _____ (if positive) or _____ (if negative) to divide that segment into the number of equal-sized divisions equal to the denominator. Then draw a dot on the tick mark indicated by the numerator. **To graph an improper fraction on a number line**, write it as a _____ and follow the process for a mixed number. **To graph a mixed number on a number line**, draw evenly spaced tick marks between the integer and the _____ (if positive) or _____ (if negative) to divide that segment into the number of equal-sized divisions equal to the denominator. Then draw a dot on the tick mark indicated by the numerator.	Graph $\dfrac{7}{10}$ on a number line. **Solution:** 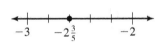 Graph $-\dfrac{13}{5}$ on a number line. **Solution:** $-\dfrac{13}{5} = -2\dfrac{3}{5}$

[5.1] For Exercises 9 and 10, graph each number on a number line.

9. $\dfrac{1}{5}$

10. $-2\dfrac{5}{8}$

Definitions/Rules/Procedures	Key Example(s)
Equivalent fractions are fractions that _____. **To write an equivalent fraction**, _____ both the numerator and denominator by the same nonzero number.	Fill in the blank so that the fractions are equivalent. **a.** $\dfrac{4}{5} = \dfrac{?}{15}$ **b.** $-\dfrac{12}{15} = -\dfrac{4}{?}$ **Solution:** **Solution:** $\dfrac{4}{5} = \dfrac{4\cdot 3}{5\cdot 3} = \dfrac{12}{15}$ $-\dfrac{12}{15} = -\dfrac{12\div 3}{15\div 3} = -\dfrac{4}{5}$

For Exercises 11 and 12, fill in the blank so that the fractions are equivalent.

11. $\dfrac{24}{30} = \dfrac{4}{?}$

12. $-\dfrac{9}{11} = -\dfrac{?}{44}$

Definitions/Rules/Procedures	Key Example(s)
A **multiple** is a number that is _____ by a given number. **To compare two fractions:** 1. Write equivalent fractions that have a(n) _____. 2. Compare the _____ of the equivalent fractions.	Use $<$, $>$, or $=$ to write a true statement. $$\dfrac{5}{8} \; ? \; \dfrac{4}{7}$$ Use cross products to write equivalent fractions with a common denominator. $$\dfrac{35}{56} = \dfrac{5 \cdot 7}{8 \cdot 7} = \dfrac{5}{8} \; ? \; \dfrac{4}{7} = \dfrac{4 \cdot 8}{7 \cdot 8} = \dfrac{32}{56}$$ $$\dfrac{35}{56} > \dfrac{32}{56}; \text{ therefore, } \dfrac{5}{8} > \dfrac{4}{7}.$$

[5.1] For Exercises 13 and 14, use $<$, $>$, or $=$ to write a true statement.

13. $\dfrac{5}{9} \; ? \; \dfrac{7}{13}$

14. $-\dfrac{5}{24} \; ? \; -\dfrac{3}{16}$

5.2 Simplifying Fractions and Rational Expressions

Definitions/Rules/Procedures	Key Example(s)
A fraction is in **lowest terms** when the greatest common factor of its numerator and denominator is ____. **To simplify a fraction to lowest terms**, divide the numerator and denominator by their _____. **To simplify a fraction to lowest terms using primes:** 1. Replace the numerator and denominator with their prime factorizations. 2. Divide out _____. 3. Multiply the remaining factors. A **rational expression** is a fraction in which the numerator and denominator are _____.	Simplify $\frac{70}{182}$ to lowest terms. $$\dfrac{70}{182} = \dfrac{70 \div 14}{182 \div 14} = \dfrac{5}{13} \quad \text{or}$$ $$\dfrac{70}{182} = \dfrac{\overset{1}{2} \cdot 5 \cdot \overset{1}{7}}{\underset{1}{2} \cdot \underset{1}{7} \cdot 13} = \dfrac{5}{13}$$ Simplify $-\dfrac{18x^3}{24x^2}$ to lowest terms. $$-\dfrac{18x^3}{24x^2} = -\dfrac{\overset{1}{2} \cdot \overset{1}{3} \cdot 3 \cdot \overset{1}{x} \cdot \overset{1}{x} \cdot x}{\underset{1}{2} \cdot 2 \cdot 2 \cdot \underset{1}{3} \cdot \underset{1}{x} \cdot \underset{1}{x}} = -\dfrac{3x}{4}$$

[5.2] Simplify to lowest terms.

15. $\dfrac{15}{35}$

16. $-\dfrac{84}{105}$

17. $-\dfrac{8m^3n}{26n^2}$

18. $\dfrac{6xy^2}{20y^5}$

5.3 Multiplying Fractions, Mixed Numbers, and Rational Expressions

Definitions/Rules/Procedures	Key Example(s)
To multiply fractions:	Multiply.
Method 1: 1. _____ the numerators and _____ the denominators.	**a.** $-\dfrac{14}{15}\cdot\dfrac{9}{16} = -\dfrac{\overset{1}{2}\cdot 7}{3\cdot\underset{1}{5}}\cdot\dfrac{\overset{1}{3}\cdot 3}{2\cdot 2\cdot 2\cdot\underset{1}{2}} = -\dfrac{21}{40}$
2. Simplify the product.	
Method 2: 1. _____ factors of any numerator with identical factors of any denominator.	**b.** $\dfrac{9a}{10b}\cdot\dfrac{-16}{21a^2} = \dfrac{\overset{1}{3}\cdot 3\cdot\overset{1}{a}}{2\cdot 5\cdot\underset{1}{b}}\cdot\dfrac{-2\cdot 2\cdot 2\cdot 2}{\underset{1}{3}\cdot 7\cdot\underset{1}{a}\cdot\underset{1}{a}}$
2. _____ the remaining numerator factors and multiply the remaining denominator factors.	$= -\dfrac{24}{35ab}$
To multiply mixed numbers:	**c.** $-6\dfrac{3}{8}\cdot\left(-1\dfrac{1}{3}\right) = -\dfrac{51}{8}\cdot\left(-\dfrac{4}{3}\right)$
1. Write the mixed numbers as _____.	$= -\dfrac{\overset{1}{3}\cdot 17}{2\cdot 2\cdot\underset{1}{2}}\cdot\left(-\dfrac{\overset{1}{2}\cdot\overset{1}{2}}{\underset{1}{3}}\right)$
2. Divide out factors of any numerator with identical factors of any denominator.	
3. Multiply the remaining factors (numerator by numerator and denominator by denominator).	$= \dfrac{17}{2}$
4. Simplify. If the product is an improper fraction, write it as a mixed number in simplest form.	$= 8\dfrac{1}{2}$

[5.3] For Exercises 19–22, multiply. Write the product in lowest terms.

19. $\dfrac{4}{9}\cdot\dfrac{12}{20}$

20. $\dfrac{10n}{6p^3}\cdot\dfrac{14np}{8n}$

21. $\dfrac{h^3}{4k}\cdot\left(-\dfrac{12k}{15h}\right)$

22. $-4\dfrac{1}{2}\cdot 2\dfrac{2}{3}$

Definitions/Rules/Procedures	Key Example(s)
To evaluate a fraction raised to a power, evaluate both the numerator and denominator raised to that power.	Simplify.
In math language:	**a.** $\left(\dfrac{2}{3}\right)^4 = \dfrac{2^4}{3^4} = \dfrac{2\cdot 2\cdot 2\cdot 2}{3\cdot 3\cdot 3\cdot 3} = \dfrac{16}{81}$
$\left(\dfrac{a}{b}\right)^n = \dfrac{?}{?}$ when $b \neq 0$	**b.** $\left(\dfrac{-2x^2}{5}\right)^3 = \dfrac{(-2x^2)^3}{5^3}$
	$= \dfrac{(-2x^2)(-2x^2)(-2x^2)}{5\cdot 5\cdot 5}$
	$= \dfrac{-8x^6}{125}$

[5.3] For Exercises 23 and 24, simplify.

23. $\left(\dfrac{1}{2}\right)^6$

24. $\left(-\dfrac{3}{5}xy^3\right)^2$

5.4 Dividing Fractions, Mixed Numbers, and Rational Expressions

Definitions/Rules/Procedures	Key Example(s)
To divide fractions: 1. Write an equivalent multiplication by changing the operation symbol from division to multiplication and changing _____. 2. Multiply. Use either method described in the procedure for multiplying fractions. (Section 5.3) 3. Simplify. If the result is an improper fraction, write it as a mixed number in simplest form. **To divide mixed numbers:** 1. Write the mixed numbers as improper fractions. 2. Write an equivalent multiplication by _____ _____ _____. 3. Multiply. Use either method described in the procedure for multiplying fractions. (Section 5.3) 4. Simplify. If the product is an improper fraction, write it as a mixed number in simplest form.	Divide. a. $\dfrac{9}{16} \div \dfrac{3}{10} = \dfrac{9}{16} \cdot \dfrac{10}{3}$ $= \dfrac{\overset{1}{3} \cdot 3}{\underset{1}{2} \cdot 2 \cdot 2 \cdot 2} \cdot \dfrac{\overset{1}{2} \cdot 5}{\underset{1}{3}}$ $= \dfrac{15}{8}$ $= 1\dfrac{7}{8}$ b. $-\dfrac{5k}{16m^2} \div \dfrac{25k}{2m} = -\dfrac{5k}{16m^2} \cdot \dfrac{2m}{25k}$ $= -\dfrac{\overset{1}{5} \cdot \overset{1}{k}}{\underset{1}{2} \cdot 2 \cdot 2 \cdot 2 \cdot \underset{1}{m} \cdot m} \cdot \dfrac{\overset{1}{2} \cdot \overset{1}{m}}{\underset{1}{5} \cdot 5 \cdot \underset{1}{k}}$ $= -\dfrac{1}{40m}$ c. $-2\dfrac{2}{5} \div 5\dfrac{1}{4} = -\dfrac{12}{5} \div \dfrac{21}{4}$ $= -\dfrac{12}{5} \cdot \dfrac{4}{21}$ $= -\dfrac{2 \cdot 2 \cdot \overset{1}{3}}{5} \cdot \dfrac{2 \cdot 2}{\underset{1}{3} \cdot 7}$ $= -\dfrac{16}{35}$

[5.4] For Exercises 25–28, divide. Write the quotient in simplest form.

25. $-\dfrac{5}{6} \div \left(\dfrac{-15}{28}\right)$

26. $\dfrac{20a}{12ab^3} \div \dfrac{10}{8b}$

27. $-\dfrac{8x^6y}{25x^3} \div \dfrac{2xy}{5z^2}$

28. $7\dfrac{1}{5} \div 2\dfrac{4}{5}$

Definitions/Rules/Procedures	Key Example(s)
To find the square root of a fraction, try the following: 1. Find the square root of the _____ and _____ separately; then simplify. 2. _____ the fraction; then find the square root of the quotient.	Simplify. a. $\sqrt{\dfrac{49}{100}} = \dfrac{\sqrt{49}}{\sqrt{100}} = \dfrac{7}{10}$ b. $\sqrt{\dfrac{75}{3}} = \sqrt{25} = 5$

[5.4] For Exercises 29 and 30, evaluate the square root.

29. $\sqrt{\dfrac{100}{36}}$

30. $\sqrt{\dfrac{50}{2}}$

5.5 Least Common Multiple

Definitions/Rules/Procedures	Key Example(s)
The **least common multiple (LCM)** of a set of numbers is _____ _____.	Find the LCM. **a.** 18 and 24 Listing method:

To find the least common multiple (LCM) by listing, ... gives this content:

Definitions/Rules/Procedures	Key Example(s)
The **least common multiple (LCM)** of a set of numbers is _____ _____. **To find the least common multiple (LCM) by listing,** _____ _____. **To find the LCM of a set of numbers using prime factorization:** 1. Find the prime factorization of each given number. 2. Write a factorization that contains _____ _____. 3. Multiply to get the LCM.	Find the LCM. **a.** 18 and 24 Listing method: 24 48 72 is not is not is divisible divisible divisible by 18. by 18. by 18. $\text{LCM}(18, 24) = 72$ Prime factorization method: $$18 = 2 \cdot 3^2$$ $$24 = 2^3 \cdot 3$$ $$\text{LCM}(18, 24) = 2^3 \cdot 3^2$$ $$= 2 \cdot 2 \cdot 2 \cdot 3 \cdot 3$$ $$= 72$$ **b.** $18x^3y$ and $24xz^2$ $$\text{LCM}(18x^3y, 24xz^2) = 2^3 \cdot 3^2 x^3 y z^2$$ $$= 8 \cdot 9x^3yz^2$$ $$= 72x^3yz^2$$

[5.5] For Exercises 31 and 32, find the LCM.

31. 28 and 24

32. $15x$ and $20x^2y$

5.6 Adding and Subtracting Fractions, Mixed Numbers, and Rational Expressions

Definitions/Rules/Procedures	Key Example(s)
To add or subtract fractions with the same denominator: 1. Add or subtract the _____ and keep the same _____. 2. Simplify. **Note:** With rational expressions, we can combine only like terms in the numerators. For unlike terms, we must express the sum or difference as a polynomial.	Add or subtract. **a.** $\dfrac{2}{9} + \dfrac{1}{9} = \dfrac{3}{9} = \dfrac{1}{3}$ **b.** $\dfrac{5h}{2} - \dfrac{1}{2} = \dfrac{5h - 1}{2}$ **c.** $\dfrac{2x - 9}{5y} - \dfrac{x + 1}{5y}$ $$= \dfrac{(2x - 9) - (x + 1)}{5y}$$ $$= \dfrac{(2x - 9) + (-x - 1)}{5y}$$ $$= \dfrac{x - 10}{5y}$$

For Exercises 35–36, add or subtract.

33. $\dfrac{5}{9} + \dfrac{1}{9}$

34. $\dfrac{4}{15} - \dfrac{1}{15}$

35. $\dfrac{3n}{8} + \dfrac{n}{8}$

36. $\dfrac{7x-1}{y} - \dfrac{2x+3}{y}$

Definitions/Rules/Procedures	Key Example(s)
To add or subtract fractions with different denominators: 1. Write the fractions as _____ _____. 2. Add or subtract the numerators and keep the common denominator. 3. Simplify.	Add or subtract. **a.** $\dfrac{1}{6} + \dfrac{3}{4} = \dfrac{1(2)}{6(2)} + \dfrac{3(3)}{4(3)}$ $\qquad\quad = \dfrac{2}{12} + \dfrac{9}{12}$ $\qquad\quad = \dfrac{11}{12}$ **b.** $\dfrac{1}{2} - \dfrac{5}{8} = \dfrac{1(4)}{2(4)} - \dfrac{5}{8}$ $\qquad\quad = \dfrac{4}{8} - \dfrac{5}{8}$ $\qquad\quad = -\dfrac{1}{8}$ **c.** $\dfrac{3}{5x^2} - \dfrac{9}{10xy} = \dfrac{3(2y)}{5x^2(2y)} - \dfrac{9(x)}{10xy(x)}$ $\qquad\qquad = \dfrac{6y}{10x^2y} - \dfrac{9x}{10x^2y}$ $\qquad\qquad = \dfrac{6y - 9x}{10x^2y}$

[5.6] For Exercises 37–40, add or subtract.

37. $\dfrac{5}{6} + \dfrac{1}{5}$

38. $\dfrac{9}{15} - \dfrac{4}{5}$

39. $\dfrac{7}{6h} - \dfrac{3}{4h}$

40. $\dfrac{5}{8a} - \dfrac{7}{12}$

Definitions/Rules/Procedures	Key Example(s)
To add or subtract mixed numbers: **Method 1:** Write as _____; then add or subtract those _____. **Method 2:** Add or subtract the _____ separately.	Add or subtract. **a.** $2\dfrac{3}{5} + 5\dfrac{3}{4}$ **Method 1:** $2\dfrac{3}{5} + 5\dfrac{3}{4} = \dfrac{13}{5} + \dfrac{23}{4}$ $\qquad\qquad\qquad = \dfrac{13(4)}{5(4)} + \dfrac{23(5)}{4(5)}$ $\qquad\qquad\qquad = \dfrac{52}{20} + \dfrac{115}{20}$ $\qquad\qquad\qquad = \dfrac{167}{20}$ $\qquad\qquad\qquad = 8\dfrac{7}{20}$

Definitions/Rules/Procedures	Key Example(s)
	b. $8\frac{1}{4} - 5\frac{2}{3}$ **Method 2:** $8\frac{1}{4} - 5\frac{2}{3} = 8\frac{1(3)}{4(3)} - 5\frac{2(4)}{3(4)}$ $= 8\frac{3}{12} - 5\frac{8}{12}$ $= 7\frac{15}{12} - 5\frac{8}{12}$ $= 2\frac{7}{12}$

For Exercises 41 and 42, add or subtract.

41. $4\frac{5}{8} + 6\frac{2}{3}$

42. $-5\frac{1}{6} - \left(-2\frac{1}{3}\right)$

5.7 Order of Operations; Evaluating and Simplifying Expressions

Definitions/Rules/Procedures	Key Example(s)
Perform operations in the following order: 1. _____. These include parentheses ( ), brackets [ ], braces { }, absolute value | |, radicals $\sqrt{}$, and fraction bars ——. 2. _____ in order as they occur from left to right 3. _____ in order as they occur from left to right 4. _____ in order as they occur from left to right **Note:** Use G $\overrightarrow{\text{ER}}$ $\overrightarrow{\text{MD}}$ $\overrightarrow{\text{AS}}$ to remember the order.	Simplify. $\left(\frac{3}{4}\right)^2 - \frac{1}{6} \div \frac{2}{3}$ $= \frac{9}{16} - \frac{1}{6} \div \frac{2}{3}$ Simplify the exponential form. $= \frac{9}{16} - \frac{1}{6} \cdot \frac{3}{2}$ Rewrite the division as multiplication. $= \frac{9}{16} - \frac{1}{\underset{2}{6}} \cdot \frac{\overset{1}{3}}{2}$ Divide out the common factor. $= \frac{9}{16} - \frac{1}{4}$ Multiply. $= \frac{9}{16} - \frac{1(4)}{4(4)}$ Rewrite with the LCD, 16. $= \frac{9}{16} - \frac{4}{16}$ $= \frac{5}{16}$ Subtract.

[5.7] For Exercises 43 and 44, simplify.

43. $3\frac{1}{2} + \frac{1}{4}\left(\frac{2}{3} - \frac{1}{6}\right)$

44. $\left(4 + \frac{1}{12}\right) - 8 \div \frac{3}{4}$

Definitions/Rules/Procedures	Key Example(s)

To evaluate an expression:

1. _____ with the corresponding given values.

2. _____ using the order of operations agreement.

To combine like terms, add the _____ and keep the _____ and their _____ the same.

To add polynomials, _____.

To subtract polynomials:

1. Write the subtraction expression as an equivalent addition expression.

   a. Change the operation symbol from a − to a +.

   b. Change the subtrahend (second polynomial) to its additive inverse by _____.

2. Combine like terms.

To multiply monomials:

1. Multiply the _____.

2. _____ of the like variables.

To multiply two polynomials:

1. Multiply _____ in the second polynomial by _____ in the first polynomial.

2. _____ like terms.

Evaluate $\frac{2}{3}x - 5xy$ when $x = -6$ and $y = \frac{1}{10}$.

$\frac{2}{3}(-6) - 5(-6)\left(\frac{1}{10}\right)$ Replace variables with values.

$= \frac{2}{3}\left(-\frac{\overset{2}{6}}{1}\right)$ Write as improper fractions and divide out common factors.

$-\frac{\overset{1}{5}}{1}\left(-\frac{\overset{3}{6}}{1}\right)\left(\frac{1}{\underset{1}{10}}\right)$

$= -4 + 3$ Multiply.

$= -1$ Add.

Combine like terms.

$\frac{7}{12}x^2 + \frac{1}{4}x - \frac{2}{3}x^2 + \frac{1}{2}x$

$= \frac{7}{12}x^2 - \frac{2}{3}x^2 + \frac{1}{4}x + \frac{1}{2}x$ Collect the like terms.

$= \frac{7}{12}x^2 - \frac{2(4)}{3(4)}x^2 +$ Rewrite the fractions with their LCD.
$\frac{1}{4}x + \frac{1(2)}{2(2)}x$

$= \frac{7}{12}x^2 - \frac{8}{12}x^2 +$ Add the coefficents and keep the variables and their exponents the same.
$\frac{1}{4}x + \frac{2}{4}x$

$= -\frac{1}{12}x^2 + \frac{3}{4}x$

Add or subtract.

a. $\left(\frac{1}{6}x + \frac{4}{5}\right) + \left(\frac{1}{6}x - \frac{1}{5}\right)$

$= \frac{1}{6}x + \frac{1}{6}x + \frac{4}{5} - \frac{1}{5}$ Collect the like terms.

$= \frac{2}{6}x + \frac{3}{5}$ Combine the like terms.

$= \frac{1}{3}x + \frac{3}{5}$ Simplify.

Definitions/Rules/Procedures	Key Example(s)
	b. $\left(\dfrac{5}{9}x - \dfrac{3}{4}\right) - \left(\dfrac{1}{9}x - \dfrac{1}{12}\right)$

$= \left(\dfrac{5}{9}x - \dfrac{3}{4}\right) + \left(-\dfrac{1}{9}x + \dfrac{1}{12}\right)$ Rewrite as addition.

$= \dfrac{5}{9}x - \dfrac{1}{9}x - \dfrac{3}{4} + \dfrac{1}{12}$ Collect the like terms.

$= \dfrac{5}{9}x - \dfrac{1}{9}x - \dfrac{3(3)}{4(3)} + \dfrac{1}{12}$ Rewrite the fractions with their LCD.

$= \dfrac{5}{9}x - \dfrac{1}{9}x - \dfrac{9}{12} + \dfrac{1}{12}$

$= \dfrac{4}{9}x - \dfrac{8}{12}$ Combine the like terms.

$= \dfrac{4}{9}x - \dfrac{2}{3}$ Simplify.

Multiply.

a. $\left(\dfrac{5}{6}x^2\right)\left(\dfrac{7}{10}xy\right) = \dfrac{5}{6} \cdot \dfrac{\overset{1}{7}}{\underset{2}{10}}x^{2+1}y = \dfrac{7}{12}x^3y$

b. $\left(\dfrac{1}{4}m - 1\right)\left(\dfrac{2}{3}m + 4\right)$

$= \dfrac{1}{\underset{2}{4}}m \cdot \dfrac{\overset{1}{2}}{3}m + \dfrac{1}{\underset{1}{4}}m \cdot \dfrac{\overset{1}{4}}{1} - 1 \cdot \dfrac{2}{3}m - 1 \cdot 4$

$= \dfrac{1}{6}m^2 + m - \dfrac{2}{3}m - 4$ Multiply.

$= \dfrac{1}{6}m^2 + \dfrac{3}{3}m - \dfrac{2}{3}m - 4$ Write with the LCD.

$= \dfrac{1}{6}m^2 + \dfrac{1}{3}m - 4$ Combine like terms.

---

45. Evaluate $\dfrac{3}{4}m - 2n$ when $m = -\dfrac{5}{9}$ and $n = 1\dfrac{3}{4}$.

[5.7] For Exercises 46–50, simplify and write the resulting polynomial in descending order of degree.

46. $\dfrac{3}{5}x^2 + 9x - \dfrac{1}{2}x^2 - 2 - 11x$

47. $\left(\dfrac{1}{4}n^2 - \dfrac{2}{3}n - 3\right) + \left(\dfrac{3}{8}n^2 + 1\right)$

48. $\left(12y^3 - \dfrac{5}{6}y + \dfrac{2}{5}\right) - \left(\dfrac{1}{2}y^3 + 2y^2 + \dfrac{1}{3}\right)$

49. $\left(\dfrac{2}{7}b^3\right)\left(-\dfrac{7}{8}b\right)$

50. $\left(\dfrac{1}{4}x + 2\right)\left(5x - \dfrac{3}{2}\right)$

5.8 Solving Equations

Definitions/Rules/Procedures	Key Example(s)

To solve a linear equation in one variable:
 	
1. Simplify each side of the equation.
 	
  a. Distribute to eliminate parentheses.
 	
  b. Eliminate fractions by _____ both sides of the equation by the _____ of all of the fractions. (Optional)
 	
  c. Combine like terms.
 	
2. Use the addition principle of equality so that all variable terms are on one side of the equation and all constants are on the other side. Then combine like terms.
 	
  Note Eliminate the variable term with the lesser coefficient to avoid negative coefficients.
 	
3. Use the multiplication principle of equality to eliminate the coefficient of the variable.
	Solve.

Solve.

$$\frac{2}{3}(x - 1) = \frac{1}{2}x - \frac{2}{5}$$

Eliminate the fractions by multiplying both sides by the LCD, 30.

$$30\left(\frac{2}{3}x - \frac{2}{3}\right) = \left(\frac{1}{2}x - \frac{2}{5}\right)30$$

$$\frac{\overset{10}{\cancel{30}}}{1} \cdot \frac{2}{\underset{1}{\cancel{3}}}x - \frac{\overset{10}{\cancel{30}}}{1} \cdot \frac{2}{\underset{1}{\cancel{3}}} = \frac{\overset{15}{\cancel{30}}}{1} \cdot \frac{1}{\underset{1}{\cancel{2}}}x - \frac{\overset{6}{\cancel{30}}}{1} \cdot \frac{2}{\underset{1}{\cancel{5}}}$$

$$20x - 20 = 15x - 12$$
$$\underline{-15x \qquad\qquad -15x}$$
$$5x - 20 = \quad 0 - 12$$

Subtract 15x from both sides

$$5x - 20 = -12$$
$$\underline{+20 \qquad +20}$$
$$5x - \quad 0 = \quad\; 8$$

$$\frac{5x}{5} = \frac{8}{5}$$

Add 20 to both sides.

$$1x = \frac{8}{5}$$

Divide both sides by 5.

$$x = 1\frac{3}{5}$$

Write the answer as a mixed number.

[5.8] For Exercises 51–54, solve and check.

51. $y - 3\frac{4}{5} = \frac{1}{3}$

52. $-\frac{3}{8}n = \frac{4}{3}$

53. $\frac{1}{2}m - 5 = \frac{3}{4}$

54. $\frac{1}{2}\left(n - \frac{2}{3}\right) = \frac{3}{4}n + 2$

Learning Strategy

Writing formulas in different colors or sizes next to a corresponding picture or diagram on a study sheet can help you memorize them.

—Stefanie P.

Geometry

Circles

[5.3] A **circle** is a collection of points that are _____ from a central point, called the *center*.

[5.3] The **radius** of a circle is the distance from the _____ to any point on the circle.

[5.3] The **diameter** of a circle is the distance _____ a circle along a straight line through the center.

[5.3] The **circumference** of a circle is the _____ a circle.

[5.3] An **irrational number** is a number that _____ be expressed in the form $\frac{a}{b}$, where a and b are integers and $b \neq 0$.

Geometry

[5.3] π is an irrational number that is the ratio of the _____ of a circle to its _____.

[5.3] Diameter of a circle: $d = $ _____ r

[5.3] Radius of a circle: $r = $ _____ d

[5.3] Circumference of a circle given its radius: $C = $ _____

[5.3] Circumference of a circle given its diameter: $C = $ _____

[5.7] Area of a circle: $A = $ _____

Triangles

[5.3] Area of a triangle: $A = $ _____

Trapezoids

A **trapezoid** is a four-sided figure with _____ pair of parallel sides.

Area of a trapezoid: $A = $ _____

For Exercises 55–70, solve.

55. [5.3] At a clothing store, $\frac{5}{9}$ of the merchandise is discounted. $\frac{3}{10}$ of these discounted items are shirts. What fraction of all items in the store are discounted shirts?

56. [5.3] Find the area of the triangle.

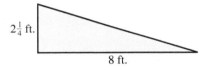

57. [5.3] The circular top of a can of beans has a diameter of $3\frac{1}{2}$ inches. Find the radius.

58. [5.3] A wheel on a piece of exercise equipment has a radius of $14\frac{1}{5}$ centimeters. Find the diameter.

59. [5.3] A tree has a diameter of $2\frac{1}{2}$ feet. What is the circumference? Use $\frac{22}{7}$ for π.

60. [5.4] A box of rice contains $4\frac{1}{2}$ cups of uncooked rice. The label indicates that a single serving is $\frac{1}{4}$ cup of uncooked rice. How many servings are in the box?

61. [5.4] The circumference of a Frisbee is 33 inches. Find the radius.

62. [5.6] Respondents to a survey had three ways to respond: agree, disagree, or no opinion. The following graph shows the fraction of respondents that agreed or disagreed. What fraction of the respondents had no opinion?

Survey

63. [5.7] The following table shows the circumference of six cantaloupes taken from a particular garden. Find the mean and median circumference.

$18\frac{1}{2}$ in.	$17\frac{3}{4}$ in.	$18\frac{3}{4}$ in.	$19\frac{1}{4}$ in.	$16\frac{3}{4}$ in.	$18\frac{1}{4}$ in.

64. [5.7] Find the area of the trapezoid.

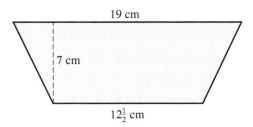

65. [5.7] Find the surface area of a circular tabletop with a diameter of 4 feet. Use $\frac{22}{7}$ for π.

66. [5.7] Find the area of the shaded region.

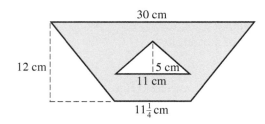

67. [5.8] $\frac{1}{4}$ of the sum of n and 2 is the same as $\frac{3}{5}$ less than $\frac{3}{4}$ of n. Translate to an equation; then solve.

68. [5.8] Find the length b in the trapezoid shown assuming that the area is 19 square meters.

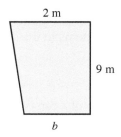

69. [5.8] A section of PVC pipe is $2\frac{1}{3}$ times the length of a second section. The two pipes together have a length of 35 feet. What are the lengths of the two sections?

70. [5.8] Two pipes are welded together forming two angles as shown. The smaller angle measures $\frac{3}{4}$ of the measure of the larger angle. Find the angle measurements.

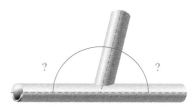

Chapter 5 Practice Test

For Extra Help — CHAPTER Test Prep VIDEOS. Step-by-step test solutions are found on the Chapter Test Prep Videos available via the Video Resources on DVD, in MyMathLab®, and on YouTube (search "CarsonPrealgebra" and click on "Channels").

[5.1] **1.** Name the fraction in lowest terms represented by the shaded region.

1._____

[5.1] **2.** Graph $-3\frac{1}{4}$ on a number line.

2._____

$\begin{bmatrix}5.1\\5.2\end{bmatrix}$ **3.** Use $<$, $>$, or $=$ to write a true statement.

   a. $\frac{3}{4}$? $\frac{9}{12}$ **b.** $\frac{14}{15}$? $\frac{5}{6}$

3. a._____

b._____

[5.1] **4. a.** Write $\frac{37}{6}$ as a mixed number.
   b. Write $-4\frac{5}{8}$ as an improper fraction.

4. a._____

b._____

[5.1] **5.** Simplify.

   a. $\frac{-16}{1}$ **b.** $\frac{0}{14}$ **c.** $\frac{27}{27}$

5. a._____

b._____

c._____

[5.2] **6.** Simplify to lowest terms.

   a. $\frac{24}{40}$ **b.** $-\frac{9x^5y}{30x^3}$

6. a._____

b._____

[5.3] **7.** Multiply. Write the product in lowest terms.

   a. $-2\frac{1}{4}\cdot 5\frac{1}{6}$ **b.** $\frac{2a}{7b^3}\cdot\frac{14b}{3a}$

7. a._____

b._____

[5.3] **8.** Evaluate.

   a. $\left(\frac{2}{5}\right)^2$ **b.** $\left(\frac{1}{4}\right)^3$

8. a._____

b._____

[5.4] **9.** Divide. Write the quotient in lowest terms.

   a. $3\frac{1}{8}\div\frac{5}{4}$ **b.** $\frac{-2m^4n}{9n^2}\div\frac{5m}{12n^2}$

9. a._____

b._____

[5.5] **10.** Find the LCM of $18t^3$ and $12t^2u$.

10._____

[5.6] **11.** Add or subtract.

   a. $\frac{3}{10}+\frac{1}{4}$ **b.** $-3\frac{1}{8}-\left(-1\frac{1}{2}\right)$

11. a._____

b._____

[5.6] **12.** Add or subtract.

   a. $-\frac{4x}{15}+\left(-\frac{x}{15}\right)$ **b.** $\frac{7}{4a}-\frac{5}{6a}$

12. a._____

b._____

13._____

[5.7] **13.** Simplify. $2\dfrac{1}{4} + \dfrac{5}{8}\left(\dfrac{4}{5} - \dfrac{3}{10}\right)$

14. a._____

  b._____

[5.7] **14.** Simplify.

  a. $\left(\dfrac{1}{4}n^2 - \dfrac{1}{2}n + 2\right) - \left(\dfrac{1}{4}n^2 + n + 5\right)$ **b.** $\left(m + \dfrac{3}{4}\right)\left(2m - \dfrac{1}{2}\right)$

15._____

[5.8] **15.** Solve and check. $\dfrac{1}{4}m - \dfrac{2}{3} = \dfrac{5}{6}$

16._____

[5.3] **16.** In a new housing development, $\frac{3}{4}$ of the lots were sold in the first year. $\frac{5}{6}$ of these lots had finished houses on them by the end of the first year. What fraction of all lots in the development had finished houses on them by the end of the first year?

17._____

[5.3] **17.** Find the area of the triangle.

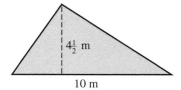

$4\frac{1}{2}$ m

10 m

18. _____

[5.3] **18.** A compact disc has a diameter of $4\frac{3}{4}$ inches. What is the circumference? Use $\frac{22}{7}$ for π.

19. _____

[5.4] **19.** Pam is going to cut a 35-inch strip of ribbon into pieces that are $4\frac{3}{8}$ inches long. How many pieces will she have?

20. _____

[5.6] **20.** A testing company examines the results of a certain multiple-choice test question that had three possible responses: A, B, or C. The following graph shows the fraction of people who answered either A or B. What fraction chose C?

Test Question Responses

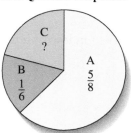

21._____

[5.7] **21.** The following table shows the height of four siblings. Find the mean, median, and mode height.

$69\frac{1}{2}$ in.	70 in.	$70\frac{3}{4}$ in.	$71\frac{1}{2}$ in.

[5.7]   **22.** Find the area of the trapezoid.

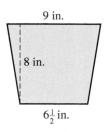

9 in.

8 in.

$6\frac{1}{2}$ in.

22._____

[5.7]   **23.** Find the area of a circle with a diameter of 5 inches. Use $\frac{22}{7}$ for π.

23._____

[5.8]   **24.** $\frac{2}{3}$ of the sum of n and 5 is the same as $\frac{3}{4}$ less than $\frac{1}{2}$ of n. Translate to an equation; then solve.

24._____

[5.8]   **25.** Catrina owns two chocolate shops, one downtown and one in a shopping mall. In one week, the downtown shop earned twice as much revenue as the shop in the mall. If the total revenue that week from both shops was $15,600, how much did each shop earn?

25._____

Chapters 1–5 Cumulative Review Exercises

For Exercises 1–4, answer true or false.

[1.1] **1.** -1 is a whole number.

[2.2] **2.** The sum of two numbers that have the same sign is always positive.

[4.2] **3.** $2x = 5x^2 + 8$ is a linear equation.

[5.3] **4.** $\left(\dfrac{3}{4}\right)^2 = \dfrac{3^2}{4^2}$

[3.6] **5.** Explain how to find the greatest common factor of a set of numbers using prime factorization.

[5.6] **6.** Explain in your own words how to add or subtract fractions.

[1.1] **7.** Write the word name for 4,582,601.

[1.3] **8.** Estimate $461 \cdot 72$ by rounding.

[3.2] **9.** What is the coefficient and degree of $-x$?

[5.5] **10.** Find the LCM of 24 and 30.

For Exercises 11–17, simplify.

[2.5] **11.** $-3^4 - 5\sqrt{100 - 64}$

[2.5] **12.** $18 - 5(6) \div [9 + 3(-7)]$

[5.2] **13.** $\dfrac{36}{40}$

[5.4] **14.** $4\dfrac{1}{3} \div 5\dfrac{1}{9}$

[5.3] **15.** $-\dfrac{5x^3}{9} \cdot \dfrac{27}{40x}$

[5.6] **16.** $7\dfrac{1}{6} - 3\dfrac{1}{2}$

[5.6] **17.** $\dfrac{4}{5x} + \dfrac{2}{3}$

[5.7] **18.** Evaluate $mn - 2n^3$ when $m = \frac{5}{6}$ and $n = -4$.

[5.7] **19.** Combine like terms and write the resulting polynomial in descending order of degree.

$12b^2 - \dfrac{2}{3}b + 3b^2 - b^3 - \dfrac{1}{4}b$

[3.3] **20.** Subtract. $(x^4 - 5x^3 - 10x + 18) - (x^4 + 2x^3 - 3x - 8)$

[3.4] **21.** Multiply. $(x - 7)(x + 7)$

[3.6] **22.** Factor. $18m^3 + 24m^2 - 30m$

For Exercises 23 and 24, solve and check.

[4.3] **23.** $3(y - 4) - 8 = 7y - 16$

[5.8] **24.** $\dfrac{3}{4}n - 1 = -\dfrac{2}{5}$

For Exercises 25–30, solve.

[1.5] **25.** The following table shows the address and listed price for five houses for sale in a particular neighborhood. Find the mean, median, and mode price.

Address	Price
112 Oak St.	$175,000
204 Sycamore St.	$182,000
715 Maple St.	$215,000
512 Maple St.	$245,000
313 Pine Ave.	$255,000

[2.3] **26.** The table below lists the assets and debts for the Krueger family. Calculate their net worth.

Assets	Debts
Savings = $1260	Credit card balance = $872
Checking = $945	Mortgage = $57,189
Furniture = $13,190	Automobile 1 = $3782
	Automobile 2 = $12,498

[3.7] **27.** a. Write an expression in simplest form for the perimeter of the rectangle shown.

   b. Find the perimeter if y is 8 feet.

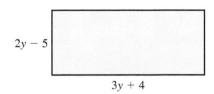

$2y - 5$

$3y + 4$

[4.3] **28.** A storage chest has a volume of 24 cubic feet. If the length is 4 feet and the width is 3 feet, find the height.

[5.4] **29.** A tree has a circumference of 44 inches. What is the diameter?

$\begin{bmatrix} 4.5 \\ 5.8 \end{bmatrix}$ **30.** A support beam is to be angled so that the angle made between the beam and the ground on one side of the beam is 35° less than twice the adjacent angle on the other side of the beam. What are the angle measurements?

CHAPTER

6

Decimals

Chapter Overview

In this chapter, we will study decimal notation, which is another way to represent the fractions and mixed numbers that we learned in Chapter 5. Like Chapter 5, we will review Chapters 1–4 as we study decimal numbers as follows:

▶ Positive and negative decimal numbers

▶ Expressions with decimal numbers

▶ Equations with decimal numbers

After completing this chapter, we will have explored all of the different types of numbers in the set of real numbers.

6.1 Introduction to Decimal Numbers

Warm-up

[5.1] 1. Graph $3\frac{5}{8}$ on a number line.

[5.2] 2. Simplify $-\frac{45}{60}$ to lowest terms.

Objective 1 Write decimals as fractions or mixed numbers.

In this chapter, we explore numbers written in **decimal notation.**

Definition **Decimal notation:** A base-10 notation for expressing fractions.

By *base-10 notation,* we mean that each place value is a power of 10. As the following table shows, a decimal point separates the whole number place values from the fractional place values.

Decimal Point

Whole numbers				Fractions			
◄ ...	Hundreds	Tens	Ones	Tenths	Hundredths	Thousandths	Ten thousandths	... ►

To see the connection between decimals and fractions, let's look at a decimal number written in expanded form. Consider the number 23.791.

Decimal Point

Whole numbers		Fractions		
Tens	Ones	Tenths	Hundredths	Thousandths
2	3	7	9	1

Expanded form: $23.791 = 2 \cdot 10 + 3 \cdot 1 + 7 \cdot \frac{1}{10} + 9 \cdot \frac{1}{100} + 1 \cdot \frac{1}{1000}$

We can determine the equivalent fraction by multiplying and adding.

$$= 20 + 3 + \frac{7}{10} + \frac{9}{100} + \frac{1}{1000}$$ **Multiply.**

$$= 23 + \frac{7(100)}{10(100)} + \frac{9(10)}{100(10)} + \frac{1}{1000}$$ **Write equivalent fractions with their LCD, 1000.**

Note The last place value determines the common denominator.

Objectives

1 Write decimals as fractions or mixed numbers.

2 Write a word name for a decimal number.

3 Graph decimals on a number line.

4 Use < or > to write a true statement.

5 Round decimal numbers to a specified place.

Connection Notice that the fraction names follow the same pattern as the whole number side of the decimal point. We simply place *th* in each whole place name. *Tens* becomes *tenths*, *hundreds* becomes *hundredths*, and so on.

Answers to Warm-up

1.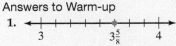

2. $-\frac{3}{4}$

Connection We read $23\frac{791}{1000}$ as "twenty-three and seven hundred ninety-one thousandths"; so 23.791 is read the same way.

$$= 23 + \frac{700}{1000} + \frac{90}{1000} + \frac{1}{1000}$$

$$= 23 + \frac{791}{1000} \qquad \text{Add the fractions.}$$

$$= 23\frac{791}{1000} \qquad \text{Write as a mixed number.}$$

▲

Note The whole number part becomes the integer part of a mixed number. The numerator of the fraction contains all of the decimal digits, and the denominator is the last place value.

Our example suggests a fast way to write a decimal number as a fraction or mixed number.

Procedure

To write a decimal number as a fraction or mixed number in simplest form:
1. Write all digits to the left of the decimal point as the integer part of a mixed number.
2. Write all digits to the right of the decimal point in the numerator of a fraction.
3. Write the denominator indicated by the last place value.
4. Simplify to lowest terms.

▶ **Your Turn 1**

Write as a fraction or mixed number in simplest form.
 a. 0.79
 b. −0.08
 c. 2.6
 d. −14.675

Example 1 Write as a fraction or mixed number in simplest form.

a. 0.84

Solution: Write all digits to the right of the decimal point in the numerator. Write the denominator indicated by the last place value. Then simplify to lowest terms.

$$0.\overline{84} = \frac{84}{100} = \frac{84 \div 4}{100 \div 4} = \frac{21}{25}$$

Hundredths

Because $0.84 = \frac{84}{100}$, we read 0.84 as "eighty-four hundredths."

b. −217.5

Solution: Write the digits to the left of the decimal point as the integer part of a mixed number. Write the digit to the right of the decimal point in the numerator. Write the denominator indicated by the last place value. Then simplify.

$$-217.\overline{5} = -217\frac{5}{10} = -217\frac{5 \div 5}{10 \div 5} = -217\frac{1}{2}$$

Tenths

Connection Because the denominators will always be powers of 10 and the primes that divide powers of 10 are 2 and 5, these fractions will reduce only if the numerators are divisible by 2 or 5.

We read −217.5 as "negative two hundred seventeen and five tenths," or, in simplest form, "negative two hundred seventeen and one-half." Note that it is also common to say "negative two hundred seventeen point five."

◀ **Do Your Turn 1**

Answers to Your Turn 1
a. $\frac{79}{100}$ **b.** $-\frac{2}{25}$ **c.** $2\frac{3}{5}$ **d.** $-14\frac{27}{40}$

In Chapter 5, we learned that a rational number is a number that can be written in the form $\frac{a}{b}$ where a and b are integers and $b \neq 0$. So 23.791, 0.84, and -217.5 are rational numbers because they can be written as fractions whose numerators and denominators are integers.

Objective 2 Write a word name for a decimal number.

In Example 1, we learned to read a decimal number using its equivalent fraction or mixed number. Similarly, when we write a word name for a decimal number, we write the name of its equivalent fraction or mixed number.

$0.39 = \frac{39}{100}$ and is written *thirty-nine hundredths*

$19.8571 = 19\frac{8571}{10,000}$ and is written *nineteen and eight thousand five hundred seventy-one ten-thousandths*

Notice that the word *and* separates the integer and fraction parts of both mixed numbers and decimal numbers; so *and* takes the place of the decimal point in the word name for a decimal number that has an integer amount.

Procedure

To write the word name for a decimal number with no integer part (no digits other than 0 to the left of the decimal point):

1. Write the word name for digits as if they represented a whole number.
2. Write the name of the last place value.

To write a word name for a decimal number with both integer and fractional parts:

1. Write the name of the integer number part.
2. Write the word *and* for the decimal point.
3. Write the name of the fractional part.

Example 2 Write the word name.

a. 0.0000489

Answer: four hundred eighty-nine ten-millionths

b. 98.10479

Answer: ninety-eight and ten thousand four hundred seventy-nine hundred-thousandths

▶ **Do Your Turn 2**

Objective 3 Graph decimals on a number line.

Decimal numbers can be graphed on a number line. The denominator of the equivalent fraction describes the number of divisions between integers. The numerator of the equivalent fraction describes how many of those divisions we are interested in.

▶ **Your Turn 2**

Write the word name.

a. 0.91
b. −0.602
c. 2.7
d. 124.90017

Answers to Your Turn 2
a. ninety-one hundredths
b. negative six hundred two thousandths
c. two and seven tenths
d. one hundred twenty-four and ninety thousand seventeen hundred-thousandths

▶ **Your Turn 3**

Graph on a number line.

a. 0.7

b. 12.45

c. −6.21

d. −19.106

Note 14 and 14.0 are the same, just as 14.5 and 14.50 are the same. Zero digits are understood to continue on forever beyond the last nonzero digit in a decimal number. Writing those zeros is optional, but it is simpler not to write them. ▶

Example 3 Graph on a number line.

a. 3.6

Solution: Because 3.6 means $3\frac{6}{10}$, we divide the distance between 3 and 4 into 10 equal-size divisions and draw a dot on the sixth division mark.

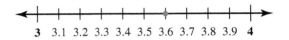

b. 14.517

Solution: Because 14.517 means $14\frac{517}{1000}$, we should divide the space between 14 and 15 into 1000 divisions and count to the 517th mark. Of course, this is rather tedious; so we'll gradually zoom in on smaller and smaller sections of the number line.

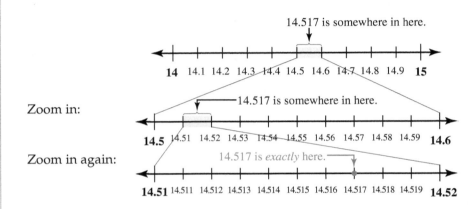

Rather than going through all of the "zoom" stages, we could go directly to the last number line. Notice that the end points on the last number line are determined by the next-to-last place, in this case the hundredths place. We then draw nine marks to divide the space between the end points into ten equal divisions and count to the seventh mark.

c. −2.75

Solution: Let's zoom in on tenths to graph the hundredths. Remember, on a number line, negative numbers decrease to the left. (Their absolute values increase to the left.)

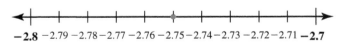

◀ **Do Your Turn 3**

Answers to Your Turn 3

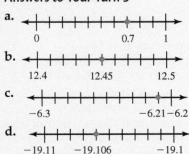

Objective 4 Use < or > to write a true statement.

Remember that we can use a number line to compare two numbers. Let's compare 13.7 and 13.2 by graphing them on a number line.

Because 13.7 is farther to the right than 13.2 is, we can say that 13.7 > 13.2.

We can also compare the numbers without graphing them. Look at the digits in 13.7 and 13.2 from left to right. They match until we get to the tenths place, and it is this place with its different digits that tells us which number is greater. The greater digit, 7, is in the greater number.

Procedure

To determine which of two positive decimal numbers is greater, compare the digits in the corresponding place values from left to right until you find two different digits in the same place. The greater number contains the greater of those two digits.

Note If both numbers are nega-
◄ tive, the greater number is closer to zero and contains the lesser of the two different digits.

Example 4   Use $<$ or $>$ to write a true statement.

a. 35.619 ? 35.625

Answer: $35.619 < 35.625$

Explanation: Compare the digits in the corresponding places from left to right.

$$35.619$$
$$\updownarrow$$
$$35.625$$

Note The digits match
◄ from left to right until the hundredths place.

In the hundredths places, because 2 is greater than 1, we can say that 35.625 is greater than 35.619. On a number line, 35.625 is farther to the right.

b. −0.00981 ? −0.009806

Answer: $-0.00981 < -0.009806$

Explanation: Compare the digits in the corresponding places from left to right.

$$-0.00981$$
$$\updownarrow$$
$$-0.009806$$

Note The digits match from
◄ left to right until the hundred-thousandths place.

Because both numbers are negative, the number closest to zero is the greater number and contains the lesser of the two different digits. Because 0 is less than 1, −0.009806 is the greater number.

► **Do Your Turn 4**

Objective 5   Round decimal numbers to a specified place.

Rounding decimal numbers is much like rounding whole numbers. Suppose we want to round 14.517 to the nearest hundredth. Let's use the graph of 14.517 from Example 3b.

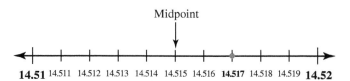

Notice that the nearest hundredths to 14.517 are 14.51 and 14.52. Because 14.517 is closer to 14.52, we say that it rounds up to 14.52.

► **Your Turn 4**

Use $<$ or $>$ to write a true statement.

a. 0.61 ? 0.65

b. 45.192 ? 45.092

c. 0.002 ? 0.00056

d. −5.71 ? −5.8

e. −0.04701 ? −0.0471

Answers to Your Turn 4
a. $0.61 < 0.65$
b. $45.192 > 45.092$
c. $0.002 > 0.00056$
d. $-5.71 > -5.8$
e. $-0.04701 > -0.0471$

Rounding to a different place value would require a different scale. For example, to round 14.517 to the nearest tenth, we would use a number line with the nearest tenths, 14.5 and 14.6, as end points.

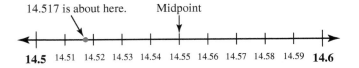

14.517 is about here.   Midpoint

14.5   14.51   14.52   14.53   14.54   14.55   14.56   14.57   14.58   14.59   14.6

Because 14.517 is closer to 14.5, it rounds down to 14.5.

If we had to round to the nearest whole (ones place), we would use 14 and 15 as our end points and the tenths place would determine whether we round up or down.

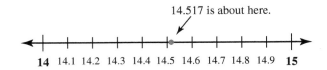

14.517 is about here.

14   14.1   14.2   14.3   14.4   14.5   14.6   14.7   14.8   14.9   15

Because 14.517 is slightly greater than the midpoint on this scale, it rounds up to 15. Even if it were exactly halfway, we agree to round up. Notice that like rounding whole numbers, the digit in the place to the right of the place to be rounded can be used to determine whether to round a decimal number up or down.

Procedure

To round a number to a given place value:
1. Identify the digit in the given place value.
2. If the digit to the right of the given place value is 5 or greater, round up by increasing the digit in the given place value by 1. If the digit to the right of the given place value is 4 or less, round down by keeping the digit in the given place value the same.
3. Change all digits to the right of the rounded place value to zeros. Decimal digits to the right of the rounded place can be eliminated.

▶ **Your Turn 5**

Round 59.61538 to the specified place.
a. Tenths
b. Hundredths
c. Ten-thousandths
d. Whole number

Example 5 Round 917.28053 to the specified place.

a. Tenths

Solution: The nearest tenths are 917.2 and 917.3. Rounding to the nearest tenth means that we consider the digit in the hundredths place, which is 8. Because 8 is greater than 5, we round up to 917.3.

Answer: 917.3

b. Thousandths

Solution: The nearest thousandths are 917.280 and 917.281. Looking at the ten-thousandths place, we see a 5; so we round up to 917.281.

Answer: 917.281

c. Whole number

Solution: The nearest whole numbers are 917 and 918. The 2 in the tenths place indicates that we round down to 917.

Answer: 917

Answers to Your Turn 5
a. 59.6
b. 59.62
c. 59.6154
d. 60

◀ **Do Your Turn 5**

6.1 Exercises

Objective 1

Prep Exercise 1 Complete the table of place values.

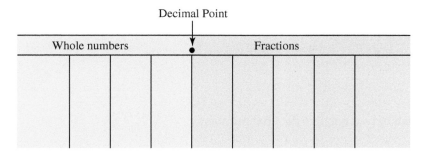

Prep Exercise 2 Explain how to write a decimal number
as a fraction or mixed number in simplest form.

For Exercises 1–20, write as a fraction or mixed number in simplest form.

1. 0.2	**2.** 0.6	**3.** 0.25	**4.** 0.75
5. 0.375	**6.** 0.125	**7.** 0.24	**8.** 0.48
9. 1.5	**10.** 4.2	**11.** 18.75	**12.** 6.4
13. 9.625	**14.** 14.375	**15.** 7.36	**16.** 9.54
17. −0.008	**18.** −0.0005	**19.** −13.012	**20.** −219.105

Objective 2

Prep Exercise 3 When writing the word name for a decimal number greater than 1,
as in 2.56, what word takes the place of the decimal point?

For Exercises 21–34, write the word name.

21. 0.097	**22.** 0.415
23. 0.002015	**24.** 0.030915
25. 31.98	**26.** 15.73
27. 521.608	**28.** 343.406
29. 4159.6	**30.** 2109.003
31. −107.99	**32.** −2106.1
33. −0.50092	**34.** −0.000982

Objective 3

Prep Exercise 4 To graph 0.8 on a number line, divide the space between 0 and 1
into _____ equal-size spaces and place a dot on the eighth mark.

For Exercises 35–46, graph on a number line.

35. 0.8	**36.** 0.4

37. 1.3

38. 6.9

39. 4.25

40. 13.75

41. 8.06

42. 11.05

43. −3.21

44. −6.45

45. −19.017

46. −15.002

Objective 4

Prep Exercise 5 Explain how to determine which of two positive decimal numbers is greater.

For Exercises 47–58, use < or > to make a true statement.

47. 0.81 ? 0.8

48. 0.15 ? 0.5

49. 2.891 ? 2.8909

50. 153.77 ? 153.707

51. 0.001983 ? 0.001985

52. 0.090194 ? 0.091094

53. 0.0245 ? 0.00963

54. 0.007 ? 0.0602

55. −1.01981 ? −1.10981

56. −18.1095 ? −18.1085

57. −145.7183 ? −14.57183

58. −26.71803 ? −267.1803

Objective 5

Prep Exercise 6 When rounding a decimal number, if the place value to be rounded is a decimal place value, all digits to the right of that place value can be _____ .

For Exercises 59–64, round 610.28315 to the specified place.

59. tenths

60. hundredths

61. ten-thousandths

62. whole number

63. tens

64. hundreds

For Exercises 65–70, round 0.95106 to the specified place.

65. whole number

66. tens

67. tenths

68. hundredths

69. thousandths

70. ten-thousandths

For Exercises 71–76 round − 408.06259 to the specified place.

71. tenths

72. hundredths

73. ten-thousandths

74. thousandths

75. tens

76. whole number

Review Exercises

[5.6] **1.** Add. $\dfrac{37}{100} + \dfrac{4}{100}$

[2.3] **2.** Subtract. $-29 - (-16)$

For Exercises 3 and 4, simplify.

[3.2] **3.** $7a^2 - a + 15 - 10a^2 + a - 7a^3$

[3.3] **4.** $(5x^2 + 9x - 18) - (x^2 - 12x - 19)$

[4.2] **5.** Solve and check. $x - 19 = -36$

[5.6] **6.** Find the perimeter of a $12\frac{1}{2}$-foot-wide by 16-foot-long room.

6.2 Adding and Subtracting Decimal Numbers

Objectives

1 Add decimal numbers.

2 Subtract decimal numbers.

3 Add and subtract signed decimal numbers.

4 Simplify, add, or subtract polynomials containing decimal numbers.

5 Solve equations using the addition principle.

6 Solve applications.

Warm-up

[5.6] 1. Estimate; then find the actual difference. $14\frac{1}{4} - 7\frac{5}{6}$

[6.1] 2. Write 37.08 as a mixed number in simplest form.

Objective 1 Add decimal numbers.

In Section 6.1, we learned that decimal numbers are equivalent to mixed numbers; so adding decimal numbers must be related to adding mixed numbers. With mixed numbers, we add the integers and fractions separately; the same idea must apply to their decimal equivalents. Look at the following mixed number addition and its equivalent decimal addition.

$$2\frac{82}{100}$$
$$+43\frac{1}{100}$$
$$\overline{45\frac{83}{100}}$$

$$2.82$$
$$+43.01$$
$$\overline{45.83}$$

Note The decimal point separates the whole and fraction place values; so by aligning the decimal point vertically, we align the corresponding place values.

Connection Aligning the place values aligns decimal digits that are the numerators of fractions with common denominators (tenths with tenths, hundredths with hundredths, and so on). By adding those digits, we are adding numerators of fractions with common denominators.

Procedure

To add decimal numbers:

1. Stack the numbers so that the corresponding place values align. (Vertically align the decimal points.)
2. Add the digits in the corresponding place values.
3. Place the decimal point in the sum so that it aligns with the decimal points in the addends.

Note Remember, an estimate is intended to be a quick prediction of what we can expect the actual answer to be. If the actual answer reasonably agrees with the estimate, we can be confident that the actual calculation is correct. In each of our examples, the actual answer agrees reasonably with the estimate.

Example 1   Estimate. Then find the actual sum.

a. 51.092 + 23.64

Solution: $\begin{array}{r} 51.092 \\ +23.64 \\ \end{array}$ Stack the numbers so that the corresponding place values align. (Vertically align the decimal points.)

Estimate: Use the same procedure as for estimating a sum of whole numbers.

$\begin{array}{r} 51.092 \longrightarrow 50 \\ +23.64 \longrightarrow +20 \\ \hline 70 \end{array}$ Round the smallest number, 23.64, to one nonzero digit, which is to the nearest ten: 20. Round 51.092 to the nearest ten as well: 50. Then add the rounded numbers.

Actual: $\begin{array}{r} \overset{1}{51.092} \\ +23.640 \\ \hline 74.732 \end{array}$ **Note** Some people write the understood 0s beyond the last nonzero digit instead of having blank spaces.

Answers to Warm-up

1. Estimate: 6; Actual: $6\frac{5}{12}$

2. $37\frac{2}{25}$

► **Your Turn 1**

Estimate. Then find the actual sum.

a. $14.103 + 7.035$
b. $0.1183 + 0.094$
c. $6.981 + 0.57 + 23.1$
d. $0.46 + 781 + 34$

b. $32.09 + 4.103 + 19.6423$

Solution:

Estimate:		
32.09	$\longrightarrow$	32
4.103	$\longrightarrow$	4
+ 19.6423	$\longrightarrow$	+ 20
		56

Round the smallest number, 4.103, to one nonzero digit, which is to the nearest one (whole) = 4. Round the other two numbers to the nearest one as well. Then add.

$$
\begin{array}{r}
\overset{1\ \ 1}{32.0900} \\
4.1030 \\
+\ 19.6423 \\
\hline
55.8353
\end{array}
$$

Actual:

c. $76 + 0.0189 + 3.99$

Solution:

Estimate:		
76	$\longrightarrow$	$\overset{1\ 1}{76.00}$
0.0189	$\longrightarrow$	0.02
+ 3.99	$\longrightarrow$	+ 3.99
		80.01

Note Estimates are supposed to be easy to perform mentally. In this case, our method leads to a calculation that may not be easy to perform mentally. If so, consider rounding to easier numbers, such as the nearest whole numbers.

Actual:

$$
\begin{array}{r}
\overset{11\ 1}{76.0000} \\
0.0189 \\
+\ 3.9900 \\
\hline
80.0089
\end{array}
$$

$$
\begin{array}{r}
76 \\
0 \\
+\ 4 \\
\hline
80
\end{array}
$$

◄ **Do Your Turn 1**

Objective 2 Subtract decimal numbers.

We subtract decimal numbers the same way we add, except that we subtract the digits in the corresponding place values.

> **Procedure**
>
> To subtract decimal numbers:
> 1. Stack the number with the greater absolute value on top so that the corresponding place values align. (Vertically align the decimal points.)
> 2. Subtract the digits in the corresponding place values.
> 3. Place the decimal point in the difference so that it aligns with the decimal points in the minuend and subtrahend.

Example 2 Subtract.

a. $58.941 - 2.54$

Solution: Stack the number with the greater absolute value on top.

Estimate:		Actual:	
59			58.941
− 3			− 2.540
56			56.401

Answers to Your Turn 1
a. Estimate: 21
   Actual: 21.138
b. Estimate: 0.2
   Actual: 0.2123
c. Estimate: 30.7
   Actual: 30.651
d. Estimate: 815.5
   Actual: 815.46

b. $179 - 48.165$

Solution:

Estimate:
$$\begin{array}{r} 180 \\ -\ 50 \\ \hline 130 \end{array}$$

Actual:
$$\begin{array}{r} \overset{8\ \ 9\,9\,10}{179.000} \\ -\ 48.165 \\ \hline 130.835 \end{array}$$

◄ **Note** Because 179 is an integer, the decimal point is understood to be to the right of the 9 digit. We can write as many 0s to the right of the decimal point as we like.

c. $98.036 - 9.482$

Solution:

Estimate:
$$\begin{array}{r} \overset{8\,18}{98} \\ -\ 9 \\ \hline 89 \end{array}$$

Actual:
$$\begin{array}{r} \overset{17}{\overset{8\ 7\ \ 9\,13}{98.036}} \\ -\ \ 9.482 \\ \hline 88.554 \end{array}$$

► **Do Your Turn 2**

► **Your Turn 2**

Estimate. Then find the actual difference.

a. $146.79 - 35.14$

b. $809 - 162.648$

c. $0.80075 - 0.06791$

d. $10.302 - 9.8457$

Objective 3 Add and subtract signed decimal numbers.

To add or subtract signed decimal numbers, we follow the same rules we developed in Chapter 2 for integers.

Example 3 | Add.

a. $-15.79 + 8.4$

Solution: Because the two numbers have different signs, we subtract their absolute values and keep the sign of the number with the greater absolute value.

Note To subtract, stack the ► number with the greater absolute value on top and align the decimal points.

$$\begin{array}{r} \overset{0\,15}{1\cancel{5}.79} \\ -\ 8.40 \\ \hline 7.39 \end{array}$$

◄ **Note** To avoid confusion in the calculation, we left out the negative signs for -15.79 and -7.39. Although it is okay to leave signs out of a calculation, we must be careful to write them properly in our answer statement.

Answer: -7.39 Because -15.79 has the greater absolute value, our sum is negative.

Note In terms of credits and debts, we have a debt of $15.79 and make a payment of $8.40. Because our payment is not enough to pay off the debt, we still have a debt of $7.39.

b. $-68.95 + (-14.50)$

Solution: Because the two numbers have the same sign, we add and keep the same sign.

$$\begin{array}{r} \overset{1\ 1}{68.95} \\ +\ 14.50 \\ \hline 83.45 \end{array}$$

◄ **Note** Again, to avoid confusion, we left the negative signs out of the calculation, then put them back in the answer statement.

Answer: -83.45 Because both addends are negative, the sum is negative.

Note In terms of credits and debts, we have a debt of $68.95 and add another debt of $14.50; so we are increasing our debt to $83.45.

► **Do Your Turn 3**

► **Your Turn 3**

Add.

a. $90.56 + (-59.8)$

b. $-104 + (-54.88)$

c. $-5.14 + 0.75 + (-4.50) + 2.95$

d. $54.6 + (-14.50) + (-6.78) + (-1.3)$

Answers to Your Turn 2

a. Estimate: 110
   Actual: 111.65

b. Estimate: 600
   Actual: 646.352

c. Estimate: 0.73
   Actual: 0.73284

d. Estimate: 0
   Actual: 0.4563

Answers to Your Turn 3

a. 30.76

b. -158.88

c. -5.94

d. 32.02

Recall from Chapter 2 that to subtract signed numbers, we can write an equivalent addition statement. To write an equivalent addition, we change the subtraction sign to an addition sign and change the subtrahend to its additive inverse.

▶ **Your Turn 4**

Subtract.

  a. $1.987 - 10.002$

  b. $4.1 - (-0.951)$

  c. $-213.4 - 90.9$

  d. $-0.017 - (-0.056)$

Example 4   Subtract.

  a. $0.08 - 5$

Solution: Write an equivalent addition statement.

Change the minus sign to a plus sign. $\underline{\qquad 0.08 - 5 \qquad}$ Change the subtrahend to its additive inverse.

$$= 0.08 + (-5)$$

Because we are adding numbers with different signs, we subtract their absolute values and keep the sign of the number with the greater absolute value.

$$\begin{array}{r} \overset{4\ 9\,10}{5.\cancel{0}\cancel{0}} \\ -0.08 \\ \hline 4.92 \end{array}$$ Subtract absolute values.

Answer: -4.92 In the equivalent addition, -5 has the greater absolute value; so the result is negative.

> **Note** In terms of credits and debts, we have $0.08 in a bank account (asset) and write a check for $5. This means that we are overdrawn by $4.92.

  b. $-16.2 - (-9.07)$

Solution: $-16.2 - (-9.07) = -16.2 + 9.07$ Write the equivalent addition statement.

Because we are adding two numbers with different signs, we subtract their absolute values and keep the sign of the number with the greater absolute value.

$$\begin{array}{r} \overset{0\,16\ \ 1\,10}{1\cancel{6}.\cancel{2}\cancel{0}} \\ -9.07 \\ \hline 7.13 \end{array}$$ Subtract absolute values.

Answer: -7.13 In the equivalent addition, -16.2 has the greater absolute value; so the result is negative.

◀ **Do Your Turn 4**

▶ **Your Turn 5**

Simplify.

  a. $5.2x^2 - 9.87x + 15x^2 - 2.35x - 10x^3 + 1.6$

  b. $4.2a^4 - 2.67a + 1 + 15.7a^2 + a^4 - 5.1 - 3.9a$

Objective 4 Simplify, add, or subtract polynomials containing decimal numbers.

In Section 3.2, we learned that we can simplify expressions by combining like terms. Recall that like terms have the same variables raised to the same exponents. To combine like terms, add the coefficients and keep the variables and their exponents the same.

Example 5   Simplify. $4.2n^2 - 6.1n + 9 + 2.35n - 10n^2 + 2.09$

Solution: Combine like terms.

> **Note** Remember, collecting like terms is an optional step.

$$4.2n^2 - 6.1n + 9 + 2.35n - 10n^2 + 2.09$$

$$= \underbrace{4.2n^2 - 10n^2} - \underbrace{6.1n + 2.35n} + \underbrace{9 + 2.09}$$ Collect like terms.

$$= \quad -5.8n^2 \qquad\qquad -3.75n \qquad +11.09$$ Combine like terms.

◀ **Do Your Turn 5**

Answers to Your Turn 4

a. -8.015

b. 5.051

c. -304.3

d. 0.039

Answers to Your Turn 5

a. $-10x^3 + 20.2x^2 - 12.22x + 1.6$

b. $5.2a^4 + 15.7a^2 - 6.57a - 4.1$

In Section 3.3, we learned how to add and subtract polynomials. To add polynomials, we combine like terms. To subtract, we write an equivalent addition expression using the additive inverse of the subtrahend, which is the second polynomial.

Example 6 Add or subtract.

a. $(8.2y^3 - 6.1y^2 + 10.65) + (y^3 - y - 12)$

Solution: Combine like terms.

$$(8.2y^3 - 6.1y^2 + 10.65) + (y^3 - y - 12)$$
$$= \underbrace{8.2y^3 + y^3}\ - 6.1y^2 - y + \underbrace{10.65 - 12} \qquad \text{Collect like terms.}$$
$$= \quad 9.2y^3 \quad\ -6.1y^2 - y \qquad\quad -1.35 \qquad \text{Combine like terms.}$$

Note Remember, if a term has no visible coefficient, as in y^3, the coefficient is understood to be 1; so $8.2y^3 + y^3 = 9.2y^3$. Also, because $-6.1y^2$ and $-y$ have no like terms, we write them unchanged in the result.

b. $(19.15x^2 + 8x - 7) - (10.1x^2 + x - 9.1)$

Solution: Write the subtraction as an equivalent addition. Then combine like terms.

$$(19.15x^2 + 8x - 7) - (10.1x^2 + x - 9.1)$$
$$= (19.15x^2 + 8x - 7) + (-10.1x^2 - x + 9.1) \qquad \text{Write an equivalent addition.}$$
$$= \underbrace{19.15x^2 - 10.1x^2}\ + \underbrace{8x - x}\ \underbrace{-7 + 9.1} \qquad \text{Collect like terms.}$$
$$= \quad 9.05x^2 \qquad\quad +7x \qquad +2.1 \qquad \text{Combine like terms.}$$

▶ **Do Your Turn 6**

▶ **Your Turn 6**

Add or subtract.

a. $(t^3 - 4.15t^2 + 7.3t - 16) + (0.4t^3 + 3.9t^2 + 1.73)$

b. $(8.1m^4 + 2.78m^2 - 5.15m + 10.3n) + (1.7m^4 - 5.2m^2 - 3.1m - 15n)$

c. $(15.7n^2 + 0.2n - 1.8) - (2.17n^2 + 0.97)$

d. $(0.14x^2 + 2.9x + 4.8) - (x^2 - 9.35x + 12.8)$

Note Remember, to get the additive inverse of a polynomial, change the sign of each term in the polynomial; so the additive inverse of $10.1x^2 + x - 9.1$ is $-10.1x^2 - x + 9.1$.

Objective 5 Solve equations using the addition principle.

In Section 4.2, we learned to solve equations using the addition principle of equality. The addition principle states that the same number may be added or subtracted on both sides of an equation without affecting its solution(s).

Example 7 Solve and check. $n + 4.78 = -5.62$

Solution:

Note Because $+4.78$ and -4.78 are additive inverses, their sum is 0.

$$n + 4.78 = -5.62$$
$$\underline{-4.78 \qquad -4.78} \qquad \text{Subtract 4.78 from both sides.}$$
$$n + \quad 0 = -10.40$$
$$n = -10.4$$

Check:

$$n + 4.78 = -5.62$$
$$-10.4 + 4.78 \overset{?}{=} -5.62 \qquad \text{In the original equation, replace } n \text{ with } -10.4.$$
$$-5.62 = -5.62 \qquad \text{True; so } -10.4 \text{ is the solution.}$$

▶ **Do Your Turn 7**

▶ **Your Turn 7**

Solve and check.

a. $-10.15 + x = 2.8$

b. $-14.6 = y - 2.99$

Objective 6 Solve applications.

Many applications of adding and subtracting decimals can be found in the field of accounting. We developed some principles of accounting in Chapters 1 and 2. Let's extend those principles to decimal numbers.

Answers to Your Turn 6

a. $1.4t^3 - 0.25t^2 + 7.3t - 14.27$

b. $9.8m^4 - 2.42m^2 - 8.25m - 4.7n$

c. $13.53n^2 + 0.2n - 2.77$

d. $-0.86x^2 + 12.25x - 8$

Answers to Your Turn 7

a. $x = 12.95$

b. $y = -11.61$

▶ **Your Turn 8**

Solve.

a. The Jacksons will be closing on their new house in a few days. Below is a summary of their debts and credits for the closing. What should be the amount of the check they bring to closing?

Gross amount from borrower:
(Debits)
Contract Sales Price   89,980.00
Settlement Charges   1,025.12

Amount paid by the borrower:
(Credits)
Earnest Money   1,000.00
Principle Amount of
    New Loan(s)   85,300.00
Portion of Appraisal   125.10
County Taxes   188.36

b. The closing price of a stock one day was $25.50. The next day the closing price of the same stock was $28.00. Let n represent the amount the stock's closing price changed. Write an equation that can be used to find the amount the closing price changed; then solve the equation.

c. The perimeter of the following figure is 84.5 meters. Write an equation that can be used to find the unknown side length; then solve the equation.

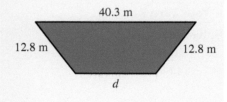
40.3 m
12.8 m   12.8 m
d

Example 8 Solve.

a. Find the final balance in the following checkbook register.

Date	No.	Transaction Description	Debits	Credits	Balance
3 - 30		Deposit		1254 88	1380 20
3 - 30	2341	National Mortgage	756 48		
3 - 30	2342	Electric & Gas	125 56		
3 - 30	2343	Bell South	43 70		
3 - 30	2344	Visa	150 00		
3 - 30	2345	Domino's	12 50		
4 - 01		Transfer from savings		200 00	

Solution: Subtract the sum of all check amounts (debits) from the initial balance and then add the $200 transfer (credit).

$$\text{Balance} = \text{initial balance} - \text{sum of all debits} + \text{credit}$$

$$B = 1380.20 - (756.48 + 125.56 + 43.70 + 150.00 + 12.50) + 200.00$$

$$B = 1380.20 - 1088.24 + 200.00$$

$$B = 291.96 + 200.00$$

$$B = 491.96$$

Answer: The final balance is $491.96.

b. The closing price of a stock one day was $40.00. The next day the closing price of the same stock was $36.50. Let n represent the amount the stock's closing price changed. Write an equation that can be used to find the amount the closing price changed; then solve the equation.

Solution: The price on any given day is determined by adding the amount of change to the previous day's closing price.

$$\text{Previous closing price} + \text{amount of change} = \text{current closing price}$$

Equation: $40.00 + n = 36.50$

Solution: $40.00 + n = 36.50$

     $\underline{-40.00}$ $\underline{-40.00}$ **Subtract 40.00 from both sides.**

                 $n = -3.50$

Answer: The closing price decreased $3.50.

◀ **Do Your Turn 8**

Answer to Your Turn 8
a. $4391.66   **b.** $25.50 + n = 28.00$; $2.50 increase
c. $12.8 + 40.3 + 12.8 + d = 84.5$; 18.6 m

6.2 Exercises   (For Extra Help)   MyMathLab®

Objective 1

Prep Exercise 1 To add decimal numbers, stack the numbers so that the place values _____ .

Prep Exercise 2 Place the decimal point correctly in the sum shown.

    24.5
$\underline{+ 2.81}$
    2731

For Exercises 1–6, estimate. Then find the actual sum.

1. $34.51 + 125.2$

2. $9.167 + 12.32$

3. $58.915 + 0.8361$

4. $0.5082 + 1.946$

5. $312.98 + 6.337 + 14$

6. $0.183 + 75 + 245.9$

Objective 2

Prep Exercise 3 When stacking two decimal numbers to subtract them, which of the two numbers is placed on top?

Prep Exercise 4 Place the decimal point correctly in the difference shown.

$$\begin{array}{r} 497.6 \\ -\ \ 34.2 \\ \hline 4634 \end{array}$$

For Exercises 7–12, estimate. Then find the actual difference.

7. $864.55 - 23.1$

8. $1964.86 - 213.5$

9. $809.06 - 24.78$

10. $6.002 - 4.135$

11. $0.1005 - 0.08261$

12. $5.0002 - 0.19873$

Objective 3

For Exercises 13–24, add or subtract.

13. $25 + (-14.89)$

14. $-20 + 8.007$

15. $-0.16 + (-4.157)$

16. $-7.09 + (-10.218)$

17. $0.0015 - 1$

18. $0.0209 - 1$

19. $-30.75 - 159.27$

20. $-980.4 - 19.62$

21. $-90 - (-16.75)$

22. $-100 - (-63.912)$

23. $-0.008 - (-1.0032)$

24. $-0.005 - (-2.0031)$

For Exercises 25–28, simplify.

25. $5y - 2.81y^2 + 7.1 - 10y^2 - 8.03$

26. $0.8x^3 - 7x + 2.99 - 1.2x - 9 + 0.91x^3$

27. $a^3 - 2.5a + 9a^2 + a - 12 + 3.5a^3 + 1.4$

28. $4.2m^2 + m^3 - 9m + 0.6 - 1.02m - m^3 + 0.5$

Objective 4

For Exercises 29–38, add or subtract.

29. $(x^2 + 5.2x + 3.4) + (9.2x^2 - 6.1x + 5)$

30. $(0.4y^3 - 8.2y + 2.1) + (3.4y^3 - y^2 - 3)$

31. $(3.1a^3 - a^2 + 7.5a - 0.01) + (6.91a^3 - 3.91a^2 + 6)$

32. $(20.5m^4 + 16m^2 - 18.3m - 0.77) + (0.9m^4 - 2.08m^2 + 1)$

33. $(1.4n^2 + 8.3n - 0.6) - (n^2 - 2.5n + 3)$

34. $(3.1y^3 - y^2 + 0.7) - (2.3y^3 + 1.7y^2 - 4)$

35. $(5k^3 + 2.5k^2 - 6.2k - 0.44) - (2.2k^3 - 6.2k - 0.5)$

36. $(1.2n^3 - 7.1n + 0.4) - (3.1n^3 - 6n^2 + 8.9n + 1.3)$

37. $(12.91x^3 - 16.2x^2 - x + 7) - (10x^3 - 5.1x^2 - 3.6x + 11.45)$

38. $(6a^4 - 3.1a^2 + 1.2a - 1) - (5.4a^4 + 1.7a^2 - 2.9a - 1.32)$

Objective 5

For Exercises 39–44, solve and check.

39. $m + 3.67 = 14.5$

40. $x - 5.8 = 4.11$

41. $2.19 = 8 + y$

42. $6.09 = n + 7.1$

43. $k - 4.8 = -8.02$

44. $-16.7 = h - 7.59$

Objective 6

For Exercises 45–58, solve.

45. Following is Adam's checkbook register. Find his final balance.

Date	No.	Transaction Description	Debits		Credits		Balance	
2-24		Deposit			428	45	682	20
2-25	952	Food Lion	34	58				
2-25	953	Cash	30	00				
2-25	954	Green Hills Apartments	560	00				
2-25	955	Electric & Gas	58	85				
2-26	956	American Express	45	60				

46. Following is Jada's checkbook register. Find her balance.

Date	No.	Transaction Description	Debits		Credits		Balance	
10-26	1011	Sears	36	70			250	20
10-27	1012	Frank's Costumes	25	00				
10-30	1013	Credit Union Mortgage	858	48				
10-30	1014	Electric Co.	119	45				
10-30	1015	Water Co.	17	50				
10-30	1016	Food Mart	84	95				
10-31		Deposit			1355	98		

47. Tim runs a landscaping company. Following is a balance sheet that shows the charges for a client. What is the client's final balance?

Charges		Payments Received	$200
Trencher rental	$75.00		
PVC pipe	$15.50		
Sprinkler heads	$485.65		
Back flow prevention	$45.79		

48. Following is a settlement sheet for the Smiths, who will be closing on their new house in a few days. What should be the amount of the check they bring to closing?

Meyer, Lewis, Callahan, Garrity and Pederson, Esq.
One Financial Center
Boston, Massachusetts 02111

700 Philadelphia Avenue
Washington, D.C. 20004
Stephen L. Meyer

Direct Dial Number
617.555.0000

October 23, 2010

Gross amount from borrower: (Debts)		Amounts paid by the borrower: (Credits)	
		Earnest Money	$2000
Contract Sales Price	$179,900	Principal Amount of	
Settlement Charges	$2054.75	New Loan(s)	$161,000
		Portion of Appraisal	$245.50
		County Taxes	$359.80

49. Following is Catherine's receipt from the grocery store. If she gave the cashier $20, how much did she get back in change?

Milk	$1.59
Yogurt	$2.49
Bread	$1.79
Soup	$0.79
Cereal	$3.45
Tax	$0.51

50. Following is Selina's receipt from a clothing store. She has a $50 bill. Is this enough? If not, how much more does she need?

Tank T	$12.95
Jeans	$35.95
Tax	$ 2.45

51. Maria gets paid once a month. Following is a list of deductions from her paycheck. Find Maria's net monthly pay.

CURRENT GROSS:	2904.17		
Taxes		Other Deductions	
FICA	174.25	MED SPEND	45.00
MEDICARE	39.64	DENT	12.74
FED W/H	235.62	BLUE CROSS	138.15
STATE W/H	131.56	401K	200.00
		STD LIFE	7.32

52. Tony gets paid twice a month. Following is a list of deductions from his paycheck. Find Tony's net pay.

CURRENT GROSS:	1067.08		
Taxes		Other Deductions	
FICA	62.27	HMO BLUE	210.96
MEDICARE	14.57	DENT	7.54
FED W/H	128.05	STATE RET	64.02
STATE W/H	53.35	STATE OP LIFE	6.45

53. The closing price of a stock one day was $35.25. The next day the closing price of the same stock was $32.50. Let n represent the amount the stock's closing price changed. Write an equation that can be used to find the amount the closing price changed; then solve the equation.

54. The closing price of a stock one day was $27.90. The next day the closing price of the same stock was $30.45. Let n represent the amount the stock's closing price changed. Write an equation that can be used to find the amount the closing price changed; then solve the equation.

55. The perimeter of the following trapezoid is 96.1 meters. Write an equation to find the unknown side length. Then solve the equation.

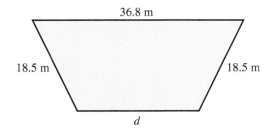

56. The perimeter of the following triangle is 6.9 miles. Write an equation to find the unknown side length. Then solve the equation.

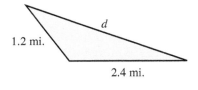

57. Use the following line graph, which shows the closing price of Hewlett-Packard stock each day from July 19, 2010, to July 23, 2010.

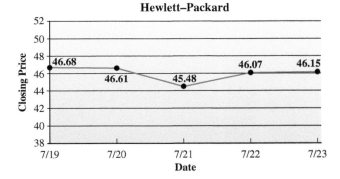

a. Complete the following table. In the row labeled "Equation," for each day, write an equation that can be used to find the amount Hewlett-Packard's stock closing price changed from the previous day's closing price. Use the variable n to represent the change in closing price. Solve the equation and write the amount the price changed in the row labeled "Price change."

	July 20	July 21	July 22	July 23
Equation				
Price change				

b. Which day had the greatest change in closing price?

58. Use the following line graph, which shows the closing price of Whole Foods Market stock each day from August 16, 2010, to August 20, 2010.

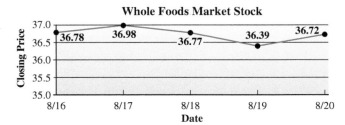

a. Complete the following table. In the row labeled "Equation," for each day, write an equation that can be used to find the amount Whole Food Market's stock closing price changed from the previous day's closing price. Use the variable n to represent the change in closing price. Solve the equation and write the amount the price changed in the row labeled "Price change."

	August 17	August 18	August 19	August 20
Equation				
Price change				

b. Which day had the greatest change in closing price?

Puzzle Problem At a bookstore, a calculator and a pen cost a total of $15.50. The calculator costs $15 more than the pen. What is the cost of each item?

Review Exercises

For Exercises 1 and 2, multiply.

[2.4] **1.** $-12 \cdot 25$

[5.3] **2.** $\dfrac{3}{10} \cdot \dfrac{7}{10}$

For Exercises 3 and 4, divide.

[1.4] **3.** $10{,}404 \div 34$

[5.4] **4.** $9\dfrac{1}{6} \div 3\dfrac{1}{4}$

[2.4] **5.** Evaluate. $\sqrt{121}$

For Exercises 6 and 7, multiply.

[3.4] **6.** $-2x^2 \cdot 9x^5$

[3.4] **7.** $(y + 3)(y - 5)$

$\begin{bmatrix} 2.4 \\ 4.3 \end{bmatrix}$ **8.** Solve and check. $-8x = 72$

6.3 Multiplying Decimal Numbers; Exponents with Decimal Bases

Objectives

1 Multiply decimal numbers.
2 Multiply signed decimal numbers.
3 Evaluate exponential forms with decimal bases.
4 Write a number in scientific notation in standard form.
5 Write standard form numbers in scientific notation.
6 Multiply monomials.
7 Multiply polynomials.
8 Solve applications.

Objective 1 Multiply decimal numbers.

Because decimal numbers are base-10 numbers, we should be able to multiply them in the same way we multiply whole numbers. But we must be careful to place the decimal point correctly in the product.

Consider $(0.3)(0.7)$. Because we know how to multiply fractions, let's write these decimal numbers as equivalent fractions to see what we can discover about placing the decimal point in the product.

$$(0.3)(0.7) = \frac{3}{10} \cdot \frac{7}{10}$$

◄ **Note** 0.3 is equivalent to $\frac{3}{10}$, and 0.7 is equivalent to $\frac{7}{10}$.

$$= \frac{21}{100}$$

Note There are two decimal places in the product because we multiplied tenths times tenths to create hundredths.

$$= 0.21$$ ◄

What if the problem were $0.03 \cdot 0.7$?

$$0.03 \cdot 0.7 = \frac{3}{100} \cdot \frac{7}{10}$$

◄ **Note** 0.03 is equivalent to $\frac{3}{100}$.

$$= \frac{21}{1000}$$

Note There are now three decimal places in the product because we multiplied hundredths times tenths to create thousandths.

$$= 0.021$$ ◄

Our examples suggest the following conclusion and procedure:

Conclusion: Because the numerators of the fraction equivalents are whole numbers, when we multiply decimal numbers, we can multiply the decimal digits just as we would multiply whole numbers. Also, because we multiply the denominators, the number of decimal places in the product is equal to the total number of decimal places in the factors.

Procedure

To multiply decimal numbers:

1. Multiply as if the decimal numbers were whole numbers.
2. Place the decimal point in the product so that it has the same number of decimal places as the total number of decimal places in the factors.

Note When multiplying decimal numbers, we do not need to align the decimal points because the process does not depend on decimal alignment.

Example 1 Estimate. Then find the actual product.

a. $(2.31)(1.7)$

Solution:

Estimate: Use the same procedure as for estimating a product of whole numbers.

$$
\begin{array}{r}
2.31 \rightarrow \quad 2 \\
\times \ 1.7 \rightarrow \times 2 \\
\hline
4
\end{array}
$$

Round *each* number so that it has one nonzero digit. Then multiply the rounded numbers.

▶ **Your Turn 1**

Estimate. Then find the actual product.

a. $2.04 \cdot 0.09$ b. $1.005 \cdot 34.6$
c. $6.8 \cdot 12.5$ d. $15 \cdot 0.007$

Actual:
$$
\begin{array}{r}
\overset{2}{2.31} \leftarrow \text{2 decimal places} \\
\times\ 1.7 \leftarrow +\text{1 decimal place} \\
\hline
1617 \\
+\ 231 \\
\hline
3.927 \leftarrow \text{3 decimal places}
\end{array}
$$

Note The estimate helps check our decimal placement. Our estimate, 4, is reasonably close to the actual product, 3.927. Incorrect decimal placements such as 39.27 and 0.3927 disagree with the estimate.

b. $(3.04)(0.35)$

Solution:

Estimate:

$$
\begin{array}{r}
3.04 \rightarrow 3 \leftarrow \text{0 decimal places} \\
\times\ 0.35 \rightarrow 0.4 \leftarrow +\text{1 decimal place} \\
\hline
1.2 \leftarrow \text{1 decimal place}
\end{array}
$$

Actual:
$$
\begin{array}{r}
\overset{1}{\overset{2}{3.04}} \leftarrow \text{2 decimal places} \\
\times\ 0.35 \leftarrow +\text{2 decimal places} \\
\hline
1520 \\
+\ 912 \\
\hline
1.0640 \leftarrow \text{4 decimal places}
\end{array}
$$

Answer: 1.064   ◀ **Note** After placing the decimal point, 0 digits to the right of the last nonzero decimal digit can be deleted. We can see why using fractions.

$$(3.04)(0.35) = \frac{304}{100} \cdot \frac{35}{100} = \frac{10{,}640}{10{,}000} = \frac{1064}{1000} = 1.064$$

Dividing out the common factor of 10 deletes the last 0 in the numerator and denominator.

◀ **Do Your Turn 1**

Connection

$$
\begin{aligned}
(2.31)(1.7) &= \left(2\tfrac{31}{100}\right)\left(1\tfrac{7}{10}\right) \\
&= \tfrac{231}{100} \cdot \tfrac{17}{10} \\
&= \tfrac{3927}{1000} \\
&= 3\tfrac{927}{1000} \\
&= 3.927
\end{aligned}
$$

Adding the total number of places in the factors corresponds to adding exponents of the powers of 10 in the denominators. We add the exponents because we are multiplying exponential forms that have the same base.

$$\frac{231}{100} \cdot \frac{17}{10} = \frac{231}{10^2} \cdot \frac{17}{10^1} = \frac{3927}{10^{2+1}} = \frac{3927}{10^3} = 3.927$$

Number of decimal places in 2.31	Number of decimal places in 1.7	Number of decimal places in the product, 3.927

Objective 2 Multiply signed decimal numbers.

Recall that the product of two numbers that have the same sign is positive and the product of two numbers that have different signs is negative.

Answers to Your Turn 1

a. Estimate: 0.18
   Actual: 0.1836
b. Estimate: 30
   Actual: 34.773
c. Estimate: 70
   Actual: 85
d. Estimate: 0.14
   Actual: 0.105

Example 2 Multiply. $(0.09)(-0.051)$

Solution:

Note To avoid confusion, we took out the negative signs while calculating but put them back in the answer statement.

$$
\begin{array}{r}
0.051 \leftarrow \quad \text{3 decimal places} \\
\times \quad 0.09 \leftarrow \quad \underline{+2} \text{ decimal places} \\
\hline
0459 \\
0000 \\
+0000 \\
\hline
0.00459 \leftarrow \quad \text{5 decimal places}
\end{array}
$$

Answer: -0.00459 Because the two factors have different signs, the product is negative.

► **Do Your Turn 2**

Your Turn 2

Multiply.
  a. $-1.5 \cdot 2.91$
  b. $(-6.45)(-8)$

Objective 3 Evaluate exponential forms with decimal bases.

Because an exponent indicates to multiply the base repeatedly, we must be careful how we place the decimal in the result.

Example 3 Evaluate. $(-0.04)^3$

Solution: Use -0.04 as a factor three times.

$$(-0.04)^3 = (-0.04)(-0.04)(-0.04)$$
$$= -0.000064$$

Note The product is negative because we have an odd number of negative factors.

Note We can quickly determine the number of decimal places in the product by multiplying the number of decimal places in the base by the exponent.

$$(-0.04)^3 = -0.000064$$

number of · exponent = number of
decimal   decimal
places in   places in the
the base   product
  2   ·   3   =   6

Answer: -0.000064

► **Do Your Turn 3**

Your Turn 3

Evaluate.
  a. $(0.008)^2$
  b. $(0.32)^0$
  c. $(-0.12)^2$
  d. $(-0.6)^3$

Objective 4 Write a number in scientific notation in standard form.

Scientists, mathematicians, and engineers often work with very large numbers. When these very large numbers have many 0 digits, it is convenient to express the numbers in **scientific notation.**

Definition Scientific notation: A notation in the form $a \times 10^n$ where a is a decimal number whose absolute value is greater than or equal to 1 but less than 10 and n is an integer.

The number 2.37×10^4 is in scientific notation because the absolute value of 2.37 is 2.37, which is greater than or equal to 1 but less than 10, and the power of 10 is 4, which is an integer. The number 42.5×10^3 is not in scientific notation because the absolute value of 42.5 is greater than 10.

Answers to Your Turn 2
a. -4.365 **b.** 51.6

Answers to Your Turn 3
a. 0.000064   **b.** 1   **c.** 0.0144
d. -0.216

What does 2.37×10^4 mean?

$$2.37 \times 10^4 = 2.37 \times 10 \times 10 \times 10 \times 10$$

2.37	23.7	237	2370
$\times\ 10$	$\times\ 10$	$\times\ 10$	$\times\ 10$
000	000	000	0000
$+\ 237$	$+\ 237$	$+\ 237$	$+\ 2370$
23.70	237.0	2370	23,700

Notice that each multiplication by 10 moves the decimal point one place to the right. Because we multiply by four factors of 10, the decimal point moves a total of four places to the right from where it started in 2.37.

Conclusion: The number of 10s is determined by the exponent; so the exponent actually determines the number of places the decimal point moves.

Procedure

To change from scientific notation with a positive integer exponent to standard notation for a given number, move the decimal point to the right the number of places indicated by the exponent.

Connection Because 10^6 is 1,000,000, the number 8.23×10^6 is often read as 8.23 million. Similarly, 10^9 is 1,000,000,000; so -5.1782×10^9 is often read as -5.1782 billion.

▶ **Your Turn 4**

Write each number in standard form; then write the word name.

a. 7.8×10^3

b. 9.102×10^9

c. 4.1518×10^{12}

d. 5.4×10^7

Answers to Your Turn 4

a. 7800; seven thousand, eight hundred

b. 9,102,000,000; nine billion, one hundred two million

c. 4,151,800,000,000; four trillion, one hundred fifty-one billion, eight hundred million

d. 54,000,000; fifty-four million

Example 4   Write the number in standard form; then write its word name.

a. 8.23×10^6

Solution: Multiplying 8.23 by 10^6 causes the decimal point to move six places to the right from its original position in 8.23.

$$8.23 \times 10^6 = 8\,2\,3\,0\,0\,0\,0$$
$$1\,2\,3\,4\,5\,6$$

◀ **Note** Moving two of the six places puts the decimal point after the 23; so we must place four 0s after the 3 digit to account for all six places.

$$= 8{,}230{,}000$$

Word name: Eight million, two hundred thirty thousand

b. -5.1782×10^9

Solution: Multiplying -5.1782 by 10^9 causes the decimal point to move nine places to the right from its original position in -5.1782.

$$-5.1782 \times 10^9 = -5\,1\,7\,8\,2\,0\,0\,0\,0\,0$$
$$1\,2\,3\,4\,5\,6\,7\,8\,9$$

◀ **Note** Moving four of the nine places puts the decimal point after the 1782; so we must place five 0s after the 2 digit to account for all nine places.

$$= -5{,}178{,}200{,}000$$

Word name: Negative five billion, one hundred seventy-eight million, two hundred thousand.

◀ **Do Your Turn 4**

Objective 5 Write standard form numbers in scientific notation.

Suppose we want to write 62,910,000 in scientific notation. First, we must determine the decimal position.

$$62{,}910{,}000$$
$$\phantom{62{,}9}\uparrow$$

The decimal must go between the 6 and 2 to express a number whose absolute value is greater than or equal to 1 but less than 10.

Next, we determine the power of 10.

$$62{,}910{,}000 = 6.2910000 \times 10^7$$

Because there are seven places between the new decimal position and the original decimal position, the power of 10 is 7.

Finally, we can simplify by eliminating unnecessary 0s.

$$62{,}910{,}000 = 6.2910000 \times 10^7 = 6.291 \times 10^7$$

Because 0s to the right of the last nonzero decimal digit are not needed, we can delete these four 0s.

Our example suggests the following procedure:

Procedure

To write a standard form number whose absolute value is greater than 1 in scientific notation:

1. Move the decimal point so that the number's absolute value is greater than or equal to 1 but less than 10. (Place the decimal point to the right of the first nonzero digit.)
2. Write the decimal number multiplied by 10^n where n is the number of places between the new decimal position and the original decimal position.
3. Delete 0s to the right of the last nonzero digit.

Example 5 Write each number in scientific notation.

a. 417,000,000,000

Solution:

$$417{,}000{,}000{,}000 = 4.17 \times 10^{11}$$

Move the decimal here to express a number whose absolute value is greater than or equal to 1 but less than 10.

There are 11 places between the new decimal position and the original decimal position. After placing the decimal point, the nine 0s can be deleted.

b. −2,917,300,000

Solution:

$$-2{,}917{,}300{,}000 = -2.9173 \times 10^9$$

Move the decimal here to express a number whose absolute value is greater than or equal to 1 but less than 10.

There are 9 places between the new decimal position and the original decimal position. After placing the decimal point, the five 0s can be deleted.

▶ Do Your Turn 5

Objective 6 Multiply monomials.

We learned how to multiply monomials in Section 3.4. Recall that we multiply the coefficients and add exponents of the like variables.

▶ **Your Turn 5**

Write each number in scientific notation.

a. 91,000
b. 108,300,000
c. −6,390,000
d. −40,913,000,000,000

Answers to Your Turn 5
a. 9.1×10^4
b. 1.083×10^8
c. -6.39×10^6
d. -4.0913×10^{13}

▶ **Your Turn 6**

Multiply.
a. $(1.7n^3)(6n^2)$
b. $(-4.03a)(-2.1a^3)$

Example 6 Multiply.

a. $(2.1y^7)(3.43y^8)$

Solution:

$$(2.1y^7)(3.43y^8) = 2.1 \cdot 3.43 \cdot y^{7+8}$$ Multiply the coefficients and add the exponents of the like variables.

$$= 7.203y^{15}$$

Connection Multiplying monomials is like multiplying numbers expressed in scientific notation. For example, $(2.1y^7)(3.43y^8)$ corresponds to $(2.1 \times 10^7)(3.43 \times 10^8)$.

$$(2.1y^7)(3.43y^8) \qquad (2.1 \times 10^7)(3.43 \times 10^8)$$
$$= 2.1 \cdot 3.43 \cdot y^{7+8} \qquad = 2.1 \times 3.43 \times 10^{7+8}$$
$$= 7.203y^{15} \qquad = 7.203 \times 10^{15}$$

b. $(6.13x^2)(-4.5x^4)$

Solution:

$$(6.13x^2)(-4.5x^4) = 6.13 \cdot (-4.5) \cdot x^{2+4}$$ Multiply the coefficients and add the exponents of the like variables.

$$= -27.585x^6$$

◀ Do Your Turn 6

In Section 3.4, we learned how to simplify monomials raised to a power. We evaluate the coefficient raised to the power and multiply each variable's exponent by the power.

▶ **Your Turn 7**

Simplify.
a. $(1.2y^4)^2$
b. $(-0.5m^3)^3$

Example 7 Simplify. $(0.2a^2)^3$

Solution:

$$(0.2a^2)^3 = (0.2)^3 a^{2\cdot3}$$ Evaluate the coefficient, 0.2, raised to the power, 3, and multiply the variable's exponent, 2, by the power, 3.

$$= 0.008a^6$$

◀ Do Your Turn 7

Objective 7 Multiply polynomials.

In Section 3.4, we learned to multiply polynomials using the distributive property.

Example 8 Multiply.

a. $4.2x^2(5x^3 + 9x - 7)$

Solution:

$$4.2x^2(5x^3 + 9x - 7)$$

$$= 4.2x^2 \cdot 5x^3 + 4.2x^2 \cdot 9x - 4.2x^2 \cdot 7$$ Using the distributive property, multiply each term in $5x^3 + 9x - 7$ by $4.2x^2$.

$$= 21x^5 + 37.8x^3 - 29.4x^2$$ Simplify.

Answers to Your Turn 6
a. $10.2n^5$ b. $8.463a^4$

Answers to Your Turn 7
a. $1.44y^8$ b. $-0.125m^9$

b. $(3.1n - 2)(1.7n + 6.5)$

Solution:

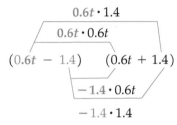

$$= 3.1n \cdot 1.7n + 3.1n \cdot 6.5 - 2 \cdot 1.7n - 2 \cdot 6.5 \qquad \text{Using the distributive property, multiply each term in } 1.7n + 6.5 \text{ by each term in } 3.1n - 2 \text{ (FOIL).}$$

$$= 5.27n^2 + 20.15n - 3.4n - 13 \qquad \text{Simplify.}$$

$$= 5.27n^2 + 16.75n - 13 \qquad \text{Combine like terms.}$$

c. $(0.6t - 1.4)(0.6t + 1.4)$

Solution:

$$= 0.6t \cdot 0.6t + 0.6t \cdot 1.4 - 1.4 \cdot 0.6t - 1.4 \cdot 1.4 \qquad \text{Using the distributive property, multiply each term in } 0.6t + 1.4 \text{ by each term in } 0.6t - 1.4 \text{ (use FOIL).}$$

$$= 0.36t^2 + 0.84t - 0.84t - 1.96 \qquad \text{Simplify.}$$

$$= 0.36t^2 - 1.96 \qquad \text{Combine like terms.}$$

▶ **Do Your Turn 8**

Objective 8 Solve applications.

A common situation involving multiplication is when we purchase or sell more than one unit of a product at the same price, which is called its **unit price.**

Definition Unit price: The price of each unit.

For example, suppose gasoline is sold at a unit price of $2.89 per gallon, which means that we pay $2.89 for every 1 gallon of gasoline that we purchase. The unit price can be expressed as a fraction.

$$\$2.89 \text{ per gallon} = \$2.89/\text{gallon} = \frac{\$2.89}{1 \text{ gallon}}$$

Because the unit price is $2.89 for every 1 gallon, if we buy more or less than 1 gallon, we can multiply the unit price by the quantity purchased to find the total price. For example, if we buy 10 gallons, we multiply 10 by 2.89 to get the total price.

$$\text{Total price} = \text{quantity} \cdot \text{unit price}$$

$$= 10 \text{ gal.} \cdot \frac{\$2.89}{1 \text{ gal.}}$$

▶ **Your Turn 8**

Multiply.
a. $-2.8y^2(3.5y^3 - 4y + 5.1)$
b. $(7x - 4.1)(3.2x - 2)$
c. $(1.9y + 5)(1.9y - 5)$

TACTILE VISUAL AUDITORY

Learning Strategy
Recall that we can use FOIL to remember how to multiply the terms in two binomials. We multiply the *first* terms, then the *outer* terms, then the *inner* terms, and finally the *last* terms.

Note Because the binomials $0.6t - 1.4$ and $0.6t + 1.4$ are conjugates, the like terms are additive inverses resulting in a product that is a difference of squares.

Note The word *per* indicates division and translates to the symbol /, or a fraction line.

Answers to Your Turn 8
a. $-9.8y^5 + 11.2y^3 - 14.28y^2$
b. $22.4x^2 - 27.12x + 8.2$
c. $3.61y^2 - 25$

$$= \frac{10 \text{ gal.}}{1} \cdot \frac{\$2.89}{1 \text{ gal.}}$$
$$= \$28.90$$

Connection We left the units in to illustrate how the units of measurement can be *divided out* just like numbers and variables. Because *gal.* appears in the numerator and denominator, we can *divide out* this common factor. Notice that the only unit of measurement left is $. This unit analysis, as it is called, helps clarify how unit pricing works. We will discuss unit analysis in more detail when we develop rates more fully in Chapter 7.

▶ **Your Turn 9**

Solve.

a. The long-distance phone company that Janice uses charges $0.09 per minute. If she talks for 76 minutes, how much will the call cost?

b. A broker purchases 150 shares of a stock that costs $4.25 per share. What is the total cost of the purchase?

Answers to Your Turn 9
a. $6.84   **b.** $637.50

Example 9 A seafood market has salmon priced at $9.89 per pound. Find the total price of 1.75 pounds of salmon.

Solution:

$$\text{Total price} = 1.75 \times 9.89 \qquad \text{Multiply 1.75 pounds by the unit price, \$9.89 per pound.}$$

$$= 17.3075 \qquad \blacktriangleleft \text{ Note The result is in dollars; so we round to two decimal places.}$$

Answer: $17.31

◀ **Do Your Turn 9**

6.3 Exercises For Extra Help MyMathLab®

Objectives 1 and 2

Prep Exercise 1 When multiplying two decimal numbers, how do you determine the position of the decimal point in the product?

Prep Exercise 2 Place the decimal correctly in the product shown.

$$\begin{array}{r} 5.3 \\ \times \ \ 0.4 \\ \hline 212 \end{array}$$

Prep Exercise 3 When multiplying a decimal number by a power of ten, such as 10, 100, or 1000, how do you determine where to place the decimal point in the product?

For Exercises 1–6, estimate. Then find the actual product.

1. $(0.6)(0.9)$

2. $(0.7)(0.4)$

3. $(2.9)(3.65)$

4. $(6.81)(4.2)$

5. $(9.81)(71.62)$

6. $(10.4)(12.31)$

For Exercises 7–16, multiply.

7. $(0.01)(619.45)$

8. $145(0.1)$

9. $19.6(10)$

10. $(1000)5.413$

11. $(0.1508)(-100)$

12. $(-10,000)(0.052)$

13. $152(-0.001)$

14. $-0.01(6)$

15. $(-0.029)(-0.15)$

16. $(-0.18)(-0.234)$

Objective 3

For Exercises 17–24, evaluate.

17. $(0.9)^2$ **18.** $(0.13)^2$ **19.** $(2.1)^3$ **20.** $(3.5)^3$

21. $(-0.03)^4$ **22.** $(-0.2)^5$ **23.** $(-0.4)^3$ **24.** $(-1.43)^2$

Objective 4

Prep Exercise 4 Is 24.7×10^2 in scientific notation? Explain.

Prep Exercise 5 Explain how to write 4.85×10^7 in standard form.

For Exercises 25–30, write each number in standard form and write the word name.

25. The speed of light is approximately 3×10^8 meters per second.

26. A light-year (lt-yr.) is the distance light travels in 1 year and is approximately 5.76×10^{12} miles.

27. The approximate distance to the Andromeda galaxy is 2.14×10^6 light-years.

28. The national debt is about $\$1.379 \times 10^{13}$. (*Source:* Bureau of Public Debt, November 29, 2010)

29. There are approximately 3.07×10^8 people in the United States. (*Source:* U.S. Bureau of the Census, 2009)

30. Earth is about 9.292×10^7 miles from the sun.

Objective 5

Prep Exercise 6 Suppose we are writing 456,000 in scientific notation. So far we have $4.56 \times 10^?$. What will be the unknown exponent? Why?

For Exercises 31–36, write each number in scientific notation.

31. There are 2,300,000 blocks of stone in Khufu's pyramid.

32. There are approximately 191,000 words in this textbook.

33. The distance from the sun to Neptune is approximately 2,795,000,000 miles.

34. The nearest star to ours (the sun) is Proxima Centauri at a distance of about 24,700,000,000,000 miles.

35. Human DNA consists of 5,300,000,000 nucleotide pairs.

36. An FM radio station broadcasts at a frequency of 102,300,000 hertz.

Objective 6

For Exercises 37–42, multiply.

37. $(3.2x^5)(1.8x^3)$ **38.** $(9.15y^2)(7.2y^4)$ **39.** $(0.2a^2)(-0.08a)$

40. $(-1.67n^3)(2.4n^2)$ **41.** $(0.5t^3)^2$ **42.** $(-0.2z^5)^3$

Objective 7

For Exercises 43–54, multiply.

43. $0.5(1.8y^2 + 2.6y + 38)$ **44.** $0.2(6.8m^3 + 0.44m - 7)$

45. $-6.8(0.02a^3 - a + 1.9)$ **46.** $-0.14(0.9n^2 - 40n - 8.7)$

47. $(x + 1.12)(x - 0.2)$ **48.** $(p - 1.5)(p + 7.9)$

49. $(0.4x - 1.22)(16x + 3)$ **50.** $(6.1y + 2)(0.8y - 5)$

51. $(5.8a + 9)(5.8a - 9)$ **52.** $(0.1m - 0.41)(0.1m + 0.41)$

53. $(k - 0.1)(14.7k - 2.8)$ **54.** $(9t - 0.5)(2t - 6.01)$

Objective 8

For Exercises 55 and 56, solve.

55. According to AAA, the highest average price of gasoline on record in the United States occurred on July 7, 2008, when it reached $4.11 per gallon. At that price, how much did it cost to fill the 32-gallon fuel tank in a Hummer H2 SUV? (Source: AAA Fuel Gauge Report.)

56. At the peak price of $4.11 per gallon, how much did it cost to fill the 10.6-gallon fuel tank in a Honda Insight?

For Exercises 57 and 58, use the following table that shows the pricing schedule for a phone company's long-distance rates.

Time	Rate
8 A.M.–5 P.M.	$0.12/minute
5 P.M.–10 P.M.	$0.10/minute
10 P.M.–8 A.M.	$0.09/minute

57. Joanne makes a long-distance call at 4:12 P.M. and talks for 32 minutes. How much does the call cost?

58. Joanne makes another long-distance call at 10:31 P.M. and talks for 21 minutes. How much does the call cost?

For Exercises 59–74, solve.

59. Stan buys 4.75 pounds of onions at $0.79 per pound. What is the total cost of the onions?

60. Nadine buys 5.31 pounds of chicken at $1.19 per pound. What is the total cost of the chicken?

61. A gas company charges $1.3548 per therm used. If Ruth uses 76 therms during a particular month, what will be the cost?

62. An electric company charges $0.1316 per kilowatt-hour. Margaret uses 698 kilowatt-hours one particular month. What will be the cost?

63. A van is assessed to have a value of $6400. Property tax on vans in the owner's county is $0.0285 per dollar of the assessed value. What will be the tax on the van?

64. One year after the purchase of a new sedan, it is assessed at a value of $9850. Property tax for the sedan in the county where the owner lives is $0.0301 per dollar of assessed value. What is the property tax on the sedan?

65. The current in a circuit is measured to be -3.5 amperes. If the resistance is 0.8 ohm, what is the voltage? (Use $V = ir$.)

66. The current in a circuit is measured to be 24.5 amperes. If the resistance is 1.2 ohms, what is the voltage? (Use $V = ir$.)

67. In a test, the speed of a car is measured to be 12.5 feet per second. How far will the car travel in 4.5 seconds? (Use $d = rt$.)

68. The speed of a 747 aircraft is 565 miles per hour. How far will the aircraft fly in 3.5 hours? (Use $d = rt$.)

69. A square top covers the components in an electronic device. The top is 4.8 centimeters on each side. What is the area of the top?

70. A rectangular computer processor is 3.4 centimeters by 3.6 centimeters. What is the area of the processor?

71. Find the area of the lot shown.

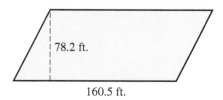

78.2 ft.

160.5 ft.

72. Find the area of the brace shown.

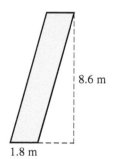

8.6 m

1.8 m

73. Find the volume.

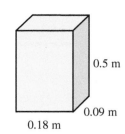

0.5 m

0.09 m

0.18 m

74. Find the volume.

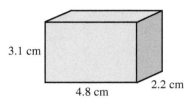

3.1 cm

4.8 cm

2.2 cm

Review Exercises

For Exercises 1 and 2, evaluate.

[1.4] 1. $3280 \div 16$

[2.4] 2. $\sqrt{169}$

[5.2] 3. Divide. $\dfrac{12x^7}{2x^3}$

[4.3] *For Exercises 4–6, solve and check.*

4. $9t = -81$

5. $3x + 5 = 32$

6. $-2a + 9 = 3a - 6$

[5.3] 7. Find the radius and circumference of a circle with a 15-inch diameter. Use $\dfrac{22}{7}$ for π.

6.4 Dividing Decimal Numbers; Square Roots with Decimals

Objectives

1 Divide decimal numbers.
2 Write fractions and mixed numbers as decimal numbers.
3 Evaluate square roots.
4 Divide monomials with decimal coefficients.
5 Solve equations using the multiplication principle.
6 Solve applications.

Warm-up

[1.4] **1.** Divide. $6503 \div 16$
[6.3] **2.** Multiply. $(-24.3)(4.7)$

Objective 1 Divide decimal numbers.

Suppose we must divide $4.8 \div 2$. We will see that we can use long division. However, to gain some insight into the process, we will divide by writing 4.8 as an equivalent mixed number.

$$4.8 \div 2 = 4\frac{8}{10} \div 2 \quad \text{Write 4.8 as a mixed number.}$$

$$= \frac{48}{10} \div 2 \quad \text{Write the mixed number as a fraction.}$$

$$= \frac{\overset{24}{48}}{10} \cdot \frac{1}{\underset{1}{2}}$$

Note After writing the equivalent multiplication, we can divide 48 by 2. This suggests that we can divide the original digits as if they were whole numbers.

Note The denominator, 10, indicates that the quotient is in terms of tenths.

$$= \frac{24}{10}$$

$$= 2\frac{4}{10}$$

Note Although we could have simplified the fraction, we left it in terms of tenths so that writing the equivalent decimal number would be easy.

$$= 2.4$$

Using this lengthy method, we can see that $4.8 \div 2 = 2.4$.

Notice that if we set up the problem in long division form and write the decimal in the quotient directly above its position in the dividend, we get the same result.

$$\begin{array}{r} 2.4 \\ 2\overline{)4.8} \\ \underline{-4} \\ 08 \\ \underline{-8} \\ 0 \end{array}$$

Our example suggests the following procedure:

Procedure

To divide a decimal number by a whole number using long division:
1. Divide the divisor into the dividend as if both numbers were whole numbers. Align the digits in the quotient properly.
2. Write the decimal in the quotient directly above its position in the dividend.
3. If after bringing down the last digit in the dividend the remainder is not 0, write a 0 digit to the right of the last nonzero digit in the dividend, bring that 0 down, and divide. Continue adding 0 digits and dividing until the remainder is 0 or a digit or group of digits repeats without end in the quotient. To indicate repeated digits in the quotient, write a repeat bar over the digit or group of digits that repeat.

Example 1 Estimate. Then find the actual quotient. $9 \div 2$

Solution:

Estimate: Rounding both numbers to single nonzero digit leaves them unchanged. Think of the division as a fraction and then write it as a mixed number.

$$9 \div 2 = \frac{9}{2} = 4\frac{1}{2}$$

Actual:
```
        4.5
    2)9.0
     -8
      10
     -10
       0
```
Use long division with 9 written as 9.000... Continue to divide until the remainder is 0. Write the decimal in the quotient directly above its position in 9.0.

Note Our decimal answer 4.5 is the same as our estimate $4\frac{1}{2}$.

$$4.5 = 4\frac{5}{10} = 4\frac{1}{2}$$

Answer: 4.5 ◄

▶ **Do Your Turn 1**

Example 2 Estimate. Then find the actual quotient. $3.2 \div 3$

Solution:

Estimate: $3 \div 3 = 1$

Actual:
```
        1.066 . . .
    3)3.200
     -3
      02
      -0
       20
      -18
        20
       -18
         2
```

◄ **Note** The 6 digit will continue in an unending pattern. To indicate this repetitive pattern, we use a repeat bar over the 6.

Warning Place the repeat bar carefully. For example, $1.\overline{06}$ means 1.060606..., which is not our quotient.

Answer: $1.0\overline{6}$

▶ **Do Your Turn 2**

Now let's explore division when the divisor is a decimal number, as in $1.78 \div 0.5$. The approach is to rewrite the division so that the divisor is a whole number. To make 0.5 a whole number, the decimal point needs to move one place to the right. So we multiply 0.5 by 10. To keep the division expression equivalent to the original expression, we must multiply 1.78 by 10 as well. We are using the rule for writing equivalent fractions. To see how this works, we will write the division in fraction form.

$$1.78 \div 0.5 = \frac{1.78}{0.5}$$

$$= \frac{1.78(10)}{0.5(10)}$$ **Multiply the numerator (dividend) and denominator (divisor) by 10.**

$$= \frac{17.8}{5}$$ ◄ **Note** Multiplying the numerator (dividend) and denominator (divisor) by 10 gives an equivalent division with a whole number divisor.

In long division format, the process is as follows:

$$0.5\overline{)1.7\,8} \longrightarrow \begin{array}{r} 3.56 \\ 5\overline{)17.80} \\ -15 \\ \hline 28 \\ -25 \\ \hline 30 \\ -30 \\ \hline 0 \end{array}$$

Note We move the decimal point 1 place to the right in the divisor to make it a whole number (multiply it by 10). Because we moved the decimal point 1 place to the right in the divisor, we must also move it 1 place to the right in the dividend (multiply it by 10 as well).

Our example suggests the following procedure:

> ### Procedure
>
> To divide using long division when the divisor is a decimal number:
> 1. Move the decimal point in the divisor to the right enough places to make the divisor an integer (multiply by an appropriate power of 10).
> 2. Move the decimal point in the dividend the same number of places (multiply it by the same power of 10).
> 3. Follow the same steps as for dividing a decimal number by a whole number.

▶ **Your Turn 3**

Estimate. Then find the actual quotient.

a. $0.768 \div 0.6$
b. $13.49 \div 1.42$
c. $-605.3 \div 0.02$
d. $16 \div -0.03$

▲
Note For Your Turn 3d, after moving the decimal point, we have $1600 \div 3$. Rounding these numbers for the estimate gives $2000 \div 3$. However, it is easier to bend the rules for rounding and think of 1600 as 1500 so that we have $1500 \div -3 = -500$. This is not only easier, it's more accurate.

Answers to Your Turn 3
a. Estimate: $1.\overline{6}$
   Actual: 1.28
b. Estimate: 10
   Actual: 9.5
c. Estimate: $-30,000$
   Actual: $-30,265$
d. Estimate: -500
   Actual: $-533.\overline{3}$

Example 3   Estimate. Then find the actual quotient. $24.1 \div 0.04$

Solution:

$$0.04\overline{)24.1\,0}$$

Note When we move the decimal point two places to the right in 24.1, we move past an understood 0 digit that is to the right of 1 so that the result is 2410.

Move the decimal point 2 places to the right in 0.04 to make it an integer (multiply it by 100). Then move the decimal point 2 places to the right in 24.1 as well (multiply it by 100).

Estimate: Round the numbers after moving the decimal point. $2000 \div 4 = 500$

$$\text{Actual:}\quad \begin{array}{r} 602.5 \\ 4\overline{)2410.0} \\ -24 \\ \hline 01 \\ -0 \\ \hline 10 \\ -8 \\ \hline 20 \\ -20 \\ \hline 0 \end{array}$$

Divide. Remember to write the decimal point in the quotient directly above its position in the dividend.

▲
Note Rounding 2410 to 2400 is still easy and makes for a more accurate estimate.
$2400 \div 4 = 600$

Answer: 602.5

◀ **Do Your Turn 3**

Objective 2 Write fractions and mixed numbers as decimal numbers.

Because a fraction is a notation for division, we can write fractions as decimals by dividing the denominator into the numerator. We follow the same procedure as for dividing decimal numbers when the divisor is a whole number.

Procedure

To write a fraction as a decimal number, divide the denominator into the numerator.

Example 4 Write $\dfrac{5}{8}$ as a decimal number.

Solution:

$$
\begin{array}{r}
0.625 \\
8\overline{)5.000} \\
-48 \\
\hline
20 \\
-16 \\
\hline
40 \\
-40 \\
\hline
0
\end{array}
$$

Divide the denominator into the numerator.

Connection We can verify that $0.625 = \frac{5}{8}$ by writing 0.625 as a fraction in lowest terms.

$$0.625 = \frac{625}{1000} = \frac{625 \div 125}{1000 \div 125} = \frac{5}{8}$$

Answer: 0.625

▶ **Do Your Turn 4**

▶ **Your Turn 4**

Write as a decimal number.

a. $\dfrac{3}{4}$

b. $-\dfrac{5}{12}$

How can we write mixed numbers as decimal numbers? We saw in Section 6.1 that $23.791 = 23\frac{791}{1000}$. In reverse, if we want to write a mixed number as a decimal number, the integer part of the mixed number, in this case 23, is the integer part of the decimal number and the fraction part, $\frac{791}{1000}$, is equivalent to the decimal digits.

Procedure

To write a mixed number as a decimal:

1. Write the integer part of the mixed number to the left of a decimal point.
2. Divide the denominator into the numerator to determine the decimal digits. Write those digits to the right of the decimal point.

VISUAL

Learning Strategy

If you are a visual learner, imagine the space between the integer and fraction part of a mixed number as the decimal point in its decimal number equivalent.

$$-17\frac{2}{11} = -17.\overline{18}$$

Example 5 Write $-17\dfrac{2}{11}$ as a decimal number.

Solution: The integer part of the mixed number, -17, is written to the left of the decimal point. The decimal equivalent of $\frac{2}{11}$ is written to the right of the decimal point. To find the decimal digits, we divide 11 into 2.

Note The decimal digits repeat in an unending pattern; so we use a repeat bar over those digits that repeat.

$$
\begin{array}{r}
0.1818\ldots \\
11\overline{)2.0000} \\
-11 \\
\hline
90 \\
-88 \\
\hline
20 \\
-11 \\
\hline
90
\end{array}
$$

Connection Notice that $-17.\overline{18}$ is a rational number because it is equivalent to $-17\frac{2}{11}$, which we can express as the fraction $-\frac{189}{11}$. It turns out that all nonterminating decimal numbers with repeating digits can be expressed as fractions; so they are all rational numbers. But there are nonterminating decimal numbers without repeating digits, which means that there are decimal numbers that are not rational numbers. We will learn about these numbers in the next objective.

Answer: $-17.\overline{18}$

▶ **Do Your Turn 5**

▶ **Your Turn 5**

Write as a decimal number.

a. $25\dfrac{1}{8}$

b. $-19\dfrac{5}{6}$

Answers to Your Turn 4
a. 0.75   b. $-0.41\overline{6}$

Answers to Your Turn 5
a. 25.125
b. $-19.8\overline{3}$

Objective 3 Evaluate square roots.

In previous chapters, we discussed square roots of perfect squares, as in $\sqrt{64}$, and square roots of fractions with perfect square numerators and denominators, as in $\sqrt{\frac{100}{121}}$. Recall that the radical sign indicates the principal square root, which is either 0 or a positive number whose square is the radicand.

$\sqrt{64} = 8$ because $(8)^2 = 64$

$\sqrt{\frac{100}{121}} = \frac{10}{11}$ because $\left(\frac{10}{11}\right)^2 = \frac{100}{121}$

Note The numbers 64, 100, and 121 are perfect squares because their square roots are whole numbers. (For a list of perfect squares, see Table 1.1.)

We now consider square roots of numbers that are not perfect squares, such as $\sqrt{14}$. To evaluate $\sqrt{14}$, we must find a positive number that can be squared to equal 14. Because 14 is between the perfect squares 9 and 16, we can conclude that it must be true that $\sqrt{14}$ is between $\sqrt{9}$ and $\sqrt{16}$. Therefore, $\sqrt{14}$ is between 3 and 4.

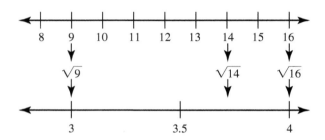

Notice that $\sqrt{14}$ is not an integer. Using our number line, we can approximate $\sqrt{14}$ to be about 3.7. We can test our approximation by squaring it.

$$(3.7)^2 = 13.69$$

Because 13.69 is less than 14, our approximation is too small. Let's try squaring 3.8.

$$(3.8)^2 = 14.44$$

So $\sqrt{14}$ is between 3.7 and 3.8. We need to go to hundredths to refine further.

We could continue guessing and refining indefinitely because $\sqrt{14}$ is a nonterminating decimal number that does not have a pattern of unending repeating digits. Because we cannot express the exact value of $\sqrt{14}$ as a decimal or fraction, we say that it is an *irrational number*. In fact, the square root of any whole number that is not a perfect square is an irrational number.

We can use the radical sign to express the exact value of the square root of a whole number that is not a perfect square. For example, $\sqrt{14}$ is the exact expression for the square root of 14. If the radical form is not desirable, we can approximate an irrational square root with a rational number. To approximate, we can guess, verify, and refine as we did earlier or use a calculator.

Approximating $\sqrt{14}$ using a calculator gives

$$\sqrt{14} \approx 3.741657387.$$

Note Remember that this symbol ▲ means *is approximately equal to.*

Connection We defined irrational numbers in Section 5.3 when we discussed the irrational number π.

▶ **Your Turn 6**

Approximate the square root to the nearest hundredth.

a. $\sqrt{50}$
b. $\sqrt{94}$
c. $\sqrt{128}$
d. $\sqrt{212}$

Answers to Your Turn 6
a. ≈ 7.07 b. ≈ 9.70 c. ≈ 11.31
d. ≈ 14.56

Example 6   Approximate $\sqrt{29}$ to the nearest hundredth.

Answer: ≈ 5.39

◀ **Do Your Turn 6**

What if we must find the square root of a decimal number? Some decimal numbers have exact decimal square roots; others are irrational.

$$\sqrt{0.09} = 0.3 \quad \text{because} \quad (0.3)^2 = 0.09$$
$$\sqrt{1.21} = 1.1 \quad \text{because} \quad (1.1)^2 = 1.21$$
$$\sqrt{0.0064} = 0.08 \quad \text{because} \quad (0.08)^2 = 0.0064$$

Notice that each of the radicands is a perfect square if we take out the decimal point.

0.09 is similar to 9,

1.21 is similar to 121,

and 0.0064 is similar to 64.

Also notice that the square roots have the same digits as the square roots of the corresponding perfect squares. The tricky part is placing the decimal point. If we square a decimal number, the product will always have twice as many places as the original decimal number.

$$(0.3)^2 = 0.09 \quad (0.03)^2 = 0.0009 \quad (0.003)^2 = 0.000009$$

1 place   2 places   2 places 4 places   3 places   6 places

This means that when we find the square root of a decimal number that corresponds to a perfect square, the decimal number must have an *even number of decimal places* if it is to have an exact decimal square root. The exact square root will have half the number of decimal places and the same digits as the square root of the corresponding perfect square. If the radicand has an odd number of decimal digits, the square root is irrational.

$$\sqrt{0.0081} = 0.09 \quad \text{whereas} \quad \sqrt{0.081} \text{ is irrational}$$

4 places   2 places

◄ **Note** Using a calculator, we can approximate the irrational root.

$$\sqrt{0.081} \approx 0.284604989$$

Rule

The square root of a decimal number that corresponds to a perfect square and has an even number of decimal places will be a decimal number with half the number of decimal places and the same digits as the square root of the corresponding perfect square.

Example 7   Find the square root. $\sqrt{0.000196}$

Solution:

$$\sqrt{0.000196} = 0.014$$

6 places   3 places

196 is a perfect square, and 0.000196 has an even number of decimal places; so its square root has half the number of decimal places and the same digits as the square root of 196, which is 14.

► **Do Your Turn 7**

► **Your Turn 7**

Evaluate the square root.
a. $\sqrt{0.01}$
b. $\sqrt{0.0016}$
c. $\sqrt{0.25}$
d. $\sqrt{0.000169}$

The rational and irrational numbers together form the set of **real numbers.** If we travel the entire number line from end to end, every number we encounter is a real number, either rational or irrational.

Definition   Real numbers: The set of all rational and irrational numbers.

Answers to Your Turn 7
a. 0.1   b. 0.04   c. 0.5   d. 0.013

┌─────────────────── **Real Numbers** ───────────────────┐

| **Rational Numbers:** Real numbers that can be expressed in the form $\frac{a}{b}$, where a and b are integers and $b \neq 0$, such as $-4\frac{3}{4}$, $-\frac{2}{3}$, 0.018, $0.\overline{3}$, and $\frac{5}{8}$. | **Irrational Numbers:** Real numbers that are not rational, such as $-\sqrt{2}$, $-\sqrt{3}$, $\sqrt{0.8}$, and π. |

Integers: $\ldots, -3, -2, -1, 0, 1, 2, 3, \ldots$

Whole Numbers: $0, 1, 2, 3, \ldots$

Natural Numbers: $1, 2, 3, \ldots$

Objective 4 Divide monomials with decimal coefficients.

We learned how to divide monomials in Section 3.6. In Section 5.2, we learned to simplify rational expressions. From what we learned in Sections 3.6 and 5.2, we can say that to divide monomial expressions, we divide the coefficients, subtract exponents for the like variables, and write any unlike variables as they are.

▶ **Your Turn 8**

Divide.

a. $1.28y^5 \div 0.4y$

b. $\dfrac{-0.8tu^7}{-1.6u^3}$

Example 8   Divide.

a. $8.2n^6 \div 2.5n^2$

Solution:

$$8.2n^6 \div 2.5n^2 = (8.2 \div 2.5)n^{6-2}$$ Divide the coefficients and subtract exponents for the like bases.

$$= 3.28n^4$$

b. $\dfrac{-3.85x^5yz}{0.5x^4y}$

Solution:

$$\frac{-3.85x^5yz}{0.5x^4y} = \frac{-3.85}{0.5}x^{5-4}y^{1-1}z$$

Note Remember, ▶ y^0 simplifies to 1.

$$= -7.7x^1y^0z$$ Divide the coefficients, subtract exponents for the like bases, and write the unlike bases as they are.

$$= -7.7xz$$

Connection We could use the approach we learned in Section 5.2 for simplifying rational expressions to simplify the expressions in Example 8.

a. $8.2n^6 \div 2.5n^2 = \dfrac{8.2n^6}{2.5n^2} = \dfrac{8.2 \cdot \overset{1}{\cancel{n}} \cdot \overset{1}{\cancel{n}} \cdot n \cdot n \cdot n \cdot n}{2.5 \cdot \underset{1}{\cancel{n}} \cdot \underset{1}{\cancel{n}}} = 3.28n^4$

b. $\dfrac{-3.85x^5yz}{0.5x^4y} = \dfrac{-3.85 \cdot \overset{1}{\cancel{x}} \cdot \overset{1}{\cancel{x}} \cdot \overset{1}{\cancel{x}} \cdot \overset{1}{\cancel{x}} \cdot x \cdot \overset{1}{\cancel{y}} \cdot z}{0.5 \cdot \underset{1}{\cancel{x}} \cdot \underset{1}{\cancel{x}} \cdot \underset{1}{\cancel{x}} \cdot \underset{1}{\cancel{x}} \cdot \underset{1}{\cancel{y}}} = -7.7xz$

◀ **Do Your Turn 8**

Answers to Your Turn 8

a. $3.2y^4$ b. $0.5tu^4$

Objective 5 Solve equations using the multiplication principle.

In Section 4.3, we learned to solve equations using the multiplication principle of equality, which says that we can multiply (or divide) both sides of an equation by the same nonzero amount without affecting its solution(s). Let's apply the principle to equations that contain decimal numbers.

Example 9   Solve and check. $-14.2x = 39.05$

Solution: $\dfrac{-14.2x}{-14.2} = \dfrac{39.05}{-14.2}$ Divide both sides by -14.2.

$\qquad\qquad 1x = -2.75$

$\qquad\qquad\ x = -2.75$

Check:

$\qquad\qquad -14.2x = 39.05$

$\qquad -14.2(-2.75) \stackrel{?}{=} 39.05$ In the orginal equation, replace x with -2.75.

$\qquad\qquad 39.05 = 39.05$ True; so -2.75 is the solution.

▶ **Do Your Turn 9**

Objective 6 Solve applications.

Example 10   Laura owes a hospital $1875.84. She arranges to pay the debt in monthly installments over one year. How much is each monthly payment?

Solution: Because there are 12 months in a year, she will make 12 payments. Therefore, to calculate the amount of each payment (P), we divide $1875.84 by 12.

$$P = 1875.84 \div 12$$
$$P = 156.32$$

Answer: Each payment is $156.32.

▶ **Do Your Turn 10**

In Chapter 5, we learned about circles. We learned that the circumference of a circle is the distance around its edge and is found using the formula $C = \pi d$ or $C = 2\pi r$. Remember, π is an irrational number, and in Chapter 5, we approximated its value with the fraction $\frac{22}{7}$. We can now find a decimal equivalent to $\frac{22}{7}$ by dividing 22 by 7.

$$\begin{array}{r} 3.14\dots \\ 7)\overline{22.00} \\ -21 \\ \hline 10 \\ -7 \\ \hline 30 \\ -28 \\ \hline 2 \end{array}$$

◀ **Note** The π key on a calculator gives a more accurate decimal approximation, 3.141592654.

When solving problems involving π, if we are given no instructions about which approximation to use, we choose the approximation that seems best suited to the problem. Or we can express the exact answer in terms of π (similar to using the radical sign to express the exact answer when a radicand is not a perfect square).

Connection Although it seems contradictory, the *multiplication* principle of equality allows us to *divide* both sides of an equation by the same nonzero number because division can be written as multiplication using the reciprocal of the divisor.

▶ **Your Turn 9**

Solve and check.
  a. $37.994 = 3.14d$
  b. $-9.8m = 158.76$

▶ **Your Turn 10**

Solve.
  a. To avoid interest charges, Mark takes a 90-day same-as-cash option to pay off the puchase of a DVD player. The cash amount would be $256.86. If he designs his budget to make three equal payments, how much should each payment be?
  b. A patient is to receive 1.5 liters of saline over 60 minutes. How much should the patient receive every minute?

Answers to Your Turn 9
a. $d = 12.1$ **b.** $m = -16.2$

Answers to Your Turn 10
a. $85.62   **b.** 0.025 L

▶ **Your Turn 11**

Solve. Use 3.14 to approximate π.

a. The radius of an engine cylinder is 1.78 centimeters. What is the circumference?

b. The circumference of a crater is 16.8 kilometers. What is the diameter?

Example 11 Solve. Use 3.14 to approximate π.

a. The radius of Earth is about 6378 kilometers at the equator. Find the circumference of Earth at the equator.

Solution:
$$C = 2\pi(6378) \quad \text{In } C = 2\pi r, \text{ replace } r \text{ with 6378.}$$
$$C = 12{,}756\pi \quad \text{Multiply 6378 by 2 to get the exact answer in terms of } \pi.$$
$$C \approx 12{,}756(3.14) \quad \text{Replace } \pi \text{ with the approximate value 3.14.}$$
$$C \approx 40{,}053.84 \quad \text{Multiply.}$$

Answer: The circumference of Earth is approximately 40,053.84 kilometers at the equator.

Connection Earlier in this section, we learned to express the exact value of an irrational square root by leaving the radical sign in the expression. Similarly, we can express the exact value of a calculation involving π by leaving the symbol π in the expression.

Exact	Approximation
$\sqrt{17}$	≈ 4.123105626
$12{,}756\pi$	$\approx 40{,}053.84$

Of Interest

The official diameter of a basketball rim is exactly 18 inches; so the circumference of 56.52 inches is an approximation calculated using 3.14.

Answers to Your Turn 11

a. ≈ 11.1784 cm

b. ≈ 5.35 km

b. Find the diameter of an official basketball rim, which has a circumference of approximately 56.52 inches.

Solution:
$$56.52 \approx (3.14)d \quad \text{In } C = \pi d, \text{ replace } C \text{ with 56.52 and } \pi \text{ with 3.14.}$$
$$\frac{56.52}{3.14} \approx \frac{3.14d}{3.14} \quad \text{Divide both sides by 3.14.}$$
$$18 \approx d \quad \text{Simplify.}$$

Answer: The diameter of an official basketball rim is 18 inches.

◀ **Do Your Turn 11**

6.4 Exercises

For Extra Help

MyMathLab®

Objective 1

Prep Exercise 1 Given the long division shown, place the decimal point correctly in the quotient.

$$
\begin{array}{r}
62 \\
4\overline{)24.8} \\
-24 \\
\hline
08 \\
-8 \\
\hline
0
\end{array}
$$

Prep Exercise 2 To divide $65.76 \div 0.02$ using long division, how many places should the decimal point be moved in the dividend and divisor to complete the long division?

For Exercises 1–20, estimate. Then find the actual quotient.

1. $286.2 \div 12$

2. $194.16 \div 24$

3. $0.48 \div 20$

4. $3.392 \div 32$

5. $5640 \div 9.4$ **6.** $93.96 \div 10.8$ **7.** $0.288 \div 0.06$ **8.** $0.7208 \div 0.08$

9. $-6.4 \div 0.16$ **10.** $28.5 \div (-0.15)$ **11.** $1 \div 0.008$ **12.** $0.1 \div 0.002$

13. $20.6 \div (-1000)$ **14.** $8.93 \div (-10,000)$ **15.** $-8.145 \div (-0.01)$ **16.** $-21.7 \div (0.001)$

17. $19.6 \div 0.11$ **18.** $30.13 \div 0.99$ **19.** $-0.0901 \div (-0.085)$ **20.** $-339.2 \div (-0.106)$

Objective 2

Prep Exercise 3 How do you write a fraction as a decimal number?

For Exercises 21–36, write as a decimal number.

21. $\dfrac{3}{5}$ **22.** $\dfrac{1}{8}$ **23.** $\dfrac{9}{20}$ **24.** $\dfrac{7}{25}$

25. $-\dfrac{7}{16}$ **26.** $-\dfrac{5}{32}$ **27.** $\dfrac{13}{30}$ **28.** $\dfrac{1}{6}$

29. $13\dfrac{1}{4}$ **30.** $25\dfrac{1}{2}$ **31.** $-17\dfrac{5}{8}$ **32.** $-5\dfrac{4}{5}$

33. $104\dfrac{2}{3}$ **34.** $76\dfrac{5}{6}$ **35.** $-216\dfrac{4}{7}$ **36.** $-99\dfrac{5}{13}$

Objective 3

Prep Exercise 4 The square root of a decimal number that corresponds to a perfect square and has an even number of decimal places will be a decimal number with _____ the number of decimal places and the same digits as the square root of the corresponding perfect square.

For Exercises 37–48, evaluate the square root. If the root is irrational, approximate the square root to the nearest hundredth.

37. $\sqrt{0.0016}$ **38.** $\sqrt{2.25}$ **39.** $\sqrt{0.25}$ **40.** $\sqrt{0.0081}$
41. $\sqrt{24}$ **42.** $\sqrt{108}$ **43.** $\sqrt{200}$ **44.** $\sqrt{78}$
45. $\sqrt{1.69}$ **46.** $\sqrt{0.0256}$ **47.** $\sqrt{0.009}$ **48.** $\sqrt{2.5}$

Objective 4

For Exercises 49–54, divide.

49. $10.2x^6 \div 0.4x^4$ **50.** $5.44y^4 \div 1.7y$ **51.** $\dfrac{-0.96m^3n}{0.15m}$

52. $\dfrac{3.42hk^7}{-3.8k^2}$ **53.** $\dfrac{-3.03a^5bc}{-20.2a^4b}$ **54.** $\dfrac{-2.7t^4u^3v^2}{-0.75uv^2}$

Objective 5

For Exercises 55–62, solve and check.

55. $2.1b = 12.642$ **56.** $81.5y = 21.19$ **57.** $-0.88h = 1.408$ **58.** $4.066 = -0.38k$

59. $-2.28 = -3.8n$ **60.** $-2.89 = -8.5p$ **61.** $-28.7 = 8.2x$ **62.** $10.8t = -9.072$

Objective 6

For Exercises 63–70, solve.

63. Carla purchases a computer on a 90-day same-as-cash option. The purchase price is $1839.96. If she plans to make three equal payments to pay off the debt, how much should each payment be?

64. Brad has a student loan balance of $1875.96. If he agrees to make equal monthly payments over a three-year period, how much will each payment be?

65. A patient is to receive radiation treatment for cancer. The treatment calls for a total of 324 rads administered in 8 bursts of focused radiation. How many rads should each burst be?

66. The FDA recommends that for a healthy diet, a person consuming 2000 calories per day should limit saturated fat intake to about 20 grams per day. If a person splits the recommended saturated fat intake equally among three meals, how much saturated fat is allowed in each meal? (*Source:* U.S. Food and Drug Administration, 2005.)

67. The voltage in a circuit is 12 volts. The current is measured to be 0.03 amp. What is the resistance? (Use $V = ir$.)

68. The voltage in a circuit measures -40 volts. If the resistance is 500 ohms, what is the current? (Use $V = ir$.)

69. A plane is flying at 350 miles per hour. How long will it take the plane to reach a city that is 600 miles away? (Use $d = rt$.)

70. The area of a rectangular painting is 358.875 square inches. If the length is 16.5 inches, what is the width? (Use $A = bh$.)

For Exercises 71–76, use 3.14 to approximate π.

71. The radius of the moon is about 1087.5 miles. Calculate the circumference of the moon at its equator.

72. The solid rocket boosters used to propel the space shuttle into orbit have O-rings that fit around the boosters at connections. If the diameter of one rocket booster is 12.17 feet, what is the circumference of the O-ring?

Of Interest

On January 28, 1986, the space shuttle *Challenger* exploded 73 seconds after launch, killing all seven crew members. The cause of the disaster was a weakness in an attach ring on one of the rocket boosters. At the time, the rings did not completely circle the cases of the rocket boosters. After the disaster, the rings were redesigned to completely circle the rocket boosters.

73. Haleakala crater on Maui island, Hawaii, is the largest inactive volcanic crater in the world and has a circumference of about 21 miles at its rim. What is the diameter at the rim?

74. A giant sequoia is measured to have a circumference of 26.4 meters. What is the diameter?

Of Interest

The giant sequoia tree is found along the western slopes of the Sierra Nevada range. The trees reach heights of nearly 100 meters, and some are over 4000 years old.

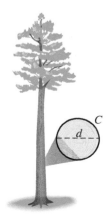

75. To manufacture a light fixture, a company bends 6.5 feet of metal tubing into a circle. What is the radius of this circle?

76. The circumference of Earth along the equator is approximately 24,888.98 miles. What is Earth's equatorial radius?

Review Exercises

For Exercises 1 and 2, simplify.

[2.5] 1. $16 - 9[12 + (8 - 13)] \div 3$

[5.3] 2. $\dfrac{3}{5} \cdot 30$

[5.7] 3. Evaluate $\dfrac{1}{2}at^2$ when $a = -10$ and $t = 6$.

For Exercises 4 and 5, find the area.

[5.3] 4.

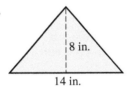

8 in.

14 in.

[5.7] 5.

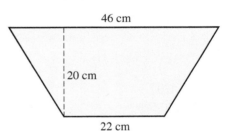

46 cm

20 cm

22 cm

6.5 Order of Operations and Applications in Geometry

Objectives

1 Simplify numerical expressions using the order of operations agreement.

2 Simplify expressions containing fractions and decimals.

3 Solve application problems requiring more than one operation.

4 Find weighted means.

5 Evaluate expressions.

6 Find the area of a triangle, trapezoid, and circle.

7 Find the volume of a cylinder, pyramid, cone, and sphere.

8 Find the area and volume of a composite form.

▶ **Your Turn 1**

Simplify.

a. $6.5 - 1.2(14.4 + 0.6)$

b. $1.8 + 0.4(9.02 - 10.52)$

c. $-0.2 \div 1.6 - 20(0.2)^3$

d. $(1.5)^2 + 2.52 \div 0.4(-0.6)$

e. $[(12.8 - 15.3) \div 10] - 0.6\sqrt{0.16}$

f. $\sqrt{1.69} + 0.2[(0.011 + 0.013) \div (0.5 - 0.8)]$

Warm-up

[2.5] **1.** Simplify. $(-4)^3 - 36 \div (-9)(2) - \sqrt{100}$

[5.7] **2.** Evaluate $\frac{4}{9}x^2 + y$ when $x = \frac{3}{4}$ and $y = \frac{2}{3}$.

Objective 1 Simplify numerical expressions using the order of operations agreement.

Example 1   Simplify.

a. $2.75 - 8.2(3.4 + 4)$

Solution: $2.75 - 8.2(3.4 + 4)$

$= 2.75 - 8.2(7.4)$ Add in the parentheses. $3.4 + 4 = 7.4$

$= 2.75 - 60.68$ Multiply. $8.2(7.4) = 60.68$

$= -57.93$ Subtract. $2.75 - 60.68 = -57.93$

b. $12.6 \div 0.2 - 6.5(0.4)^2$

Solution: $12.6 \div 0.2 - 6.5(0.4)^2$

$= 12.6 \div 0.2 - 6.5(0.16)$ Evaluate the exponential form. $(0.4)^2 = 0.16$

$= 63 - 1.04$ Multiply and divide from left to right. $12.6 \div 0.2 = 63$ and $6.5(0.16) = 1.04$

$= 61.96$ Subtract. $63 - 1.04 = 61.96$

c. $[(1.2 + 0.3) \div 5] \div 2\sqrt{0.09}$

Solution: $[(1.2 + 0.3) \div 5] \div 2\sqrt{0.09}$

$= [1.5 \div 5] \div 2\sqrt{0.09}$ Add in the parentheses. $1.2 + 0.3 = 1.5$

$= 0.3 \div 2\sqrt{0.09}$ Divide in the brackets. $1.5 \div 5 = 0.3$

$= 0.3 \div 2(0.3)$ Evaluate the square root. $\sqrt{0.09} = 0.3$

$= 0.15(0.3)$ Divide. $0.3 \div 2 = 0.15$

$= 0.045$ Multiply. $0.15(0.3) = 0.045$

◀ **Do Your Turn 1**

Objective 2 Simplify expressions containing fractions and decimals.

We will consider three methods for simplifying expressions that contain both decimals and fractions.

Answers to Your Turn 1
a. -11.5 **b.** 1.2 **c.** -0.285
d. -1.53 **e.** -0.49 **f.** 1.284

Answers to Warm-up
1. -66 **2.** $\frac{11}{12}$

Procedure

To simplify numerical expressions that contain both fractions and decimals:

1. **Method 1:** Write each decimal number as a fraction. or
   Method 2: Write each fraction as a decimal number. or
   Method 3: Write each decimal number over 1 in fraction form.
2. Follow the order of operations agreement.

Consider an example using each of the three methods.

Example 2   Simplify. $\dfrac{2}{3}(6.09) - \dfrac{3}{8}$

Method 1: Write each decimal number as a fraction.

$$\frac{2}{3}(6.09) - \frac{3}{8} = \frac{2}{3} \cdot \frac{609}{100} - \frac{3}{8} \qquad \text{Write 6.09 as } \tfrac{609}{100}.$$

$$= \frac{2}{\underset{1}{3}} \cdot \frac{\overset{203}{609}}{\underset{50}{100}} - \frac{3}{8} \qquad \text{Divide out the common factors.}$$

$$= \frac{203}{50} - \frac{3}{8} \qquad \text{Multiply.}$$

$$= \frac{203(4)}{50(4)} - \frac{3(25)}{8(25)} \qquad \text{The LCD for 50 and 8 is 200.}$$

$$= \frac{812}{200} - \frac{75}{200} \qquad \text{Write equivalent fractions with 200 as the denominator.}$$

$$= \frac{737}{200} \qquad \text{Subtract numerators and keep the common denominator.}$$

$$= 3\frac{137}{200}, \text{ or } 3.685 \qquad \text{Write as a mixed number or decimal number.}$$

Working with fractions can be tedious; so let's explore two other methods.

Method 2: Write each fraction as a decimal number.

$$\frac{2}{3}(6.09) - \frac{3}{8} = 0.\overline{6}(6.09) - 0.375 \qquad \text{Write } \tfrac{2}{3} \text{ as } 0.\overline{6} \text{ and } \tfrac{3}{8} \text{ as 0.375.}$$

Note To complete the calculation, we must round $0.\overline{6}$, which makes the answer an approximation.

$$\blacktriangleright \approx 0.667(6.09) - 0.375 \qquad \text{Round } 0.\overline{6} \text{ to the nearest thousandth.}$$

$$\approx 4.06203 - 0.375 \qquad \text{Multiply.}$$

$$\approx 3.68703 \qquad \text{Subtract.}$$

You may want to avoid this method if any of the fractions' decimal equivalents have nonterminating digits.

Method 3: Write each decimal number over 1 in fraction form.

$$\frac{2}{3}(6.09) - \frac{3}{8} = \frac{2}{3} \cdot \frac{6.09}{1} - \frac{3}{8} \qquad \text{Write 6.09 over 1.}$$

$$= \frac{12.18}{3} - \frac{3}{8} \qquad \text{Multiply. } 2 \cdot 6.09 = 12.18 \text{ and } 3 \cdot 1 = 3$$

$$= 4.06 - 0.375 \qquad \text{Divide denominators into numerators to get decimals. } 12.18 \div 3 = 4.06 \text{ and } 3 \div 8 = 0.375$$

$$= 3.685 \qquad \text{Subtract.}$$

▶ **Your Turn 2**

Simplify.

a. $\dfrac{3}{4}(24.6)$

b. $(-5.052)\dfrac{5}{6}$

c. $\dfrac{1}{2}(7.9) - \dfrac{2}{5}$

d. $\dfrac{1}{4} + (-2.7) \div \dfrac{9}{2}$

◀ **Note** We can write the decimal number over 1 because the quotient of a number and 1 is the number.

Answers to Your Turn 2
a. 18.45   b. −4.21
c. 3.55   d. −0.35

This method avoided the nonterminating decimal digits because the product of 6.09 and 2 is divisible by the denominator, 3. This won't always work, but the method is worth a try if you see that a fraction's decimal equivalent has nonterminating decimal digits.

◄ Do Your Turn 2

Objective 3 Solve application problems requiring more than one operation.

Example 3 The following tables show the pricing schedule and long-distance calls made by Glenda.

Long-distance rate schedule

Time of call	Rate
8 A.M.–5 P.M.	$0.12/minute
5 P.M.–10 P.M.	$0.10/minute
10 P.M.–8 A.M.	$0.09/minute

Glenda's long-distance summary

Time of call	City	Minutes
7:05 A.M.	Los Angeles, Calif.	37
9:52 A.M.	Columbus, Ohio	21
1:12 A.M.	Boston, Mass.	13
4:47 P.M.	Atlanta, Ga.	24
7:35 P.M.	Los Angeles, Calif.	20
10:05 P.M.	Richmond, Va.	13

a. Find the mean, median, and mode duration of Glenda's long-distance calls.

Solution:

$$\text{Mean} = \frac{37 + 21 + 13 + 24 + 20 + 13}{6} = \frac{128}{6} = 21.\overline{3} \text{ min.}$$

Divide the sum of the minutes by the number of calls.

For the median, we arrange the call durations in order from least to greatest.

13, 13, 20, 21, 24, 37

$$\text{Median} = \frac{20 + 21}{2} = \frac{41}{2} = 20.5 \text{ min.}$$

Because there is an even number of calls, the median duration is the mean of the middle two.

$\text{Mode} = 13 \text{ min.}$ The mode duration is the duration that occurs most often, in this case 13 minutes.

b. Find the total cost of the calls.

Solution: The unit price depends on the time of the call. Each call except the call to Atlanta occurred at a time so that a single unit price applies.

Because the call to Atlanta began at 4:47 P.M. and lasted until 5:11 P.M., the first 13 minutes of the call cost $0.12 per minute and the last 11 minutes, which occurred after 5:00 P.M., cost $0.10 per minute.

Let's add a column to the table of the cost of each call found by multiplying the number of minutes by the unit price. The total cost is found by adding all of those costs.

Time of call	City	Minutes	Cost
7:05 A.M.	Los Angeles, Calif.	37	$37(0.09) = \$3.33$
9:52 A.M.	Columbus, Ohio	21	$21(0.12) = \$2.52$
1:12 P.M.	Boston, Mass.	13	$13(0.12) = \$1.56$
4:47 P.M.	Atlanta, Ga.	24	$13(0.12) + 11(0.10) = \$2.66$
7:35 P.M.	Los Angeles, Calif.	20	$20(0.10) = \$2.00$
10:05 P.M.	Richmond, Va.	13	$13(0.09) = \$1.17$

Total cost = $13.24

▶ **Do Your Turn 3**

Objective 4 Find weighted means.

Sometimes when we calculate a mean, some scores in the data set have more weight than others. An example of a weighted mean is **grade point average (GPA)**, where each letter grade has a **grade point** value.

Definitions **Grade point average (GPA):** The sum of the total grade points earned divided by the total number of credit hours.

Grade point: A numerical value assigned to a letter grade.

The following list shows commonly used grade point equivalents for each letter grade.

$$
\left.
\begin{array}{l}
A = 4 \\
B+ = 3.5 \\
B = 3 \\
C+ = 2.5 \\
C = 2 \\
D+ = 1.5 \\
D = 1 \\
F = 0
\end{array}
\right\}
$$

◀ Number of grade points per credit hour of the course

Note Some schools also use − grades such as A−, B−, C−, and D−. Other schools do not use + or − grades at all.

We say that GPA is a *weighted* mean because we determine the total number of grade points for a course by multiplying the letter grade's equivalent grade point value by the course's number of credit hours. Suppose you receive an A in a 3-credit-hour course. Because an A is worth 4 grade points for each credit of the course, we multiply $4 \cdot 3$ to equal 12 grade points for that course.

To calculate the GPA, we must calculate the total grade points for all courses, then divide by the total number of credit hours.

$$\text{GPA} = \frac{\text{total grade points}}{\text{total credits}}$$

Connection We can use dimensional analysis.

$$\frac{4 \text{ grade points}}{\text{credit hour}} \cdot \frac{3 \text{ credit hours}}{1} = 12 \text{ grade points}$$

Example 4 A student's grade report is shown below. Calculate the student's GPA rounded to the nearest thousandth.

Course	Credits	Grade
MATH 100	5	B+
ENG 101	3	C+
PSY 101	3	C
BIO 101	4	A

▶ **Your Turn 3**

Use the rate schedule in Example 3 to solve the following.

a. Find the mean, median, and mode duration of the calls made in the following table.

b. Find the total cost of the calls.

Long-distance summary

Time of call	City	Minutes
8:25 P.M.	Dallas, Tex.	12
9:45 A.M.	New York, N.Y.	34
6:35 P.M.	Chicago, Il.	17
4:52 P.M.	Denver, Colo.	22

Answers to Your Turn 3
a. mean: 21.25 min.; median: 19.5 min; no mode
b. $9.34

A student's grade report is shown below. Calculate the student's GPA rounded to the nearest thousandth.

Course	Credits	Grade
MATH 100	5	C
ENG 101	3	A
PSY 101	3	B+
BIO 101	4	C+

Solution: First, calculate each course's grade points by multiplying the course credits by the value of the letter grade received. Then add all of those results to find the total grade points.

Course	Credits	Grade	Grade points
MATH 100	5	B+	$5(3.5) = 17.5$
ENG 101	3	C+	$3(2.5) = 7.5$
PSY 101	3	C	$3(2.0) = 6.0$
BIO 101	4	A	$4(4.0) = 16.0$
Totals	15		47.0

Note
◀ B+ = 3.5 grade points/credit
C+ = 2.5 grade points/credit
C = 2.0 grade points/credit
A = 4.0 grade points/credit

$$\text{GPA} = \frac{\text{total grade points}}{\text{total credits}} = \frac{47}{15} = 3.1\overline{3}$$

Divide the total grade points by the total number of credit hours.

Answer: Rounded to the nearest thousandth, this student's GPA is 3.133.

◀ **Do Your Turn 4**

Objective 5 Evaluate expressions.

Remember, to evaluate a variable expression, replace each variable in the expression with the corresponding value and then calculate following the order of operations agreement.

Evaluate each expression using the given values.

a. $mv^2; m = 10, v = 0.5$

b. $\dfrac{Mm}{d^2}; M = 3.5 \times 10^5,$
  $m = 4.8, d = 2$

c. $\left(1 + \dfrac{r}{n}\right)^{nt}; r = 0.06,$
  $n = 2, t = 0.5$

Example 5   Evaluate each expression using the given values.

a. $vt + \dfrac{1}{2}at^2; v = 30, t = 0.4, a = 12.8$

Solution:

$(30)(0.4) + \dfrac{1}{2}(12.8)(0.4)^2$ Replace v with 30, t with 0.4, and a with 12.8.

$= (30)(0.4) + 0.5(12.8)(0.4)^2$ Write $\frac{1}{2}$ as 0.5.

$= (30)(0.4) + 0.5(12.8)(0.16)$ Evaluate the exponential form. $(0.4)^2 = 0.16$

$= 12 + 1.024$ Multiply. $(30)(0.4) = 12$ and $0.5(12.8)(0.16) = 1.024$

$= 13.024$ Add. $12 + 1.024 = 13.024$

b. $\left(1 + \dfrac{r}{n}\right)^{nt}; r = 0.08, n = 4, t = 0.5$

Solution: $\left(1 + \dfrac{0.08}{4}\right)^{(4)(0.5)}$ Replace r with 0.08, n with 4, and t with 0.5.

Note When an exponent contains operations, we must ◀ simplify those operations before evaluating the exponential form.

$= (1 + 0.02)^2$ Divide 0.08 by 4 in the parentheses and multiply $(4)(0.5)$ to simplify the exponent.

$= (1.02)^2$ Add in the parentheses.

$= 1.0404$ Square 1.02.

◀ **Do Your Turn 5**

Objective 6 Find the area of a triangle, trapezoid, and circle.

Now that we have seen how to simplify expressions containing both fractions and decimal numbers, let's revisit finding the area of a triangle, a trapezoid, and a circle. Below we list the formulas and the sections where we learned them.

Area of a triangle: $A = \dfrac{1}{2}bh$ Section 5.3

Area of a trapezoid: $A = \dfrac{1}{2}h(a + b)$ Section 5.7

Area of a circle: $A = \pi r^2$ Section 5.7

Example 6

a. Find the area of the triangle shown.

Solution: Use the formula $A = \dfrac{1}{2}bh$.

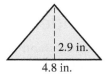
2.9 in.
4.8 in.

$A = \dfrac{1}{2}(4.8)(2.9)$ **Replace b with 4.8 and h with 2.9.**

$A = \dfrac{1}{2}(13.92)$ **Multiply 4.8 and 2.9.**

$A = \dfrac{1}{2} \cdot \dfrac{13.92}{1}$ **Write 13.92 over 1.**

$A = \dfrac{13.92}{2}$ **Multiply.**

$A = 6.96$ **Divide 13.92 by 2.**

Note Although we chose to keep $\frac{1}{2}$ in fraction form, we could have written it as a decimal and then multiplied.
$A = 0.5(4.8)(2.9)$
$A = 2.4(2.9)$
$A = 6.96$

Answer: The area is 6.96 in.^{2}.

b. Find the area of the trapezoid shown.

Solution: Use the formula $A = \frac{1}{2}h(a + b)$.

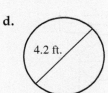
5.34 m
9 m
4.86 m

$A = \dfrac{1}{2}(9)(4.86 + 5.34)$ **Replace h with 9, a with 4.86, and b with 5.34.**

$A = \dfrac{1}{2}(9)(10.2)$ **Add in the parentheses.**

$A = \dfrac{1}{2}(91.8)$ **Multiply 9 by 10.2.**

$A = \dfrac{1}{2} \cdot \dfrac{91.8}{1}$ **Write 91.8 over 1.**

$A = \dfrac{91.8}{2}$ **Multiply.**

$A = 45.9$ **Divide 91.8 by 2.**

Answer: The area is 45.9 m^2.

c. Find the area of the circle shown. Use 3.14 for π.

Solution: Use the formula $A = \pi r^2$. We are given the diameter of 5.6 centimeters, but the formula requires the radius. Because radius is half the diameter, the radius is 2.8 centimeters.

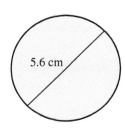
5.6 cm

Your Turn 6

Find the area. For parts c and d, use 3.14 for π.

a.
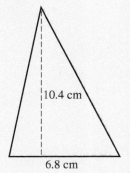
10.4 cm
6.8 cm

b.
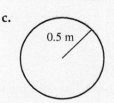
0.4 km
0.6 km
0.8 km

c.

0.5 m

d.
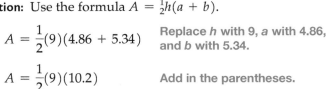
4.2 ft.

Answers to Your Turn 6
a. 35.36 cm^2 b. 0.36 km^2
c. 0.785 m^2 d. 13.8474 ft.^{2}

$$A = \pi(2.8)^2 \quad \text{Replace } r \text{ with 2.8.}$$
$$A = \pi(7.84) \quad \text{Square 2.8.}$$
$$A = 7.84\pi \quad \text{Express the exact answer with } \pi.$$
$$A \approx 7.84(3.14) \quad \text{Replace } \pi \text{ with 3.14.}$$
$$A \approx 24.6176 \quad \text{Multiply.}$$

Note Because π is irrational, 7.84π is the exact representation of the area. When we use an approximate value for π, the calculation becomes an approximation. This is why we use $\approx$.

Answer: The area is approximately 24.6176 cm². ◄ **Note** Remember, area is always expressed in square units. We write square centimeters as cm².

◄ **Do Your Turn 6**

Objective 7 Find the volume of a cylinder, pyramid, cone, and sphere.

Cylinder

To understand how to derive the formula for the volume of a cylinder, let's look more closely at the formula for the volume of a box, $V = lwh$. Notice that the base of a box is a rectangle, and we calculate the area of that base by multiplying l by w. So the formula for the volume of a box is a result of multiplying the area of the base, lw, by the height of the box, h.

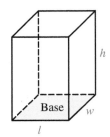

$$\text{Volume} = \text{area of the base} \cdot \text{height}$$
$$V = \quad lw \quad \cdot \quad h$$

We can apply this same approach to deriving the volume of any three-dimensional object that is formed by repeatedly stacking a shape. To understand what we mean by repeatedly stacking a shape, think about a box as a stack of papers. A single sheet of paper is very thin when viewed along its edge but is a rectangle when viewed from the top or bottom. When we stack many of these thin rectangles on top of each other, we form a box shape.

A cylinder is formed by repeatedly stacking a circular base. Imagine coins stacked on top of each other. Following the same approach that we used for a box, we find the volume of a cylinder by multiplying the area of its base, which is πr^2, by its height, h.

$$\text{Volume} = \text{area of the base} \cdot \text{height}$$
$$V = \quad \pi r^2 \quad \cdot \quad h$$
$$V = \pi r^2 h$$

► Your Turn 7

Find the volume. Use 3.14 for π.

a. 6.2 in. 1.2 in.

b. 1.5 cm 4.8 cm

Answers to Your Turn 7
a. ≈ 144.84 in.³
b. ≈ 33.912 cm³

Example 7   Find the volume of the cylinder. Use 3.14 for π.

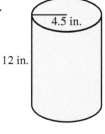

Solution: Use the formula $V = \pi r^2 h$.

$$V = \pi(4.5)^2(12) \quad \text{Replace } r \text{ with 4.5 and } h \text{ with 12.}$$
$$V = \pi(20.25)(12) \quad \text{Square 4.5.}$$
Exact answer → $V = 243\pi$ Multiply 20.25 by 12.
$$V \approx 243(3.14) \quad \text{Replace } \pi \text{ with 3.14.}$$
Approximation → $V \approx 763.02$ Multiply.

Note Remember, volume is always expressed in cubic units. We write cubic inches as in.³.

Answer: The volume is approximately 763.02 in.³.

◄ **Do Your Turn 7**

Pyramid

To help you understand the formula for the volume of a four-sided pyramid, notice that one would fit inside a box that has a base of the same length and width. The pyramid's volume occupies $\frac{1}{3}$ the volume of the box.

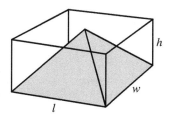

$$V = \frac{1}{3} \cdot \text{volume of a box with the same length, width, and height}$$

$$V = \frac{1}{3}lwh$$

Example 8   Find the volume of the pyramid.

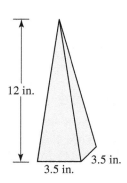

Solution: Use the formula $V = \frac{1}{3}lwh$.

$V = \frac{1}{3}(3.5)(3.5)(12)$ Replace *l* with 3.5, *w* with 3.5, and *h* with 12.

$V = \frac{1}{3}(147)$ Multiply 3.5, 3.5, and 12.

$V = \frac{1}{3} \cdot \frac{147}{1}$ Write 147 over 1.   ◄

$V = \frac{147}{3}$ Multiply.

$V = 49$ Divide 147 by 3.

Note We avoid the decimal equivalent of $\frac{1}{3}$ because it is a nonterminating decimal and its use would lead to an approximate answer.

Answer: The volume is 49 in.^{3}.

► Do Your Turn 8

► Your Turn 8

Find the volume.

a.

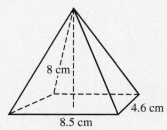

b.

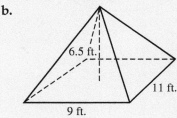

Cone

A cone is much like a pyramid except that the base is a circle. Also, like a pyramid, which has $\frac{1}{3}$ the volume of a box with the same-size base and height, a cone has $\frac{1}{3}$ the volume of a cylinder with the same-size base and height.

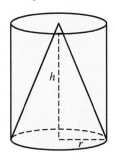

$$V = \frac{1}{3} \cdot \text{volume of a cylinder with the same-size} \\ \text{base and height}$$

$$V = \frac{1}{3}\pi r^2 h$$

Example 9   Find the volume of the cone. Use 3.14 for π.

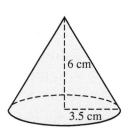

Solution: Use the formula $V = \frac{1}{3}\pi r^2 h$.

$$V = \frac{1}{3}\pi (3.5)^2 (6)$$ Replace *r* with 3.5 and *h* with 6.

Answers to Your Turn 8
a. $104.2\overline{6}$ cm^3 **b.** 214.5 ft.^{3}

► **Your Turn 9**

Find the volume. Use 3.14 for π.

a.

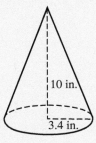

10 in.

3.4 in.

b.

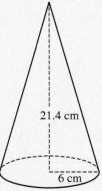

21.4 cm

6 cm

► **Your Turn 10**

Solve. Use 3.14 for π.

a. Find the volume of a sphere with a radius of 3 feet.

b. Find the volume of a sphere with a radius of 8 centimeters.

Answers to Your Turn 9
a. ≈ 120.99 in.^{3}
b. ≈ 806.352 cm^3

Answers to Your Turn 10
a. ≈ 113.04 ft.^{3}
b. $\approx 2143.57\overline{3}$ cm^3

$$V = \frac{1}{3}\pi(12.25)(6)$$ Square 3.5.

$$V = \frac{1}{3}(73.5)\pi$$ Multiply 12.25 by 6.

$$V = \frac{1}{3} \cdot \frac{73.5}{1}\pi$$ Write 73.5 over 1.

$$V = \frac{73.5}{3}\pi$$ Multiply.

Exact answer → $V = 24.5\pi$ Divide 73.5 by 3.

$V \approx 24.5(3.14)$ Replace π with 3.14.

Approximation → $V \approx 76.93$ Multiply.

Answer: The volume is approximately 76.93 cm^3.

◄ **Do Your Turn 9**

Sphere

A ball is a sphere. Every point on the surface of a sphere is equally distant from the center. It turns out that the volume of a sphere is equal to four cones that have the same radius as the sphere and height equal to the radius.

$$\text{Volume of one such cone} = \frac{1}{3}\pi r^2 \cdot r = \frac{1}{3}\pi r^{2+1} = \frac{1}{3}\pi r^3$$

▲

Note Because the height of the cone is equal to the radius of the sphere, we use r in place of h.

Because the volume of a sphere is equal to four of these cones, we must multiply the expression we found previously by 4.

$$\text{Volume of a sphere} = 4 \cdot \frac{1}{3}\pi r^3 = \frac{4}{1} \cdot \frac{1}{3}\pi r^3 = \frac{4}{3}\pi r^3$$

The volume of a sphere can be found using the formula $V = \frac{4}{3}\pi r^3$.

Example 10   Find the volume of the sphere. Use 3.14 for π.

Solution: Use the formula $V = \frac{4}{3}\pi r^3$.

$$V = \frac{4}{3}\pi(0.2)^3$$ Replace r with 0.2.

$$V = \frac{4}{3}\pi(0.008)$$ Cube 0.2.

$$V = \frac{4}{3}\left(\frac{0.008}{1}\right)\pi$$ Write 0.008 over 1.

$$V = \frac{0.032}{3}\pi$$ Multiply.

$V = 0.010\overline{6}\pi$ Divide 0.032 by 3.

$V \approx 0.010\overline{6}(3.14)$ Replace π with 3.14.

$V \approx 0.033493$ Multiply.

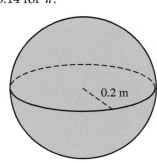

0.2 m

Answer: The volume is approximately $0.03349\overline{3}$ m^3.

◄ **Do Your Turn 10**

Objective 8 Find the area and volume of a composite form.

Composite forms are geometric forms that combine several basic forms. The key to finding the area of a composite form is to determine whether two or more forms have been put together or whether a form or forms have been removed from a larger form.

Example 11   Find the area. Use 3.14 for π.

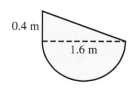

0.4 m

1.6 m

Understand This shape is a combination of a half-circle and a triangle. The area of a full circle is found by $A = \pi r^2$; so for a half-circle, the area formula is $A = \frac{1}{2}\pi r^2$. Note that 1.6 meters is the diameter; so the radius is 0.8 meter. Also, the formula for the area of a triangle is $A = \frac{1}{2}bh$.

Plan Add the area of the half-circle and the area of the triangle.

Execute $A = $ area of half-circle + area of triangle

$$A = \qquad \frac{1}{2}\pi r^2 \quad + \quad \frac{1}{2}bh$$

$A = \frac{1}{2}\pi(0.8)^2 \; + \; \frac{1}{2}(1.6)(0.4)$ **Replace r with 0.8, b with 1.6, and h with 0.4.**

$A = \frac{1}{2}\pi(0.64) \; + \; \frac{1}{2}(1.6)(0.4)$ **Square 0.8.**

$A = 0.5(0.64)\pi + 0.5(1.6)(0.4)$ **Write each $\frac{1}{2}$ as 0.5 and use the commutative property to exchange 0.64 and π.**

Exact answer → $A = 0.32\pi + 0.32$ **Multiply.**

$A \approx 0.32(3.14) + 0.32$ **Replace π with 3.14.**

$A \approx 1.0048 + 0.32$ **Multiply.**

Approximation → $A \approx 1.3248$ **Add.**

Answer The area of the shape is approximately 1.3248 m².

Check We can check by reversing each calculation. The check will be left to the reader.

▶ **Do Your Turn 11**

Example 12   The tub inside a washing machine is a small cylinder within a larger cylinder. The space between the cylinders is where the clothes are placed. Find the volume of the space for clothes inside the washing machine shown.

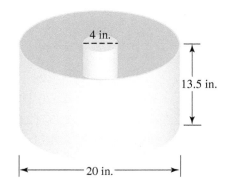

4 in.

13.5 in.

20 in.

Understand The volume of a cylinder is found by $V = \pi r^2 h$. The radius of the smaller cylinder is 2 inches. The radius of the larger cylinder is 10 inches. The height of both cylinders is 13.5 inches.

Plan Subtract the volume of the inner cylinder from the volume of the outer cylinder.

▶ **Your Turn 11**

Find the area. Use 3.14 for π.

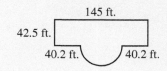

145 ft.

42.5 ft.

40.2 ft.   40.2 ft.

Answer to Your Turn 11

≈ 7800.47 ft.²

▶ **Your Turn 12**

Find the amount of storage space under the countertop. (Assume that the sink is half of a sphere.)

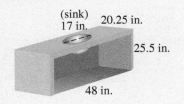

(sink) 17 in. 20.25 in.

25.5 in.

48 in.

Execute V = volume of outer cylinder − volume of inner cylinder

$$V = \qquad \pi r^2 h \qquad\qquad - \qquad\qquad \pi r^2 h$$

$V = \pi(10)^2(13.5) - \pi(2)^2(13.5)$ **Replace each r with the corresponding radius.**

$V = \pi(100)(13.5) - \pi(4)(13.5)$ **Square 10 and 2.**

$V = 1350\pi - 54\pi$ **Multiply.**

$V = 1296\pi$ **Combine like terms.**

$V \approx 1296(3.14)$ **Replace π with 3.14.**

$V \approx 4069.44$ **Multiply.**

◀ **Note** Because 1350π and -54π are both in terms of π, they are like terms. We can combine by subtracting the coefficients and keeping the result in terms of π.

Answer The space for clothes inside the washing machine is about 4069.44 in.³

◀ **Do Your Turn 12**

Of Interest

In any list of the greatest mathematicians, the name Archimedes is usually at the top. Archimedes was born in 287 B.C. in the city of Syracuse in what is now Sicily. In his youth, he studied in Alexandria, Egypt, where he met and befriended Eratosthenes (famous for his sieve method for finding primes).

In pure mathematics, Archimedes developed methods for finding the volume of many shapes and objects, including the cylinder, cone, and sphere. He developed a method for calculating π. He placed the value as being between $3\frac{10}{71}$ and $3\frac{1}{7}$. He even developed some of the ideas involved in calculus, preceding Isaac Newton (the accepted inventor of calculus) by 1800 years.

Answer to Your Turn 12
$\approx 23{,}500.43$ in.³

6.5 Exercises

For Extra Help   MyMathLab®

Objective 1

Prep Exercise 1 To simplify $\frac{2}{3}(4.8) + \frac{1}{4}$, which is better—writing the decimals as fractions or writing the fractions as decimals? Why?

For Exercises 1–14, simplify.

1. $0.64 + 2.5(0.8)$

2. $9.28 - 0.56(12)$

3. $-5.1(3.4 - 5.7)$

4. $-0.25(0.7 + 5.2)$

5. $88.2 - 2.2(0.45 + 20.1)$

6. $9.5(2.6 - 0.99) - 4.33$

7. $(0.4)^2 - 2.8 \div 0.2(1.6)$

8. $16.1 \div 0.4(0.3) + (1.3)^2$

9. $10.7 - 18\sqrt{1.96} + (74.6 - 88.1)$

10. $7.5 + 2.2\sqrt{0.25} - 36.8 \div 8$

11. $\sqrt{0.0081} + 120.8 \div 4(2.5 - 6.4)$

12. $40.1 - 6.9 \div 4.6(1.4)^2 + \sqrt{0.36}$

13. $[-5.53 \div (0.68 + 0.9)] + 0.7\sqrt{2.25}$

14. $[(6.2 - 10) \div 8] - 2\sqrt{1.21}$

Objective 2

For Exercises 15–22, simplify.

15. $\dfrac{3}{5}(-0.85)$

16. $\dfrac{7}{10}(12.88 - 4.38)$

17. $\dfrac{1}{4} + \dfrac{2}{3}(0.06)$

18. $(-12.9)\dfrac{5}{6} + \dfrac{3}{4}$

19. $-2\dfrac{4}{9}(1.8) - \dfrac{3}{8}$

20. $3\dfrac{1}{8} - (0.45)\left(5\dfrac{1}{3}\right)$

21. $\dfrac{2}{5} \div (-0.8) + \left(\dfrac{1}{3}\right)^2$

22. $\dfrac{1}{6} - \dfrac{1}{24}(0.4 + 1.2)$

Objective 3

For Exercises 23 and 24, the following table shows the pricing schedule for a phone company's long-distance rate.

Time	Rate
8 A.M.–5 P.M.	$0.12/minute
5 P.M.–10 P.M.	$0.10/minute
10 P.M.–8 A.M.	$0.09/minute

23. Below is Vicki's long-distance call summary.

    a. Find the mean, median, and mode duration of the calls.
    b. Find the total cost of the calls.

Time of call	City	Number of minutes
1:09 P.M.	Denver, Colo.	15
12:32 A.M.	Austin, Tex.	38
7:57 A.M.	Phoenix, Ariz.	12
8:47 P.M.	Portland, Oreg.	26

24. Below is a portion of Teshieka's long-distance call summary.

    a. Find the mean, median, and mode duration of the calls.
    b. Find the total cost of the calls.

Time of call	City	Number of minutes
5:01 P.M.	Boston, Mass.	22
9:07 A.M.	New York, N.Y.	64
7:32 A.M.	Washington, D.C.	31
3:08 P.M.	Orlando, Fla.	12

25. The following table contains the height in inches of the nine most recent presidents. Find the mean, median, and mode height.

President	Height	President	Height
Barack Obama	73	Jimmy Carter	69.5
George W. Bush	71.5	Gerald Ford	72
Bill Clinton	74	Richard Nixon	71.5
George H. W. Bush	74	Lyndon Johnson	75.5
Ronald Reagan	73		

Of Interest

The tallest president was Abraham Lincoln, who was 76 inches tall (6′ 4″). The second tallest was Lyndon Johnson at 75.5 inches tall (6′ 3½″). The shortest president was James Madison at 64 inches tall (5′ 4″).

26. The following table lists the height in inches of the first nine presidents. Find the mean, median, and mode height.

President	Height	President	Height
George Washington	74	John Quincy Adams	67
John Adams	67	Andrew Jackson	67
Thomas Jefferson	74.5	Martin Van Buren	66
James Madison	64	William Henry Harrison	68
James Monroe	72		

(*Source:* Wikipedia.)

27. The following graph shows the closing price of Walmart stock each day for eight consecutive days. Find the mean, median, and mode closing price.

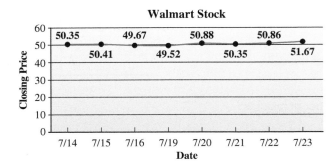

28. The following graph shows the closing price of Sony Corporation stock each day for eight consecutive days. Find the mean, median, and mode closing price.

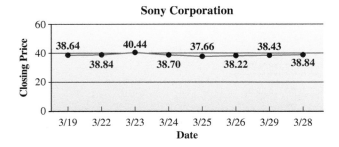

For Exercises 29 and 30, use the following table of a company's reimbursements for business travel.

Reimbursements for Business Travel
$0.35 per mile when using own vehicle
$0.15 per mile when using company vehicle
$10 breakfast when traveling between 12 A.M. and 11 A.M.
$12 lunch when traveling between 11 A.M. and 3 P.M.
$15 dinner when traveling between 3 P.M. and 12 A.M.

29. Aimee uses her own vehicle on a business trip. She leaves on a Tuesday at 11:30 A.M. and notes that the odometer reads 75,618.4. She returns on Thursday and arrives at 6:45 P.M. At the conclusion of her trip, she finds that the odometer reads 76,264.1. How much should she be reimbursed for mileage and food?

30. Miguel uses a company vehicle and leaves at 7 A.M. on a Monday. He notes that the odometer reads 45,981.6. He returns on Friday, arriving at 2:15 P.M. At the conclusion of the trip, he notes that the odometer reads 46,610.8. How much should he be reimbursed for mileage and food?

31. Chan receives the following printout of charges that he plans to pay on a 90-day same-as-cash option. He wants to split the charges into three equal payments. How much is each payment?

Description	Amount
1 EPS keyboard	$1645.95
2 cables	$14.95 each
Sales tax	$83.79

32. Shannon makes the following purchases for her new apartment, using a special credit agreement with the store. The credit line is interest-free as long as she pays off the balance in six months. After six months, all six months' worth of interest will be added to the remaining balance. To avoid the interest, she decides to split the charges into six equal payments. How much is each payment?

Description	Amount
1 table	$354.95
4 chairs	$74.95 each
Sales tax	$32.74

Objective 4

For Exercises 33–36, a student's grade report is shown. Calculate the student's GPA rounded to the nearest thousandth.

33.

Course	Credits	Grade
MATH 101	3.0	B
ENG 101	3.0	C
HIS 102	3.0	B+
CHM 101	4.0	A

34.

Course	Credits	Grade
MATH 100	5.0	B
ENG 100	3.0	C+
PSY 101	3.0	F
BIO 101	4.0	C

35.

Course	Credits	Grade
MATH 100	5.0	B
ENG 100	3.0	B+
PSY 101	3.0	D+
BIO 101	4.0	C
COL 101	2.0	A

36.

Course	Credits	Grade
MATH 101	3.0	A
ENG 100	3.0	C
SPA 101	3.0	A
PHY 101	4.0	D+
COL 101	2.0	B

Objective 5

For Exercises 37–46, evaluate each expression using the given values.

37. $\frac{1}{2}mv^2$; $m = 1.25$, $v = -8$

38. $\frac{1}{2}mv^2$; $m = 84.6$, $v = 0.2$

Of Interest

$\frac{1}{2}mv^2$ is an expression used to calculate the energy of a moving object, where m is the mass of the object and v is the velocity. As a formula, we would write $E = \frac{1}{2}mv^2$.

39. $\dfrac{w}{h^2} \cdot 705; w = 150, h = 67$

40. $\dfrac{w}{h^2} \cdot 705; w = 185, h = 70$

Of Interest

$\dfrac{w}{h^2} \cdot 705$ is used to calculate a person's body mass index (BMI), where w is the person's weight in pounds and h is the person's height in inches. According to the National Institute of Health, people with a BMI of 25 or more have an increased risk for cardiovascular and other diseases.

41. $mc^2; m = 2.5 \times 10^6, c = 3 \times 10^8$

42. $mc^2; m = 3.6 \times 10^4, c = 3 \times 10^8$

Of Interest

mc^2 is an expression developed by Albert Einstein to calculate the energy of a particle, where m is the mass of the particle and c is the speed of light. It appears in possibly the most famous formula of physics: $E = mc^2$.

43. $vt + \dfrac{1}{2}at^2; v = 20, t = 0.6, a = -12.5$

44. $vt + \dfrac{1}{2}at^2; v = 35, t = 1.2, a = 10$

Of Interest

$vt + \frac{1}{2}at^2$ is an expression used to calculate the distance an object travels that has an initial velocity, v, and acceleration or deceleration, a, over a period of time, t. As a formula, we would write $d = vt + \frac{1}{2}at^2$.

45. $\left(1 + \dfrac{r}{n}\right)^{nt}; r = 0.12, n = 4, t = 0.5$

46. $\left(1 + \dfrac{r}{n}\right)^{nt}; r = 0.04, n = 8, t = 0.25$

Of Interest

$\left(1 + \frac{r}{n}\right)^{nt}$ is a part of the formula used for calculating the balance in an account that has earned interest at an interest rate of r compounded n times per year in t years. We will study the formula for compound interest in Section 8.5.

Objective 6

For Exercises 47–52, find the area. Use 3.14 for π.

47.

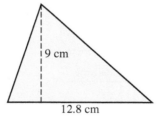

9 cm

12.8 cm

48.

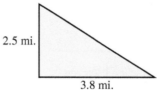

2.5 mi.

3.8 mi.

49.

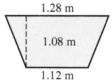

1.28 m

1.08 m

1.12 m

50.

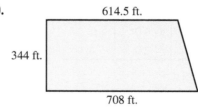

614.5 ft.

344 ft.

708 ft.

51.

10.5 in.

52.

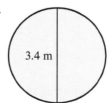

3.4 m

Objective 7

Prep Exercise 2 What are the formulas for the volume of a cylinder and a cone? How are they related?

Prep Exercise 3 What is the formula for the volume of a pyramid? How is it related to the volume of a box?

Prep Exercise 4 What is the formula for the volume of a sphere? What do its symbols represent?

For Exercises 53–60, find the volume. Use 3.14 for π.

53. A quarter 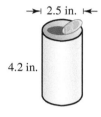 0.2 cm
← 2.4 cm →

54. Canned drink ↦ 2.5 in. ↤

4.2 in.

55. The Castillo at Chichen Itza in the Yucatan forest in Mexico is 90 feet tall.

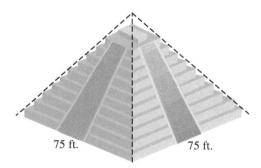

75 ft.   75 ft.

Of Interest

The Castillo at Chichen Itza is a pyramid temple built by the Mayans around A.M. 500. The pyramid is a calendar. There are four sets of steps, one set on each side of the pyramid, with 91 steps in each. At the top is a small temple with 1 step on all four sides for a total of 365 steps, one for each day of the year. At the bottom of each set of steps are serpent heads that stare out from either side of the steps. At the equinoxes, the sunlight hits the ridged sides of the pyramid at the perfect angle to cast a shadow that connects with the serpent head, giving the appearance of a serpent snaking its way down the pyramid.

56. Menkaure's pyramid at Giza, Egypt, is 203 feet tall.

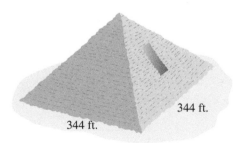

344 ft.

344 ft.

Of Interest

Menkaure was king of Egypt from 2490 to 2472 B.C. His pyramid at Giza is the smallest of the three pyramids. The original height of the pyramid was to be 215 feet, but it is only 203 feet because the smooth outer case of granite was never completed.

57. Onyx decorative cone

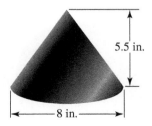

5.5 in.

← 8 in. →

58. Afterburner fire cone from a jet engine

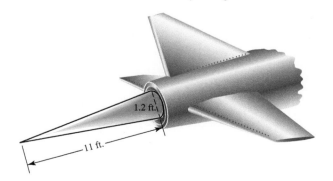

1.2 ft.

11 ft.

59. If Earth is a sphere with a radius of 3884.3 miles, find the volume of Earth.

60. If the sun is a sphere with a radius of 423,858 miles, find the volume of the sun.

Objective 8

For Exercises 61 and 62, find the area of the composite shape. Use 3.14 for π.

61.

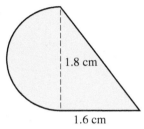

1.8 cm

1.6 cm

62.

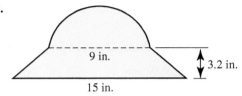

9 in.

3.2 in.

15 in.

For Exercises 63 and 64, find the area of the shaded region. Use 3.14 for π.

63.

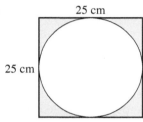

25 cm

25 cm

64.

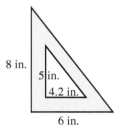

8 in.

5 in.

4.2 in.

6 in.

For Exercises 65 and 66, find the volume of the composite object. Use 3.14 for π.

65. Water tower

64 ft.

45.5 ft.

20 ft.

66. Minisub

1.8 m

2 m   2.1 m   0.9 m

67. A pyramid built with cut blocks of stone has an inner chamber that is 5.8 meters long by 5.2 meters wide by 10.8 meters high. The height of the pyramid is 146.5 meters. Find the volume of stone used to build the pyramid.

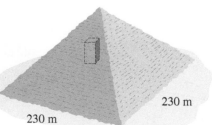

230 m

230 m

68. The diameter of a bowling ball is 10 inches. Each hole is a cylinder 3 inches deep with a 1-inch diameter. Find the volume of material used to make the ball. Use 3.14 for π.

69. a. A grain of salt is a cube that is 0.03 inch along each side. What is the volume of a grain of salt?

    b. About how many grains of salt would be in 1 cubic inch?

70. A grain of salt is a cube that is 0.03 inch along each side. Calculate the surface area of a grain of salt.

Puzzle Problem Estimate the volume of our galaxy. It is a spiral galaxy 100,000 light-years across and about 1000 light-years thick at the center. Use 3.14 for π. (*Hint*: Imagine two identical cones joined at the base.)

Review Exercises

For Exercises 1 and 2, solve and check.

[4.3] 1. $4(x - 6) + 1 = 9x - 2 - 8x$

[5.8] 2. $\frac{1}{3}y - \frac{3}{4} = 6$

[5.7] 3. Find the area of the composite shape.

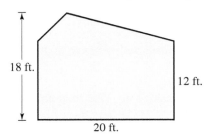

18 ft.
12 ft.
20 ft.

[5.7] 4. Find the area of the shaded region.

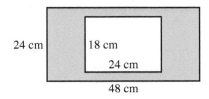

24 cm
18 cm
24 cm
48 cm

[5.8] 5. Translate to an equation; then solve. $\frac{1}{2}$ the sum of x and 12 is equal to 9 less than $\frac{3}{4}$ of x.

[4.5] 6. Marla has $160 in ten- and five-dollar bills. If she has 4 more tens than fives, how many of each bill does she have?

6.6 Solving Equations and Problem Solving

Objectives

1 Solve equations using the addition and multiplication principles of equality.

2 Eliminate decimal numbers from equations using the multiplication principle.

3 Solve problems involving one unknown.

4 Solve problems using the Pythagorean theorem.

5 Solve problems involving two unknowns.

▶ **Your Turn 1**

Solve and check.
a. $4.5y - 21.99 = 6$
b. $0.08n + 0.4 = 0.244$

Warm-up

[6.5] 1. Simplify. $\frac{2}{3}(16.8) + 10.9$

[5.8] 2. Solve. $\frac{5}{6}x + \frac{2}{3} = \frac{3}{4}$

Objective 1 Solve equations using the addition and multiplication principles of equality.

Example 1 Solve and check. $6 - 4.8t = 3.552$

Solution:

$$6 - 4.8t = 3.552$$
$$\underline{-6 \qquad\qquad -6} \qquad \text{Subtract 6 from both sides to isolate } -4.8t.$$
$$0 - 4.8t = -2.448$$
$$\frac{-4.8t}{-4.8} = \frac{-2.448}{-4.8} \qquad \text{Divide both sides by } -4.8 \text{ to isolate } t.$$
$$t = 0.51$$

Check:

$$6 - 4.8t = 3.552$$
$$6 - 4.8(0.51) \stackrel{?}{=} 3.552 \qquad \text{In the original equation, replace } t \text{ with 0.51.}$$
$$6 - 2.448 \stackrel{?}{=} 3.552$$
$$3.552 = 3.552 \qquad \text{True; so 0.51 is the solution.}$$

◀ **Do Your Turn 1**

Objective 2 Eliminate decimal numbers from equations using the multiplication principle.

In Chapter 5, we learned to eliminate fractions from an equation by multiplying both sides by the LCD (lowest common denominator). We can use the same method to eliminate decimal numbers from an equation, except that we multiply both sides by an appropriate power of 10. Remember, multiplying a decimal number by a power of 10 moves the decimal point to the right; so by using a large enough power of 10, the resulting equation will contain only integers. The number with the most decimal places determines the minimum power of 10 needed. For example, in $2.5x - 0.64 = 8$, because 0.64 has the most decimal places (two), we need to multiply both sides of the equation by at least 100.

$$100(2.5x - 0.64) = 100 \cdot 8 \qquad \text{Multiply both sides by 100.}$$
$$100 \cdot 2.5x - 100 \cdot 0.64 = 100 \cdot 8 \qquad \text{Distribute 100.}$$
$$250x - 64 = 800$$

▲

Note Multiplying both sides by 100 transforms the equation from a decimal equation to an equation that contains only integers.

With decimal numbers, the LCD will be a power of 10 with the same number of 0 digits as decimal places in the number with the most decimal places. We can amend our procedure for solving equations to include this technique.

Connection Writing the fraction equivalents of the decimal numbers in $2.5x - 0.64 = 8$ gives $\frac{25}{10}x - \frac{64}{100} = 8$. Notice that the LCD is 100; so this really is the same method as presented in Chapter 5.

Answers to Your Turn 1
a. $y = 6.22$ b. $n = -1.95$

Answers to Warm-up
1. 22.1
2. $\frac{1}{10}$

Procedure

To solve a linear equation in one variable:
1. Simplify each side of the equation.
   a. Distribute to eliminate parentheses.
   b. Eliminate fractions and/or decimal numbers by multiplying both sides of the equation by the LCD. In the case of decimal numbers, the LCD is the power of 10 with the same number of 0 digits as decimal places in the number with the most decimal places.
   c. Combine like terms.
2. Use the addition principle of equality so that all variable terms are on one side of the equation and all constants are on the other side. Then combine like terms.

> **Note** Eliminate the variable term with the lesser coefficient to avoid negative coefficients.

3. Use the multiplication principle of equality to eliminate the coefficient of the variable.

Let's look at Example 1 again using this technique.

Example 2 Solve and check. $6 - 4.8t = 3.552$

Solution:

$$1000 \cdot (6 - 4.8t) = 1000 \cdot 3.552$$ Multiply both sides by 1000 to eliminate the decimal numbers.

$$1000 \cdot 6 - 1000 \cdot 4.8t = 1000 \cdot 3.552$$ Distribute 1000.

$$6000 - 4800t = 3552$$ Multiply.

$$\underline{-6000 \qquad\qquad -6000}$$ Subtract 6000 from both sides to isolate $-4800t$.

$$0 - 4800t = -2448$$

$$\frac{-4800t}{-4800} = \frac{-2448}{-4800}$$ Divide both sides by -4800 to isolate t.

$$t = 0.51$$

Check:

$$6 - 4.8t = 3.552$$

$$6 - 4.8(0.51) \overset{?}{=} 3.552$$ In the original equation, replace t with 0.51.

$$6 - 2.448 \overset{?}{=} 3.552$$

$$3.552 = 3.552$$ True; so 0.51 is the solution.

▶ **Do Your Turn 2**

Example 3 Solve and check. $0.3(x - 1.2) = 1.46 - (2.1x + 0.22)$

Solution:

$$0.3(x - 1.2) = 1.46 - (2.1x + 0.22)$$

$$0.3x - 0.36 = 1.46 - 2.1x - 0.22$$ Distribute to eliminate parentheses.

> **Note** 3.552 has the most decimal places with three. To transform to an integer, we need its decimal point to move at least three places to the right; so 1000 is the minimum power of ten we can use. Because all of the other numbers have fewer decimal places, they too will become or remain integers after being multiplied by 1000.

▶ **Your Turn 2**

Solve and check.
 a. $4.5y - 21.99 = 6$
 b. $0.08n + 0.4 = 0.244$

Answers to Your Turn 2
a. $y = 6.22$ b. $n = -1.95$

▶ **Your Turn 3**

Solve and check.

a. $18.6k - 1.25 = 9.4k + 1.326$

b. $11 - 2(5.6 + x) = 0.7x + 1.3$

$$100 \cdot (0.3x - 0.36) = 100 \cdot (1.46 - 2.1x - 0.22)$$ **Multiply both sides by 100 to eliminate the decimal numbers.**

$$100 \cdot 0.3x - 100 \cdot 0.36 = 100 \cdot 1.46 - 100 \cdot 2.1x - 100 \cdot 0.22$$ **Distribute.**

$$30x - 36 = 146 - 210x - 22$$ **Multiply.**

$$30x - 36 = 124 - 210x$$ **Combine like terms.**

$$\underline{+ 210x \qquad\qquad\qquad + 210x}$$ **Add 210x to both sides so that the x terms are on the same side of the equation.**

$$240x - 36 = 124 + 0$$

$$240x - 36 = 124$$

$$\underline{+ 36 \quad + 36}$$ **Add 36 to both sides to isolate 240x.**

$$240x + 0 = 160$$

$$\frac{240x}{240} = \frac{160}{240}$$ **Divide both sides by 240 to isolate x.**

$$x = 0.\overline{6}$$

Check: Because $0.\overline{6}$ is a nonterminating decimal number, we must round to check with it. Consequently, the result on each side will be close to the same value but not exactly the same. We will round $0.\overline{6}$ to the nearest thousandth, 0.667.

$$0.3(x - 1.2) = 1.46 - (2.1x + 0.22)$$

$$0.3(0.667 - 1.2) \stackrel{?}{=} 1.46 - (2.1[0.667] + 0.22)$$ **Replace x with 0.667.**

$$0.3(-0.533) \stackrel{?}{=} 1.46 - (1.4007 + 0.22)$$

$$-0.1599 \stackrel{?}{=} 1.46 - 1.6207$$

$$-0.1599 \approx -0.1607$$

Note As predicted, the values are not equal, but they are reasonably close. If we had rounded to ten-thousandths or hundred-thousandths, the results would be even closer.

Note Using the fraction equivalent of $0.\overline{6}$, which is $\frac{2}{3}$, would give the same result on both sides. However, we have not learned how to find the equivalent fraction of a decimal number with repeating digits. For now, unless you happen to remember the equivalent fraction, round the repeating decimal to check.

◀ **Do Your Turn 3**

Objective 3 Solve problems involving one unknown.

Let's consider some problems where we translate key words directly to an equation.

Example 4   5.9 more than 0.2 times n is equal to 6.7 less than the product of 1.4 and n. Translate to an equation; then solve for n.

Solution: Translate:

<u>5.9 more than 0.2 times n</u> <u>is equal to</u> <u>6.7 less than the product of 1.4 and n</u>.

$$\underset{5.9 + 0.2n}{\downarrow} \qquad\qquad \underset{=}{\downarrow} \qquad\qquad \underset{1.4n - 6.7}{\downarrow}$$

Note When we translate *less than* to subtraction, the subtrahend appears before the words *less than* and the minuend appears after.

Solve: $10(5.9 + 0.2n) = 10(1.4n - 6.7)$ **Multiply both sides by 10 to eliminate decimal numbers.**

$$10 \cdot 5.9 + 10 \cdot 0.2n = 10 \cdot 1.4n - 10 \cdot 6.7$$

$$59 + 2n = 14n - 67$$ **Subtract 2n from both sides.**

$$\underline{-2n \qquad -2n}$$

$$59 + 0 = 12n - 67$$

$$59 = 12n - 67$$

$$\underline{+ 67 \qquad\qquad + 67}$$ **Add 67 to both sides.**

$$126 = 12n + 0$$

$$\frac{126}{12} = \frac{12n}{12}$$ **Divide both sides by 12.**

$$10.5 = n$$

Answers to Your Turn 3

a. $k = 0.28$ b. $x = -0.\overline{5}$

Check:

$$5.9 + 0.2n = 1.4n - 6.7$$

$$5.9 + 0.2(10.5) \stackrel{?}{=} 1.4(10.5) - 6.7 \qquad \text{In the original equation, replace } n \text{ with 10.5.}$$

$$5.9 + 2.1 \stackrel{?}{=} 14.7 - 6.7$$

$$8 = 8 \qquad\qquad \text{It checks.}$$

▶ **Do Your Turn 4**

▶ **Your Turn 4**

Solve.

a. 18.6 less than y is the same as 1.25 times y plus 2.18. Translate to an equation; then solve for y.

b. 0.4 times the difference of m and 1.8 is equal to -0.8 times m minus 9.36. Translate to an equation; then solve for m.

Example 5   Karen uses a cell phone service that charges $39.99 per month for 450 minutes. After 450 minutes, it costs $0.45 for each additional minute. If Karen's bill is $82.29, how many total minutes did she use?

Understand We must calculate Karen's total cell phone use. The total charges are $82.29. To calculate the total charges, the company charged a flat fee of $39.99 for the first 450 minutes and $0.45 for each additional minute.

$$\text{flat fee} + \text{cost of additional minutes} = \text{total charges}$$

Because $0.45/minute is a unit price, the cost of additional minutes is found by multiplying the additional minutes by 0.45. If we let m be the number of additional minutes, $0.45m$ describes the cost of those additional minutes.

Plan Write an equation and then solve.

Execute flat fee + cost of additional minutes = total charges

$$39.99 \quad + \qquad\quad 0.45m \qquad\quad = 82.29$$

$$100(39.99 + 0.45m) = 100(82.29) \qquad \text{Multiply both sides by 100 to eliminate decimals.}$$

$$100 \cdot 39.99 + 100 \cdot 0.45m = 100 \cdot 82.29 \qquad \text{Distribute 100.}$$

$$3999 + 45m = 8229 \qquad\qquad \text{Subtract 3999 from both sides.}$$

$$\underline{-3999 \qquad\qquad -3999}$$

$$0 + 45m = 4230$$

$$\frac{45m}{45} = \frac{4230}{45} \qquad\qquad \text{Divide both sides by 45.}$$

$$m = 94$$

Answer Karen used her cell phone 94 additional minutes. This is in addition to the 450 minutes she paid for with $39.99; so she used a total of

$$450 + 94 = 544 \text{ minutes}$$

Check Verify that 544 minutes at $39.99 for the first 450 minutes and $0.45 per minute for the additional 94 minutes comes to a total of $82.29.

$$\text{total charges} = 39.99 + 0.45(94)$$

$$= 39.99 + 42.30$$

$$= 82.29 \qquad\qquad \text{It checks.}$$

▶ **Do Your Turn 5**

▶ **Your Turn 5**

Solve.

a. A plumber charges a flat fee of $75 plus $7.50 for every quarter of an hour spent working. Dina suspects that the $127.50 he charged is too much. How long should he have worked?

b. A cell phone company charges $59.99 for up to 900 minutes of use plus $0.40 for each additional minute. Ron's bill comes to a total of $81.59. How many minutes did he spend using his cell phone?

Answers to Your Turn 4

a. $y - 18.6 = 1.25y + 2.18$; $y = -83.12$

b. $0.4(m - 1.8) = -0.8m - 9.36$; $m = -7.2$

Answers to Your Turn 5

a. 7 quarter hours, or 1.75 hr.

b. 954 min.

Objective 4 Solve problems using the Pythagorean theorem.

One of the most popular theorems in mathematics is the Pythagorean theorem. The theorem is named after the Greek mathematician Pythagoras. The theorem describes a relationship that exists among the lengths of the sides of any **right triangle.**

Definition Right triangle: A triangle that has one right angle.

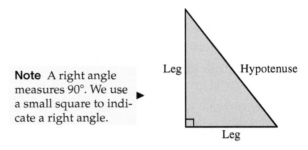

Note A right angle measures 90°. We use a small square to indicate a right angle. ▶

Leg

Hypotenuse

Leg

In a right triangle, the two sides that form the 90° angle are called **legs.** The side directly across from the 90° angle is called the **hypotenuse.**

Definitions Legs: The sides that form the 90° angle in a right triangle.

Hypotenuse: The side directly across from the 90° angle in a right triangle.

Pythagoras saw that the sum of the areas of the squares on the legs of a right triangle is the same as the area of the square on the hypotenuse. Consider a right triangle with side lengths of 3 feet, 4 feet, and 5 feet.

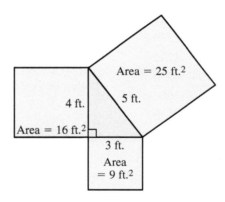

Area = 25 ft.^{2}

4 ft.

5 ft.

Area = 16 ft.^{2}

3 ft.

Area = 9 ft.^{2}

If we add the areas of the squares on the legs, $9 + 16$, we get the same area as the square on the hypotenuse, 25.

$$16 \text{ ft.}^2 + 9 \text{ ft.}^2 = 25 \text{ ft.}^2$$

Pythagoras proved that this relationship is true for every right triangle, which is why it is called the Pythagorean theorem.

Rules

The Pythagorean Theorem
Given a right triangle, where a and b represent the lengths of the legs and c represents the length of the hypotenuse, $a^2 + b^2 = c^2$.

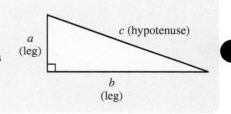

c (hypotenuse)

a (leg)

b (leg)

The Pythagorean theorem can be used to find a missing length in a right triangle when the other two lengths are known.

Example 6

a. To steady a telephone pole, a wire is to be attached 21 feet up on the pole and then attached to a stake 20 feet out from the base of the pole. What length of wire will be needed between the pole and the stake? (Ignore the wire needed for attachment.)

Understand Draw a picture. We assume that the pole makes a 90° angle with the ground. The wire connects to the pole to form a right triangle.

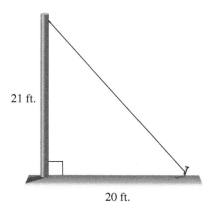

21 ft.

20 ft.

Plan Use the Pythagorean theorem. The missing length is the hypotenuse, which is c in the formula.

$$a^2 + b^2 = c^2$$
$$(21 \text{ ft.})^2 + (20 \text{ ft.})^2 = c^2 \quad \text{Replace } a \text{ with 21 and } b \text{ with 20.}$$
$$441 \text{ ft.}^2 + 400 \text{ ft.}^2 = c^2 \quad \text{Square 21 and 20.}$$
$$841 \text{ ft.}^2 = c^2 \quad \text{Add.}$$
$$\sqrt{841} \text{ ft.} = c$$
$$29 \text{ ft.} = c$$

◄ **Note** The equation $841 = c^2$ means that the square of the value c is 841; so the value of c must be the square root of 841.

Note We have placed the units in the equation as a reminder that 441 ft.² and 400 ft.² are the areas of the squares on the legs. Their sum, 841 ft.², is the area of the square on the hypotenuse. To get the length from the area, we must use a square root.

Answer The wire must be at least 29 ft. to connect the pole and the stake.

Check Verify that $a^2 + b^2 = c^2$ is true when a is 21, b is 20, and c is 29.

$$(21)^2 + (20)^2 \stackrel{?}{=} (29)^2$$
$$441 + 400 \stackrel{?}{=} 841$$
$$841 = 841 \quad \text{It checks.}$$

b. How high up on the side of a building will a 20-foot ladder reach if the base is placed 8 feet from the base of the building?

Understand Draw a picture of the situation.

When the ladder is placed against the building, it forms a right triangle. The 20-foot ladder is the hypotenuse, and the 8-foot distance from the base of the building is one of the legs. We must find the vertical distance, which is the other leg.

Plan Use the Pythagorean theorem.

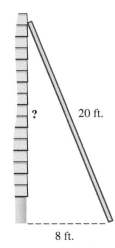

?

20 ft.

8 ft.

▶ **Your Turn 6**

Solve.

a. In the construction of a roof, three boards are used to form a right triangle frame. An 8-foot board and a 6-foot board are brought together to form a 90° angle. How long must the third board be?

b. A wire attached to a telephone pole is pulled taut and attached to a stake in the ground 4 meters from the base of the pole. If the length of wire is 10 meters, how high up the pole is the wire attached?

Execute $a^2 + b^2 = c^2$

$$(8)^2 + b^2 = (20)^2$$ 　Replace a with 8 and c with 20.

$$64 + b^2 = 400$$ 　Square 8 and 20.

$$\underline{-64 \qquad\quad -64}$$ 　Subtract 64 from both sides to isolate b^2.

$$0 + b^2 = 336$$

$$b^2 = 336$$ 　◀ **Note** This equation says that the square of our unknown value is 336; so the unknown value, b, must be the square root of 336.

$$b = \sqrt{336}$$ 　◀ Because $\sqrt{336}$ is irrational, this is the exact answer.

$$b \approx 18.33$$ 　◀ This is the approximate answer rounded to the nearest hundredth.

Answer The ladder would reach approximately 18.33 ft. up the side of the building.

Check Verify that the sum of the squares of the legs is the same as the square of the hypotenuse.

$$(8)^2 + (18.33)^2 \overset{?}{=} (20)^2$$

$$64 + 335.9889 \overset{?}{=} 400$$

$$399.9889 \approx 400$$ 　◀

Note Because 18.33 is an approximation, we cannot expect the results to match exactly, but they should be close. 399.9889 is close enough to 400 for us to accept that 18.33 is a good approximation.

◀ **Do Your Turn 6**

Of Interest

Pythagoras was a Greek mathematician who lived in the town of Croton in what is now Italy. He is believed to have lived from 569 B.C. to 500 B.C. The Pythagorean theorem is named after Pythagoras not because he discovered the relationship, as commonly thought, but because he and his followers are the first to have proved that the relationship is true for all right triangles.

The Pythagoreans also believed in the mystical power of numbers and that the universe was built on whole numbers. The very theorem that he and his followers proved wreaked havoc on their beliefs, however. It quickly became obvious that the legs could be measured by whole numbers as desired, but the hypotenuse would have to be a value other than a whole number. For example, if the legs are both 1 unit in length, we have the following:

$$1^2 + 1^2 = c^2$$
$$1 + 1 = c^2$$
$$2 = c^2$$
$$\sqrt{2} = c$$

$\sqrt{2}$ is an irrational number; so it cannot be expressed exactly as a decimal or fraction. It is said that this distressed Pythagoras tremendously because he was so sure that the universe was based on whole numbers.

Objective 5 Solve problems involving two unknowns.

In Section 4.5, we developed the use of a table to help organize information in problems involving two unknowns. The table we used had four columns:

Categories	Value	Number	Amount

Answers to Your Turn 6
a. 10 ft.　b. ≈9.17 m

The two unknowns are written in the categories column. The value column contains the given values of each category. The number column will describe the number of items in each category. The amount column will be found by the following relationship:

$$\text{value} \cdot \text{number} = \text{amount}$$

Example 7   Betty tells her son that she has 15 coins in her change purse that total $2.85. If she keeps only quarters and dimes in her change purse, how many of each coin does she have?

Understand There are two unknowns in the problem: the number of quarters and the number of dimes. Given that there are 15 coins, if n represents the number of one coin type, then $15 - n$ represents the number of the other type. Let's let n represent the number of quarters so that $15 - n$ represents the number of dimes.

Categories	Value	Number	Amount
Quarters	0.25	n	$0.25n$
Dimes	0.10	$15 - n$	$0.10(15 - n)$

▲

Because value · number = amount, we multiply the expressions in the number column by the expressions in the value column to get the expressions in the amount column.

We are also given the total amount of money, which is $2.85. This total is calculated by adding the amount in quarters plus the amount in dimes.

$$\text{amount in quarters} + \text{amount in dimes} = \text{total amount of money}$$

Plan Write an equation; then solve.

Execute   amount in quarters + amount in dimes = total amount of money

$$0.25n \qquad + \qquad 0.10(15 - n) = 2.85$$

$$0.25n \qquad + \qquad 1.5 - 0.1n \quad = 2.85 \qquad \text{Distribute 0.10 to clear parentheses.}$$

$$100 \cdot 0.25n + 100 \cdot 1.5 - 100 \cdot 0.10n = 100 \cdot 2.85 \qquad \text{Multiply both sides by 100 to eliminate decimals.}$$

$$25n + 150 - 10n = 285 \qquad \text{Multiply.}$$

$$15n + 150 = 285 \qquad \text{Combine like terms.}$$

$$\underline{\quad -150 \quad -150} \qquad \text{Subtract 150 from both sides.}$$

$$15n + 0 = 135$$

$$\frac{15n}{15} = \frac{135}{15} \qquad \text{Divide both sides by 15.}$$

$$n = 9$$

Answer Because n represents the number of quarters, there are 9 quarters. To find the number of dimes, we use the expression $15 - n$ from the table: $15 - 9 = 6$ dimes.

Check Verify that 9 quarters and 6 dimes are 15 coins that total $2.85.

$$9 + 6 = 15 \text{ coins} \quad 0.25(9) + 0.10(6) \overset{?}{=} 2.85$$

$$2.25 + 0.60 \overset{?}{=} 2.85$$

$$2.85 = 2.85 \qquad \text{Everything checks.}$$

Note We let n represent the number of quarters because quarters have greater value than do dimes. As we will see, this choice will avoid negative coefficients in the last steps of solving the equation.

Learning Strategy

To see how to relate the number of coins, make up some numeric examples. Suppose 5 of the 15 coins are quarters; then there must be $15 - 5 = 10$ dimes. Or if 6 of the 15 coins are quarters, there must be $15 - 6 = 9$ dimes. So if there are n quarters, there must be $15 - n$ dimes.

Note When we combine $25n$ and $-10n$, we get $15n$, which has a positive coefficient. Choosing to let n represent the greater-valued coin, quarters, leads to this positive coefficient. We'll see that if we let n be the number of dimes, we get a negative coefficient.

> **Your Turn 7**

Solve.

a. Jack sold a mixture of 6-string electric and 12-string acoustic guitar strings to a customer. He remembers that the customer bought 20 packs but doesn't remember how many of each type. The price for a pack of 6-string electric strings is $6.95 and $10.95 for a pack of 12-string acoustic strings. If the total sale was $159.00, how many of each pack did the customer buy?

b. Kelly has only nickels and dimes in her pocket. If she has 12 more dimes than nickels and the total is $3.90, how many of each coin does she have?

Answers to Your Turn 7
a. 5 packs of 12-string, 15 packs of 6-string
b. 18 nickels, 30 dimes

We could have let n be the number of dimes. If so, then $15 - n$ would have described the number of quarters. The solution would have proceeded as follows:

Categories	Value	Number	Amount
Quarters	0.25	$15 - n$	$0.25(15 - n)$
Dimes	0.10	n	$0.10n$

amount in quarters + amount in dimes = total amount of money

$$0.25(15 - n) \quad + \quad 0.10n \quad = \quad 2.85$$

Note We chose to solve this equation without clearing the decimals.

$$3.75 - 0.25n \quad + \quad 0.10n \quad = \quad 2.85$$

Note Now when we combine $-0.25n$ and $0.10n$, the result, $-0.15n$, has a negative coefficient. Choosing to let n represent the number of the smaller-valued coin leads to this negative coefficient.

$$3.75 - 0.15n = 2.85$$
$$-3.75 \qquad\qquad -3.75$$
$$0 - 0.15n = -0.90$$
$$\frac{-0.15n}{-0.15} = \frac{-0.90}{-0.15}$$
$$n = 6$$

There are 6 dimes and $15 - 6 = 9$ quarters.

Notice that we get the same final answer regardless of which number of coins we let n represent.

◄ **Do Your Turn 7**

6.6 Exercises   For Extra Help   MyMathLab®

Objectives 1 and 2

Prep Exercise 1 Explain how to use the multiplication principle to eliminate decimal numbers from an equation.

Prep Exercise 2 To eliminate the decimal numbers in the equation $0.4x + 1.2 = 0.25x - 3$, what is the least power of 10 that you could use?

For Exercises 1–18, solve and check.

1. $4.5n + 7 = 7.9$

2. $3x - 0.8 = 1.6$

3. $15.5y + 11.8 = 21.1$

4. $12.1x + 5.6 = 10.924$

5. $16.7 - 3.5t = 17.12$

6. $0.15 = 0.6 - 1.2n$

7. $0.62k - 12.01 = 0.17k - 14.8$

8. $0.72p - 1.21 = 1.37p - 3.29$

9. $0.8n + 1.22 = 0.408 - 0.6n$

10. $5.1m + 7.5 = 4.05 - 4.9m$

11. $4.1 - 1.96x = 4.2 - 1.99x$

12. $12.14 - 2.42g = 4.76 - 0.78g$

13. $0.4(8 + t) = 5t + 0.9$

14. $8k - 2.98 = 2.6(k - 0.8)$

15. $1.5(x + 8) = 3.2 + 0.7x$

16. $0.06(m - 11) = 22 - 0.05m$

17. $20(0.2n + 0.28) = 3.98 - (0.3 - 3.92n)$

18. $4(2.55 - x) - 5.8x = 12.2 - (8 + 11.4x)$

Objective 3

For Exercises 19–26, solve.

19. 7 more than 3.2 times p is equal to 9.56.

20. Twice m decreased by 9.9 is the product of m and -6.25.

21. 18.75 more than 3.5 times t is equal to 1.5 minus t.

22. The product of 3.2 and n is the same as the sum of n and 1.43.

23. 0.48 less than 0.2 times y is the same as 0.1 times the difference of 2.76 and y.

24. 0.6 times the sum of k and 1.5 is equal to 0.42 plus the product of 1.2 and k.

25. There is a huge crater on Mimas, one of Saturn's moons. The crater is circular and has an area of 7850 square kilometers. What is the diameter of the crater? Use 3.14 for π.

26. Crater Lake in southern Oregon is actually the top of an inactive volcano (Mount Mazama). The lake is circular and has an area of 20 square miles. What is the diameter of the lake? Use 3.14 for π.

Objective 4

Prep Exercise 3 What is a right triangle?

Prep Exercise 4 What are the legs of a right triangle?

Prep Exercise 5 What is the hypotenuse?

Prep Exercise 6 What is the formula for the Pythagorean theorem, and what does each variable in it represent?

For Exercises 27–38, use the Pythagorean theorem.

27. Rebekah is a scientist studying plant life in the rain forests of South America. From her base camp, she hikes 4 miles straight south then heads east for 3 miles. Because it is getting late in the day, she decides to head straight back to camp from her present location rather than backtrack the way she came. What is the distance back to camp?

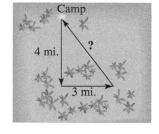

28. A helicopter flies east 9 miles then south 12 miles. How far is the helicopter from its original position?

29. A screen on a small notebook computer is 6 inches wide by 8 inches long. What is the distance along the diagonal?

30. An 8-foot-wide by 15-foot-long rectangular steel frame is to be fitted with a diagonal beam welded to the corners of the frame. How long must the diagonal beam be?

31. A softball diamond is a square with bases at each corner. The distance between bases is 65 feet. What is the distance from home plate to second base?

32. A baseball diamond is a square with bases at each corner. The distance between bases is 90 feet. What is the distance from home plate to second base?

33. A 25-foot ladder is placed 7 feet from the base of a wall. How high up the wall does the ladder reach?

34. A rope 12 feet long is tied to a tree, pulled tight, and secured 9 feet from the base of the tree. How high up the tree is the rope tied?

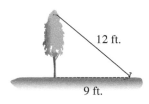

35. A 14.5-foot beam is to be attached to a vertical beam 3.5 feet above a horizontal support beam. How long must the horizontal support beam be so that it can be attached to the bottom of the 14.5-foot beam? (Ignore the width of the beams.)

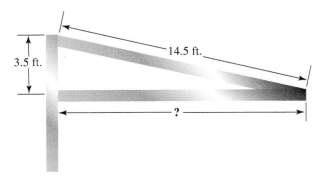

36. A 50-foot zip line is strung across a lake so that it begins in a tower 20 feet above the ground on one side of a pond and is secured at the ground on the other side of the pond. How far from the base of the tower is the point on the opposite bank where the zip line is secured?

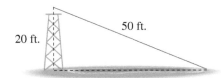

37. A fire truck has a ladder that can extend to 60 feet. The bottom of the ladder sits atop the truck so that the bottom of the ladder is 8 feet above the ground.

   a. If the truck parks 30 feet from the bottom of a building, what is the highest point the ladder can reach?

   b. If a person is in a seventh-story window awaiting rescue, can the ladder reach if each story is 10.5 feet?

38. The base of the pyramid in front of the Louvre in Paris, France, is a square measuring 35 meters along each side. The length of each face of the pyramid to the pinnacle is 27 meters. How tall is the pyramid?

Of Interest

The pyramid in front of the Louvre was constructed in 1989. The pyramid contains 673 glass segments. (603 are rhombus-shaped, and 70 are triangular.)

Objective 5

For Exercises 39–42, complete the table. Do not solve.

39. At a yard sale, Barry sold paperback books for $0.50 and hardcover books for $1.50. At the conclusion of the sale, he found that he sold 14 more paperback books than hardcover books.

Categories	Selling price	Number of books sold	Income
Paperback			
Hardcover			

40. A grocery store sells two different boxes of a particular cereal. The large box sells for $4.59; the small box, for $3.79. In one day, the store sold 32 more small boxes than large boxes.

Categories	Selling price	Number of boxes sold	Income
Small box			
Large box			

41. A department store has a special sale on T-shirts. Children's T-shirts sell for $8.99; adult shirts, for $14.99. In one particularly busy hour, the store sold 45 T-shirts.

Categories	Selling price	Number of T-shirts sold	Income
Children's T-shirts			
Adult T-shirts			

42. A restaurant sells two sizes of pizza. The large one-topping pizza sells for $16.50, and the small one-topping pizza sells for $13.50. In one day, the restaurant sold 42 one-topping pizzas.

Categories	Selling price	Number of pizzas sold	Income
Large			
Small			

For Exercises 43–46, use a four-column table to solve.

43. Latisha has a change purse in which she keeps only quarters and half-dollars. She has 12 more quarters than half-dollars. If the total in the purse is $6.75, how many of each coin does she have?

44. Jose has 22 coins totaling $4.10. If all of the 22 coins are either nickels or quarters, how many of each coin does he have?

45. Bernice sells drinks at college football games and gets paid by the number of each size drink she sells. There are two drink sizes, 12 ounces and 16 ounces. The 12 ounce sells for $1.50; the 16 ounce, for $2.00. She knows she sold 65 drinks but cannot remember how many of each size. If her total sales are $109, how many of each size did she sell?

46. Arturo sells two different home security systems. He receives $225 commission for each A system and $275 commission for each B system that he sells. In one month, he sells 14 systems. If he receives a total commission of $3400, how many of each system did he sell?

Puzzle Problem A cube has a surface area of 10.14 square feet. What are the dimensions of the cube?

Review Exercises

[1.1] 1. Write the word name for 24,915,000,204.

[1.1] 2. Round 46,256,019 to the nearest hundred-thousand.

[5.6] 3. Add. $4\frac{5}{8} + 8\frac{5}{6}$

[5.6] 4. Subtract. $12\frac{1}{4} - 7\frac{4}{5}$

[3.6] 5. Divide. $\frac{28x^9}{7x^3}$

[4.3] 6. Explain the mistake; then work the problem correctly.

$$-3x - 18 = -24$$
$$\underline{+\ 18\quad +18}$$
$$-3x + 0 = -6$$
$$\frac{-3x}{3} = \frac{-6}{3}$$
$$x = -2$$

Chapter 6   Summary and Review Exercises

6.1 Introduction to Decimal Numbers

Definitions/Rules/Procedures	Key Example(s)
Decimal notation is _____ notation for expressing fractions. **To write a decimal number as a fraction or mixed number in simplest form:** 1. Write all digits to the left of the decimal point as the integer part of a mixed number. 2. Write all the digits to the right of the decimal point in the _____ of a fraction. 3. Write the denominator indicated by the last place value. 4. Simplify to lowest terms.	Write as a fraction or mixed number in simplest form. **a.** $0.24 = \dfrac{24}{100} = \dfrac{6}{25}$ **b.** $16.082 = 16\dfrac{82}{1000} = 16\dfrac{41}{500}$

Complete the following table of place values.

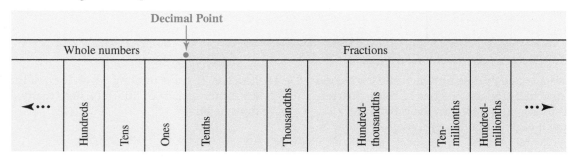

[6.1] For Exercises 1–3, write as a fraction or mixed number in simplest form.

1. 0.8   **2.** 0.024   **3.** −2.65

Definitions/Rules/Procedures	Key Example(s)
To write the word name for a decimal number with no integer part (no digits other than 0 to the left of the decimal point): 1. Write the word name for the digits to the right of the decimal point as if they represented a whole number. 2. Write the name of the _____. **To write the word name for a decimal number with both integer and fractional parts:** 1. Write the name of the integer number part. 2. Write the word _____ for the decimal point. 3. Write the name of the fractional part.	Write the word name for the given number. **a.** 0.0351 **Answer:** three hundred fifty-one ten-thousandths **b.** 147.29 **Answer:** one hundred forty-seven and twenty-nine hundredths

[6.1] For Exercises 4–6, write the word name.

4. 24.39   **5.** 581.459   **6.** 0.2917

Definitions/Rules/Procedures	Key Example(s)
To determine which of two positive decimal numbers is greater, _____ _____ until you find two different digits in the same place value. The greater number contains the greater of those two digits. If both numbers are negative, the greater number is _____ and contains the lesser of the two different digits.	Use < or > to write a true sentence. **a.** 0.517 ? 0.516 **Answer:** 0.517 > 0.516 **b.** −0.4912 ? −0.4812 **Answer:** −0.4912 < −0.4812

[6.1] *For Exercises 7 and 8, use < or > to write a true statement.*

7. 2.001 ? 2.0009

8. −0.016 ? −0.008

Definitions/Rules/Procedures	Key Example(s)
To round a decimal number to a given place value: 1. Identify the digit in the given place value. 2. If the digit to the right of the given place value is 5 or greater, round _____ by increasing the digit in the given place value by 1. If the digit to the right of the given place value is 4 or less, round down by keeping the digit in the given place value the same. 3. Change all digits to the right of the rounded place value to _____. Decimal digits to the right of the rounded place can be eliminated.	Round each number to the specified place. **a.** 4.619 to the nearest tenth. **Answer:** 4.6 **b.** 18.5141 to the nearest whole. **Answer:** 19 **c.** 0.00127 to the nearest ten-thousandth. **Answer:** 0.0013

[6.1] *For Exercises 9–12, round 31.8052 to the specified place.*

9. whole number   **10.** tenths   **11.** hundredths   **12.** thousandths

6.2 Adding and Subtracting Decimal Numbers

Definitions/Rules/Procedures	Key Example(s)
To add decimal numbers: 1. Stack the numbers so that the corresponding place values align. (Vertically align the decimal points.) 2. Add the digits in the corresponding place values. 3. Place the decimal point in the sum so that _____ _____. **To subtract decimal numbers:** 1. Stack the number with the greater absolute value _____ so that the corresponding place values align. (Vertically align the decimal points.) 2. Subtract the digits in the corresponding place values. 3. Place the decimal point in the difference so that it aligns with the decimal points in the minuend and subtrahend.	Estimate. Then find the actual sum. 2.035 + 15.98 **Estimate:** 2 +16 / 18 **Actual:** 2.035 +15.980 / 18.015 Estimate. Then find the actual difference. 60.043 − 4.798 **Estimate:** 60 − 5 / 55 **Actual:** 60.043 − 4.798 / 55.245 Simplify. $$5.1x^3 + 2.91x - 7x^2 + 9.8 - 3.4x - 3.8x^2$$ $$= 5.1x^3 - 10.8x^2 - 0.49x + 9.8$$

[6.2] For Exercises 13 and 14, estimate. Then find the actual sum or difference.

13. $24.81 + 54.2 + 9.005$ **14.** $9.005 - 1.089$

[6.2] For Exercises 15–17, add or subtract.

15. $6.7 + (-12.9)$ **16.** $(-20) - (-13.88)$ **17.** $-1.082 + 29.1 - 4.9$

[6.2] For Exercises 18–20, simplify.

18. $9y^4 + 8y - 12.2y^2 + 2.1y^4 - 16y + 3.8$ **19.** $(0.3a^2 + 8.1a + 6) + (5.1a^2 - 2.9a + 4.33)$

20. $(2.6x^3 - 98.1x^2 + 4) - (5.3x^3 - 2.1x + 8.2)$

6.3 Multiplying Decimal Numbers

Definitions/Rules/Procedures	Key Example(s)
To multiply decimal numbers: 1. Multiply as if the decimal numbers were whole numbers. 2. Place the decimal point in the product so that it has _____ _____.	Estimate. Then find the actual product. $(6.45)(2.8)$ **Estimate:** $\begin{array}{r} 6 \\ \times\ 3 \\ \hline 18 \end{array}$ **Actual:** $\begin{array}{r} {\scriptstyle 3\ 4} \\ 6.45 \\ \times\ 2.8 \\ \hline 5\ 160 \\ 12\ 90 \\ \hline 18.060 \end{array}$

[6.3] For Exercises 21 and 22, estimate. Then find the actual product.

21. $(3.44)(5.1)$ **22.** $(1.5)(-9.67)$

[6.3] For Exercises 23–28, multiply.

23. $(1.2)^2$ **24.** $(-0.9x^3y)(4.2x^5y)$ **25.** $(-0.3x^3)^3$

26. $0.2(0.6n^3 - 1.4n^2 + n - 18)$ **27.** $(4.5a - 6.1)(2a - 0.5)$ **28.** $(1.3x + 4.2)(1.3x - 4.2)$

Definitions/Rules/Procedures	Key Example(s)
Scientific notation is a notation in the form $a \times 10^n$ where a is a decimal number whose absolute value is _____ and n is an integer. **To change from scientific notation with a positive exponent to standard notation for a given number,** move the decimal point to the _____ the number of places indicated by the exponent. **To write a standard form number whose absolute value is greater than 1 in scientific notation:** 1. Move the decimal point so that the number's absolute value is greater than or equal to 1 but less than 10. (Place the decimal point to the right of the first nonzero digit.) 2. Write the decimal number multiplied by 10^n where n is the number of places between the _____ decimal position and the _____ decimal position. 3. Delete 0s to the right of the last nonzero digit.	Write in standard form. **a.** $3.59 \times 10^7 = 35{,}900{,}000$ **b.** $-4.2 \times 10^5 = -420{,}000$ Write in scientific notation. **a.** $4{,}791{,}000{,}000 = 4.791 \times 10^9$ **b.** $-87{,}050{,}000{,}000{,}000 = -8.705 \times 10^{13}$

[6.3] **29.** Write 3.58×10^9 in standard form.

30. Write $-103,000$ in scientific notation.

6.4 Dividing Decimal Numbers; Square Roots with Decimals

Definitions/Rules/Procedures	Key Example(s)

To divide a decimal number by a whole number using long division:

1. Divide the divisor into the dividend as if both numbers were whole numbers. Align the digits in the quotient properly.

2. Write the decimal point in the quotient _____ _____.

3. If after you bring down the last digit in the dividend the remainder is not 0, write a 0 digit to the right of the last nonzero digit in the dividend, bring that 0 down, and divide. Continue adding 0 digits and dividing until the remainder is 0 or a digit or group of digits repeats without end in the quotient. To indicate repeated digits in the quotient, write a repeat bar over the digit or group of digits that repeat.

To divide using long division when the divisor is a decimal number:

1. Move the decimal point in the divisor to the right enough places to make the divisor

   _____.

2. Move the decimal point in the dividend

   _____.

3. Follow the same steps as for dividing a decimal number by a whole number.

Estimate. Then find the actual quotient.

a. $20.4 \div 6$

Estimate: $20 \div 6 = \dfrac{20}{6}$

$= 3\dfrac{2}{3}$

$= 3.\overline{6}$

$$\begin{array}{r} 3.4 \\ 6\overline{)20.4} \\ -18 \\ \hline 24 \\ -24 \\ \hline 0 \end{array}$$

Actual:

Answer: 3.4

b. $3.68 \div 11$

Estimate: $4 \div 10 = \dfrac{4}{10}$

$= 0.4$

$$\begin{array}{r} 0.334545\ldots \\ 11\overline{)3.680000} \\ -33 \\ \hline 38 \\ -33 \\ \hline 50 \\ -44 \\ \hline 60 \\ -55 \\ \hline 50 \end{array}$$

Actual:

Answer: $0.3\overline{345}$

c. $50.8 \div 0.02 = 5080 \div 2$

Estimate: $5000 \div 2 = 2500$

Actual:

$$\begin{array}{r} 2540 \\ 2\overline{)5080} \\ -4 \\ \hline 10 \\ -10 \\ \hline 08 \\ -8 \\ \hline 0 \end{array}$$

Answer: 2540

d. $0.694 \div 0.22 = 69.4 \div 22$

Estimate: Instead of $70 \div 20$, it is easier to bend the rounding rules and use $60 \div 20 = 3$.

Actual:

$$\begin{array}{r} 3.154545\ldots \\ 22\overline{)69.400000} \\ -66 \\ \hline 34 \\ -22 \\ \hline 120 \\ -110 \\ \hline 100 \\ -88 \\ \hline 120 \end{array}$$

Answer: $3.1\overline{54}$

[6.4] For Exercises 31-33, estimate. Then find the actual quotient.

31. $21 \div 4$ **32.** $3.6 \div 9$ **33.** $-54 \div (-0.09)$

[6.4] 34. Divide.

34. $\dfrac{-2.38a^4bc}{6.8ab}$

Definitions/Rules/Procedures	Key Example(s)
To write a fraction as a decimal number, divide the _____. **To write a mixed number as a decimal number:** 1. Write the integer part of the mixed number _____ _____. 2. _____ to determine the decimal digits. Write those digits to the right of the decimal point.	Write $\frac{5}{8}$ as a decimal number. $\begin{array}{r} 0.625 \\ 8\overline{)5.000} \\ -48 \\ \hline 20 \\ -16 \\ \hline 40 \\ -40 \\ \hline 0 \end{array}$ **Answer:** $\frac{5}{8} = 0.625$ Write $14\frac{2}{3}$ as a decimal number. $\begin{array}{r} 0.66\ldots \\ 3\overline{)2.00} \\ -18 \\ \hline 20 \\ -18 \\ \hline 20 \end{array}$ **Answer:** $14.\overline{6}$

[6.4] For Exercises 35 and 36, write as a decimal number.

35. $\dfrac{3}{5}$ **36.** $-4\dfrac{2}{11}$

Definitions/Rules/Procedures	Key Example(s)
The square root of a decimal number that corresponds to a perfect square and has an even number of decimal places will be a decimal number with _____ the number of decimal places and the same digits as the square root of the corresponding perfect square. The set of **real numbers** contains all _____ and _____ numbers.	Simplify. $\sqrt{0.0049} = 0.07$

[6.4] **37.** Find the square root. $\sqrt{0.64}$ *[6.4]* **38.** Approximate $\sqrt{148}$ to the nearest hundredth.

6.5 Order of Operations and Applications in Geometry

Definitions/Rules/Procedures	Key Example(s)

To simplify numerical expressions that contain both fractions and decimals:
 	
1. **Method 1:** Write each decimal number as a fraction.
  or
  Method 2: Write each fraction as a decimal number.
  or
  Method 3: Write each decimal number over 1 in fraction form.
 	
2. Follow the order of operations agreement.
	Simplify.
	(Method 3 illustrated here)
	
	$$\frac{2}{3}(6.09) - \frac{3}{8} = \frac{2}{3} \cdot \frac{6.09}{1} - \frac{3}{8}$$
	
	$$= \frac{12.18}{3} - \frac{3}{8}$$
	
	$$= 4.06 - 0.375$$
	
	$$= 3.685$$

[6.5] For Exercises 39 and 40, simplify.

39. $3.6 + (2.5)^2 - 1.5(9 - 0.4)$

40. $\frac{1}{6}(0.24) + \frac{2}{3}(-3.3)$

41. [6.5] Evaluate $\frac{1}{2}at^2$ when $a = 9.8$ and $t = 2$.

6.6 Solving Equations and Problem Solving

Definitions/Rules/Procedures	Key Example(s)

To solve a linear equation in one variable:
 	
1. Simplify each side of the equation.
  a. Distribute to eliminate parentheses.
  b. Eliminate fractions and/or decimal numbers by multiplying both sides of the equation by the LCD. In the case of decimal numbers, the LCD is the power of 10 with the same number of 0 digits as _____ in the number with the most decimal places
  c. Combine like terms.
 	
2. Use the addition principle of equality so that all variable terms are on one side of the equation and all constants are on the other side. Then combine like terms.
 	
    Note Eliminate the variable term with the lesser coefficient to avoid negative coefficients.
 	
3. Use the multiplication principle of equality to eliminate the coefficient of the variable.
	Solve and check.
	
	$8.3 - (2.4x + 0.7) = 14.6x - 5(x + 6.16)$
	$8.3 - 2.4x - 0.7 = 14.6x - 5x - 30.8$
	$10(8.3 - 2.4x - 0.7) = (14.6x - 5x - 30.8)10$
	$83 - 24x - 7 = 146x - 50x - 308$
	$76 - 24x = 96x - 308$
	$\underline{ + 24x + 24x}$
	$76 + 0 = 120x - 308$
	$76 = 120x - 308$
	$\underline{+ 308 + 308}$
	$384 = 120x + 0$
	$\dfrac{384}{120} = \dfrac{120x}{120}$
	$3.2 = 1x$
	$3.2 = x$
	
	Check:
	$8.3 - (2.4x + 0.7) = 14.6x - 5(x + 6.16)$
	$8.3 - [2.4(3.2) + 0.7] \overset{?}{=} 14.6(3.2) - 5(3.2 + 6.16)$
	$8.3 - (7.68 + 0.7) \overset{?}{=} 46.72 - 5(9.36)$
	$8.3 - 8.38 \overset{?}{=} 46.72 - 46.8$
	$-0.08 = -0.08$
	
	True; so 3.2 is the solution.

[6.6] For Exercises 42–46, solve and check.

42. $y - 0.58 = -1.22$

43. $-1.6n = 0.032$

44. $4.5k - 2.61 = 0.99$

45. $2.1x - 12.6 = 1.9x - 10.98$

46. $3.6n - (n - 13.2) = 10(0.4n + 0.62)$

Formulas

[6.3] **Unit price** is the price of _____ unit.

[6.3] Total price = _____ · unit price

[6.4] Circumference of a circle given its radius: $C =$ _____

[6.4] Circumference of a circle given its diameter: $C =$ _____

[6.5] Area of a circle: $A =$ _____

[6.5] Area of a triangle: $A =$ _____

[6.5] Area of a trapezoid: $A =$ _____

[6.5] Volume of a cylinder: $V =$ _____

[6.5] Volume of a pyramid: $V =$ _____

[6.5] Volume of a cone: $V =$ _____

[6.5] Volume of a sphere: $V =$ _____

[6.6] A **right triangle** is a triangle that has one _____.

[6.6] The **legs** of a right triangle form the _____ angle.

[6.6] The **hypotenuse** of a right triangle is the side _____ from the 90° angle.

[6.6] **The Pythagorean theorem**
Given a right triangle, where a and b represent the lengths of the legs and c represents the length of the hypotenuse, _____.

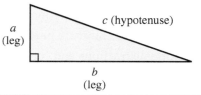

[6.4]
[6.5] **47.** Find the radius, circumference, and area of the circle. Use 3.14 for π.

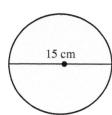

15 cm

[6.5] **48.** Find the area.

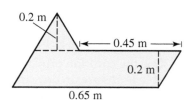

0.2 m

0.45 m

0.2 m

0.65 m

[6.5] **49.** Find the area of the shaded region. Use 3.14 for π.

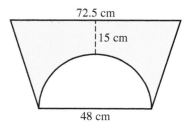

72.5 cm

15 cm

48 cm

[6.5] For Exercises 50–54, find the volume. Use 3.14 for π.

50.

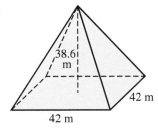

51.

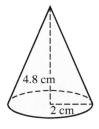

52.

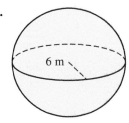

53.

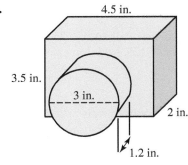

54.

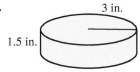

For Exercises 55–62, solve.

[6.2] **55.** Sonja's checkbook register is shown below. Find her final balance.

Date	No.	Transaction Description	Debits		Credits		Balance	
3 - 21	1201	Papa John's	16	70			150	20
3 - 22	1202	Food Lion	84	61				
3 - 25	1203	Ford Motor Credit	245	16				
3 - 25	1204	Electric & Gas	67	54				
3 - 25	1205	Fleet Mortgage	584	95				
3 - 26		Deposit			1545	24		

[6.3] **56.** Caroline buys 3.4 pounds of flounder at $5.99 per pound. What is the total cost of the flounder?

[6.4] **57.** Naja has a student loan balance of $2854.80. If he agrees to make equal monthly payments over a five-year period, how much will each payment be?

[6.5] **58.** The following graph shows the closing price of Sears stock each day from October 25 to November 5.

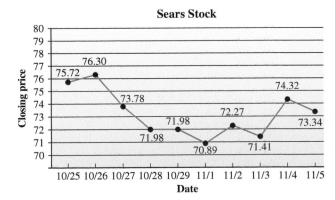

a. Find the mean closing price for those days.

b. Find the median closing price for those days.

c. Find the mode closing price for those days.

[6.5] **59.** A student's grade report is shown below. Calculate the student's GPA rounded to the nearest thousandth.

Course	Credits	Grade
ENG 101	3.0	C
BIO 101	4.0	D
MAT 102	3.0	B
PHY 101	3.0	C

[6.6] **60.** A 12-foot board is resting against a tree. The bottom of the board is 3 feet from the base of the tree. How far is the top of the board from the base of the tree? (Assume that the tree forms a right angle with the ground.)

[6.6] **61.** An electrician charges $75 for each service call and $6.25 for every 15 minutes spent working. If Brittany has $125, how long will the electrician work?

For Exercise 62, use a four-column table to solve.

[6.6] **62.** Elissa has a change purse that contains only nickels and dimes. She has 9 more dimes than nickels. If she has a total of $2.40 in the purse, how many of each coin does she have?

Chapter 6 Practice Test

For Extra Help CHAPTER **Test Prep** VIDEOS

Step-by-step test solutions are found on the Chapter Test Prep Videos available via the Video Resources on DVD, in **MyMathLab**", and on You Tube (search "CarsonPrealgebra" and click on "Channels").

[6.1] **1.** Write the word name for 56.789.

[6.1] **2.** Write 0.68 as a fraction in lowest terms.

[6.1] **3.** Use $<$ or $>$ to write a true statement.
-0.0059 ? -0.0058

[6.1] **4.** Round 2.0915 to the specified place.
a. whole number **b.** tenths **c.** hundredths **d.** thousandths

For Exercises 5–9, estimate. Then find the actual sum, difference, product, or quotient.

[6.2] **5.** $4.591 + 34.6 + 2.8$

[6.2] **6.** $27.004 - 3.098$

[6.3] **7.** $(4.5)(-8.61)$

[6.4] **8.** $4.8 \div 16$

[6.4] **9.** $(-55.384) \div (-9.2)$

[6.4] **10.** Evaluate. $\sqrt{0.81}$

[6.4] **11.** Write $3\frac{5}{6}$ as a decimal number.

[6.3] **12.** Write 2.97×10^6 in standard form.

[6.3] **13.** Write $-35,600,000$ in scientific notation.

[6.5] **14.** Simplify. $3.2 + (0.12)^2 - 0.6(2.4 - 5.6)$

[6.5] **15.** Evaluate $\frac{1}{2}mv^2$ when $m = 12.8$ and $v = 0.2$.

[6.2] **16.** Simplify. $(7.6x^3 - 2.3x^2 - 9) - (6.7x^3 + 4.7x - 12.8)$

For Exercises 17 and 18, multiply.

[6.3] **17.** $2.5x^3 \cdot 0.6x$

18. $(3.2n + 1.5)(0.4n - 6)$

[6.6] **19.** Solve and check. $6.5t - 12.8 = 13.85$

1. _____
2. _____
3. _____
4. a. _____
   b. _____
   c. _____
   d. _____
5. _____
6. _____
7. _____
8. _____
9. _____
10. _____
11. _____
12. _____
13. _____
14. _____
15. _____
16. _____
17. _____
18. _____
19. _____

20. _____

21. _____

22. _____

23. _____

24. a._____

b._____

c._____

25. a._____

b._____

[6.6]   **20.** Solve and check. $1.8(k - 4) = -3.5k - 4.02$

[6.2]   **21.** Following is a copy of Dedra's bill at a restaurant. She pays with a $20 bill. How much change should she receive?

Guest Check	
Club sandwich	7.95
Salad	2.95
Beverage	1.50
Tax	0.57

[6.3]   **22.** A fish market has shrimp on sale for $7.99 per pound. How much does 5 pounds cost?

[6.4]   **23.** Bill receives the following printout of charges that he plans to pay on a 90-day same-as-cash option. He wants to split the charges into three equal payments. How much is each payment?

Description	Amount
1 TV	$385.80
2 speakers	$59.90 each
Sales tax	$25.28

[6.5]   **24.** The following graph shows the closing price of Xerox Corporation stock from July 14 to July 19, 2010.
   a. Find the mean closing price for those days.
   b. Find the median closing price for those days.
   c. Find the mode closing price for those days.

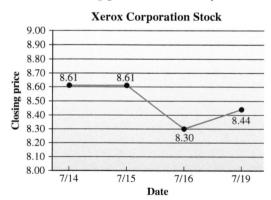

$\begin{bmatrix} 6.4 \\ 6.5 \end{bmatrix}$ **25.** A circle has a radius of 12 feet. Use 3.14 for π.
   a. Find the circumference.
   b. Find the area.

[6.5] For Exercises 26 and 27, find the volume. Use 3.14 for π.

26.

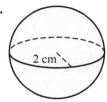

2 cm

27.

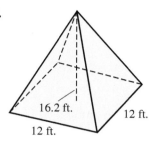

16.2 ft.

12 ft.

12 ft.

26. _____

27. _____

[6.5] **28.** Find the area.

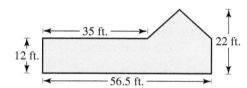

35 ft.

12 ft.

22 ft.

56.5 ft.

28. _____

[6.6] **29.** Carlton has a 20-foot ladder that he places 5 feet from the base of his house. How high up the side of the house will the ladder reach?

29. _____

[6.6] **30.** Yolanda sells two different kinds of cakes, round-layered cakes and sheet cakes. The round-layered cakes sell for $14, and the sheet cakes sell for $12.50. In one month, she sold 12 cakes and made a total of $162. How many of each cake did she sell?

30. _____

Chapters 1-6 Cumulative Review Exercises

For Exercises 1–4, answer true or false.

[3.2] **1.** $4x^3 - 7$ is a binomial.

[3.4] **2.** The conjugate of $x - 9$ is $-x + 9$.

[3.5] **3.** 221 is a composite number.

[5.3] **4.** π is a rational number.

For Exercises 5 and 6, complete the rule.

[1.4] **5.** If the divisor is 0 and the dividend is any number other than 0, the quotient is _____.

[3.2] **6.** When combining like terms, we add or subtract the _____ and keep the _____ the same.

[3.2] **7.** What is the degree of $3b^2 - b^4 + 9b + 4b^6 - 17$?

[5.1] **8.** Graph $-2\frac{3}{4}$ on a number line.

[6.1] **9.** Round 2.0185 to the nearest hundredth.

[6.3] **10.** Write 7.68×10^8 in standard form; then write the word name.

For Exercises 11–21, simplify.

[2.5] **11.** $\{9 - 4[6 + (-16)]\} \div 7$

[5.3] **12.** $-\frac{1}{9} \cdot 5\frac{1}{6}$

[5.4] **13.** $-\frac{12n^2}{13m} \div \frac{-15n}{26m}$

[5.6] **14.** $4\frac{5}{8} + 7\frac{1}{2}$

[5.6] **15.** $\frac{x}{6} + \frac{3}{4}$

[6.5] **16.** $0.48 + 0.6\left(4\frac{1}{3}\right) + \sqrt{0.81}$

[6.2] **17.** Simplify and write the result in descending order of degree.

$$2.6x^3 - \frac{1}{4}x^2 + 3.8 - x^3 + x^2 - \frac{1}{4}$$

[6.2] **18.** Subtract and write the result in descending order of degree. $(14m^3 - 5.7m^2 + 7) - (9.1m^3 + m^2 - 11.6)$

[6.5] **19.** Evaluate $4h^2 - k^3$ when $h = \frac{3}{4}$ and $k = -0.2$.

[3.4] **20.** Multiply. $(4y - 3)(y + 2)$

[3.6] **21.** Factor. $30n^5 + 15n^2 - 25n$

For Exercises 22–24, solve and check.

[6.6] **22.** $-0.06k = 0.408$

[5.8] **23.** $\frac{2}{3}x + \frac{1}{4} = \frac{5}{6}$

[4.3] **24.** $2n + 8 = 5(n - 3) + 2$

For Exercises 25–30, solve.

[2.6] **25.** The annual financial report for a business indicates that the total revenue was $2,609,400 and the total costs were $1,208,500. What was the net? Did the business have a profit or loss?

[3.7] **26. a.** Write an expression for the volume of the box shown.
b. Find the volume of the box if w is 6 centimeters.

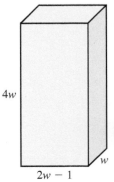

$4w$

$2w - 1$

w

[4.5] **27.** An entrance to an amusement park is to be an isosceles triangle. The length of each equal-length side is 12 feet longer than twice the length of the base. The perimeter is 89 feet. What are the lengths of the base and the sides of equal length?

[5.6] **28.** A question on a survey allows for a response of yes, no, or no opinion. $\frac{1}{6}$ of the respondents said yes, and $\frac{3}{5}$ said no. What fraction had no opinion?

[6.5] **29.** Find the area of the shape. Use 3.14 for π.

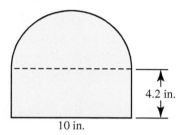

4.2 in.

10 in.

[6.6] **30.** Cedrick makes two different sizes of clay bowls. He sells the larger size for $12.50, and the smaller size sells for $8.50. If he sells a total of 20 bowls at a fair and makes a total of $206, how many of each size did he sell? (Use a four-column table.)

CHAPTER

7

Ratios, Proportions, and Measurement

Chapter Overview

Now that the real number system has been developed in Chapters 1–6, we will consider applications containing all types of numbers. In this chapter, we will consider the following topics:

▶ Ratios

▶ Proportions

▶ Unit conversion

As we will see, ratios are like fractions with any type of number in the numerator or denominator. Those ratios can be used to solve proportions or convert units of measurement.

7.1 Ratios, Probability, and Rates

Warm-up

[5.2] 1. Out of 120 participants in a poll about a new product, 45 said that they liked the product. What fraction of the participants liked the product? Write the fraction in lowest terms.

[6.4] 2. Divide. $\dfrac{4.68}{0.4}$

Objective 1 Write ratios in simplest form.

In this section, we will solve problems involving **ratios.**

Definition **Ratio:** A comparison between two quantities using a quotient.

When a ratio is written in words, the word *to* separates the quantities. A ratio can be written with a colon or in fraction form. The following table shows two ratios for a committee of 12 males and 18 females.

In words	Colon notation	Fraction form
the ratio of males *to* females	12:18	$\dfrac{12}{18}$
the ratio of females *to* males	18:12	$\dfrac{18}{12}$

Note In fraction form, the quantity to the left of the word *to* goes in the numerator and the quantity to the right of the word *to* goes in the denominator.

In this text, we will translate ratios to fraction form. The following rule shows how to translate words to a ratio in fraction form.

Rules

For two quantities, a and b, the ratio of a to b is $\dfrac{a}{b}$.

Example 1 The pitch, or degree of slant, of a roof is to be 8 inches of vertical rise for every 12 inches of horizontal length. Write the ratio of vertical rise to horizontal length in simplest form.

Solution: ratio of vertical rise to horizontal length $= \dfrac{8}{12} = \dfrac{2}{3}$

This means that measuring 12 inches out then 8 inches up is equivalent to measuring 3 inches out then 2 inches up. The pitch of the roof is the same using either ratio.

Connection The ratio of the vertical rise to the horizontal length is known as the *slope* of an incline. Slope is often referred to as the ratio of *rise to run* or *rise over run*.

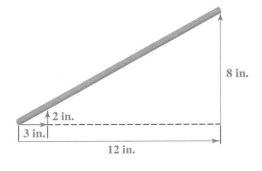

2 in.

3 in.

8 in.

12 in.

▶ **Do Your Turn 1**

Objectives

1 Write ratios in simplest form.
2 Determine probabilities.
3 Calculate unit ratios.
4 Write rates in simplest form.
5 Use unit price to determine the better buy.

▶ **Your Turn 1**

Stacie measures 40 inches from her feet to her navel and 24 inches from her navel to the top of her head. Write the ratio of the smaller length to the larger length in simplest form.

Answer to Your Turn 1

$\dfrac{3}{5}$

Answers to Warm-up

1. $\dfrac{3}{8}$ 2. 11.7

A ratio can have fractions or decimals in its numerator or denominator. This can make simplifying more challenging.

▶ **Your Turn 2**

A recipe calls for $2\frac{1}{2}$ cups of milk and $\frac{3}{4}$ of a cup of sugar. Write the ratio of milk to sugar in simplest form.

Of Interest

Gears in machines operate on the principle of ratios. In an automobile, the number of revolutions of the driveshaft is converted by a gear to a number of tire revolutions. In a lower gear, the engine is turning at many more revolutions than the tire, while in a higher gear, the opposite is true.

Example 2   For every $2\frac{1}{2}$ rotations of the pedals on a bicycle, the back wheel rotates $3\frac{1}{4}$ times. Write the ratio of the number of pedal rotations to the number of back wheel rotations.

Solution: number of pedal rotations to number of back wheel rotations $= \dfrac{2\frac{1}{2}}{3\frac{1}{4}}$

This is a complex fraction, which we learned to simplify in Chapter 5. To simplify, we divide the mixed numbers.

$$\frac{2\frac{1}{2}}{3\frac{1}{4}} = 2\frac{1}{2} \div 3\frac{1}{4} \quad \text{Rewrite using a division sign.}$$

$$= \frac{5}{2} \div \frac{13}{4} \quad \text{Write as improper fractions.}$$

$$= \frac{5}{\overset{}{2}} \cdot \frac{\overset{2}{4}}{13} \quad \text{Write as an equivalent multiplication and divide out the common factor 2.}$$

$$= \frac{10}{13} \quad \text{Multiply.}$$

This means that every 10 rotations of the pedals result in 13 rotations of the back wheel.

Example 2 suggests the following procedure:

Procedure

To simplify a ratio of two fractions or mixed numbers, divide the fractions or mixed numbers.

◀ **Do Your Turn 2**

Now let's try simplifying ratios of decimal numbers.

Example 3   The diameter of a piece of pipe is 0.42 meter. The length is 3.5 meters. Write the ratio of the diameter to the length in simplest form.

Solution: diameter to length $= \dfrac{0.42}{3.5}$

Note Because we multiply the numerator and denominator by the same power of 10, the decimal will move to the right the same number of places in the numerator and denominator.

To write in simplest form, we must eliminate the decimals so that the numerator and denominator are integers. In Chapter 6, we learned to eliminate decimals by multiplying by an appropriate power of 10. In Chapter 5, we learned that we write equivalent fractions by multiplying the numerator and denominator by the same number. Because 0.42 has the most decimal places, it determines the power of 10 that we use. Therefore, in this case, we multiply both the numerator and denominator by 100.

$$\frac{0.42}{3.5} = \frac{0.42(100)}{3.5(100)} = \frac{42}{350} = \frac{3}{25}$$

Answer to Your Turn 2
$\dfrac{10}{3}$

Example 3 suggests the following procedure:

Procedure

To eliminate decimals in a ratio, multiply the numerator and denominator by an appropriate power of 10 as determined by the decimal number with the most decimal places.

▶ **Do Your Turn 3**

Of Interest

The golden ratio is an irrational number with an approximate value of 0.618 to 1. Various parts of the human body are in the golden ratio. Some examples are shown below.

$$\frac{\text{navel to top of head}}{\text{bottom of foot to navel}} \qquad \frac{\text{center of eyes to top of head}}{\text{bottom of chin to center of eyes}}$$

$$\frac{\text{shoulders to top of head}}{\text{navel to shoulders}} \qquad \frac{\text{bottom to tip of middle finger}}{\text{wrist to bottom of middle finger}}$$

Note that these ratios are 0.618 to 1 for humans on average. Because individuals differ, it is normal to see slight variation from one person to another. The golden ratio also appears in plant growth, art, architecture, and musical instruments. The front of the Greek building the Parthenon fits inside a rectangle with a ratio of width to length that is in the golden ratio. The ratio of the length of the neck of a violin to the length of its body is in the golden ratio.

Objective 2 Determine probabilities.

Ratios are used in calculating probabilities. Suppose a friend approached you with a coin and suggested the following game: If the coin lands heads up, you win; if it lands tails up, your friend wins. What is the likelihood that you will win? Because there are two possible outcomes, both of which are equally likely to occur, heads or tails, and you win with one of those possibilities, we say that you have a 1 in 2 chance of winning.

Note that we can write *1 in 2* as the fraction $\frac{1}{2}$. The fraction $\frac{1}{2}$ is the **theoretical probability** that you will win.

Definition **Theoretical probability of equally likely outcomes:** The ratio of the number of favorable outcomes to the total number of possible outcomes.

$$\text{the probability of a coin landing heads up} = \frac{1}{2} \begin{array}{l} \text{1 side is heads} \\ \text{out of} \\ \text{2 possible outcomes.} \end{array}$$

Notice that your friend has an equal chance of winning because there is 1 tail out of 2 possible outcomes.

$$\text{the probability of a coin landing tails up} = \frac{1}{2}$$

Procedure

To determine a probability:
1. Write the number of favorable outcomes in the numerator.
2. Write the total number of possible outcomes in the denominator.
3. Simplify.

▶ **Your Turn 3**

An ant weighing 0.004 gram can carry an object weighing up to 0.02 gram. Write the ratio of the object's weight to the ant's weight in simplest form.

Answer to Your Turn 3
$\frac{5}{1}$

Note that this does not mean that if your friend wins one time, you are guaranteed to win the next time. In fact, your friend could have a streak of several wins before you win, if at all. On any given day, the number of times you win may be more or less than $\frac{1}{2}$ of the time. However, over a lifetime, if you were to look at how many times you won, it would be close to $\frac{1}{2}$ of the time. Taken to an extreme, if you could play the game forever, you would win exactly $\frac{1}{2}$ of the time.

Some of the problems we will explore involve dice and cards. A six-sided game die is a cube with dots numbering each of the six sides.

A standard deck of cards contains 52 cards with four suits of 13 cards. There are two red suits—diamonds and hearts—and two black suits—spades and clubs. The 13 cards are as follows: Ace, 2, 3, 4, 5, 6, 7, 8, 9, 10, Jack, Queen, and King. The Jack, Queen, and King are referred to as *face cards*.

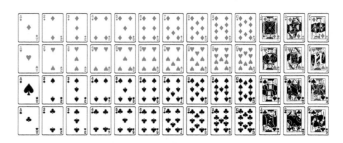

▶ Your Turn 4

Determine the probability. Write each probability in simplest form.

a. What is the probability of selecting a 2 from a standard deck of cards?

b. Suppose 28 males and 30 females put their names in a hat for a drawing. If the winner's name is selected from the hat at random, what is the probability that the winner will be male?

Example 4   Determine the probability. Write each probability in simplest form.

a. What is the probability of rolling a 2 on a six-sided die?

Solution: Write the ratio of the number of 2s to the total number of outcomes on a six-sided die. There is only one 2 on a six-sided die.

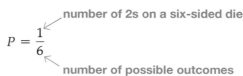

number of 2s on a six-sided die

$$P = \frac{1}{6}$$

number of possible outcomes

b. What is the probability of tossing a 7 on a six-sided die?

Solution: Write the ratio of the number of 7s to the total number of outcomes on a six-sided die. There are no 7s on a normal six-sided die.

$$P = \frac{0}{6} = 0$$

c. What is the probability of tossing a 1, 2, 3, 4, 5, or 6 on a six-sided die?

Solution: There are six favorable outcomes out of six possible outcomes.

$$P = \frac{6}{6} = 1$$ ◀ **Note** An event with a probability of 1 is certain to occur.

d. What is the probability of drawing a red King out of a standard deck of 52 cards?

Solution: Write the ratio of the number of red Kings to the total number of cards in the deck. There are two red Kings—the King of diamonds and the King of hearts—in a standard deck of 52 cards.

$$P = \frac{2}{52} = \frac{1}{26}$$

◀ Do Your Turn 4

Answers to Your Turn 4

a. $\frac{1}{13}$ b. $\frac{14}{29}$

Objective 3 Calculate unit ratios.

Sometimes it is more useful to write a ratio as a **unit ratio.**

Definition **Unit ratio:** A ratio in which the denominator is 1.

For example, a debt-to-income ratio is more informative when expressed as a unit ratio. Suppose a person pays \$200 toward debt out of every \$1000 of income.

$$\text{debt-to-income ratio} = \frac{200}{1000} = \frac{1}{5}$$

The simplified ratio indicates that the person has \$1 of debt for every \$5 of income. Writing the ratio as a unit ratio will tell us how much debt the person pays for every \$1 of income. Because a unit ratio has a denominator of 1, we divide the denominator into the numerator.

$$\text{debt-to-income ratio} = \frac{200}{1000} = \frac{1}{5} = \frac{0.2}{1}$$

The unit ratio $\frac{0.2}{1}$, or 0.2, means that the person has a debt of \$0.2 for every \$1 of income.

Procedure

To write a ratio as a unit ratio, divide the denominator into the numerator.

Example 5   Write each ratio as a unit ratio and interpret the result.

a. A family has total debt of \$5865 and a gross annual income of \$45,200. What is its debt-to-income ratio?

Solution: $\text{debt-to-income ratio} = \dfrac{5865}{45,200}$ Write the ratio with debt in the numerator and income in the denominator.

$\approx \dfrac{0.13}{1}$, or 0.13   Divide to find the unit ratio.

Answer: The family has a debt-to-income ratio of approximately 0.13, which means that it owes about \$0.13 for every \$1 of income.

b. A certain college has 9752 students and 460 faculty members. What is the student-to-faculty ratio?

Solution: $\text{student-to-faculty ratio} = \dfrac{9752}{460}$ Write the ratio with the number of students in the numerator and the number of faculty in the denominator.

$= \dfrac{21.2}{1}$, or 21.2   Divide to find the unit ratio.

Answer: The student-to-faculty ratio of 21.2 means that there are about 21 students for every faculty member at the college.

▶ **Do Your Turn 5**

Objective 4 Write rates in simplest form.

We can now build on what we've learned about ratios to discuss **rates.**

Definition **Rate:** A ratio comparing two different measurements.

▶ **Your Turn 5**

Write each ratio as a unit ratio and interpret the result.

a. A family has a total debt of \$7842 and a gross income of \$34,240. What is its debt-to-income ratio?

b. A certain college has a total of 14,280 students and 788 faculty. What is the student-to-faculty ratio?

Answers to Your Turn 5
a. ≈ 0.23; the family owes about \$0.23 for every \$1.00 of income.
b. ≈ 18.1; there are about 18 students for every faculty member.

For example, suppose a car travels 180 miles in 3 hours and we want to find the car's rate in miles per hour. The desired units, miles per hour, tell us how to set up the ratio. The measurement to the left of the word *per* goes in the numerator. The measurement to the right of the word *per* goes in the denominator.

$$\text{rate} = \frac{180 \text{ miles}}{3 \text{ hours}}$$

◄ **Note** Because the rate is to be in miles per hour, we place 180 miles in the numerator and 3 hours in the denominator.

A rate is usually easier to interpret when it is written as a **unit rate**.

Definition Unit rate: A rate in which the denominator is 1.

We can write $\frac{180 \text{ miles}}{3 \text{ hours}}$ as a unit by dividing 180 by 3 so that the denominator is 1. Because the units miles and hours are different, they cannot be divided out; so we leave them in fraction form.

$$\frac{180 \text{ miles}}{3 \text{ hours}} = \frac{60 \text{ miles}}{1 \text{ hour}} = 60\frac{\text{miles}}{\text{hour}} \text{ or } 60 \text{ mi./hr.}$$

The unit rate 60 miles per hour means that the car travels 60 miles for every 1 hour. Our exploration suggests the following procedure for finding a unit rate:

Note The simplified unit rate 60 mi./hr. is often abbreviated as 60 mph where the *p* in *mph* stands for "per" and takes the place of the slash.

Procedure

To write a rate as a unit rate, divide the denominator into the numerator. Because the units are different, leave the units in fraction form.

► Your Turn 6

Solve.

a. A cyclist rides 20 miles in 48 minutes. What is the cyclist's average rate in miles per minute?

b. Sasha drives 45 miles in $\frac{3}{4}$ hour. After a break, she drives another 60 miles in $\frac{4}{5}$ hour. What was her average rate for the entire trip in miles per hour?

Example 6 An athlete is clocked running the 40-yard dash in 4.4 seconds. What is the athlete's average rate in yards per second?

Solution: $r = \frac{40 \text{ yd.}}{4.4 \text{ sec.}}$

The rate needs to be in yards per second; so place the number of yards in the numerator and seconds in the denominator.

$$= 9.\overline{09} \text{ yd./sec.} \approx 9.1 \text{ yd./sec.} \quad \text{Divide.}$$

◄ Do Your Turn 6

Another common rate is **unit price,** which relates price to quantity.

Definition Unit price: A unit rate relating price to quantity.

Procedure

If q units of a product have a total price of p, the unit price is found by the formula $u = \frac{p}{q}$.

Note Units written for unit price always have a money unit over a quantity unit. For example, dollars per pound is written \$/lb. and cents per ounce is written ¢/oz.

Answers to Your Turn 6
a. ≈ 0.42 mi./min.
b. ≈ 67.7 mi./hr.

Example 7 A 15-ounce box of cereal costs $3.56. What is the unit price in cents per ounce?

Solution: The unit price needs to be in cents per ounce; so $3.56 needs to be converted to cents. Because there are 100 cents in every dollar, we multiply 3.56 by 100. Multiplying by 100 moves the decimal point to the right 2 places; so $3.56 = 356¢.

$$u = \frac{356¢}{15 \text{ oz.}}$$

The unit price needs to be in cents per ounce; so write the number of cents in the numerator and the number of ounces in the denominator.

$$= 23.7\overline{3}¢/\text{oz.} \approx 23.73¢/\text{oz.}$$ Divide.

▶ **Do Your Turn 7**

Objective 5 Use unit price to determine the better buy.

As consumers, we are often faced with a choice between two different sizes of the same item. Unit prices can be used to determine which of two different-size quantities is a better buy.

Example 8 Determine the better buy.

a. 15-ounce box of bran cereal for $2.45 or 20-ounce box of bran cereal for $3.59

Solution: Let F represent the unit price of the 15-ounce box and T represent the unit price of the 20-ounce box. Calculate the unit price for each box; then compare unit prices. The box with the smaller unit price is the better buy.

15-oz. box
$$F = \frac{\$2.45}{15 \text{ oz.}} = \$0.16\overline{3}/\text{oz.}$$

20-oz. box
$$T = \frac{\$3.59}{20 \text{ oz.}} = \$0.1795/\text{oz.}$$

Answer: Because $0.16\overline{3}/oz. is the smaller unit price, the 15-ounce box is the better buy. In other words, we would pay less for each ounce of bran cereal by buying the 15-ounce box than we would pay if we bought the 20-ounce box.

b. two 10.5-ounce containers of name-brand yogurt for $1.40   or   a 32-ounce container of store-brand yogurt for $2.49

Solution: Let N represent the unit price of the name-brand yogurt and S represent the unit price of the store-brand yogurt. Calculate the unit prices; then compare.

Name brand
$$N = \frac{\$1.40}{21 \text{ oz.}} = \$0.0\overline{6}/\text{oz.}$$

Store brand
$$S = \frac{\$2.49}{32 \text{ oz.}} \approx \$0.078/\text{oz.}$$

▲

Note Buying two containers of the 10.5-ounce yogurt means that we are buying 2(10.5) = 21 ounces of yogurt for $1.40.

Answer: Because $0.0\overline{6}/oz. is the smaller unit price, buying two containers of the 10.5-ounce name-brand yogurt is a better buy than buying 32 ounces of the store brand.

▶ **Do Your Turn 8**

▶ **Your Turn 7**

Solve.

a. A 50-ounce bottle of detergent costs $4.69. What is the unit price in cents per ounce?

b. 4.5 pounds of chicken cost $6.70. What is the unit price in dollars per pound?

▶ **Your Turn 8**

Determine the better buy.

a. a 10.5-ounce can of soup for 79¢
or
a 16-ounce can of soup for $1.29

b. three 2-pound bags of store-brand french fries for $5.67
or
a 2-pound bag of name-brand french fries for $2.69

Answers to Your Turn 7
a. 9.38¢/oz.   b. ≈$1.49/lb.

Answers to Your Turn 8
a. 10.5 oz. can
b. three 2 lb. bags of store-brand french fries

7.1 Exercises (For Extra Help) MyMathLab®

Objective 1

Prep Exercise 1 What is a ratio?

Prep Exercise 2 For two quantities a and b, the ratio of a to b is _____.

For Exercises 1–12, write each ratio in simplest form.

1. There are 62 males and 70 females attending a conference.

   a. What is the ratio of males to total attendance?
   b. What is the ratio of females to total attendance?
   c. What is the ratio of males to females?
   d. What is the ratio of females to males?

2. The following table shows the number of mature trees in a region of forest.

Tree species	Number of mature trees
Pine	488
Maple	264
Oak	114
Other	295

   a. What is the ratio of pine trees to maple trees?
   b. What is the ratio of maple trees to oak trees?
   c. What is the ratio of pine trees to total trees?
   d. What is the ratio of oak trees to total trees?

3. Calcium nitride is a chemical compound that is composed of three calcium atoms and two nitrogen atoms.

   a. What is the ratio of calcium to nitrogen?
   b. What is the ratio of nitrogen to calcium?
   c. What is the ratio of calcium to total atoms in the compound?
   d. What is the ratio of nitrogen to total atoms in the compound?

4. Hydrogen phosphate is a chemical compound composed of three atoms of hydrogen, one atom of phosphorous, and four atoms of oxygen.

   a. What is the ratio of hydrogen to oxygen?
   b. What is the ratio of phosphorous to oxygen?
   c. What is the ratio of hydrogen to total atoms in the compound?
   d. What is the ratio of oxygen to total atoms in the compound?

5. A roof is to have a pitch, or slant, of 14 inches vertically to 16 inches horizontally. Write the ratio of vertical distance to horizontal distance in simplest form.

6. The steps on a staircase rise 8 inches and are 10 inches wide. In simplest form, write the ratio of the rise to the width of each step.

7. The back wheel of a bicycle rotates $3\frac{1}{2}$ times with $2\frac{1}{4}$ rotations of the pedals. In simplest form, write the ratio of back wheel rotations to pedal rotations.

8. At 20 miles per hour in second gear, a car engine is running about 3000 revolutions per minute (rpm). The drive tire is rotating about 260 revolutions per minute. What is the ratio of the engine revolutions per minute to the tire revolutions per minute?

 Learning Strategy

When quizzing someone or watching someone work a problem, be careful that you don't offer tips or hints while the person speaks or works. Let him or her say or do the entire procedure, even if you note a mistake. Correct only after the person has completed the procedure. Remember, nobody will be there during a test to give hints or clues that something is wrong.

Of Interest

Chemical compounds are written with subscripts that indicate the number of atoms of each element in the compound. Calcium nitride, for example, is written as Ca_3N_2.

Of Interest

Hydrogen phosphate is written as H_3PO_4. Note that there is no subscript written for P. The subscript is an understood 1, just as when there is no exponent written for a particular base in exponential form.

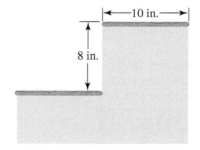

9. In one week, Nina receives 7 bills, 3 letters, 17 advertisements, and 3 credit card offers in her mail.

   a. What is the ratio of bills to total pieces of mail?

   b. What is the ratio of advertisements to total pieces of mail?

   c. What is the ratio of letters to credit card offers?

   d. What is the ratio of bills and letters to advertisements and credit card offers?

10. Nutrition facts for a cereal bar are shown in the table below.

Total bar weight	37 g
Total fat	3 g
Saturated fat	0.5 g
Total carbohydrates	27 g
Fiber	1 g
Sugars	13 g
Protein	2 g

   a. What is the ratio of fiber to total carbohydrates?

   b. What is the ratio of protein to total fat?

   c. What is the ratio of total carbohydrates to the weight of the bar?

   d. What is the ratio of saturated fat to total fat?

11. The following table contains the population of each county in Rhode Island in 2009. (*Source:* U.S. Bureau of the Census, Department of Commerce.)

County	Population (in thousands)
Bristol	50
Kent	169
Newport	80
Providence	628
Washington	127

   a. What is the ratio of the population of Bristol to the population of Newport?

   b. What is the ratio of the population of Newport to the population of Providence?

   c. What is the ratio of the population of Kent to the total population of Rhode Island?

   d. What is the ratio of the population of Providence to the total population of Rhode Island?

12. The bridge and nut on a standard guitar are set so that the length of each string between them is $25\frac{1}{2}$ inches. The seventh fret is placed on the neck of the guitar exactly 17 inches from the bridge.

   a. What is the ratio of the distance between the bridge and the seventh fret to the total length of the strings?

   b. What is the ratio of the distance between the bridge and the seventh fret to the distance between the seventh fret and the nut?

   c. What is the ratio of the distance between the seventh fret and the nut to the total length of the string?

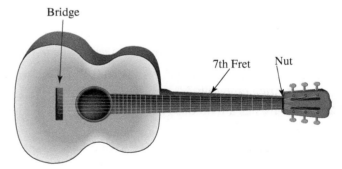

Of Interest

Musical tones are created on stringed instruments by changing the string length. Shortening the string length causes the pitch to get higher. With guitars, violins, and basses, this is achieved by placing a finger on a string. By placing a finger at the seventh fret on a guitar, the string length is shortened from 25.5 inches to 17 inches, which creates a musical tone that is higher in pitch than the tone created by plucking the string with no finger placement (open). In fact, the musical tone created at the seventh fret is a musical fifth from the open string tone. Musical notes are named after the letters of the alphabet; so if the tone of an open string is named A, the note at the seventh fret is named E, which is the fifth letter from A.

Objective 2

Prep Exercise 3 The theoretical probability of equally likely outcomes is _____
_____.

For Exercises 13–22, determine the probability. Write each probability in simplest form.

13. What is the probability of selecting a King in a standard deck of cards?

14. What is the probability of selecting a red Ace in a standard deck of cards?

15. What is the probability of selecting an Ace or a King in a standard deck of cards?

16. What is the probability of selecting a Queen or a Jack in a standard deck of cards?

17. What is the probability of tossing a 1 or a 6 on a six-sided game die?

18. What is the probability of tossing a 1, 2, or 3 on a six-sided game die?

19. What is the probability of tossing a 10 on a six-sided game die?

20. What is the probability of selecting a 1 in a standard deck of cards?

21. Suppose 12 males and 14 females put their names in a hat for a drawing. What is the probability that the winner will be female?

22. Miwa submits 14 entries in a random-drawing contest. If the company receives 894,264 entries, what is the probability that Miwa will win?

23. The following table shows the number of candies of each color in a bag of M&M's®.

Green	42
Red	28
Brown	30
Blue	25

a. What is the probability that a candy chosen at random will be blue?
b. What is the probability that a candy chosen at random will be red?
c. What is the probability that a candy chosen at random will be green or brown?
d. What is the probability that a candy chosen at random will be a color other than red?

24. The following table shows the home states for people attending a conference.

Georgia	46
South Carolina	42
North Carolina	38
Tennessee	18
Florida	14

a. What is the probability that a person chosen at random is from North Carolina?
b. What is the probability that a person chosen at random is from Florida?
c. What is the probability that a person chosen at random is from Georgia or South Carolina?
d. What is the probability that a person chosen at random is from a state other than Tennessee?

Objective 3

Prep Exercise 4 What is a unit ratio?

Prep Exercise 5 Explain how to write a ratio as a unit ratio.

For Exercises 25–32, write each ratio as a unit ratio and interpret the result.

25. A family has a total debt of $4798 and a gross income of $29,850. What is its debt-to-income ratio?

26. A family has a total debt of $9452 and a gross income of $42,325. What is its debt-to-income ratio?

27. A family is seeking a mortgage to purchase a new home. If the mortgage is approved, the new payment will be $945. If the gross monthly income is $3650, what is the payment-to-income ratio?

28. A family pays $550 in rent and has a gross monthly income of $2240. What is the rent-to-income ratio?

29. At a certain college campus, there are 12,480 students and 740 faculty. What is the student-to-faculty ratio?

30. A certain college has 4212 students and 310 faculty. What is the student-to-faculty ratio?

31. At the close of the stock market on a particular day, each share of a computer company's stock was selling for $54\frac{1}{8}$. The annual earnings per share were $4.48. What is the ratio of selling price to annual earnings per share?

32. At the close of the stock market on a particular day, each share of a textile company's stock was selling for $12\frac{5}{16}$. The annual earnings per share were $3.52. What is the ratio of selling price to annual earning per share?

Of Interest

Debt-to-income ratios are one of the determining factors in qualifying for a loan. More specifically, lenders consider the ratio of monthly debt to gross monthly income. This ratio is called the *back-end ratio*. We'll say more about this in Section 7.6.

Of Interest

The ratio of mortgage payment to gross monthly income is another determining factor in qualifying for a loan. This ratio is called the *front-end* ratio.

Of Interest

The ratio of selling price to annual earnings for a stock is called its *price-to-earnings ratio,* or *P/E ratio* for short. Companies with higher (15+) P/E ratios generally show a trend of increasing profits. Companies with lower (less than 5) P/E ratios are considered high-risk investments.

Objective 4

Prep Exercise 6 What is a rate?

Prep Exercise 7 What is a unit price?

For Exercises 33–38, solve.

33. Tara drove 306.9 miles in 4.5 hours. What was her average rate in miles per hour?

34. Li drove 130.2 kilometers in $2\frac{1}{3}$ hours. What was Li's average rate in kilometers per hour?

35. A long-distance phone call lasting 23 minutes costs $2.76. What is the unit price in cents per minute?

36. 16.8 gallons of gas cost $39.48. What is the unit price in dollars per gallon?

37. Gina paid $5.92 for $\frac{3}{4}$ pound of shrimp. What was the unit price in dollars per pound?

38. A 32-ounce bag of frozen french fries costs $2.59. What is the unit price in cents per ounce?

Objective 5

Prep Exercise 8 Brand A of a certain product has a unit price of $0.45 per ounce, whereas brand B has a unit price of $0.42 per ounce. Which is the better buy? Why?

For Exercises 39–44, determine the better buy.

39. a 10.5-ounce can of soup for $1.79   or
    a 16-ounce can of soup for $2.19

40. a 15-ounce can of fruit cocktail for $1.29   or
    a 30-ounce can of fruit cocktail for $2.19

41. a bag containing 24 diapers for $10.99   or
    $15.99 for a bag containing 40 diapers

42. a 14-ounce box of rice for $2.59   or
    $3.69 for a 32-ounce box of rice

43. two 15.5-ounce boxes of store-brand whole grain cereal for $5.00 or a 20-ounce box of name-brand whole grain cereal for $3.45

44. three 15-ounce cans of mixed vegetables for $1.98   or
    two 12.5-ounce cans of mixed vegetables for $1.19

45. Use the following graph, which shows the number of people per 10,000 that were victims of the indicated crimes. (*Source:* National Crime Victimization Survey, U.S. Department of Justice.)

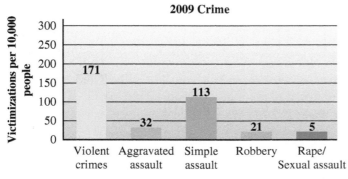

2009 Crime

a. Complete the following table.

	Violent crimes	Aggravated assault	Simple assault	Robbery	Rape/Sexual assault
Ratio per 10,000 people					
Ratio per 1000 people					
Ratio per 100 people					

b. In the row labeled "Ratio per 10,000 people," interpret the simplified ratio in the column labeled "Aggravated assault."

c. How many violent crimes occurred per 1000 people?
d. How many robberies occurred per 100 people?

46. Use the following graph, which shows that the average hourly wage for education and health services is increasing at nearly a constant rate. (*Source:* U.S. Bureau of Labor Statistics.)

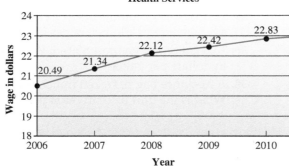

We can use the following formula to determine the rate of increase between any two years.

$$\text{rate of increase} = \frac{\text{greater wage} - \text{lesser wage}}{\text{later year} - \text{earlier year}}$$

a. Complete the following table.

	2006 to 2007	2007 to 2008	2008 to 2009	2009 to 2010	2006 to 2010
Rate of Increase					

b. Round the rate of increase for 2006 to 2010 to the nearest hundredth.
c. The number you found in part b is a rough approximation of the rate of increase from year to year. Using that approximation, complete the following table by calculating the expected hourly wage for the indicated years.

	2011	2012	2013	2014
Expected hourly wage				

Puzzle Problem According to the Bureau of the Census, every 7 seconds a baby is born, every 31 seconds an immigrant arrives, and every 13 seconds a person dies. Determine the time it takes the population to increase 1 person. (*Hint:* Use the LCM.)

Review Exercises

For Exercises 1–3, multiply.

[5.3] 1. $5\frac{1}{4} \cdot \frac{6}{7}$

[6.3] 2. $(9.6)(12.5)$

[6.5] 3. $4.5\left(2\frac{3}{4}\right)$

For Exercises 4 and 5, solve and check.

[5.8] 4. $\frac{2}{3}y = 5$

[6.4] 5. $10.4x = 89.44$

7.2 Proportions

Warm-up *Solve.*

[5.4] 1. $\dfrac{5}{6}x = \dfrac{7}{8}$

[7.1] 2. Margaret spends $2.32 for 2.61 pounds of sweet potatoes. Find the unit price.

Objective 1 Determine whether two ratios are proportional.

Now that we have learned about ratios in Section 7.1, we can solve problems involving equal ratios in a **proportion.**

Definition Proportion: An equation in the form $\dfrac{a}{b} = \dfrac{c}{d}$, where $b \neq 0$ and $d \neq 0$.

For example, $\dfrac{5}{8} = \dfrac{10}{16}$ is a proportion. We can see why because $\dfrac{5}{8} = \dfrac{5 \cdot 2}{8 \cdot 2} = \dfrac{10}{16}$ or $\dfrac{10}{16} = \dfrac{10 \div 2}{16 \div 2} = \dfrac{5}{8}$. In Chapter 5, we also learned that the cross products of equivalent fractions are equal.

$5 \cdot 16 = 80 \qquad 8 \cdot 10 = 80$

$$\dfrac{5}{8} \bowtie \dfrac{10}{16}$$

◄ **Note** The cross products of equivalent fractions are equal.

We can show that the cross products are equal for any proportion. We begin with the general form of a proportion from the definition $\dfrac{a}{b} = \dfrac{c}{d}$, where $b \neq 0$ and $d \neq 0$. We can use the multiplication principle of equality to eliminate the fractions by multiplying both sides of the proportion by the LCD, which is bd.

$$\dfrac{\overset{1}{\cancel{bd}}}{1} \cdot \dfrac{a}{\cancel{b}} = \dfrac{\overset{1}{\cancel{bd}}}{1} \cdot \dfrac{c}{\cancel{d}}$$
$$ad = bc$$

Note that ad and bc are the cross products.

$$\overset{ad}{} \qquad \overset{bc}{}$$
$$\dfrac{a}{b} \bowtie \dfrac{c}{d}$$

We can write the following rule for proportions:

Rule

If two ratios are proportional, their cross products are equal.

In math language: If $\dfrac{a}{b} = \dfrac{c}{d}$, where $b \neq 0$ and $d \neq 0$, then $ad = bc$.

Example 1 Determine whether the ratios are proportional.

a. $\dfrac{3.5}{8} \overset{?}{=} \dfrac{1.4}{3.2}$

Note Although the Commutative property of multiplication allows us to write the factors in any order, we will write them in the order indicated by the arrows: denominator × numerator.

Solution:

$$(3.2)(3.5) = 11.2 \qquad (8)(1.4) = 11.2$$

Find the cross products.

Answer: The cross products are equal; so the ratios are proportional.

Objectives

1 Determine whether two ratios are proportional.

2 Solve for an unknown value in a proportion.

3 Solve proportion problems.

4 Use proportions to solve for unknown lengths in similar figures.

▶ **Your Turn 1**

Determine whether the ratios are proportional.

a. $\dfrac{14}{16} \overset{?}{=} \dfrac{18}{20}$

b. $\dfrac{3.6}{5} \overset{?}{=} \dfrac{18}{25}$

c. $\dfrac{3\frac{1}{4}}{16} \overset{?}{=} \dfrac{6}{29\frac{7}{13}}$

b. $\dfrac{\frac{3}{4}}{\frac{3}{10}} \overset{?}{=} \dfrac{\frac{5}{12}}{\frac{1}{2}}$

Solution: $\dfrac{1}{2}\cdot\dfrac{3}{4} = \dfrac{3}{8}$ $\dfrac{\overset{1}{3}}{\underset{2}{10}}\cdot\dfrac{\overset{1}{5}}{\underset{4}{12}} = \dfrac{1}{8}$

$$\dfrac{\frac{3}{4}}{\frac{3}{10}} \overset{?}{=} \dfrac{\frac{5}{12}}{\frac{1}{2}}$$

Find the cross products.

Answer: The cross products are not equal; so the ratios are not proportional.

◀ **Do Your Turn 1**

Objective 2 Solve for an unknown value in a proportion.

Cross products can be used to solve for an unknown value in a proportion.

Learning Strategy

If you are a visual learner, draw the arrows as in Example 2 to help visualize the pattern of cross products.

Procedure

To solve a proportion using cross products:

1. Find the cross products.
2. Set the cross products equal to each other.
3. Solve the resulting equation.

▶ **Your Turn 2**

Solve.

a. $\dfrac{5}{9} = \dfrac{n}{27}$

b. $\dfrac{y}{16} = \dfrac{7}{10}$

Example 2 Solve. $\dfrac{5}{12} = \dfrac{x}{8}$

Solution: Because this is a proportion, the cross products must be equal.

$8\cdot 5 = 40 \qquad\qquad 12\cdot x = 12x$ Calculate the cross products.

$$\dfrac{5}{12} \overset{?}{=} \dfrac{x}{8}$$

$40 = 12x$ Set the cross products equal to each other.

$\dfrac{40}{12} = \dfrac{12x}{12}$ Divide both sides by 12 to isolate x.

$\dfrac{10}{3} = x$

$3\dfrac{1}{3} = x$ ◀ **Note** We could have expressed the answer as $3.\overline{3}$.

◀ **Do Your Turn 2**

Answers to Your Turn 1
a. no   b. yes   c. yes

Answers to Your Turn 2
a. $n = 15$ b. $y = 11.2$

Example 3 Solve. $\dfrac{2\frac{1}{4}}{8} = \dfrac{-9}{b}$

Solution: Because this is a proportion, the cross products must be equal.

$b \cdot 2\frac{1}{4} = 2\frac{1}{4}b$ $\qquad$ $(8)(-9) = -72$ $\qquad$ Calculate the cross products.

$$\dfrac{2\frac{1}{4}}{8} \mathbin{\diagup\!\!\!\!=\!\!\!\!\diagup} \dfrac{-9}{b}$$

$2\frac{1}{4}b = -72$ $\qquad$ Set the cross products equal to each other.

$\dfrac{9}{4}b = -72$ $\qquad$ Write the mixed number as an improper fraction.

$\overset{1}{\underset{1}{\dfrac{4}{9}}} \cdot \overset{1}{\underset{1}{\dfrac{9}{4}}}b = \dfrac{\overset{8}{-72}}{1} \cdot \dfrac{4}{\underset{1}{9}}$ $\qquad$ Multiply both sides by $\frac{4}{9}$ to isolate b.

$b = -32$

▶ **Do Your Turn 3**

▶ **Your Turn 3**
Solve.
a. $\dfrac{4\frac{2}{3}}{m} = \dfrac{\frac{3}{4}}{-9}$

b. $\dfrac{-6.5}{y} = \dfrac{-13}{14}$

Objective 3 Solve proportion problems.

In application problems that can be solved using a proportion, a ratio will be given. You will be asked to find an unknown part of an equivalent ratio when the numerator or denominator is increased or decreased. The following problem is a typical proportion problem.

The label on a bag of frozen vegetables indicates that 16 ounces serve 5 people. At that rate, how many ounces are needed to serve 8 people?

Given ratio: 16 ounces serve 5 people. $\qquad$ Translation: $\dfrac{16 \text{ ounces}}{5 \text{ people}}$

Unknown: How many ounces are needed to serve 8 people? Translation: $\dfrac{n \text{ ounces}}{8 \text{ people}}$

At that rate indicates that the ratios are equal; so we can write a proportion.

$$\dfrac{16 \text{ ounces}}{5 \text{ people}} = \dfrac{n \text{ ounces}}{8 \text{ people}}$$

Notice that in the proportion, the information from the problem is paired both horizontally (straight across) *and* vertically (up and down).

$\dfrac{16 \text{ ounces}}{5 \text{ people}} = \dfrac{n \text{ ounces}}{8 \text{ people}}$ $\qquad$ ◀ **Horizontal pairs:** Ounces are in both numerators, ◀ and people are in both denominators.

Vertical pairs: 16 ounces pairs with 5 people, and the unknown number of ounces pairs with 8 people.

Here is another correct way to pair the same information horizontally *and* vertically:

$\dfrac{16 \text{ ounces}}{n \text{ ounces}} = \dfrac{5 \text{ people}}{8 \text{ people}}$ $\qquad$ ◀ **Horizontal pairs:** 16 ounces pairs with 5 people, and ◀ the unknown number of ounces pairs with 8 people.

Vertical pairs: The units match up and down, ounces with ounces and people with people.

Answers to Your Turn 3
a. $m = -56$ b. $y = 7$

Our example suggests the following procedure:

Procedure

To translate an application problem to a proportion, write two ratios set equal to each other so that the information from the problem pairs both horizontally (straight across) and vertically (up and down).

▶ **Your Turn 4**

Solve.

a. Erica drove 303.8 miles using 12.4 gallons of gasoline. At this rate, how much gasoline would it take to drive 1000 miles?

b. Xion can mow a 2500-square-foot lawn in 20 minutes. At this rate, how long will it take him to mow a 12,000-square-foot lawn?

Example 4 Bobbie can read 40 pages in 50 minutes. At that rate, how long will it take her to read 220 pages?

Solution: We are given a rate of 40 pages in 50 minutes and asked to find the time it takes to read 220 pages at the same rate. Because the rates are equivalent, we can translate to a proportion.

$$\frac{40 \text{ pages}}{50 \text{ min.}} = \frac{220 \text{ pages}}{x \text{ min.}}$$

$$x \cdot 40 = 40x \qquad \qquad 50 \cdot 220 = 11{,}000$$

$$\frac{40}{50} \times \frac{220}{x}$$

$$40x = 11{,}000$$

$$\frac{40x}{40} = \frac{11{,}000}{40}$$

$$x = 275$$

Note Before solving the proportion, a quick estimate can show what to expect as an answer. Notice that 220 pages is a little over 5 times the 40 pages Bobbie can read in 50 minutes. Therefore, it should take a little over 5 times the number of minutes. Because $5 \times 50 = 250$ minutes, we should expect the answer to be a little more than 250 minutes.

Answer: It will take Bobbie 275 minutes to read 220 pages.

Our answer reasonably agrees with our estimate. If you set up the proportion incorrectly, you will get an unreasonable answer.

For example, suppose we had set up Example 4 as shown to the right.

$$\frac{40 \text{ pages}}{50 \text{ min.}} = \frac{x \text{ min.}}{220 \text{ pages}}$$

Warning This setup is incorrect because the information is not paired horizontally.

Solving this incorrect setup, we get

$$220 \cdot 40 = 8800 \qquad \qquad 50 \cdot x = 50x$$

$$\frac{40}{50} \times \frac{x}{220}$$

$$8800 = 50x$$

$$\frac{8800}{50} = \frac{50x}{50}$$

$$176 = x$$

Warning 176 minutes is unreasonable because it is a little over three times the amount of time ($3 \cdot 50 = 150$ minutes) while she is reading more than five times the number of pages ($5 \cdot 40 = 200$ pages).

◀ **Do Your Turn 4**

Answers to Your Turn 4
a. ≈ 40.8 gal.　b. 96 min.

Objective 4 Use proportions to solve for unknown lengths in similar figures.

Consider triangles *ABC* and *DEF*.

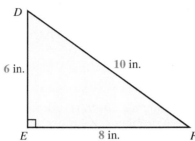

Connection In Section 4.5, we learned that congruent angles have the same measure.

Although the triangles are different in size, their corresponding angle measurements are congruent. In symbols, we write $\angle A \cong \angle D$, $\angle B \cong \angle E$, and $\angle C \cong \angle F$.

Besides having all angles congruent, notice that each side of triangle *DEF* is twice as long as the corresponding side of triangle *ABC*. Or we could say that each side of triangle *ABC* is half as long as the corresponding side of triangle *DEF*. In other words, the corresponding side lengths are proportional.

$$\begin{array}{c} \text{triangle } DEF \to \\ \text{to} \\ \text{triangle } ABC \to \end{array} \quad \frac{6}{3} = \frac{10}{5} = \frac{8}{4} = \frac{2}{1}$$

$$\begin{array}{c} \text{triangle } ABC \to \\ \text{to} \\ \text{triangle } DEF \to \end{array} \quad \frac{3}{6} = \frac{5}{10} = \frac{4}{8} = \frac{1}{2}$$

Because the corresponding angles in the two triangles are congruent and the side lengths are proportional, we say that these are **similar figures.**

Definition **Similar figures:** Figures with the same shape, congruent corresponding angles, and proportional side lengths.

Example 5 Find the unknown length in the similar triangles.

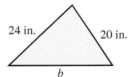

Solution: Because the triangles are similar, the lengths of the corresponding sides are proportional. We can write ratios of the lengths in the smaller triangle to the corresponding lengths in the larger triangle as shown below.

$$\begin{array}{c} \text{smaller triangle} \to \\ \text{to} \\ \text{larger triangle} \to \end{array} \quad \frac{18}{24} = \frac{15}{20} = \frac{21}{b}$$

◄ **Note** We can use either $\frac{18}{24}$ or $\frac{15}{20}$ as the given ratio in our proportion.

$$\frac{18}{24} = \frac{21}{b}$$

$b \cdot 18 = 18b \qquad 24 \cdot 21 = 504$

$$\frac{18}{24} \diagup \frac{21}{b}$$

$$18b = 504$$

$$\frac{18b}{18} = \frac{504}{18}$$

$$b = 28$$

or

$$\frac{15}{20} = \frac{21}{b}$$

$b \cdot 15 = 15b \qquad 20 \cdot 21 = 420$

$$\frac{15}{20} \diagup \frac{21}{b}$$

$$15b = 420$$

$$\frac{15b}{15} = \frac{420}{15}$$

$$b = 28$$

Answer: The unknown length is 28 in.

▶ **Your Turn 5**

Find the unknown length in the similar figure.

a.

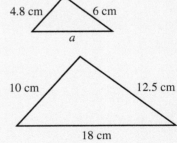

b.

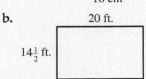

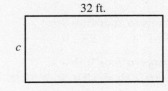

Connection Here is another correct setup where we relate the corresponding sides horizontally.

$$\text{the upper left side} \rightarrow \frac{18}{21} = \frac{24}{b} \leftarrow \text{the base of each triangle}$$

◀ **Do Your Turn 5**

Sometimes more than one length is unknown in problems involving similar figures. To solve these problems, we must be given the length of at least one pair of corresponding sides.

▶ **Your Turn 6**

Find the unknown lengths in the similar figures.

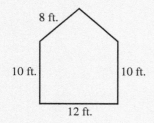

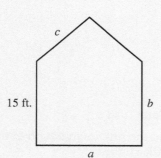

Example 6 Find the unknown lengths in the similar figures.

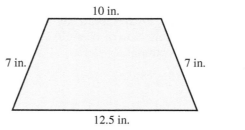

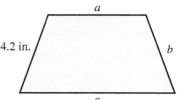

Solution: Because the figures are similar, the ratios of the lengths of the corresponding sides are proportional. The corresponding sides can be related as follows:

$$\text{larger trapezoid} \rightarrow \frac{7}{4.2} = \frac{10}{a} = \frac{7}{b} = \frac{12.5}{c} \leftarrow \text{smaller trapezoid}$$

Note It is not necessary to write a proportion to find b. Because two sides in the larger trapezoid measure 7 inches, their corresponding sides in the smaller trapezoid must be equal in length; so we can conclude that $b = 4.2$ inches.

To find a: $\dfrac{7}{4.2} = \dfrac{10}{a}$

$a \cdot 7 = 7a \qquad 4.2 \cdot 10 = 42$

$\dfrac{7}{4.2} \diagup\hspace{-1.2em}\diagdown \dfrac{10}{a}$

$7a = 42$

$\dfrac{7a}{7} = \dfrac{42}{7}$

$a = 6$

To find c: $\dfrac{7}{4.2} = \dfrac{12.5}{c}$

$c \cdot 7 = 7c \qquad 4.2 \cdot 12.5 = 52.5$

$\dfrac{7}{4.2} \diagup\hspace{-1.2em}\diagdown \dfrac{12.5}{c}$

$7c = 52.5$

$\dfrac{7c}{7} = \dfrac{52.5}{7}$

$c = 7.5$

Answer: The missing lengths are $a = 6$ in., $b = 4.2$ in., and $c = 7.5$ in.

Connection We can verify the reasonableness of the answers by doing quick estimates of the missing side lengths. Comparing the smaller to the larger trapezoid, we see that the 4.2-inch side is a little more than half the corresponding 7-inch side in the larger trapezoid. Each of the corresponding sides should follow the same rule. The 10-inch side of the larger trapezoid should correspond to a length of a little more than 5 for side a of the smaller trapezoid, and it does at 6 inches. The 12.5-inch side should correspond to a length of a little more than 6.25 for side c on the smaller trapezoid, and it does at 7.5 inches.

Answers to Your Turn 5
a. 8.64 cm **b.** 23.2 ft.

Answers to Your Turn 6
$a = 18$ ft., $b = 15$ ft., $c = 12$ ft.

◀ **Do Your Turn 6**

7.2 Exercises For Extra Help MyMathLab®

Objective 1

Prep Exercise 1 What is a proportion?

Prep Exercise 2 If two ratios are proportional, their _____ are equal.

For Exercises 1–14, determine whether the ratios are proportional.

1. $\dfrac{3}{4} \overset{?}{=} \dfrac{9}{12}$

2. $\dfrac{15}{25} \overset{?}{=} \dfrac{3}{5}$

3. $\dfrac{15}{4} \overset{?}{=} \dfrac{7}{2.4}$

4. $\dfrac{7}{4} \overset{?}{=} \dfrac{15}{6.2}$

5. $\dfrac{15}{24} \overset{?}{=} \dfrac{2.5}{3.2}$

6. $\dfrac{10}{14} \overset{?}{=} \dfrac{2.5}{3.5}$

7. $\dfrac{9.5}{14} \overset{?}{=} \dfrac{28.5}{42}$

8. $\dfrac{6.5}{14} \overset{?}{=} \dfrac{10.5}{42}$

9. $\dfrac{16.8}{40.2} \overset{?}{=} \dfrac{17.5}{42.5}$

10. $\dfrac{15.2}{40.2} \overset{?}{=} \dfrac{60.8}{120.4}$

11. $\dfrac{\frac{4}{5}}{\frac{7}{10}} \overset{?}{=} \dfrac{12}{10\frac{1}{2}}$

12. $\dfrac{\frac{2}{5}}{\frac{3}{10}} \overset{?}{=} \dfrac{15}{10\frac{1}{2}}$

13. $\dfrac{2\frac{1}{3}}{3\frac{1}{2}} \overset{?}{=} \dfrac{5\frac{1}{2}}{7\frac{1}{3}}$

14. $\dfrac{3\frac{1}{4}}{5\frac{1}{2}} \overset{?}{=} \dfrac{4\frac{2}{3}}{6\frac{1}{2}}$

Objective 2

Prep Exercise 3 Explain how to solve for an unknown amount in a proportion.

For Exercises 15–30, solve.

15. $\dfrac{x}{8} = \dfrac{20}{32}$

16. $\dfrac{28}{42} = \dfrac{y}{6}$

17. $\dfrac{-3}{8} = \dfrac{n}{20}$

18. $\dfrac{m}{12} = \dfrac{2}{-5}$

19. $\dfrac{2}{n} = \dfrac{7}{21}$

20. $\dfrac{5}{k} = \dfrac{20}{4}$

21. $-\dfrac{4}{28} = \dfrac{5}{m}$

22. $-\dfrac{26}{4} = \dfrac{130}{t}$

23. $\dfrac{18}{h} = \dfrac{21.6}{30}$

24. $\dfrac{20}{32.8} = \dfrac{8}{k}$

25. $\dfrac{-14}{b} = \dfrac{3.5}{6.25}$

26. $\dfrac{4.9}{6.4} = \dfrac{-24.5}{c}$

27. $\dfrac{4\frac{3}{4}}{5\frac{1}{2}} = \dfrac{d}{16\frac{1}{2}}$

28. $\dfrac{\frac{3}{5}}{6\frac{1}{2}} = \dfrac{1\frac{2}{3}}{j}$

29. $\dfrac{-9.5}{22} = \dfrac{-6\frac{1}{3}}{t}$

30. $\dfrac{u}{-\frac{3}{8}} = \dfrac{-4.2}{5}$

Objective 3

Prep Exercise 4 The following application problem has been correctly translated to a proportion. A 5-pound bag of fertilizer covers 2000 square feet. How many pounds are needed to cover 5000 square feet?

$$\text{Translation: } \frac{5 \text{ lb.}}{2000 \text{ ft.}^2} = \frac{n \text{ lb.}}{5000 \text{ ft.}^2}$$

a. Why is the horizontal pairing correct?

b. Why is the vertical pairing correct?

For Exercises 31–42, solve.

31. William drove 358.4 miles using 16.4 gallons of gasoline. At this rate, how much gasoline would he use to drive 750 miles?

32. A carpet cleaning company claims that it will clean 400 square feet for $54.95. At this rate, how much should it charge to clean 600 square feet?

33. Gunther estimates that a 12.5-pound turkey will provide 30 servings. At this rate, how many servings would a 15-pound turkey provide?

34. A 20-pound bag of fertilizer covers 5000 square feet. How many pounds of fertilizer should be used for a lawn that is 12,000 square feet? If the fertilizer comes only in 20-pound bags, how many bags must be purchased?

35. Nathan notes that he consumes 16 ounces of yogurt in five days. At this rate, how many ounces will he consume in a year? (1 year $= 365\frac{1}{4}$ days)

36. In taking a patient's pulse, a nurse counts 19 heartbeats in 15 seconds. How many heartbeats is this in 1 minute? (1 minute $= 60$ seconds)

37. A recipe for bran muffins calls for $1\frac{1}{2}$ cups of bran flakes. The recipe yields 8 muffins. How much bran flakes should be used to make a dozen muffins?

38. The recipe for bran muffins from Exercise 37 calls for $2\frac{1}{2}$ teaspoons of baking powder. How much baking powder should be used to make a dozen muffins?

39. A $42\frac{1}{2}$-foot wall measures $8\frac{1}{2}$ inches on a scale drawing. How long should a $16\frac{1}{4}$-foot wall measure on the drawing?

40. Corrine is drafting the blueprints for a house. The scale for the drawing is $\frac{1}{4}$ inch $= 1$ foot. How long should she draw the line representing a wall that is $13\frac{1}{4}$ feet long?

41. A building company estimates that construction on a high-rise building will move at a pace of 18 vertical feet every 30 days. At this rate, how long will it take to complete a building that is 980 feet tall?

42. A water treatment plant can treat 40,000 gallons of water in 21 hours. How many gallons does the plant treat in a week? (1 week $= 168$ hours)

Objective 4

Prep Exercise 5   What are similar figures?

For Exercises 43–46, find the unknown lengths in the similar figures.

43.

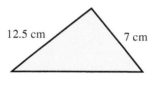
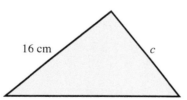

44.

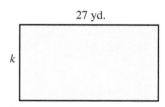

45.

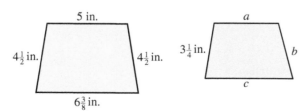

46.

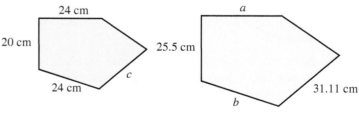

47. Yury wants to estimate the height of his house using a small mirror. He places the mirror on the ground 16.5 meters from the house and walks back (away from the mirror and the house) for 3 meters until he can see the reflection of the top of the house in the mirror. If the distance from the ground to Yury's eyes is 1.8 meters, how tall is the house? (*Note:* The two triangles are similar.)

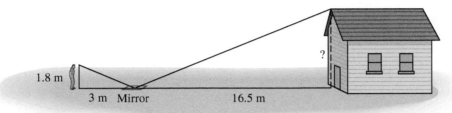

48. A kite is held out on a line that is nearly straight. When 7 meters of line have been let out, the kite is 6 meters off the ground. How high is the kite if 100 meters of line have been let out and the angle of the line has not changed? (*Hint:* Use similar triangles.)

49. To estimate the height of a tree, a forester uses the concept of similar triangles. A particular tree is found to have a shadow measuring 84 feet in length. The forester has her own shadow measured at the same time. Her shadow measures $7\frac{1}{2}$ feet. If she is $5\frac{1}{2}$ feet tall, how tall is the tree?

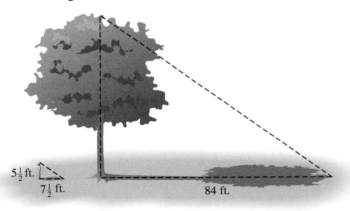

50. To estimate the height of the great pyramid in Egypt, the Greek mathematician Thales performed a procedure similar to that in Exercise 49, except that he supposedly used a staff instead of his body. Suppose the shadow from the staff measured 6 feet and, at the same time, the shadow of the pyramid measured $524\frac{2}{3}$ feet from the center of the pyramid to the tip of the shadow. If the staff was $5\frac{1}{2}$ feet tall, what was the height of the pyramid?

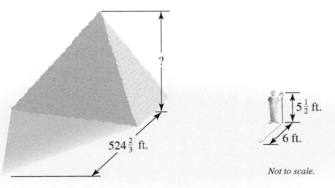

Not to scale.

Of Interest

Thales was a Greek mathematician who lived from 640 B.C. to 546 B.C. He was born in Miletus, a city on the coast of what is now Turkey. Thales is considered to be the father of science and mathematics in the Greek culture. He is the first to have proposed the importance of proof. It is said that he asserted that water is the origin of all things and that matter is infinitely divisible.

51. To estimate the distance across a river, an engineer creates similar triangles. The idea is to create two similar right triangles, as shown in the figure. Calculate the width of the river.

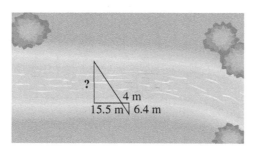

52. The method described in Exercise 51 could be used to determine the width of the Grand Canyon at a given point. Consider the measurements shown. Calculate the width of the canyon.

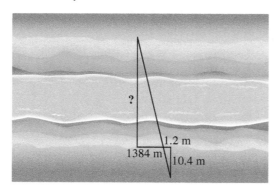

Of Interest

Located in northwest Arizona, the Grand Canyon is about 277 miles (446 kilometers) long and ranges in width from 4 to 18 miles (6.4 to 29 kilometers). The canyon reaches depths of up to 1 mile (1.6 kilometers).

| Puzzle Problem It takes a clock 5 seconds to strike six o'clock. How many seconds will it take for the clock to strike twelve o'clock?

Review Exercises

For Exercises 1–5, evaluate.

[5.3] 1. $\dfrac{4}{9} \cdot \dfrac{5}{12}$

[5.3] 2. $9\dfrac{1}{3}(144)$

[5.4] 3. $60\dfrac{3}{4} \div 3$

[6.4] 4. $70.5 \div 12$

[6.5] 5. $59.2 \cdot \dfrac{1}{2} \cdot \dfrac{1}{2} \cdot \dfrac{1}{4}$

7.3 American Measurement; Time

Warm-up

[6.5] 1. Multiply. $(52.4)\frac{1}{12}$

[7.2] 2. If Ian can mow a 4000-square-foot lawn in 1.5 hours, how long will it take him to mow a 6000-square-foot lawn?

Objectives

1. Convert American units of length.
2. Convert American units of area.
3. Convert American units of capacity.
4. Convert American units of weight.
5. Convert units of time.
6. Convert American units of speed.

Objective 1 Convert American units of length.

To convert from one unit of measurement to another, we will use *dimensional analysis*, which is also called the *factor-label method*. First, consider units of length. Following are some facts about American units of length.

$$1 \text{ foot} = 12 \text{ inches}$$
$$1 \text{ yard} = 3 \text{ feet}$$
$$1 \text{ mile} = 5280 \text{ feet}$$

Each fact can be written as a ratio. In each of those ratios, the numerator is equal to the denominator; so each ratio can be inverted without changing its value, which is 1.

$$\frac{1 \text{ ft.}}{12 \text{ in.}} = \frac{12 \text{ in.}}{1 \text{ ft.}} = 1 \qquad \frac{1 \text{ yd.}}{3 \text{ ft.}} = \frac{3 \text{ ft.}}{1 \text{ yd.}} = 1 \qquad \frac{1 \text{ mi.}}{5280 \text{ ft.}} = \frac{5280 \text{ ft.}}{1 \text{ mi.}} = 1$$

In the factor-label method, we multiply by these ratios; so each is called a **conversion factor**.

Definition Conversion factor: A ratio used to convert from one unit of measurement to another by multiplication.

Because a conversion factor is equivalent to 1, multiplying a measurement by one of them gives an equivalent measurement with different units. For example, multiplying 4 feet by $\frac{12 \text{ in.}}{1 \text{ ft.}}$ causes the unit feet to divide out so that the result is equivalent to 4 feet but now in inches.

$$4 \text{ ft.} = \frac{4 \text{ ft.}}{1} \cdot \frac{12 \text{ in.}}{1 \text{ ft.}} = 48 \text{ in.} \quad \blacktriangleleft$$

Note The unit feet appears in the numerator of one factor and in the denominator of the other factor; so it divides out, leaving the unit inches.

Procedure

To convert units using dimensional analysis, multiply the given measurement by conversion factors so that the undesired units divide out, leaving the desired units.

Notes If an undesired unit is in a numerator, that unit should appear in the denominator of a conversion factor.

If an undesired unit is in the denominator, that unit should appear in the numerator of a conversion factor.

Example 1 Convert using dimensional analysis.

a. 3.5 miles to feet

Solution: Because 3.5 miles can be written over 1, the undesired unit, miles, is in the numerator. Therefore, we need a conversion factor with miles in the

Answers to Warm-up

1. $4.3\overline{6}$ **2.** 2.25 hr. or $2\frac{1}{4}$ hr.

▶ **Your Turn 1**

Convert using dimensional analysis.

a. 20 feet to inches

b. $4\frac{3}{10}$ miles to feet

c. 62 feet to yards

denominator. The desired unit is feet. The conversion factor with feet in the numerator and miles in the denominator is $\frac{5280 \text{ ft.}}{1 \text{ mi.}}$.

$$3.5 \text{ mi.} = \frac{3.5 \text{ mi.}}{1} \cdot \frac{5280 \text{ ft.}}{1 \text{ mi.}} = 18{,}480 \text{ ft.}$$

b. $8\frac{2}{3}$ yards to feet

Solution: Because $8\frac{2}{3}$ yards can be written over 1, the undesired unit, yards, is in the numerator. Therefore, we need a conversion factor with yards in the denominator. The desired unit is feet. The conversion factor with feet in the numerator and yards in the denominator is $\frac{3 \text{ ft.}}{1 \text{ yd.}}$.

$$8\frac{2}{3} \text{ yd.} = \frac{26 \text{ yd.}}{3} = \frac{\overset{}{26 \text{ yd.}}}{\underset{1}{3}} \cdot \frac{\overset{1}{3 \text{ ft.}}}{1 \text{ yd.}} = 26 \text{ ft.}$$

c. 78 inches to feet

Solution: Because 78 inches can be written over 1, the undesired unit, inches, is in the numerator. Therefore, we need a conversion factor with inches in the denominator. The desired unit is feet. The conversion factor with feet in the numerator and inches in the denominator is $\frac{1 \text{ ft.}}{12 \text{ in.}}$.

$$78 \text{ in.} = \frac{\overset{13}{78 \text{ in.}}}{1} \cdot \frac{1 \text{ ft.}}{\underset{2}{12 \text{ in.}}} = \frac{13}{2} \text{ ft.} = 6\frac{1}{2} \text{ ft., or } 6.5 \text{ ft.}$$

◀ **Do Your Turn 1**

Sometimes we may not know a direct relation between the undesired unit and the desired unit. In such cases, we must convert to one or more intermediate units to get to the desired unit. This means that we will use more than one conversion factor.

Example 2   Convert using dimensional analysis.

a. 10 yards to inches

Solution: If we do not know a direct relation of yards to inches, we can convert yards to feet, then feet to inches. To convert from yards to feet, use $\frac{3 \text{ ft.}}{1 \text{ yd.}}$. Then to convert from feet to inches, use $\frac{12 \text{ in.}}{1 \text{ ft.}}$.

$$10 \text{ yd.} = \frac{10 \text{ yd.}}{1} \cdot \frac{3 \text{ ft.}}{1 \text{ yd.}} \cdot \frac{12 \text{ in.}}{1 \text{ ft.}} = 360 \text{ in.}$$

Note Multiplying these two ratios gives the direct fact that there are 36 inches to 1 yard.

b. 4400 yards to miles

Solution: Because we do not have a direct relation of yards to miles, we can convert yards to feet, using $\frac{3 \text{ ft.}}{1 \text{ yd.}}$ and then convert feet to miles using $\frac{1 \text{ mi.}}{5280 \text{ ft.}}$.

$$4400 \text{ yd.} = \frac{4400 \text{ yd.}}{1} \cdot \frac{3 \text{ ft.}}{1 \text{ yd.}} \cdot \frac{1 \text{ mi.}}{5280 \text{ ft.}} = \frac{13{,}200}{5280} \text{ mi.} = 2\frac{1}{2} \text{ mi., or } 2.5 \text{ mi.}$$

Answers to Your Turn 1
a. 240 in.
b. 22,704 ft.
c. $20\frac{2}{3}$ yd. or $20.\overline{6}$ yd.

If we had known the direct relations in Example 2, we would not have had to use several conversion factors. The more relations you know, the fewer conversion factors you have to use. However, this means more to remember. So there's a trade-off: memorize more facts and use fewer conversion factors or memorize fewer facts and use more conversion factors. Each has its merits, but in the end, it's up to you to choose which is better.

▶ **Do Your Turn 2**

▶ **Your Turn 2**

Convert using dimensional analysis.
 a. 0.8 yard to inches
 b. 7392 yards to miles

Objective 2 Convert American units of area.

Now let's use dimensional analysis to convert units of area. Recall that area is always represented in square units, like square inches (in.^{2}), square feet (ft.^{2}), square yards (yd.^{2}), and square miles (mi.^{2}).

1 ft. = 12 in.

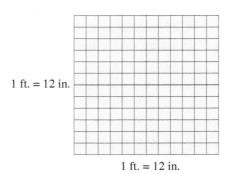

1 ft. = 12 in.

A square foot is a 1-foot by 1-foot square. If we subdivide that square into inches, we have a 12-inch by 12-inch square.

Calculating the area of a 1-foot by 1-foot square as a 12-inch by 12-inch square, we find that it contains 144 square inches. Note that if we square both sides of the relation for length, we get the relation for area.

$$1 \text{ ft.} = 12 \text{ in.} \quad \leftarrow \text{relation for length}$$
$$(1 \text{ ft.})^2 = (12 \text{ in.})^2 \quad \text{Square both sides.}$$
$$1 \text{ ft.}^2 = 144 \text{ in.}^2 \quad \leftarrow \text{relation for area}$$

Conclusion: We can derive the relations for area by squaring the relations for length.

Because 1 yd. = 3 ft., we can conclude that 1 yd.^{2} = 9 ft.^{2}.

Because 1 mi. = 5280 ft., we can conclude that 1 mi.^{2} = 27,878,400 ft.^{2}.

| **Example 3** | Convert using dimensional analysis.

a. 8 square feet to square inches

Solution: To convert square feet to square inches, we need a conversion factor with square feet in the denominator and square inches in the numerator; so we use $\dfrac{144 \text{ in.}^2}{1 \text{ ft.}^2}$.

$$8 \text{ ft.}^2 = \frac{8 \text{ ft.}^2}{1} \cdot \frac{144 \text{ in.}^2}{1 \text{ ft.}^2} = 1152 \text{ in.}^2$$

b. 150 square feet to square yards

Solution: To convert square feet to square yards, we need a conversion factor with square feet in the denominator and square yards in the numerator; so we use $\dfrac{1 \text{ yd.}^2}{9 \text{ ft.}^2}$.

$$150 \text{ ft.}^2 = \frac{\overset{50}{150} \text{ ft.}^2}{1} \cdot \frac{1 \text{ yd.}^2}{\underset{3}{9} \text{ ft.}^2} = \frac{50}{3} \text{ yd.}^2 = 16\frac{2}{3} \text{ yd.}^2, \text{ or } 16.\overline{6} \text{ yd.}^2$$

▶ **Do Your Turn 3**

▶ **Your Turn 3**

Convert using dimensional analysis.
 a. 12 square yards to square feet
 b. 6000 square inches to square feet
 c. 8 square yards to square inches

Answers to Your Turn 2
a. 28.8 in.   b. 4.2 mi.

Answers to Your Turn 3
a. 108 ft.^{2}
b. $41\frac{2}{3}$ ft.^{2} or $41.\overline{6}$ ft.^{2}
c. 10,368 in.^{2}

> ▶ **Your Turn 4**
>
> A room is 24 feet long by $16\frac{1}{2}$ feet wide. Find its area in square yards.

Example 4 A room that is 25.5 feet long by 18 feet wide is to be carpeted. Calculate the area of the room in square yards.

Understand We must calculate the area of a rectangular room. The formula for the area of a rectangle is $A = lw$.

The given measurements are in terms of feet. If we calculate the area with these measurements, the area unit will be square feet. The desired unit is square yards. There are two approaches to this problem:

1. Calculate the area in terms of square feet; then convert square feet to square yards.
2. Convert feet to yards and calculate the area with the measurements in terms of yards so that the outcome is in square yards.

Plan Let's use the first approach. We will calculate the area using the given measurements so that the area will be in square feet. We will then convert square feet to square yards.

Execute $A = lw$

$A = (25.5)(18)$ Replace *l* with 25.5 and *w* with 18.

$A = 459$ Multiply.

Now convert 459 square feet to square yards.

$$459 \text{ ft.}^2 = \frac{459 \text{ ft.}^2}{1} \cdot \frac{1 \text{ yd.}^2}{9 \text{ ft.}^2} = \frac{459}{9} \text{ yd.}^2 = 51 \text{ yd.}^2$$

Answer The room will require 51 square yards of carpet.

Check Let's use the second method we described as a check. We will convert the given measurements to yards, then calculate the area.

Length: $25.5 \text{ ft.} = \frac{25.5 \text{ ft.}}{1} \cdot \frac{1 \text{ yd.}}{3 \text{ ft.}} = \frac{25.5}{3} \text{ yd.} = 8.5 \text{ yd.}$

Width: $18 \text{ ft.} = \frac{18 \text{ ft.}}{1} \cdot \frac{1 \text{ yd.}}{3 \text{ ft.}} = \frac{18}{3} \text{ yd.} = 6 \text{ yd.}$

Area: $A = (8.5)(6)$ Replace *l* with 8.5 and *w* with 6.

$A = 51 \text{ yd.}^2$

◀ **Do Your Turn 4**

Objective 3 Convert American units of capacity.

Now consider **capacity.**

Definition **Capacity:** A measure of the amount a container holds.

The American system uses units such as gallons (gal.), quarts (qt.), pints (pt.), cups (c.), and ounces (oz.) to measure the capacity of a container. Following are some facts about capacity units.

1 cup = 8 ounces
1 pint = 2 cups
1 quart = 2 pints
1 gallon = 4 quarts

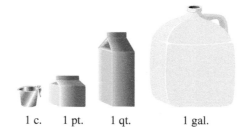

1 c.   1 pt.   1 qt.   1 gal.

Answer to Your Turn 4
44 yd.^{2}

Example 5   Convert using dimensional analysis.

a. 5 pints to ounces

Solution: Because we do not have a direct relation of pints to ounces, we convert pints to cups and then cups to ounces. To convert pints to cups, use $\frac{2\,c.}{1\,pt.}$. To convert cups to ounces, use $\frac{8\,oz.}{1\,c.}$.

$$5\,pt. = \frac{5\,pt.}{1}\cdot\frac{2\,c.}{1\,pt.}\cdot\frac{8\,oz.}{1\,c.} = 80\,oz.$$

b. 240 ounces to gallons

Solution: Because we do not have a direct relation of ounces to gallons, we convert ounces to cups, cups to pints, pints to quarts, and then quarts to gallons.

$$240\,oz. = \frac{240\,oz.}{1}\cdot\frac{1\,c.}{8\,oz.}\cdot\frac{1\,pt.}{2\,c.}\cdot\frac{1\,qt.}{2\,pt.}\cdot\frac{1\,gal.}{4\,qt.} = \frac{240}{128}\,gal. = 1\frac{7}{8}\,gal.,\ or\ 1.875\,gal.$$

▶ **Do Your Turn 5**

Objective 4 Convert American units of weight.

Now let's convert units of **weight**.

Definition Weight: A measure of the downward force due to gravity.

In the American system, weight can be expressed in terms of ounces (oz.), pounds (lb.), and tons (T). Following is a list of two facts about weight.

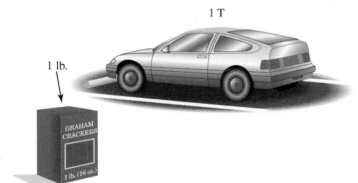

1 pound = 16 ounces
1 ton = 2000 pounds

Example 6   Convert using dimensional analysis.

a. 12 ounces to pounds

Solution: To convert ounces to pounds, use $\frac{1\,lb.}{16\,oz.}$.

$$12\,oz. = \frac{\overset{3}{12}\,oz.}{1}\cdot\frac{1\,lb.}{\underset{4}{16}\,oz.} = \frac{3}{4}\,lb.,\ or\ 0.75\,lb.$$

b. 0.2 ton to ounces

Solution: We do not have a direct relation of tons to ounces; so we will convert tons to pounds, and then pounds to ounces.

$$0.2\,T = \frac{0.2\,T}{1}\cdot\frac{2000\,lb.}{1\,T}\cdot\frac{16\,oz.}{1\,lb.} = 6400\,oz.$$

▶ **Do Your Turn 6**

▶ **Your Turn 5**

Convert using dimensional analysis.
  a. 6 quarts to cups
  b. 20 ounces to pints
  c. 3 gallons to cups

▶ **Your Turn 6**

Convert using dimensional analysis.
  a. 12 pounds to ounces
  b. 5000 pounds to tons

Answers to Your Turn 5
a. 24 c.   b. 1.25 pt.   c. 48 c.

Answers to Your Turn 6
a. 192 oz.   b. 2.5 T

Objective 5 Convert units of time.

Now let's consider units of time. Time is measured in units of seconds (sec.), minutes (min.), hours (hr.), days (d.), weeks (wk.), months (mo.), and years (yr.).

$$1 \text{ minute} = 60 \text{ seconds}$$
$$1 \text{ hour} = 60 \text{ minutes}$$
$$1 \text{ day} = 24 \text{ hours}$$
$$1 \text{ year} = 365\frac{1}{4} \text{ or } 365.25 \text{ days}$$

◄ **Note** Unit abbreviations for time vary. Years can be y. or yr. Days can be d. or da. Hours can be h. or hr. Seconds can be s. or sec. However, the only abbreviation for minutes is min. because mi. is miles and m is meters.

▶ Your Turn 7

Convert using dimensional analysis.

a. 8.5 minutes to seconds

b. 5 days to minutes

c. 1 year to hours

Example 7 Convert using dimensional analysis.

a. $4\frac{1}{2}$ hours to seconds

Solution: We do not have a direct relation of hours to seconds; so we will convert hours to minutes and then minutes to seconds.

$$4\frac{1}{2}\text{ hr.} = \frac{9\text{ hr.}}{\underset{1}{2}}\cdot\frac{\overset{30}{60}\text{ min.}}{1\text{ hr.}}\cdot\frac{60\text{ sec.}}{1\text{ min.}} = 16{,}200\text{ sec.}$$

b. 15 minutes to days

Solution: We do not have a direct relation of minutes to days; so we will convert minutes to hours and then hours to days.

$$15\text{ min.} = \frac{\overset{1}{15}\text{ min.}}{1}\cdot\frac{1\text{ hr.}}{\underset{4}{60}\text{ min.}}\cdot\frac{1\text{ d.}}{24\text{ hr.}} = \frac{1}{96}\text{ d. or } 0.01041\overline{6}\text{ d.}$$

◄ **Do Your Turn 7**

Objective 6 Convert American units of speed.

The final type of measurement that we will consider is speed. Speed is a rate describing the distance traveled in an amount of time. Some units of speed are feet per second (ft./sec.) and miles per hour (mi./hr.). We can convert the distance unit, the time unit, or both units within the speed unit.

Example 8 Convert using dimensional analysis.

a. 90 feet per second to feet per minute

Solution: We need to convert only the time part of the speed unit from seconds to minutes. Because seconds is in the denominator of the given unit, we need a conversion factor with seconds in the numerator. Because there are 60 seconds to 1 minute, we use 60 sec./1 min.

$$90\text{ ft./sec.} = \frac{90\text{ ft.}}{1\text{ sec.}}\cdot\frac{60\text{ sec.}}{1\text{ min.}} = \frac{5400\text{ ft.}}{1\text{ min.}} = 5400\text{ ft./min.}$$

b. 60 miles per hour to feet per second

Solution: We need to convert the distance units from miles to feet and the time units from hours to seconds. Because the given speed has miles in the numerator, we use a conversion factor with miles in the denominator. Because the given speed has hours in the denominator, we use a conversion factor with hours in the numerator.

Answers to Your Turn 7

a. 510 sec.

b. 7200 min.

c. 8766 hr.

We can carry out the two conversions in two stages.

$$60 \text{ mi./hr.} = \frac{60 \text{ mi.}}{1 \text{ hr.}} \cdot \frac{5280 \text{ ft.}}{1 \text{ mi.}} = \frac{316{,}800 \text{ ft.}}{1 \text{ hr.}} = 316{,}800 \text{ ft./hr.}$$

$$316{,}800 \text{ ft./hr.} = \frac{316{,}800 \text{ ft.}}{1 \text{ hr.}} \cdot \frac{1 \text{ hr.}}{60 \text{ min.}} \cdot \frac{1 \text{ min.}}{60 \text{ sec.}} = \frac{316{,}800 \text{ ft.}}{3600 \text{ sec.}} = 88 \text{ ft./sec.}$$

Or we can perform both conversions in a single line.

$$60 \text{ mi./hr.} = \frac{60 \text{ mi.}}{1 \text{ hr.}} \cdot \frac{5280 \text{ ft.}}{1 \text{ mi.}} \cdot \frac{1 \text{ hr.}}{60 \text{ min.}} \cdot \frac{1 \text{ min.}}{60 \text{ sec.}} = \frac{316{,}800 \text{ ft.}}{3600 \text{ sec.}} = 88 \text{ ft./sec.}$$

► **Do Your Turn 8**

► Your Turn 8

Convert using dimensional analysis.

a. 500 feet per minute to feet per second
b. 28 feet per second to miles per hour
c. 9 yards per second to feet per minute

Answers to Your Turn 8
a. $8\frac{1}{3}$ ft./sec. or $8.\overline{3}$ ft./sec.
b. $19\frac{1}{11}$ mi./hr. or $19.\overline{09}$ mi./hr.
c. 1620 ft./min.

7.3 Exercises For Extra Help MyMathLab®

Objectives 1 and 2

Prep Exercise 1 What is a conversion factor?

Prep Exercise 2 Explain how to convert units using dimensional analysis.

Prep Exercise 3 Complete the relations for American units of length.

1 foot = _____ inches
1 yard = _____ feet
1 mile = _____ feet

Prep Exercise 4 Complete the relations for American units of area.

$1 \text{ ft.}^2 =$ _____ in.^2
$1 \text{ yd.}^2 =$ _____ ft.^2

For Exercises 1–20, convert using dimensional analysis.

1. 52 yards to feet
2. 6 feet to inches
3. 90 inches to feet
4. 45 feet to yards
5. 6.5 miles to feet
6. 7392 feet to miles
7. $5\frac{1}{2}$ feet to inches
8. $15\frac{2}{3}$ yards to feet
9. 20.8 miles to yards
10. 0.2 yard to inches
11. 99 inches to yards
12. 3168 yards to miles
13. 30 square yards to square feet
14. 648 square inches to square feet
15. $10\frac{1}{2}$ square yards to square inches

16. 40 square feet to square yards

17. Blueprint paper can be purchased in a roll that is 36 inches wide. If unrolled, the paper would extend 150 feet in length. What is the total area of the blueprint paper in square feet?

18. Blueprint paper also comes in a roll that is 42 inches wide. If unrolled, the paper would extend 150 feet in length. What is the total area of the blueprint paper in square feet?

19. A room that is 13.5 feet long by 12.5 feet wide is to be carpeted. Calculate the area in square yards.

20. A room that is 16.5 feet long by 14 feet wide is to be carpeted. Calculate the area in square yards.

Objective 3

Prep Exercise 5 Complete the relations for American units of capacity.

1 cup = _____ ounces

1 pint = _____ cups

1 quart = _____ pints

1 gallon = _____ quarts

For Exercises 21–32, convert using dimensional analysis.

21. 9 pints to cups

22. 6 quarts to pints

23. 2.5 cups to ounces

24. 16.4 gallons to quarts

25. 240 ounces to pints

26. 12 cups to pints

27. $8\frac{1}{2}$ pints to quarts

28. 60 ounces to quarts

29. 0.4 gallon to ounces

30. 5.2 gallons to pints

31. 40.8 pints to gallons

32. $6\frac{1}{3}$ cups to pints

Objective 4

Prep Exercise 6 Complete the relations for American units of weight.

1 pound = _____ ounces

1 ton = _____ pounds

For Exercises 33–40, convert using dimensional analysis.

33. $12\frac{1}{4}$ pounds to ounces

34. $9\frac{1}{2}$ pounds to ounces

35. 180 ounces to pounds

36. 124 ounces to pounds

37. 1200 pounds to tons

38. 6200 pounds to tons

39. 4.2 tons to pounds

40. 6.4 tons to pounds

Objective 5

Prep Exercise 7 Complete the relations for American units of time.

1 minute = _____ seconds

1 hour = _____ minutes

1 day = _____ hours

1 year = _____ days

For Exercises 41–56, convert using dimensional analysis.

41. 150 minutes to hours

42. 200 minutes to hours

43. 90 seconds to minutes

44. 20 seconds to minutes

45. $10\frac{1}{2}$ minutes to seconds

46. 12.2 minutes to seconds

47. $2\frac{1}{4}$ hours to minutes

48. $8\frac{1}{2}$ hours to minutes

49. 0.2 hour to seconds

50. 1.5 hours to seconds

51. 30 days to years

52. 90 days to years

53. 1 day to minutes

54. 7 days to seconds

55. 72 years to minutes

56. 85 years to seconds

Objective 6

For Exercises 57–64, solve using dimensional analysis.

57. The speed of sound in dry air at 0°C is 1088 feet per second. Calculate the speed of sound in miles per minute.

58. Use the information in Exercise 57 to calculate the speed of sound in miles per hour.

59. An investigating officer examining skid marks at the scene of an accident estimates that the speed of the vehicle was 80 feet per second. The driver of the vehicle claims to have been going 40 miles per hour. Are they in agreement?

60. A defendant in a collision case claims to have been traveling at 35 miles per hour around a blind curve and did not have time to swerve to miss rear-ending the plaintiff, who had just pulled into the road. The investigating officer determines that any speed less than 60 feet per second would have allowed time to brake or swerve to avoid the collision. How would you judge the defendant?

61. The escape velocity for Earth is 36,687.6 feet per second. Calculate this speed in miles per hour.

62. The minimum speed required to achieve orbit around Earth is 25,957 feet per second. Calculate this speed in miles per hour.

63. While driving along a highway at 60 miles per hour, a driver sees a sign indicating that the lane ends 500 feet ahead. After passing the sign, if she maintains her speed, how much time does she have to merge into the other lane before the lane ends?

64. While driving along a highway at 70 miles per hour, a driver sees a sign indicating uneven pavement 500 feet ahead. After passing the sign, if he maintains his speed, how much time does he have before he reaches the uneven pavement?

Puzzle Problem   A light-year (lt-yr.) is the distance light travels in one year. If light travels at a rate of 186,282 miles per second, how far is 1 light-year in miles? *Proxima Centauri* is the next nearest star to our star (the sun) at a distance of about 4.2 light-years. What is this distance in miles?

Of Interest

Escape velocity is the speed that an object must achieve to escape the gravitational pull of a planet. Any object traveling slower than the escape velocity will return to the planet or orbit it. An object traveling less than the minimum speed to achieve orbit will return to the planet. Objects traveling at speeds greater than the minimum speed to achieve orbit but less than the escape velocity will orbit the planet. For Earth, objects traveling at exactly 25,957 feet per second will orbit in a circle. As speed is increased, the path becomes more elliptical.

Review Exercises

For Exercises 1–6, perform the indicated operation.

[6.3] 1. 8.4(100)

[6.3] 2. 0.95(1000)

[6.3] 3. (0.01)(32.9)

[6.4] 4. 45.6 ÷ 10

[6.4] 5. 9 ÷ 1000

[6.4] 6. 4800 ÷ 1000

7.4 Metric Measurement

Objectives

1 Convert units of metric length.
2 Convert units of metric capacity.
3 Convert units of metric mass.
4 Convert units of metric area.

Warm-up

[6.3] 1. Multiply. (0.65) 1000
[7.3] 2. Convert 132 inches to yards.

The difficulty with the American system of measurement is that each set of facts contains different numbers. The metric system is easier because to convert from one unit to the next, we multiply or divide by the same number, 10. Multiplying or dividing by 10 is easy because we just move the decimal point to the right or left.

Each type of measurement has a **base unit.**

Definition Base unit: A basic unit; other units are named relative to it.

To the right are the metric base units for length, capacity, and mass.

Type of measurement	Base unit
Length	meter (m)
Capacity	liter (L)
Mass	gram (g)

To further simplify the system, prefixes are used with the base unit to indicate the size of the measurement relative to the base unit. The prefixes we will learn are listed below in order from largest size to smallest size.

Note These three units are larger than the base unit. ▼

Note These three units are smaller than the base unit. ▼

| kilo- (k) | hecto- (h) | deka- (da) | base unit | deci- (d) | centi- (c) | milli- (m) |

Of Interest

Originally, a meter was defined to be 1/40,000,000 of a circle on Earth's surface passing through the poles. Now we define a meter to be the distance traveled by light in empty space in 1/299,792,458 of a second.

Objective 1 Convert units of metric length.

As we've mentioned, the base unit for length is the meter (m), which is a little longer than a yard. Attaching the base unit, meter, to each prefix is shown in the following table:

kilometer (km)	hectometer (hm)	dekameter (dam)	meter (m)	decimeter (dm)	centimeter (cm)	millimeter (mm)
1 km = 1000 m	1 hm = 100 m	1 dam = 10 m	1 m	1 m = 10 dm	1 m = 100 cm	1 m = 1000 mm

Here are some comparisons to help you visualize the size of these units.

A kilometer is about 0.6 mile.

A hectometer is about 110 yards, which is the length of a football field including one end zone.

A dekameter is about the length of a school bus.

Answers to Warm-up
1. 650
2. $3.\overline{6}$ yd. or $3\frac{2}{3}$ yd.

A decimeter is about the width of your hand across your palm (include your thumb).

A centimeter is about the width of your pinky finger at the tip.

A millimeter is about the thickness of your fingernail.

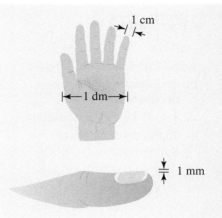

We can use these facts to convert metric length using dimensional analysis. Suppose we want to convert 2.8 meters to centimeters. We use the fact that there are 100 centimeters to 1 meter.

$$2.8 \text{ m} = \frac{2.8 \text{ m}}{1} \cdot \frac{100 \text{ cm}}{1 \text{ m}} = 280 \text{ cm}$$

Note that we multiplied by 100, which causes the decimal point to move 2 places to the right of its position in 2.8. Also notice that centimeter is 2 places to the right of meter in the table of units.

kilometer hectometer dekameter meter decimeter centimeter millimeter
  (km)   (hm)   (dam)   (m)   (dm)   (cm)   (mm)

Suppose we want to convert 2.8 meters to kilometers. We use the fact that 1 kilometer is equal to 1000 meters. Based on what we just saw, what do you expect to happen?

$$2.8 \text{ m} = \frac{2.8 \text{ m}}{1} \cdot \frac{1 \text{ km}}{1000 \text{ m}} = \frac{2.8}{1000} \text{ km} = 0.0028 \text{ km}$$

Note that we divided by 1000, which causes the decimal point to move 3 places to the left. Also notice that kilometer is 3 places to the left of meter in the table of units.

kilometer hectometer dekameter meter decimeter centimeter millimeter
  (km)   (hm)   (dam)   (m)   (dm)   (cm)   (mm)

Our examples suggest the following fast method for converting units in the metric system:

Procedure

To convert units in the metric system:
1. Using the table of metric units, count the number of units from the undesired unit to the desired unit and note the direction (right or left).

kilo-   hecto-   deka-   base unit   deci-   centi-   milli-
 (k)   (h)   (da)   (m)   (d)   (c)   (m)

2. Move the decimal point the same number of places and in the same direction as determined in step 1.

Learning Strategy

Many people remember the order of units by making up a sentence with the initial letters of each unit. Here's an example:

King **H**ect**or**'s **Da**ta **Ba**se **De**ci**dedly Cent**ralized **Milli**ons

| **Example 1** | Convert.

a. 0.15 meter to decimeters

Solution: In the table of units, decimeter is the first unit to the right of meter; so move the decimal point 1 place to the right of its position in 0.15.

kilometer hectometer dekameter meter decimeter centimeter millimeter
  (km)   (hm)   (dam)   (m)   (dm)   (cm)   (mm)

$$0.15 \text{ m} = 1.5 \text{ dm}$$

Answer: 1.5 dm

▶ **Your Turn 1**

Convert.
a. 9.74 meters to millimeters
b. 0.61 dekameter to decimeters
c. 456 meters to hectometers
d. 20,800 centimeters to kilometers

b. 56.1 dekameters to centimeters

Solution: In the table of units, centimeter is the third unit to the right of dekameter (including the base unit); so move the decimal point 3 places to the right of its position in 56.1.

kilometer (km)	hectometer (hm)	dekameter (dam)	meter (m)	decimeter (dm)	centimeter (cm)	millimeter (mm)

56.1 dam = 56, 1 0 0 cm

Answer: 56,100 cm

c. 2500 millimeters to meters

Solution: In the table of units, meter is the third unit to the left of millimeter; so move the decimal point 3 places to the left of its position in 2500.

kilometer (km)	hectometer (hm)	dekameter (dam)	meter (m)	decimeter (dm)	centimeter (cm)	millimeter (mm)

2500 mm = 2.500 m

Note In 2500, the decimal point is to the right of the last 0. After we move the decimal point 3 places to the left to get 2.500, those 0s are to the right of the last nonzero decimal digit; so they are no longer needed.

Answer: 2.5 m

d. 970 centimeters to hectometers

Solution: Using the table of units, we see that hectometer is the fourth unit to the left of centimeter; so move the decimal point 4 places to the left of its position in 970.

kilometer (km)	hectometer (hm)	dekameter (dam)	meter (m)	decimeter (dm)	centimeter (cm)	millimeter (mm)

970 cm = 0.0970 hm

Answer: 0.097 hm

◀ **Do Your Turn 1**

Of Interest

The fact that $1 \, ml = 1 \, cm^3$ is extremely important for people in health-related fields. Medications are often prescribed in units of milliliters, yet the packaging for the medication is expressed in cubic centimeters or vice versa. The person administering the medication needs to recognize that milliliters and cubic centimeters are used interchangeably.

Objective 2 Convert units of metric capacity.

Consider metric capacity. The base unit for metric capacity is the liter (L). A liter is the amount of liquid that would fill a 10-centimeter by 10-centimeter by 10-centimeter cube. In other words, a liter of liquid occupies the same space as 1000 cubic centimeters.

We need to make an important connection. There are 1000 milliliters to 1 liter. If there are 1000 milliliters and 1000 cubic centimeters to 1 liter, we can conclude that 1 milliliter must occupy 1 cubic centimeter of space. The common abbreviation for cubic centimeter is cc.

Conclusion: $1 \, ml = 1 \, cm^3$, or 1 cc

We can slightly alter the table of units for capacity by including the cubic centimeters with milliliters.

kiloliter (kl)	hectoliter (hl)	dekaliter (dal)	liter (L)	deciliter (dl)	centiliter (cl)	milliliter or cubic centimeter (ml or cm³ or cc)
1 kl = 1000 L	1 hl = 100 L	1 dal = 10 L	1 L	1 L = 10 dl	1 L = 100 cl	1 L = 1000 ml
						1 L = 1000 cm³ or cc

Answers to Your Turn 1
a. 9740 mm   b. 61 dm
c. 4.56 hm   d. 0.208 km

A liter is slightly more than a quart.

A milliliter (or cubic centimeter) is a little less than a quarter of a teaspoon.

Warning Make sure you understand the difference between centiliter and cubic centimeter.

Example 2 Convert.

a. 680 deciliters to hectoliters

Solution: Hectoliter is the third unit to the left of deciliter; so move the decimal point 3 places to the left in 680.

$$680 \text{ dl} = 0.68 \text{ hl}$$

b. 0.0184 dekaliter to milliliters

Solution: Milliliter is the fourth unit to the right of dekaliter; so move the decimal point 4 places to the right in 0.0184.

$$0.0184 \text{ dal} = 184 \text{ ml}$$

c. 0.0085 liter to cubic centimeters

Solution: Cubic centimeters and milliliters are interchangeable; so this is the same as converting from liters to milliliters. Because milliliter is the third unit to the right of liter, move the decimal point 3 places to the right in 0.0085.

$$0.0085 \text{ L} = 8.5 \text{ cc}$$

▶ Do Your Turn 2

Objective 3 Convert units of metric mass.

Now let's convert units of **mass**.

Definition **Mass:** A measure of the amount of matter that makes up an object.

Mass is different from weight. Weight is a measure of the force due to gravity and is dependent on the mass of an object and the gravitational pull on that object by a planet or another object.

The base unit for metric mass is the gram (g).

kilogram (kg)	hectogram (hg)	dekagram (dag)	gram (g)	decigram (dg)	centigram (cg)	milligram (mg)
1 kg=1000 g	1 hg = 100 g	1 dag = 10 g	1 g	1 g = 10 dg	1 g =100 cg	1 g = 1000 mg

A large paper clip has a mass of about 1 gram. Because this is quite small, most everyday objects are measured in terms of kilograms: 1 liter of water has a mass of 1 kilogram; so a 2-liter bottle of water has a mass of 2 kilograms. You are probably familiar with the milligram as well. Medicines in tablet form are usually measured in milligrams. Common over-the-counter pain relief medicines are usually in doses of around 200 milligrams.

Example 3 Convert.

a. 250 milligrams to grams

Solution: Gram is the third unit to the left of milligram; so move the decimal point 3 places to the left in 250.

$$250 \text{ mg} = 0.25 \text{ g}$$

▶ **Your Turn 2**

Convert.

a. 1.2 kiloliters to dekaliters

b. 7800 centiliters to liters

c. 0.17 centiliter to cubic centimeters

Connection 1 liter of water fills a 10-centimeter by 10-centimeter by 10-centimeter cube and has a mass of 1 kilogram.

Answers to Your Turn 2
a. 120 dal   b. 78 L   c. 1.7 cc

▶ **Your Turn 3**

Convert.

a. 6500 milligrams to dekagrams

b. 0.15 kilogram to grams

c. 2.6 centigrams to decigrams

Note Remember, the abbreviation for the American ton is T.

Connection We have said that a 10-centimeter by 10-centimeter by 10-centimeter cube of water has a capacity of 1 liter and a mass of 1 kilogram. A metric ton would be the mass of a cube of water 1000 times the volume of the 10-centimeter by 10-centimeter by 10-centimeter cube. This cube would be 100 centimeters by 100 centimeters by 100 centimeters, which is actually 1 meter by 1 meter by 1 meter. In other words, 1 metric ton is the mass of a 1-meter by 1-meter by 1-meter cube of water. The volume of this cube is 1 cubic meter, and the capacity is 1000 liters, or 1 kiloliter.

▶ **Your Turn 4**

Convert.

a. 2090 kilograms to metric tons

b. 0.06 metric ton to kilograms

c. 0.00591 metric ton to grams

Answers to Your Turn 3
a. 0.65 dag   b. 150 g
c. 0.26 dg

Answers to Your Turn 4
a. 2.09 t   b. 60 kg   c. 5910 g

b. 0.062 hectograms to grams

Solution: Gram is the second unit to the right of hectogram; so move the decimal point 2 places to the right in 0.062.

$$0.062 \, \text{hg} = 6.2 \, \text{g}$$

◀ **Do Your Turn 3**

For very large measurements of mass, we use the metric ton (t).

$$1 \, \text{t} = 1000 \, \text{kg}$$

Let's see if we can discover a way to use the table of units to convert to metric tons. Suppose we want to convert 4500 kilograms to metric tons. We could use the preceding fact in a dimensional analysis setup.

$$4500 \, \text{kg} = \frac{4500 \, \text{kg}}{1} \cdot \frac{1 \, \text{t}}{1000 \, \text{kg}} = \frac{4500}{1000} \text{t} = 4.5 \, \text{t}$$

Note that we divided by 1000 to convert kilograms to metric tons, which caused the decimal point to move 3 places to the left in 4500. This corresponds to moving three units to the left on the chart.

Conclusion: We can think of the metric ton as being three units to the left of kilogram. The two units in between are unnamed.

metric ton (t)	unnamed unit	unnamed unit	kilogram (kg)	hectogram (hg)	dekagram (dag)	gram (g)

Example 4   Convert.

a. 64,000 kilograms to metric tons

Solution: We can think of metric ton as being the third unit to the left of kilogram; so we move the decimal point 3 places to the left in 64,000.

$$64,000 \, \text{kg} = 64 \, \text{t}$$

b. 0.0058 metric ton to kilograms

Solution: If metric ton is the third unit to the left of kilogram, kilogram is the third unit to the right of metric ton; so we move the decimal point 3 places to the right in 0.0058.

$$0.0058 \, \text{t} = 5.8 \, \text{kg}$$

c. 240,000 dekagrams to metric tons

Solution: Kilogram is two units to the left of dekagram, and metric ton is three units to the left of kilogram. Therefore, metric ton is a total of five units to the left of dekagram; so we move the decimal point 5 places to the left in 240,000.

$$240,000 \, \text{dag} = 2.4 \, \text{t}$$

◀ **Do Your Turn 4**

Objective 4 Convert units of metric area.

To discover the facts about area in the American system, we determined how many square inches fit in 1 square foot. We can perform the same analysis with the metric system. Let's find the number of square decimeters that fit in 1 square meter. Because there are 10 decimeters to 1 meter, a 1-meter by 1-meter square is a 10-decimeter by 10-decimeter square.

If we calculate the area of this square meter in terms of decimeters, we have $10 \cdot 10 = 100$ square decimeters.

$$1 \text{ m}^2 = 100 \text{ dm}^2$$

We concluded in our discussion of area facts for the American system that we could simply square both sides of the length relations. Notice that the same applies to the metric relations.

$$1 \text{ m} = 10 \text{ dm} \qquad \leftarrow \text{ relation for length}$$
$$(1 \text{ m})^2 = (10 \text{ dm})^2 \qquad \text{Square both sides.}$$
$$1 \text{ m}^2 = 100 \text{ dm}^2 \qquad \leftarrow \text{ relation for area}$$

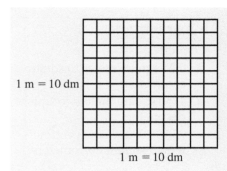

$1 \text{ m} = 10 \text{ dm}$

$1 \text{ m} = 10 \text{ dm}$

Deriving the other relations in this way, we have the following table:

square kilometer (km^2)	square hectometer (hm^2)	square dekameter (dam^2)	square meter (m^2)	square decimeter (dm^2)	square centimeter (cm^2)	square millimeter (mm^2)
1 km^2 $= 1{,}000{,}000 \text{ m}^2$	1 hm^2 $= 10{,}000 \text{ m}^2$	1 dam^2 $= 100 \text{ m}^2$	1 m^2	1 m^2 $= 100 \text{ dm}^2$	1 m^2 $= 10{,}000$ cm^2	1 m^2 $= 1{,}000{,}000$ mm^2

We can still use the table to convert units of metric area, but we have to alter our thinking a bit. Consider the following conversions and look for a pattern:

Suppose we want to convert 2.4 square meters to square decimeters.

$$2.4 \text{ m}^2 = \frac{2.4 \text{ m}^2}{1} \cdot \frac{100 \text{ dm}^2}{1 \text{ m}^2} = 240 \text{ dm}^2$$

Now let's convert 2.4 square meters to square centimeters.

$$2.4 \text{ m}^2 = \frac{2.4 \text{ m}^2}{1} \cdot \frac{10{,}000 \text{ cm}^2}{1 \text{ m}^2} = 24{,}000 \text{ cm}^2$$

Do you see a pattern? Consider converting 2.4 square meters to square millimeters.

$$2.4 \text{ m}^2 = \frac{2.4 \text{ m}^2}{1} \cdot \frac{1{,}000{,}000 \text{ mm}^2}{1 \text{ m}^2} = 2{,}400{,}000 \text{ mm}^2$$

Note To convert square meters to square decimeters, we multiply by 100, or 10^2, which moves the decimal point to the right 2 places. That's twice the number of places as for converting meters to decimeters.

Note To convert square meters to square centimeters, we multiply by 10,000, or 10^4, which moves the decimal point to the right 4 places. That's twice the number of places as for converting meters to centimeters.

Note To convert square meters to square millimeters, we multiply by 1,000,000, or 10^6, which moves the decimal point to the right 6 places. That's twice the number of places as for converting meters to millimeters.

Procedure

To convert metric units of area, move the decimal point twice the number of places as units from the initial area unit to the desired area unit on the table of units.

Example 5 Convert.

a. 0.0081 square meter to square millimeters

Solution: Square millimeter is the third unit to the right of square meter. With area units, we move the decimal point twice the number of places; so the decimal point moves 6 places to the right.

$$0.0081 \text{ m}^2 = 8100 \text{ mm}^2$$

b. 45,100 square meters to square hectometers

Solution: Square hectometer is the second unit to the left of square meter. Moving the decimal point twice that number of places means that it moves 4 places to the left.

$$45{,}100 \text{ m}^2 = 4.51 \text{ hm}^2$$

VISUAL

Learning Strategy

Use the exponent of 2 in the area units as a reminder to move the decimal point twice the number of units from the initial unit to the desired unit on the chart.

▶ **Your Turn 5**

Convert.

a. 0.25 square meter to square centimeters

b. 4750 square meters to square decimeters

c. 95,100,000 square centimeters to square hectometers

c. 0.000075 square kilometer to square decimeters

Solution: Square decimeter is the fourth unit to the right of square kilometer. Moving the decimal point twice that number of places means that it moves 8 places to the right.

$$0.000075 \text{ km}^2 = 7500 \text{ dm}^2$$

◀ **Do Your Turn 5**

Connection The metric system has more prefixes than we have studied so far. Two such prefixes are mega and giga, which you have probably heard in describing computer memory (megabytes and gigabytes). We can use a power of 10 to indicate the size of a unit with a particular prefix relative to the base unit. For example, because a kilometer is 1000 times the size of 1 meter, we can say that $1 \text{ km} = 1000 \text{ m} = 10^3 \text{ m}$. Or because there are 1000 millimeters in 1 meter, we can say that 1 millimeter is 1/1000, or $\frac{1}{10^3}$, of 1 meter. Below is a list of the prefixes, their size relative to the base unit, and the common word name.

Prefix	Factor	Word name
Tera- (T)	$= 10^{12}$	Trillion
Giga- (G)	$= 10^9$	Billion
Mega- (M)	$= 10^6$	Million
Kilo- (k)	$= 10^3$	Thousand
Hecto- (h)	$= 10^2$	Hundred
Deka- (da)	$= 10^1$	Ten
Base unit	$= 10^0 = 1$	
Deci- (d)	$= \frac{1}{10}$	Tenth
Centi- (c)	$= \frac{1}{10^2}$	Hundredth
Milli- (m)	$= \frac{1}{10^3}$	Thousandth
Micro- (μ)	$= \frac{1}{10^6}$	Millionth
Nano- (n)	$= \frac{1}{10^9}$	Billionth
Pico- (p)	$= \frac{1}{10^{12}}$	Trillionth

We can connect these powers of 10 with scientific notation. For example, 6.5 gigabytes can be expressed as follows:

$$6.5 \text{ GB} = 6.5 \text{ billion bytes}$$
$$= 6.5 \times 10^9 \text{ bytes}$$
$$= 6,500,000,000 \text{ bytes}$$

On the smaller side, X-rays have a wavelength of about 3 nanometers.

$$3 \text{ nm} = 3 \text{ billionths of a meter} = 3 \times \frac{1}{10^9} \text{ m}$$
$$= \frac{3}{1,000,000,000} \text{ m} = 0.000000003 \text{ m}$$

Answers to Your Turn 5
a. 2500 cm²
b. 475,000 dm²
c. 0.951 hm²

7.4 Exercises   For Extra Help   MyMathLab®

Objective 1

Prep Exercise 1 What is a base unit?

Prep Exercise 2 List the six basic prefixes in the metric system in order from largest to smallest.

Prep Exercise 3 What is the base unit for length in the metric system?

For Exercises 1–12, convert.

1. 4.5 meters to centimeters

2. 0.81 meter to millimeters

3. 0.07 kilometer to meters

4. 1.3 hectometers to meters

5. 3800 meters to kilometers

6. 125 centimeters to meters

7. 9500 millimeters to meters

8. 3.48 dekameters to meters

9. 1540 decimeters to dekameters

10. 0.45 centimeter to millimeters

11. 112 meters to hectometers

12. 79 millimeters to decimeters

Objective 2

Prep Exercise 4 What is the base unit for capacity in the metric system?

Prep Exercise 5 A cubic centimeter is equivalent to which capacity unit?

For Exercises 13–24, convert.

13. 0.12 liter to milliliters

14. 0.05 liter to milliliters

15. 0.005 kiloliter to liters

16. 145 deciliters to liters

17. 9.5 centiliters to dekaliters

18. 0.248 hectoliter to liters

19. 0.08 liter to cubic centimeters

20. 6 cubic centimeters to liters

21. 12 cubic centimeters to milliliters

22. 0.085 dekaliter to cubic centimeters

23. 0.4 liter to cubic centimeters

24. 145 centiliters to cubic centimeters

Objective 3

Prep Exercise 6 What is the base unit for mass in the metric system?

Prep Exercise 7 How many kilograms are in a metric ton?

For Exercises 25–36, convert.

25. 12 grams to milligrams

26. 4500 grams to kilograms

27. 0.009 kilogram to grams

28. 900 milligrams to grams

29. 0.05 dekagram to centigrams

30. 0.045 hectogram to milligrams

31. 3600 kilograms to metric tons

32. 850 kilograms to metric tons

33. 0.106 metric ton to kilograms

34. 5.07 metric tons to kilograms

35. 65,000 hectograms to metric tons

36. 0.008 metric ton to dekagrams

Objective 4

Prep Exercise 8 To convert metric units of area, move the decimal point _____ the number of places as units from the initial area unit to the desired area unit on the table of units.

For Exercises 37–44, convert.

37. 1.8 square meters to square centimeters

38. 580,000 square meters to square kilometers

39. 2200 square centimeters to square meters

40. 19,000 square meters to square hectometers

41. 0.05 square centimeter to square millimeters

42. 2.4 square decimeters to square centimeters

43. 44,300 square decimeters to square hectometers

44. 0.068 square kilometer to square meters

Puzzle Problem A light-year (lt-yr.) is the distance light travels in one year. The speed of light is 3×10^8 meters per second. Calculate the distance of 1 light-year in kilometers. The Magellanic Cloud is the nearest galaxy to our own Milky Way galaxy at a distance of 150,000 light-years. Calculate this distance in kilometers.

Review Exercises

For Exercises 1 and 2, evaluate.

[5.7] **1.** $6\frac{1}{2} \div 3 \cdot 10$

[6.5] **2.** $8.4(12) \div 100$

[6.2] **3.** Combine like terms. $6.5x^2 - 8x + 14.4 - 10.2x - 20$

[6.3] **4.** Distribute. $100(4.81y^2 - 2.8y + 7)$

[6.6] **5.** Solve and check. $7x - 3.98 = 2.4x + 8.9$

7.5 Converting Between Systems; Temperature

Objectives

1 Convert units of length.
2 Convert units of capacity.
3 Convert units of mass/weight.
4 Convert units of temperature.

Warm-up

[6.5] 1. Simplify. $\dfrac{5}{9}(56 - 32)$

[7.4] 2. Convert 2750 milliliters to liters.

Objective 1 Convert units of length.

When converting between systems of measurement, we use dimensional analysis. The key is to have at least one relation that acts as a bridge from one system to the other. However, the more relations we know, the more likely we'll be able to convert directly to the desired unit.

Let's consider units of length. The table to the right lists common relations for converting between metric and American units of length. Note that most conversions are approximations.

American to metric	Metric to American
1 in. = 2.54 cm	1 m ≈ 3.281 ft.
1 yd. ≈ 0.914 m	1 m ≈ 1.094 yd.
1 mi. ≈ 1.609 km	1 km ≈ 0.621 mi.

▶ **Your Turn 1**

Convert.

a. 60 miles to kilometers
b. 42 meters to feet
c. $9\frac{1}{2}$ inches to centimeters

Example 1   Convert.

a. 100 yards to meters

Solution: The undesired unit is yards, and the desired unit is meters; so we need a conversion factor with meters in the numerator and yards in the denominator. We can use $\dfrac{0.914\,\text{m}}{1\,\text{yd.}}$ or $\dfrac{1\,\text{m}}{1.094\,\text{yd.}}$.

$$100\,\text{yd.} \approx \frac{100\,\text{yd.}}{1} \cdot \frac{0.914\,\text{m}}{1\,\text{yd.}} = 91.4\,\text{m} \quad\text{or}$$

$$100\,\text{yd.} \approx \frac{100\,\text{yd.}}{1} \cdot \frac{1\,\text{m}}{1.094\,\text{yd.}} = \frac{100}{1.094}\,\text{m} = 91.4\,\text{m}$$

Note Because both conversion relations are approximations, using either produces a result that is an approximation of the original measurement. Therefore, 100 yd. ≈ 91.4 m. The equal signs that appear in the conversion steps indicate that a result is equal to only the previous expression, not the original measurement.

b. 5 kilometers to miles

Solution: The undesired unit is kilometers, and the desired unit is miles; so we need a conversion factor with miles in the numerator and kilometers in the denominator. We can use $\dfrac{0.621\,\text{mi.}}{1\,\text{km}}$ or $\dfrac{1\,\text{mi.}}{1.609\,\text{km}}$.

$$5\,\text{km} \approx \frac{5\,\text{km}}{1} \cdot \frac{0.621\,\text{mi.}}{1\,\text{km}} \approx 3.1\,\text{mi.} \quad\text{or}$$

$$5\,\text{km} \approx \frac{5\,\text{km}}{1} \cdot \frac{1\,\text{mi.}}{1.609\,\text{km}} = \frac{5}{1.609}\,\text{mi.} \approx 3.1\,\text{mi.}$$

◀ **Do Your Turn 1**

If we do not know a direct relation from the initial unit to the desired unit, we select a known relation to a unit that is close to the desired unit in the target system, then convert from that unit to the desired unit.

Answers to Your Turn 1
a. ≈ 96.54 km
b. ≈ 137.8 ft.
c. 24.13 cm

Answers to Warm-up
 1. $13\frac{1}{3}$ or $13.\overline{3}$ **2.** 2.75 L

Example 2 Convert.

a. 0.3 mile to meters

Solution: The relations that we have relate miles to kilometers; so we will convert miles to kilometers and then use the metric unit table to convert kilometers to meters.

$$0.3 \text{ mi.} \approx \frac{0.3 \text{ mi.}}{1} \cdot \frac{1.609 \text{ km}}{1 \text{ mi.}} = 0.4827 \text{ km}$$

On the metric table of units, meter is the third unit to the right of kilometer; so we move the decimal point 3 places to the right in 0.4827.

$$0.4827 \text{ km} = 482.7 \text{ m}$$ ◀ **Note** Because we used 1.609 km ≈ 1 mi. initially, this result is an approximation; so 0.3 mi. ≈ 482.7 m.

b. 64 millimeters to inches

Solution: The only relation we have to inches is 1 inch = 2.54 centimeters. To use this relation, we need the initial unit to be in centimeters. Therefore, we first convert 64 millimeters to centimeters.

On the metric table of units, centimeter is one unit to the left of millimeter; so we move the decimal point 1 place to the left in 64.

$$64 \text{ mm} = 6.4 \text{ cm}$$

Now we can use the fact that 1 inch = 2.54 cm.

$$6.4 \text{ cm} = \frac{6.4 \text{ cm}}{1} \cdot \frac{1 \text{ in.}}{2.54 \text{ cm}} = \frac{6.4}{2.54} \text{ in.} \approx 2.52 \text{ in.}$$

▶ **Do Your Turn 2**

Objective 2 Convert units of capacity.

Although the units and numbers change, the method for converting units of capacity does not change.

American to metric	Metric to American
1 qt. ≈ 0.946 L	1 L ≈ 1.057 qt.

Example 3 Convert.

a. 2 gallons to liters

Solution: The relations that we have relate quarts to liters. Because the initial unit is gallons, we must convert gallons to quarts using the fact that 4 quarts = 1 gallon. We can then use the fact that 0.946 liter ≈ 1 quart.

$$2 \text{ gal.} \approx \frac{2 \text{ gal.}}{1} \cdot \frac{4 \text{ qt.}}{1 \text{ gal.}} \cdot \frac{0.946 \text{ L}}{1 \text{ qt.}} = 7.568 \text{ L}$$ ◀ **Note** Because we used the approximation 0.946 L ≈ 1 qt., the result is an approximation of the original measurement; so 2 gal. ≈ 7.568 L.

b. 40 milliliters to ounces

Solution: The relations that we have relate quarts to liters. Because the initial unit is milliliters, we must first convert milliliters to liters. Because liter is three units to the left of milliliter, we move the decimal point 3 places to the left.

$$40 \text{ ml} = 0.04 \text{ L}$$

▶ **Your Turn 2**

Convert.

a. 412 feet to decimeters

b. 0.5 meter to inches

c. 20 hectometers to yards

▶ **Your Turn 3**

Convert.

a. 6 quarts to liters

b. 30 gallons to liters

c. 60 milliliters to ounces

d. 20 cubic centimeters to ounces

Answers to Your Turn 2

a. ≈ 1255.78 dm

b. ≈ 19.7 in.

c. ≈ 2188 yd.

Note Because milliliters and cubic centimeters are the same size, we would follow the same procedure as in Example 3b to convert cubic centimeters to ounces.

We can now use the fact that 1.057 quarts $\approx$ 1 liter. However, our target unit is ounces; so we'll need to convert quarts to ounces.

$$0.04 \text{ L} \approx \frac{0.04 \text{ L}}{1} \cdot \frac{1.057 \text{ qt.}}{1 \text{ L}} \cdot \frac{2 \text{ pt.}}{1 \text{ qt.}} \cdot \frac{2 \text{ c.}}{1 \text{ pt.}} \cdot \frac{8 \text{ oz.}}{1 \text{ c.}} \approx 1.35 \text{ oz.}$$

◀ **Do Your Turn 3**

Objective 3 Convert units of mass/weight.

The primary conversion relations that link American weight to metric mass are shown to the right.

American to metric	Metric to American
1 lb. $\approx$ 0.454 kg	1 kg $\approx$ 2.2 lb.
1 T $\approx$ 0.907 t	1 t $\approx$ 1.1 T

▶ **Your Turn 4**

Convert.

a. 50 pounds to kilograms

b. 4.2 metric tons to American tons

c. 40 ounces to grams

Example 4   Convert.

a. 180 pounds to kilograms

Solution: We have a direct relation of pounds to kilograms:
1 pound $\approx$ 0.454 kilogram.

$$180 \text{ lb.} \approx \frac{180 \text{ lb.}}{1} \cdot \frac{0.454 \text{ kg}}{1 \text{ lb.}} = 81.72 \text{ kg}$$

b. $6\frac{1}{2}$ tons to metric tons

Solution: We have a direct relation of American tons to metric tons:
1 American ton $\approx$ 0.907 metric ton.

$$6\frac{1}{2} \text{ T} \approx \frac{6\frac{1}{2} \text{ T}}{1} \cdot \frac{0.907 \text{ t}}{1 \text{ T}} \approx 5.9 \text{ t}$$

c. 250 grams to ounces

Solution: The relations that we have relate kilograms to pounds. Because the initial unit is grams, we must first convert grams to kilograms. Because kilogram is three units to the left of gram, we move the decimal point 3 places to the left.

$$250 \text{ g} = 0.25 \text{ kg}$$

We can now use the fact that 1 kilogram $\approx$ 2.2 pounds. However, our target unit is ounces; so we must convert pounds to ounces.

$$0.25 \text{ kg} \approx \frac{0.25 \text{ kg}}{1} \cdot \frac{2.2 \text{ lb.}}{1 \text{ kg}} \cdot \frac{16 \text{ oz.}}{1 \text{ lb.}} = 8.8 \text{ oz.}$$

Note Remember, there are 16 ounces (oz.) in 1 pound (lb.).

◀ **Do Your Turn 4**

Objective 4 Convert units of temperature.

Temperature is a measure of the heat of a substance. In the American system, temperature is measured in degrees Fahrenheit (°F). In the metric system, temperature is measured in degrees Celsius (°C). To relate the two systems, we need two reference points. For simplicity, we will use the freezing point and the boiling point of water.

Answers to Your Turn 3
a. $\approx$ 5.676 L   b. $\approx$ 113.52 L
c. $\approx$ 2.03 oz.   d. $\approx$ 0.676 oz.

Answers to Your Turn 4
a. $\approx$ 22.7 kg
b. $\approx$ 4.62 T
c. $\approx$ 1135 g

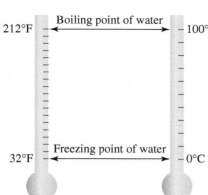

On the Fahrenheit scale, the boiling point of water is 212°F and the freezing point is 32°F. The difference between these two points is 212°F − 32°F = 180°F.

On the Celsius scale, the boiling point of water is 100°C. and the freezing point is 0°C. The difference between these two points is 100°C − 0°C= 100°C.

Notice that a change of 180°F is the same as a change of 100°C. We can write this relation as a ratio, which we can use to convert between the two temperature scales.

$$\begin{array}{l} \text{degrees Fahrenheit} \rightarrow \\ \text{to} \\ \text{degrees Celsius} \rightarrow \end{array} \quad \frac{180°\text{F}}{100°\text{C}} = \frac{9°\text{F}}{5°\text{C}}$$

Let's try using our ratio to convert 100°C to degrees Fahrenheit. We should get 212°F.

$$100°\text{C} = \frac{100°\cancel{\text{C}}}{1} \cdot \frac{9°\text{F}}{5°\cancel{\text{C}}} = 180°\text{F}$$ ◄ But 180°F isn't correct. We should get 212°F.

The reason multiplying by $\frac{9}{5}$ alone does not produce the correct temperature is because the two scales do not have the same reference point for 0. The Fahrenheit scale is shifted by 32°F. Therefore, to convert degrees Celsius to degrees Fahrenheit, after multiplying by $\frac{9}{5}$, we need to *add* 32°F.

$$100°\text{C} = \frac{100°\cancel{\text{C}}}{1} \cdot \frac{9°\text{F}}{5°\cancel{\text{C}}} + 32°\text{F}$$
$$= 180°\text{F} + 32°\text{F}$$
$$= 212°\text{F}$$

We can write a formula to perform this conversion. To convert from degrees Celsius to degrees Fahrenheit, multiply degrees Celsius by $\frac{9}{5}$, then add 32.

Conclusion: $F = \dfrac{9}{5}C + 32$

| **Example 5** | Convert 45°C to degrees Fahrenheit.

Solution: Because the desired unit is degrees Fahrenheit, we use the formula $F = \frac{9}{5}C + 32$.

$$F = \frac{9}{5}(45) + 32 \qquad \text{Replace } C \text{ with 45.}$$

$$F = \frac{9}{\cancelto{1}{5}}\left(\frac{\cancelto{9}{45}}{1}\right) + 32 \qquad \text{Divide out the common factor, 5.}$$

$$F = 81 + 32 \qquad \text{Multiply.}$$
$$F = 113°\text{F} \qquad \text{Add.}$$

► **Do Your Turn 5**

► **Your Turn 5**

Convert.
a. 60°C to degrees Fahrenheit
b. 120.5°C to degrees Fahrenheit
c. 20°C to degrees Fahrenheit
d. −5°C to degrees Fahrenheit

Answers to Your Turn 5
a. 140°F b. 248.9°F
c. 68°F d. 23°F

Now let's convert degrees Fahrenheit to degrees Celsius. To derive a formula, let's convert 86°F to degrees Celsius using $F = \frac{9}{5}C + 32$. We will replace F with 86 and then solve for C. Note the order of operations as we solve.

$$86 = \frac{9}{5}C + 32 \qquad \text{Replace } F \text{ with 86.}$$

$$86 = \frac{9}{5}C + 32$$

$$\underline{-32 \qquad\qquad -32} \qquad \text{Subtract 32 from both sides.}$$

$$54 = \frac{9}{5}C + 0$$

$$\frac{5}{\underset{1}{\cancel{9}}} \cdot \frac{\overset{6}{\cancel{54}}}{1} = \frac{\overset{1}{\cancel{5}}}{\underset{1}{\cancel{9}}} \cdot \frac{\overset{1}{\cancel{9}}}{\underset{1}{\cancel{5}}}C \qquad \text{Multiply both sides by } \frac{5}{9} \text{ to isolate } C.$$

$$30°C = C$$

In the solution, we subtracted 32 from 86 to get 54 and then multiplied that difference by $\frac{5}{9}$. This suggests that to convert degrees Fahrenheit to degrees Celsius, we subtract 32 from the given degrees Fahrenheit and then multiply that difference by $\frac{5}{9}$.

Conclusion: $C = \frac{5}{9}(F - 32)$

Note The parentheses are needed so that the subtraction occurs *before* the multiplication.

► Your Turn 6

Convert.
a. 77°F to degrees Celsius
b. 40.1°F to degrees Celsius
c. −28°F to degrees Celsius
d. 0°F to degrees Celsius

Answers to Your Turn 6
a. 25°C   b. 4.5°C
c. −33.$\overline{3}$°C   d. −17.$\overline{7}$°C

Example 6   Convert 80°F to degrees Celsius.

Solution: Because the desired unit is degrees Celsius, we use the formula $C = \frac{5}{9}(F - 32)$.

$$C = \frac{5}{9}(80 - 32) \qquad \text{Replace } F \text{ with 80.}$$

$$C = \frac{5}{9}(48) \qquad \text{Subtract.}$$

$$C = \frac{5}{\underset{3}{\cancel{9}}}\left(\frac{\overset{16}{\cancel{48}}}{1}\right) \qquad \text{Divide out the common factor, 3.}$$

$$C = 26\frac{2}{3}°C, \text{ or } 26.\overline{6}°C \qquad \text{Multiply.}$$

◄ Do Your Turn 6

7.5 Exercises   For Extra Help MyMathLab®

Objectives 1–3

Prep Exercise 1 What conversion fact correctly completes the following dimensional analysis?

$$9 \text{ in.} \times \frac{?}{1 \text{ in.}} = 22.86 \text{ cm}$$

Prep Exercise 2 What conversion fact correctly completes the following dimensional analysis?

$$2.5 \text{ kg} \times \frac{?}{1 \text{ kg}} \approx 5.5 \text{ lb.}$$

For Exercises 1–28, convert.

1. A football field measures 100 yards. Convert the length of the field to meters.

2. A circular table has a diameter of 44 inches. Convert the diameter to centimeters.

3. A marathon is 26.2 miles. Convert to kilometers.

4. A mechanic uses a $\frac{3}{4}$-inch wrench. Convert to millimeters.

5. A carpenter uses a $\frac{3}{8}$-inch drill bit. Convert to millimeters.

6. A family builds a 4-foot fence. Convert to centimeters.

7. An athlete prepares for the 400-meter dash. Convert to yards.

8. A swimmer prepares for the 100-meter freestyle. Convert to feet.

9. A sign in Germany indicates that it is 40 kilometers to Berlin. Convert to miles.

10. A sign in France indicates that road construction is 0.4 kilometer ahead. Convert to feet.

11. An oceanographer measures a fish to be 45 centimeters long. What is the length in inches?

12. The width of a pencil is about 7 millimeters. What is the width in inches?

13. How many liters are in a 16-ounce bottle of soda?

14. How many milliliters are in an 8-ounce can of soda?

15. A recipe calls for $2\frac{1}{2}$ cups of milk. Convert to milliliters.

16. How many liters are in a half-gallon of milk?

17. How many gallons are in a 3-liter cola?

18. How many ounces are in a 2-liter cola?

19. The instructions on a pain reliever for children indicate to give the child 20 milliliters. How many ounces is this?

20. A child is to receive 40 milliliters of liquid medicine. How many ounces is this?

21. A woman weighs 135 pounds. How many kilograms is this?

22. A man weighs 168 pounds. How many kilograms is this?

23. How many grams is a 12-ounce steak?

24. How many grams is a 2.5-ounce serving of rice?

25. A pill contains 500 milligrams of medicine. How many ounces is this?

26. A pill contains 300 milligrams of aspirin. How many ounces is this?

27. A person from Russia claims that her mass is 55 kilograms. What is her weight in pounds?

28. A person from Japan claims to have a mass of 68 kilograms. What is his weight in pounds?

Objective 4

Prep Exercise 3 What is the formula for converting degrees Celsius to degrees Fahrenheit?

Prep Exercise 4 What is the formula for converting degrees Fahrenheit to degrees Celsius?

For Exercises 29–40, convert.

29. Temperatures in the Sahara desert can reach $54\frac{4}{9}°$C. What is this in degrees Fahrenheit?

30. Temperatures on the moon can reach 127°C at noon. What is this in degrees Fahrenheit?

31. On a cold day in December, the temperature is reported to be 6°C. What is this in degrees Fahrenheit?

32. The lowest temperatures on the moon can reach -173°C just before dawn. What is this in degrees Fahrenheit?

33. Absolute zero is the temperature at which molecular energy is at a minimum. This temperature is -273.15°C. What is this in degrees Fahrenheit?

34. Liquid oxygen boils at a temperature of -182.96°C. What is this in degrees Fahrenheit?

Of Interest

Absolute zero refers to 0 on the Kelvin scale. The Kelvin scale is what most scientists use to measure temperatures. Increments on the Kelvin scale are equivalent to increments on the Celsius scale; so an increase or decrease of 1 Kelvin (K) is the same as an increase or a decrease of 1 degree Celsius.

35. Normal body temperature is 98.6°F. What is normal body temperature in degrees Celsius?

36. A recipe calls for the oven to be preheated to 350°F. What is this temperature in degrees Celsius?

37. On a hot summer day, the temperature is reported to be 102°F. What is this in degrees Celsius?

38. During the summer, it is suggested to keep the thermostat at 78°F. What is this in degrees Celsius?

39. During January, the temperature is reported to be −5°F. What is this in degrees Celsius?

40. At the South Pole, the temperature can get down to −40°F. What is this in degrees Celsius?

Puzzle Problem   A common unit of antiquity was the cubit, which was the length of the forearm from the tip of the middle finger to the elbow. In the Biblical story of Noah's ark, the ark was to be built 300 cubits long, 50 cubits wide, and 30 cubits high. If the cubit was about 1.5 feet in length, what were the dimensions of the ark in feet? Yards? Meters?

Review Exercises

For Exercises 1 and 2, evaluate.

[6.4] **1.** $896 \div 3200$

[6.5] **2.** $(90 + 110 + 84) \div 2400$

[5.7] **3.** Combine like terms. $-\frac{1}{3}x + \frac{2}{5}y - x - \frac{3}{4}y$

For Exercises 4 and 5, solve and check.

[5.8] **4.** $2(t + 1) = 5t$

[7.2] **5.** $\frac{m}{20.4} = \frac{8}{15}$

7.6 Applications and Problem Solving

Objectives

1 Use debt-to-income ratios to decide loan qualification.

2 Calculate the maximum monthly PITI payment that meets the front-end ratio qualification for a loan.

3 Calculate the maximum debt that meets the back-end ratio qualification for a loan.

4 Calculate medical dosages.

Warm-up

[7.1] 1. A family has a total debt of $14,880 and a gross monthly income of $62,000. Find the family's debt-to-income ratio as a unit ratio.

[7.5] 2. Convert 175 pounds to kilograms.

Objective 1 Use debt-to-income ratios to decide loan qualification.

In Section 7.1, we considered ratios of debt to income. Lenders consider debt-to-income ratios to determine whether an applicant qualifies for a loan. The two ratios commonly used are the **front-end ratio** and the **back-end ratio.**

Definitions Front-end ratio: The ratio of the total monthly house payment to gross monthly income.

Back-end ratio: The ratio of the total monthly debt payments to gross monthly income.

The total monthly house payment includes the principal and interest for the mortgage, property taxes, and homeowners insurance premiums. It is usually referred to as the PITI payment.

Suppose a family with a gross monthly income of $4000 is trying to qualify for a loan. If it were to get the loan, the monthly PITI payment would be approximately $1000.

$$\text{front-end ratio} = \frac{1000}{4000} = 0.25$$

Because the general guideline for lenders on a conventional loan is for the front-end ratio not to be greater than 0.28, this family would meet the qualification for the front-end ratio.

Now let's consider the back-end ratio. The total monthly debt includes the monthly PITI payment, minimum payments for all credit cards, all auto loan payments, and student loan payments. Sometimes student loans and car loans are not considered if they are to be paid off within a certain time frame. Suppose the family has the following monthly debts:

$$\text{monthly PITI payment} = \$1000$$
$$\text{credit card 1 minimum payment} = \$50$$
$$\text{credit card 2 minimum payment} = \$40$$
$$\text{car loan} = \$253$$
$$\underline{\text{student loan} = \$104}$$
$$\textbf{total monthly debt} = \textbf{\$1447}$$

Now we can calculate the back-end ratio.

$$\text{back-end ratio} = \frac{1447}{4000} = 0.36175$$

The general guideline for lenders on a conventional loan is that the back-end ratio should not be greater than 0.36. This family's back-end ratio is just slightly greater than the maximum allowed value for the back-end ratio.

In summary, the front-end ratio for this family tells us that it has enough income to make the monthly PITI payment, but the back-end ratio indicates that it has a bit too much total debt.

Answers to Warm-up
1. 0.24 **2.** 79.45 kg

These two ratios are only part of what lenders consider in deciding whether a person qualifies for a loan or mortgage. Two other factors considered are credit score and the type of loan.

Credit scores are calculated by national credit bureaus. Generally, lenders will request credit reports from several different bureaus. For example, it is common to request three reports and use the median of the three scores in the decision process. Credit scores range from about 450 up to 900. A score of 650 is generally considered acceptable, greater than 650 is good, and less than 650 is considered risky.

Of Interest

Following are some factors, in order of importance, that impact a credit score.

1. Payment history or unresolved collections (Nonpayment and late payment have a negative impact.)
2. Excessive lines of credit (More than three different lines of credit has a negative impact.)
3. Balances (A balance at or near the maximum has a negative impact.)

The type of loan affects the factors considered as well. For example, with VA (Veteran Affairs) loans, the front-end ratio and credit score are not considered. The back-end ratio can be up to 0.41. Of course, the primary qualification with a VA loan is that the applicant must be a veteran. With an FHA (Federal Housing Authority) loan, the credit score is not considered. The front-end ratio can be up to 0.29 and back-end ratio up to 0.41.

Table 7.1 Factors for loan qualification

	Conventional	VA	FHA
1. Front-end ratio should not exceed	0.28	N/A	0.29
2. Back-end ratio should not exceed	0.36	0.41	0.41
3. Credit score should be	650 or higher	N/A	N/A

▶ **Your Turn 1**

Use Table 7.1 to solve.

a. The Cicone family is trying to qualify for a conventional loan. The gross monthly income is $2870. If the family gets the loan, the monthly PITI payment will be approximately $897. The total monthly debt, excluding the monthly PITI payment, is $223. The family has a credit score of 700. Does the family qualify?

b. The Mathias family is trying to qualify for an FHA loan. Its gross monthly income is $2580. If the family gets the loan, the monthly PITI payment will be approximately $710. The family has the following monthly debt payments:

credit card 1 = $25

car loan 1 = $265

student loan = $78

The family's credit score is 750. Does the family qualify?

Example 1   The Jones family is trying to qualify for a conventional loan. Its gross monthly income is $4200. If it gets the loan, the monthly PITI payment will be approximately $985. The total monthly debt, excluding the PITI payment, is $535. The credit score is 720. Does the family qualify?

Understand We are to decide whether the Jones family qualifies for the loan. Because the family is applying for a conventional loan, we must consider the front-end ratio, back-end ratio, and credit score.

Plan 1. Calculate the front-end ratio. The front-end ratio should not exceed 0.28.

$$\text{Front-end ratio} = \frac{\text{monthly PITI payment}}{\text{gross monthly income}}$$

2. Calculate the back-end ratio. The back-end ratio should not exceed 0.36.

$$\text{Back-end ratio} = \frac{\text{total monthly debt payments}}{\text{gross monthly income}}$$

3. Consider the credit score. The credit score should be 650 or higher.

Execute front-end ratio $= \dfrac{985}{4200} \approx 0.235$

back-end ratio $= \dfrac{1520}{4200} \approx 0.362$

credit score $= 720$

Note The total monthly debt payment is found by adding the monthly PITI payment to the current monthly debt.

$985 + 535 = 1520$

Answer The back-end ratio is just slightly greater than 0.36. However, the Jones family has a good credit score at 720. The good credit score should be taken into consideration to offset the 0.362 back-end ratio. The family would most likely qualify for the loan.

Check Verify the ratio calculations by reversing the process. This will be left to the reader.

▶ Do Your Turn 1

Objective 2 Calculate the maximum monthly PITI payment that meets the front-end ratio qualification for a loan.

When buying a home, it is helpful to consider the maximum monthly PITI payment for which you can qualify. Knowing that maximum monthly payment will help you set a price limit on the homes you consider.

Example 2 Suppose a family has a gross monthly income of $3000. What is the maximum monthly PITI payment that would meet the front-end ratio qualification for a conventional loan?

Understand We are to calculate the maximum monthly PITI payment that would meet the front-end ratio qualification for a conventional loan. The general guideline is that the front-end ratio should not exceed 0.28. The front-end ratio is calculated as follows:

$$\text{Front-end ratio} = \frac{\text{monthly PITI payment}}{\text{gross monthly income}}$$

Plan Let p represent the monthly PITI payment. Replace each part of the front-end ratio formula with appropriate values and solve for p.

Execute $$0.28 = \frac{p}{3000}$$

$$\frac{3000}{1} \cdot \frac{0.28}{1} = \frac{p}{3000} \cdot \frac{3000}{1} \qquad \text{Multiply both sides by 3000 to isolate } p.$$

$$840 = p$$

Answer The maximum PITI payment that would meet the front-end ratio qualification for a conventional loan with a gross monthly income of $3000 is $840.

Check Verify that the front-end ratio with a monthly PITI payment of $840 and gross monthly income of $3000 is in fact 0.28.

$$\text{Front-end ratio} = \frac{840}{3000} = 0.28$$

▶ Do Your Turn 2

Objective 3 Calculate the maximum debt that meets the back-end ratio qualification for a loan.

In addition to finding the maximum monthly PITI payment to qualify with the front-end ratio, it also is helpful to calculate the maximum debt you can have yet still satisfy the back-end ratio qualification.

▶ **Your Turn 2**

Suppose a person has a gross monthly income of $1680. What is the maximum monthly PITI payment that would meet the front-end ratio qualification for a FHA loan?

Answers to Your Turn 1
a. front-end = 0.313; back-end = 0.39; the family does not qualify.
b. front-end = 0.275; back-end = 0.418; although the back-end ratio is just slightly greater than the required 0.41, the other factors warrant qualification.

Answer to Your Turn 2
$487.20

▶ **Your Turn 3**

Suppose a family has a gross monthly income of $2350. What is the maximum debt that would meet the back-end ratio qualification for a VA loan?

Example 3 Suppose a family has a gross monthly income of $3200. What is the maximum debt that would meet the back-end ratio qualification for an FHA loan?

Understand We are to calculate the maximum debt that would meet the back-end ratio qualification for an FHA loan. The general guideline is that the back-end ratio should not exceed 0.41. The back-end ratio is calculated as follows:

$$\text{Back-end ratio} = \frac{\text{total monthly debt payments}}{\text{gross monthly income}}$$

Plan Let d represent the total monthly debt. Replace each part of the back-end ratio formula with appropriate values and solve for d.

Execute
$$0.41 = \frac{d}{3200}$$

$$\frac{3200}{1} \cdot \frac{0.41}{1} = \frac{d}{3200} \cdot \frac{3200}{1}$$

$$1312 = d$$

Answer The maximum monthly debt that would meet the back-end ratio qualification for an FHA loan with a gross monthly income of $3200 is $1312.

Check Verify that the back-end ratio with a total monthly debt of $1312 and gross monthly income of $3200 is in fact 0.41.

$$\text{Back-end ratio} = \frac{1312}{3200} = 0.41$$

◀ **Do Your Turn 3**

Objective 4 Calculate medical dosages.

We can use dimensional analysis to determine appropriate medical dosages. Recall that the main idea is to note the initial unit and the desired unit. Then you multiply by appropriate unit fractions so that the undesired unit divides out, leaving the desired unit.

▶ **Your Turn 4**

A patient with a mass of 68 kilograms is to receive isoniazid, an antitubercular drug. The order is to administer 10 milligrams per kilogram. How much isoniazid should be given?

Example 4 A patient with a mass of 42 kilograms is to receive an antibiotic. The order is to administer 20 milligrams per kilogram. How much of the antibiotic should be given?

Understand The order of 20 milligrams per kilogram means that the patient should receive 20 milligrams of the antibiotic for every 1 kilogram of the patient's mass.

Plan Multiply the patient's mass by the dosage per unit mass.

Execute total dose = patient's mass · dosage per unit of mass

$$\text{total dose} = \frac{42 \text{ kg}}{1} \cdot \frac{20 \text{ mg}}{1 \text{ kg}} = 840 \text{ mg}$$

Answer The patient should receive 840 milligrams of the antibiotic.

Check Verify that a 42-kilogram patient receiving 840 milligrams of antibiotic receives 20 milligrams of antibiotic per kilogram of mass.

$$\frac{840 \text{ mg}}{42 \text{ kg}} = 20 \text{ mg/kg}$$

◀ **Do Your Turn 4**

Answer to Your Turn 3
$963.50

Answer to Your Turn 4
680 mg

Many medications are administered in intravenous (IV) solutions. A regulator is used to regulate the number of drops of the solution the patient receives over a period of time (usually 1 minute). Medical orders are usually not written in terms of the number of drops per minute; so for the regulator to be set properly, they must be converted to number of drops per minute.

Example 5   A patient is to receive 500 milliliters of 5% D/W solution IV over 4 hours. The label on the IV bag indicates that 10 drops dissipate 1 milliliter of the solution. How many drops should the patient receive each minute?

Understand The order is for 500 milliliters over 4 hours. With this particular IV, 10 drops of the solution is equal to 1 milliliter. We can write these rates as follows:

$$\frac{500 \text{ ml}}{4 \text{ hr.}} \quad \text{and} \quad \frac{10 \text{ drops}}{1 \text{ ml}}$$

We want to end up with the number of drops per minute; so milliliters and hours are undesired units. We need to convert hours to minutes and milliliters to drops.

Plan Use the fact that 1 hour is 60 minutes to convert hours to minutes and that 10 drops is 1 ml to convert milliliters to drops.

Execute $$\frac{500 \text{ ml}}{4 \text{ hr.}} \cdot \frac{10 \text{ drops}}{1 \text{ ml}} \cdot \frac{1 \text{ hr.}}{60 \text{ min}} = \frac{5000 \text{ drops}}{240 \text{ min.}} = 20.8\overline{3} \text{ drops/min.}$$

Answer The patient should receive about 20 to 21 drops per minute.

Check Verify that a rate of 21 drops per minute with every 10 drops equal to 1 milliliter is a total of 500 milliliters after 4 hours.

$$\frac{4 \text{ hr.}}{1} \cdot \frac{60 \text{ min.}}{1 \text{ hr.}} \cdot \frac{21 \text{ drops}}{1 \text{ min.}} = 5040 \text{ drops}$$

If the patient receives 5040 drops in 4 hours and every 10 drops is 1 milliliter, we can calculate the total number of milliliters the patient receives.

$$\frac{5040 \text{ drops}}{1} \cdot \frac{1 \text{ ml}}{10 \text{ drops}} = \frac{5040}{10} \text{ ml} = 504 \text{ ml}$$

Because we rounded the number of drops to a whole number, this result does not match 500 milliliters exactly. However, it is reasonable.

▶ Do Your Turn 5

▶ **Your Turn 5**

A patient is to receive 250 cubic centimeters of 5% D/W solution IV over 4 hours. The label on the IV bag indicates that 15 drops dissipate 1 cubic centimeter of the solution. How many drops should the patient receive each minute?

Answer to Your Turn 5
15 to 16 drops per min.

7.6 Exercises For Extra Help MyMathLab®

Objective 1

Prep Exercise 1 What is a front-end ratio?

Prep Exercise 2 What is a back-end ratio?

Prep Exercise 3 What is the PITI payment?

For Exercises 1–16, use the table to solve.

Factors for loan qualification

	Conventional	VA	FHA
1. Front-end ratio should not exceed	0.28	N/A	0.29
2. Back-end ratio should not exceed	0.36	0.41	0.41
3. Credit score should be	650 or higher	N/A	N/A

1. The Wu family is trying to qualify for a conventional loan. Its gross monthly income is $3850. If the family gets the loan, the monthly PITI payment will be approximately $845. The family's monthly debt, excluding the monthly PITI payment, is $435. The family has a credit score of 700. Does the Wu family qualify?

2. The Deas family is trying to qualify for a conventional loan. Its gross monthly income is $3240. If the family gets the loan, the monthly PITI payment will be approximately $925. The total monthly debt, excluding the monthly PITI payment, is $495. The credit score is 680. Does the Deas family qualify?

3. The Rivers family is trying to qualify for a VA loan. The gross monthly income is $2845. If the family gets the loan, the monthly PITI payment will be approximately $756. The credit score is 660. The family has the following monthly debt payments:

   credit card 1 = $30
   credit card 2 = $45
   car loan 1 = $279
   student loan = $93

   Does the Rivers family qualify?

4. The Santana family is trying to qualify for a VA loan. The gross monthly income is $4200. If the family gets the loan, the monthly PITI payment will be approximately $987. The credit score is 650. The family has the following monthly debt payments:

   credit card 1 = $35
   credit card 2 = $85
   car loan 1 = $305
   car loan 2 = $289

   Does the Santana family qualify?

5. The Bishop family is trying to qualify for an FHA loan. The gross monthly income is $2530. If the family gets the loan, the monthly PITI payment will be approximately $785. The family has the following monthly debt payments:

   credit card 1 = $40
   car loan = $249
   student loan = $65

   The credit score is 700. Does the Bishop family qualify?

6. The Nicks family is trying to qualify for an FHA loan. The gross monthly income is $2760. If the family gets the loan, the monthly PITI payment will be approximately $805. The family has the following monthly debt payments:

   credit card 1 = $25
   credit card 2 = $35
   car loan 1 = $279

   The credit score is 650. Does the Nicks family qualify?

Objective 2

7. Suppose a family has a gross monthly income of $4800. What is the maximum monthly PITI payment that would meet the front-end ratio qualification for a conventional loan?

8. Suppose a family has a gross monthly income of $2980. What is the maximum monthly PITI payment that would meet the front-end ratio qualification for a conventional loan?

9. Suppose a person has a gross monthly income of $1840. What is the maximum monthly PITI payment that would meet the front-end ratio qualification for an FHA loan?

10. Suppose a person has a gross monthly income of $1580. What is the maximum monthly PITI payment that would meet the front-end ratio qualification for an FHA loan?

Objective 3

11. Suppose a family has a gross monthly income of $3480. What is the maximum debt that would meet the back-end ratio qualification for a conventional loan?

12. Suppose a family has a gross monthly income of $2240. What is the maximum debt that would meet the back-end ratio qualification for a conventional loan?

13. Suppose a family has a gross monthly income of $1675. What is the maximum debt that would meet the back-end ratio qualification for a VA loan?

14. Suppose a family has a gross monthly income of $2110. What is the maximum debt that would meet the back-end ratio qualification for a VA loan?

15. Suppose a family has a gross monthly income of $2375. What is the maximum debt that would meet the back-end ratio qualification for an FHA loan?

16. Suppose a family has a gross monthly income of $2680. What is the maximum debt that would meet the back-end ratio qualification for an FHA loan?

Objective 4

Prep Exercise 4 Identify the mistake in the following calculation for an IV drip rate.

$$\frac{300 \text{ ml}}{4 \text{ hr.}} \times \frac{1 \text{ hr.}}{60 \text{ min.}} \times \frac{1 \text{ ml}}{20 \text{ drops}} = 0.0625 \text{ drops/min.}$$

For Exercises 17–24, use dimensional analysis.

17. A patient with a mass of 55 kilograms is to receive an antibiotic. The order is to administer 9.5 milligrams per kilogram. How much of the antibiotic should be given?

18. A patient with a mass of 81 kilograms is to receive 1.4 milligrams per kilogram of metroprolol tartrate (an antihypertensive drug). How much of the drug should the patient receive?

19. A patient weighing 132 pounds is to receive 0.04 milligram per kilogram of clonazepam (an anticonvulsant drug). If each tablet contains 0.5 milligram, how many tablets should the patient receive?

20. A patient weighing 165 pounds is to receive 7.2 milligrams per kilogram of azithromycin. If each capsule contains 300 milligrams, how many capsules should the patient receive?

21. A patient is to receive 250 milliliters of 5% D/W solution IV over 6 hours. The label on the box of the IV indicates that 15 drops dissipate 1 milliliter of the solution. How many drops should the patient receive each minute?

22. A patient is to receive 500 milliliters of 5% D/W solution IV over 2 hours. The label on the box of the IV indicates that 15 drops dissipate 1 milliliter of the solution. How many drops should the patient receive each minute?

23. 500 milliliters of 5% D/W solution IV contains 20,000 units of heparin. If the patient receives 10 milliliters per hour, how many units is the patient receiving each hour?

24. 500 milliliters of 5% D/W solution IV contains 20 units of pitocin. If the patient receives 0.002 unit per minute, how many milliliters is she receiving in an hour?

Review Exercises

[7.1] 1. Simplify the ratio 0.14 to 0.5.

[7.1] 2. What is the probability of randomly selecting a King or a Jack from a shuffled standard deck of cards?

[7.1] 3. A 16-ounce bottle of apple juice costs $2.29. What is the unit price in dollars per ounce?

[7.3] 4. Convert 75 feet to yards.

[7.4] 5. Convert 8.5 grams to milligrams.

Chapter 7 Summary and Review Exercises

Learning Strategies

Spread out study time. Don't do it all the night before because you retain more when you study the same material over several sessions.

—Sindhu P.

Set goals for how much you will complete in a specific time frame. After completing that goal, reward yourself with a break and do something fun.

—Cara S.

7.1 Ratios

Definitions/Rules/Procedures	Key Example(s)
A **ratio** is a comparison between two quantities using a(n) _____. **For two quantities, a and b, the ratio** of a to b is _____. **To simplify a ratio of two fractions or mixed numbers,** _____ the fractions or mixed numbers. **To eliminate decimals in a ratio**, multiply the numerator and denominator by _____ as determined by the decimal number with the most decimal places.	A roof has a vertical rise of 8 inches for every 13 inches of horizontal length. What is the ratio of vertical rise to horizontal length? **Answer:** $\dfrac{8}{13}$ A recipe calls for $\frac{1}{2}$ cup of flour to $2\frac{1}{4}$ cups of milk. Write the ratio of flour to milk in simplest form. $$\frac{\frac{1}{2}}{2\frac{1}{4}} = \frac{1}{2} \div 2\frac{1}{4}$$ $$= \frac{1}{2} \div \frac{9}{4}$$ $$= \frac{1}{\underset{1}{2}} \cdot \frac{\overset{2}{4}}{9}$$ $$= \frac{2}{9}$$ A small plastic tube has a diameter of 0.08 centimeter and a length of 12.5 centimeters. What is the ratio of length to diameter? $$\frac{12.5}{0.08} = \frac{12.5(100)}{0.08(100)} = \frac{1250}{8} = \frac{625}{4}$$

For Exercises 1 and 2, write each ratio in simplest form.

[7.1] 1. A company has 32 male employees and 22 female employees. Write the ratio of female to male employees.

[7.1] 2. The shortest side on a right triangle measures 2.5 inches. The hypotenuse measures 10.8 inches. Write the ratio of the length of the shortest side to the length of the hypotenuse.

Definitions/Rules/Procedures	Key Example(s)
The **theoretical probability of equally likely outcomes** is the ratio of the number of _____ outcomes to the total number of _____ outcomes. **To determine a probability:** 1. Write the number of _____ outcomes in the numerator. 2. Write the total number of _____ outcomes in the denominator. 3. Simplify.	In simplest form write the probability of drawing a red 10 from a standard deck of cards. There are two red 10s in a standard deck of 52 cards. $$P = \frac{2}{52} = \frac{1}{26}$$

[7.1] For Exercises 3 and 4, determine the probability. Write each probability in simplest form.

3. What is the probability of randomly selecting a Queen or a King from a shuffled standard deck of cards?

4. What is the probability of rolling a 1 or a 2 on a six-sided die?

Definitions/Rules/Procedures	Key Example(s)
A **unit ratio** is a ratio in which the denominator is ____. **To write a ratio as a unit ratio,** divide _____ _____.	A college has 12,495 students and 525 faculty members. Write the unit ratio of students to faculty members. $$\frac{12,495}{525} = 23.8$$

[7.1] For Exercises 5 and 6, find the unit ratio.

5. A college has 4500 students and 275 faculty. Write a unit ratio of students to faculty.

6. The price for each share of stock for a communications company closed at $12\frac{5}{8}$. The annual per share earning for the stock was $1.78. Calculate the price-to-earnings ratio.

Definitions/Rules/Procedures	Key Example(s)
A **rate** is a ratio comparing two _____ measurements. A **unit rate** is a rate in which the _____ is 1. **To write a rate as a unit rate,** _____ _____. Because the units are different, leave the units in fraction form. A **unit price** is a unit rate relating _____. If q units of a product have a total price of p, the unit price is found by the formula $u = $ _____.	A 17.6-ounce container of yogurt costs $4.89. Find the unit price in cents per ounce. Round to the nearest tenth of a cent per ounce. **Solution:** $4.89 = (100\cent)4.89 = 489\cent$ Convert dollars to cents. $u = \dfrac{489\cent}{17.6 \text{ oz.}}$ In $u = \dfrac{p}{q}$, replace p with $489\cent$ and q with 17.6 ounces. $u = 27.8\cent/\text{ounce}$ Divide.

[7.1] For Exercises 7–10, solve.

7. Sekema drove 210.6 miles in 3 hours. What was her average rate in miles per hour?

8. A phone call lasting 28 minutes costs $2.52. What is the unit price in cents per minute?

9. A 32-ounce bag of sugar costs $1.44. What is the unit price in cents per ounce?

10. A 64-ounce bottle of juice costs $2.49. What is the unit price in cents per ounce?

[7.1] For Exercises 11 and 12, determine which is the better buy.

11. $0.55 for a 24-ounce cup of cola

or

$0.79 for a 32-ounce cup of cola

12. two 12-ounce cans of store-brand frozen juice for $2.58

or

four 8-ounce cans of name-brand frozen juice for $3.90

7.2 Proportions

Definitions/Rules/Procedures	Key Example(s)
A **proportion** is an equation in the form _____, where $b \neq 0$ and $d \neq 0$. **When two ratios are proportional,** their _____ are equal. **In math language,** if $\dfrac{a}{b} = \dfrac{c}{d}$, where $b \neq 0$ and $d \neq 0$, _____.	Determine whether the ratios are proportional. $9.6 \cdot 8 = 76.8 \qquad 6.4 \cdot 12 = 76.8$ $\dfrac{8}{6.4} \overset{?}{=} \dfrac{12}{9.6}$ Because the cross products are equal, the ratios are proportional.

[7.2] For Exercises 13 and 14, determine whether the ratios are proportional.

13. $\dfrac{5}{9} \overset{?}{=} \dfrac{7}{12}$

14. $\dfrac{3}{7.2} \overset{?}{=} \dfrac{0.75}{1.8}$

Definitions/Rules/Procedures	Key Example(s)
To solve a proportion using cross products: 1. Find the cross products. 2. _____. 3. Solve the resulting equation.	Solve. $\dfrac{2.8}{-9} = \dfrac{m}{15}$ $15(2.8) = 42 \qquad\qquad -9 \cdot m = -9m$ $\dfrac{2.8}{-9} = \dfrac{m}{15}$ $42 = -9m$ $\dfrac{42}{-9} = \dfrac{-9m}{-9}$ $-4.\overline{6} = m$

[7.2] For Exercises 15 and 16, solve.

15. $\dfrac{1.8}{-5} \overset{?}{=} \dfrac{k}{12.5}$

16. $\dfrac{2}{m} = \dfrac{3\frac{1}{4}}{15}$

Definitions/Rules/Procedures	Key Example(s)
To translate an application problem to a proportion, write two ratios set equal to each other so that the information from the problem pairs both _____ (straight across) and _____ (up and down). **Similar figures** are figures with the same shape, _____ corresponding angles, and _____ side lengths.	A garden plan shows 30 plants covering 96 square feet. How many plants would be needed to cover a garden area that is 140 square feet? $\dfrac{30 \text{ plants}}{96 \text{ ft.}^2} = \dfrac{n \text{ plants}}{140 \text{ ft.}^2}$ $4200 = 96n$

Definitions/Rules/Procedures	Key Example(s)
	$$\frac{4200}{96} = \frac{96n}{96}$$ $$43.75 = n$$ **Answer:** Only whole plants can be purchased; so the larger garden would require about 44 plants.

[7.2] **17.** The instructions on the back of a can of wood sealant indicate that 3.78 liters will cover about 25 square meters of mildly porous wood. If the can contains 4.52 liters of sealant, how many square meters of mildly porous wood will it cover?

[7.2] **18.** On a map, $\frac{1}{4}$ inch represents 20 miles. Suppose two cities are $3\frac{1}{2}$ inches apart on the map. How far apart are the two cities in miles?

[7.2] *For Exercises 19 and 20, find the unknown length in the similar figures.*

19.

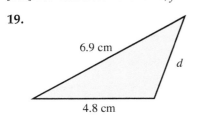 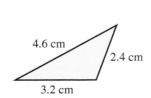

20.

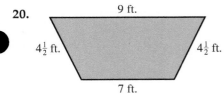 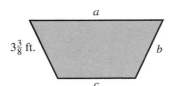

7.3 American Measurement; Time

Definitions/Rules/Procedures	Key Example(s)
A **conversion factor** is a(n) _____ used to convert from one unit of measurement to another by multiplication. **To convert units using dimensional analysis**, multiply the given measurement by conversion factors so that the undesired units _____, leaving the desired units. **Note:** If an undesired unit is in a numerator, that unit should appear in the _____ of a conversion factor. If an undesired unit is in a denominator, that unit should appear in the _____ of a conversion factor. **Capacity** is a measure of the amount a container _____. **Weight** is a measure of the downward _____ due to gravity.	Convert. **a.** 4.5 feet to inches $$4.5 \text{ ft.} = \frac{4.5 \text{ ft.}}{1} \cdot \frac{12 \text{ in.}}{1 \text{ ft.}} = 54 \text{ in.}$$ **b.** 72 square feet to square yards $$72 \text{ ft.}^2 = \frac{72 \text{ ft.}^2}{1} \cdot \frac{1 \text{ yd.}^2}{9 \text{ ft.}^2} = \frac{72}{9} \text{ yd.}^2$$ $$= 8 \text{ yd.}^2$$ **c.** 20 cups to quarts $$20 \text{ c.} = \frac{20 \text{ c.}}{1} \cdot \frac{1 \text{ pt.}}{2 \text{ c.}} \cdot \frac{1 \text{ qt.}}{2 \text{ pt.}} = \frac{20}{4} \text{ qt.}$$ $$= 5 \text{ qt.}$$

Definitions/Rules/Procedures	Key Example(s)

Complete the table.

	Conversion fact
Length	1 ft. = _____ in.
	1 yd. = _____ ft.
	1 mi. = _____ ft.
Capacity	1 c. = _____ fl. oz.
	1 pt. = _____ c.
	1 qt. = _____ pt.
	1 gal. = _____ qt.
Weight	1 lb. = _____ oz.
	1 T = _____ lb.
Time	1 min. = _____ sec.
	1 hr. = _____ min.
	1 day = _____ hr.
	1 yr. = _____ days

d. 0.4 ton to pounds

$$0.4\,T = \frac{0.4\,T}{1} \cdot \frac{2000\ \text{lb.}}{1\,T} = 800\ \text{lb.}$$

e. 3960 minutes to days

$$3960\ \text{min.} = \frac{3960\ \text{min.}}{1} \cdot \frac{1\ \text{hr.}}{60\ \text{min.}} \cdot$$

$$\frac{1\text{d.}}{24\ \text{hr.}} = \frac{3960}{1440}\,\text{d.} = 2.75\ \text{d.}$$

[7.3] For Exercises 21–30, use dimensional analysis to convert.

21. 12 yards to inches

22. 9152 yards to miles

23. 40 ounces to pounds

24. 4800 pounds to tons

25. 6 quarts to ounces

26. 49.6 pints to gallons

27. $2\frac{1}{4}$ hours to minutes

28. 4 years to days

29. A room that is 14 feet wide by 15.5 feet long is to be carpeted. Calculate the area of the room in square yards.

30. A meteor streaks through Earth's upper atmosphere at a speed of 116,160 feet per second. Calculate the speed in miles per hour.

7.4 Metric Measurement

Definitions/Rules/Procedures	Key Example(s)

A **base unit** is a(n) _____ unit; other units are named relative to it.

In the metric system, the base unit for **length** is the _____, the base unit for **capacity** is the _____, and the base unit for **mass** is the _____.

To convert units in the metric system:

1. Using the table of metric units, count the number of units from the undesired unit to the desired unit and note the direction (right or left).

_____	hecto-	deka-	base unit	deci-	_____	milli-
(k)	(h)	(da)	(m)	(d)	(c)	(m)

Convert.

a. 2.8 meters to centimeters

Because centimeter is the second unit to the right of meter in the table, we move the decimal point 2 places to the right in 2.8.

$$2.8\ \text{m} = 280\ \text{cm}$$

b. 0.085 liter to milliliters

Because milliliter is the third unit to the right of liter in the table, we move the decimal point 3 places to the right in 0.085.

$$0.085\ \text{L} = 85\ \text{ml}$$

Definitions/Rules/Procedures	Key Example(s)
2. Move the decimal point the same number of _____ and in the same direction as determined in step 1. Memory tip: **King Hector's Data Base Deci**dedly **Cent**ralized **Milli**ons **Mass** is a measure of the amount of _____ that makes up an object. **Additional Relations** _____ ml = 1 cm^3, or 1 cc 1 t = _____ kg **To convert metric units of area**, move the decimal point _____ the number of places as units from the initial area unit to the desired area unit on the table of units.	**c.** 97,500 decigrams to kilograms Because kilogram is the fourth unit to the left of decigram in the table, we move the decimal point 4 places to the left in 97,500. $$97{,}500 \text{ dg} = 9.75 \text{ kg}$$ **d.** 0.005 square centimeter to square millimeter Square millimeter is one unit to the right of square centimeter. Moving the decimal point twice that number of places means that we move it 2 places to the right in 0.005. $$0.005 \text{ cm}^2 = 0.5 \text{ mm}^2$$

[7.4] For Exercises 31–36, convert.

31. 5 meters to decimeters

32. 375 millimeters to meters

33. 0.26 kilogram to grams

34. 950,000 grams to metric tons

35. 285 deciliters to liters

36. 0.075 liter to cubic centimeters

7.5 Converting between Systems; Temperature

Definitions/Rules/Procedures	Key Example(s)
Complete the table.	Convert. **a.** 50 yards to meters

	American to metric	Metric to American
Length	1 in. = _____ cm	1 m ≈ _____ ft.
	1 yd. ≈ _____ m	1 m ≈ _____ yd.
	1 mi. ≈ _____ km	1 km ≈ _____ mi.
Capacity	1 qt. ≈ _____ L	1 L ≈ _____ qt.
Mass	1 lb. ≈ _____ kg	1 kg ≈ _____ lb.
	1 T ≈ _____ t	1 t ≈ _____ T

Key Example(s) continued:

a. 50 yards to meters

$$50 \text{ yd.} \approx \frac{50 \text{ yd.}}{1} \cdot \frac{0.914 \text{ m}}{1 \text{ yd.}} = 45.7 \text{ m}$$

b. 3 liters to pints

$$3 \text{ L} \approx \frac{3 \text{ L}}{1} \cdot \frac{1.057 \text{ qt.}}{1 \text{ L}} \cdot \frac{2 \text{ pt.}}{1 \text{ qt.}}$$
$$= 6.342 \text{ qt.}$$

c. 192 ounces to kilograms

$$192 \text{ oz.} \approx \frac{192 \text{ oz.}}{1} \cdot \frac{1 \text{ lb.}}{16 \text{ oz.}} \cdot \frac{0.454 \text{ kg}}{1 \text{ lb.}}$$
$$= 5.448 \text{ kg}$$

To convert degrees Fahrenheit to degrees Celsius, use the following formula:

$$C = \underline{\hspace{2cm}}$$

To convert degrees Celsius to degrees Fahrenheit, use the following formula:

$$F = \underline{\hspace{2cm}}$$

[7.5] For Exercises 37–42, convert.

37. 20 feet to meters

38. 12 miles to kilometers

39. 145 pounds to kilograms

40. 3 liters to gallons

41. 90°F to degrees Celsius

42. −4°C to degrees Fahrenheit

7.6 Applications and Problem Solving

Definitions/Rules/Procedures	Key Example(s)
A **front-end-ratio** is the ratio of the total monthly house payment to _____. A **back-end-ratio** is the ratio of the total monthly _____ to gross monthly income.	The Jackson family is trying to qualify for an FHA loan. The gross monthly income is $3600. If approved, the monthly PITI payment will be about $950. The total monthly debt, excluding the PITI payment, is $480. The credit score is 640. Does the family qualify?

Factors for loan qualification

	Conventional	VA	FHA
Front-end ratio should not exceed	0.28	N/A	0.29
Back-end ratio should not exceed	0.36	0.41	0.41
Credit score should be	650 or higher	N/A	N/A

Solution: This is an FHA loan; so only the front-end and back-end ratios apply.

$$\text{Front-end ratio} = \frac{950}{3600} \approx 0.264$$

$$\text{Back-end ratio} = \frac{950 + 480}{3600} = \frac{1430}{3600} \approx 0.397$$

Both ratios are less than the required values; so the Jackson family qualifies for the loan.

[7.6] For Exercises 43–48, solve.

43. A patient is to receive 450 milliliters of NS (normal saline) IV in 6 hours. If the IV dissipates 1 milliliter every 10 drops, what is the drip rate in drops per minute?

44. A patient is to receive an antibiotic. The patient has a mass of 32 kilograms. The order for the antibiotic is 2.9 milligrams per kilogram. How much should the patient receive?

45. A family applying for a loan has a gross monthly income of $3450. If the monthly PITI payment will be $978, what is the ratio of monthly PITI payment to gross monthly income (front-end ratio)?

46. A family has the following monthly debts:

$$\begin{aligned} \text{monthly PITI} &= \$565 \\ \text{credit card} &= \$45 \\ \text{car loan} &= \$225 \\ \text{student loan} &= \$125 \end{aligned}$$

If the gross monthly income is $2420, what is the ratio of monthly debt to gross monthly income (back-end ratio)?

47. A family with a gross monthly income of $1970 is applying for an FHA loan. To qualify, the front-end ratio must not exceed 0.29. What is the maximum monthly PITI payment the family can have and qualify for the loan?

48. A family with a gross monthly income of $4200 is applying for a VA loan. To qualify, the back-end ratio must not exceed 0.41. What is the maximum monthly debt the family can have and qualify for the loan?

Chapter 7 Practice Test

[7.1]   **1.** A computer screen is 9.5 inches wide by 13 inches high. Write the ratio of the width to the height in simplest form.

1. _____

[7.1]   **2.** Dianne has 22 entries in a random-drawing contest that has 4578 total entries. What is the probability that Dianne will win? Write the probability in simplest form.

2. _____

[7.1]   **3.** What is the probability of rolling a 9 or a 10 on a die that has 10 sides? Write the probability in simplest form.

3. _____

[7.1]   **4.** A college has 3850 students and 125 faculty. Write a unit ratio of students to faculty.

4. _____

[7.1]   **5.** Laurence drove 158.6 miles in 2.25 hours. What was his average rate in miles per hour?

5. _____

[7.1]   **6.** A 15-ounce can of mixed vegetables costs $0.65. What is the unit price in cents per ounce?

6. _____

[7.1]   **7.** Determine the better buy: $1.19 for a 20-ounce can of pineapple or $0.79 for an 8-ounce can of pineapple?

7. _____

[7.2]   **8.** Solve. $\dfrac{-6.5}{12} = \dfrac{n}{10.8}$

8. _____

[7.2]   **9.** Find the unknown length in the similar figures.

9. _____

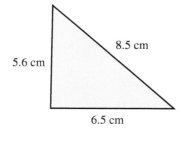
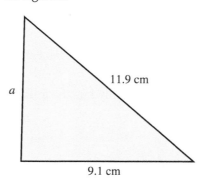

[7.2]   **10.** A company charges $49 to install 20 square feet of hardwood floor. How much would it charge to install the same floor in a 210-square-foot room?

10. _____

[7.2]   **11.** On a map, $\frac{1}{4}$ inch represents 50 miles. Suppose two cities are $4\frac{1}{2}$ inches apart on the map. How far apart are the two cities in miles?

11. _____

[7.3] For Exercises 12–16, convert using dimensional analysis.

12. _____

12. 14 feet to inches

13. _____

13. 20 pounds to ounces

14. _____

14. 36 pints to gallons

15. _____

15. 150 minutes to hours

16. _____

[7.3] **16.** A room that is 20 feet long by 12.5 feet wide is to be carpeted. Calculate the area of the room in square yards.

For Exercises 17–23, convert.

17. a. _____
    b. _____

[7.4] **17. a.** 0.058 meter to centimeters
       b. 420 meters to kilometers

18. a. _____
    b. _____

[7.4] **18. a.** 24 grams to kilograms
       b. 0.091 metric ton to grams

19. a. _____
    b. _____

[7.4] **19. a.** 280 square centimeters to square meters
       b. 0.8 square meter to square decimeters

20. a. _____
    b. _____

[7.4] **20. a.** 6.5 deciliters to liters
       b. 0.8 liter to cubic centimeters

21. _____

[7.5] **21.** 75 feet to meters

22. _____

[7.5] **22.** 30°F to degrees Celsius

23. _____

[7.5] **23.** −12°C to degrees Fahrenheit

24. _____

[7.6] **24.** A family has the following monthly debts:

> monthly PITI = $714
> credit card 1 = $128
> credit card 2 = $78
> car payment 1 = $295
> car payment 2 = $324

If the gross monthly income is $3420, what is the ratio of monthly debt to gross monthly income (back-end ratio)?

25. _____

[7.6] **25.** A patient is to receive 750 milliliters of NS IV in 8 hours. If the IV dissipates 1 milliliter every 10 drops, what is the drip rate in drops per minute?

Chapters 1–7 Cumulative Review Exercises

For Exercises 1–4, answer true or false.

[1.3, 3.4] **1.** $5^2 \cdot 5^6 = 5^{12}$

[5.4] **2.** $\frac{4}{9}x = 12$ is a linear equation.

[6.3] **3.** $-2.4 \times 10^6 > -2.4 \times 10^5$

[6.4] **4.** $\sqrt{5}$ is a rational number.

For Exercises 5 and 6, fill in the blank.

[2.4] **5.** The product of two negative numbers is a(n) _____ number.

[5.8] **6.** To clear fractions from an equation, we can use the multiplication principle of equality and multiply both sides by the _____ of the denominators.

[3.5] **7.** Find the prime factorization of 2400.

[5.1] **8.** Graph $-3\frac{2}{3}$ on a number line.

[5.5] **9.** Find the LCM of $8k^2$ and $10k$.

[6.3] **10.** Write 914,000,000 in scientific notation.

For Exercises 11–16, simplify.

[2.5] **11.** $[10 - 3(7)] - [16 \div 2(7 + 3)]$

[5.4] **12.** $5\frac{2}{3} \div \left(-2\frac{1}{6}\right)$

[5.6] **13.** $10\frac{1}{2} - 4\frac{3}{5}$

[5.4] **14.** $\sqrt{\frac{45}{5}}$

[6.5] **15.** $\left(8\frac{2}{3}\right)(-3.6)$

[6.3] **16.** $(-0.3)^4$

[6.5] **17.** Evaluate $\frac{1}{2}mv^2$ when $m = 38$ and $v = 0.4$.

[6.2] **18.** Subtract. $(12.4y^3 - 8.2y^2 + 9) - (4.1y^3 + y - 1.2)$

[5.7] **19.** Multiply. $\left(\frac{2}{3}a - 1\right)\left(\frac{3}{4}a - 7\right)$

[5.3] **20.** Simplify. $\frac{9x^2}{20y} \cdot \frac{5y}{12x^6}$

For Exercises 21–24, solve and check.

[5.8] **21.** $\frac{3}{4}x - 5 = \frac{1}{5}x + 3$

[7.2] **22.** $\frac{10.2}{b} = \frac{-12.5}{20}$

[7.3] **23.** Convert 12.5 miles to feet.

[7.4] **24.** Convert 0.48 kilogram to grams.

For Exercises 25–30, solve.

[1.6] **25.** Find the volume.

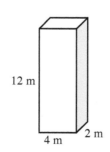

12 m

4 m

2 m

[3.7] **26.** Write an expression in simplest form for the area of the following parallelogram.

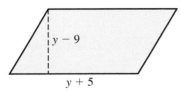

[4.5] **27.** The width of a rectangle is 3 less than the length. If the perimeter is 50 meters, find the length and width.

[5.7] **28.** Find the area.

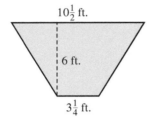

[6.5] **29.** The following graph shows a company's quarterly revenue for one year. Find the mean, median, and mode quarterly revenue.

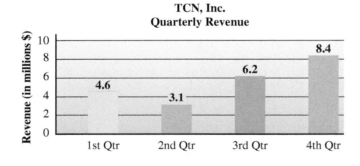

[7.1] **30.** Determine the better buy: 16 ounces of beans for $0.49 or 28 ounces of beans for $1.19?

Chapter Overview

In Chapter 8, we will learn about a special type of ratio called a percent. We will begin the chapter by converting fractions and decimals to percent notation and vice versa. We will also use what we have learned about solving equations and proportions to solve problems involving percents.

8.1 Introduction to Percent

Objectives

1 Write a percent as a fraction or a decimal number.

2 Write a fraction or a decimal number as a percent.

3 Draw circle graphs.

Warm-up

[6.4] 1. Write $\frac{3}{8}$ as a decimal number.

[6.4] 2. Divide. $4.8 \div 100$

Objective 1 Write a percent as a fraction or a decimal number.

Think about the word *percent*. It is a compound word made from the prefix *per,* which means "for each" or "divide," and the suffix *cent,* coming from the Latin word *centum,* which means 100. Therefore, **percent** means for each 100 or divide by 100.

Definition **Percent:** A ratio representing some part out of 100.

The symbol for percent is %. For example, 20% is read *twenty percent* and means "20 out of 100." The shaded region in the figure to the left represents 20% because 20 out of the 100 squares are shaded. A percent can also be written as a fraction or decimal number.

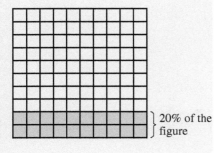
} 20% of the figure

$$20\% = 20 \;\underbrace{\text{out of } 100}$$

$$= \frac{20}{100}$$

$$= \frac{1}{5} \text{ or } 0.2$$

◀ **Note** Using the definition of percent, we can replace the percent sign with *out of 100* or *per 100,* which translates to a fraction with a denominator of 100.

Procedure

To write $n\%$ as a fraction or a decimal number:

1. Write $n\%$ as $\frac{n}{100}$.
2. Simplify to the desired form.

First, let's consider writing percents as fractions in lowest terms.

Example 1 Write each percent as a fraction in lowest terms.

a. 12%

Solution: $12\% = \frac{12}{100} = \frac{3}{25}$ Write 12 over 100; then simplify. ◀

Note Recall from Chapter 5 that to simplify a fraction, we divide the numerator and denominator by their GCF. In this case, the GCF of 12 and 100 is 4.

b. 20.5%

Solution: $20.5\% = \frac{20.5}{100}$ Write 20.5 over 100.

$$= \frac{20.5(10)}{100(10)}$$ Clear the decimal by multiplying the numerator and denominator by 10.

$$= \frac{205}{1000}$$ Multiply.

$$= \frac{41}{200}$$ Simplify to lowest terms.

Note In Section 7.1, we learned to clear decimals in a ratio by multiplying the numerator and denominator by an appropriate power of 10, as determined by the decimal number with the greatest number of decimal places. In this case, 20.5 ▶ has the most decimal places. To clear 1 place, we multiply by 10.

c. $30\frac{1}{4}\%$

Solution: $30\frac{1}{4}\% = \dfrac{30\frac{1}{4}}{100}$ Write $30\frac{1}{4}$ over 100.

$= 30\frac{1}{4} \div 100$ Write the complex fraction in division form.

$= \dfrac{121}{4} \cdot \dfrac{1}{100}$ Write an equivalent multiplication with improper fractions.

$= \dfrac{121}{400}$ Multiply.

▶ **Do Your Turn 1**

▶ **Your Turn 1**

Write each percent as a fraction in lowest terms.

 a. 48%

 b. 6.25%

 c. $10\frac{1}{3}\%$

Now let's consider writing percents as decimal numbers. Because a percent translates to a ratio with a denominator of 100, we will be dividing by 100, which causes a decimal point to move 2 places to the left. If the percent contains a mixed number or fraction, we will write the mixed number or fraction as a decimal number before dividing by 100.

Example 2   Write each percent as a decimal number.

a. 92%

Solution: $92\% = \dfrac{92}{100} = 0.92$ Write 92 over 100; then divide, which causes the decimal point to move 2 places to the left.

b. 125%

Solution: $125\% = \dfrac{125}{100} = 1.25$ Write 125 over 100; then divide.

c. 9.2%

Solution: $9.2\% = \dfrac{9.2}{100} = 0.092$ Write 9.2 over 100; then divide.

d. $21\frac{2}{3}\%$

Solution:

$21\frac{2}{3}\% = \dfrac{21\frac{2}{3}}{100}$ Write $21\frac{2}{3}$ over 100.

$= \dfrac{21.\overline{6}}{100}$ Write $21\frac{2}{3}$ as a decimal number.

$= 0.21\overline{6}$ Divide by 100.

Note Remember, to write a fraction as a decimal number, we ◀ divide the denominator into the numerator.

▶ **Do Your Turn 2**

▶ **Your Turn 2**

Write each percent as a decimal number.

 a. 14%

 b. 250%

 c. 1.7%

 d. $83\frac{1}{4}\%$

Objective 2 Write a fraction or a decimal number as a percent.

We have seen that $20\% = \frac{20}{100} = \frac{1}{5} = 0.2$. Note that to write a percent as a fraction or as a decimal number, we divide by 100. To reverse this process, we multiply by 100 and attach a percent sign.

Answers to Your Turn 1

a. $\dfrac{12}{25}$ b. $\dfrac{1}{16}$ c. $\dfrac{31}{300}$

Answers to Your Turn 2

a. 0.14 b. 2.5
c. 0.017 d. 0.8325

Procedure

To write a fraction or a decimal number as a percent:

1. Multiply by 100%.
2. Simplify.

First, we consider writing fractions as percents.

▶ **Your Turn 3**

Write each fraction as a percent.

a. $\dfrac{3}{5}$

b. $\dfrac{5}{16}$

c. $\dfrac{4}{9}$

Connection Notice that $100\% = \dfrac{100}{100} = 1$. Therefore, when we multiply a number by 100%, we are multiplying by 1, which is why the resulting percent is equal to the given number.

| **Example 3** | Write each fraction as a percent. |

a. $\dfrac{1}{2}$

Solution: $\dfrac{1}{2} = \dfrac{1}{\underset{1}{2}} \cdot \dfrac{\overset{50}{100}}{1}\%$ Multiply by 100%.

$= 50\%$ Simplify.

b. $\dfrac{5}{8}$

Solution: $\dfrac{5}{8} = \dfrac{5}{\underset{2}{8}} \cdot \dfrac{\overset{25}{100}}{1}\%$ Multiply by 100%.

$= \dfrac{125}{2}\%$ Simplify.

$= 62\dfrac{1}{2}\%$, or 62.5%

c. $\dfrac{2}{3}$

Solution: $\dfrac{2}{3} = \dfrac{2}{3} \cdot \dfrac{100}{1}\%$ Multiply by 100%.

$= \dfrac{200}{3}\%$ Simplify.

$= 66\dfrac{2}{3}\%$, or $66.\overline{6}\%$

◀ **Do Your Turn 3**

Now let's consider writing decimal numbers as percents. Multiplying a decimal number by 100 causes the decimal point to move 2 places to the right.

| **Example 4** | Write each decimal number as a percent. |

a. 0.7

Solution: $0.7 = (0.7)(100\%) = 70\%$ Multiply by 100%, which causes the decimal point to move 2 places to the right.

b. 0.018

Solution: $0.018 = (0.018)(100\%) = 1.8\%$ Multiply by 100%.

c. 2

Solution: $2 = (2)(100\%) = 200\%$ Multiply by 100%.

Answers to Your Turn 3
a. 60%

b. $31\dfrac{1}{4}\%$ or 31.25%

c. $44\dfrac{4}{9}\%$ or $44.\overline{4}\%$

d. $0.\overline{25}$

Solution: $0.\overline{25} = (0.\overline{25})(100\%) = 25.\overline{25}\%$ **Multiply by 100%.**

▲

Note Remember, the repeat bar means that the decimal digits beneath it repeat without end; so $0.\overline{25} = 0.\overline{252525}\ldots$.

▶ **Do Your Turn 4**

Objective 3 Draw circle graphs.

Example 5 The results of a state reading comprehension assessment show that at a particular elementary school, 24% of the students read below grade level, 58% read on grade level, and 18% read above grade level. Draw a circle graph that represents the results of the test.

Solution: Notice that $24\% + 58\% + 18\% = 100\%$; so the three percents together form the whole circle graph. We will draw wedges to represent each percent. In a circle, lightly draw 10 equal-size wedges so that each wedge represents 10%.

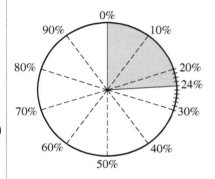

The wedge for 24% goes from 0% (straight up) to a point on the circle 4 tick marks past 20%. Shade this wedge.

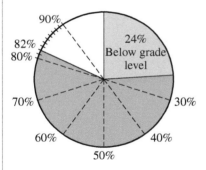

The wedge representing 58% is added to the wedge for 24%. Because $24\% + 58\% = 82\%$, the wedge representing 58% extends from 24% to 82%. Draw a radius line to the second tick mark past 80% and shade this wedge a different color.

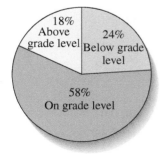

Because 82% of the circle has been shaded so far, the remaining wedge must represent 18% ($100\% - 82\% = 18\%$). Shade it in a third color.

▶ **Do Your Turn 5**

▶ **Your Turn 4**

Write each decimal number as a percent.
a. 0.49
b. 0.883
c. 1.5
d. $0.\overline{13}$

▶ **Your Turn 5**

In a survey, 42% of the people rated a product as excellent, 34% rated it good, 14% rated it fair, and 10% rated it poor. Draw a circle graph that represents the results of the survey.

Answers to Your Turn 4
a. 49% b. 88.3%
c. 150% d. $13.\overline{13}\%$

Answers to Your Turn 5

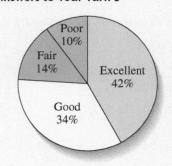

8.1 Exercises (For Extra Help) MyMathLab®

Learning Strategy

Whether or not homework is assigned, I always go through the text and do as many exercises that have answers available as possible. Having the answers is very important for me because I like to practice but am doubtful of my work until I know I have the correct answer.

—Deanna C.

Objective 1

Prep Exercise 1 A percent is a ratio representing some part out of _____.

Prep Exercise 2 To write $n\%$ as a fraction or decimal number:
1. Write $n\%$ as _____.
2. Simplify to the desired form.

For Exercises 1–12, write each percent as a fraction in lowest terms.

1. 20%

2. 30%

3. 15%

4. 85%

5. 12.4%

6. 14.8%

7. 3.75%

8. 1.25%

9. $45\frac{1}{2}\%$

10. $65\frac{1}{4}\%$

11. $33\frac{1}{3}\%$

12. $66\frac{2}{3}\%$

For Exercises 13–24, write each percent as a decimal number.

13. 75%

14. 58%

15. 135%

16. 210%

17. 12.9%

18. 11.6%

19. 1.65%

20. 100.5%

21. $53\frac{2}{5}\%$

22. $70\frac{3}{4}\%$

23. $16\frac{1}{6}\%$

24. $6\frac{2}{3}\%$

Objective 2

Prep Exercise 3 To write a fraction or decimal number as a percent:
1. Multiply by _____.
2. Simplify.

For Exercises 25–36, write each fraction as a percent.

25. $\frac{1}{4}$

26. $\frac{3}{4}$

27. $\frac{2}{5}$

28. $\frac{1}{5}$

29. $\frac{3}{8}$

30. $\frac{7}{8}$

31. $\frac{1}{6}$

32. $\frac{5}{6}$

33. $\frac{1}{3}$

34. $\frac{1}{9}$

35. $\frac{4}{9}$

36. $\frac{5}{11}$

For Exercises 37–52, write each decimal number as a percent.

37. 0.96

38. 0.42

39. 0.8

40. 0.6

41. 0.09

42. 0.01

43. 1.2

44. 3.58

45. 0.028

46. 0.065

47. 0.004

48. 0.007

49. $0.\overline{6}$

50. $0.\overline{3}$

51. $1.\overline{63}$

52. $0.\overline{47}$

Objective 3

Prep Exercise 4 To draw a circle graph containing percents that total 100%, start by dividing the circle into _____ equal-size wedges each representing _____%.

For Exercises 53–56, draw a circle graph showing each percent.

53. A poll shows that 36% of participants agree with a proposed bill, 48% disagree, and the rest have no opinion.

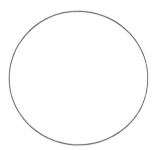

54. A company report shows that 67% of the company's expenses go to salaries, 23% to facilities, and the rest to miscellaneous expenses.

55. Students in an English course can earn an A, a B, a C, or an F. The following table shows the results of a class.

A	12%
B	25%
C	45%
F	18%

56. After listening to a speaker at a conference, participants are asked to rate the presentation as excellent, good, fair, or poor. The following table shows the response.

Excellent	32%
Good	48%
Fair	12%
Poor	8%

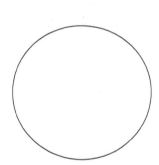

Review Exercises

[6.3] *For Exercises 1–3, evaluate.*

    1. 0.6(24)   **2.** 0.45(3600)   **3.** 0.05(76.8)

[6.4] *For Exercises 4 and 5, solve and check.*

    4. $0.15x = 200$

    5. $0.75y = 28.5$

[5.8] 6. Three-fourths of a number is twenty-four. Find the number.

8.2 Solving Basic Percent Sentences

Objectives

1 Identify the unknown amount in a basic percent sentence.

2 Solve for the unknown part in a basic percent sentence.

3 Solve for the unknown whole in a basic percent sentence.

4 Solve for the unknown percent in a basic percent sentence.

Warm-up

[6.4] **1.** Solve. $0.4n = 25$

[7.2] **2.** Solve. $\dfrac{p}{100} = \dfrac{5.4}{12}$

Objective 1 Identify the unknown amount in a basic percent sentence.

To solve an application problem involving a percent, it is helpful to reduce the problem to a simple percent sentence. Every basic sentence has the following structure with three numbers: *percent, whole,* and *part*.

A percent of a whole is a part.

The following table shows how the basic sentence might be written with each unknown.

	A percent	of	a whole	is	a part.
Unknown part:	20%	of	80	is	what number?
Unknown whole:	32%	of	what number	is	12?
Unknown percent:	What percent	of	240	is	60?

Phrases around the word *is* can exchange position; so the basic sentence can also be written as follows:

A part is a percent of a whole.

Note To rewrite the basic sentence, the words on either side of the word *is* exchange places.

A percent of a whole is a part.

A part is a percent of a whole.

Each question in the table can be rewritten using the alternative form of the sentence.

	A part	is	a percent	of	a whole.
Unknown part	What number	is	20%	of	80?
Unknown whole	12	is	32%	of	what number?
Unknown percent	60	is	what percent	of	240?

Notice that no matter how the question is written, the percent is easy to identify because it has a percent sign or the words *what percent* are used. The whole amount always follows the word *of*. If you can identify those two pieces, the third piece must be the part.

Example 1 For each basic percent sentence, identify the unknown amount as the percent, whole, or part.

a. 30% of what number is 64?

Answer: whole

Explanation: The whole amount always follows the word *of*.

b. What number is 30% of 80?

Answer: part

Explanation: 30% has a percent sign; so it is the percent. The number 80 follows the word *of*; so it is the whole amount. Because the percent and whole amount are given, the part must be the unknown amount.

c. What percent of 72 is 12?

Answer: percent

Explanation: The words *what percent* indicate that the percent is unknown.

▶ **Do Your Turn 1**

Objective 2 Solve for the unknown part in a basic percent sentence.

We will learn two methods for solving for an unknown number in a basic percent sentence. To learn how to use both methods, consider the following basic percent sentence:

$$20\% \text{ of } 50 \text{ is } 10.$$
$$\text{percent} \quad \text{whole} \quad \text{part}$$

Method 1: Word-for-word translation to an equation

As we learned in Section 8.1, the percent is a ratio with a denominator of 100; so 20% means $\frac{20}{100}$, or 0.2.

Because the word *of* is preceded by the percent, which is a fraction, it translates to multiplication. The word *is* translates to an equal sign.

$$\begin{array}{cccccc} 20\% & \text{of} & 50 & \text{is} & 10. \\ \downarrow & \downarrow & & \downarrow & \\ 0.2 & \cdot & 50 & = & 10 \end{array}$$

Note To multiply 20% and 50, 20% must be written as a fraction or decimal. Most people ◀ prefer to use the decimal form.

Therefore, multiplying the whole by the percent, written as a fraction or decimal, gives the part.

Method 2: Proportion

The percent is a ratio, $20\% = \frac{20}{100}$. The part, 10, is a part of the whole 50; so those two numbers form a second ratio, $\frac{10}{50}$. Simplifying $\frac{20}{100}$ and $\frac{10}{50}$ to lowest terms shows that they are equal ratios.

$$\frac{20}{100} = \frac{\overset{1}{2} \cdot \overset{1}{2} \cdot \overset{1}{5}}{\underset{1}{2} \cdot \underset{1}{2} \cdot \underset{1}{5} \cdot 5} = \frac{1}{5} \qquad \frac{10}{50} = \frac{\overset{1}{2} \cdot \overset{1}{5}}{\underset{1}{2} \cdot \underset{1}{5} \cdot 5} = \frac{1}{5}$$

Because $\frac{20}{100}$ and $\frac{10}{50}$ are equivalent, we can write the following proportion:

$$\text{percent} \begin{cases} \dfrac{20}{100} = \dfrac{10}{50} \end{cases} \begin{matrix} \leftarrow \text{part} \\ \leftarrow \text{whole} \end{matrix}$$

As a proportion, the percent, written as a ratio with a denominator of 100, is equal to the ratio of the part to the whole.

▶ **Your Turn 1**

For each basic percent sentence, identify the unknown amount as the percent, whole, or part.

  a. 70% of 40 is what number?
  b. 25% of what number is 60?
  c. 6.4 is what percent of 18?
  d. What number is 5% of 46.5?

TACTILE   VISUAL
Learning Strategy

If you are a visual or tactile learner, to understand the proportion, imagine a container filled with liquid. If we divide the container into 100 equal-size divisions, the liquid comes to the mark for 20 units. If we divide the container into 50 equal-size divisions, the liquid comes to the mark for 10 units. Either way, 1/5 of the container is filled.

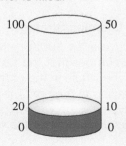

Answers to Your Turn 1
a. part
b. whole
c. percent
d. part

The two methods are summarized in the following procedure:

Procedure

To translate a basic percent sentence to an equation, use one of the following methods:

Method 1: Write the percent as a fraction or decimal number, *of* as multiplication, and *is* as an equal sign.

$$\text{A } \textbf{percent} \text{ of a } \textbf{whole} \text{ is a } \textbf{part.}$$
$$\text{percent} \;\cdot\; \text{whole} \;=\; \text{part}$$

Method 2: Write a proportion in the following form:

$$\text{Percent} = \frac{\text{part}}{\text{whole}}$$

Note In the proportion, the percent is written as a ratio with a denominator of 100.

As we explore the two methods, think about which method you prefer. You may find one method preferable over the other depending on what is unknown in the sentence.

Example 2 Translate to an equation and solve.

a. 60% of 40 is what number?

Solution: 60% is the percent. Because 40 follows the word *of*, it is the whole amount. Therefore, the part is unknown. We will use n to represent the unknown part.

Method 1: 60%   of   40 is what number?

$$0.6 \;\cdot\; 40 = n \qquad \text{Write 60\% as a decimal number, } \textit{of} \text{ as } \cdot,$$
$$\textit{is} \text{ as } =, \text{ and } \textit{what number} \text{ as } n.$$
$$24 = n \qquad \text{Multiply.}$$

Method 2:

60% written as a ratio

$$\frac{60}{100} = \frac{n}{40} \;\begin{matrix}\leftarrow \text{ part} \\ \leftarrow \text{ whole}\end{matrix} \qquad \text{Use percent} = \frac{\text{part}}{\text{whole}}.$$

$$40 \cdot 60 = 2400 \qquad 100 \cdot n = 100n \qquad \text{Cross multiply.}$$

$$\frac{60}{100} \diagup\!\!\!\!\times\!\!\!\!\diagdown \frac{n}{40}$$

$$2400 = 100n \qquad\qquad\qquad \text{Equate the cross products.}$$

$$\frac{2400}{100} = \frac{100n}{100} \qquad\qquad \text{Divide both sides by 100.}$$

$$24 = n$$

b. What number is 110% of 40?

Solution: 110% is the percent. Because 40 follows the word *of*, it is the whole amount. Therefore, the part is unknown.

Method 1: What number is 110% of 40?

$$n = 1.1 \;\cdot\; 40 \qquad \text{Write } \textit{what number} \text{ as } n, \textit{ is} \text{ as } =, 110\%$$
$$\text{as a decimal number, and } \textit{of} \text{ as } \cdot.$$
$$n = 44 \qquad\qquad \text{Multiply.}$$

AUDITORY

Learning Strategy

If you are an auditory learner, to remember that the part goes over the whole in the proportion, try saying *is over of*. The part is always near the word *is*, and the whole always follows the word *of*.

Method 2:

110% written as a ratio $\begin{cases} \dfrac{110}{100} = \dfrac{n}{40} \end{cases}$ ← part ← whole

Use percent $= \dfrac{\text{part}}{\text{whole}}$.

$110 \cdot 40 = 4400 \qquad 100 \cdot n = 100n$ Cross multiply.

$\dfrac{110}{100} \bowtie \dfrac{n}{40}$

$4400 = 100n$ Equate the cross products.

$\dfrac{4400}{100} = \dfrac{100n}{100}$ Divide both sides by 100.

$44 = n$

c. What number is 15.2% of $20\frac{1}{3}$?

Solution: 15.2% is the percent. Because $20\frac{1}{3}$ follows the word *of*, it is the whole amount. Therefore, the part is unknown.

Method 1: What number is 15.2% of $20\frac{1}{3}$?

$n = 0.152 \cdot 20\frac{1}{3}$ Write *what number* as *n*, *is* as =, 15.2% as a decimal number, and *of* as ·.

$n = \dfrac{0.152}{1} \cdot \dfrac{61}{3}$ Write 0.152 over 1 and $20\frac{1}{3}$ as an improper fraction.

$n = \dfrac{9.272}{3}$ Multiply.

$n = 3.090\overline{6}$ Divide.

Method 2:

15% written as a ratio $\begin{cases} \dfrac{15.2}{100} = \dfrac{n}{20\frac{1}{3}} \end{cases}$ ← part ← whole

Use percent $= \dfrac{\text{part}}{\text{whole}}$.

$\dfrac{15.2}{1} \cdot \dfrac{61}{3} = \dfrac{927.2}{3} = 309.0\overline{6} \qquad 100 \cdot n = 100n$ Write the mixed number as a fraction and cross multiply.

$\dfrac{15.2}{100} \bowtie \dfrac{n}{\frac{61}{3}}$

$309.0\overline{6} = 100n$ Equate the cross products.

$\dfrac{309.0\overline{6}}{100} = \dfrac{100n}{100}$ Divide both sides by 100.

$3.090\overline{6} = n$

▶ **Do Your Turn 2**

Objective 3 Solve for the unknown whole in a basic percent sentence.

Now let's translate sentences in which the whole is unknown.

Connection 110% = 100% + 10%. Because 100% = 1, 100% of a number means to multiply by 1, which gives the number. So 100% of 40 is 40. 10% of a number means $\frac{1}{10}$ of the number; so 10% of 40 is 4. Therefore, 110% of 40 means to take all of 40 and add 10% to it; so 40 + 4 = 44. This is what happens when we multiply by the decimal equivalent of 110%, which is 1.1.

$\begin{array}{r} 4.0 \\ \times 1.1 \\ \hline 4.0 \\ +40. \\ \hline 44.0 \end{array}$ ← 10% of 40 ← +100% of 40 ← 110% of 40

▶ **Your Turn 2**

Translate to an equation and solve.

a. 30% of 148 is what number?

b. 120% of 48.5 is what number?

c. What number is 8% of $46\frac{1}{2}$?

Answers to Your Turn 2

a. $0.3 \cdot 148 = n; \dfrac{30}{100} = \dfrac{n}{148}; 44.4$

b. $1.2 \cdot 48.5 = n; \dfrac{120}{100} = \dfrac{n}{48.5}; 58.2$

c. $n = 0.08 \cdot 46\frac{1}{2}; \dfrac{8}{100} = \dfrac{n}{46\frac{1}{2}}; 3.72$

► Your Turn 3

Translate to an equation and solve.

a. 45% of what number is 95.4?

b. $80\frac{1}{4}$ is 5% of what number?

Note For consistency, we will always use w to represent an unknown whole amount. ►

Example 3 Translate to an equation and solve.

a. 32% of what number is 12?

Solution: 32% is the percent. Because *what number* follows the word *of*, the whole amount is unknown. Therefore, 12 must be the part. We will use w to represent the unknown whole.

Method 1:   32% of what number is 12?

$$0.32 \quad \cdot \quad w \qquad = 12$$

Write 32% as a decimal number, *of* as ·, *what number* as w, and *is* as =.

$$\frac{0.32w}{0.32} = \frac{12}{0.32}$$

Divide both sides by 0.32.

$$w = 37.5$$

Method 2:

32% written as a ratio $\begin{cases} \dfrac{32}{100} = \dfrac{12}{w} \end{cases}$ ← part ← whole

Use percent = $\dfrac{\text{part}}{\text{whole}}$.

$$32 \cdot w = 32w \qquad 100 \cdot 12 = 1200$$

Cross multiply.

$$\frac{32}{100} \diagdown \frac{12}{w}$$

$$32w = 1200$$

Equate the cross products.

$$\frac{32w}{32} = \frac{1200}{32}$$

Divide both sides by 32.

$$w = 37.5$$

b. $16\frac{1}{2}$ is 2% of what number?

Solution: 2% is the percent. Because *what number* follows the word *of*, the whole amount is unknown. Therefore, $16\frac{1}{2}$ must be the part.

Method 1: $16\frac{1}{2}$ is 2% of what number?

$$16\frac{1}{2} = 0.02 \cdot w$$

Write *is* as =, 2% as a decimal number, *of* as ·, and *what number* as w.

$$\frac{16.5}{0.02} = \frac{0.02w}{0.02}$$

Write $16\frac{1}{2}$ as a decimal number and divide both sides by 0.02.

$$825 = w$$

Method 2:

2% written as a ratio $\begin{cases} \dfrac{2}{100} = \dfrac{16\frac{1}{2}}{w} \end{cases}$ ← part ← whole

Use percent = $\dfrac{\text{part}}{\text{whole}}$.

$$2 \cdot w = 2w \qquad 100 \cdot 16.5 = 1650$$

Cross multiply.

$$\frac{2}{100} \diagdown \frac{16.5}{w}$$

$$2w = 1650$$

Equate the cross products.

$$\frac{2w}{2} = \frac{1650}{2}$$

Divide both sides by 2.

$$w = 825$$

Answers to Your Turn 3

a. $0.45w = 95.4; \dfrac{45}{100} = \dfrac{95.4}{w}; 212$

b. $80\frac{1}{4} = 0.05w;$

$\dfrac{5}{100} = \dfrac{80\frac{1}{4}}{w}; 1605$

◄ Do Your Turn 3

Objective 4 Solve for the unknown percent in a basic percent sentence.

Now let's translate sentences in which the percent is unknown.

| **Example 4** | Translate to an equation and solve.

a. What percent of 240 is 60?

Solution: *What percent* means that the percent is unknown. Because 240 follows the word *of*, it is the whole amount. Therefore, 60 must be the part. In Method 1, we will use p for the unknown because the answer will be a decimal number and not a percent. In Method 2, we will use P because the result will be the percent value.

Method 1: What percent of 240 is 60?

$$p \quad \cdot \; 240 = 60 \qquad \text{Write } \textit{what percent} \text{ as } p, \textit{of} \text{ as } \cdot,$$
$$\qquad\qquad\qquad\qquad\qquad \text{and } \textit{is} \text{ as } =.$$

Note The result is a decimal number, not a percent. To write it as a percent, we must multiply it by 100%.

$$240p = 60 \qquad \text{Simplify.}$$
$$\frac{240p}{240} = \frac{60}{240} \qquad \text{Divide both sides by 240.}$$
$$p = 0.25$$

Answer: $(0.25)(100\%) = 25\%$ \qquad Multiply by 100% to write the result as a percent.

Method 2:

$P\%$ written as a ratio $\left\{ \dfrac{P}{100} = \dfrac{60}{240} \begin{array}{l} \leftarrow \text{part} \\ \leftarrow \text{whole} \end{array} \right.$ \qquad Use percent $= \dfrac{\text{part}}{\text{whole}}$.

$$240 \cdot P = 240P \qquad\qquad 100 \cdot 60 = 6000 \qquad \text{Cross multiply.}$$

$$\frac{P}{100} \diagup\!\!\!\!\diagdown \frac{60}{240}$$

$$240P = 6000 \qquad\qquad\qquad \text{Equate the cross products.}$$
$$\frac{240P}{240} = \frac{6000}{240} \qquad\qquad\qquad \text{Divide both sides by 240.}$$
$$P = 25 \qquad\qquad \blacktriangleleft \textbf{Note} \text{ In the original proportion, } \dfrac{P}{100} \text{ means}$$

Answer: 25% \qquad Attach the % sign. \qquad $P\%$. Therefore, if the value of P is 25, all we need to do is attach the percent sign.

b. 17.64 is what percent of 84?

Solution: *What percent* means that the percent is unknown. Because 84 follows the word *of*, it is the whole amount. Therefore, 17.64 must be the part.

Method 1: 17.64 is what percent of 84?

$$17.64 = \qquad p \quad \cdot \; 84 \qquad \text{Write } \textit{is} \text{ as } =, \textit{what percent} \text{ as } p,$$
$$\qquad\qquad\qquad\qquad\qquad\qquad \text{and } \textit{of} \text{ as } \cdot.$$

$$17.64 = 84p \qquad\qquad \text{Simplify.}$$
$$\frac{17.64}{84} = \frac{84p}{84} \qquad\qquad \text{Divide both sides by 84.}$$
$$0.21 = p$$

Answer: $(0.21)(100\%) = 21\%$ \qquad Multiply by 100% to write the result as a percent.

▶ **Your Turn 4**

Translate to an equation and solve.

  a. What percent of 400 is 50?

  b. 15.37 is what percent of 14.5?

  c. What percent is 19 out of 40?

  d. 32 of 72 is what percent?

Method 2:

$P\%$ written as a ratio $\begin{cases} \dfrac{P}{100} = \dfrac{17.64}{84} \end{cases} \begin{array}{l} \leftarrow \text{part} \\ \leftarrow \text{whole} \end{array}$ Use $\text{percent} = \dfrac{\text{part}}{\text{whole}}$.

$$84 \cdot P = 84P \qquad 100 \cdot 17.64 = 1764 \qquad \text{Cross multiply.}$$

$$\dfrac{P}{100} \diagup\!\!\!\times\!\!\!\diagdown \dfrac{17.64}{84}$$

$$84P = 1764 \qquad\qquad\qquad \text{Equate the cross products.}$$

$$\dfrac{84P}{84} = \dfrac{1764}{84} \qquad\qquad \text{Divide both sides by 84.}$$

$$P = 21$$

Answer: 21%   Attach the % sign.

Note The phrase *15 out of 40* can also be written as *15 of 40*.

c. What percent is 15 out of 40? ◀

Solution: *What percent* means that the percent is unknown. Because 40 follows the word *of*, it is the whole amount. Therefore, 15 must be the part. More specifically, *15 out of 40* translates to the fraction $\dfrac{15}{40}$.

Method 1: What percent is 15 out of 40?

$$\qquad\qquad \downarrow \qquad\qquad \downarrow$$

$$p \quad = \quad \dfrac{15}{40} \qquad \begin{array}{l}\text{Write } \textit{what percent} \text{ as } p, \textit{ is as} \\ =, \text{ and } \textit{15 out of 40} \text{ as } \dfrac{15}{40}.\end{array}$$

$$p = 0.375 \qquad \text{Divide.}$$

Answer: $(0.375)(100\%) = 37.5\%$ Multiply by 100% to write the result as a percent.

Method 2:

$P\%$ written as a ratio $\begin{cases} \dfrac{P}{100} = \dfrac{15}{40} \end{cases} \begin{array}{l} \leftarrow \text{part} \\ \leftarrow \text{whole} \end{array}$ Use $\text{percent} = \dfrac{\text{part}}{\text{whole}}$.

$$40 \cdot P = 40P \qquad 100 \cdot 15 = 1500 \qquad \text{Cross multiply.}$$

$$\dfrac{P}{100} \diagup\!\!\!\times\!\!\!\diagdown \dfrac{15}{40}$$

$$40P = 1500 \qquad\qquad\qquad \text{Equate the cross products.}$$

$$\dfrac{40P}{40} = \dfrac{1500}{40} \qquad\qquad \text{Divide both sides by 40.}$$

$$P = 37.5$$

Answer: 37.5%   Attach the % sign.

◀ **Do Your Turn 4**

Answers to Your Turn 4

a. $p \cdot 400 = 50$;

$\dfrac{P}{100} = \dfrac{50}{400}$; 12.5%

b. $15.37 = p \cdot 14.5$;

$\dfrac{P}{100} = \dfrac{15.37}{14.5}$; 106%

c. $p = \dfrac{19}{40}$; $\dfrac{P}{100} = \dfrac{19}{40}$; 47.5%

d. $\dfrac{32}{72} = p$; $\dfrac{32}{72} = \dfrac{P}{100}$; $44.\overline{4}\%$

8.2 Exercises For Extra Help **MyMathLab®**

Objective 1

Prep Exercise 1 Complete the basic percent sentence.

A percent of a(n) _____ is a(n) _____.

For Exercises 1–12, identify the unknown amount as the percent, part, or whole.

  1. 45% of 30 is what number?

  2. 10% of 80 is what number?

3. 20% of what number is 56?

4. 15% of what number is 72?

5. What percent of 90 is 15?

6. What percent of 60 is 24?

7. 8 is 15% of what number?

8. 12 is 25% of what number?

9. What number is 6.5% of 84?

10. What number is 7.5% of 28?

11. 3.6 is what percent of 15?

12. 4.8 is what percent of 12?

Objective 2

Prep Exercise 2   Complete the translation for the following basic percent sentence. Do not solve.

$$25\% \text{ of } 60 \text{ is what number?}$$

Word for word to an equation: _____ · _____ = _____

$$\text{Proportion: } \frac{\rule{1cm}{0.4pt}}{100} = \frac{\rule{1cm}{0.4pt}}{\rule{1cm}{0.4pt}}$$

Prep Exercise 3   When solving for an unknown part in a basic percent sentence, why might translating word for word to an equation be preferable over using the proportion method?

For Exercises 13–24, translate to an equation and solve.

13. 40% of 350 is what number?

14. 60% of 400 is what number?

15. What number is 15% of 78?

16. What number is 5% of 44?

17. 150% of 60 is what number?

18. What number is 105% of 84?

19. What number is 16% of $30\frac{1}{2}$?

20. 95% of $40\frac{3}{5}$ is what number?

21. What number is 6.5% of 22,800?

22. What number is 3.5% of 24,500?

23. What number is $5\frac{1}{2}$% of 280?

24. $10\frac{1}{4}$% of 68 is what number?

Objective 3

Prep Exercise 4   Complete the translation for the following basic percent sentence. Do not solve.

$$40\% \text{ of what number is } 15?$$

Word for word to an equation: _____ · _____ = _____

$$\text{Proportion: } \frac{\rule{1cm}{0.4pt}}{100} = \frac{\rule{1cm}{0.4pt}}{\rule{1cm}{0.4pt}}$$

For Exercises 25–34, translate to an equation and solve.

25. 35% of what number is 77?

26. 70% of what number is 45.5?

27. 4800 is 40% of what number?

28. 1950 is 65% of what number?

29. 605 is 2.5% of what number?

30. 645 is 12.9% of what number?

31. 105% of what number is 49.14?

32. 120% of what number is 101.52?

33. 47.25 is $10\frac{1}{2}$% of what number?

34. 2000 is $6\frac{1}{4}$% of what number?

Objective 4

Prep Exercise 5 Complete the translation for the following basic percent sentence. Do not solve.

What percent of 24 is 5?

Word for word to an equation:_____ · _____ = _____

Proportion: $\dfrac{}{100} = \dfrac{}{}$

Prep Exercise 6 When solving for an unknown percent in a basic percent sentence, why might translating to a proportion be preferable over using word-for-word translation to an equation?

For Exercises 35–46, translate to an equation and solve.

35. What percent of 68 is 17?

36. What percent of 120 is 48?

37. What percent of 75 is 80?

38. What percent of 90 is 110?

39. 2.142 is what percent of 35.7?

40. 25.12 is what percent of 125.6?

41. $7\frac{7}{20}$ is what percent of $24\frac{1}{2}$?

42. $5\frac{5}{8}$ is what percent of $9\frac{3}{8}$?

43. What percent is 21 out of 60?

44. What percent is 114 out of 200?

45. What percent is 43 of 65?

46. What percent is 29 of 30?

For Exercises 47–58, translate to an equation and solve.

47. 65% of what number is 52?

48. 90% of what number is 133.2?

49. What number is 15% of $9\frac{1}{2}$?

50. 98% of $60\frac{1}{4}$ is what number?

51. What percent of $30\frac{1}{3}$ is $18\frac{1}{5}$?

52. $5\frac{1}{20}$ is what percent of $12\frac{5}{8}$?

53. What number is $12\frac{1}{2}$% of 440?

54. $5\frac{1}{4}$% of 164 is what number?

55. 120% of what number is 65.52?

56. 110% of what number is 251.35?

57. 16.9% of 2450 is what number?

58. 14.8% of 160 is what number?

59. 17 of 19 is what percent?

60. What percent is 169 of 180?

For Exercises 61–64, explain the mistake, then solve the problem correctly.

61. What number is 2% of 70?

$$n = 0.2 \cdot 70$$
$$n = 14$$

62. What percent of 60 is 15?

$$60p = 15$$
$$\frac{60p}{60} = \frac{15}{60}$$
$$p = 0.25$$

63. What percent is 8 of 20?

$$\frac{P}{20} = \frac{8}{100}$$
$$\frac{100P}{100} = \frac{160}{100}$$
$$P = 1.6\%$$

64. 5% of what number is 24?

$$\frac{5}{100} = \frac{n}{24}$$
$$\frac{120}{100} = \frac{100n}{100}$$
$$1.2 = n$$

Review Exercises

For Exercises 1–3, evaluate.

[5.3] **1.** $\frac{9}{16} \cdot \frac{20}{27}$

[6.3] **2.** 0.48(560)

[6.3] **3.** 0.05(24.8)

[5.3] **4.** $\frac{5}{8}$ of the respondents in a survey disagreed with a given statement. If 400 people were surveyed, how many disagreed with the given statement?

[5.3] **5.** $\frac{3}{4}$ of the respondents in a survey agreed with a given statement. Of these, $\frac{4}{5}$ were female. What fraction of all respondents were females who agreed with the particular statement?

8.3 Solving Percent Problems (Portions)

Objectives

1 Solve for the part in percent problems.
2 Solve for the whole amount in percent problems.
3 Calculate the percent.

Warm-up *[8.2] Translate word for word to an equation or a proportion; then solve.*

1. 45% of 82 is what number?
2. What percent of 20 is 5.4?

Objective 1 Solve for the part in percent problems.

Now that we have seen how to translate basic percent sentences, we can solve percent problems that are more complex. Our strategy with these more complex problems is to use basic sentences as a guide to translate those problems to an equation (word for word or proportion) that we can solve.

The problems in this section will be phrased in terms of whole amounts and portions (parts) of those whole amounts.

Procedure

To solve problems involving percents:

1. Identify the percent, whole, and part, noting which is unknown.
2. Write a basic percent sentence (if needed).
3. Translate to an equation (word for word or proportion).
4. Solve for the unknown.

First, we consider problems in which the part is unknown.

▶ **Your Turn 1**

Translate to an equation; then solve.

Sabrina scored 80% on a test with 60 total questions. How many questions did she answer correctly?

Example 1   30% of a 400-milliliter solution is acetone. How many milliliters of the solution is acetone?

Solution: The unknown is the *part* of the solution that is acetone. We are given the *percent* (30%) and the capacity of the solution (400 ml), which is the *whole* amount.

Note When the part is unknown, using word-for-word translation requires fewer steps than the proportion method does.

A percent of a whole is a part.

30% of 400 is what number?

$0.3 \cdot 400 = n$ Translate to an equation.

$120 = n$ Multiply.

Answer: The solution contains 120 ml of acetone.

◀ **Do Your Turn 1**

Salespeople often earn a **commission** based on sales. The commission can be a percent of the total sales amount or a percent of the profit.

Definition Commission: The portion of sales earnings that a salesperson receives.

Answer to Your Turn 1

$0.8 \cdot 60 = n$ or $\dfrac{80}{100} = \dfrac{n}{60}$; 48

Answers to Warm-up

1. $0.45 \cdot 82 = n$; $\dfrac{45}{100} = \dfrac{n}{82}$; 36.9
2. $p \cdot 20 = 5.4$; $\dfrac{P}{100} = \dfrac{54}{20}$; 27%

Example 2   Carrie earns 15% of total sales in commission. If she sells $2485 in merchandise over a two-week period, what is her commission?

Solution: The unknown is Carrie's commission, which is the *part*. We are given her commission rate (15%) and her total sales ($2485), which is the *whole*.

Her commission is 15% of her total sales.

What number is 15% of 2485?

$$n = 0.15 \cdot 2485 \quad \text{Translate to an equation.}$$
$$n = 372.75 \quad \text{Multiply.}$$

Answer: Carrie's commission is $372.75.

▶ **Do Your Turn 2**

Objective 2 Solve for the whole amount in percent problems.

Now let's consider problems in which the whole amount is unknown.

Example 3 Translate to an equation; then solve.

a. Tedra sells real estate and receives a commission of $5445.60. If her commission rate is 6%, what was the total sale?

Solution: We must find Tedra's total sale, which is the *whole*. We are given her commission rate (6%) and her commission ($5445.60), which is the *part*.

5445.60 is 6% of the total sale.

5445.60 is 6% of what number?

$$5445.60 = 0.06 \cdot w \quad \text{Translate to an equation.}$$
$$\frac{5445.60}{0.06} = \frac{0.06w}{0.06} \quad \text{Divide both sides by 0.06 to isolate } w.$$
$$90{,}760 = w$$

Answer: Tedra's total sale was $90,760.

b. A reporter states that 456 people in a survey indicated that they believed their schools were safe. The reporter went on to say that this was 48% of the respondents. How many respondents were involved in the survey?

Solution: We must find the total respondents in a survey, which is the *whole*. We are given that 456 people represent 48% of the total respondents.

456 is 48% of the total respondents.

456 is 48% of what number?

$$456 = 0.48 \cdot w \quad \text{Translate to an equation.}$$
$$\frac{456}{0.48} = \frac{0.48w}{0.48} \quad \text{Divide both sides by 0.48 to isolate } w.$$
$$950 = w$$

Answer: 950 respondents were involved in the survey.

▶ **Do Your Turn 3**

Objective 3 Calculate the percent.

Finally, we will solve problems in which the percent is unknown. Recall that if we use the proportion method to find the percent, the answer will be in percent form.

▶ **Your Turn 2**

Translate to an equation; then solve.

Brin sells scanning equipment. She earns 20% of the net profit in commission. If the net profit was $16,148, what was her commission?

▶ **Your Turn 3**

Translate to an equation; then solve.

a. Todd earned 20% commission on the sale of medical equipment. If Todd received $3857.79 as a commission, what was his total sale?

b. A power amplifier is operating at 80% of its full power capability. If its current output is 400 W (watts), what is its full power capability?

Answer to Your Turn 2

$0.2 \cdot 16{,}148 = n$ or $\dfrac{20}{100} = \dfrac{n}{16{,}148}$;

$3229.60

Answers to Your Turn 3

a. $3857.79 = 0.2 \cdot w$ or

$\dfrac{20}{100} = \dfrac{3857.79}{w}$; $19,288.95

b. $400 = 0.8 \cdot w$ or $\dfrac{80}{100} = \dfrac{400}{w}$;

500 W

▶ **Your Turn 4**

Translate to an equation; then solve.

a. Out of the 286 students taking foreign language at a certain school, 234 passed. What percent passed?

b. After four months of construction, 26 floors of the 55 floors in a skyscraper are complete. What percent remains to be completed?

Answers to Your Turn 4

a. $p = \dfrac{234}{286}$ or $\dfrac{P}{100} = \dfrac{234}{286}$;

81.$\overline{81}$%

b. $p = \dfrac{29}{55}$ or $\dfrac{P}{100} = \dfrac{29}{55}$; 52.$\overline{72}$%

Example 4 Translate to an equation; then solve.

a. Latasha answered 52 questions correctly on a test with a total of 60 questions. What percent of the questions did she answer correctly?

Solution: We must find the percent of the questions that Latasha answered correctly given that she answered 52 correctly (*part*) out of 60 total questions (*whole*).

What percent is 52 out of 60?

P% written as a ratio $\left\{ \dfrac{P}{100} = \dfrac{52}{60} \right.$ ← number correct (part)
← total questions (whole)   Use percent = $\dfrac{\text{part}}{\text{whole}}$.

$60P = 5200$ Equate the cross products.

$\dfrac{60P}{60} = \dfrac{5200}{60}$ Divide both sides by 60 to isolate P.

$P = 86.\overline{6}$

Answer: Latasha answered 86.$\overline{6}$% of the questions correctly. Her score on the test would most likely be rounded to 87%.

b. A manufacturing plant produces 485 units of its product in one week. If three units are found to be defective, what percent of the units produced are *without* defect?

Solution: We must find the percent of the units produced that are without defect given that 3 units are defective (*part*) out of 485 units produced (*whole*).

First, we need to find the number of units that are without defect by subtracting the number of defective units from the total number of units.

Number of units produced without defect = $485 - 3 = 482$

What percent is 482 out of 485?

P% written as a ratio $\left\{ \dfrac{P}{100} = \dfrac{482}{485} \right.$ ← units without defect (part)
← total units produced (whole)   Use percent = $\dfrac{\text{part}}{\text{whole}}$.

$485P = 48{,}200$ Equate the cross products.

$\dfrac{485P}{485} = \dfrac{48{,}200}{485}$ Divide both sides by 485 to isolate P.

$P \approx 99.4$

Answer: Approximately 99.4% of the products manufactured in a week are without defect.

◀ **Do Your Turn 4**

8.3 Exercises For Extra Help MyMathLab®

Objective 1

Prep Exercise 1 What is the general approach to solving problems involving percents?

Prep Exercise 2 Identify the percent, whole amount, and part.

20% of the students in a class received an A on their first test. If there are 24 students in the class, how many received an A?

Percent: _____

Whole amount: _____

Part: _____

For Exercises 1–10, translate to an equation and solve.

1. 60% of an 800-milliliter mixture is HCl (hydrochloric acid). How many milliliters of the solution are HCl?

2. The label on a bottle of rubbing alcohol indicates that it is 70% isopropyl alcohol. If the bottle contains 473 milliliters, how many milliliters of isopropyl alcohol does it contain?

Of Interest

Here's a case where 1 + 1 does *not* equal 2. Measure 1 cup of water and mix with 1 cup of rubbing alcohol. This will not equal exactly 2 cups of liquid. The reason is that the water and alcohol molecules are different sizes and shapes, allowing the molecules to *squeeze* together so that the total volume is less than 2 cups. The purer the alcohol, the more evident this effect.

3. Sabrina scored 90% on a test with 30 total questions.
   a. How many questions did she answer correctly?

   b. What percent did she answer incorrectly?
   c. How many questions did she answer incorrectly?

4. Willis scored 85% on a test with 40 questions.
   a. How many questions did he answer correctly?

   b. What percent did he answer incorrectly?
   c. How many questions did he answer incorrectly?

5. Wright earns 10% of total sales in commission. If he sells $3106 in merchandise over a two-week period, what is his commission?

6. Kera is a salesperson at a car dealership and earns 25% of the profit in commission. If in one month the dealership made a profit of $9680 from her sales, how much will she receive as commission?

7. Cush sells cars and earns 25% of the profit in commission. In one month, the dealership grossed $78,950. If the total cost to the company was $62,100, what was Cush's commission?

8. A writer earns 15% of the publishing company's net profit on the sale of her book in royalties (commission). If the company sells 13,000 copies of the book and makes a profit of $6 on each book, what is the writer's royalty?

9. The following graph shows where the money spent on a gallon of gas went in 2009.

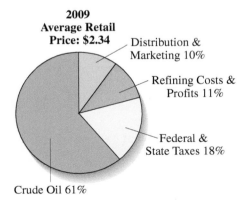

2009 Average Retail Price: $2.34

Distribution & Marketing 10%

Refining Costs & Profits 11%

Federal & State Taxes 18%

Crude Oil 61%

What We Pay For in a Gallon of Regular Grade Gasoline
(*Source*: U.S. Energy Information Administration. http://www.eia.doe.gov/oog/info/gdu/gasdiesel.asp)

Complete the following table showing how much of the price of a gallon of gas goes toward each expense.

Expense	Percent	Equation	Amount spent
Taxes			
Distribution & Marketing			
Refining			
Crude Oil			

10. The following graph shows where the money spent on a gallon of gas went in 2010.

**What We Pay For in a Gallon of
Regular Gasoline (July 2010)
Retail Price: $2.73/gallon**

Taxes 15%

Distribution &
Marketing 10%

Refining 9%

Crude Oil 66%

(*Source*: U.S. Energy Information Administration. http://www.eia.doe.gov/oog/info/gdu/gasdiesel.asp)

Complete the following table showing how much of the price of a gallon of gas goes toward each expense.

Expense	Percent	Equation	Amount spent
Taxes			
Distribution & Marketing			
Refining			
Crude Oil			

Objective 2

Prep Exercise 3 Identify the percent, whole amount, and part.

Gary earns a commission rate of 12%. If he earned $4800 in commission, what were his total sales?
Percent: _____
Whole amount: _____
Part: _____

For Exercises 11–16, translate to an equation and solve.

11. Rosita sells real estate and receives a commission of $750. If her commission rate is 6% of the seller's net, what is the seller's net?

12. Andre earns 20% of the net profit in commission on the sale of cash registers. If he receives $2675.92 as a commission, what is the net profit on his sales?

13. A basketball team in one game made 42% of the total number of attempted shots. If the players made a total of 38 baskets, how many shot attempts did they have in the game?

14. A football quarterback in one game completes 34% of his pass attempts. If he completed 14 passes, how many pass attempts did he have in the game?

15. A company spends $245,000 on advertising during the first quarter of the year. If this represents 49% of the company's advertising budget for the year, what is the company's advertising budget for the year?

16. $5,710,900 of the budget at a school comes from the state. If this represents 65% of the school's total budget, what is the school's total budget?

Objective 3

Prep Exercise 4 Identify the percent, whole amount, and part.

In a survey, 4000 people were asked if they agreed or disagreed with a legislative decision. 2400 agreed with the decision. What percent agreed?
Percent: _____
Whole amount: _____
Part: _____

17. Hunter answered 48 questions correctly out of a total of 54 questions. What percent of the total questions did Hunter answer correctly?

18. Corrina answered 41 questions correctly out of a total of 45 questions. What percent of the total questions did she answer correctly?

19. A baseball player has 36 hits in a season where he was at bat 110 times. What percent of his times at bat did he get a hit?

20. In one season, a football kicker scores 33 extra points in 35 attempts. What percent of his attempts did he score?

Of Interest

In baseball, a player's batting average is the ratio of the number of hits to the number of times at bat. It is expressed as a decimal number to the nearest thousandth. The batting average can be expressed as a percent by multiplying by 100. For example, a player with a batting average of 0.345 gets a hit 34.5% (a little more than $\frac{1}{3}$) of the times he goes to bat.

21. Suppose in a 24-hour period a person spends 8 hours sleeping, 7 hours working, 1 hour commuting, and 1 hour eating. What percent of the day is left?

22. If Mozart composed his first music at the age of 4 and continued composing and performing until he died at the age of 35, what percent of his life was spent composing music?

Of Interest

Wolfgang Amadeus Mozart (1756–1791) is considered to be one of the greatest classical composers. By the age of 6, he was an accomplished performer and had already composed several pieces of high quality. It is said that he completed his compositions in his head and then merely copied the completed work to paper with few or no corrections. He composed nearly 600 pieces in his lifetime and performed throughout Europe for royalty and common citizens.

23. According to a report by the Nielsen Company, the average American spends 31.5 hours per week watching television. What percent of the week does the average American spend watching television? (*Source*: Nielsen 2010 Media Industry Fact Sheet.)

24. According to a report by the Nielsen Company, kids aged 6–11 watch 28 hours of live television per week. What percent of the week do kids aged 6–11 spend watching live television? (*Source*: Nielsen 2010 Media Industry Fact Sheet.)

25. Below are the results of a survey in which respondents could agree, disagree, or say that they have no opinion.

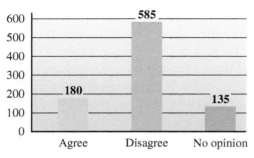

a. Complete the following table.

	Agreed	Disagreed	No opinion
Percent of Respondents			

b. Draw lines in the circle below to create a circle graph with the appropriate percentages.

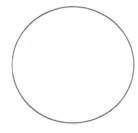

c. What percent of all respondents either agreed or disagreed?
d. What percent of all respondents either disagreed or said that they have no opinion?

26. Use the following graph to answer the questions.

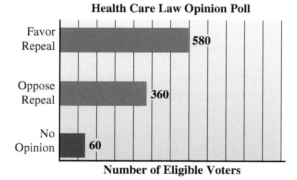

(*Source*: Rasmussen Reports, August 2010.)

a. Complete the following table.

	In favor of repeal	Oppose repeal	No opinion
Percent of those polled			

b. Draw lines in the circle below to create a circle graph with the appropriate percentages.

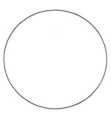

c. What percent of all polled either favor or oppose repeal?
d. What percent of all polled either oppose repeal or have no opinion?

27. The following table contains the Smith family's net monthly income and expenses.

Income	Expenses
Mr. Smith: $3480.25	Mortgage: $1580
Mrs. Smith: $1879.75	Car Loan 1: $420
	Car Loan 2: $345
	Credit Card Payments: $280
	Utilities: $540
	Groceries: $600

a. Complete the following table with the percent of the total monthly income that each expense represents. (Round each percent to one decimal place.)

	Mortgage	Car loan 1	Car loan 2	Credit card payments	Utilities	Groceries
Percent of total monthly income						

b. What is the total percent of the family's income paid toward expenses?
c. What percent of the family's income is left after all expenses are paid?
d. Draw a circle graph showing the percent of the total monthly income that each expense represents, along with the percent remaining after all expenses are paid.

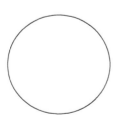

28. The following table contains the Jones family's net monthly income and expenses.

Income	Expenses
Mr. Jones: $2252.70	Mortgage: $1680
Mrs. Jones: $2597.30	Car Loan 1: $380
	Car Loan 2: $465
	Credit Card Payments: $220
	Utilities: $485
	Groceries: $450

a. Complete the following table with the percent of the total monthly income that each expense represents. (Round each percent to one decimal place.)

	Mortgage	Car loan 1	Car loan 2	Credit card payments	Utilities	Groceries
Percent of total monthly income						

b. What is the total percent of the family's income paid toward expenses?

c. What percent of the family's income is left after all expenses are paid?

d. Draw a circle graph showing the percent of the total monthly income that each expense represents, along with the percent remaining after all expenses are paid.

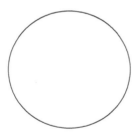

For Exercises 29–40, translate to an equation and solve. (mixed)

29. The nutrition label on a cereal bar shows that one bar has 3.5 grams of total fat and that this is 6% of the recommended daily value. Based on the label, how many grams of fat are recommended?

30. The nutrition label on a jar of peanut butter shows that a serving has 3 grams of saturated fat and that this represents 16% of the recommended daily value. Based on the label, how many grams of saturated fat are recommended?

31. According to the Center for Disease Control, 35% of emergency room visits are injury-related. If 119 million visits occurred one year, how many were injury-related? (*Source:* Center for Disease Control and Prevention.)

32. According to the Bureau of Justice Statistics, in 2009 93% of America's inmates were men. If there were about 1.6 million inmates in America, how many were men? (*Source:* Bureau of Justice Statistics.)

33. The sun is a sphere with a radius of about 695,990 kilometers. The core of the sun is a sphere with a radius of about 170,000 kilometers. What percent of the sun's total volume is made up of the core?

Of Interest

The sun is so large that if Earth were placed at its center, the moon would be about halfway to the surface of the sun.

34. In a 1860-square-foot house, the living room is 16 feet wide by 20 feet long. What percent of the total area of the house is the living room area?

35. Liz earns 8% of total sales in commission and $6 per hour. If she works a total of 60 hours in two weeks and her total sales are $4249, what will her gross pay be for the two-week period?

36. Vastine earns 10% of total sales in commission and $5 per hour. If he works a total of 80 hours over two weeks and has total sales of $5827, what is his gross pay for the two-week period?

37. Of the world's 6.9 billion people, about 1.4 billion live without electricity. What percent of the world's population live without electricity? (*Source: World Energy Outlook, 2010.*)

38. The 2011 federal budget has $0.9 trillion allocated for the Department of Health and Human Services. If the total budget was $3.8 trillion, what percent of the total budget went toward the Department of Health and Human Services? (*Source*: Budget of the United States Government, Fiscal Year 2011.)

39. The Ramos family has $45,000 in home equity. If this represents 18% of the value of the home, how much is the home worth?

40. After two months of fund-raising, an organization raised 72% of its goal. If the current amount raised is $61,200, what is the goal amount?

Puzzle Problem There are approximately 2×10^{11} stars in our galaxy. Suppose 10% of those stars have planets orbiting them. Suppose 1% of those planetary systems have a planet with a climate that could support life. Suppose that 1% of those planets that could support life actually have living organisms. Suppose that 1% of those planets with life have intelligent life. Suppose that 1% of the intelligent life have developed civilizations with communication devices. Suppose that 1% of those civilizations with communication devices are within a 500 light-year radius from Earth. Based on these suppositions, how many stars might we expect to have planets with intelligent life that have communication devices within a 500 light-year radius?

Review Exercises

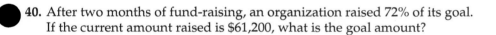

[6.2] 1. Combine like terms. $5.1x + 9.8 + x - 12.4$

For Exercises 2–4, solve and check.

[6.6] 2. $14.7 = x + 0.05x$

[6.6] 3. $24.48 = y - 0.2y$

[7.2] 4. $\dfrac{m}{2.4} = \dfrac{5.6}{8.4}$

[1.2] 5. Amelia receives a raise so that her new salary is $28,350. If her former salary was $25,800, what was the amount of the raise?

8.4 Solving Problems Involving Percent of Increase or Decrease

Objectives

1 Use a percent of increase to find an increase amount and the final amount after the increase.

2 Use a percent of decrease to find a decrease amount and the final amount after the decrease.

3 Find an unknown initial amount in a percent of increase or decrease problem.

4 Find a percent of increase or decrease.

Connection To calculate the amount of increase or decrease mentally when the percent is a multiple of 5%, calculate the 10% amount, then adjust by an appropriate factor. Because 10% is 0.1, when we multiply a decimal number by 10%, the decimal point moves one place to the left.

> 10% of 35 is 3.5 (Move the decimal point left one place in 35.)

Because 5% is half of 10%, the 5% amount is half of the 10% amount. Because 10% of 35 is 3.5, 5% of 35 will be half of 3.5, which is 1.75.

We could also calculate percents that are multiples of 10%. For example, 20% is twice the 10% amount. Because 10% of 35 is 3.5, 20% is twice 3.5, which is 7.

Answers to Warm-up

1. $0.2 \cdot 420 = n; \dfrac{20}{100} = \dfrac{n}{420};$

84 employees

2. $p \cdot 150 = 48; \dfrac{P}{100} = \dfrac{48}{150}; 32\%$

Warm-up *Translate to an equation; then solve.*

[8.3] 1. In one week, 20% of a company's employees were out with flu. If the company has 420 employees, how many were out with flu that week?

[8.3] 2. A theater has 150 seats. If 48 people attend a show, what percent of the seats are occupied?

In this section, we consider situations where, after calculating the part, we add it to or subtract it from the whole to arrive at a final amount. For example, with sales tax, we add the part (sales tax) to the whole (initial price) to find a final price. The percent used to find the sales tax is a percent of increase. With discount, we subtract the part (discount) from the whole (initial price) to find a final price. The percent used to find the discount is a percent of decrease. For these problems, it is helpful to rephrase the basic percent sentence and think of the whole as an *initial amount* and the part as an *increase or decrease amount*.

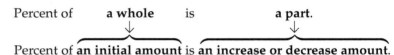

Percent of **an initial amount** is **an increase or decrease amount**.

The sentence is the same whether the situation involves a percent of increase or decrease. However, after calculating the increase or decrease amount, we either add or subtract.

Objective 1 Use a percent of increase to find an increase amount and the final amount after the increase.

First, we will consider situations such as salary raises and sales tax that use a percent of increase.

> #### Procedure
>
> Given a percent of increase and an initial amount, to find the final amount after the increase:
>
> 1. Find the increase amount.
> 2. Add the increase amount to the initial amount.

Example 1

a. A software package costs $35. If the sales tax rate is 5%, find the sales tax and the total amount of the purchase.

Solution: Use the *percent*, 5%, and the *initial amount*, $35, to find the sales tax (the *increase amount*.) Then add the sales tax to the initial price to find the final price.

A **percent** of **an initial amount** is **an increase amount**.

5%	of	35		is what number?
0.05	•	35		= n
			1.75	= n

Note Because the increase amount (part) is unknown, we use word-for-word translation to an equation.

Write 5% as 0.05, *of* as • , *is* as =, and *what number* as n.

Multiply.

Answer: Sales tax = $1.75

Final price = $35 + $1.75 = $36.75   Add the sales tax to the initial price.

Alternative solution: Consider the fact that we pay the entire price of the item, which is 100% of the price, plus tax. So if we pay 100% of $35 plus 5% of $35 in tax, the total is 105% of $35.

The final price is 105% of $35.

final price = 1.05(35)   Translate to an equation.

final price = 36.75   Multiply.

Look at the multiplication process to see why this works.

$$
\begin{array}{r}
35 \\
\times\ 1.05 \\
\hline
175 \\
000 \\
+\ 35 \\
\hline
36.75
\end{array}
$$

Multiplying 35 by the 5 digit in ◄ 1.05 places the tax in the sum.

Multiplying 35 by the digit 1 in the ones place of 1.05 places the initial price in the sum. ▶

The tax and initial price are added within the multiplication pro-◄ cedure to end up with the total amount.

Conclusion: In general, to calculate the total amount given the initial amount and the percent of increase, we can multiply the initial amount by the sum of 100% and the percent of increase.

b. Carlita's current annual salary is $23,500. If she gets a 4% raise, what is her new salary?

Solution: Let's use the alternative solution we learned in Example 1a. If the 4% raise is added to 100% of her current salary to get her new salary, the new salary is 104% of the current salary (4% + 100% = 104%).

Carlita's new salary is 104% of her current salary, $23,500.

$S = 1.04 \cdot 23{,}500$ Translate to an equation.

$S = 24{,}440$ Multiply.

▲

Note The variable is often chosen to be the first letter of the word that describes the unknown in the problem. In this case, we chose S to represent *salary*.

Answer: Carlita's new salary is $24,440.

▶ **Do Your Turn 1**

Objective 2 Use a percent of decrease to find a decrease amount and the final amount after the decrease.

Now let's consider situations such as pay deductions and discount that use a percent of decrease.

Procedure

Given a percent of decrease and an initial amount, to find the final amount after the decrease:

1. Find the decrease amount.
2. Subtract the decrease amount from the initial amount.

▶ **Your Turn 1**

Solve.

a. A suit is priced at $128.95. Find the sales tax and total amount if the tax rate is 6%.

b. Benjamin has a current salary of $22,600. If he receives a 3% raise, what is his new salary?

Note If we need the amount of the raise, we can subtract the initial salary from the new salary.

raise
◄ amount = $24,440 − $23,500
        = $940

Answers to Your Turn 1
a. $7.74; $136.69
b. $23,278

Solve.

a. A coat with an initial price of $88.95 is marked 40% off. Find the discount amount and final price.

b. Sandra's total deductions come to about 31% of her gross pay. If her gross pay is $1337.50, what is her net pay?

Example 2   Solve.

a. A 20% discount is to be applied to a sweater with an initial price of $49.95. What are the discount amount and final price?

Solution: Use the *percent*, 20%, and the *initial amount*, $49.95, to find the discount (the *decrease amount*). Then subtract the discount from the initial price to find the final price.

A **percent** of **an initial amount** is **a decrease amount**.

20%	of	49.95	is what number?	Write 20% as 0.2, *of* as $\cdot$, *is* as $=$, and *what number* as n.

$$0.2 \cdot 49.95 = n$$
$$9.99 = n \qquad \text{Multiply.}$$

Answer: Discount = $9.99;

Final price = $49.95 − $9.99 = $39.96   Subtract the discount from the initial price.

Alternative solution: Consider the fact that with discounts, we are subtracting the discount amount from the initial price. In our example, we subtracted 20% of the initial price from 100% of the initial price so that the final price is 80% of the initial price.

The final price is 80% of the initial price.

final price = $0.8 \cdot 49.95$ Translate to an equation.
final price = 39.96   Multiply.

Conclusion: In general, to calculate the final amount given the initial amount and the percent of decrease, we can multiply the initial amount by a percent found by subtracting the percent of decrease from 100%.

b. Approximately 28% of Deanne's gross monthly salary is deducted for Social Security, federal withholdings, state withholdings, and insurance. If Deanne's gross monthly salary is $2111.67, what is her net pay after the deductions?

Solution: Let's use the alternative solution we learned in Example 2a. If 28% of her gross monthly salary is deducted from her gross monthly salary, her net pay will be 72% of her gross monthly salary (100% − 28% = 72%).

Deanne's net pay is 72% of her gross monthly salary, $2111.67.

$$n = 0.72 \cdot 2111.67 \qquad \text{Translate to an equation.}$$
$$n \approx 1520.40 \qquad \text{Multiply.}$$

Answer: Deanne's net pay is $1520.40.

◀ **Do Your Turn 2**

Note If we need the amount deducted, we can subtract the net pay from the gross montly salary.

amount deducted = $2111.67 − $1520.40
= $591.27

▶

Objective 3 Find an unknown initial amount in a percent of increase or decrease problem.

Now let's consider percent of increase or decrease problems in which the initial amount is unknown. We must look carefully at the given information as there are two cases to consider.

Case 1: Given the percent of increase/decrease and the amount of increase/decrease

Answers to Your Turn 2
a. $35.58; $53.37   **b.** $922.88

Example 3 Solve.

a. Carmine receives a raise of $1950, which is a 6% increase to his salary. What was his former salary?

Solution:

A **percent** of **an initial amount** is **an increase amount**.

6% of the former salary is the raise amount, $1950.

$$0.06 \cdot f = 1950 \quad \text{Translate to an equation.}$$
$$\frac{0.06f}{0.06} = \frac{1950}{0.06} \quad \text{Divide both sides by 0.06 to isolate } f.$$
$$f = 32{,}500$$

Answer: Carmine's former salary was $32,500.

b. If 32% of Darlene's gross monthly salary results in total deductions of $918.40, what is her gross monthly salary?

Solution:

A **percent** of **an initial amount** is **a decrease amount**.

32% of her gross monthly salary is total deductions, $918.40.

$$0.32 \cdot g = 918.4 \quad \text{Translate to an equation.}$$
$$\frac{0.32g}{0.32} = \frac{918.4}{0.32} \quad \text{Divide both sides by 0.32 to isolate } g.$$
$$g = 2870$$

Answer: Darlene's gross monthly salary is $2870.

▶ **Do Your Turn 3**

Case 2: Given the percent of increase/decrease and the final amount after the increase/decrease

Example 4

a. After an 8% raise, Angela's new salary is $45,360. What was her former salary?

Solution: Angela's new salary is her former salary plus the raise amount, which is 8% of her former salary.

new salary = former salary + raise amount

new salary = former salary + 8% · former salary

Note The expression $f + 0.08f$ is the same as $1f + 0.08f$; so when we combine those two like terms, we add the coefficients 1 and 0.08 to get $1.08f$.

$$45{,}360 = f + 0.08f \quad \text{Translate to an equation with } f \text{ representing her former salary.}$$
$$45{,}360 = 1.08f \quad \text{Combine like terms.}$$
$$\frac{45{,}360}{1.08} = \frac{1.08f}{1.08} \quad \text{Divide both sides by 1.08 to isolate } f.$$
$$42{,}000 = f$$

Answer: Angela's former salary was $42,000.

b. After deductions, Rodney's net pay is $2516. If the total of the deductions was 26% of his gross monthly salary, what was his gross monthly salary?

Solution: Rodney's net pay is his gross monthly salary minus the total deductions, which are 26% of his gross monthly salary.

▶ **Your Turn 3**

Solve.

a. Julia indicates that she received a 5% raise in the amount of $2260. What was her former salary?

b. Clayton purchased a suitcase on sale for 20% off. If the discount amount was $37.09, what was the initial price?

Connection The equation in Example 4a agrees with our alternative method for finding the final amount given the initial amount and the percent of increase. The equation $45{,}360 = f + 0.08f$ indicates that 45,360 is the sum of 100% of the former salary and 8% of the former salary. After combining like terms, we have the equation $45{,}360 = 1.08f$, which indicates that 45,360 is 108% of the former salary.

Answers to Your Turn 3
a. $45,200   b. $185.45

Connection The equation in Example 4b agrees with our alternative method for finding the final amount given an initial amount and a percent of decrease. The equation $2516 = g - 0.26g$ indicates that 2516 is the amount left after subtracting 26% of the gross monthly salary from 100% of the gross monthly salary. After combining like terms, we have the equation $2516 = 0.74g$, which indicates that 2516 is 74% of Rodney's gross monthly salary.

▶ **Your Turn 4**

a. Josh recalls paying a total of $16.75 for a music CD. He knows that the sales tax rate in his state is 5%. What was the initial price of the CD?

b. Clarece purchased a robe discounted 25%. If the price after the discount was $59.96, what was the initial price?

---

net pay = gross monthly salary − total deductions

net pay = gross monthly salary − 26% · gross monthly salary

Note The expression $g - 0.26g$ is the same as $1g - 0.26g$; so when we combine those two like terms, we subtract the coefficients 1 and 0.26 to get $0.74g$.

▶ $2516 = g - 0.26g$ Translate to an equation with g representing his gross monthly salary.

$2516 = 0.74g$ Combine like terms.

$\dfrac{2516}{0.74} = \dfrac{0.74g}{0.74}$ Divide both sides by 0.74 to isolate g.

$3400 = g$

Answer: Rodney's gross monthly salary was $3400.

◀ **Do Your Turn 4**

Objective 4 Find a percent of increase or decrease.

To find the percent of increase or decrease, we need the amount of the increase or decrease and the initial amount. Because we are finding a percent, we prefer the proportion method.

Procedure

To find a percent of increase or decrease:

1. Find the amount of the increase or decrease.
   a. For an increase situation, subtract the initial amount from the final amount.
   b. For a decrease situation, subtract the final amount from the initial amount.
2. Translate to an equation with an unknown percent. For the proportion method, use the following form:

$$\frac{P}{100} = \frac{\text{increase or decrease amount}}{\text{initial amount}}$$

In Example 5, the amount of the increase or decrease is given; so step 1 of the procedure will not be needed.

Example 5 Solve.

a. Tiera is visiting a different state and purchases a shirt that is priced $24.95. She notes that the tax on her receipt is $2.25. What is the sales tax rate?

Solution: The *increase amount* $2.25 is an unknown percent of the *initial amount*, $24.95.

A **percent** of **an initial amount** is **an increase amount**.

What percent of   24.95   is   2.25?

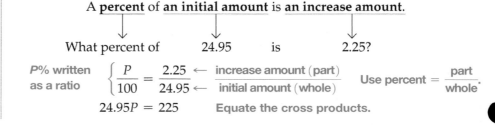

P% written as a ratio $\left\{ \dfrac{P}{100} = \dfrac{2.25}{24.95} \right.$ ← increase amount (part)   Use percent $= \dfrac{\text{part}}{\text{whole}}$.
         ← initial amount (whole)

$24.95P = 225$ Equate the cross products.

Answers to Your Turn 4
a. $15.95   b. $79.95

$$\frac{24.95P}{24.95} = \frac{225}{24.95}$$ Divide both sides by 24.95.

$$P \approx 9$$

Answer: The tax rate is 9%.   Attach a percent sign to the result.

b. By using coupons, a family reduced grocery expenses by $45 one month. If the family spent $600 the previous month, what was the percent of decrease?

Solution: The *decrease amount* $45 is an unknown percent of the *initial amount* $600.

A **percent** of **an initial amount** is **a decrease amount**.

What percent of   600   is   45?

P% written as a ratio $\begin{cases} \dfrac{P}{100} = \dfrac{45}{600} \end{cases}$ ← decrease amount (part) ← initial amount (whole)   Use percent = $\dfrac{part}{whole}$.

$$600P = 4500$$ Equate the cross products.

$$\frac{600P}{600} = \frac{4500}{600}$$ Divide both sides by 600.

$$P = 7.5$$

Answer: The percent of decrease was 7.5%.   Attach a percent sign to the result.

▶ **Do Your Turn 5**

In Example 6, the initial amount and final amount are given; so step 1 of the procedure is needed to find the amount of increase or decrease.

Example 6 Solve.

a. Jerod was making $12.50 per hour. He is given a raise so that now he is making $14.00 per hour. What was the percent of increase?

Solution: We know the *initial amount*, $12.50, and the *final amount*, $14.00, after the increase. But to find the percent of the increase, we need the increase amount, which we find by subtracting.

increase amount = $14.00 − $12.50 = $1.50   Subtract the initial amount from the final amount.

Now we can find the percent of increase just as we did in Example 5.

A **percent** of **an initial amount** is **an increase amount**.

What percent of   12.50   is   1.50?

P% written as a ratio $\begin{cases} \dfrac{P}{100} = \dfrac{1.50}{12.50} \end{cases}$ ← increase amount (part) ← initial amount (whole)   Use percent = $\dfrac{part}{whole}$.

$$12.50P = 150$$ Equate the cross products.

$$\frac{12.50P}{12.50} = \frac{150}{12.50}$$ Divide both sides by 12.50.

$$P = 12$$

Answer: The percent of increase was 12%.   Attach a percent sign to the result.

b. Andrew purchases a stock at $56.75 per share. Two months later he sees that the stock price is down to $48.50 per share. What is the percent of decrease in price?

Note Most likely the actual tax rate is exactly 9%, and when the tax amount was calculated, it was rounded to the nearest cent.

▶ **Your Turn 5**
Solve.
a. Marvin received a raise of $1728. If his former salary was $28,800, what was the percent of increase?
b. A salesperson says that she can take $1280 off the price of a car. If the initial price is $25,600, what is the percent of decrease?

Answers to Your Turn 5
a. 6%   b. 5%

▶ **Your Turn 6**

Solve.

a. A college increases tuition from $4800 to $6000 per semester. What is the percent of increase in tuition?

b. Marianne purchases a small rug on sale. The initial price was $185.79. The price after the discount was $157.92. What was the percent of decrease (discount rate)?

Solution: We know the *initial amount*, $56.75, and the *final amount*, $48.50, after the decrease. But to find the percent of the decrease, we need the decrease amount, which we find by subtracting.

decrease amount = $56.75 - 48.50 = 8.25$ Subtract the final amount from the initial amount.

Now we can find the percent of the decrease.

A **percent** of **an initial amount** is **a decrease amount.**
What percent of   56.75   is   8.25?

P% written as a ratio $\begin{cases} \dfrac{P}{100} = \dfrac{8.25}{56.75} \leftarrow \text{decrease amount (part)} \\ \phantom{\dfrac{P}{100} = \dfrac{8.25}{56.75}} \leftarrow \text{initial amount (whole)} \end{cases}$ Use percent $= \dfrac{\text{part}}{\text{whole}}$.

$56.75P = 8.25$ Equate the cross products.

$\dfrac{56.75P}{56.75} = \dfrac{8.25}{56.75}$ Divide both sides by 56.75.

$P \approx 14.5$

Answer: The percent of decrease is about 14.5%.   Attach a percent sign to the result.

Answers to Your Turn 6
a. 25%   **b.** 15%

◀ **Do Your Turn 6**

8.4 Exercises   For Extra Help   MyMathLab®

Objective 1

Prep Exercise 1 Identify the percent, initial amount, and amount of increase or decrease.

Sharon's current salary is $34,000. If she receives a 3% raise, how much will the raise be?

Percent: _____
Initial amount: _____
Amount of increase or decrease: _____

Prep Exercise 2 Given a percent of increase and an initial amount, explain how to find the final amount after the increase.

For Exercises 1–36, solve.

1. A desk is priced at $185.95. The sales tax rate is 5%. Find the sales tax and the total amount of the purchase.

2. A guitar is priced at $585.95. Find the sales tax and the total amount of the purchase if the tax rate is 6%.

Of Interest

Many states have a law that sets the maximum sales tax that can be charged on a purchase. It may be beneficial to shop around in different states for large purchases.

3. The bill in a restaurant comes to a total of $48.75. The patrons decide to tip the server 20%. What tip amount should be left, and how much do they pay including the tip?

4. The bill in a restaurant comes to a total of $32.85. The patrons decide to tip the server 15%. What tip amount should be left, and how much do they pay including the tip?

Connection When tipping, most people round to whole dollar amounts. A convenient scale for 15% is to recognize that every $10 of the bill corresponds to a tip of $1.50. The rest of the scale would be as follows:

Total Bill	Tip
$5	$0.75
$10	$1.50
$20	$3.00

In our example of a total bill of $48.75, we would round to $50. $50 is two $20s and one $10. Each $20 is a $3 tip and the $10 is a $1.50 tip for a total tip of $7.50.

$$\text{Total bill} = \$20 + \$20 + \$10 = \$50$$
$$\downarrow \qquad \downarrow \qquad \downarrow \qquad \downarrow$$
$$\text{Tip} = \$3 + \$3 + \$1.50 = \$7.50$$

5. Michele's current annual salary is $29,600. If she gets a 3.5% raise, what is her new salary?

6. Lou's current annual salary is $23,800. If he receives a 6.5% raise, what is his new salary?

Objective 2

Prep Exercise 3 Given a percent of decrease and an initial amount, explain how to find the final amount after the decrease.

7. A 30% discount is to be applied to a dress with an initial price of $65.99. What are the discount amount and final price?

8. A sweater with an initial price of $42.95 is marked 25% off. What are the discount amount and final price?

9. Approximately 27% of Sheila's gross monthly income is deducted for Social Security, federal withholdings, state withholdings, and insurance. If her gross monthly salary is $2708.33, what is her net pay?

10. Approximately 29% of Natron's gross monthly income is deducted for Social Security, federal withholdings, state withholdings, and insurance. If his gross monthly salary is $2487.50, what is his net pay?

11. In July, a company's total expenses were $12,250. In August, the company cut costs by 4%. How much were the costs reduced? What amount did the company spend in August?

12. A large company has 4200 employees. As a result of a declining economy, the company must downsize and let 6% of its employees go. How many employees are let go? How many are left?

13. Jack received a 4% raise that amounts to $1008. What was his former annual salary?

14. Lucia received a 5.5% raise in the amount of $1793. What was her former salary?

15. Rachel paid a total of $45.10 for a pair of shoes. She knows that the sales tax rate in her state is 5%. What was the initial price of the shoes?

16. Paul paid $635.95 for a new stereo system. If the sales tax rate in his state is 6%, what was the initial price?

Objective 3

Prep Exercise 4 Identify the percent, initial amount, and amount of increase or decrease.

A 25% discount is applied to the price of an appliance, resulting in a discount of $80. What was the original price of the appliance?

Percent: _____
Initial amount: _____
Amount of increase or decrease: _____

17. If 26% of Gordon's gross monthly salary results in total deductions of $674.96, what is his gross monthly salary?

18. Marcelle's retirement deduction from her monthly salary is $187.50. If the retirement deduction is 6% of her gross monthly salary, what is her gross monthly salary?

19. Patricia purchased a table on sale for 30% off. If the discount amount was $44.09, what was the initial price?

20. A lamp is on sale for 15% off. If the discount amount is $13.49, what is the initial price?

Objective 4

Prep Exercise 5 One month a business earns $40,000 in revenue. The next month the business increases the revenue $5000.

a. Is $40,000 an initial amount, a final amount, or an increase amount?

b. Is $5000 an initial amount, a final amount, or an increase amount?

c. With the information given, to find the percent of increase, do you need to subtract prior to setting up the proportion? Why or why not?

d. Write a proportion that can be used to find the percent of increase.

21. Melliah purchases a tennis racquet that is priced $54.95. She notes that the tax on her receipt is $3.85. What is the sales tax rate?

22. The property tax on a vehicle is $286.45. If the vehicle is assessed at a value of $8500, what is the tax rate?

23. Tina received a raise of $849.10. If her former salary was $24,260, what was the percent of the increase?

24. Jana notes that her July electricity bill is $8.96 higher than last year's amount. If last year's bill was $148.75, what is the percent of increase?

25. A salesperson says that he can take $1512 off the price of a car. If the initial price is $18,900, what is the percent of decrease?

26. A company's cost of production for one unit of a product was $328.50. After taking some cost-cutting measures, the company was able to reduce the cost of production by $14.50 per unit. What is the percent of decrease in cost?

Prep Exercise 6 By exercising and changing his diet, Marcus drops from 220 pounds to 180 pounds.

a. Is 220 an initial amount, a final amount, or an decrease amount?

b. Is 180 an initial amount, a final amount, or an decrease amount?

c. With the information given, to find the percent of decrease, do you need to subtract prior to setting up the proportion? Why or why not?

d. Write a proportion that can be used to find the percent of decrease.

27. Lena was making $8.50 per hour. She is given a raise so that now she is making $10.00 per hour. What was the percent of increase?

28. Gloria notes that her water bill was $18.50 in March. Her April water bill is $32.40. What is the percent of increase?

29. Van purchases a DVD player on sale. The initial price was $175.90. The price after the discount was $149.52. What was the percent of decrease (discount rate)?

30. Risa's gross two-week salary is $1756.25. If her net pay is $1334.75, what is the percent of decrease?

For Exercises 31–36, use the following graph, which shows the number of new homes sold in the United States each month in 2009.

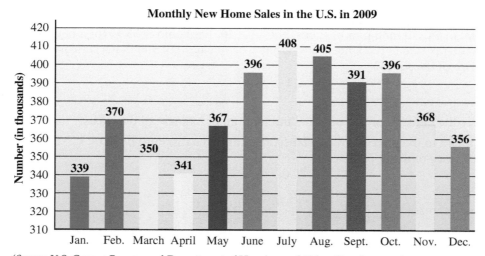

Monthly New Home Sales in the U.S. in 2009

(*Source*: U.S. Census Bureau and Department of Housing and Urban Development)

31. What was the percent of increase in the number of new homes sold from January to February?

32. What was the percent of increase in the number of new homes sold from April to May?

33. What was the percent of decrease in the number of new homes sold from March to April?

34. What was the percent of decrease in the number of new homes sold from August to September?

35. What was the percent of decrease in the number of new homes sold from February to April?

36. What was the percent of decrease in the number of new homes sold from July to December?

For Exercises 37–42, use the following graph, which shows the average price for regular gasoline at the beginning of January of each year.

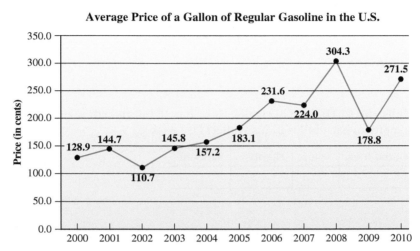

Average Price of a Gallon of Regular Gasoline in the U.S.

(*Source*: U.S. Department of Energy, Energy Information Administration)

37. What was the percent of decrease in the average price of a gallon of gasoline from January 2001 to January 2002?

38. What was the percent of decrease in the average price of a gallon of gasoline from January 2008 to January 2009?

39. What was the percent of increase in the average price of a gallon of gasoline from January 2007 to January 2008?

40. What was the percent of increase in the average price of a gallon of gasoline from January 2009 to January 2010?

41. What was the percent of increase in the average price of a gallon of gasoline from January 2002 to January 2008?

42. What was the percent of increase in the average price of a gallon of gasoline from January 2000 to January 2010?

For Exercises 43–56, solve. (Mixed)

43. A sofa with an initial price of $699.99 is marked 35% off. What are the discount amount and final price?

44. A necklace with an initial price of $349.90 is marked 45% off. What are the discount amount and final price?

45. A social worker worked a total of 1259 cases in one year. The next year he worked 1540 cases. What was the percent of increase in caseload?

46. At the end of the first year of operation for a business, the business recorded a profit of $15,280. Two years later the business recorded a profit of $80,450. What was the percent of increase in profit?

47. A car is priced at $26,450. If the sales tax rate is 6%, calculate the sales tax and the total amount of the purchase.

48. A computer is priced at $1998.95. Find the sales tax and total amount of the purchase if the tax rate is 7%.

49. After deductions, Luther's net pay is $1850.40. If the total of the deductions was 28% of his gross monthly salary, what was his gross monthly salary?

50. After deductions, Wendy's net pay is $1016.60. If the total of the deductions was 32% of her grosstwo-week salary, what was her gross two-week salary?

51. By insulating their water heater, adding a storm door, and installing new double-pane windows, the Goto family decreased their electric and gas bills from a total of $1536.98 in one year to $1020.57 in the next year. What was the percent of decrease in the Goto family's electric and gas charges?

52. Glenda is an office manager. Through her suggestions for improving efficiency and cutting excessive spending, the cost of operation dropped from $65,886.35 in one year to $60,245.50 in the next year. What was the percent of decrease in operation cost?

53. Huang received a 4% raise. His new salary is $30,420. What was his former salary?

54. Denzel received a 9% raise. His new salary is $39,828.60. What was his former salary?

55. Hannah purchased a curio stand discounted 35%. If the price after the discount was $246.68, what was the initial price?

56. Vladimir purchased a set of golf clubs discounted 15%. If the price after the discount was $577.96, what was the initial price?

Review Exercises

For Exercises 1 and 2, simplify.

[5.7] **1.** $\left(\dfrac{1}{5}\right)^3$

[6.5] **2.** $(1 + 0.08)^2$

For Exercises 3 and 4, solve and check.

[4.3] **3.** $20 = 9(y + 2) - 3y$

[6.4] **4.** $0.12x = 450$

[4.5] **5.** Will has some $5 bills and $10 bills in his wallet. If he has 16 bills worth a total of $105, how many of each bill is in his wallet? Use a table.

Categories	Value	Number	Amount

8.5 Solving Problems Involving Interest

Warm-up *[8.4]* *Solve.*

1. Kenny purchases a coat for $49.95. If the sales tax rate is 7%, find the sales tax and the final price.
2. Yolanda's current salary is $56,000. If she receives a 2.5% raise, what is her new salary?

Objective 1 Solve problems involving simple interest.

In this section, we explore a type of percent increase situation in which the percent is called an **interest rate**, the initial amount is called the **principal**, and the increase amount is called **interest**.

Definitions **Interest rate:** A percent used to calculate interest.

Principal: An initial amount of money invested or borrowed.

Interest: An amount of money that is a percent of the principal.

In general, a principal amount of money is deposited or borrowed and an interest rate is used to calculate the interest that is added to the principal after a certain amount of time. We will learn two ways to calculate interest. First, we will consider **simple interest**.

Definition **Simple interest:** Interest calculated using only the interest rate, the original principal, and the amount of time in which the principal earns interest.

To derive the formula for simple interest, we can use dimensional analysis. Interest rates are *annual percentage rates*, which means that they determine the amount of interest per year. So if $100 is invested in an account earning 5% simple interest, 5% of $100 is added to the account each year as interest.

$$\text{Simple interest after 1 year} = \$100 \cdot \frac{0.05}{1 \text{ yr.}} \cdot 1 \text{ yr.} = \$5$$

$$\text{Simple interest after 2 years} = \$100 \cdot \frac{0.05}{1 \text{ yr.}} \cdot 2 \text{ yr.} = \$10$$

The dimensional analysis indicates that to calculate simple interest, we multiply the original principal by the interest rate and the time in years.

Rule

To calculate simple interest, use the following formula:

$$I = Prt$$

I represents the simple interest.
P represents the principal.
r represents the simple interest rate.
t represents the time in years.

Because interest is an increase amount, after the interest is calculated, we find the final balance by adding the interest to the principal.

Objectives

1 Solve problems involving simple interest.

2 Solve problems involving compound interest.

3 Solve problems involving amortization.

Note The interest rate means 5% per year, or 0.05/yr. Multiplying the interest rate by the time in years causes the unit year to divide out so that the result is in dollars.

Answers to Warm-up
1. sales tax = $3.50;
   final price = $53.45
2. $57,400

Procedure

To find the final balance after calculating simple interest, add the interest to the principal.

$$\text{Final balance} = \text{principal} + \text{interest}$$

▶ **Your Turn 1**

Find the simple interest and the final balance.

a. $640 at 8% simple interest rate for 1 year

b. $4500 at 12% simple interest for 1 quarter ($\frac{1}{4}$ of a year)

c. $900 at 5.5% simple interest for 30 days

Example 1   Find the simple interest and final balance.

a. $800 at 3% for 1 year

Solution:

$I = Prt$

$I = (800)(0.03)(1)$ Replace P with 800, r with 0.03, and t with 1.

$I = 24$ Multiply.

Answer: Interest = $24

Final balance = $800 + $24 = $824   Add the interest to the principal.

b. $1200 at 4.5% for 5 years

Solution:

$I = Prt$

$I = (1200)(0.045)(5)$ Replace P with 1200, r with 0.045, and t with 5.

$I = 270$ Multiply.

Answer: Interest = $270

Final balance = $1200 + $270 = $1470   Add the interest to the principal.

c. $2000 at 15% for 90 days

Solution: The time must be in years; so we convert 90 days to years.

$$90 \text{ days} = \frac{\overset{18}{\cancel{90 \text{ days}}}}{1} \cdot \frac{1 \text{ yr.}}{\underset{73}{\cancel{365 \text{ days}}}} = \frac{18}{73}\text{yr.}$$ Convert 90 days to years.

Now find the simple interest.

$I = Prt$

$I = (2000)(0.15)\left(\dfrac{18}{73}\right)$ Substitute for $P, r,$ and t.
Remember that $15\% = 0.15$.

$I = \dfrac{2000}{1} \cdot \dfrac{0.15}{1} \cdot \dfrac{18}{73}$ Write 2000 and 0.15 over 1.

$I = \dfrac{5400}{73}$ Multiply.

$I \approx 73.97$ Divide and round to the nearest hundredth.

Note We round to the nearest hundredth because interest is money.

Answer: Interest = $73.97

Final balance = $2000 + $73.97 = $2073.97   Add the interest to the principal.

◀ **Do Your Turn 1**

Answers to Your Turn 1
a. $51.20; $691.20
b. $135; $4635
c. $4.07; $904.07

What if the principal is unknown? Like the problems involving percent of increase, our approach will depend on the given information. If we are given the percent, interest rate, and time, we can use $I = Prt$ and solve for P.

Example 2   Sandy notes that $6.80 was deposited in her savings account after 3 months. The account earns 5.5% simple interest. What was the principal?

Solution: The time must be in years; so we convert 3 months to years.

$$3 \text{ mo.} = \frac{\overset{1}{\cancel{3 \text{ mo}}}}{1} \cdot \frac{1 \text{ year}}{\underset{4}{\cancel{12 \text{ mo}}}} = \frac{1}{4}\text{yr.} \qquad \text{Convert 3 months to years.}$$

Because $6.80 is simple interest, we use the simple interest formula $I = Prt$.

$$I = Prt$$

$$6.80 = P(0.055)\left(\frac{1}{4}\right) \qquad \text{Replace } I \text{ with 6.80, } r \text{ with 0.055, and } t \text{ with } \frac{1}{4}.$$

$$6.80 = P\left(\frac{0.055}{1} \cdot \frac{1}{4}\right) \qquad \text{Write 0.055 over 1.}$$

$$6.80 = P\left(\frac{0.055}{4}\right) \qquad \text{Multiply.}$$

$$6.80 = 0.01375P \qquad \text{Divide 0.055 by 4.}$$

$$\frac{6.80}{0.01375} = \frac{0.01375P}{0.01375} \qquad \text{Divide both sides by 0.01375 to isolate } P.$$

$$494.55 \approx P$$

Answer: The principal was $494.55.

◄ **Note** We round to the nearest hundredth because principal is money.

▶ **Do Your Turn 2**

Now suppose the final balance is given instead of the interest. Because the final balance is a combination of principal and interest, we can write the following equation:

$$\text{Final balance} = \text{principal} + \text{interest}$$

Because interest is calculated by the relationship Prt, we can write this equation as follows:

$$B = P + Prt$$

We can rewrite this formula by factoring out P.

$$B = P(1 + rt)$$

Example 3   Pedro notes that his new balance at the end of the year is $1441.60. If the account earns 6% simple interest, what was the principal?

Solution: Because $1441.60 is a final balance, we use the final balance formula $B = P(1 + rt)$.

$$B = P(1 + rt)$$

$$1441.60 = P[1 + 0.06(1)] \qquad \text{Replace } B \text{ with 1441.60, } r \text{ with 0.06, and } t \text{ with 1.}$$

$$1441.60 = 1.06P \qquad \text{In the bracket, multiply and then add.}$$

$$\frac{1441.60}{1.06} = \frac{1.06P}{1.06} \qquad \text{Divide both sides by 1.06 to isolate } P.$$

$$1360 = P$$

Answer: The principal was $1360.

▶ **Do Your Turn 3**

Objective 2 Solve problems involving compound interest.

Where simple interest is calculated based on the original principal only, **compound interest** uses the principal *and* prior earned interest.

▶ **Your Turn 2**

Solve.

a. Ingrid notes that her savings account earned $42.32 in 1 year. If the account earns 6.2% simple interest, what was her principal?

b. Lawrence notes that his account earned $59.50 in half a year. If the account earns 14% simple interest, what was his principal?

Connection Example 3 is the same type of problem as Example 4a in Section 8.4. Consider the meaning of the equation $1441.60 = 1.06P$. Because 1.06 is 106%, this equation is saying that 1441.60 is 106% of the principal. This should make sense because 1441.60 is 100% (all) of the principal combined with the interest, where the interest is 6% of the principal.

▶ **Your Turn 3**

Solve.

a. The total amount that Nala must repay on a 1-year loan with 8% simple interest is $12,960. What was the principal?

b. Ron notes that his new balance at the end of 6 months is $1221. If the account earns 4% simple interest, what was the principal?

Answers to Your Turn 2
a. $682.58   b. $850

Answers to Your Turn 3
a. $12,000   b. $1197.06

Definition **Compound interest:** Interest calculated using the principal and prior earned interest.

To *compound* interest means that after interest is calculated and added to the principal, the process is repeated with the new balance. When interest is compounded annually, the final balance at the end of each year is used to find the final balance for the next year. For example, let's find the final balance if $5000 is invested at 4% compounded annually for 2 years.

First Year: $B = P(1 + rt)$

$B = (5000)[1 + (0.04)(1)]$ Replace P with 5000, r with 0.04, and t with 1.

$B = 5000[1.04]$ Simplify in the brackets.

$B = \$5200$ Multiply.

Second year: $B = 5200[1.04]$ Replace P with the first year's balance, 5200.

$B = \$5408$ Multiply.

Notice that we repeadedly multiply by 1.04; so the following expression describes the entire calculation.

$B = 5000\underbrace{(1.04)(1.04)}$

$B = 5000(1.04)^2$ ◀ **Note** Because 1.04 is used as a factor twice, we could use exponential form. The exponent corresponds to the number of times the interest is compounded.

$B = \$5408$

Now suppose the interest is compounded more frequently, such as *semiannually*, which means every six months (half of a year). Because 4% is an annual percentage rate, it must be converted to a semiannual rate.

$$\text{Semiannual rate} = \frac{4\%}{1 \text{ year}} \cdot \frac{1}{2}\text{year} = \frac{4\%}{2} = 2\%$$

This means that every six months, 2% of the principal is added to the account. Also, every six months means that the interest is compounded twice per year; so in two years, 4 compound periods occur.

$B = 5000\underbrace{(1.02)(1.02)(1.02)(1.02)}$

$B = 5000(1.02)^4$ ◀ **Note** We divide the interest rate by the number of compounds per year. Also, the exponent is the total number of compound periods. It is found by multiplying the time by the number of compounds per year.

$B \approx \$5412.16$

Our exploration suggests the following procedure for the finding the final balance if interest is compounded.

Procedure

To find the final balance if interest is compounded, use the following formula:

$$B = P\left(1 + \frac{r}{n}\right)^{nt}$$

B represents the final balance.
P represents the principal.
r represents the annual percentage rate.
n represents the number of compound periods per year.
t represent the time in years.

Note Common compound periods and corresponding values of n:

◀ **Annually:** once per year, $n = 1$
Semiannually: twice per year, $n = 2$
Quarterly: four times per year, $n = 4$
Daily: 365 times per year, $n = 365$

Because the final balance is the principal and interest combined, if we needed to know the interest, we could subtract the principal from the final balance.

$$\text{Interest} = \text{final balance} - \text{principal}$$
$$= \$5412.16 \quad - \$5000 = \$412.16$$

Procedure

To find the interest if the principal and the final balance are known, subtract the principal from the final balance.

$$\text{Interest} = \text{final balance} - \text{principal}$$

Example 4 Find the final balance and interest.

a. $15,000 at 12% compounded annually for 4 years

Solution: $B = P\left(1 + \dfrac{r}{n}\right)^{nt}$

$B = 15{,}000\left[1 + \dfrac{0.12}{1}\right]^{1\cdot4}$ — Replace P with 15,000, r with 0.12, n with 1, and t with 4.

$B = 15{,}000[1 + 0.12]^4$ — Divide 0.12 by 1 and multiply 4 by 1.

$B = 15{,}000[1.12]^4$ — Add 1 and 0.12.

$B \approx \$23{,}602.79$ — Evaluate the exponential expression; then multiply.

$\text{Interest} = \$23{,}602.79 - \$15{,}000 = \$8602.79$ — Subtract the principal from the final balance.

b. $15,000 at 12% compounded semiannually for 4 years

Solution: $B = P\left(1 + \dfrac{r}{n}\right)^{nt}$

$B = 15{,}000\left[1 + \dfrac{0.12}{2}\right]^{2\cdot4}$ — Replace P with 15,000, r with 0.12, n with 2, and t with 4.

$B = 15{,}000[1 + 0.06]^8$ — Divide 0.12 by 2 and multiply $2\cdot4$.

$B = 15{,}000[1.06]^8$ — Add 1 and 0.06.

$B \approx \$23{,}907.72$ — Evaluate the exponential expression; then multiply.

$\text{Interest} = \$23{,}907.72 - \$15{,}000 = \$8907.72$ — Subtract the principal from the final balance.

▶ **Do Your Turn 4**

Example 5 Paula has a balance of $685.72 on a credit card that has a 14.9% interest rate. The interest with this particular card is compounded daily. If she does not charge any money to this account until after she receives her next bill (30 days), what will her balance be on the next bill?

Understand We must calculate the final balance given the principal, the interest rate, and the fact that the interest is compounded daily for 30 days. Compounding daily means that the interest is compounded 365 times per year; so $n = 365$.

▶ **Your Turn 4**

Find the final balance and interest.

a. $12,000 at 14% compounded annually for 2 years

b. $12,000 at 14% compounded semiannually for 2 years

c. $8000 at 8% compounded semiannually for 3 years

d. $8000 at 8% compounded quarterly for 3 years

Connection Comparing Example 4b with 4a, we see that if the same principal is invested at the same rate but compounded more frequently, more interest is earned.

Answers to Your Turn 4
a. $B = \$15{,}595.20; I = \$3595.20$
b. $B = \$15{,}729.55; I = \$3729.55$
c. $B = \$10{,}122.55; I = \$2122.55$
d. $B = \$10{,}145.93; I = \$2145.93$

Marla has a balance of $426.85 on a credit card that has a 12.9% interest rate. The interest with this particular card is compounded daily. If she does not charge any money to this account until after she receives her next bill (30 days), what will her balance be on the next bill?

Plan Use the compound interest formula $B = P\left(1 + \dfrac{r}{n}\right)^{nt}$.

$P = \$685.72$

$r = 14.9\% = 0.149$

$n = 365$

$t = \dfrac{30}{365}$

◀

Note Because t must be in years, we must convert 30 days to years.

$$30 \text{ days} \cdot \frac{1 \text{ year}}{365 \text{ days}} = \frac{30}{365} \text{ years}$$

We did not simplify to lowest terms because we will use a calculator to calculate the final balance.

Execute $B = P\left(1 + \dfrac{r}{n}\right)^{nt}$

$B = 685.72\left[1 + \dfrac{0.149}{365}\right]^{365 \cdot \frac{30}{365}}$

Replace P with 685.72, r with 0.149, n with 365, and t with $\frac{30}{365}$.

$B = 685.72[1 + 0.000408219]^{30}$

Divide 0.149 by 365 and multiply $365 \cdot \frac{30}{365}$.

$B = 685.72[1.000408219]^{30}$

Add.

$B \approx 694.17$

Evaluate the exponential expression; then multiply.

Answer Paula's balance will be $694.17 on the next bill.

◀ **Do Your Turn 5**

Of Interest

Read the fine print of a credit card statement to find out how the interest is calculated for the card. Sometimes several methods are listed for the same type of card. The method used will be indicated somewhere on the statement. In addition to the daily compounding method used in Example 5, here are a few other methods.

Method 1: Apply the monthly percentage rate to the average daily balance. Applying this method to Example 5 gives the following:

$$\text{monthly percentage rate} = 14.9\% \div 12 = 1.242\% = 0.01242$$
$$\text{finance charge (interest)} = 685.72(0.01242) \approx \$8.52$$
$$\text{final balance} = 685.72 + 8.52 = \$694.24$$

Method 2: Apply the daily percentage rate to the average daily balance. Then multiply the result times the number of days in the billing cycle (simple interest). Applying this method to Example 5 gives the following:

$$\text{daily percentage rate} = 14.9\% \div 365 = 0.04082\% = 0.0004082$$
$$\text{finance charge} = 685.72(0.0004082)(30) \approx \$8.40 \quad \text{There were 30 days in the billing cycle.}$$

$$\text{final balance} = 685.72 + 8.40 = \$694.12$$

Objective 3 Solve problems involving amortization.

In Example 5, notice that the interest charged during the month was $8.45 ($694.17 − 685.72 = 8.45$). If Paula paid just $8.45, her balance would go back to the initial balance, $685.72, and the next month's balance would be back to $694.17. Thus, if all she does is pay the finance charge (interest), she will never pay off the debt.

To pay off a debt, we must pay more than the interest. Any amount that we pay in addition to the interest will decrease the principal. This will then decrease the amount of interest for the next payment.

For example, suppose we borrow $1000 at 12% and make a $200 payment each month. We will need to calculate the balance at the end of each month. Because the interest rate is 12%, the monthly rate will be 1%.

First month balance: $B = 1000\left[1 + 0.12\left(\frac{1}{12}\right)\right] = 1000(1.01) = \$1010$

Applying the $200 payment: $1010 - \$200 = \$810$

> **Note** $10 of the $200 payment is going toward interest while the other $190 is going toward the principal. In our next payment, we will owe less interest because the principal is now less.

Second month balance: $B = 810(1.01) = \$818.10$

Applying the $200 payment: $818.10 - \$200 = \$618.10$

> **Note** Now only $8.10 is interest; so the rest of the $200 payment, which amounts to $191.90, goes toward principal. In the next payment, we will pay even less interest and more of the $200 goes toward principal.

This process continues until we pay off the loan. The payments we make to pay off a loan are called *installments*. When we pay off a loan in installments, we **amortize** the loan.

Definition **Amortize:** To pay off a loan or debt in installments.

In amortizing a loan, we are interested in the payment we must make. The factors that affect this payment are the principal, percentage rate, number of payments per year, and agreed-upon time to pay off the loan.

Because the calculations can be rather involved, it is common to use amortization tables to determine the payment. Most bookstores carry amortization books that list tables by percentage rate. Table 8.1 (next page) shows the monthly payment for various principal amounts borrowed at a rate of 9% and paid off in 3, 4, 5, 15, or 30 years.

Notice that under the 30-year column and the row for 90,000, the payment is $724.16. Note that the table does not list every possible principal amount. However, we can use combinations of the listed principal values to determine the payment for a principal that is not listed.

Example 6 Use Table 8.1 to find the monthly payment for $72,300 at 9% for 15 years.

Understand We must determine the monthly payment using the amortization table. Because 72,300 is not listed, we must use combinations of principals that are listed. We can use any of the principals whose sum is 72,300.

Plan Locate principal amounts whose sum is 72,300, find each of their monthly payments in the 15-year column, and add all of the monthly payments to get the total monthly payment for 72,300.

Execute Write 72,300 as a sum of values in the amount column of Table 8.1.

$$72,300 = 70,000 + 2000 + 200 + 100$$

Now using the 15-year column, we find the monthly payment for each of our listed principals and add those monthly payments.

$$72,300 = 70,000 + 2000 + 200 + 100$$
$$709.99 + 20.29 + 2.03 + 1.01 = 733.32$$

Answer The monthly payment will be $733.32.

▶ **Do Your Turn 6**

▶ **Your Turn 6**

Use Table 8.1 to find the monthly payment.
a. $15,000 for 5 years
b. $125,000 for 30 years
c. $97,500 for 15 years

Answers to Your Turn 6
a. $311.38
b. $1005.78
c. $988.91

Table 8.1 9% APR Monthly Payment Schedule

Amount	3 Years	4 Years	5 Years	15 Years	30 Years
100	3.18	2.49	2.08	1.01	0.80
200	6.36	4.98	4.15	2.03	1.61
500	15.90	12.44	10.38	5.07	4.02
1,000	31.80	24.89	20.76	10.14	8.05
2,000	63.60	49.77	41.52	20.29	16.09
5,000	159.00	124.43	103.79	50.71	40.23
6,000	190.80	149.31	124.55	60.86	48.28
7,000	222.60	174.20	145.31	71.00	56.32
8,000	254.40	199.08	166.07	81.14	64.37
9,000	286.20	223.97	186.83	91.28	72.42
10,000	318.00	248.85	207.58	101.43	80.46
15,000	477.00	373.28	311.38	152.14	120.69
20,000	635.99	497.70	415.17	202.85	160.92
25,000	794.99	622.13	518.96	253.57	201.16
30,000	953.99	746.55	622.75	304.28	241.39
35,000	1112.99	870.98	726.54	354.99	281.62
36,000	1144.79	895.86	747.30	365.14	289.66
37,000	1176.59	920.75	768.06	375.28	297.71
38,000	1208.39	945.63	788.82	385.42	305.76
39,000	1240.19	970.52	809.58	395.56	313.80
40,000	1271.99	995.40	830.33	405.71	321.85
41,000	1303.79	1020.29	851.09	415.85	329.90
42,000	1335.59	1045.17	871.85	425.99	337.94
43,000	1367.39	1070.06	892.61	436.13	345.99
44,000	1399.19	1094.94	913.37	446.28	354.03
45,000	1430.99	1119.83	934.13	456.42	362.08
46,000	1462.79	1144.71	954.88	466.56	370.13
47,000	1494.59	1169.60	975.64	476.71	378.17
48,000	1526.39	1194.48	996.40	486.85	386.22
49,000	1558.19	1219.37	1017.16	496.99	394.27
50,000	1589.99	1244.25	1037.92	507.13	402.31
51,000	1621.79	1269.14	1058.68	517.28	410.36
52,000	1653.59	1294.02	1079.43	527.42	418.40
53,000	1685.39	1318.91	1100.19	537.56	426.45
54,000	1717.19	1343.79	1120.95	547.70	434.50
55,000	1748.99	1368.68	1141.71	557.85	442.54
56,000	1780.79	1393.56	1162.47	567.99	450.59
57,000	1812.58	1418.45	1183.23	578.13	458.63
58,000	1844.38	1443.33	1203.98	588.27	466.68
59,000	1876.18	1468.22	1224.74	598.42	474.73
60,000	1907.98	1493.10	1245.50	608.56	482.77
65,000	2066.98	1617.53	1349.29	659.27	523.00
70,000	2225.98	1741.95	1453.08	709.99	563.24
75,000	2384.98	1866.38	1556.88	760.70	603.47
80,000	2543.98	1990.80	1660.67	811.41	643.70
85,000	2702.98	2115.23	1764.46	862.13	683.93
90,000	2861.98	2239.65	1868.25	912.84	724.16
95,000	3020.97	2364.08	1972.04	963.55	764.39
100,000	3179.97	2488.50	2075.84	1014.27	804.62
105,000	3338.97	2612.93	2179.63	1064.98	844.85
110,000	3497.97	2737.35	2283.42	1115.69	885.08
120,000	3815.97	2986.21	2491.00	1217.12	965.55
130,000	4133.97	3235.06	2698.59	1318.55	1046.01
140,000	4451.96	3483.91	2906.17	1419.97	1126.47
150,000	4769.96	3732.76	3113.75	1521.40	1206.93
175,000	5564.95	4354.88	3632.71	1774.97	1408.09
200,000	6359.95	4977.01	4151.67	2028.53	1609.25
225,000	7154.94	5599.13	4670.63	2282.10	1810.40
250,000	7949.93	6221.26	5189.59	2535.67	2011.56

8.5 Exercises   For Extra Help   MyMathLab®

Objective 1

Prep Exercise 1 What is principal?

Prep Exercise 2 What is interest?

Prep Exercise 3 What formula is used to calculate simple interest? What does each variable represent?

Prep Exercise 4 After calculating simple interest, how do you determine the final balance?

For Exercises 1–10, calculate the simple interest and final balance.

1. $4000 at 3% for 1 year

2. $500 at 4% for 1 year

3. $350 at 5.5% for 2 years

4. $1200 at 6.5% for 3 years

5. $12,250 at 12.9% for 5 years

6. $18,000 at 14.5% for 4 years

7. $2400 at 8% for six months $\left(\dfrac{1}{2}\text{year}\right)$

8. $840 at 12% for 3 months $\left(\dfrac{1}{4}\text{year}\right)$

9. $2000 at 6.9% for 60 days

10. $600 at 4.9% for 30 days

For Exercises 11–14, solve.

11. Nuno notes that his savings account earned $331.12 in 1 year. If the account earns 6.9% simple interest, what was his principal?

12. Sharice notes that her savings account earned $21.28 in 1 year. If the account earns 5.5% simple interest, what was her principal?

13. Kelly notes that her savings account earned $15.57 in 3 months. If the simple interest rate for her account is 3.2%, what was her principal?

14. Roger notes that his savings account earned $24.56 in 9 months. If the simple interest rate for his account is 4.8%, what was his principal?

Objective 2

Prep Exercise 5 What formula is used to calculate the final balance when interest is compounded? What does each variable represent?

Prep Exercise 6 Suppose interest is compounded semiannually. What is the value of n in the compound interest formula?

Prep Exercise 7 After using the compound interest formula to find a final balance, how do you determine the interest?

For Exercises 15–30, find the final balance and interest.

15. $5000 at 8% compounded annually for 2 years

16. $6400 at 12% compounded annually for 2 years

17. $840 at 6% compounded annually for 2 years

18. $1450 at 10% compounded annually for 2 years

19. $400 at 9% compounded annually for 3 years

20. $650 at 5% compounded annually for 3 years

21. $14,000 at 8% compounded semiannually for 1 year

22. $20,000 at 10% compounded semiannually for 1 year

23. $1600 at 6% compounded semiannually for 3 years

24. $960 at 14% compounded semiannually for 3 years

25. $290 at 9% compounded semiannually for 4 years

26. $500 at 11% compounded semiannually for 5 years

27. $1300 at 12% compounded quarterly for 2 years

28. $900 at 8% compounded quarterly for 3 years

29. $450 at 6% compounded quarterly for 4 years

30. $2400 at 10% compounded quarterly for 4 years

For Exercises 31–34, solve.

31. Kevin has a balance of $860.20 on a credit card that has a 16.9% interest rate. The interest is compounded daily. If he does not charge any money to this account until after he receives his next bill (30 days), what will his balance be on the next bill?

32. Phyllis has a balance of $285.32 on a credit card that has a 7.9% interest rate. The interest is compounded daily. If she does not charge any money to this account until after she receives her next bill (30 days), what will her balance be on the next bill?

33. Lindsey has a balance of $694.75 on a credit card that has a 19.8% interest rate. The interest is compounded daily. If she does not charge any money to this account until after she receives her next bill (30 days), what will her balance be on the next bill?

34. Scott has a balance of $1248.56 on a credit card that has a 21.9% interest rate. The interest is compounded daily. If he does not charge any money to this account until after he receives his next bill (30 days), what will his balance be on the next bill?

Objective 3

Prep Exercise 8   What does amortize mean?

For Exercises 35–44, use Table 8.1 to find the monthly payment.

35. $7000 for 5 years

36. $9000 for 3 years

37. $17,000 for 4 years

38. $21,000 for 5 years

39. $92,500 for 30 years

40. $78,000 for 15 years

41. $84,100 for 15 years

42. $103,500 for 30 years

43. $115,200 for 30 years

44. $147,200 for 15 years

Review Exercises

[5.1] 1. Graph $-6\frac{3}{4}$ on a number line.

[6.4] 2. Write $\frac{5}{6}$ as a decimal number.

[6.4] 3. Simplify. $\dfrac{65 + 90 + 72 + 84}{4}$

[3.1] 4. Evaluate $2x + 3y$ when $x = -3$ and $y = 4$.

[5.4] 5. Solve and check. $\dfrac{3}{8}x = -\dfrac{21}{30}$

Chapter 8 Summary and Review Exercises

8.1 Introduction to Percent

Definitions/Rules/Procedures	Key Example(s)
Percent is a ratio representing some part out of _____. **To write $n\%$ as a fraction or a decimal number:** 1. Write $n\%$ as _____. 2. Simplify to the desired form.	Write each percent as a fraction in lowest terms. a. $45\% = \dfrac{45}{100} = \dfrac{9}{20}$ b. $16.2\% = \dfrac{16.2}{100} = \dfrac{16.2(10)}{100(10)} = \dfrac{162}{1000} = \dfrac{81}{500}$ c. $9\frac{1}{2}\% = \dfrac{9\frac{1}{2}}{100} = \dfrac{19}{2} \div 100 = \dfrac{19}{2}\cdot\dfrac{1}{100} = \dfrac{19}{200}$ Write each percent as a decimal number. a. $62\% = \dfrac{62}{100} = 0.62$ b. $4.5\% = \dfrac{4.5}{100} = 0.045$ c. $33\frac{1}{3}\% = 33.\overline{3}\% = \dfrac{33.\overline{3}}{100} = 0.\overline{3}$

[8.1] *For Exercises 1–4, write each percent as a fraction in lowest terms.*

1. 40%   **2.** 26%   **3.** 6.5%   **4.** $24\frac{1}{2}\%$

[8.1] *For Exercises 5–8, write each percent as a decimal number.*

5. 16%   **6.** 150%   **7.** 3.2%   **8.** $40\frac{1}{3}\%$

Definitions/Rules/Procedures	Key Example(s)
To write a fraction or a decimal number as a percent: 1. Multiply by _____. 2. Simplify.	Write each fraction as a percent. a. $\dfrac{3}{5} = \dfrac{3}{5}\cdot\dfrac{\overset{20}{100}}{1}\% = 60\%$ b. $\dfrac{2}{3} = \dfrac{2}{3}\cdot\dfrac{100}{1}\% = \dfrac{200}{3}\% = 66\frac{2}{3}\%$ Write each decimal number as a percent. a. $0.49 = (0.49)(100\%) = 49\%$ b. $0.067 = (0.067)(100\%) = 6.7\%$ c. $1.03 = (1.03)(100\%) = 103\%$

[8.1] *For Exercises 9–12, write each number as a percent.*

9. $\dfrac{3}{8}$ **10.** $\dfrac{4}{9}$ **11.** 0.54   **12.** 1.3

8.2 Solving Basic Percent Sentences

Definitions/Rules/Procedures	Key Example(s)
To translate a basic percent sentence to an equation, use one of the following methods: **Method 1**: Write the percent as a fraction or decimal number, *of* as _____, and *is* as a(n) _____. A **percent** of a **whole is** a **part**. percent · whole = part **Method 2**: Write a proportion in the following form: Percent = _____ **Note** Either method works in all cases. However, we suggest using Method 1 when the part or whole is unknown and Method 2 when the percent is unknown.	Translate to an equation and solve. **a.** 40% of 70 is what number? Translation: $40\% \cdot 70 = n$ Solution: $(0.4)(70) = n$ $28 = n$ **b.** 15% of what number is 60? Translation: $0.15w = 60$ Solution: $\dfrac{0.15w}{0.15} = \dfrac{60}{0.15}$ $w = 400$ **c.** What percent is 20 out of 30? (Or what percent is 20 of 30?) Translation: $\dfrac{P}{100} = \dfrac{20}{30}$ Solution: $30P = 2000$ $\dfrac{30P}{30} = \dfrac{2000}{30}$ $P = 66.\overline{6}\%$

[8.2] For Exercises 13–16, translate to an equation and solve.

13. What number is 15% of 90?

14. 12.8% of what number is 5.12?

15. 12.5 is what percent of 20?

16. What percent is 40 of 150?

8.3 Solving Percent Problems (Portions)

Definitions/Rules/Procedures	Key Example(s)
To solve application problems involving percents: 1. Identify the _____, _____, and _____, noting which is unknown. 2. Write a basic percent sentence (if needed). 3. Translate to an equation (word for word or proportion). 4. Solve for the unknown. **Commission** is the portion of _____ earnings that a salesperson receives.	Brianna earns 20% commission. One week her sales totaled $4500. Find her commission. **Solution:** The *percent* is 20%. The $4500 in total sales is the *whole*. The unknown commission is the *part*. A **percent** of a **whole** is a **part**. 20% of 4500 is what number? 0.2 · 4500 = n **Translate to an equation.** $900 = n$ **Multiply.** **Answer**: Brianna's commission is $900.

[8.3] For Exercises 17–20, translate to an equation and solve.

17. Use the graph at right, which shows the results of a survey of 680 people.

    a. How many people agreed?

    b. How many had no opinion?

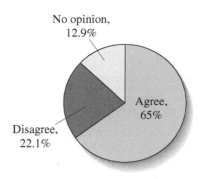

No opinion, 12.9%

Agree, 65%

Disagree, 22.1%

18. Trent earns 15% commission. One week his sales totaled $7800. Find his commission.

19. After traveling 180 miles, Boris completed only 35% of a trip. What was the total distance of the trip?

20. In a survey of 554 adults, 321 supported raising the minimum wage. What percent of those surveyed supported raising the minimum wage?

8.4 Solving Problems Involving Percent of Increase or Decrease

Definitions/Rules/Procedures	Key Example(s)
Given a percent of increase and an initial amount, to find the final amount after the increase: 1. Find the _____ amount. 2. Add the increase amount to the _____ amount.	A book costs $6.95. If the tax rate is 5%, find the sales tax and the final price. **Solution:** A **percent** of **an initial amount** is **an increase amount**. 5% of 6.95 is what number? $0.05 \cdot 6.95 = n$ Write 5% as 0.05, *of* as ·, *is* as =, and what number as *n*. $0.35 \approx n$ Multiply. Sales tax = $0.35 Final price = $6.95 + $0.35 = $7.30
Given a percent of decrease and an initial amount, to find the final amount after the decrease: 1. Find the _____ amount. 2. Subtract the decrease amount from the _____ amount.	A table is discounted 30%. If the original price was $275, what is the discount and the price after the discount? **Solution:** A **percent** of **an initial amount** is **a decrease amount**. 30% of 275 is what number? $0.3 \cdot 275 = n$ Write 30% as 0.3, *of* as ·, *is* as =, and what number as *n*. $82.5 = n$ Multiply. Discount = $82.50 Final price = $275 − $82.50 = $192.50

[8.4] For Exercises 21–24, translate to an equation and solve.

21. A home theater system is priced at $495.95. The sales tax rate is 6%. Calculate the sales tax and total amount of the purchase.

22. A dress is on sale for 25% off. If the initial price is $56.95, what is the price after the discount?

23. Kat received a 3.5% raise that amounted to $1102.50. What was her former salary?

24. Gaven lost 4% of his body weight so that he now weighs 168 pounds. What did he weigh before he lost weight?

Definitions/Rules/Procedures	Key Example(s)

To find a percent of increase or decrease:
 	
1. Find the amount of the increase or decrease.
  a. For an increase situation, subtract _____ _____.
  b. For a decrease situation, subtract _____ _____.
 	
2. Translate to an equation with an unknown percent. For the proportion method, use the following form:
 	
$$\frac{P}{100} = \frac{\text{increase or decrease amount}}{\text{initial amount}}$$
	Marcus notices that his electric bill increased from $120 in January of one year to $140 in January of the next year. What was the percent of the increase?
	
	Increase amount: $140 − $120 = $20
	
	Basic sentence: What percent of 120 is 20?
	
	P% written as a ratio $\left\{ \dfrac{P}{100} = \dfrac{20}{120} \right.$ ← increase amount
	← initial amount
	
	$$120P = 2000$$
	$$\frac{120P}{120} = \frac{2000}{120}$$
	$$P = 16.\overline{6}$$
	
	Answer: The percent of increase was $16.\overline{6}$%.

[8.4] For Exercises 25 and 26, translate to an equation and solve.

25. The Johnsons' power bill increased $27.68 from last month. If the bill was $86.50 last month, what was the percent of increase?

26. Dave's hourly wage is raised from $12.00 per hour to $15.50 per hour. What is the percent of increase?

8.5 Solving Problems Involving Interest

Definitions/Rules/Procedures	Key Example(s)
An **interest rate** is a(n) _____ used to calculate interest. **Principal** is a(n) _____ amount of money invested or borrowed. **Interest** is an amount of money that is a percent of the _____. **Simple interest** is interest calculated using only the interest rate, the _____, and the amount of time in which the principal earns interest. **To calculate simple interest, use the following formula:** $$I = Prt$$ I represents the simple _____. P represents the _____. r represents the simple interest _____. t represents the _____ in years.	Find the simple interest and final balance if $600 is invested at 4% for 9 months. $$I = Prt$$ $$I = (600)(0.04)\left(\frac{3}{4}\right)$$ $$I = 18$$ Interest = $18 Final balance = $600 + $18 = $618 Bill notes that his new balance at the end of the year is $1310.40. If the account earns 4% interest, what was the principal? $$B = P(1 + rt)$$ $$1310.40 = P[1 + 0.04(1)]$$ $$1310.40 = 1.04P$$ $$\frac{1310.40}{1.04} = \frac{1.04P}{1.04}$$ $$1260 = P$$ **Answer:** The principal was $1260.

Definitions/Rules/Procedures	Key Example(s)
To find the final balance after calculating simple interest, add the _____ to the _____. Final balance = _____ + _____	

[8.5] *For Exercises 27 and 28, find the simple interest and final balance.*

27. $480 at 6% for 1 year

28. $8000 at 5.2% for 3 months.

29. After 1 year at 8%, the simple interest added to a savings account was $62.80. What was the principal?

30. After 1 year at 4.5% simple interest, the balance in an account is $627. If no deposits or withdrawals were made during the year, what was the principal?

Definitions/Rules/Procedures	Key Example(s)
To find the final balance if interest is compounded, use the following formula: $$B = P\left(1 + \frac{r}{n}\right)^{nt}$$ B represents the _____. P represents the _____. r represents the _____. n represents the number of _____ per year. t represents the _____ in years. **To find the interest if the principal and the final balance are known, subtract the _____ from the _____.** Interest = _____ − _____	Find the final balance and interest if $5000 is invested at 8% compounded quarterly for 2 years. $$B = P\left(1 + \frac{r}{n}\right)^{nt}$$ $$B = 5000\left(1 + \frac{0.08}{4}\right)^{4\cdot 2}$$ $$B = 5000(1.02)^8$$ $$B \approx 5858.30$$ Final balance = $5858.30 Interest = $5858.30 − $5000 = $858.30

[8.5] *For Exercises 31 and 32, find the final balance and interest.*

31. $5000 at 8% compounded annually for 3 years

32. $1800 at 12.4% compounded quarterly for 1 year

Definitions/Rules/Procedures	Key Example(s)
To amortize a loan means to _____ a loan or debt in installments.	Use Table 8.1 to find the monthly payment if 92,500 is to be amortized at 9% APR over 30 years. 92,500 = 90,000 + 2000 + 500 ↓ ↓ ↓ 724.16 + 16.09 + 4.02 = 744.27 **Answer:** $744.27

[8.5] *For Exercises 33 and 34, use Table 8.1 to find the monthly payment.*

33. $90,500 for 15 years

34. $162,000 for 30 years

Chapter 8 Practice Test

For Extra Help

Step-by-step test solutions are found on the Chapter Test Prep Videos available via the Video Resources on DVD, in MyMathLab®, and on You Tube▪ (search "CarsonPrealgebra" and click on "Channels").

1. _____

2. _____

3. _____

4. _____

5. _____

6. _____

7. _____

8. _____

9. _____

10._____

11._____

12._____

13._____

[8.1] *For Exercises 1 and 2, write each percent as a fraction in lowest terms.*

1. 24%

2. $40\frac{3}{4}\%$

[8.1] *For Exercises 3 and 4, write each percent as a decimal number.*

3. 4.2%

4. $12\frac{1}{2}\%$

[8.1] *For Exercises 5–8, write each number as a percent.*

5. $\frac{2}{5}$

6. $\frac{5}{9}$

7. 0.26

8. 1.2

[8.2] *For Exercises 9–12, translate to an equation and solve.*

9. What number is 15% of 76?

10. 6.5% of what number is 8.32?

11. 14 is what percent of 60?

12. What percent is 12 of 32?

For Exercises 13–19, translate to an equation and solve.

[8.3]   **13.** Use the graph at right, which shows the grade distribution in a class of 50 people. How many people earned an A?

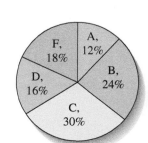

[8.3] **14.** Carolyn earns 25% of total sales in commission. If she sells a total of $1218 in merchandise in one week, what is her commission?

14. _____

[8.3] **15.** The Morgan family has a net monthly income of $2786.92. If the family has a $345.75 car payment, what percent of the net monthly income goes toward paying for the car?

15. _____

[8.4] **16.** A microwave oven is priced at $295.75. The sales tax rate is 5%. Calculate the sales tax and total amount of the purchase.

16. _____

[8.4] **17.** A pair of boots is on sale for 30% off. If the initial price of the boots is $84.95, what is the price after the discount?

17. _____

[8.4] **18.** Barbara received a 2.5% raise that amounted to $586.25. What was her former salary?

18. _____

[8.4] **19.** Andre's hourly wage is raised from $8.75 per hour to $10.50 per hour. What is the percent of increase?

19. _____

[8.5] *For Exercises 20 and 21, find the simple interest and final balance.*

20. $5000 at 6% for 1 year

20. _____

21. $800 at 4% for 6 months

21. _____

[8.5] **22.** After 1 year at 6% simple interest rate, the balance in an account is $2544. If no deposits or withdrawals were made during the year, what was the principal?

22. _____

[8.5] *For Exercises 23 and 24, find the final balance and interest.*

23. $2000 at 9% compounded annually for 3 years

23. _____

24. $1200 at 4% compounded quarterly for 1 year

24. _____

25. A family borrows $112,000 at 9% APR to be amortized over 30 years. Use Table 8.1 to find the monthly payment.

25. _____

Chapters 1-8 Cumulative Review Exercises

For Exercises 1–4, answer true or false.

[2.4] **1.** There are two solutions for $(?)^2 = 81$: $+9$ and -9.

[3.5] **2.** 91 is a composite number.

[6.4] **3.** π is an irrational number.

[8.1] **4.** $9\% > 0.1$

[5.6] **5.** Explain how to add fractions.

[6.6] **6.** Explain how to clear decimal numbers from an equation.

[1.3] **7.** Estimate $7805 \cdot 246$ by rounding.

[3.2] **8.** What is the degree of $-5y$?

[5.5] **9.** Find the LCM of $24y^2$ and $30x$.

[6.1] **10.** Graph -5.6 on a number line.

For Exercises 11–14, simplify.

[2.5] **11.** $\left[12 + 4(6 - 8)\right] - (-2)^3 + \sqrt{100 - 36}$

[5.7] **12.** $10\frac{5}{8} \div \left(-\frac{3}{16}\right) - \frac{1}{2}$

[6.5] **13.** $4.86 \div 0.8 + 58.9$

[6.5] **14.** $\frac{3}{4} - (0.6)^2$

[5.4] **15.** Simplify. $\frac{10m}{9n^2} \div \frac{5}{12n}$

[5.6] **16.** Simplify. $\frac{3}{5} - \frac{y}{4}$

$\begin{bmatrix} 5.7 \\ 6.2 \end{bmatrix}$ **17.** Subtract.

$$\left(x^4 - \frac{1}{3}x^2 + 9.6\right) - \left(4x^3 + \frac{3}{5}x^2 - 14.6\right)$$

[6.3] **18.** Multiply. $(6.2x - 1)(4x + 5)$

[8.1] **19.** Write $12\frac{1}{2}\%$ as a decimal number.

[8.1] **20.** Write 0.71 as a percent.

For Exercises 21 and 22, solve and check.

[5.8] **21.** $\frac{5}{8}y - 3 = \frac{3}{4}$

[6.6] **22.** $3.5x - 12.1 = 6.8x + 3.08$

[7.3] **23.** Convert 9.5 pounds to ounces.

[7.4] **24.** Convert 80 milliliters to liters.

For Exercises 25–30, solve.

[1.5] **25.** Use the following graph to find the mean, median, and mode income of the actors listed. Write the answers in standard form.

Highest Paid Actors of 2010
(in millions)

[bar chart]

Johnny Depp $75
Sandra Bullock $56
Ben Stiller $53
Tom Hanks $45
Adam Sandler $40
Reese Witherspoon $32
Cameron Diaz $32
Leonardo DiCaprio $28
Jennifer Aniston $27
Daniel Radcliffe $25

(*Source*: Forbes.com)

[3.4] **26.** Write an expression in simplest form for the area of the rectangle shown. Calculate the area if x is 3 inches.

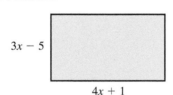

3x − 5

4x + 1

[4.5] **27.** The sum of two integers is 123. One integer is three more than twice the other integer. What are the integers?

[6.5] **28.** Find the area of the shaded region. Use 3.14 for π.

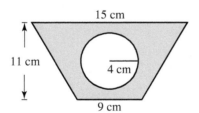

15 cm

11 cm

4 cm

9 cm

[7.2] **29.** The following triangles are similar. Find the unknown side lengths.

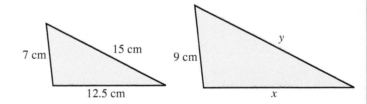

7 cm   15 cm

12.5 cm

9 cm y

x

[8.4] **30.** Of the 450 people in attendance at a retirement seminar, 384 are over 50. What percent of the people in attendance are over 50?

9

More with Geometry and Graphs

Chapter Overview

In Chapter 9, additional topics in geometry will be explored. After formally defining some of the basic terminology, we will explore angles more extensively. Then we will add to what we have learned about plotting points on a number line and explore coordinate geometry.

9.1 Points, Lines, and Angles

Warm-up *[4.3] Solve.*
1. $n + 3n - 20 = 180$
2. $4a + 36 = 7a - 90$

Objective 1 Identify points, lines, line segments, rays, planes, and parallel lines.

In this section, we will explore more concepts in geometry. First, we need to define some terms.

Definitions **Point:** A position in space having no length or width.

Line: A straight, one-dimensional figure extending forever in both directions.

Line segment: A straight, one-dimensional figure extending between two end points.

Ray: A straight, one-dimensional figure extending forever in one direction from a single point.

Plane: A flat, two-dimensional surface extending forever in both dimensions.

Parallel lines: Lines lying in the same plane that do not intersect.

Usually, the geometrical figures defined above are named using letters as shown in Example 1.

Objectives

1 Identify points, lines, line segments, rays, planes, and parallel lines.

2 Identify angles.

3 Solve problems involving angles formed by intersecting lines.

4 Solve problems involving angles in a triangle.

5 Solve problems involving congruent triangles.

Example 1 Identify the figure.

a. $A \cdot$

Answer: point A

b.

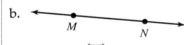

Answer: line $\overleftrightarrow{MN}$

Note The two points, M and N, fix the line in place and are used to name the line. A line can also be named using a lower-case letter, as in part f below.

c.

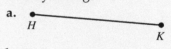

Answer: ray $\overrightarrow{XY}$

d.

Answer: line segment $\overline{AB}$

e.

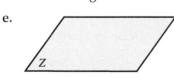

Answer: plane Z

Note When drawing a plane, we draw a parallelogram to give the impression of a flat surface. Although we have drawn boundaries, it is assumed that the plane has no boundaries.

f.

Answer: parallel lines l and m; to indicate that they are parallel, we can write $l \| m$.

▶ **Do Your Turn 1**

▶ **Your Turn 1**

Identify the figure.

a.

H

K

b. $T \cdot$

c.

p

n

Answers to Your Turn 1
a. line segment $\overline{HK}$
b. point T c. $p \| n$

Answers to Warm-up
1. $n = 50$ 2. $a = 42$

Objective 2 Identify angles.

In Section 4.5, we introduced angles. In this section, we will explore angles in a little more detail. When two lines, line segments, rays, or planes intersect, they form an **angle.**

Definition Angle: A figure formed by the intersection of two lines, line segments, rays, or planes.

To simplify our discussion, we will speak of angles as being formed by two rays. The common end point for the two rays is called the vertex of the angle. For example, the vertex of the following angle is point *B*.

An angle is named using the vertex, or the vertex along with a point on each ray forming the angle, as follows:

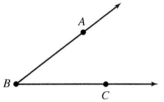

$$\angle B \quad \text{or} \quad \angle ABC \quad \text{or} \quad \angle CBA$$

Notice that if three points are used, as in $\angle ABC$ or $\angle CBA$, the vertex is the letter written in the middle.

Of Interest

Angles can also be measured in radians.

Usually, angles are measured in degrees. Imagine a greeting card. When the card is closed, the measure of the angle between the front and back is 0°. As the card is opened, the angle increases. When the card is opened to the point where it can lie flat, the measure of the angle is 180°.

An angle can open beyond 180°. Imagine the greeting card being opened past flat.

We will focus on angles measuring between 0° and 180°. Those angles can be categorized according to their measure.

Definitions Acute angle: An angle measuring between 0° and 90°.

Right angle: An angle measuring 90°.

Obtuse angle: An angle measuring between 90° and 180°.

Straight angle: An angle measuring 180°.

Example 2   State whether the angle is acute, right, obtuse, or straight.

a.

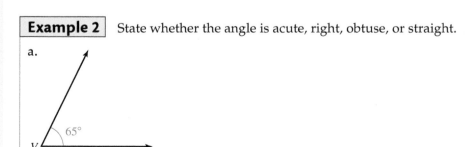

Answer: Angle *V* is acute because its measure is between 0° and 90°.

b.

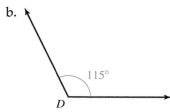

Answer: Angle D is obtuse because its measure is between 90° and 180°.

c.

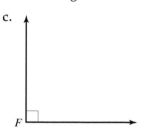

> **Note** Remember, the small square indicates that the measure of the angle is 90°.

Answer: Angle F is a right angle because its measure is 90°.

d.

Answer: Angle C is a straight angle because its measure is 180°.

▶ **Do Your Turn 2**

Also, in Section 4.5, we defined **congruent angles, complementary angles,** and **supplementary angles.** Let's review those definitions.

Definitions **Congruent angles:** Angles that have the same measurement.

Complementary angles: Two angles whose measurements sum to 90°.

Supplementary angles: Two angles whose measurements sum to 180°.

Congruent Angles

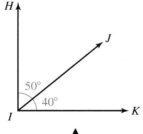

Note $\angle ABC$ and $\angle DEF$ are congruent because their measurements are equal. Using the symbol for congruence, $\cong$, we can write $\angle ABC \cong \angle DEF$.

Complementary Angles

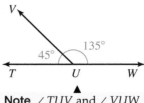

Note $\angle HIJ$ and $\angle JIK$ are complementary because 50° + 40° = 90°.

Supplementary Angles

V

45° 135°

T U W

Note $\angle TUV$ and $\angle VUW$ are supplementary because 45° + 135° = 180°.

▶ **Your Turn 2**

State whether the angle is acute, right, obtuse, or straight.

a.

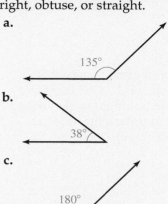

135°

b.

38°

c.

180°

d.

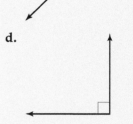

Answers to Your Turn 2
a. obtuse   **b.** acute
c. straight   **d.** right

Now we can explore some properties of the angles formed from intersecting lines. If two lines intersect, two pairs of congruent angles are formed, called **vertical angles.**

Definition **Vertical angles:** A pair of congruent angles formed by the intersection of two lines.

We see vertical angles in the following figures.

Connection Notice that ∠*ABC* and ∠*CBE* together form a straight angle; so they are supplementary, as are ∠*ABD* and ∠*DBE*.

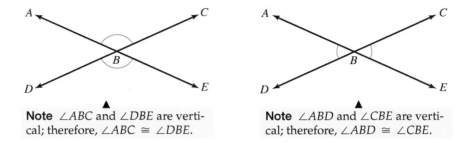

Note ∠*ABC* and ∠*DBE* are vertical; therefore, ∠*ABC* ≅ ∠*DBE*.

Note ∠*ABD* and ∠*CBE* are vertical; therefore, ∠*ABD* ≅ ∠*CBE*.

If the two lines that intersect form a 90° angle, the two lines are **perpendicular.**

Definition **Perpendicular lines:** Two lines that intersect to form a 90° angle.

For example, in the following figure, lines $\overleftrightarrow{AB}$ and $\overleftrightarrow{CD}$ are perpendicular.

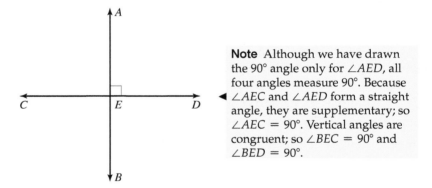

Note Although we have drawn the 90° angle only for ∠*AED*, all four angles measure 90°. Because ∠*AEC* and ∠*AED* form a straight angle, they are supplementary; so ∠*AEC* = 90°. Vertical angles are congruent; so ∠*BEC* = 90° and ∠*BED* = 90°.

When two lines are intersected by a third line (called a *transversal*), eight angles are created, as shown in the following figure.

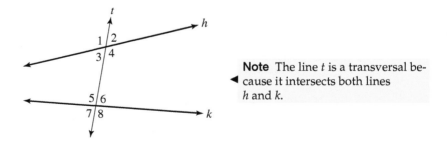

Note The line *t* is a transversal because it intersects both lines *h* and *k*.

Pairs of those eight angles have special names: **corresponding angles, alternate interior angles, alternate exterior angles, consecutive interior angles,** and **consecutive exterior angles.**

Definitions Corresponding angles: In a figure formed by a transversal intersecting two lines, a pair of angles in the same relative position.

Examples: $\angle 1$ and $\angle 5$, $\angle 2$ and $\angle 6$, $\angle 3$ and $\angle 7$, or $\angle 4$ and $\angle 8$

Alternate interior angles: In a figure formed by a transversal intersecting two lines, a pair of angles between the lines but on opposite sides of the transversal.

Examples: $\angle 3$ and $\angle 6$ or $\angle 4$ and $\angle 5$

Alternate exterior angles: In a figure formed by a transversal intersecting two lines, a pair of angles outside the lines but on opposite sides of the transversal.

Examples: $\angle 1$ and $\angle 8$ or $\angle 2$ and $\angle 7$

Consecutive interior angles: In a figure formed by a transversal intersecting two lines, a pair of angles between the lines on the same side of the transversal.

Examples: $\angle 3$ and $\angle 5$ or $\angle 4$ and $\angle 6$

Consecutive exterior angles: In a figure formed by a transversal intersecting two lines, a pair of angles outside the lines on the same side of the transversal.

Examples: $\angle 1$ and $\angle 7$ or $\angle 2$ and $\angle 8$

If two parallel lines are intersected by a transversal, the following rules apply:

Rules

If two parallel lines, m and n, are intersected by a transversal, t, all of the following apply:

The corresponding angles are congruent.

$$\angle 1 \cong \angle 5, \angle 2 \cong \angle 6, \angle 3 \cong \angle 7, \text{ and } \angle 4 \cong \angle 8$$

The alternate interior angles are congruent.

$$\angle 3 \cong \angle 6 \text{ and } \angle 4 \cong \angle 5$$

The alternate exterior angles are congruent.

$$\angle 1 \cong \angle 8 \text{ and } \angle 2 \cong \angle 7$$

The consecutive interior angles are supplementary.

$$\angle 3 + \angle 5 = 180° \text{ and } \angle 4 + \angle 6 = 180°$$

The consecutive exterior angles are supplementary.

$$\angle 1 + \angle 7 = 180° \text{ and } \angle 2 + \angle 8 = 180°$$

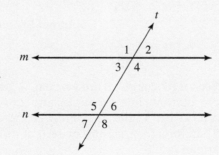

Example 3   In the following figure, lines l and m are parallel. State whether each pair of angles is congruent or supplementary and explain why.

a. $\angle 2$ and $\angle 6$

Answer: congruent; they are corresponding angles.

b. $\angle 3$ and $\angle 6$

Answer: congruent; they are alternate interior angles.

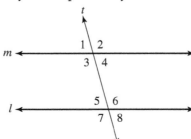

▶ **Your Turn 3**

In the following figure, lines *i* and *j* are parallel. State whether each pair of angles is congruent or supplementary and explain why.

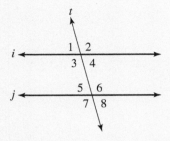

a. ∠1 and ∠5
b. ∠4 and ∠6
c. ∠5 and ∠8
d. ∠3 and ∠4

▶ **Your Turn 4**

In the following figure, lines *h* and *k* are parallel. If ∠1 = 42°, find the measure of each of the other angles.

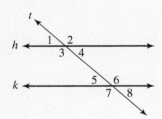

Answers to Your Turn 3
a. congruent; they are corresponding angles.
b. supplementary; they are consecutive interior angles.
c. congruent; they are vertical angles.
d. supplementary; they form a straight angle.

Answer to Your Turn 4

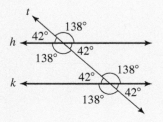

c. ∠1 and ∠7

Answer: supplementary; they are consecutive exterior angles.

d. ∠4 and ∠6

Answer: supplementary; they are consecutive interior angles.

e. ∠1 and ∠4

Answer: congruent; they are vertical angles.

f. ∠5 and ∠7

Answer: supplementary; they form a straight angle.

◀ Do Your Turn 3

Example 4   In the following figure, lines *p* and *q* are parallel. If ∠1 = 57°, find the measure of each of the other angles.

Solution: Because ∠1 and ∠4 are vertical, ∠4 and ∠5 are alternate interior angles and ∠5 and ∠8 are vertical; all of those angles are congruent. Therefore, the measure of each of those angles is 57°.

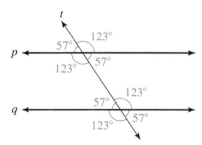

$$∠1 \cong ∠4 \cong ∠5 \cong ∠8 = 57°$$

Because ∠1 and ∠2 form a straight angle, they are supplementary; so ∠1 + ∠2 = 180°.

$57° + ∠2 = 180°$ **In ∠1 + ∠2 = 180°, replace ∠1 with 57°.**

$$\underline{\begin{array}{rcl} 57° + ∠2 &=& 180° \\ -57° \qquad\qquad &-&57° \end{array}}$$ **Subtract 57 from both sides.**

$$0 + ∠2 = 123°$$
$$∠2 = 123°$$

Because ∠2 and ∠3 are vertical, ∠3 and ∠6 are alternate interior angles and ∠6 and ∠7 are vertical angles; they are all congruent. Therefore, the measure of each of those angles is 123°.

$$∠2 \cong ∠3 \cong ∠6 \cong ∠7 = 123°$$

The figure to the right contains each angle's measurement.

◀ Do Your Turn 4

Objective 3 Solve problems involving angles formed by intersecting lines.

In Section 4.5, we solved problems with two unknown angles given that the two angles were congruent, complementary, or supplementary. Now we will solve similar problems using what we just learned about angles formed by intersecting lines.

Example 5   In the figure to the right, four angles are formed by the intersecting lines. Find the measure of each angle.

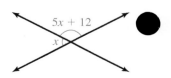

Understand Because the two angles with variable expressions form a straight angle, they are supplementary; so the sum of their measures is 180°.

Plan Translate to an equation; then solve. The sum of x and $5x + 12$ is 180.

Execute $x + 5x + 12 = 180$

$6x + 12 = 180$ **Combine like terms.**

$6x + 12 = 180$

$\underline{-12 \quad -12}$ **Subtract 12 from both sides.**

$6x + 0 = 168$

$\dfrac{6x}{6} = \dfrac{168}{6}$ **Divide both sides by 6.**

$x = 28$

The measure of the smaller angle is 28°. To find the measure of the larger angle, we replace x in $5x + 12$ with 28.

$5(28) + 12$ **Replace x in $5x + 12$ with 28.**

$= 140 + 12$

$= 152$

Answer The two angles are 28° and 152°. Because the two intersecting lines form two pairs of vertical angles, the measures of the other angles are also 28° and 152°.

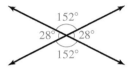

Check The sum of the measures of the supplementary angles is 28° + 152° = 180°.

▶ **Do Your Turn 5**

Example 6 In the figure to the right, lines m and n are parallel. The transversal t forms eight angles. Find the measure of each angle.

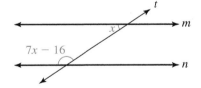

Understand The variable expressions represent the measures of angles that are consecutive interior angles; so they are supplementary.

Plan Translate to an equation; then solve. The sum of x and $7x - 16$ is 180. After finding the values of x and $7x - 16$, we can use the rules about the angles formed by a transversal intersecting two parallel lines to find the other angles.

Execute $x + 7x - 16 = 180$

$8x - 16 = 180$ **Combine like terms.**

$8x - 16 = 180$

$\underline{+16 \quad +16}$ **Add 16 to both sides.**

$8x + 0 = 196$

$\dfrac{8x}{8} = \dfrac{196}{8}$ **Divide both sides by 8.**

$x = 24.5$

The measure of the smaller angle is 24.5°. To find the measure of the larger angle, we replace x in $7x - 16$ with 24.5.

$7(24.5) - 16$ **Replace x in $7x - 16$ with 24.5.**

$= 171.5 - 16$

$= 155.5$

▶ **Your Turn 5**

In the following figure, four angles are formed by the intersecting lines. Find the measure of each angle.

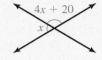

▶ **Your Turn 6**

In the following figure, m and n are parallel lines. The transversal t forms eight angles. Find the measure of each angle.

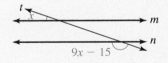

Answer to Your Turn 5

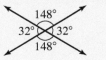

Answer to Your Turn 6

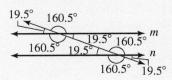

Connection Look at the final figure in Example 6. Notice that the vertical angles, corresponding angles, alternate interior angles, and alternate exterior angles are congruent. Also, the angle pairs forming straight angles, consecutive interior angles, and consecutive exterior angles are supplementary.

Answer The two angles are 24.5° and 155.5°. We can find all of the other angles using the rules concerning angles formed by a transversal intersecting two parallel lines.

Check The sum of the measures of the two consecutive interior angles is 24.5° + 155.5° = 180°.

◄ **Do Your Turn 6**

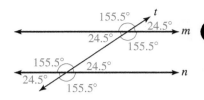

Objective 4 Solve problems involving angles in a triangle.

In Section 4.5, we learned about equilateral and isosceles triangles, which are classifications related to the lengths of the sides. Triangles can also be classified according to their angles. We will consider **right, acute,** and **obtuse** triangles.

Definitions Right triangle: A triangle with one angle measuring 90°.

Acute triangle: A triangle in which each angle measures less than 90°.

Obtuse triangle: A triangle in which one angle measures greater than 90°.

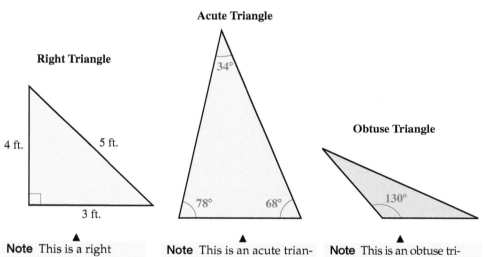

Note This is a right triangle because the measure of one angle is 90°.

Note This is an acute triangle because the measure of each angle is less than 90°.

Note This is an obtuse triangle because the measure of one angle is greater than 90°.

► **Your Turn 7**

For each triangle, indicate whether it is a right triangle, an acute triangle, or an obtuse triangle and explain why.

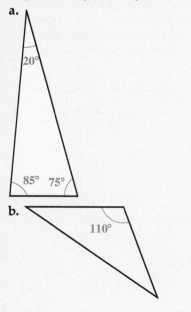

Example 7 For each triangle, indicate whether it is a right triangle, an acute triangle, or an obtuse triangle. Explain why.

a.

Answer: obtuse triangle; the measure of one angle is greater than 90°.

b.

Answer: right triangle; the measure of one angle is 90°.

◄ **Do Your Turn 7**

Answers to Your Turn 7
a. acute triangle; the measure of each angle is less than 90°.
b. obtuse triangle; the measure of one angle is greater than 90°.

We can use what we learned about parallel lines intersected by a transversal to derive a property about all triangles. In the figure to the right, lines l and m are parallel and we have numbered the angles in the triangle, $\angle 1$, $\angle 2$, and $\angle 3$. Using the rules that vertical angles and corresponding angles are congruent, we can conclude that the three angles along the top are congruent to $\angle 1$, $\angle 2$, and $\angle 3$, as shown below.

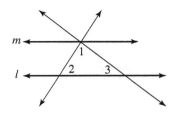

Note Along the top, $\angle 1$, $\angle 2$, and $\angle 3$ together form a straight angle; so their sum is 180°. Therefore, the sum of the measures of the angles in any triangle is 180°.

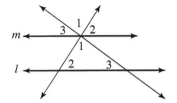

Rule
For any triangle, the sum of the measures of the angles is 180°.

Example 8 Find the measure of each angle in the following triangle.

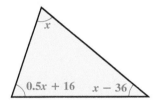

Understand The sum of the measures of the angles is 180°.

Plan Translate to an equation; then solve. The sum of x, $0.5x + 16$, and $x - 36$ is 180.

Execute $x + 0.5x + 16 + x - 36 = 180$

$2.5x - 20 = 180$ Combine like terms.

$2.5x - 20 = 180$

$\underline{+20 \quad +20}$ Add 20 to both sides.

$2.5x + 0 = 200$

$\dfrac{2.5x}{2.5} = \dfrac{200}{2.5}$ Divide both sides by 2.5.

$x = 80$

The measure of the angle labeled x is 80°. To find the measure of the other two angles, we replace x in $0.5x + 16$ and $x - 36$ with 80.

$0.5(80) + 16$ Replace x in $0.5x + 16$ with 80.
$= 40 + 16$
$= 56$

$80 - 36$ Replace x in $x - 36$ with 80.
$= 44$

Answer The measures of the angles are 80°, 56°, and 44°.

Check The sum of the measures of the angles in the triangle is $80° + 56° + 44° = 180°$.

▶ **Do Your Turn 8**

Connection In Section 4.5, we learned that the three sides of an equilateral triangle are of equal length. It can be shown that all three angles also have equal measure. Because the sum of the angles in any triangle is 180°, in any equilateral triangle, each angle must measure 60°.

▶ **Your Turn 8**
Find the measure of each angle in the following triangle.

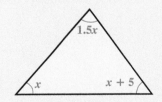

Answer to Your Turn 8
50°, 75°, 55°

Objective 5   Solve problems involving congruent triangles.

In Section 7.2, we learned that similar triangles have the same shape but not necessarily the same size. *Congruent* triangles, on the other hand, have the same shape and the same size, which means that all of the sides are equal length and all of the angles are congruent. The two triangles below are congruent.

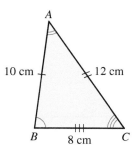

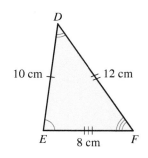

◀ **Note** Angles with the same number of arcs are congruent. For example, $\angle BAC$ and $\angle EDF$ have two arcs; so they are congruent. Sides with the same number of slashes are equal length.

The following four rules can be used to determine whether two triangles are congruent.

Rules

The Side-Side-Side (SSS) Postulate
If the lengths of the three sides of one triangle are equal to the lengths of the three sides of another triangle, the triangles are congruent.

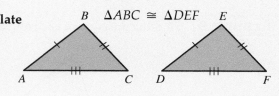

$\triangle ABC \cong \triangle DEF$

The Side-Angle-Side (SAS) Postulate
If two sides and the included angle of one triangle are equal to two sides and the included angle of another triangle, the triangles are congruent.

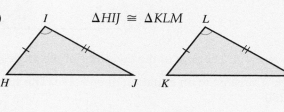

$\triangle HIJ \cong \triangle KLM$

The Angle-Side-Angle (ASA) Postulate
If two angles and the included side of one triangle are equal to two angles and the included side of another triangle, the triangles are congruent.

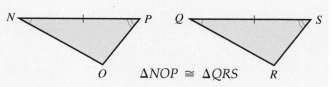

$\triangle NOP \cong \triangle QRS$

The Angle-Angle-Side (AAS) Postulate
If two angles and a nonincluded side of one triangle are equal to two angles and the corresponding nonincluded side of another triangle, the triangles are congruent.

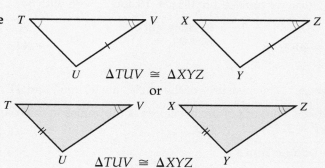

$\triangle TUV \cong \triangle XYZ$
or
$\triangle TUV \cong \triangle XYZ$

Example 9 Determine whether the following triangles are congruent. If they are congruent, state the rule that explains why.

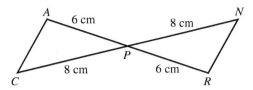

Solution: $\overline{AP} \cong \overline{RP}$ and $\overline{CP} \cong \overline{NP}$. Also, $\angle APC$ and $\angle RPN$ are vertical angles. Therefore, they are congruent.

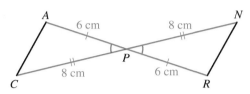

Because the corresponding congruent sides form congruent angles, we have side-angle-side.

Answer: The triangles are congruent because of SAS.

▶ **Do Your Turn 9**

Example 10 The following triangles are congruent. Find the measure of the angles.

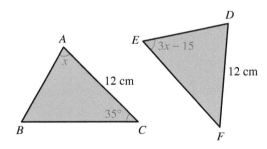

Solution: Because the triangles are congruent, their angles are congruent; so $\angle ABC \cong \angle DEF = 3x - 15$, $\angle BAC \cong \angle EDF = x$, and $\angle ACB \cong \angle DFE = 35°$. We also know that the sum of the measures of the angles in a triangle is $180°$; so we can write the following equation:

$$\angle ABC \text{ or } \angle DEF + \angle BAC \text{ or } \angle EDF + \angle ACB \text{ or } \angle DFE = 180$$
$$3x - 15 \qquad + \qquad x \qquad + \qquad 35 \qquad = 180$$

▶ **Your Turn 9**

Determine whether the following triangles are congruent. If they are congruent, state the rule that explains why.

a.

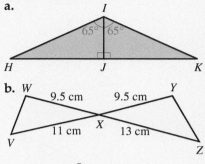

b.

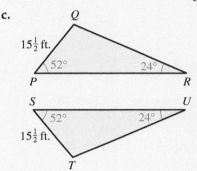

c.

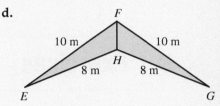

d.

Answers to Your Turn 9
a. congruent; ASA
b. not congruent
c. congruent; AAS
d. congruent; SSS

▶ **Your Turn 10**

The following triangles are congruent. Find the measure of the angles.

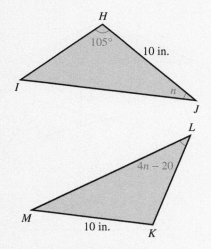

Answers to Your Turn 10
$\angle IHJ \cong \angle LKM = 105°$;
$\angle IJH \cong \angle LMK = 19°$;
$\angle HIJ \cong \angle KLM = 56°$

$$4x + 20 = 180 \qquad \text{Combine like terms.}$$

$$\begin{array}{r} 4x + 20 = 180 \\ \underline{-20 \qquad -20} \\ 4x + 0 = 160 \end{array} \qquad \text{Subtract 20 from both sides.}$$

$$\frac{4x}{4} = \frac{160}{4} \qquad \text{Divide both sides by 4.}$$

$$x = 40°$$

Answers: $\angle BAC \cong \angle EDF = x = 40°$

$\qquad\qquad \angle ABC \cong \angle DEF = 3x - 15$

$\qquad\qquad\qquad\qquad\qquad = 3(40) - 15 \qquad$ In $3x - 15$, replace x with 40.

$\qquad\qquad\qquad\qquad\qquad = 120 - 15 \qquad$ Multiply.

$\qquad\qquad\qquad\qquad\qquad = 105° \qquad$ Subtract.

$\qquad \angle ACB \cong \angle DFE = 35°$

◀ **Do Your Turn 10**

9.1 Exercises For Extra Help MyMathLab®

Objective 1

Prep Exercise 1 What is a point?

Prep Exercise 2 What are parallel lines?

For Exercises 1–12, identify the figure.

1.
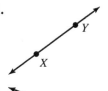

2.

3.
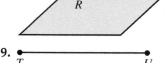

4.

5. $J \bullet$

6. $P \bullet$

7.
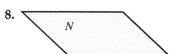

8.

9.

$T \qquad\qquad\qquad U$

10.

11.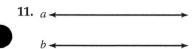

12.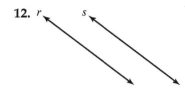

Objective 2

Prep Exercise 3 What is the difference between an acute angle and an obtuse angle?

For Exercises 13–20, state whether the angle is acute, right, obtuse, or straight.

13.

14.

15.

16.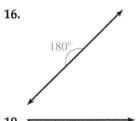

17.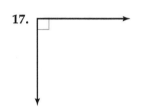

18.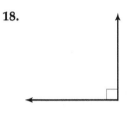

19.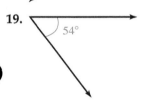

20.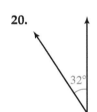

Objective 3

Prep Exercise 4 In the figure to the right, lines *l* and *m* are parallel. What rule applies to the alternate interior angles?

Prep Exercise 5.   In the figure to the right, lines *l* and *m* are parallel. What rule applies to the consecutive interior angles?

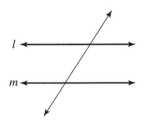

For Exercises 21–34, use the following figure. Lines l and m are parallel. State whether each pair of angles is congruent or supplementary and explain why.

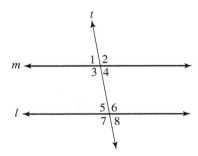

21. ∠1 and ∠4

22. ∠6 and ∠7

23. ∠4 and ∠5

24. ∠3 and ∠6

25. ∠2 and ∠8

26. ∠1 and ∠7

27. $\angle 3$ and $\angle 4$ **28.** $\angle 6$ and $\angle 8$ **29.** $\angle 2$ and $\angle 7$ **30.** $\angle 1$ and $\angle 8$

31. $\angle 3$ and $\angle 5$ **32.** $\angle 4$ and $\angle 6$ **33.** $\angle 3$ and $\angle 7$ **34.** $\angle 2$ and $\angle 6$

For Exercises 35–38, use the following figure. Lines h and k are parallel.

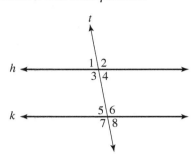

35. If $\angle 1 = 52°$, find the measure of each of the following.
    a. $\angle 3$ **b.** $\angle 4$ **c.** $\angle 5$ **d.** $\angle 6$

36. If $\angle 8 = 64°$, find the measure of each of the following.
    a. $\angle 7$ **b.** $\angle 6$ **c.** $\angle 4$ **d.** $\angle 1$

37. If $\angle 2 = 110°$, find the measure of each of the following.
    a. $\angle 4$ **b.** $\angle 5$ **c.** $\angle 7$ **d.** $\angle 8$

38. If $\angle 6 = 114°$, find the measure of each of the following.
    a. $\angle 7$ **b.** $\angle 3$ **c.** $\angle 2$ **d.** $\angle 1$

For Exercises 39–42, four angles are formed by the intersecting lines. Find the measure of each angle.

39.
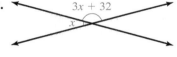

40.
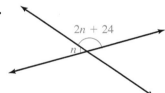

41.
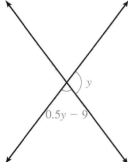

42.
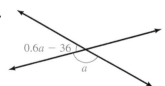

43. In the following figure, lines r and s are parallel. The transversal t forms eight angles. Find the measure of each angle.

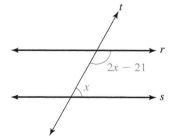

44. In the following figure, lines *l* and *m* are parallel. The transversal *t* forms eight angles. Find the measure of each angle.

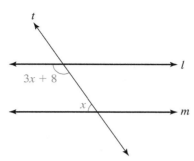

45. In the following figure, lines *i* and *j* are parallel. The transversal *t* forms eight angles. Find the measure of each angle.

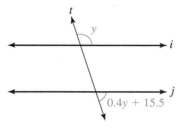

46. In the following figure, lines *l* and *m* are parallel. The transversal *t* forms eight angles. Find the measure of each angle.

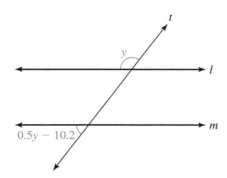

Objective 4

Prep Exercise 6 For any triangle, what is the sum of the measures of the angles?

For Exercises 47–52, indicate whether each triangle is a right triangle, an acute triangle, or an obtuse triangle. Explain why.

47.

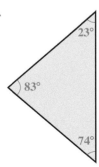

48.

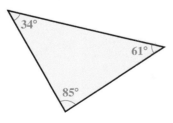

49.

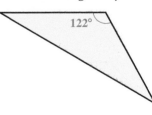

50.

51.

52.

For Exercises 53–56, find the measure of each angle in the triangle.

53.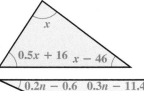

54.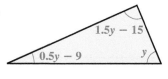

55.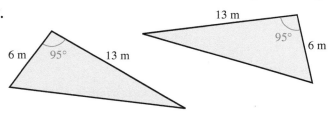

56.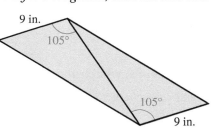

Objective 5

Prep Exercise 7 If two triangles are congruent, what is true about their angles and side lengths?

Prep Exercise 8 Which postulate—SSS, SAS, ASA, or AAS—explains why the triangles to the right are congruent?

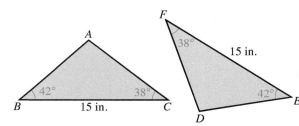

For Exercises 57–68, determine whether the triangles are congruent. If they are congruent, state the rule that explains why.

57.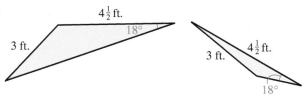

58.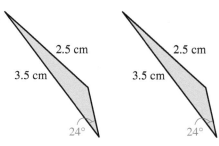

59.

60.

61.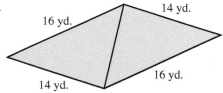

62.

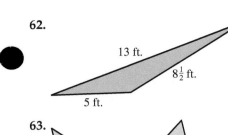
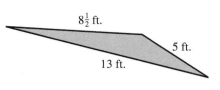

63.

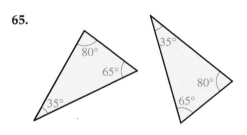

25.5 cm

25.5 cm

45°

45°

64.

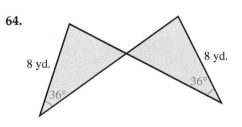

8 yd.

8 yd.

36°

36°

65.

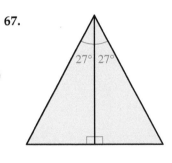

80°

65°

35°

35°

80°

65°

66.

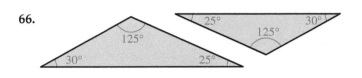

125°

30°

25°

25°

125°

30°

67.

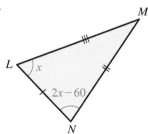

27° 27°

68.

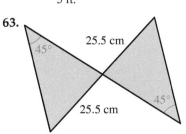

40° 40°

For Exercises 69–72, find the measures of the angles in the congruent triangles.

69.

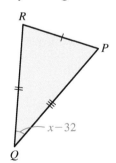

M

R

P

L

x

$2x-60$

N

$x-32$

Q

70.

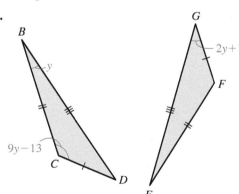

G

$2y+1$

B

y

F

$9y-13$

C

D

E

71.

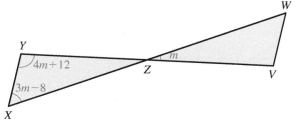

72.

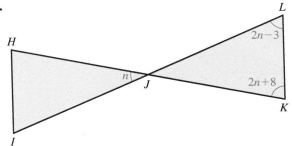

Puzzle Problem In the figure to the right, lines m and n are parallel. Triangle ABC is an equilateral triangle. Find the measure of each angle indicated.

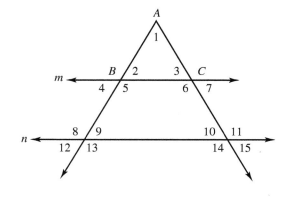

Review Exercises

[1.5] **1.** Find the mean and median of the following set of scores.

$$84, 98, 46, 72, 88, 98$$

[2.1] For Exercises 2 and 3, graph on a number line.

2. 3

3. -4

[3.1] For Exercises 4 and 5, evaluate the expression $x + 2y$ using the indicated values.

4. $x = 3, y = -5$

5. $x = -1, y = -2$

9.2 The Rectangular Coordinate System

Warm-up

[2.1] 1. Graph -3 on a number line.

[9.1] 2. In the figure shown, find the measure of each angle.

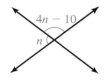

Objective 1 Determine the coordinates of a given point.

In 1619, René Descartes, the French philosopher and mathematician, recognized that positions of points in a plane could be described using two number lines that intersect at a right angle. Each number line is called an **axis.**

Definition   Axis: A number line used to locate a point in a plane.

Two perpendicular axes form the *rectangular*, or *Cartesian, coordinate system*, named in honor of René Descartes. Usually, we call the horizontal axis the x-axis and the vertical axis the y-axis.

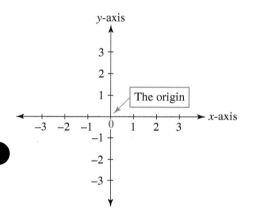

The point where the axes intersect is 0 for both the x-axis and y-axis. This position is called the *origin*. The positive numbers are to the right and up from the origin, whereas negative numbers are to the left and down from the origin.

Any point in the plane can be described using two numbers, one number from each axis. To avoid confusion, these two numbers are written in a specific order. The number representing a point's horizontal distance from the origin is given first, and the number representing a point's vertical distance from the origin is given second. Because the order in which we say or write these two numbers matters, we say that they form an *ordered pair*. Each number in an ordered pair is called a *coordinate* of the ordered pair. We write an ordered pair as follows:

(horizontal coordinate, vertical coordinate)

Consider the point labeled A in the coordinate plane below. The point is drawn at the intersection of the third line to the right of the origin and the fourth line up from the origin. The ordered pair that describes point A is $(3, 4)$.

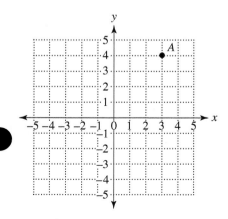

Connection Think of the intersecting lines in the grid as avenues and streets in a city such as New York, where avenues run north-south and streets run east-west. To describe an intersection, a person would say the avenue first, then the street. Point A on the figure is at the intersection of 3rd Avenue and 4th Street—or as a New Yorker would say, "3rd and 4th."

Objectives

1 Determine the coordinates of a given point.

2 Plot points in the coordinate plane.

3 Determine the quadrant for a given ordered pair.

4 Find the midpoint of two points in the coordinate plane.

Of Interest

Born in 1596 near Tours, France, René Descartes became skeptical of the teachings of his day and began formulating his own methods for reasoning. One of his most famous philosophical conclusions is "I think, therefore I am." In 1637, after some urging by his friends, he allowed one work known as the *Method* to be printed. It was in this book that the rectangular coordinate system and analytical geometry were given to the world.

Answers to Warm-up

1.

2.

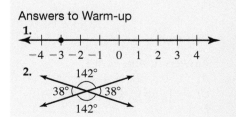

Procedure

To determine the coordinates of a given point in the rectangular coordinate system:

1. Follow a vertical line from the point to the x-axis (horizontal axis). The number at this position on the x-axis is the first coordinate.
2. Follow a horizontal line from the point to the y-axis (vertical axis). The number at this position on the y-axis is the second coordinate.

▶ **Your Turn 1**

Determine the coordinates of each point shown.

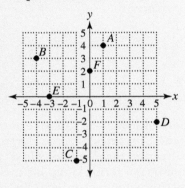

Example 1 Determine the coordinates of each point shown.

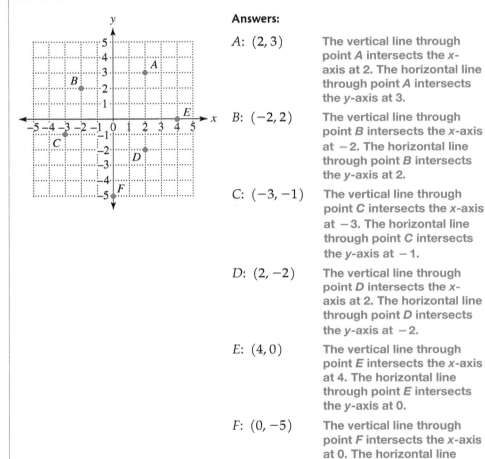

Answers:

A: $(2, 3)$ The vertical line through point A intersects the x-axis at 2. The horizontal line through point A intersects the y-axis at 3.

B: $(-2, 2)$ The vertical line through point B intersects the x-axis at -2. The horizontal line through point B intersects the y-axis at 2.

C: $(-3, -1)$ The vertical line through point C intersects the x-axis at -3. The horizontal line through point C intersects the y-axis at -1.

D: $(2, -2)$ The vertical line through point D intersects the x-axis at 2. The horizontal line through point D intersects the y-axis at -2.

E: $(4, 0)$ The vertical line through point E intersects the x-axis at 4. The horizontal line through point E intersects the y-axis at 0.

F: $(0, -5)$ The vertical line through point F intersects the x-axis at 0. The horizontal line through point F intersects the y-axis at -5.

◀ **Do Your Turn 1**

Objective 2 Plot points in the coordinate plane.

Now let's learn how to plot a point in the rectangular coordinate system given its coordinates in an ordered pair. Recall that the first coordinate in an ordered pair describes a point's horizontal distance from the origin and that the second coordinate describes that point's vertical distance from the origin. For example, to plot $(3, 4)$, we move from the origin three units to the right, then four units up.

$$\underset{\substack{\nearrow \qquad \nwarrow \\ \text{right 3} \qquad \text{up 4}}}{(3, 4)}$$

Answers to Your Turn 1
A: $(1, 4)$ B: $(-4, 3)$
C: $(-1, -5)$ D: $(5, -2)$
E: $(-3, 0)$ F: $(0, 2)$

Procedure

To plot a point given an ordered pair of coordinates:

1. Beginning at the origin $(0, 0)$, move right or left along the x-axis the amount indicated by the first coordinate.
2. From that position on the x-axis, move up or down the amount indicated by the second coordinate.
3. Draw a dot to represent the point described by the coordinates.

Example 2 Plot and label the point described by each ordered pair of coordinates.

a. $(4, -2)$ b. $(-3, -4)$ c. $(0, 5)$

Solution:

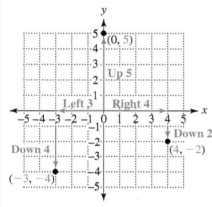

 Learning Strategy

If you are a tactile learner, when plotting an ordered pair, try using your pencil or finger to trace the path to the point. More specifically, move your pencil point or fingertip along the x-axis the amount indicated by the x-coordinate (blue arrows in Example 2). Then from that position on the x-axis, move up or down the amount indicated by the y-coordinate (red arrows in Example 2). The resulting position is the point described by the ordered pair.

▶ **Do Your Turn 2**

Objective 3 Determine the quadrant for a given ordered pair.

The two axes divide the coordinate plane into four regions called **quadrants.**

Definition Quadrant: One of four regions created by the intersection of the axes in the coordinate plane.

The quadrants are numbered using Roman numerals, as shown below.

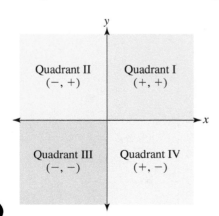

We can use the signs of the coordinates to determine the quadrant in which a point lies. For each point in quadrant I, both coordinates are positive. Each point in quadrant II has a negative first coordinate (horizontal) and a positive second coordinate (vertical). For each point in quadrant III, both coordinates are negative. Each point in quadrant IV has a positive first coordinate and a negative second coordinate.

▶ **Your Turn 2**

Plot and label the point described by each ordered pair of coordinates.

a. $(-4, 2)$
b. $(3, 1)$
c. $(0, -4)$
d. $(-2, 0)$
e. $(-1, -5)$
f. $(3, -2)$

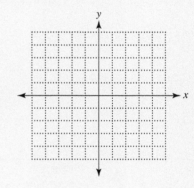

Answers to Your Turn 2

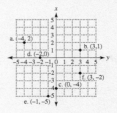

Procedure

To determine the quadrant for a given ordered pair, consider the signs of the coordinates.

$(+, +)$ means that the point is in quadrant I.

$(-, +)$ means that the point is in quadrant II.

$(-, -)$ means that the point is in quadrant III.

$(+, -)$ means that the point is in quadrant IV.

▶ **Your Turn 3**

Determine the quadrant in which each point is located.

a. $(79, 109)$

b. $(-207, -12)$

c. $(-45, 0)$

d. $(-55, 72)$

Example 3   Determine the quadrant in which each point is located.

a. $(-45, 19)$

Answer: quadrant II (upper left) because the first coordinate is negative and the second coordinate is positive

b. $(102, -68)$

Answer: quadrant IV (lower right) because the first coordinate is positive and the second coordinate is negative

c. $(0, 91)$

Answer: Because the x-coordinate is 0, this point is on the y-axis and is not in a quadrant.

◀ **Do Your Turn 3**

Objective 4 Find the midpoint of two points in the coordinate plane.

Suppose we draw a line segment with end points at $(2, 3)$ and $(6, 9)$ as shown in the graph. The point at $(4, 6)$ is the **midpoint** between the end points because it divides the line segment into two segments of equal length.

Definition Midpoint of two points: The point that lies halfway between the two points on a line segment connecting the two points.

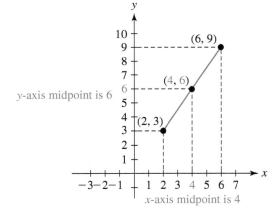

We can check that $(4, 6)$ is in fact the midpoint by constructing two right triangles as shown. Because the two triangles are identical, the distance from $(2, 3)$ to $(4, 6)$ is equal to the distance from $(4, 6)$ to $(6, 9)$. Therefore, $(4, 6)$ is exactly halfway between $(2, 3)$ and $(6, 9)$.

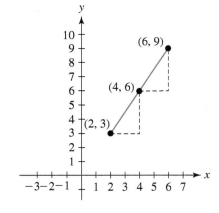

Answers to Your Turn 3

a. I b. III c. $(-45, 0)$ is not in a quadrant; it is on the x-axis. d. II

Notice that the x-coordinate of the midpoint, 4, is the mean of the x-coordinates of the two given points, $(2, 3)$ and $(6, 9)$.

$$\frac{2 + 6}{2} = \frac{8}{2} = 4$$

Similarly, the y-coordinate of the midpoint, 6, is the mean of the y-coordinates of the two given points, $(2, 3)$ and $(6, 9)$.

$$\frac{3 + 9}{2} = \frac{12}{2} = 6$$

Our example suggests the following rule:

> ## Rule
>
> **Midpoint**
> Given two points with coordinates (x_1, y_1) and (x_2, y_2), the coordinates of their midpoint are $\left(\dfrac{x_1 + x_2}{2}, \dfrac{y_1 + y_2}{2}\right)$.

Example 4   Find the midpoint of the given points.

a. $(2, 6)$ and $(-4, -14)$

Solution: Use the midpoint formula. We will consider $(2, 6)$ to be (x_1, y_1) and $(-4, -14)$ to be (x_2, y_2).

$$\text{midpoint} = \left(\frac{2 + (-4)}{2}, \frac{6 + (-14)}{2}\right) \qquad \text{Using } \left(\frac{x_1 + x_2}{2}, \frac{y_1 + y_2}{2}\right)\text{, replace } x_1 \text{ with}$$
$$2, x_2 \text{ with } -4, y_1 \text{ with 6, and } y_2 \text{ with } -14.$$
$$= \left(\frac{-2}{2}, \frac{-8}{2}\right) \qquad\qquad \text{Add in the numerators.}$$
$$= (-1, -4) \qquad\qquad \text{Simplify.}$$

b. $(-4, 8)$ and $(9, -1)$

Solution: Use the midpoint formula. We will consider $(-4, 8)$ to be (x_1, y_1) and $(9, -1)$ to be (x_2, y_2).

$$\text{midpoint} = \left(\frac{-4 + 9}{2}, \frac{8 + (-1)}{2}\right) \qquad \text{Using } \left(\frac{x_1 + x_2}{2}, \frac{y_1 + y_2}{2}\right)\text{, replace } x_1 \text{ with}$$
$$-4, x_2 \text{ with } 9, y_1 \text{ with 8, and } y_2 \text{ with } -1.$$
$$= \left(\frac{5}{2}, \frac{7}{2}\right) \qquad\qquad \text{Add in the numerators.}$$

▲

Note Athough it is not preferred, we could also express the co-ordinates as mixed numbers $\left(2\frac{1}{2}, 3\frac{1}{2}\right)$ or as decimals $(2.5, 3.5)$.

▶ **Do Your Turn 4**

Note Because of the commutative property of addition, it does not matter which point we consider to be (x_1, y_1) and which we consider to be (x_2, y_2). To illustrate, let's work through Example 4a again with $(-4, -14)$ as (x_1, y_1) and $(2, 6)$ as (x_2, y_2).

$$\text{midpoint} = \left(\frac{-4 + 2}{2}, \frac{-14 + 6}{2}\right)$$
$$= \left(\frac{-2}{2}, \frac{-8}{2}\right) = (-1, -4)$$

▶ **Your Turn 4**

Find the midpoint of the given points.

a. $(9, 1)$ and $(5, -3)$
b. $(1, -6)$ and $(8, 4)$
c. $(-2, 7)$ and $(-5, -2)$

Answers to Your Turn 4

a. $(7, -1)$ b. $\left(\dfrac{9}{2}, -1\right)$

c. $\left(-\dfrac{7}{2}, \dfrac{5}{2}\right)$

9.2 Exercises For Extra Help MyMathLab®

Objective 1

Prep Exercise 1 Draw and label the two axes that form the rectangular coordinate system. Label the origin.

Prep Exercise 2 When writing an ordered pair, which is written first, the horizontal-axis coordinate or the vertical-axis coordinate?

For Exercises 1–4, determine the coordinates of each point shown.

1.

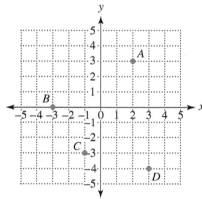

2.

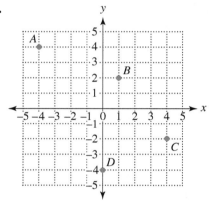

3.

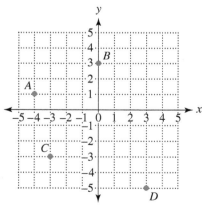

4.

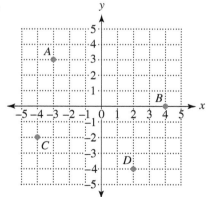

Objective 2

Prep Exercise 3 Describe how to use the coordinates in an ordered pair to plot a point in the rectangular coordinate system.

For Exercises 5–8, plot and label the point described by each ordered pair of coordinates.

5. $(4, 5), (3, -2), (-4, -1), (0, 2)$

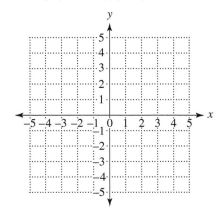

6. $(-1, 0), (2, 1), (-3, -3), (5, -2)$

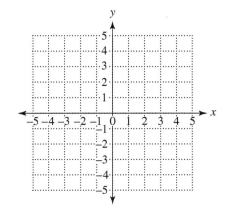

7. $(-4, 0), (1, -4), (-3, -5), (4, 4)$

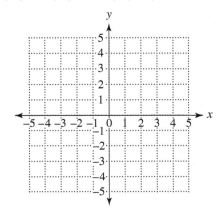

8. $(-2, -4), (0, -3), (5, -3), (-4, 2)$

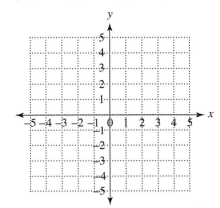

Objective 3

Prep Exercise 4 Which quadrant is the upper-left quadrant?

Prep Exercise 5 What are the signs of the coordinates of a point in the third quadrant?

For Exercises 9–20, determine the quadrant in which the point is located.

9. $(-99, 201)$

10. $(-61, -78)$

11. $(105, 50)$

12. $(-84, 56)$

13. $(42, -60)$

14. $(78, 65)$

15. $(-92, -106)$

16. $(28, -59)$

17. $(0, -34)$

18. $(0, 47)$

19. $(39, 0)$

20. $(-86, 0)$

Objective 4

Prep Exercise 6 Given two points with coordinates (x_1, y_1) and (x_2, y_2), what are the coordinates of their midpoint?

For Exercises 21–34, find the midpoint of the given points.

21. $(0, 6)$ and $(8, 0)$

22. $(1, 7)$ and $(9, -8)$

23. $(3, -1)$ and $(7, 2)$

24. $(1, -2)$ and $(10, -14)$

25. $(2, -6)$ and $(-3, 8)$

26. $(-4, 9)$ and $(-1, -1)$

27. $(0, -8)$ and $(-7, 11)$

28. $(-12, -8)$ and $(0, -3)$

29. $(-9, -3)$ and $(-5, -2)$

30. $(-11, -4)$ and $(-17, -7)$

31. $(1.5, -9)$ and $(-4.5, 2.2)$

32. $(-3.6, 12.4)$ and $(1.2, 4.8)$

33. $\left(3\frac{1}{4}, 8\right)$ and $\left(5, 2\frac{1}{2}\right)$

34. $\left(-14, 6\frac{2}{3}\right)$ and $\left(7, -2\frac{5}{6}\right)$

Review Exercises

[5.1] **1.** Graph $-3\frac{1}{2}$ on a number line.

For Exercises 2 and 3, evaluate.

[3.1] **2.** $2x - 3y$; $x = 4$, $y = 2$

[3.1] **3.** $-4x^2 - 5y$; $x = -3$, $y = 0$

For Exercises 4 and 5, solve and check.

[4.3] **4.** $2x - 8 = 10$

[5.8] **5.** $-\frac{5}{6}y = \frac{3}{4}$

9.3   Graphing Linear Equations

Warm-up

[3.1] 1. Evaluate $2x - 3y$ when $x = 4$ and $y = -1$.

[9.2] 2. In what quadrant is $(-12, -30)$ located?

Objectives

1 Determine whether a given ordered pair is a solution for a given equation with two unknowns.

2 Find three solutions for an equation in two unknowns.

3 Graph linear equations in x and y.

4 Graph horizontal and vertical lines.

5 Given an equation, find the coordinates of the x- and y-intercepts.

Objective 1 Determine whether a given ordered pair is a solution for a given equation with two unknowns.

In the past, we've considered equations with one variable. Now let's consider equations that have two variables, such as $x + y = 5$ and $y = 3x - 4$.

Recall that a solution for an equation with one variable is a number that can replace the variable and make the equation a true statement. For an equation with two variables, a solution will be a pair of numbers, one number for each variable, that can replace the corresponding variables and make the equation true. Because these solutions are ordered pairs, we can write them using coordinates. The ordered pair $(1, 4)$ is a solution for $x + y = 5$ because replacing x with 1 and y with 4 makes the equation true.

Note Recall from Section 9.2 that $(1, 4)$ means that $x = 1$ and $y = 4$.

$$x + y = 5$$
$$\downarrow \quad \downarrow$$
$$1 + 4 = 5$$

Replacing x with 1 and y with 4 makes the equation true; so the ordered pair $(1, 4)$ is a solution.

Procedure

To determine whether a given ordered pair is a solution for an equation with two variables:

1. Replace each variable in the equation with its corresponding coordinate.
2. Simplify both sides of the equation as needed. If the resulting equation is true, the ordered pair is a solution.

Example 1   Determine whether the ordered pair is a solution for the equation.

a. $(2, -5); 3x + y = 1$

Solution: $\begin{aligned} 3(2) + (-5) &\overset{?}{=} 1 \\ 6 + (-5) &\overset{?}{=} 1 \\ 1 &= 1 \end{aligned}$ Replace x with 2 and y with -5.
Simplify both sides of the equation.
True

Answer: Because the resulting equation is true, $(2, -5)$ is a solution for $3x + y = 1$.

b. $(-4, -7); y = 2x - 1$

Solution: $\begin{aligned} -7 &\overset{?}{=} 2(-4) - 1 \\ -7 &\overset{?}{=} -8 - 1 \\ -7 &= -9 \end{aligned}$ Replace x with -4 and y with -7.
Simplify both sides of the equation.
False

Answer: Because the resulting equation is false, $(-4, -7)$ is not a solution for $y = 2x - 1$.

▶ **Do Your Turn 1**

▶ **Your Turn 1**

Determine whether the ordered pair is a solution for the equation.

a. $(5, 0); 4x + y = 20$

b. $(-1, 6); 2y - 3x = -9$

c. $(-3, -14); y = x + 11$

d. $(0, -1); y = 3x - 1$

Answers to Your Turn 1
a. yes   **b.** no   **c.** no   **d.** yes

Answers to Warm-up
1. II   **2.** III

Fractions and decimal numbers can also be solutions.

▶ Your Turn 2

Determine whether the ordered pair is a solution for the equation.

a. $(3, 1.5); y = \dfrac{1}{2}x$

b. $\left(5, 2\dfrac{1}{3}\right); y = \dfrac{x}{3} + 1$

Example 2 Determine whether $\left(-\dfrac{1}{3}, 3.5\right)$ is a solution for $y = \dfrac{3}{2}x + 4$.

Solution: $3.5 \overset{?}{=} \overset{1}{\underset{1}{\dfrac{\overset{1}{3}}{2}}}\left(-\dfrac{1}{\underset{1}{3}}\right) + 4$ Replace x with $-\dfrac{1}{3}$ and y with 3.5. Divide out common factors.

$3.5 \overset{?}{=} -\dfrac{1}{2} + 4$ Multiply.

$3.5 \overset{?}{=} -0.5 + 4$ Write the fraction as an equivalent decimal number.

$3.5 = 3.5$ Add. The equation is true.

Answer: Because the resulting equation is true, $\left(-\dfrac{1}{3}, 3.5\right)$ is a solution for $y = \dfrac{3}{2}x + 4$.

◀ Do Your Turn 2

Objective 2 Find three solutions for an equation in two unknowns.

How can we find solutions to equations in two variables? Consider the equation $x + y = 5$. We've already seen that $(1, 4)$ is a solution. But there are other solutions. We list some of those solutions in the following table.

x	y	Ordered Pair
0	5	$(0, 5)$
1	4	$(1, 4)$
2	3	$(2, 3)$
3	2	$(3, 2)$
4	1	$(4, 1)$
5	0	$(5, 0)$
6	−1	$(6, -1)$

◀ **Note** Every solution for $x + y = 5$ is an ordered pair of numbers whose sum is 5.

In fact, $x + y = 5$ has an infinite number of solutions. For every x value, there is a corresponding y value that will add to the x value to equal 5 and vice versa. This gives a clue as to how to find solutions. We can simply choose a value for x or y and solve for the corresponding value of the other variable. (In this case, our equation was easy enough that we could solve for y mentally.)

Procedure

To find a solution to an equation in two variables:

1. Replace one of the variables with a value (any value).
2. Solve the equation for the other variable.

Answers to Your Turn 2
a. yes   b. no

| **Example 3** | Find three solutions for the equation $2x + y = 5$.

Solution: To find a solution, we replace one of the variables with a chosen value and then solve for the value of the other variable.

For the first solution, we will choose x to be 0.

▼

$$2x + y = 5$$
$$2(0) + y = 5$$
$$0 + y = 5$$
$$y = 5$$

Solution: $(0, 5)$

Note Choosing x (or y) to be 0 usually makes the equation easy to solve.

For the second solution, we will choose x to be 1.

▼

$$2x + y = 5$$
$$2(1) + y = 5$$
$$2 + y = 5$$
$$2 + y = 5$$
$$\underline{-2 \qquad -2}$$
$$0 + y = 3$$
$$y = 3$$

Solution: $(1, 3)$

For the third solution, we will choose x to be 2.

▼

$$2x + y = 5$$
$$2(2) + y = 5$$
$$4 + y = 5$$
$$4 + y = 5$$
$$\underline{-4 \qquad -4}$$
$$0 + y = 1$$
$$y = 1$$

Solution: $(2, 1)$

Keep in mind that there are an infinite number of solutions for a given equation in two variables; so you may get solutions that are different from those of someone else solving the same equation.

▶ **Do Your Turn 3**

| **Example 4** | Find three solutions for the equation $y = \dfrac{1}{3}x - 5$.

Solution: Notice that in this equation, y is isolated. If we select values for x, we will not have to isolate y as we did in Example 3. We will simply calculate the y value. Also notice that the coefficient for x is a fraction. Because we can choose any value for x, let's choose values such as 3 and 6 that will divide out nicely with the denominator of 3.

For the first solution, we will choose x to be 0.

▼

$$y = \dfrac{1}{3}x - 5$$
$$y = \dfrac{1}{3}(0) - 5$$
$$y = -5$$

Solution: $(0, -5)$

For the second solution, we will choose x to be 3.

▼

$$y = \dfrac{1}{3}x - 5$$
$$y = \dfrac{1}{3}(3) - 5$$
$$y = \dfrac{1}{3}\left(\dfrac{\overset{1}{3}}{1}\right) - 5$$
$$y = 1 - 5$$
$$y = -4$$

Solution: $(3, -4)$

For the third solution, we will choose x to be 6.

▼

$$y = \dfrac{1}{3}x - 5$$
$$y = \dfrac{1}{3}(6) - 5$$
$$y = \dfrac{1}{3}\left(\dfrac{\overset{2}{6}}{1}\right) - 5$$
$$y = 2 - 5$$
$$y = -3$$

Solution: $(6, -3)$

▶ **Do Your Turn 4**

▶ **Your Turn 3**

Find three solutions for each equation. (Answers may vary.)
a. $x + y = 6$
b. $3x + y = 9$
c. $2x - 3y = 12$

▶ **Your Turn 4**

Find three solutions for each equation. (Answers may vary.)
a. $y = x - 3$
b. $y = -4x$
c. $y = \dfrac{2}{3}x$
d. $y = \dfrac{1}{2}x + 5$

Answers to Your Turn 3
a. $(0, 6), (1, 5), (2, 4)$
b. $(0, 9), (1, 6), (2, 3)$
c. $(0, -4), (6, 0), \left(1, -\dfrac{10}{3}\right)$

Answers to Your Turn 4
a. $(0, -3), (1, -2), (2, -1)$
b. $(-1, 4), (0, 0), (1, -4)$
c. $(0, 0), (3, 2), (6, 4)$
d. $(0, 5), (2, 6), (4, 7)$

Objective 3 Graph linear equations in x and y.

We've learned that equations in two variables have an infinite number of solutions. Consequently, it is impossible to find all solutions for an equation in two variables. However, we can represent all of the solutions using a graph.

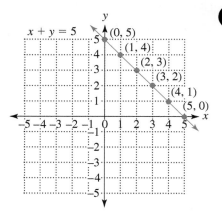

Watch what happens when we plot solutions for $x + y = 5$. Recall some of the solutions: $(0, 5), (1, 4), (2, 3), (3, 2), (4, 1),$ and $(5, 0)$.

Notice that when we plot the ordered pairs, the points lie on a straight line. In fact, all solutions for $x + y = 5$ are on that same line, which is why we say it is a *linear* equation. By connecting the points, we are graphically representing every possible solution for $x + y = 5$. Placing arrows on either end of the line indicates that the solutions continue beyond our "window" in both directions.

The graph of the solutions of every linear equation will be a straight line. Because two points determine a line in a plane, we need a minimum of two ordered pairs. This means that we must find at least two solutions in order to graph a line. However, it is wise to find three solutions, using the third solution as a check. If we plot all three points and they cannot be connected with a straight line, we know something is wrong.

▶ **Your Turn 5**

Graph each equation. (Note that you found three solutions for each of these equations in Your Turn 3.)

a. $x + y = 6$
b. $3x + y = 9$
c. $2x - 3y = 12$

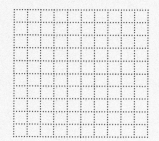

Procedure

To graph a linear equation:

1. Find at least two solutions to the equation.
2. Plot the solutions as points in the rectangular coordinate system.
3. Draw a straight line through these points.

Example 5   Graph. $2x + y = 5$

Solution: We found three solutions to this equation in Example 3. Recall those solutions: $(0, 5), (1, 3),$ and $(2, 1)$.

Now we plot each solution as a point in the rectangular coordinate system, then draw a straight line through these points.

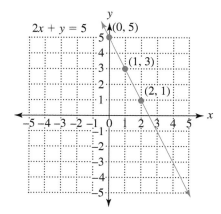

Answers to Your Turn 5

a.

b.

c.

◀ **Do Your Turn 5**

Example 6 Graph. $y = \frac{1}{3}x - 5$

Solution: We found three solutions to this equation in Example 4. Recall those solutions: $(0, -5)$, $(3, -4)$, and $(6, -3)$.

Now we plot each solution as a point in the rectangular coordinate system, then draw a straight line through these points.

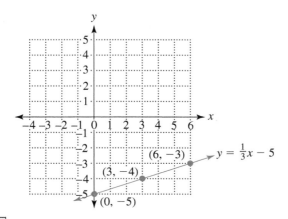

▶ Do Your Turn 6

Objective 4 Graph horizontal and vertical lines.

In Objective 3, we graphed linear equations in two variables. Now let's graph linear equations in one variable, such as $y = 3$ and $x = 2$, in which the variable is equal to a constant.

Example 7 Graph.

a. $y = 3$

Solution: The equation $y = 3$ indicates that y is equal to a constant, 3. In other words, y is always 3 no matter what value we choose for x. If we choose x to be 0, y equals 3. If we choose x to be 2, y is 3. If we choose x to be 4, y is still 3.

We now have three solutions: $(0, 3)$, $(2, 3)$, and $(4, 3)$.

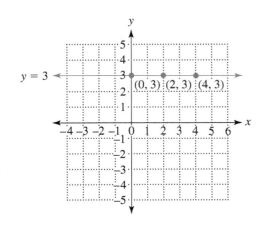

◀ **Note** The graph of $y = 3$ is a horizontal line parallel to the x-axis that passes through the y-axis at $(0, 3)$.

b. $x = 2$

Solution: The equation $x = 2$ indicates that x is equal to a constant, 2. In other words, x is always 2 no matter what value we choose for y. If we choose y to be 0, x equals 2. If we choose y to be 1, x is 2. If we choose y to be 4, x is still 2.

▶ **Your Turn 6**

Graph each equation. (Note that you found three solutions for each of these equations in Your Turn 4.)

a. $y = x - 3$ b. $y = -4x$

c. $y = \frac{2}{3}x$ d. $y = \frac{1}{2}x + 5$

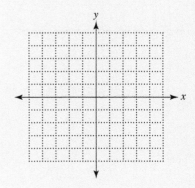

Answers to Your Turn 6
a.

b.
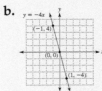

c.
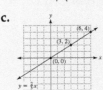

d.
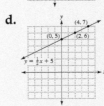

▶ **Your Turn 7**

Graph.

a. $y = -4$

b. $x = -3$

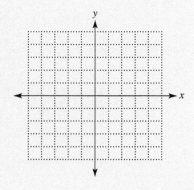

We now have three solutions: $(2, 0)$, $(2, 1)$, and $(2, 4)$.

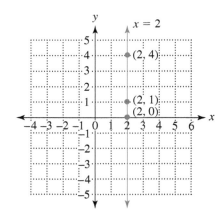

◀ **Note** The graph of $x = 2$ is a vertical line parallel to the y-axis that passes through the x-axis at $(2, 0)$.

Rules

Horizontal and Vertical Lines

The graph of $y = c$, where c is a constant, is a horizontal line parallel to the x-axis and intersects the y-axis at the point with coordinates $(0, c)$.

The graph of $x = c$, where c is a constant, is a vertical line parallel to the y-axis and intersects the x-axis at the point with coordinates $(c, 0)$.

◀ **Do Your Turn 7**

Objective 5 Given an equation, find the coordinates of the x- and y-intercepts.

Consider the following graph. Note that the line intersects the x-axis at the point with coordinates $(4, 0)$. It intersects the y-axis at the point with coordinates $(0, -2)$. These points are called *intercepts*.

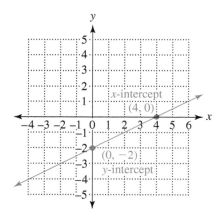

The point where a graph intersects the x-axis is called the **x-intercept.** The point where a graph intersects the y-axis is called the **y-intercept.**

Definitions **x-intercept:** A point where a graph intersects the x-axis.

y-intercept: A point where a graph intersects the y-axis.

Note that the y-coordinate of any x-intercept is always 0. Similarly, the x-coordinate of any y-intercept is always 0. These facts about intercepts suggest the following procedures:

Answers to Your Turn 7

a.

b.

Procedures

To find an x-intercept:
1. Replace y with 0 in the given equation.
2. Solve for x.

To find a y-intercept:
1. Replace x with 0 in the given equation.
2. Solve for y.

Example 8 Find the coordinates of the x- and y-intercepts.

a. $3x - 4y = 12$

Solution: **x-intercept**   **y-intercept**

For the x-intercept, replace y with 0 and solve for x.

$$3x - 4y = 12$$
$$3x - 4(0) = 12$$
$$3x - 0 = 12$$
$$\frac{3x}{3} = \frac{12}{3} \quad \text{Divide both sides by 3.}$$
$$x = 4$$

For the y-intercept, replace x with 0 and solve for y.

$$3x - 4y = 12$$
$$3(0) - 4y = 12$$
$$0 - 4y = 12$$
$$\frac{-4y}{-4} = \frac{12}{-4} \quad \text{Divide both sides by } -4.$$
$$y = -3$$

Answer: The x-intercept is $(4, 0)$, and the y-intercept is $(0, -3)$.

b. $y = \frac{3}{4}x$

Solution: **x-intercept**   **y-intercept**

$$0 = \frac{3}{4}x \quad \text{Replace } y \text{ with 0.}$$
$$\frac{4}{3} \cdot 0 = \frac{4}{3} \cdot \frac{3}{4}x \quad \text{Multiply both sides by } \frac{4}{3}.$$
$$0 = x$$

$$y = \frac{3}{4}(0) \quad \text{Replace } x \text{ with 0.}$$
$$y = 0 \quad \text{Multiply.}$$

Answer: The x- and y-intercepts are the same point, $(0, 0)$.

c. $x = -5$

Solution: The graph of $x = -5$ is a vertical line parallel to the y-axis that passes through the x-axis at the point $(-5, 0)$. Notice that this point is the x-intercept. Because the line is parallel to the y-axis, it will never intersect the y-axis. Therefore, there is no y-intercept.

d. $y = 7$

Solution: The graph of $y = 7$ is a horizontal line parallel to the x-axis that passes through the y-axis at the point $(0, 7)$. Notice that this point is the y-intercept. Because the line is parallel to the x-axis, it will never intersect the x-axis. Therefore, there is no x-intercept.

▶ Do Your Turn 8

▶ **Your Turn 8**

Find the coordinates of the x- and y-intercepts.
a. $x - 4y = 8$
b. $y = -x + 3$
c. $y = -6x$
d. $y = 9$

Answers to Your Turn 8
a. $(8, 0), (0, -2)$
b. $(3, 0), (0, 3)$
c. $(0, 0)$ for both
d. no x-intercept, $(0, 9)$

9.3 Exercises <small>For Extra Help</small> MyMathLab®

Objective 1

Prep Exercise 1 How do you determine whether an ordered pair is a solution to an equation with two variables?

For Exercises 1–12, determine whether the ordered pair is a solution for the equation.

1. $(2, 3); x + 2y = 8$

2. $(3, 1); 2x - y = 5$

3. $(-5, 2); y - 4x = 3$

4. $(4, -6); y = 3x - 1$

5. $(9, 0); y = -2x + 18$

6. $(0, -1); y = 2x + 1$

7. $(6, -2); y = -\dfrac{2}{3}x$

8. $(0, 0); y = \dfrac{2}{5}x$

9. $\left(-1\dfrac{2}{5}, 0\right); y - 3x = 5$

10. $\left(\dfrac{2}{3}, -4\dfrac{5}{6}\right); y = \dfrac{1}{4}x - 5$

11. $(2.2, -11.2); y + 6x = -2$

12. $(-1.5, -1.3); y = 0.2x - 1$

Objectives 2–3

Prep Exercise 2 How do you find a solution to an equation in two variables?

Prep Exercise 3 What will the graph of every linear equation look like?

Prep Exercise 4 Although two solutions are the minimum needed to graph a linear equation in two variables, why is it beneficial to find three solutions?

For Exercises 13–36, find three solutions for the given equation and graph. (Answers may vary for the three solutions.)

13. $x - y = 8$

14. $x + y = -5$

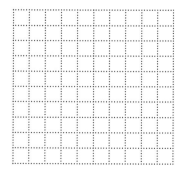

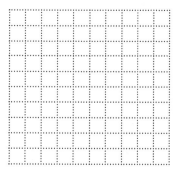

15. $2x + y = 6$

16. $3x - y = 9$

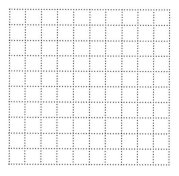

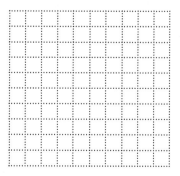

17. $y = x$

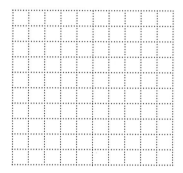

18. $y = -x$

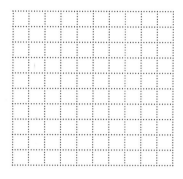

19. $y = 2x$

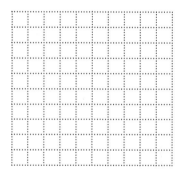

20. $y = -2x$

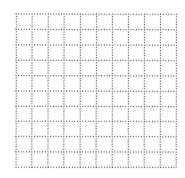

21. $y = -5x$

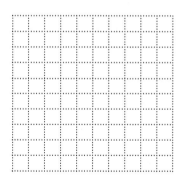

22. $y = 3x$

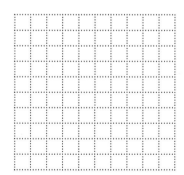

23. $y = x - 3$

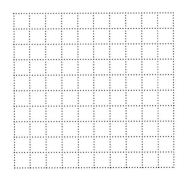

24. $y = -x + 5$

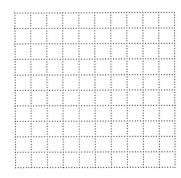

25. $y = -2x + 4$

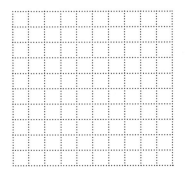

26. $y = 2x - 5$

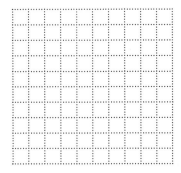

27. $y = 3x + 2$

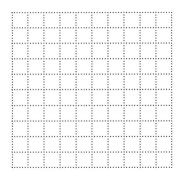

28. $y = -5x - 1$

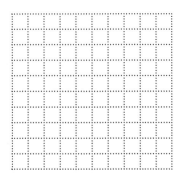

29. $y = \dfrac{1}{2}x$

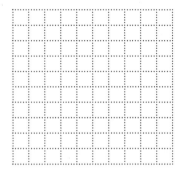

30. $y = -\dfrac{3}{4}x$

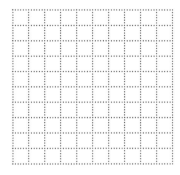

31. $y = -\dfrac{2}{3}x + 4$

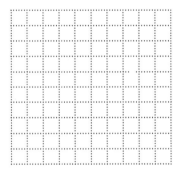

32. $y = \dfrac{4}{5}x - 1$

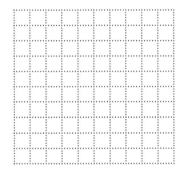

33. $x - \dfrac{1}{4}y = 2$

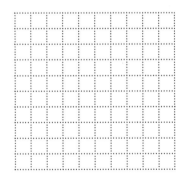

34. $\dfrac{1}{3}x + y = -1$

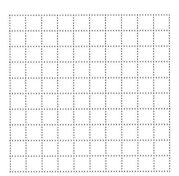

35. $y = 0.4x - 2.5$

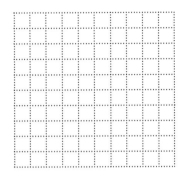

36. $1.2x + 0.5y = 6$

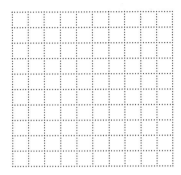

Objective 4

For Exercises 37–40, graph.

37. $y = -5$

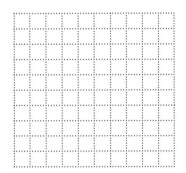

38. $x = -6$

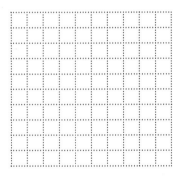

39. $x = 7$

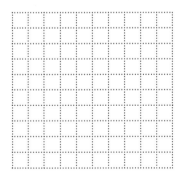

40. $y = 4$

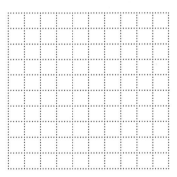

Objective 5

Prep Exercise 5 What is an x-intercept?

Prep Exercise 6 What is a y-intercept?

Prep Exercise 7 What is the y-coordinate of an x-intercept?

Prep Exercise 8 What is the x-coordinate of a y-intercept?

For Exercises 41–60, find the coordinates of the x- and y-intercepts.

41. $x + 2y = 12$

42. $x - 3y = 6$

43. $5x - 4y = 20$

44. $4x + y = 16$

45. $\dfrac{3}{4}x - y = -6$

46. $-x + \dfrac{3}{5}y = 9$

47. $6.5x + 2y = 1.3$

48. $4x - 5.6y = 2.8$

49. $y = 6x$

50. $y = -2x$

51. $y = -4x + 1$

52. $y = 2x - 3$

53. $y = \dfrac{1}{5}x - 5$

54. $y = \dfrac{2}{3}x + 6$

55. $x = -9$

56. $y = 3$

57. $y = 4$

58. $x = 2$

59. $y = -3$

60. $x = -6$

Puzzle Problem Without lifting your pencil from the paper, connect the dots below using only four straight lines.

$\cdot \quad \cdot \quad \cdot$

$\cdot \quad \cdot \quad \cdot$

$\cdot \quad \cdot \quad \cdot$

Review Exercises

$\begin{bmatrix} 3.1 \\ 6.5 \end{bmatrix}$ **1.** Evaluate $-32t + 70$ when $t = 4.2$.

[6.2] **2.** Find the perimeter of the shape.

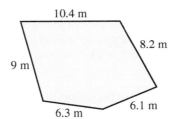

$\begin{bmatrix} 1.6 \\ 5.3 \end{bmatrix}$ **3.** Find the area of the parallelogram with a base of $6\dfrac{1}{2}$ feet and a height of 7 feet.

$\begin{bmatrix} 5.3 \\ 6.3 \end{bmatrix}$ **4.** Find the area of the triangle with a base of 4.2 inches and a height of 2.7 inches.

$\begin{bmatrix} 5.7 \\ 6.3 \end{bmatrix}$ **5.** Find the area of the trapezoid.

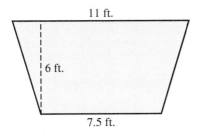

9.4 Applications with Graphing

Objectives

1 Solve problems involving linear equations in two variables.

2 Find the centroid of a figure given the coordinates of its vertices.

3 Find the area of a figure given the coordinates of its vertices.

Warm-up

[9.3] 1. Is $(-3, 10)$ a solution for $y = \dfrac{2}{3}x + 12$?

[9.3] 2. Find the x- and y-intercepts of $5x - 3y = 15$.

Objective 1 Solve problems involving linear equations in two variables.

Linear equations in two variables often describe real situations. Graphing linear equations gives us insight into the situation being described.

Example 1   The linear equation $v = -32.2t + 140$ describes the velocity of a ball thrown straight up. The variable t represents the time of the ball's flight in seconds, and v represents the velocity of the ball in feet per second (ft./sec.).

a. Find the initial velocity of the ball.

Solution: Initial velocity means the velocity at the time the ball is released, which is 0 seconds. Therefore, we must solve for v when the time is 0 ($t = 0$).

$$v = -32.2(0) + 140 \quad \text{In } v = -32.2t + 140, \text{ replace } t \text{ with } 0.$$
$$v = 0 + 140 \qquad\qquad \textbf{Multiply.}$$
$$v = 140 \qquad\qquad\quad \textbf{Add.}$$

Answer: Initially, the ball is traveling at 140 ft./sec.

Connection Values for time are usually placed on the horizontal axis and values for velocity on the vertical axis; so the t and v values can be written in an ordered pair of the form (t, v). When $t = 0$, we found that $v = 140$; so that ordered pair is $(0, 140)$. Notice that $(0, 140)$ is the intercept on the vertical axis, which is the v-axis in this case; so $(0, 140)$ is the v-intercept.

b. Find the velocity of the ball 2 seconds after being released.

Solution: Let $t = 2$ and solve for v.

$$v = -32.2(2) + 140 \quad \text{In } v = -32.2t + 140, \text{ replace } t \text{ with } 2.$$
$$v = -64.4 + 140 \qquad \textbf{Multiply.}$$
$$v = 75.6 \qquad\qquad\quad \textbf{Add.}$$

Connection When $t = 2$, $v = 75.6$; so this ordered pair is $(2, 75.6)$.

Answer: 2 sec. after being released, the ball is traveling only 75.6 ft./sec.

c. How many seconds after the ball is released does it come to a stop before descending?

Solution: When an object is at a stop, its velocity is 0; so we must solve for t when $v = 0$.

$$0 = -32.2t + 140 \quad \text{In } v = -32.2t + 140, \text{ replace } v \text{ with } 0.$$
$$\underline{-140 \qquad\qquad -140} \quad \textbf{Subtract 140 from both sides.}$$
$$-140 = -32.2t + 0$$
$$\frac{-140}{-32.2} = \frac{-32.2t}{-32.2} \qquad \textbf{Divide both sides by } -32.2.$$
$$4.3 \approx t$$

Connection In this case, $v = 0$ and we solve for t. Because t will be shown on the horizontal axis, we have just found the t-intercept. The coordinates of the t-intercept would be written this way:

$$(4.3, 0)$$

Answers to Warm-up
1. yes   2. $(3, 0), (0, -5)$

Answer: The ball will stop in midair approximately 4.3 sec. after being released.

d. Graph the equation with t shown on the horizontal axis and v on the vertical axis.

Solution: To graph, we need at least two solutions. In the answers to questions a, b, and c, we actually have three solutions.

$$(0, 140) \qquad (2, 75.6) \qquad (4.3, 0)$$

Now plot each solution in the rectangular coordinate system; then connect the points to form a straight line.

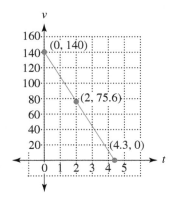

▶ **Do Your Turn 1**

Objective 2 Find the centroid of a figure given the coordinates of its vertices.

We can use coordinates to describe points that define the *corners* of a figure. A corner in a figure is called a **vertex.**

Definition **Vertex:** A point where two lines join to form an angle.

Points A, B, C, and D are the vertices of the following figure.

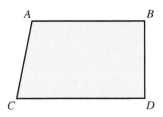

If a figure such as a triangle or parallelogram is drawn in the rectangular coordinate system, we can use the coordinates of its vertices to determine the coordinates of a point in the figure called its **centroid.**

Definition **Centroid:** The balance point, or center of gravity, of a figure.

To better understand what a centroid is, imagine cutting the figure shown above out of cardboard and balancing it on the point of a pencil. Once the figure is balanced, the tip of the pencil would be touching the figure's center of gravity, which is its centroid.

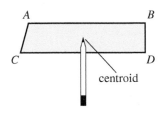

▶ **Your Turn 1**

The linear equation $p = 0.25r - 30{,}000$ describes the profit for a company, where r represents revenue.

a. Find the profit if the revenue is \$180,000.

b. Find the revenue required to break even (the point at which profit is \$0).

c. Graph the equation.

Answers to Your Turn 1
a. \$15,000 **b.** \$120,000
c.

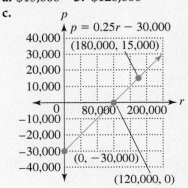

To see how we use the coordinates of the vertices of a figure in the coordinate plane to find its centroid, let's consider a line segment, which is like a stick. If we balance a stick on a pencil, the tip will touch a point halfway up the stick, which corresponds to the midpoint of a line segment. Recall from Section 9.2 that we find the coordinates of the midpoint of a line segment connecting two points in the rectangular coordinate system by finding the mean of the x-coordinates and the mean of the y-coordinates of the two points. Similarly, we find the coordinates of the centroid of a figure using the mean of the x-coordinates and the mean of the y-coordinates of the vertices.

> **Rule**
>
> Given the coordinates of the vertices of a figure in the rectangular coordinate system, the coordinates of its centroid are as follows:
>
> $$\left(\begin{array}{c} \text{mean of the } x\text{-coordinates} \\ \text{of all vertices} \end{array} , \begin{array}{c} \text{mean of the } y\text{-coordinates} \\ \text{of all vertices} \end{array} \right)$$

▶ **Your Turn 2**

Find the centroid of the triangle with vertices $(3, 4), (-4, -1)$, and $(-2, 6)$.

Note The mean of the x-coordinates is the sum of the three x-coordinates divided by 3. Similarly, the mean of the y-coordinates is the sum of the three y-coordinates divided by 3.

Example 2 Find the centroid of the triangle with vertices $(3, 2), (-4, 4)$, and $(-5, 3)$.

Solution:

$$\text{centroid} = \left(\frac{x_1 + x_2 + x_3}{3}, \frac{y_1 + y_2 + y_3}{3} \right)$$

Find the mean of the x-coordinates and the mean of the y-coordinates of the three vertices.

$$\begin{array}{ccc} (3, 2) & (-4, 4) & (-5, 3) \\ \uparrow \uparrow & \uparrow \uparrow & \uparrow \uparrow \\ x_1\, y_1 & x_2\, y_2 & x_3\, y_3 \end{array}$$

$$= \left(\frac{3 + (-4) + (-5)}{3}, \frac{2 + 4 + 3}{3} \right)$$

$$= \left(\frac{-6}{3}, \frac{9}{3} \right)$$

Add in each numerator.

$$= (-2, 3)$$

Simplify to lowest terms.

Answer: The centroid of the triangle is $(-2, 3)$.

To visualize this, let's plot the vertices and the centroid.

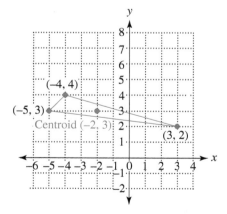

◀ **Do Your Turn 2**

Centroids of figures with four vertices can be found in much the same way.

Answer to Your Turn 2

$(-1, 3)$

Example 3 Find the centroid of the figure with vertices $(3, 2)$, $(-4, 2)$, $(3, -3)$, and $(-5, -3)$.

Solution:

$$\text{centroid} = \left(\frac{x_1 + x_2 + x_3 + x_4}{4}, \frac{y_1 + y_2 + y_3 + y_4}{4} \right)$$

Find the mean of the x-coordinates and the mean of the y-coordinates of the four vertices.

$$\underset{\substack{\uparrow \uparrow \\ x_1 \, y_1}}{(3, 2)} \quad \underset{\substack{\uparrow \uparrow \\ x_2 \, y_2}}{(-4, 2)} \quad \underset{\substack{\uparrow \uparrow \\ x_3 \, y_3}}{(3, -3)} \quad \underset{\substack{\uparrow \uparrow \\ x_4 \, y_4}}{(-5, -3)}$$

$$= \left(\frac{3 + (-4) + 3 + (-5)}{4}, \frac{2 + 2 + (-3) + (-3)}{4} \right)$$

$$= \left(\frac{-3}{4}, \frac{-2}{4} \right) \quad \text{Add in each numerator.}$$

$$= \left(-\frac{3}{4}, -\frac{1}{2} \right) \quad \text{Simplify to lowest term.}$$

Answer: The centroid is $\left(-\frac{3}{4}, -\frac{1}{2} \right)$.

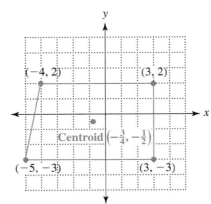

▶ **Do Your Turn 3**

▶ **Your Turn 3**

Find the centroid of the figure with vertices $(2, 3)$, $(4, -5)$, $(-4, 1)$, and $(8, 7)$.

Objective 3 Find the area of a figure given the coordinates of its vertices.

Given the coordinates of the vertices of a figure, we can also find its area.

Example 4 Find the area of the figure with vertices $(4, 3)$, $(-2, 3)$, $(-4, -2)$, and $(2, -2)$.

Understand This is a four-sided figure. We are to calculate the area.

Plan Plot the points to get a sense of the shape. Then determine the lengths of the sides that are needed to find the area. Finally, calculate the area.

Execute

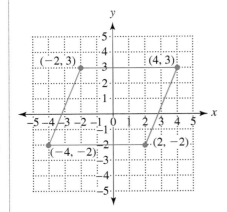

Notice that the figure is a parallelogram. The formula for the area of a parallelogram is $A = bh$; so we need the base and height. Because the base is parallel to the x-axis, we can subtract the x-coordinates.

Base: $|2 - (-4)| = |2 + 4| = |6| = 6$

Answer to Your Turn 3
$$\left(\frac{5}{2}, \frac{3}{2} \right)$$

▶ **Your Turn 4**

Find the area of the figure with the given vertices.

a. $(5, 2), (3, -1), (-1, 2),$ and $(-3, -1)$

b. $(-1, 4), (-3, 4), (-5, -5),$ and $(2, -5)$

Answers to Your Turn 4
a. 18 square units
b. 40.5 square units

Recall that height is a vertical measurement from the base to the top of a shape. The line along which height is measured is parallel to the y-axis. Therefore, we can subtract the y-coordinates.

$$\text{Height: } |3 - (-2)| = |3 + 2| = |5| = 5$$

Now we can calculate the area.

$$A = 6 \cdot 5$$
$$A = 30$$

Answer The area is 30 square units.

◀ **Do Your Turn 4**

9.4 Exercises For Extra Help MyMathLab®

Objective 1

Prep Exercise 1 The linear equation $v = -32.2t + 200$ describes the velocity of an object after being thrown straight up, where t represents the time of the object's flight in seconds and v represents the velocity in feet per second. How do we determine the amount of time it takes the object to come to a stop before descending?

For Exercises 1–10, use the given linear equation to answer the questions.

1. The linear equation $v = -32.2t + 600$ describes the velocity in feet per second of a rocket t seconds after being launched.

   a. Find the initial velocity of the rocket.
   b. Find the velocity after 3 seconds.
   c. How many seconds after launch will the rocket stop before returning to Earth?
   d. Graph the equation with t as the horizontal axis and v as the vertical axis.

Of Interest

At a velocity of 7.9 kilometers per second, an object would just overcome the effects of gravity so that it would enter the lowest possible circular orbit around Earth. An object traveling between 7.9 kilometers per second and 11.18 kilometers per second would travel in an elliptical orbit around Earth. An object traveling 11.18 kilometers per second or faster would escape Earth's gravity altogether. What are these velocities in feet per second? Miles per hour?

2. The linear equation $v = -32.2t + 86$ describes the velocity in feet per second of a ball t seconds after being thrown straight up.

   a. Find the initial velocity of the ball.
   b. Find the velocity after 1.5 seconds.
   c. How many seconds after launch will the ball stop before descending?
   d. Graph the equation with t on the horizontal axis and v on the vertical axis.

3. The equation $p = 0.24r - 45{,}000$ describes the profit for a company, where r represents revenue in dollars.

a. Find the profit if the revenue is $250,000.
b. Find the revenue required to break even (the point at which profit is $0).
c. Graph the equation with r on the horizontal axis and p on the vertical axis.

4. The equation $p = 0.18r - 36{,}000$ describes the profit for a company, where r represents revenue in dollars.

a. Find the profit if the revenue is $450,000.
b. Find the revenue required to break even (the point at which profit is $0).
c. Graph the equation with r on the horizontal axis and p on the vertical axis.

5. The equation $b = 13.5t + 300$ describes the final balance of an account t years after the initial investment is made.

a. Find the initial balance (principal). (*Hint:* $t = 0$.)

b. Find the balance after 5 years.
c. Find the balance after 20 years.
d. Graph the equation with t on the horizontal axis and b on the vertical axis.

6. The equation $b = 17.5t + 500$ describes the final balance of an account t years after the initial investment is made.

a. Find the initial balance (principal). (*Hint:* $t = 0$.)

b. Find the balance after 5 years.
c. Find the balance after 20 years.
d. Graph the equation with t on the horizontal axis and b on the vertical axis.

7. The equation $F = \dfrac{9}{5}C + 32$ is used to convert a temperature in °C to temperature in °F.

   a. What is the F-intercept?

   b. What is the C-intercept?

   c. Convert 40°C to °F.

   d. Graph the equation with C on the horizontal axis and F on the vertical axis.

8. The equation $C = \dfrac{5}{9}(F - 32)$ is used to convert a temperature in °F to temperature in °C.

   a. What is the F-intercept?

   b. What is the C-intercept?

   c. Convert 68°F to °C.

   d. Graph the equation with F on the horizontal axis and C on the vertical axis.

9. An HVAC (heating, ventilation, and air-conditioning) service technician charges $60 per hour for labor plus a $100 flat fee for the visit.

   a. Write a linear equation that describes the total cost.

   b. What would be the total cost for $2\dfrac{1}{2}$ hours of labor?

   c. How many hours of labor can the service technician work for $355?

   d. Graph the equation.

10. The weekly cost to produce a toy is $0.75 per unit plus a flat $4500 for lease, equipment, supplies, and other expenses. Let C represent the total cost and u represent the number of units produced.

    a. Write a linear equation that describes the total cost.

    b. What would be the total cost to produce 600 units?

    c. What would be the total cost to produce 800 units?

    d. Graph the equation.

Objectives 2 and 3

Prep Exercise 2 What is a vertex?

Prep Exercise 3 A figure has vertices $(2, 4)$, $(-3, 4)$, $(-3, -1)$, and $(2, -1)$. What is the figure?

Prep Exercise 4 A triangle has vertices (x_1, y_1), (x_2, y_2), and (x_3, y_3). What is the formula for calculating the coordinates of its centroid?

For Exercises 11–30, find the centroid and area of the figure with the given vertices.

11. $(0, 0), (5, 0), (4, 6)$

12. $(0, 0), (8, 0), (1, 3)$

13. $(-3, -4), (5, -4), (1, 2)$

14. $(-6, -2), (4, -2), (-4, 7)$

15. $(0, 0), (0, 6), (6, 6), (6, 0)$

16. $(-1, 0), (-1, 4), (3, 4), (3, 0)$

17. $(4, 1), (-3, 1), (-3, -5), (4, -5)$

18. $(2, 5), (-5, 5), (-5, -3), (2, -3)$

19. $(-1, 2), (-6, 2), (-6, -4), (-1, -4)$

20. $(0, -2), (8, -2), (0, -7), (8, -7)$

21. $(1, 0), (4, 6), (7, 6), (4, 0)$

22. $(2, 5), (3, -1), (7, 5), (8, -1)$

23. $(-3, 2), (4, 2), (6, -7), (-1, -7)$

24. $(-6, 5), (-4, -3), (3, 5), (5, -3)$

25. $(-3, 0), (-3, 8), (4, 8), (6, 0)$

26. $(-3, -2), (-2, 4), (6, 4), (7, -2)$

27. $(0, 3), (-3, -1), (4, 3), (5, -1)$

28. $(-2, 0), (-2, -4), (4, 0), (7, -4)$

29. $(-5, -1), (-7, -4), (3, -1), (4, -4)$

30. $(-7, -2), (-6, -5), (4, -2), (0, -5)$

Puzzle Problem Find the centroid and area of the figure with the following vertices:

$$(-2, 1), (-2, -5), (6, -5), (6, -2), (0, -2), (0, 1)$$

Review Exercises

[4.5] 1. The sum of two integers is 106. One integer is 6 more than three times the other integer. What are the integers?

[5.3] 2. A company allocates $\frac{3}{4}$ of its costs for employee wages; $\frac{2}{3}$ of the employee wages are for nonsalary wages (hourly wages). What portion of the company's total cost goes toward nonsalary wages?

[6.5] 3. Find the mean, median, and mode of the following scores.

$88, 72, 91, 84, 88, 87, 90, 56, 71, 70, 95, 82, 70, 65$

[8.3] 4. 20% of the people in a class received an A. If 45 people were in the class, how many received an A?

[8.5] 5. Find the final balance in an account if $400 earning 4.8% APR is compounded quarterly over 2 years.

Chapter 9 Summary and Review Exercises

9.1 Points, Lines, and Angles

Definitions/Rules/Procedures	Key Example(s)
A **point** is a(n) _____ in space having no length or width. A **line** is a straight, one-dimensional figure extending _____ in both directions. A **line segment** is a straight, one-dimensional figure extending between two _____. A **ray** is a straight, one-dimensional figure extending forever in _____ direction. A **plane** is a flat, two-dimensional _____ extending forever in both dimensions. **Parallel lines** are lines lying in the same place that _____.	Identify the figure. a.  **Answer:** line segment $\overline{MN}$ b.  **Answer:** plane Y

[9.1] 1. Identify the figure.

a. • P b. 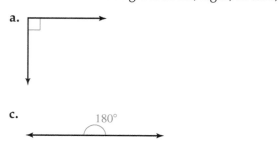   c. h ←————————→

k ←————————→

Definitions/Rules/Procedures	Key Example(s)
An **angle** is a figure formed by the _____ of two lines, line segments, rays, or planes. An **acute angle** measures _____. A **right angle** measures _____. An **obtuse angle** measures _____. A **straight angle** measures _____.	State whether the angle is acute, right, obtuse, or straight. 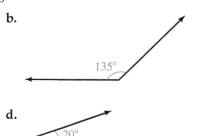 140° **Answer:** obtuse

[9.1] 2. State whether the angle is acute, right, obtuse, or straight.

a.

b.
135°

c.
180°

d.
20°

Definitions/Rules/Procedures	Key Example(s)
Congruent angles have the _____ measurement. **Complementary angles** are two angles whose measurements sum to _____. **Supplementary angles** are two angles whose measurements sum to _____. **Vertical angles** are a pair of _____ angles formed by the intersection of two lines. **Perpendicular lines** are two lines that intersect to form a(n) _____ angle. **If two parallel lines, *m* and *n*, are intersected by a transversal, *t*, all of the following apply:** 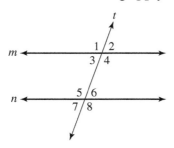 The corresponding angles are congruent. $\angle 1 \cong \angle \_\_$, $\angle 2 \cong \angle \_\_$, $\angle 3 \cong \angle \_\_$, and $\angle 4 \cong \angle \_\_$. The alternate interior angles are congruent. $\angle 3 \cong \angle \_\_$ and $\angle 4 \cong \angle \_\_$. The alternate exterior angles are congruent. $\angle 1 \cong \angle \_\_$ and $\angle 2 \cong \angle \_\_$ The consecutive interior angles are supplementary. $\angle 3 + \angle \_\_ = 180°$ and $\angle 4 + \angle \_\_ = 180°$ The consecutive exterior angles are supplementary. $\angle 1 + \angle \_\_ = 180°$ and $\angle 2 + \angle \_\_ = 180°$	In the following figure, lines *l* and *m* are parallel. The transversal *t* forms eight angles. Find the measure of each angle. 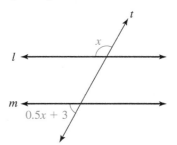 The variable expressions represent the measures of angles that are consecutive exterior angles; so they are supplementary. $$x + 0.5x + 3 = 180$$ $$1.5x + 3 = 180 \quad \text{Combine like terms.}$$ $$\frac{-3}{1.5x + 0} = \frac{-3}{177} \quad \text{Subtract 3 from both sides.}$$ $$\frac{1.5x}{1.5} = \frac{177}{1.5} \quad \text{Divide both sides by 1.5.}$$ $$x = 118$$ We determine the measure of the other angle by replacing *x* with 118 in $0.5x + 3$. $$0.5(118) + 3$$ $$= 59 + 3$$ $$= 62$$ According to the rules for angles in a figure formed by two parallel lines intersected by a transversal, the measure of each angle will be either 118° or 62°. 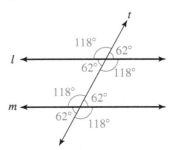

[9.1] **3.** In the figure to the right, lines *l* and *m* are parallel. State whether each pair of angles is congruent or supplementary and explain why.

    a. $\angle 1$ and $\angle 4$

    b. $\angle 4$ and $\angle 5$

    c. $\angle 2$ and $\angle 8$

    d. $\angle 6$ and $\angle 8$

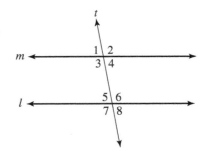

[9.1] 4. In the figure to the right, lines h and k are parallel. If $\angle 1 = 57°$, find the measure of each of the following.

   a. $\angle 3$ **b.** $\angle 4$

   c. $\angle 5$ **d.** $\angle 7$

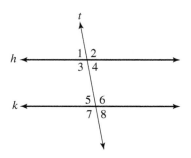

[9.1] 5. Four angles are formed by the following intersecting lines. Find the measure of each angle.

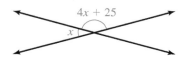

[9.1] 6. In the following figure, lines i and j are parallel. The transversal t forms eight angles. Find the measure of each angle.

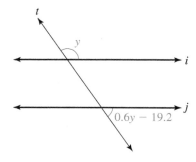

Definitions/Rules/Procedures	Key Example(s)

A **right triangle** has one angle measuring _____.
 	
 In an **acute triangle**, each angle measures _____ 90°.
 	
 In an **obtuse triangle**, one angle measures _____ 90°.
 	
 For any triangle, the sum of the measures of the angles is _____.
	Find the measure of each angle in the following triangle.
	
	 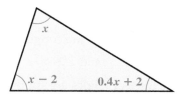
	
	 The sum of the measures of the angles is 180°.
	
	 $x + x - 2 + 0.4x + 2 = 180$
	          $2.4x = 180$ Combine like terms.
	        $\dfrac{2.4x}{2.4} = \dfrac{180}{2.4}$ Divide both sides by 2.4.
	           $x = 75$
	
	 Now determine the other measures by replacing x with 75 in $x - 2$ and $0.4x + 2$.
	
	 For $x - 2$: For $0.4x + 2$:
	    $75 - 2$ $0.4(75) + 2$
	 $= 73$ $= 30 + 2$
	                      $= 32$
	
	 Answer: 75°, 73°, 32°

[9.1] 7. Indicate whether each triangle is a right triangle, an acute triangle, or an obtuse triangle. Explain why.

a.

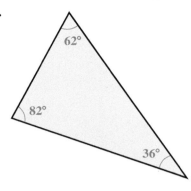

b.

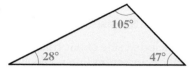

[9.1] 8. Find the measure of each angle in the triangle to the right.

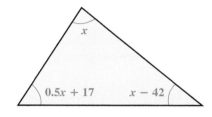

Definitions/Rules/Procedures	Key Example(s)
The Side-Side-Side (SSS) Postulate If the lengths of the three sides of one triangle are equal to the lengths of the three sides of another triangle, the triangles are congruent. 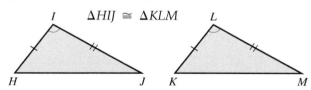 **The Side-Angle-Side (SAS) Postulate** If two sides and the included angle of one triangle are equal to two sides and the included angle of another triangle, the triangles are congruent. 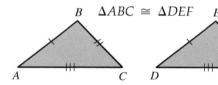 **The Angle-Side-Angle (ASA) Postulate** If two angles and the included side of one triangle are equal to two angles and the included side of another triangle, the triangles are congruent. 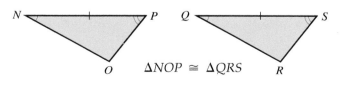	Determine whether the following two triangles are congruent. If they are congruent, state the rule that explains why. 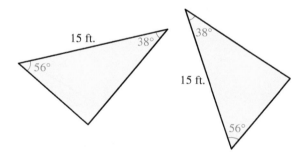 **Answer:** congruent; ASA

Definitions/Rules/Procedures	Key Example(s)
The Angle-Angle-Side (AAS) Postulate If two angles and a nonincluded side of one triangle are equal to two angles and the corresponding nonincluded side of another triangle, the triangles are congruent. 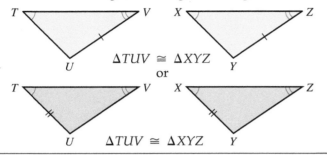 $\triangle TUV \cong \triangle XYZ$ or $\triangle TUV \cong \triangle XYZ$	

9. Determine whether the triangles are congruent. If they are congruent, state the rule that explains why.

a.
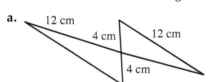

b.
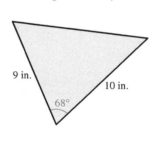

10. Find the unknown angles in the following congruent triangles.

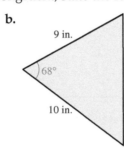

9.2 The Rectangular Coordinate System

Definitions/Rules/Procedures	Key Example(s)
An **axis** is a(n) _____ used to locate points in a plane. **To determine the coordinates of a given point in the rectangular coordinate system:** 1. Follow a vertical line from the given point to the *x*-axis (horizontal axis). The number at this position on the *x*-axis is the _____ coordinate. 2. Follow a horizontal line from the point to the *y*-axis (vertical axis). The number at this position on the *y*-axis is the _____ coordinate.	Determine the coordinates of each point shown. 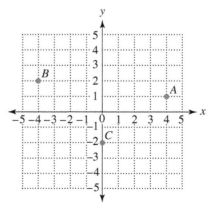 **Answers:** A: $(4, 1)$ B: $(-4, 2)$ C: $(0, -2)$

[9.2] **11.** Determine the coordinates of each point shown at right.

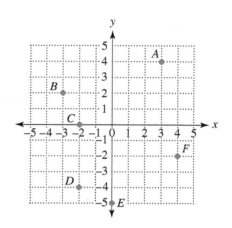

Definitions/Rules/Procedures	Key Example(s)
To plot a point given an ordered pair of coordinates:	Plot the point described by each ordered pair of coordinates.
1. Beginning at the origin, $(0, 0)$, move right or left along the x-axis the amount indicated by the _____ coordinate.	**a.** $(-3, -4)$ **b.** $(2, -3)$ **c.** $(0, 3)$
2. From that position on the x-axis, move up or down the amount indicated by the _____ coordinate.	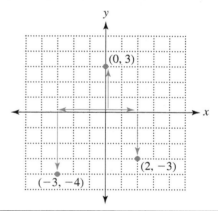
3. Draw a dot to represent the point described by the coordinates.	

[9.2] **12.** Plot and label the point described by each ordered pair of coordinates.

$A\,(5, 2)$ $B\,(2, 0)$ $C\,(-4, -3)$ $D\,(3, -5)$ $E\,(0, -4)$

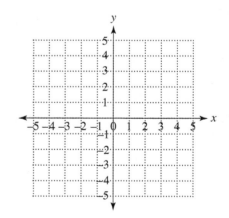

Definitions/Rules/Procedures	Key Example(s)
A **quadrant** is one of _____ regions created by the intersection of the axes in the coordinate plane.	

Definitions/Rules/Procedures	Key Example(s)
To determine the quadrant for a given ordered pair, consider the signs of the coordinates. _____ means that the point is in quadrant I. _____ means that the point is in quadrant II. _____ means that the point is in quadrant III. _____ means that the point is in quadrant IV.	Determine the quadrant in which each point is located. **a.** $(42, 95)$: quadrant I **b.** $(-65, 90)$: quadrant II **c.** $(-91, -56)$: quadrant III **d.** $(75, -102)$: quadrant IV **e.** $(-14, 0)$: not in a quadrant; this point is on the x-axis.

[9.2] **13.** Determine the quadrant in which each point is located.

      a. $(-105, 68)$ **b.** $(-47, -158)$ **c.** $(95, 72)$ **d.** $(58, 0)$

Definitions/Rules/Procedures	Key Example(s)
Given two points with coordinates (x_1, y_1) and (x_2, y_2), the coordinates of their midpoint are $$\left(\, ?, ? \, \right).$$	Find the midpoint of $(4, -9)$ and $(-12, -2)$. $$\text{midpoint} = \left(\frac{4 + (-12)}{2}, \frac{-9 + (-2)}{2} \right)$$ $$= \left(\frac{-8}{2}, -\frac{11}{2} \right)$$ $$= \left(-4, -\frac{11}{2} \right)$$

[9.2] _For Exercises 14 and 15, find the midpoint._

14. $(2, 9)$ and $(6, 1)$ **15.** $(-4, 2)$ and $(6, -3)$

9.3 Graphing Linear Equations

Definitions/Rules/Procedures	Key Example(s)
To determine whether a given ordered pair is a solution for an equation with two variables: 1. Replace each variable in the equation with its corresponding coordinate. 2. Simplify both sides of the equation as needed. If the resulting equation is _____, the ordered pair is a solution.	Determine whether the ordered pair is a solution for the equation. **a.** $(-6, 2)$; $x + 5y = 4$ $$-6 + 5(2) \overset{?}{=} 4$$ $$-6 + 10 \overset{?}{=} 4$$ $$4 = 4$$ Because the equation is true, $(-6, 2)$ is a solution. **b.** $(2.2, 4.8)$; $y = 4x - 3$ $$4.8 \overset{?}{=} 4(2.2) - 3$$ $$4.8 \overset{?}{=} 8.8 - 3$$ $$4.8 \neq 5.8$$ Because the equation is false, $(2.2, 4.8)$ is not a solution.

[9.3] _For Exercises 16–18, determine whether the ordered pair is a solution for the equation._

16. $(2, 7)$; $3x - y = -4$ **17.** $(2.5, -0.5)$; $y = -x + 2$

18. $\left(-4\frac{3}{4}, -1\frac{9}{10} \right)$; $y = \frac{2}{5}x$

Definitions/Rules/Procedures	Key Example(s)
To find a solution to an equation in two variables:	Find a solution for $2x - y = -3$.
1. Replace one of the variables with a value (any value).	We will choose x to be 1 and solve for y.
2. Solve the equation for _____.	$2(1) - y = -3$ **Replace x with 1.**
	$2 - y = -3$ **Multiply.**
	$2 - y = -3$
	$\dfrac{-2}{} \quad\quad \dfrac{-2}{}$ **Subtract 2 from both sides.**
	$0 - y = -5$
	$\dfrac{-y}{-1} = \dfrac{-5}{-1}$ **Divide both sides by -1.**
	$y = 5$ Solution: $(1, 5)$
To graph a linear equation:	Graph. $2x - y = -3$
1. Find at least _____ solutions to the equation.	In the preceding example, we found one solution, $(1, 5)$.
2. Plot the solutions as points in the rectangular coordinate system.	Two more solutions that can be found the same way are $(0, 3)$ and $(-2, -1)$.
3. Draw a(n) _____ line through these points.	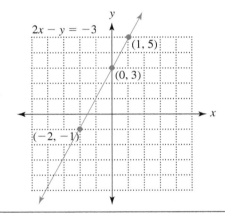

[9.3] For Exercises 19–24, find three solutions for the given equation. Then graph. (Answers may vary for the three solutions.)

19. $2x + y = 6$ **20.** $3x - 6y = 12$ **21.** $y = x + 3$

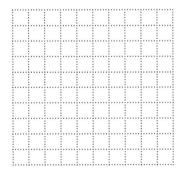   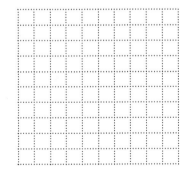   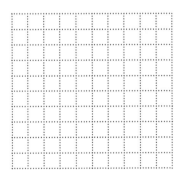

22. $y = -3x$

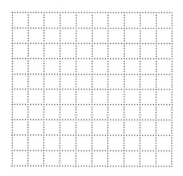

23. $y = \dfrac{2}{3}x$

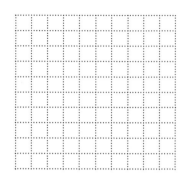

24. $y = -x - 4$

Definitions/Rules/Procedures	Key Example(s)
The graph of $y = c$, where c is a constant, is a(n) _____ line parallel to the x-axis and intersects the y-axis at the point with coordinates $(0, c)$. **The graph of $x = c$, where c is a constant,** is a(n) _____ line parallel to the y-axis and intersects the x-axis at the point with coordinates $(c, 0)$.	Graph. **a.** $y = 3$ **b.** $x = -2$ 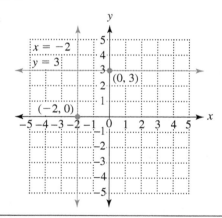

[9.3] *For Exercises 25 and 26, graph.*

25. $y = 7$

26. $x = -4$

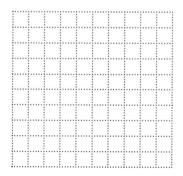

Definitions/Rules/Procedures	Key Example(s)
To find an *x*-intercept: 1. Replace _____ with 0 in the given equation. 2. Solve for _____. **To find a *y*-intercept:** 1. Replace _____ with 0 in the given equation. 2. Solve for _____.	Find the coordinates of the *x*- and *y*-intercepts for $3x + 5y = 12$. *x*-intercept: *y*-intercept: $3x + 5(0) = 12$ $3(0) + 5y = 12$ $\quad\;\; 3x = 12$ $5y = 12$ $\quad\;\; \dfrac{3x}{3} = \dfrac{12}{3}$ $\dfrac{5y}{5} = \dfrac{12}{5}$ $\quad\;\;\;\; x = 4$ *x*-intercept: $(4, 0)$ $y = \dfrac{12}{5}$ *y*-intercept: $\left(0, \dfrac{12}{5}\right)$

[9.3] For Exercises 27–31, find the coordinates of the x- and y-intercepts.

27. $5x + y = 10$ **28.** $y = 4x - 1$ **29.** $y = \dfrac{1}{5}x$

30. $x = 6$ **31.** $y = -2$

9.4 Applications with Graphing

Definitions/Rules/Procedures	Key Example(s)
A **vertex** is a(n) _____ where two lines join to form an angle. **Given the coordinates of the vertices of a figure in the rectangular coordinate system, the coordinates of its centroid are as follows:** $\left(\dfrac{\text{of the } x\text{-coordinates}}{\text{of all vertices}}, \dfrac{\text{of the } y\text{-coordinates}}{\text{of all vertices}}\right)$	Find the centroid of the triangle with vertices $(4, -3), (-2, -4),$ and $(-5, 1)$. $\text{centroid} = \left(\dfrac{4 + (-2) + (-5)}{3}, \dfrac{-3 + (-4) + 1}{3}\right)$ $= \left(\dfrac{-3}{3}, \dfrac{-6}{3}\right)$ $= (-1, -2)$

[9.4] **32.** Find the centroid of the figure with vertices $(0, -1), (0, 4), (3, 4),$ and $(7, -1)$.

[9.4] For Exercises 33 and 34, solve.

[9.4] **33.** The linear equation $p = 0.4r - 12{,}000$ describes the profit for a company, where *r* represents the revenue.

     a. Find the profit assuming that the revenue is $100,000.

     b. Find the revenue required to break even (the point at which profit is $0).

     c. Graph the equation.

[9.4] **34.** Find the area of the figure with vertices $(-2, 5), (-1, -3), (4, -3),$ and $(3, 5)$.

Chapter 9   Practice Test

For Extra Help      Step-by-step test solutions are found on the Chapter Test Prep Videos available via the Video Resources on DVD, in MyMathLab®, and on YouTube™ (search "CarsonPrealgebra" and click on "Channels").

1. a. _____

   b. _____

[9.1]   **1.** Identify the figure.

a. A B   b. R S

2. a. _____

   b. _____

[9.1]   **2.** State whether the angle is acute, right, obtuse, or straight.

a.   b.

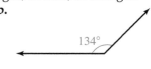

$134°$

3. _____

[9.1]   **3.** Four angles are formed by the following intersecting lines. Find the measure of each angle.

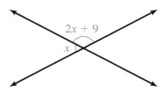

$2x + 9$

x

4. _____

[9.1]   **4.** In the following figure, lines q and r are parallel. The transversal t forms eight angles. Find the measure of each angle.

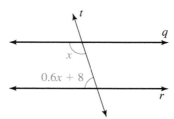

t

q

x

$0.6x + 8$

r

5. _____

[9.1]   **5.** Determine whether the following two triangles are congruent. If they are congruent, state the rule that explains why.

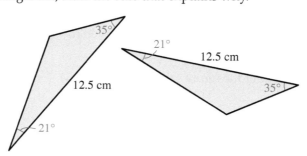

$35°$ $21°$ 12.5 cm

12.5 cm $35°$

$21°$

Learning Strategy

If a test is based on formulas or theorems, I make flash cards and separate them into piles of the ones I know and the ones I don't know until the "know" pile contains all of the cards.

—Beth M.

[9.1]   **6.** Find the measure of each angle in the following congruent triangles.

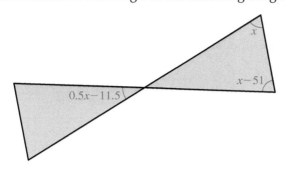

6. _____

[9.2]   **7.** Determine the coordinates of each point shown.

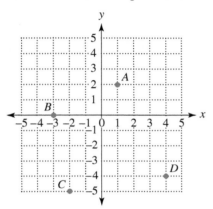

7. _____
  _____
  _____
  _____

[9.2]   **8.** In which quadrant is $(-16, 48)$ located?

8. _____

[9.2]   **9.** In which quadrant is $(0, -4)$ located?

9. _____

[9.2]   **10.** Plot and label the point described by each ordered pair of coordinates.

$A(-4, 2)$ $B(2, 1)$ $C(0, -3)$ $D(-3, -5)$

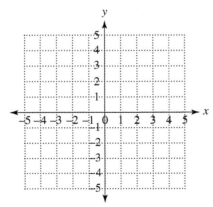

11. _____

[9.1] **11.** Find the midpoint of $(5, 3)$ and $(1, 1)$.

12. _____

[9.3] **12.** Determine whether $(5, -3)$ is a solution for $-x + 2y = -11$.

13. _____

[9.2] **13.** Determine whether $(-3, -4)$ is a solution for $y = \frac{2}{3}x + 1$.

14. _____

[9.3] **14.** Find the coordinates of the x- and y-intercepts for $3x - 5y = 6$.

15. _____

[9.3] **15.** Find the coordinates of the x- and y-intercepts for $y = 3x - 4$.

16. _____

[9.3] **16.** Find the coordinates of the x- and y-intercepts for $y = 3$.

[9.3] *For Exercises 17–22, graph.*

17. $y = -3x$

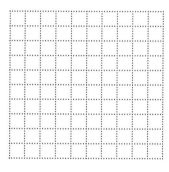

18. $y = \frac{1}{5}x - 3$

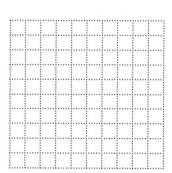

19. $x + y = 7$

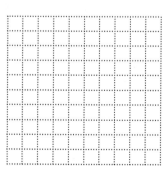

20. $3x - y = 2$

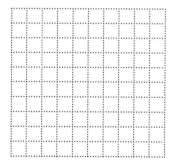

21. $y = 1$

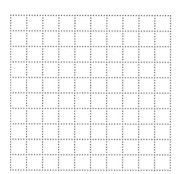

22. $x = -2$

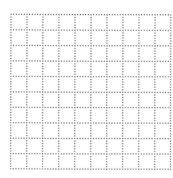

[9.4] **23.** The linear equation $v = -9.8t + 40$ describes the velocity in meters per second of a model rocket t seconds after being launched.

    a. Find the initial velocity of the rocket.

    b. Find the velocity of the rocket 3 seconds after being launched.

    c. How many seconds after launch will the rocket stop before returning to Earth?

    d. Graph the equation with t as the horizontal axis and v as the vertical axis.

23. a. _____
   b. _____
   c. _____

[9.4] **24.** Find the centroid of the figure with vertices $(4, 3)$, $(-2, 3)$, $(-2, -5)$, and $(4, -5)$.

24. _____

[9.4] **25.** Find the area of the figure with vertices $(2, 4)$, $(-5, 4)$, $(-2, -4)$, and $(5, -4)$.

25. _____

Chapters 1–9 Cumulative Review Exercises

For Exercises 1–6, answer true or false.

[2.3] **1.** $5 - 8 = 8 - 5$

[3.5] **2.** 41 is a prime number.

[4.2] **3.** $4x = 9(x + 1.5)$ is a linear equation.

[5.1] **4.** 0 is a rational number.

[8.1] **5.** $\dfrac{3}{8} > 37\%$

[9.1] **6.** Alternate interior angles are congruent.

[2.5] **7.** State the order of operations agreement.

[2.2] **8.** Explain how to add two numbers that have the same sign.

[2.2] **9.** Explain how to add two numbers that have different signs.

[2.3] **10.** Explain how to write a subtraction statement as an equivalent addition statement.

[2.4] **11.** When multiplying or dividing two numbers that have the same sign, the result is _____.

[2.4] **12.** When multiplying or dividing two numbers that have different signs, the result is _____.

$\begin{bmatrix} 1.2 \\ 1.3 \end{bmatrix}$ **13.** Estimate each calculation by rounding.
   a. $21{,}459 + 6741$
   b. $729{,}105 - 4519$
   c. 86×17
   d. $2219 \div 112$

14. Graph on a number line.
   [2.1] **a.** 3

   [5.1] **b.** $5\dfrac{1}{4}$

   [6.1] **c.** -7.2

[3.2] **15.** What is the degree of each expression?
   a. $-4y$
   b. $15x^2$
   c. $4x^3 - 9x + 10x^4 - 7$

[6.3] **17.** Write 3.53×10^6 in standard form.

[6.1] **16.** Write the word name for 4607.09.

[9.1] **18.** Is the following angle acute, right, obtuse, or straight?

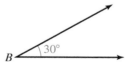

For Exercises 19–21, simplify.

[2.5] **19.** $19 - (6 - 8)^3 + \sqrt{100 - 36}$

[5.7] **20.** $-6\dfrac{1}{8} \div \dfrac{3}{16} - \dfrac{3}{4}$

[6.5] **21.** $\left(\dfrac{3}{4}\right)^2 (-8.4)$

22. Write as a decimal number.
   [6.4] **a.** $\dfrac{3}{8}$
   [6.4] **b.** $\dfrac{1}{3}$
   [8.1] **c.** 8%
   [8.1] **d.** $24\dfrac{2}{5}\%$
   [8.1] **e.** 4.5%

23. Write as a fraction.

[5.1] **a.** $5\dfrac{1}{6}$

[6.1] **b.** 0.145

[8.1] **c.** 6%

[8.1] **d.** $5\dfrac{1}{2}\%$

[8.1] **e.** 16.4%

[8.1] **24.** Write as a percent.

a. $\dfrac{1}{4}$

b. $\dfrac{4}{9}$

c. 0.65

d. 0.035

e. 2.3

[3.1] **25.** Evaluate each expression using the given values.
  a. $t^3 - 5t;\ t = -4$
  b. $-xy + 2\sqrt{x + y};\ x = -4,\ y = 20$

[6.2] **26.** Subtract.
$$\left(x^4 - \dfrac{1}{3}x^2 + 9.6\right) - \left(4x^3 + \dfrac{3}{5}x^2 - 14.6\right)$$

27. Multiply.

[5.3] **a.** $\left(\dfrac{5}{8}a^2\right)\left(-\dfrac{4}{9}a\right)$
[3.4] **b.** $(m - 9)(m + 9)$
[6.3] **c.** $(4.5t - 3)(7.1t + 6)$

[3.5] **28.** Find the prime factorization of 630.

For Exercises 29–32, solve and check.

[4.3] **29.** $-4x + 19 = -1$

[5.8] **30.** $\dfrac{3}{5}k - 7 = \dfrac{1}{4}$

[6.6] **31.** $2.4(n - 5) + 8 = 1.6n - 14$

[7.2] **32.** $\dfrac{u}{14} = \dfrac{-3}{8}$

$\begin{bmatrix} 7.3 \\ 7.4 \\ 7.5 \end{bmatrix}$ **33.** Convert.

  a. 2.4 miles to feet

  b. 4500 millimeters to meters

  c. 4 square yards to square feet

  d. 42,000 square centimeters to square meters

  e. 4500 pounds to American tons

  f. 0.23 kilogram to grams

  g. 68 cups to quarts

  h. 400 milliliters to liters

  i. 77°F to degrees Celsius

  j. −20°C to degrees Fahrenheit

For Exercises 34–50, solve.

[2.6] **34.** A family has the following assets and debts. Calculate the family's net worth.

Assets	Debts
Savings = $1282	Credit card balance = $2942
Checking = $3548	Mortgage = $83,605
Furniture = $21,781	Automobile 1 = $4269
Jewelry = $5745	

[3.7] **35.** Use the figure to the right.
  a. Write an expression in simplest form for the perimeter.
  b. Find the perimeter if h is 5 cm.
  c. Write an expression in simplest form for the area.
  d. Find the area if $h = 7$ cm.

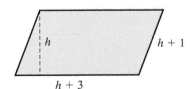

[4.5] **36.** The sum of two positive integers is 123. One integer is three more than twice the other integer. What are the integers?

[5.3] **37.** On Sandra's cell phone bill, 8 out of the 32 calls were out-of-network; $\dfrac{3}{4}$ of those calls were to her parents. What fraction of all of her cell phone calls were to her parents? How many calls were to her parents?

[5.6] **38.** A poll is taken to assess the president's approval rating. Respondents can answer four ways: excellent, good, fair, or poor. $\frac{1}{6}$ said "excellent," $\frac{3}{5}$ said "good," and $\frac{1}{8}$ said "fair."

    a. What fraction of the respondents said "excellent" or "good"?

    b. What fraction said "poor"?

39. Use the following graph, which shows the closing price for a stock each day for four days in April.

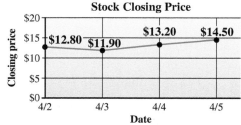

    [1.2] **a.** How much more was the closing price on April 5 than on April 3?
    [6.5] **b.** Find the mean, median, and mode closing price for the dates shown in the graph.

[6.5] **40.** Find the area of the shaded region. Use 3.14 for π.

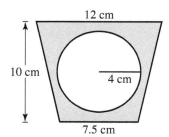

[6.5] **41.** Find the volume of a can that is 3.5 inches tall and is 3 inches in radius. Use 3.14 for π.

[6.6] **42.** A steel support beam on a tower needs to be replaced. The connecting joint is 20 feet above the ground, and the base of the support beam is 8 feet from a point directly below the connecting joint. How long is the beam?

[7.2] **43.** The triangles shown are similar. Find the unknown side lengths.

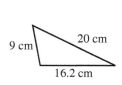 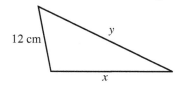

44. Use the following graph, which shows the number of students who earned the indicated grade in English 101.

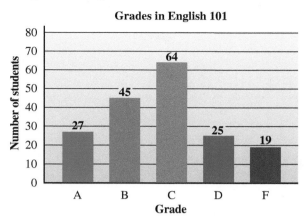

Grades in English 101

[1.2] **a.** How many students took English 101?

[5.1] **b.** What fraction, in lowest terms, of the students earned an A?

[8.4] **c.** What percent of the students earned a C?

[8.4] **d.** Draw a circle graph that shows the percent of the students who earned each grade.

[8.5] **45.** Because Coly is an employee of a store, he gets a 20% discount when he purchases merchandise. What would be his cost for an item that is normally $75.95?

[8.5] **46.** A restaurant increases its price on a menu item from $8.95 to $9.75. What is the percent of increase in price?

[8.6] **47.** How much interest is earned if $680 is invested at 4.2% simple interest rate for half a year?

[9.1] **48.** In the following figure, lines l and m are parallel. The transversal t forms eight angles. Find the measure of each angle.

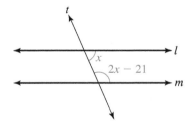

[9.2] **49.** Determine the coordinates of each point shown.

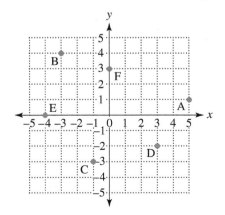

[9.3] **50.** Graph. $y = -3x + 2$

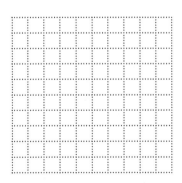

Answers

Exercise Set 1.1

Prep Exercises **1.** 0, 1, 2, 3 **2.** No, zero is not a natural number.
3.

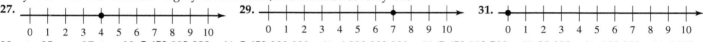

Billions period			Millions period			Thousands period			Ones period		
Hundred Billions	Ten billions	Billion	Hundred millions	Ten Millions	Millions	Hundred Thousands	Ten Thousands	Thousands	Hundreds	Tens	Ones

4. 1. Write each digit multiplied by its place value. 2. Express it all as a sum. **5.** 1. Write the name of the digits in the leftmost period. 2. Write the period name followed by a comma. 3. Repeat steps 1 and 2 until you get to the ones period. Do not follow the ones period with its name. **6.**

```
+--+--+--+--+--+-->
0  1  2  3  4  5
```

7. greater **8.** Consider the digit to the right of the desired place value. If this digit is 5 or greater, round up. If it is 4 or less, round down.

Exercises **1.** 0 **3.** 5 **5.** thousands place **7.** hundred thousands place **9.** 2 ten thousands + 4 thousands + 3 hundreds + 1 ten + 9 ones **11.** 5 millions + 2 hundred thousands + 1 ten thousand + 3 thousands + 3 hundreds + 4 ones
13. 9 ten millions + 3 millions + 1 ten thousand + 4 thousands + 8 ones **15.** 8792 **17.** 6,039,020 **19.** 40,980,109
21. seven thousand, seven hundred sixty-eight **23.** three hundred ten million, four hundred sixty-four thousand, seven hundred thirty-four **25.** one hundred eighty-six thousand, one hundred seventy-one
27.

```
+--+--+--+--•--+--+--+--+--+--+-->
0  1  2  3  4  5  6  7  8  9  10
```

29.

```
+--+--+--+--+--+--+--+--•--+--+-->
0  1  2  3  4  5  6  7  8  9  10
```

31.

```
•--+--+--+--+--+--+--+--+--+--+-->
0  1  2  3  4  5  6  7  8  9  10
```

33. < **35.** = **37.** > **39.** 5,652,992,000 **41.** 5,653,000,000 **43.** 6,000,000,000 **45.** 5,652,992,500 **47.** 30,000 **49.** 900,000 **51.** 9000
53. 142,000,000 miles; 142,000,000 because the number is reasonably accurate yet simple. (Answers may vary.) **55. a.** $64,424
b. $46,179 **c.** California, New York, Illinois, and Massachusetts **d.** $47,000 **e.** $53,000v **f.** cost of living

Exercise Set 1.2

Prep Exercises **1.** addends; sum **2.** Changing the order of the addends does not affect the sum. **3.** 1. Round the smallest number to the highest possible place value so that it has one nonzero digit. 2. Round all of the other numbers to the same place value as the smallest number. 3. Add the rounded numbers. **4.** minuend; subtrahend; difference. **5.** A constant is a symbol that does not vary in value, whereas a variable is a symbol that can vary in value. **6.** A solution to an equation is a number that makes the equation true when it replaces the variable in the equation. **7.** Write a related subtraction equation, subtracting the known addend from the sum.
8.

Addition key words/key question	Subtraction key words/ key question
(Answers may vary.) sum, plus, total, perimeter; how much in all?	(Answers may vary.) subtract, minus, difference, take away; how much is left?

Exercises **1.** Estimate: 9000 Actual: 8849 **3.** Estimate: 101,000 Actual: 100,268 **5.** Estimate: 15,800 Actual: 15,814 **7.** Estimate: 15,200 Actual: 15,246 **9.** Estimate: 181,400 Actual: 181,383 **11.** Estimate: 5400; Actual: 5352 **13.** Estimate: 34,000; Actual: 34,161 **15.** Estimate: 45,000; Actual: 45,334 **17.** Estimate: 20,000; Actual: 26,519 **19.** Estimate: 170,000; Actual: 166,488
21. $x = 4$ **23.** $t = 21$ **25.** $n = 18$ **27.** $u = 155$ **29.** $b = 225$ **31.** 1810 tickets **33.** 68 ft. **35.** $16 **37.** 1720 sq. ft. **39.** $134,650
41. 2100 K **43.** 33 A **45.** $6611 **47. a.** 69,463,000 **b.** 47,513,000 **c.** 22,000,000 **d.** 110,000 **e.** 5,000,000 **f.** 25,563,000

Review Exercises **1.** 3 hundred millions + 7 millions + 4 hundred thousands + 9 ten thousands + 1 thousand + 2 tens + 4 ones
2. one million, four hundred seventy-two thousand, three hundred fifty-nine **3.**

```
+--+--+--+--+--+--•--+--+--+--+-->
0  1  2  3  4  5  6  7  8  9  10
```

4. < **5.** 23,410,000

Exercise Set 1.3

Prep Exercises **1.** factors; product. **2.** Changing the order of the factors does not affect the product. **3.** 1. Round each factor to the highest possible place so that each has only are nonzero digit. 2. Multiply the rounded digits. **4.** $a(b + c) = ab + ac$
5. Multiply four 3s. **6.** (Answer may vary.) Multiply, times, product. **7.** A rectangular array is a rectangle formed by a pattern of neatly arranged rows and columns. **8.** The area of a shape is the total number of square units that completely fill the shape.

Exercises **1.** 352 **3.** 0 **5.** Estimate: 160; Actual: 192 **7.** Estimate: 4500; Actual: 4914 **9.** Estimate: 24,000; Actual: 22,576
11. 44,940 **13.** 995,330 **15.** 8,550,756 **17.** 241,356,618 **19.** 330,050 **21.** 16 **23.** 1 **25.** 243 **27.** 10,000,000 **29.** 42,719,296
31. 130,691,232 **33.** 9^4 **35.** 7^5 **37.** 14^6 **39.** $2 \times 10^4 + 4 \times 10^3 + 9 \times 10^2 + 2 \times 1$ **41.** $9 \times 10^6 + 1 \times 10^5 + 2 \times 10^4 + 8 \times$

$10^3 + 2 \times 10$ **43.** $4 \times 10^8 + 7 \times 10^6 + 2 \times 10^5 + 1 \times 10^4 + 9 \times 10^2 + 2 \times 10 + 5 \times 1$ **45.** 7954 intersections; 63,632 traffic lights
47. 36,792,000 heartbeats in a year; 2,759,400,000 heartbeats in a lifetime **49.** 740 mg **51.** \$41,600 **53.** 19,200,000,000 oz.
55. 112 combinations **57.** 32 binary numbers; no, it cannot recognize $26 + 10 = 36$ binary numbers. **59.** 128 locations
61. 624 ft.^{2}; no, he has overestimated. **63.** 300 ft.^{2}

Review Exercises **1.** sixteen million, five hundred seven thousand, three hundred nine **2.** 2 ten thousands + 3 thousands +
5 hundreds + 6 ones **3.** 56,988 **4.** 831,724 **5.** $n = 13$

Exercise Set 1.4
Prep Exercises 1. dividend; divisor; quotient. **2.** n **3.** 0 **4.** undefined **5.** indeterminate. **6.** 1 **7.** even **8.** the sum of the digits is divisible by 3. **9.** have 0 or 5 in the ones place. **10.** Try rounding the dividend to two nonzero digits. **11.** Write a related division equation in which the product is divided by the known factor. **12.** (Answers may vary.) divide, distribute, evenly each, split, quotient, into, per, over

Exercises 1. 1 because $26 \cdot 1 = 26$ **3.** 0 because $49 \cdot 0 = 0$ **5.** indeterminate because $x \cdot 0 = 0$ is true for any number **7.** undefined because no number can make $x \cdot 0 = 22$ true **9.** no **11.** yes **13.** yes **15.** yes **17.** no **19.** yes **21.** yes **23.** yes **25.** no
27. Estimate: 400 Actual: 403 **29.** Estimate: 200 Actual: 207 r3 **31.** Estimate: 150 Actual: 217 **33.** Estimate: 200 Actual: 246 r3
35. Estimate: 300 Actual: 304 **37.** Estimate: 40 Actual: 40 **39.** Estimate: 2000 Actual: 1900 r13 **41.** Estimate: 1500 Actual: 1601 r12
43. undefined **45.** $x = 6$ **47.** $m = 8$ **49.** $n = 7$ **51.** $t = 0$ **53.** $v = 42$ **55.** $h = 20$ **57.** \$2854 **59.** 14 ml **61.** 125 pieces of
paper **63.** 22 stamps; there are 1000 cents in \$10 and $1000 \div 44 = 22$ r32. We round down because we cannot buy part of a stamp.
65. a. 9600 boxes **b.** 1200 bundles **c.** 50 pallets **d.** 2 trucks; $50 \div 28 = 1$ r22. Because all of the pallets cannot fit on one truck, we
round up to 2. **67.** 18 ft. **69.** 22 sections

Review Exercises **1.** $t = 177$ **2.** 60 ft. **3.** Estimate: 16,000 Actual: 15,414 **4.** 625 **5.** $4 \times 10^4 + 9 \times 10^3 + 6 \times 10^2 + 2 \times 1$

Exercise Set 1.5
Prep Exercises 1. 1. Grouping symbols 2. Exponents 3. Multiplication or division from left to right 4. Addition or subtraction
from left to right **2.** Multiply 4 by 3. We multiply before subtracting. **3.** Divide 20 by 5. We multiply or divide from left to right
before adding. **4.** Subtract 4 from 10 within the parentheses. We work within grouping symbols before simplifying any other operations. **5.** describe some characteristic **6.** 1. Find the sum of the values. 2. Divide the sum by the number of values. **7.** Find
the mean of the middle two values. **8.** The value that occurs most often.

Exercises 1. 22 **3.** 45 **5.** 141 **7.** 14 **9.** 1 **11.** 35 **13.** 44 **15.** 49 **17.** 24 **19.** 5 **21.** 25 **23.** 99 **25.** 81 **27.** 37 **29.** 18
31. 83 **33.** 46 **35.** 47 **37.** 48 **39.** 30 **41.** 10 **43.** 13 **45.** 1354 **47.** Mistake: subtracted 6 from 48 instead of multiplying 5 by 6;
Correct: 18 **49.** Mistake: squared 3 and 5 instead of squaring their sum; Correct: 50 **51.** mean: 8; median: 7; mode: 13 **53.** mean:
10; median: 9; modes: 4, 17 **55.** mean: 37; median: 40; no mode **57.** mean: 62; median: 61; mode: 83 **59.** mean: 74 in.; median:
74 in.; mode: 74 in. **61.** mean: 26; median: 25; mode: 25, 21 **63.** mean: \$101,230,000; median: \$89,100,000; no mode

Review Exercises **1.** 70,000 **2.** 104,187,127 **3.** 1250 r3 **4.** 56 ft. of molding; \$224 **5.** 192 tiles

Exercise Set 1.6
Prep Exercises 1. Replace the variables with the corresponding known values; then solve for the unknown variable.
2. $p = 2l + 2w$ **3.** A parallelogram is a four-sided figure with two pairs of parallel sides. **4.** $A = bh$ **5.** Volume is the total
number of cubic units that completely fill an object. **6.** $v = lwh$ **7.** Understand the problem. **8.** Check the results.
9. Add the areas of the two smaller shapes. **10.** Subtract the area of the portion that is removed from the area of the larger shape.

Exercises 1. 94 cm **3.** 128 in. **5.** 64 km **7.** 228 m^2 **9.** 144 in.^{2} **11.** 112 km^2 **13.** 280 ft.^{3} **15.** 216 in.^{3} **17.** 210 km^3 **19.** 9 m
21. 62 in. **23.** 20 in. **25.** 2 cm **27.** 18 in. **29.** 4 ft. **31.** 21 ft. **33.** \$288 **35.** \$28 **37.** \$5525 **39. a.** 138 posts; \$276 **b.** 4360 ft.;
4 rolls of barbed wire; \$180 **c.** 552 nails; 6 boxes of nails; \$18 **d.** \$549. **e.** \$4976; He must consider how long it will take, what the
price of the tools are, and whether building a fence is something he knows how to do. **41.** 70 sheets; \$1050 **43.** 648 ft.^{2}; 2 gallons
45. 35,602 ft.^{2} **47.** 486,000 ft.^{3} **49.** 28 ft.^{3}

Review Exercises **1.** $x = 1929$ **2.** $2 \times 10^6 + 4 \times 10^5 + 8 \times 10^3 + 7 \times 10 + 3 \times 1$ **3.** Estimate: 35,000,000 Actual: 32,183,756
4. $y = 203$ **5.** 53

Chapter 1 Summary and Review Exercises
1.1 Definitions/Rules/Procedures a group of elements; a set within a set; 1, 2, 3, . . . ; 0, 1 2, 3, . . . ; Ten millions; Millions;
Hundred thousands; Thousands; Tens; digit; place-value name and a plus sign; write each digit in the place indicated by the
corresponding place value Exercises **1.** 5 millions + 6 hundred thousands + 8 ten thousands + 9 hundreds + 1 one
2. 4 ten thousands + 2 thousands + 5 hundreds + 1 ten + 9 ones **3.** 98,274 **4.** 8,020,096 Definitions/Rules/
Procedures word name of the number formed by the digits in the period Exercises **5.** nine thousand, four hundred twenty-
one **6.** one hundred twenty-three million, four hundred five thousand, six hundred Definitions/Rules/Procedures draw a dot
on the mark for that number
Exercises **7.** 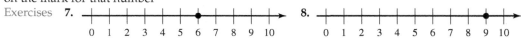 **8.**

Definitions/Rules/Procedures   a mathematical relationship that contains an equal sign ($=$); a mathematical relationship that contains an inequality symbol ($<$, $>$); $=$; greater   Exercises   **9.** $<$ **10.** $>$ Definitions/Rules/Procedures   round up by increasing the digit in the given place value by 1; round down by keeping the digit in the given place value the same   Exercises   **11.** 3,380,000   **12.** 1949–1950 to 1969–1970   **13.** 46,000,000   **14.** 2,600,000

1.2 Definitions/Rules/Procedures   combines; the corresponding place values align; the smallest number to the highest possible place value so that it has only one nonzero digit   Exercises   **15.** Estimate: 53,000; Actual: 52,721   **16.** Estimate: 39,680; Actual: 39,673   Definitions/Rules/Procedures   the total distance around the figure; add the lengths of all sides of the figure   Exercises   **17.** 40 in   **18.** 55 ft.   Definitions/Rules/Procedures   take one amount away from another; difference; unknown addend; rename the top digit; the smallest number to the highest possible place value so that it has only one nonzero digit   Exercises   **19.** Estimate: 480,000; Actual: 485,716   **20.** Estimate: 7700; Actual: 7707   Definitions/Rules/Procedures   subtraction; addend; sum   Exercises   **21.** $x = 25$ **22.** $y = 189$ Definitions/Rules/Procedures   add, plus, sum, total, increased by, in all (Answers may vary.); subtract, minus, difference, remove, decreased by, take away (Answers may vary.)   Exercises   **23.** $706   **24.** 40,839,000

1.3 Definitions/Rules/Procedures   calculates repeated addition of the same number; 0; n; the distributive property; each factor to the highest possible place value so that each has only one nonzero digit   Exercises   **25.** Estimate: 2100; Actual: 1800   **26.** Estimate: 180,000; Actual: 181,437   Definitions/Rules/Procedures   base; exponent; write the base as a factor the number of times indicated by the exponent   Exercises   **27.** 128   **28.** 125   **29.** 10^5 **30.** 7^8 Definitions/Rules/Procedures   a rectangle formed by a pattern of neatly arranged rows and columns; 1×1; square units that completely fill a shape; describes a procedure; Replace the variables with the corresponding known values; lw; multiply, times, product, each, of, by   Exercises   **31.** 64 numbers   **32.** 288 ft.^{2}

1.4 Definitions/Rules/Procedures   calculates repeated subtraction of the same number; the amount left over after dividing two whole numbers; n; undefined; 0; indeterminate; 1   Exercises   **33.** 0   **34.** undefined   **35.** 24   **36.** 1; even number   Definitions/Rules/Procedures   the sum of the digits is divisible by 3; a 0 or 5 in the ones place   Exercises   **37.** No, because $5 + 9 + 6 = 20$, which is not divisible by 3   **38.** Yes, because 1780 has 0 in the ones place; long division   Definitions/Rules/Procedures   each number to the highest possible place value so that each has one nonzero digit   Exercises   **39.** Estimate: 4000; Actual: 127   **40.** Estimate: 2000; Actual: 2109 r14   Definitions/Rules/Procedures   division; product; known factor   Exercises   **41.** $b = 24$ **42.** $k = 306$ Definitions/Rules/Procedures   divide, distribute evenly, each, split, quotient, into (Answers may vary.)   Exercises   **43.** 2 ml each minute   **44.** $3275

1.5 Definitions/Rules/Procedures   Grouping symbols; Exponents; Multiplication or division; Addition or subtraction   Exercises   **45.** 43   **46.** 58   **47.** 32   **48.** 10   Definitions/Rules/Procedures   number used to describe some characteristic; the sum of the given values and the number of values; the number of values; middle; ordered; middle value; mean of the middle two values; the value that occurs most often; most   Exercises   **49.** mean: 12; median: 10; modes 7 and 16

1.6 Definitions/Rules/Procedures   do not intersect; two pairs of parallel sides; $90°$; $1 \times 1 \times 1$; cubic units; $2l + 2w$; bh; lwh Exercises   **50.** 168 m^2 **51.** 378 ft.^{3} Definitions/Rules/Procedures   Understand; Plan; Execute; Answer; Check; add; subtract   Exercises   **52.** 89,500 ft.^{2}

Chapter 1 Practice Test
1. 5   **2.** $4 \times 10^7 + 8 \times 10^6 + 2 \times 10^5 + 1 \times 10^4 + 9 \times 10^2 + 7 \times 1$ **3.** $>$ **4.** sixty-seven million, one hundred ninety-four thousand, two hundred ten   **5.** 2,800,000   **6.** Estimate: 68,000 Actual: 68,190   **7.** Estimate: 430,000 Actual: 425,709   **8.** $x = 8$ **9.** Estimate: 2,800,000 Actual: 2,466,870   **10.** 64   **11.** Estimate: 2500 Actual: 2501 r1   **12.** $y = 18$ **13.** 25   **14.** 22   **15.** 18   **16.** 1   **17.** 2   **18. a.** $1,060,000,000 **b.** $830,000,000   **19.** $189,000,000   **20.** 83 classes   **21. a.** 93 **b.** 95   **c.** 85   **22.** 162 ft.^{2} **23. a.** 290 ft. **b.** $1740   **24.** 336 ft.^{3} **25.** 123 ft.^{2}

Chapter 1–2 Cumulative Review Exercises
1. true   **2.** false   **3.** false   **4.** true   **5.** commutative; addition   **6.** associative; multiplication   **7.** negative   **8.** Grouping symbols; Exponents or roots; Multiplication or division; Addition or subtraction   **9.** four hundred nine million, two hundred fifty-four thousand, six   **10.** 57,000   **11.** 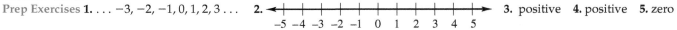   **12.** $>$ **13.** 16   **14.** 93   **15.** -50 **16.** -144 **17.** 64   **18.** 201 r2   **19.** -15 **20.** 11   **21.** -19 **22.** -48 **23.** $x = 25$ **24.** $c = -7$ **25.** perimeter: 74 m; area: 300 m^2 **26.** 36 in.^{3} **27.** mean: 80; median: 78; mode: 95   **28.** 21,325 ft.^{2} **29. a.** $36°F$ **b.** $58°F$ **30.** $-$$17,010; loss

Exercise Set 2.1
Prep Exercises 1. $\ldots -3, -2, -1, 0, 1, 2, 3 \ldots$ **2.** (number line from -5 to 5)   **3.** positive   **4.** positive   **5.** zero

Exercises   **1.** $+450$ **3.** -35 **5.** $+29,035$ **7.** $-13,200$ **9.** $+100$ **11.** -20 **13.** $-8°F$ **15.** $-20°F$ **17.** The average critical reading score of students whose parents have no high school diploma is 81 points below the overall average critical reading score.   **19.** The average math score of students whose parents have a high school diploma is 41 points below the overall average math score.   **21.** The average writing score of students whose parents have a bachelor's degree is 19 points above the overall average verbal score.   **23.** The average math score of students whose parents have a bachelor's degree is 20 points above the overall average math score.   **25.** 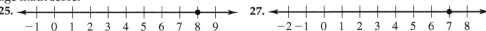   **27.** (number line from -2 to 8)

29.

31.

33.

35. <   **37.** >   **39.** <   **41.** >   **43.** >   **45.** <   **47.** 26.   **49.** 21.   **51.** 18.   **53.** 10.   **55.** 0.   **57.** 14   **59.** 18   **61.** 2004   **63.** 8   **65.** 0   **67.** 47

Review Exercises   **1.** 94,265   **2.** 864,488   **3.** 278,313   **4.** 2031 r12   **5.** $482   **6.** 46

Exercise Set 2.2

Prep Exercises **1.** Adding a debt of $6 to a debt of $24 increases the debt to $30; so the result is -30.   **2.** Making a $12 payment toward a $20 debt means that we are decreasing the debt. But because the debt exceeds the payment, we have a debt of $8 remaining; so the result is -8.   **3.** add; keep the same sign.   **4.** subtract; keep the sign of the number with the greater absolute value.   **5.** absolute value; opposite   **6.** 0

Exercises   **1.** 23   **3.** -23 **5.** -15 **7.** 46   **9.** -76 **11.** 10   **13.** -14 **15.** 14   **17.** -18 **19.** 56   **21.** 138   **23.** -72 **25.** 61   **27.** 27   **29.** -21 **31.** -21 **33.** 0   **35.** 34   **37.** 32   **39.** 0   **41.** -59 **43.** -18 **45.** 29   **47.** $-\$33$ **49.** $-\$55,551$ **51.** $29   **53.** $-78°F$ **55.** -78 ft.   **57.** 257 lb.

59. a.

	No H.S. diploma	H.S. diploma	Associate degree	Bachelor's degree	Graduate degree
Expression	$501 + (-81)$	$501 + (-37)$	$501 + (-19)$	$501 + 20$	$501 + 58$
Sum	420	464	482	521	559

b. Students whose parents achieved higher levels of education tend to perform better on the SAT.

Review Exercises   **1.** $4 \times 10^7 + 2 \times 10^6 + 5 \times 10^5 + 6 \times 10^4 + 1 \times 10^3 + 9 \times 1$ **2.** two million, four hundred seven thousand, six   **3.** 55,445   **4.** $n = 13$ **5.** 25   **6.** $7265

Exercise Set 2.3

Prep Exercises **1.** Change the operation symbol from $-$ to $+$ and change the subtrahend (second number) to its additive inverse.   **2.** The subtrahend is 8. Its additive inverse is -8.   **3.** $+(-13)$ **4.** $+(-14)$ **5.** The subtrahend is -7. Its additive inverse is 7.   **6.** $+4$ **7.** $+6$ **8.** $-9-10$

Exercises   **1.** $18 + (-25) = -7$ **3.** $-15 + (-18) = -33$ **5.** $0 + (-9) = -9$ **7.** $20 + 8 = 28$ **9.** $-14 + 18 = 4$ **11.** $-15 + 8 = -7$ **13.** $-21 + (-19) = -40$ **15.** $-4 + 19 = 15$ **17.** $31 + (-44) = -13$ **19.** $35 + 10 = 45$ **21.** $-28 + 16 = -12$ **23.** $0 + (-18) = -18$ **25.** $0 + 5 = 5$ **27.** $t = -6$ **29.** $d = 15$ **31.** $u = 13$ **33.** $m = 14$ **35.** $h = -5$ **37.** $k = -12$ **39.** $-\$119$ **41.** net: $6428 million; profit   **43.** net: $-\$21,958$; loss   **45.** 1980; $-\$3,177,000,000$ **47.** $46°F$ **49.** $55°C$ **51.** $67   **53.** 1396 ft.   **55. a.**

	Barrow, AK	Duluth, MN	Nashville, TN
Expression	$-8 - (-20)$	$18 - (-1)$	$46 - 28$
Difference	12	19	18

b. Duluth, MN

Review Exercises   **1.** 30,305   **2.** 243   **3.** 504 r5   **4.** $y = 8$ **5.** 168   **6.** 4 mi^2.

Exercise Set 2.4

Prep Exercises **1.** positive.   **2.** negative.   **3.** positive.   **4.** negative.   **5.** Multiply -8^2 by the understood -1 **6.** positive   **7.** negative   **8.** $-15 \div 3$ **9.** positive; negative   **10.** no   **11.** positive   **12.** negative

Exercises   **1.** -32 **3.** -8 **5.** 64   **7.** 168   **9.** 0   **11.** 0   **13.** -15 **15.** 380   **17.** -35 **19.** 54   **21.** -78 **23.** -13 **25.** -36 **27.** 9   **29.** -64 **31.** 81   **33.** -1024 **35.** -64 **37.** $-1,000,000$ **39.** -1 **41.** -27 **43.** -12 **45.** -3 **47.** 8   **49.** 0   **51.** -5 **53.** 31   **55.** -7 **57.** undefined   **59.** $x = 3$ **61.** $x = -2$ **63.** $t = -4$ **65.** $c = -17$ **67.** $m = 9$ **69.** $m = 0$ **71.** no solution   **73.** $d = -5$ **75.** $g = 9$ **77.** ± 4 **79.** ± 6 **81.** no real roots exist   **83.** 9   **85.** 8   **87.** 7   **89.** not a real number   **91.** -11 **93.** 0   **95.** $-\$642$ **97.** $-\$119$ **99.** $46

Review Exercises   **1.** 60 ft.   **2.** 9   **3.** 47   **4.** 18   **5.** 138 ft.^{2} **6.** 504 ft.^{3}

Exercise Set 2.5

Prep Exercises **1.** 1. Grouping symbols; 2. Exponents or roots from left to right; 3. Multiplication or division from left to right; 4. Addition or Subtraction from left to right   **2.** Multiply 5 and 2. We multiply before adding.   **3.** Subtract 8 from 3 within the parentheses. We work within grouping symbols before performing any other operations.   **4.** Divide 12 by -3. We multiply or divide from left to right.

Exercises   **1.** -2 **3.** -15 **5.** 6   **7.** -6 **9.** -1 **11.** 10   **13.** 84   **15.** -29 **17.** 7   **19.** 4   **21.** 2   **23.** 16   **25.** -32 **27.** -9 **29.** 9   **31.** -1 **33.** 21   **35.** 38   **37.** -19 **39.** -60 **41.** 53   **43.** -24 **45.** 0   **47.** -5 **49.** -32 **51.** -40 **53.** -1 **55.** -22 **57.** -23 **59.** undefined   **61.** -17 **63.** -64 **65.** no real solution   **67.** -4 **69.** -5 **71.** -1 **73.** 3   **75.** $(-2)^4$ means to find the fourth power of -2, which is $(-2)(-2)(-2)(-2) = 16 \, -2^4$ means to multiply the fourth power of 2 by an understood -1, which is

$-1 \cdot 2 \cdot 2 \cdot 2 = -16.$ **77.** The exponent is odd.   **79.** Mistake: subtracion was performed before the multiplication; Correct: 58 **81.** Mistake: the exponential form was simplified before the operations parentheses.were performed; Correct: -21 **83.** Mistake: the brackets were eliminated before the operations inside them were performed; Correct: 41   **85.** Mistake: square roots were found before subtracting in the radical; Correct: 12   **87.** mean: 17°F, median: 25°F, mode: 26°F

Review Exercises   **1.** $-\$23$ **2.** $-\$213$; loss   **3.** 73 ft   **4.** $-\$705$

Exercise Set 2.6
Prep Exercises **1.** N represents the net, R represents revenue, and C represents cost.   **2.** A positive net indicates profit, whereas a negative net indicates a loss.   **3.** V represents voltage, i represents current, and r represents resistance.   **4.** An average rate is a measure of the rate at which an object travels a total distance in a total amount of time.

Exercises   **1.** $-\$50$; loss   **3.** $-\$15,030$; loss   **5.** \$18,772; profit   **7.** $-\$10$; loss   **9.** $-63\,\text{V}$ **11.** $-44\,\text{A}$ **13.** No, the resistance is $30\,\Omega$ not $40\,\Omega$.   **15.** 22 ft./sec.   **17.** 51,180 mi.   **19.** 65 mph

Review Exercises   **1.**

-5 -4 -3 -2 -1 0 1 2 3 4 5

2. 15   **3.** $k = -110$ **4.** $m = -4$ **5.** -12 **6.** -39

Chapter 2 Summary and Review Exercises
2.1 Definitions/Rules/Procedures $\dots,-3, -2, -1, 0, 1, 2, 3, \dots$ Exercises   **1.** $-13,000$ **2.** $+212$ Definitions/Rules/Procedures   draw a dot on the mark for that number   **3.**

-9 -8 -7 -6 -5 -4 -3 -2 -1 0 1

4.

-5 -4 -3 -2 -1 0 1 2 3 4 5

Definitions/Rules/Procedures   greater than; greater than; 0   Exercises   **5.** $<$
6. $>$ Definitions/Rules/Procedures   its distance from zero; positive; positive; zero   Exercises   **7.** 41   **8.** 16

2.2 Definitions/Rules/Procedures   add their absolute values and keep the same sign; subtract their absolute values and keep the sign of the integer with the greater absolute value   Exercises   **9.** -26 **10.** 9   **11.** -12 **12.** 7   Definitions/Rules/Procedures   zero; 0; absolute value; opposite   Exercises   **13.** 27   **14.** -32

2.3 Definitions/Rules/Procedures   subtrahend (second number) to its additive inverse   Exercises   **15.** $24 + (-31) = -7$ **16.** $-20 + (-17) = -37$ **17.** $-13 + 19 = 6$ **18.** $-27 + 22 = -5$ Definitions/Rules/Procedures   known addend; sum Exercises   **19.** $x = -41$ **20.** $n = 5$

2.4 Definitions/Rules/Procedures   positive; negative; positive; negative   Exercises   **21.** 60   **22.** -54 **23.** -6 **24.** 7   **25.** 36 **26.** -16 Definitions/Rules/Procedures   positive; negative   Exercises   **27.** 81   **28.** -1000 **29.** -256 **30.** -27 Definitions/Rules/Procedures   product; known factor   Exercises   **31.** $k = -9$ **32.** $k = 8$ Definitions/Rules/Procedures   square; positive; negative; no; positive; negative   Exercises   **33.** 8 and -8 (or ± 8)   **34.** 10   **35.** -11 **36.** not a real number

2.5 Definitions/Rules/Procedures   Grouping symbols; Exponents or roots; Multiplication or division; Addition or subtraction Exercises   **37.** -12 **38.** -32 **39.** -15 **40.** 43   **41.** 16   **42.** -2 Definitions/Rules/Procedures $R - C$; ir; total; total; rt Exercises   **43.** $-\$261$ **44.** 0 lb.   Definitions/Rules/Procedures   The block is not moving   Exercises   **45.** \$719,600; profit **46.** 31°F   **47.** 70   **48.** 178   **49.** $-36\,\text{V}$ **50.** $-11\,\text{A}$ **51.** 195 mi   **52.** 60 mph

Chapter 2 Practice Test
1.

-5 -4 -3 -2 -1 0 1 2

2. 26   **3.** -12 **4.** -45 **5.** -14 **6.** -35 **7.** -16 **8.** -108 **9.** -5 **10.** 4   **11.** -84 **12.** 18 **13.** $k = 44$ **14.** $n = -9$ **15.** -64 **16.** -4 **17.** 9 and -9 (or ± 9)   **18.** 12   **19.** not a real number   **20.** -2 **21.** -1 **22.** 5 **23.** 0   **24.** undefined   **25.** -30 **26.** $-\$452$ **27.** $-\$70$ **28.** $-\$15,695$; loss   **29.** -96V **30.** 124 mi

Chapter 1–2 Cumulative Review Exercises
1. true   **2.** false   **3.** false   **4.** true   **5.** commutative; addition   **6.** associative; multiplication   **7.** negative   **8.** Grouping symbols; Exponents or roots; Multiplication or division; Addition or subtraction   **9.** four hundred nine million, two hundred fifty-four thousand, six   **10.** 57,000   **11.**

-5 -4 -3 -2 -1 0 1 2 3 4 5

12. $>$ **13.** 16   **14.** 93   **15.** -50 **16.** -144 **17.** 64 **18.** 201 r2   **19.** -15 **20.** 11   **21.** -19 **22.** -48 **23.** $x = 25$ **24.** $c = -7$ **25.** perimeter: 74 m; area: 300 m^2 **26.** 36 in.^{3} **27.** mean: 80; median: 78; mode: 95   **28.** 21,325 ft.^{2} **29. a.** 36°F   **b.** 58°F   **30.** $-\$17,010$; loss

Exercise Set 3.1
Prep Exercises **1.** An equation has an equal sign, whereas an expression does not.   **2.** Answers may vary. plus, increased by, sum, added to, more than   **3.** Answers may vary. minus, decreased by, difference, subtracted from, less than   **4.** Answers may vary. times, product, multiplied by   **5.** Answers may vary. divided by, quotient, divided into   **6.** subtracted from, less than, divided into **7.** To multiply the sum, which is the result of addition, by seven requires that the addition occur first. The parentheses are needed so that the addition occurs before the multiplication.   **8.** Replace the variables with the corresponding given values; then calculate using the order of operations.

Exercises   **1.** equation   **3.** expression   **5.** expression   **7.** equation   **9.** equation   **11.** expression   **13.** $6 + y$ **15.** $8 - n$ **17.** $n - 15$ **19.** $-7n$ **21.** $-12 \div k$ or $\dfrac{-12}{k}$ **23.** $-5 + 7n$ **25.** $20 \div n - 15$ or $\dfrac{20}{n} - 15$ **27.** xy^2 **29.** $\sqrt{h} \div k$ or $\dfrac{\sqrt{h}}{k}$ **31.** $-6(n + 3)$

33. $8(1 - n)$ **35.** $\sqrt{y} \div (x^3 + 5)$ or $\dfrac{\sqrt{y}}{x^3 + 5}$ **37.** 7 **39.** -7 **41.** -1 **43.** 14 **45.** -18 **47.** -16 **49.** -3 **51.** 40 **53.** 6 **55.** 17
57. 5 **59.** 7

61.

x	$3x - 1$
-4	-13
-2	-7
0	-1
2	5
4	11

63.

m	$m^2 - 2m + 4$
-1	7
0	4
1	3
2	4
3	7

65.

y	$|y + 2|$
-4	2
-3	1
-2	0
-1	1
0	2

67. a. $-5n - 6$; **b.** 14 **65. a.** $-3(8 - n)$; **b.** -30

Review Exercises **1.** 2 **2.** 2 hundreds + 7 tens + 9 ones **3.** 7^4 **4.** 64 **5.** 300,000 **6.** $3 + (-7) + (-9)$

Exercise Set 3.2

Prep Exercises **1.** A monomial is an expression that is a constant or a product of a constant and variables that are raised to whole number powers. Examples may vary. 12, $3x$, and $-7y^3$ **2.** The coefficient is the numerical factor in a monomial. **3.** The degree of a monomial is the sum of the exponents of all of its variables. **4.** A polynomial is an expression that is a sum of monomials. Examples may vary. $3x^2 + x - 9$ **5.** two; three **6.** The degree of a polynomial is the greatest degree of all its terms. **7.** Place the term with the greatest degree first, then the term with the next greatest degree, and so on. **8.** Like terms are monomials that have the same variables raised to the same exponents. Examples may vary. $6x$ and $-9x$ **9.** The expression is written with the fewest symbols possible. **10.** Add or subtract the coefficients and keep the variables and their exponents the same.

Exercises **1.** monomial; a product of a constant and variable raised to a whole number power **3.** not a monomial; not a product **5.** monomial; a product of a constant, -1, and a variable raised to a whole number power, 1 **7.** monomial; a constant **9.** not a monomial; not a product **11.** monomial; a product of a constant and variables raised to whole number powers **13.** coefficient: 3; degree: 8 **15.** coefficient: -9; degree: 1 **17.** coefficient: 8; degree: 0 **19.** coefficient: 1; degree: 3 **21.** coefficient: -2; degree: 7 **23.** coefficient: -1; degree: 0 **25.** first term: $5x^2$, coefficient: 5; second term: $8x$, coefficient: 8; third term: -7, coefficient: -7 **27.** first term: $6t$, coefficient: 6; second term: -1, coefficient: -1 **29.** first term: $-6x^3$, coefficient: -6; second term: x^2, coefficient: 1; third term: $-9x$, coefficient: -9; fourth term: 4, coefficient: 4 **31.** binomial **33.** monomial **35.** no special name **37.** trinomial **39.** monomial **41.** binomial **43.** 3 **45.** 6 **47.** 12 **49.** $14t^6 - 8t^4 + 9t^3 + 5t^2 - 1$ **51.** $y^5 - 18y^3 - 10y^2 + 12y + 9$ **53.** $9a^5 + 7a^3 + 2a^2 - a - 6$ **55.** like **57.** not like **59.** like **61.** like **63.** not like **65.** like **67.** $12x$ **69.** $4y$ **71.** $5m$ **73.** $-3t$ **75.** $5x^3$ **77.** $-t^2$ **79.** $-8m^5$ **81.** 0 **83.** $11x^2 + 2x + 4$ **85.** $-13a^3 + 6a + 4$ **87.** $-5c^2 - 4c$ **89.** $-7n^2 - 2n + 7$ **91.** $7m^3 + m^2 - 4$ **93.** $8x^4 - 3x^2 + 8$

Review Exercises **1.** -2 **2.** -16 **3.** -7 **4.** -26 **5.** No, because subtraction is not commutative. **6.** Yes, because addition is commutative. **7.** Yes, because multiplication is commutative. **8.** 28 m

Exercise Set 3.3

Prep Exercises **1.** Combine like terms. **2.** Change the operation sign from $-$ to $+$; then change the sign of each term in the second polynomial.

Exercises **1.** $5x + 6$ **3.** $9y - 5$ **5.** $13x - 2$ **7.** $5t + 2$ **9.** $2x^2 + 6x + 10$ **11.** $15n^2 - 12n + 3$ **13.** $7p^2 - 3p - 4$ **15.** $-8b^3 - b^2 - 9b + 11$ **17.** $3a^3 + 5a^2 + a - 2$ **19.** $8a + 4$ **21.** $4x + 7$ **23.** Mistake: combining $4y^2$ with $-2y^2$ does not equal $-2y^2$ Correct: $10y^3 + 2y^2 - 6y$ **25.** Mistake: unlike terms were combined Correct: $8x^6 + x^5 - 9x^4 + 7x^3 - 4$ **27.** $2x + 7$ **29.** $3t$ **31.** $8n - 6$ **33.** $3x - 7$ **35.** $6x^2 - x - 11$ **37.** $12a^2 - 11a - 6$ **39.** $-4x - 10$ **41.** $4m^2 - 18m + 4$ **43.** Mistake: did not change all signs in the subtrahend Correct: $2x^2 - 2x$ **45.** Mistake: adding $-9t^2$ to $-9t^2$ does not equal 0 Correct: $-5t^3 - 18t^2 + 4t + 5$ **47.**

	Expression 1	Expression 2	Sum	Difference
x	$5x + 4$	$3x - 7$	$8x - 3$	$2x + 11$
4	24	5	2	19
-2	-6	-13	-19	7

49.

	Expression 1	Expression 2	Sum	Difference
n	$n^2 + 3n - 1$	$4n^2 - n - 7$	$5n^2 + 2n - 8$	$-3n^2 + 4n + 6$
2	9	7	16	2
-3	-1	32	31	-33

Review Exercises **1.** -682 **2.** 81 **3.** -32 **4.** 128 **5.** 375 m² **6.** 20

Exercise Set 3.4

Prep Exercises **1.** add   **2.** Multiply the coefficients and add the exponents of the like variables.   **3.** multiply   **4.** Evaluate the coefficient raised to the power and multiply each variable's exponent by the power.   **5.** Use the distributive property.   **6.** Multiply each term in the first polynomial by each term in the second polynomial.   **7.** Conjugates are binomials that differ only in the sign separating the terms. Examples may vary. $x + 5$ and $x - 5$.   **8.** difference

Exercises **1.** x^7 **3.** t^{10} **5.** $21a^3$ **7.** $10n^7$ **9.** $-24u^6$ **11.** $8y^7$ **13.** $90x^7$ **15.** $-36m^4$ **17.** $60y^9$ **19.** Mistake: multiplied the exponents Correct: $45x^7$ **21.** $4x^6$ **23.** $49m^{10}$ **25.** $-32y^{30}$ **27.** $256x^{20}$ **29.** $25y^{12}$ **31.** $-125v^{18}$ **33.** $48x^{13}$ **35.** Mistake: did not raise 3 to the 4th power, added exponents of x Correct: $81x^{16}$ **37.** $8x + 12$ **39.** $15x - 35$ **41.** $-15t - 6$ **43.** $-28a + 36$ **45.** $18u^2 + 24u$ **47.** $-6a^2 - 2a$ **49.** $-10x^4 + 16x^3$ **51.** $16x^3 + 24x^2 - 32x$ **53.** $-5x^4 + 6x^3 - 9x^2$ **55.** $-6p^4 - 8p^3 + 10p^2$ **57.** $x^2 + 6x + 8$ **59.** $m^2 + 2m - 15$ **61.** $y^2 - 7y - 8$ **63.** $12x^2 - 7x - 10$ **65.** $12x^2 - 23x + 5$ **67.** $12t^2 - 17t - 5$ **69.** $2a^2 - 19a + 35$ **71.** $x + 7$ **73.** $2x - 5$ **75.** $-b - 2$ **77.** $-6x + 9$ **79.** $x^2 - 9$ **81.** $25t^2 - 36$ **83.** $36x^2 - 1$ **85.** Mistake: did not square the initial x, and 5 should be negative Correct: $6x^2 - 13x - 5$ **87.** $2x^3 + 7x^2 - 14x + 5$ **89.** $8y^3 - 10y^2 - 15y + 18$ **91.** $2a^3 + 2a^2 - 4a$ **93.** $-2x^4 - 5x^3 - 3x^2$ **95.** $-27q^4 + 3q^2$ **97.** $24y^2 + 8y$ **99.** $3x^3 + 3x^2 - 6x$

101.

	Expression 1	Expression 2	Product
x	$5x$	$-3x$	$-15x^2$
2	10	-6	-60
-3	-15	9	-135

103.

	Expression 1	Expression 2	Product
n	$n + 3$	$n - 7$	$n^2 - 4n - 21$
1	4	-6	-24
-3	0	-10	0

Review Exercises **1.** 423   **2.** -4 **3.** 11 ft.   **4.** 5 ft.   **5.** $-4x$

Exercise Set 3.5

Prep Exercises **1.** A prime number is a natural number that has exactly two different factors: 1 and the number itself. Examples may vary. 2, 3, and 5   **2.** A composite number is a natural number that has factors other than 1 and itself. Examples may vary. 4, 6, and 9   **3.** $2 \cdot 15; 3 \cdot 10; 5 \cdot 6$ **4.** The GCF of a set of a numbers is the greatest number that divides all of the given numbers with no remainder.   **5.** least

Exercises **1.** composite   **3.** prime   **5.** neither   **7.** prime   **9.** composite   **11.** composite   **13.** prime   **15.** prime   **17.** No, 2 is an even prime number.   **19.** $2^4 \cdot 5$ **21.** $2^2 \cdot 3 \cdot 13$ **23.** $2^2 \cdot 67$ **25.** $2^3 \cdot 5^2$ **27.** 7^3 **29.** $3 \cdot 5^2 \cdot 13$ **31.** $2 \cdot 3^3 \cdot 7$ **33.** $2^3 \cdot 7 \cdot 17$ **35.** 1, 2, 3, 5, 6, 9, 10, 15, 18, 30, 45, 90   **37.** 1, 3, 9, 27, 81   **39.** 1, 2, 3, 4, 5, 6, 8, 10, 12, 15, 20, 24, 30, 40, 60, 120   **41.** 24: 1, 2, 3, 4, 6, 8, 12, 24; 60: 1, 2, 3, 4, 5, 6, 10, 12, 15, 20, 30, 60 GCF = 12   **43.** 81: 1, 3, 9, 27, 81 65: 1, 5, 13, 65 GCF = 1   **45.** 72: 1, 2, 3, 4, 6, 8, 9, 12, 18, 24, 36, 72 120: 1, 2, 3, 4, 5, 6, 8, 10, 12, 15, 20, 24, 30, 40, 60, 120 GCF = 24   **47.** $140 = 2^2 \cdot 5 \cdot 7$; $196 = 2^2 \cdot 7^2$; GCF = $2^2 \cdot 7 = 28$ **49.** $130 = 2 \cdot 5 \cdot 13$; $78 = 2 \cdot 3 \cdot 13$; GCF = $2 \cdot 13 = 26$ **51.** $336 = 2^4 \cdot 3 \cdot 7$; $504 = 2^3 \cdot 3^2 \cdot 7$; GCF = $2^3 \cdot 3 \cdot 7 = 168$ **53.** $99 = 3^2 \cdot 11$; $140 = 2^2 \cdot 5 \cdot 7$; GCF = 1   **55.** $60 = 2^2 \cdot 3 \cdot 5$; $120 = 2^3 \cdot 3 \cdot 5$; $140 = 2^2 \cdot 5 \cdot 7$; GCF = $2^2 \cdot 5 = 20$ **57.** $64 = 2^6$; $160 = 2^5 \cdot 5$; $224 = 2^5 \cdot 7$; GCF = $2^5 = 32$ **59.** 10 in. by 10 in.   **61.** 8 ft; 5 in 40-ft. trench; 4 in 32-ft. trench; and 3 in 24-ft. trench   **63.** 10   **65.** $2x^2$ **67.** h^3 **69.** 1   **71.** $8x^7$ **73.** $6n^2$ **75.** $7x^3$

Review Exercises **1.** $5 \times 10^6 + 7 \times 10^5 + 8 \times 10^4 + 4 \times 10^3 + 2 \times 10^2 + 9 \times 1$ **2.** 2,019 r5   **3.** 128   **4.** No, there are only 260 possible combinations.   **5.** 6 ft. by 6 ft.

Exercise Set 3.6

Prep Exercises **1.** subtract   **2.** 1. Divide the coefficients.   2. Subtract the exponent of the divisor's variable from the exponent of the dividend's variable and keep the same variable.   **3.** Divide each term in the polynomial by the monomial.   **4.** Divide the product by the known factor.   **5.** 1. Find the GCF of the terms in the polynomial. 2. Rewrite the polynomial as a product of the GCF and the quotient of the polynomial and the GCF. Polynomial = $\text{GCF}\left(\dfrac{\text{polynomial}}{\text{GCF}}\right)$

Exercises **1.** x^7 **3.** m^6 **5.** 1   **7.** $4a^3$ **9.** $20n^8$ **11.** $-3x^4$ **13.** $-4b^2$ **15.** -14 **17.** $3a + 2$ **19.** $-7c + 4$ **21.** $3x + 2$ **23.** $-9d + 7$ **25.** $4a^3 + 6a$ **27.** $5a - 4$ **29.** $4x^3 - 2x^2 + x$ **31.** $-8c^7 - 4c^3 + 10c$ **33.** $2a^2$ **35.** $-7x^4$ **37.** $4u$ **39.** $2x + 3$ **41.** $5a - 1$ **43.** $3 - 5x$ **45.** $2t^3 - 3t^2 + 4$ **47.** $5n^2 - 4n - 2$ **49.** $9m$ **51.** $4t^2$ **53.** $4(2x - 1)$ **55.** $10(a + 2)$ **57.** $2n(n + 3)$ **59.** $x^2(7x - 3)$ **61.** $4r^3(5r^2 - 6)$ **63.** $4x(3x^2 + 5x + 8)$ **65.** $3a^3(3a^4 - 4a^2 + 6)$ **67.** $7m^5(2m^3 + 4m + 1)$ **69.** $10x^3(x^6 - 2x^2 - 4)$

Review Exercises **1.** 20 ft.   **2.** 144 m^2 **3.** 108 in.^{3} **4.** $2x^3 - 9x^2 - 17$ **5.** $x^2 - 2x - 24$

Exercise Set 3.7

Prep Exercises **1.** Add all of the side lengths.   **2.** Multiply the base by the height.   **3.** Multiply the length, width, and height.   **4.** $SA = 2lw + 2lh + 2wh$ or $SA = 2(lw + lh + wh)$ **5.** The formula is used to calculate the height of a falling object, where h represents the height in feet after the object has fallen for t seconds from an initial height of h_0 feet.   **6.** Subtract the expression for cost from the expression for revenue. ($N = R - C$)

Exercises **1.** $4x + 14$ **3.** $3n - 6$ **5.** $8b - 10$ **7. a.** 90 in.; **b.** 122 cm   **9. a.** 30 ft.; **b.** 18 yd.   **11. a.** 70 in.; **b.** 150 ft.   **13.** $2n^2 + 8n$ **15. a.** 120 km^2; **b.** 384 ft.^{2} **17.** $d^3 + 5d^2 - 6d$ **19. a.** 54 in.^{3}; **b.** 220 ft.^{3} **21.** 32 ft.^{2} of cardboard   **23.** 1536 panels   **25.** 116 ft. after 2 sec.; 36 ft. after 3 sec.   **27.** 4256 ft.   **29. a.** $35r + 75s - 245$; **b.** \$11,905 profit   **31. a.** $R = 5s + 9m + 15l$; **b.** $C = 2s + 4m + 7l$ **c.** $N = 3s + 5m + 8l$; **d.** \$87 profit

Review Exercises   **1.** 15   **2.** −1   **3.** 0   **4.** $-x^3 - 21x^2 + 14$ **5.** $-12a - 28$ **6.** 1

Chapter 3 Summary and Review Exercises

3.1 Definitions/Rules/Procedures   describes a calculation; contains an equal sign; key words; Replace the variables; Calculate
Exercises   **1.** $-2m - 8$ **2.** $5(n - 12)$ **3.** 40   **4.** −2

3.2 Definitions/Rules/Procedures   constant or a product of constants and variables; numerical factor; sum of the exponents of all variables   Exercises   **5.** Yes, it is a product of a constant and a variable raised to a whole number power.   **6.** No, it contains subtraction.   **7.** coefficient: 18; degree: 1   **8.** coefficient: 1; degree: 3   **9.** coefficient: −9; degree: 0   **10.** coefficient: −3; degree: 6
Definitions/Rules/Procedures   monomial; sum of monomials; every variable term has the same variable; more than one variable; greatest degree; greatest; next greatest   Exercises   **11.** binomial   **12.** no special name   **13.** monomial   **14.** trinomial   **15.** 6   **16.** $5a^6 + 7a^4 - a^2 - 9a + 13$ Definitions/Rules/Procedures   fewest; constant; the same variables raised to the same exponents; coefficients; variables; exponents   Exercises   **17.** They have the same variables raised to the same exponents.   **18.** The same variables are raised to different exponents.   **19.** $-a^2 + 2a - 1$ **20.** $-5x^7 + 3x^2 + 13$

3.3 Definitions/Rules/Procedures   combine like terms   Exercises   **21.** $3x + 5$ **22.** $4y^4 + 2y^3 - 10y - 4$ Definitions/Rules/Procedures   changing the sign of each of its terms   Exercises   **23.** $3a^2 - 5a$ **24.** $13h^3 - 5h^2 - 5h - 3$

3.4 Definitions/Rules/Procedures   add; keep the same base; a^{m+n}; coefficients; Add the exponents   Exercises   **25.** m^7 **26.** $-10x^5$ **27.** $-12y^5$ **28.** $6t^8$ Definitions/Rules/Procedures   multiply; keep the same base; a^{mn}; coefficient; variable's exponent   Exercises   **29.** $125x^{12}$ **30.** $4y^6$ **31.** $64t^9$ **32.** $-8a^{12}$ Definitions/Rules/Procedures   each term in the polynomial by the monomial; each term; each term; Combine; binomials; sign of the second term; difference of two squares; $a^2 - b^2$ Exercises   **33.** $6x - 8$ **34.** $-16y^2 + 20y$ **35.** $15n^3 - 3n^2 + 21n$ **36.** $a^2 - 2a - 35$ **37.** $2x^2 - x - 3$ **38.** $10y^2 - 21y + 8$ **39.** $9t^2 - 16$ **40.** $u^3 - 3u^2 - 7u + 6$ **41.** $3y + 4$ **42.** $-7x - 2$

3.5 Definitions/Rules/Procedures   natural; 1; the number itself; natural; factors other than 1 and itself; composite; prime   Exercises   **43.** composite   **44.** prime   Definitions/Rules/Procedures   prime; factors whose product is the given number   Exercises   **45.** $2^3 \cdot 3^2 \cdot 5$ **46.** $2^3 \cdot 3 \cdot 5^2 \cdot 7$ Definitions/Rules/Procedures   greatest; divides; all; greatest; common to all factorizations; least; 1   Exercises   **47.** 28   **48.** 1   **49.** $12x^5$ **50.** $6a^2$

3.6 Definitions/Rules/Procedures   subtract; keep the same base; a^{m-n}; 1; 1; coefficients; Subtract the exponent; each term   Exercises   **51.** r^6 **52.** $-4x^3$ **53.** $7t + 9$ **54.** $5x^4 - 4x^2 + 2$ Definitions/Rules/Procedures   product; known factor   Exercises   **55.** $-9m^5$ **56.** $4x^2 + 3$ **57.** $4y^4 - 3y^2 - 2$ **58.** $14b$ Definitions/Rules/Procedures   product of factors; quotient; GCF $\left(\frac{\text{polynomial}}{\text{GCF}}\right)$ Exercises   **59.** $6(5x + 2)$ **60.** $4y(3y + 5)$ **61.** $3n^3(3n - 5)$ **62.** $6x(3x^2 + 4x - 6)$

3.7 Definitions/Rules/Procedures   completely cover the outer shell; bh; lwh; $2lw + 2lh + 2wh$; $-16t^2 + h_0$; $R - C$ Exercises   **63. a.** $16x - 12$; **b.** 132 m   **64. a.** $m^2 - 2m - 15$; **b.** 33 in.^{2} **65. a.** $48n^3 - 32n^2$; **b.** 2560 ft.^{3} **66.** 52 ft.^{2} **67.** 86 ft.   **68. a.** $43n + 43b - 105$; **b.** \$31,414 profit

Chapter 3 Practice Test

1. $-7(n - 5)$ **2.** −43   **3.** coefficient: −1, degree: 4   **4.** binomial   **5.** 4   **6.** $-a^4 + 5a^2$ **7.** $5y^3 - 10y - 4$ **8.** $6y^5 + 5y^3 - y^2 - 4$ **9.** $20u^7$ **10.** $-32a^{15}$ **11.** $-2t^4 + 6t^2 + 14t$ **12.** $x^2 - 36$ **13.** $-4a^4 - 2a^3 + 2a^2$ **14.** $6x + 5$ **15.** composite   **16.** $2^2 \cdot 5 \cdot 17$ **17.** $12h^5$ **18.** m^2 **19.** $-8x^4 + 9x^2 - 11x$ **20.** $5x^3$ **21.** $4(3x - 5)$ **22.** $2y^2(5y^2 - 9y + 7)$ **23. a.** $2n^2 - 7n - 4$; **b.** 45 cm^2 **24.** 304 in.^{2} **25. a.** $84a + 91b$; **b.** \$21,217 profit

Chapters 1–3 Cumulative Review Exercises

1. false   **2.** false   **3.** false   **4.** true   **5.** negative   **6.** negative   **7.** coefficient variables exponents   **8.** a^{m+n}
9. 140,000   **10.**   **11.** −1   **12.** 5   **13.** −28   **14.** −125   **15.** 20   **16.** −15
17. 8   **18.** $4y^3 - 9y^2 + 21y - 6$ **19.** $2b^2 - 13b - 24$ **20.** $2^3 \cdot 3^2 \cdot 5$ **21.** $6n^3$ **22.** $6m^2(2m^2 - 3m + 4)$ **23.** $x = 6$
24. $x = 8$ **25.** 255 cm^2 **26.** $16w$ **27.** mean: \$122; median: \$92; mode: \$40   **28.** −\$279   **29.** 548 ft.^{2} **30. a.** $113b + 158a - 345$; **b.** \$19,711 profit

Exercise Set 4.1

Prep Exercises **1.** An expression has no equal sign, whereas an equation does.   **2.** A solution to an equation is a number that makes the equation true when it replaces the variable in the equation.   **3.** To solve an equation means to find its solution(s).   **4.** 1. Replace the variable(s) with the number. 2. Simplify both sides of the equation as needed. If the resulting equation is true, the number is a solution.

Exercises   **1.** equation   **3.** expression   **5.** expression   **7.** −5 is not a solution.   **9.** −9 is not a solution.   **11.** 3 is not a solution.   **13.** 8 is a solution.   **15.** 5 is not a solution.   **17.** −3 is a solution.   **19.** −3 is not a solution.   **21.** 0 is a solution.   **23.** 3 is not a solution.   **25.** −2 is not a solution.

Review Exercises   **1.** $5x - 12$ **2.** $-2y - 8$ **3.** $5m + 15$ **4.** $-3m + 4$ **5.** $x = 7$

Exercise Set 4.2

Prep Exercises **1.** Each variable term is a monomial of degree 1.   **2.** It contains a variable term, x^2, that is not a monomial of degree 1.   **3.** Adding the same amount to both sides of an equation does not change the equation's solution(s).   **4.** Add −9 to both sides of the equation. (Or subtract 9 from both sides of the equation.)

Exercises **1.** linear **3.** nonlinear **5.** nonlinear **7.** linear **9.** nonlinear **11.** linear **13.** $n = 6$ **15.** $c = -7$ **17.** $x = -17$ **19.** $h = 16$ **21.** $x = 8$ **23.** $x = 12$ **25.** $y = -1$ **27.** $y = -2$ **29.** $m = 6$ **31.** $y = -3$ **33.** $3295 + x = 4768$; $1473 **35.** $200 + 180 + x = 450$; 70 cc

37. a.

	'04/'05 to '05/'06	'05/'06 to '06/'07	'06/'07 to '07/'08	'07/'08 to '08/'09
Equation	$1849 + x = 1935$	$1935 + x = 2018$	$2018 + x = 2061$	$2061 + x = 2137$
Tuition increase	86	83	43	76

b. 2005/2006

39. $1600 + 4500 + 1935 + x = 10{,}500$; he must sell $2465 more. Yes, midway through the month, he has sold $8035. **41.** $3 + 4 + x = 17$; length is 10 ft. Total area of the dining room is $10 \times 8 = 80$ ft.^{2}, which is enough for the square table and chairs. **43.** $23 + 35 + l = 84$; 26 cm **45.** $2420 + 2447 + 1200 + n = 11{,}400$; $5333

Review Exercises **1.** -40 **2.** -8 **3.** -9 **4.** -2 **5.** $x = 3$

Exercise Set 4.3

Prep Exercises **1.** Multiplying (or dividing) both sides of an equation by the same amount does not change the equation's solution(s). **2.** Divide both sides of the equation by -3. **3.** Add 7 to both sides to isolate the $3x$ term. **4.** Adding $-3x$ avoids getting a negative coefficient after the like terms are combined. **5.** Distribute to eliminate parentheses. **6.** -1

Exercises **1.** $x = 7$ **3.** $a = -9$ **5.** $b = -11$ **7.** $a = 11$ **9.** $y = 6$ **11.** $x = 7$ **13.** $b = -7$ **15.** $u = -4$ **17.** $x = 11$ **19.** $k = -3$ **21.** $h = 4$ **23.** $h = -3$ **25.** $t = -8$ **27.** $x = 2$ **29.** $u = 1$ **31.** Mistake: did not write the minus sign to the left of 11 Correct: $-1 = x$ **33.** Mistake: did not change the minus sign in $x - 8$ Correct: $x = -3$ **35.** 24 in. **37.** 17 ft. **39.** 4 sec. **41.** 44 ohms **43.** 521 ft. **45.** 72 in. **47.** 4 hr. **49.** 5 sec. **51.** 615 chips

Review Exercises **1.** six million, seven hundred eighty-four thousand, two hundred nine **2.** 1 **3.** 43 **4.** $2^4 \cdot 3 \cdot 5$ **5.** $6x^3(4x^2 - 5x + 3)$

Exercise Set 4.4

Prep Exercises **1.** Answers may vary. sum, plus, added to, more than, increased by **2.** Answers may vary. difference, minus, subtracted from, less than, decreased by **3.** Answers may vary. times, product, multiplied by **4.** Answers may vary. equal, is, yields

Exercises **1.** $n + 5 = -7$; $n = -12$ **3.** $n - 6 = 15$; $n = 21$ **5.** $x + 17 = -8$; $x = -25$ **7.** $-3y = 21$; $y = -7$ **9.** $9b = -36$; $b = -4$ **11.** $5x + 4 = 14$; $x = 2$ **13.** $-6m - 16 = 14$; $m = -5$ **15.** $39 - 5x = 8x$; $x = 3$ **17.** $17 + 4t = 6t - 9$; $t = 13$ **19.** $2(b - 8) = 5 + 9b$; $b = -3$ **21.** $6x + 5(x - 7) = 19 - (x + 6)$; $x = 4$ **23.** $-8(y - 3) - 14 = -2y - (y - 5)$; $y = 1$ **25.** Mistake: incorrect subtraction order Correct: $n - 7 = 15$ **27.** Mistake: multiplied 2 times x instead of the sum Correct: $2(x + 13) = -10$ **29.** Mistake: incorrect subtraction order on the left side of the equation Correct: $16 - 6n = 2(n - 4)$

Review Exercises **1.** $6xy^5$ **2.** $9b^2(2b^3 - 3b + 6)$ **3.** $8w$ **4.** 48 ft. **5.** 9 small shrubs; $390

Exercise Set 4.5

Prep Exercises **1.** width; $v + 5$ **2.** An equilateral triangle is a triangle with all three sides of equal length. **3.** An isosceles triangle is a triangle with two sides of equal length. **4.** Complementary angles are two angles whose measurements sum to 90°. **5.** Supplementary angles are two angles whose measurements sum to 180°. **6.** second angle; $n - 20$

Exercises **1.** $4 + n + n = 34$; the numbers are 15 and 19. **3.** $3n + n = 32$; the numbers are 8 and 24. **5.** $4w + w + 4w + w = 300$; 120 ft. long by 30 ft. wide **7.** $3w - 4 + w + 3w - 4 + w = 48$; 7 in. wide by 17 in. long **9.** $2b - 11 + 2b - 11 + b = 18$; the base is 8 m, and the equal-length sides are 5 m each. **11.** $l + l + l = l + 3l - 10 + l + 3l - 10$; the length of the rectangle is 4 ft., and its width is 2 ft. The sides of the triangle are 4 ft. **13.** $a + 4a = 90$; the angles are 18° and 72°. **15.** $a + 4a + 15 = 180$; the angles are 33° and 147°. **17.** $a + 3a + 10 + a = 180$; the angles are 34°, 112°, and 34°. **19.** $43 + 2a + 2 + a = 180$; the second angle is 92°, and the third is 45°. **21.** $a + a + 10 + a - 7 = 180$; the angles are 59°, 69°, and 52°.

23.

Categories	Selling price	Number of ornaments sold	Income
Small ornament	3	$L + 8$	$3(L + 8)$
Large ornament	7	L	$7L$

25.

Categories	Selling price	Number of charms sold	Income
Small charm	4	$36 - n$	$4(36 - n)$
Large charm	6	n	$6n$

27. $8(L + 5) + 12L = 260$; 16 small bottles, 11 large bottles   **29.** $5(19 - n) + 10n = 125$; 6 \$10 bills, 13 \$5 bills
31. $12(3x) + 18x = 108$; the bills are worth \$2 and \$6.   **33.** $9(205 - x) + 6x = 1470$; hand-fed cockatiels sell for \$125, and parent-raised cockatiels sell for \$80.

Review Exercises   **1.** -3 is not a solution.   **2.** No, because it has a variable term with a degree other than 1.   **3.** $m = -7$
4. $n = -5$ **5.** $x = 3$

Chapter 4 Summary and Review Exercises

4.1 Definitions/Rules/Procedures   solution; solutions; true; Replace; true   Exercises   **1.** 1 is a solution.   **2.** -2 is not a solution.
3. 5 is not a solution.   **4.** 3 is a solution.

4.2 Definitions/Rules/Procedures   1; the same amount to both sides; additive inverse   Exercises   **5.** $n = 8$ **6.** $y = -2$
7. $-15 + t = 28$; 43°F   **8.** $-547 + p = -350$; \$197

4.3 Definitions/Rules/Procedures   the same amount; divide; Distribute; Combine; variable terms; coefficient
Exercises   **9.** $k = -9$ **10.** $m = 3$ **11.** $x = -7$ **12.** $h = -9$ **13.** $t = 3$ **14.** $y = -2$ **15.** $m = 1$ **16.** $v = -6$ **17.** $x = 3$
18. $y = -5$ **19.** 4 ft.   **20.** 8 ft.

4.4 Definitions/Rules/Procedures   unknown(s)   Exercises   **21.** $15 - 7n = 22$; $n = -1$ **22.** $5x - 4 = 3x$; $x = 2$
23. $2(n + 12) = -6n - 8$; $n = -4$ **24.** $12 - 3(x - 7) = 6x - 3$; $x = 4$

4.5 Definitions/Rules/Procedures   acted on; two; all three; the same; 90°; 180°   Exercises   **25.** $12 + n + n = 42$; the numbers are
15 and 27.   **26.** $5n + n = 36$; the numbers are 6 and 30.   **27.** $3w - 2 + w + 3w - 2 + w = 188$; 24 m wide and 70 m long
28. $b + 38 + b + 38 + b = 256$; the base is 60 in.; the equal-length sides are 98 in. each.   **29.** $16 + a + a = 180$; the angles are 98°
and 82°.   **30.** $a + 2a - 15 = 180$; the angles are 65° and 115°.   **31.** $10(n + 9) + 20n = 330$; 8 twenty-dollar bills and 17 ten-dollar
bills   **32.** $6n + 4(27 - n) = 146$; 19 large bags and 8 small bags

Chapter 4 Practice Test
1. Equation; it contains an equal sign.   **2.** -4 is not a solution.   **3.** 2 is a solution.   **4.** Nonlinear; there is a variable term with a
degree other than 1.   **5.** $n = 8$ **6.** $m = -6$ **7.** $y = 4$ **8.** $k = 1$ **9.** $x = 3$ **10.** $t = -21$ **11.** $u = -2$ **12.** $k = 9$
13. $375 + x = 458$; \$83   **14.** $96{,}800 = 242l$; 400 yd.   **15.** $4n - 9 = 23$; the number is 8.   **16.** $3(x - 5) = 4x - 9$; $x = -6$
17. $3n + n = 44$; the numbers are 11 and 33.   **18.** $2b + 9 + 2b + 9 + b = 258$; the base is 48 in., and the equal-length sides are 105 in.
each.   **19.** $a + a - 30 = 180$; the angles are 105° and 75°.   **20.** $450(12 - n) + 675n = 6300$; 8 model A guitars and 4 model B guitars

Chapters 1–4 Cumulative Review Exercises
1. true   **2.** false   **3.** false   **4.** true   **5.** positive   **6.** add   **7.**   **8.** 30   **9.** coefficient: -8;

degree: 3   **10.** 5   **11.** -61 **12.** 19   **13.** -22 **14.** -391 **15.** -7 **16.** $8t^3 + 14t^2 - 25t - 7$ **17.** $6x^3 - 7x^2 - x - 7$ **18.** $9x^2 - 25$
19. $2^2 \cdot 3 \cdot 5 \cdot 7$ **20.** $5n^2(4n^3 + 3mn - 2)$ **21.** $t = -6$ **22.** $b = 22$ **23.** $y = 4$ **24.** $n = -7$ **25.** 216 in.²   **26.** $-\$67{,}181$
27. $x^2 - 4x - 12$ **28.** 5 in.   **29.** $15 + 6n = 3$; $n = -2$ **30.** $w + 4 + w + w + 4 + w = 44$; 9 ft. wide and 13 ft. long

Exercise Set 5.1
Prep Exercises **1.** numerator: 3; denominator: 5   **2.** In an improper fraction, the value of the numerator is greater than or equal to
the value of the denominator, whereas in a proper fraction, the value of the numerator is less than the value of the denominator.
3. The numerator is the number of shaded divisions.   **4.** 1. Divide the denominator into the numerator. 2. Write the result in the

following form: quotient $\dfrac{\text{remainder}}{\text{original denominator}}$ **5.** 1. Multiply the integer by the denominator. 2. Add the resulting product to the

numerator. The sum will be the numerator of the improper fraction. 3. Keep the same denominator.   **6.** One less tick mark than the
denominator   **7.** multiply; divide   **8.** numerators

Exercises   **1.** $\frac{1}{3}$ **3.** $\frac{5}{8}$ **5.** $\frac{2}{3}$ **7. a.** $\frac{5}{16}$; **b.** $\frac{15}{16}$; **c.** The sense of smell seems to contribute more information about foods than does taste
alone. **d.** If she is trying the same foods in the same order, she may be guessing better the second time.   **9.** $\frac{17}{800}$ survive; $\frac{783}{800}$ do not
survive.   **11.** $\frac{249}{258}$ passed; $\frac{9}{258}$ did not pass. Very few of Mrs. Jones's students do not pass the course.   **13.** $\frac{451}{600}$ **15.** $\frac{149}{600}$ **17.** 23   **19.** 1
21. 0   **23.** undefined   **25.** $4\frac{2}{7}$ **27.** $21\frac{1}{4}$ **29.** $-12\frac{4}{5}$ **31.** $-12\frac{7}{8}$ **33.** $9\frac{5}{6}$ **35.** $12\frac{8}{11}$ **37.** $-104\frac{7}{8}$ **39.** $402\frac{1}{14}$ **41.** $\frac{31}{6}$ **43.** $\frac{11}{1}$
45. $-\frac{79}{8}$ **47.** $-\frac{29}{20}$ **49.**   **51.**   **53.**

55.   **57.**   **59.**   **61.** 15   **63.** 7   **65.** 30   **67.** -30

69. >   **71.** <   **73.** =   **75.** >   **77.** =   **79.** <

Review Exercises   **1.** 4   **2.** $2^3 \cdot 3 \cdot 5 \cdot 7$ **3.** 12   **4.** $8x^2$ **5.** $8x^2(5x^3 - 7)$

Exercise Set 5.2
Prep Exercises **1.** greatest common factor, 1   **2.** Divide the numerator and denominator by their greatest common factor. Or divide
out all prime factors that are common to both the numerator and denominator.

Exercises 1. $\dfrac{5}{6}$ **3.** $\dfrac{-2}{5}$ **5.** $\dfrac{1}{2}$ **7.** $-\dfrac{3}{5}$ **9.** $\dfrac{3}{4}$ **11.** $\dfrac{-3}{4}$ **13.** $\dfrac{6}{7}$ **15.** $\dfrac{7}{9}$ **17.** $\dfrac{-11}{17}$ **19.** $\dfrac{-3}{4}$ **21.** $\dfrac{7}{12}$ **23.** $\dfrac{3}{8}$ **25. a.** $\dfrac{12}{31}$; **b.** 12 out of every 31 women can be expected to use the product. **c.** 248 may not be a large enough sample. **27.** The teller made a mistake in $\dfrac{12}{125}$ of the transactions. The teller did not make a mistake in $\dfrac{113}{125}$ of the transactions. **29. a.** \$9842; **b.** $\dfrac{514}{4921}$; **c.** $\dfrac{175}{703}$; **d.** $\dfrac{11}{38}$; **e.** $\dfrac{5}{14}$ **31.** $3\dfrac{3}{4}$ **33.** $-3\dfrac{1}{3}$ **35.** $4\dfrac{1}{7}$ **37.** $-5\dfrac{1}{6}$ **39.** $\dfrac{5x}{16}$ **41.** $\dfrac{-x^2}{y}$ **43.** $-\dfrac{2n}{5m^3}$ **45.** $\dfrac{1}{4t^3u}$ **47.** $\dfrac{3ac}{5b}$ **49.** $\dfrac{-b^4}{2a^6c^5}$

Review Exercises 1. $-15{,}015$ **2.** 72 **3.** $35x^5$ **4.** $-32x^{15}$ **5.** $99\ \text{m}^2$

Exercise Set 5.3

Prep Exercises 1. 1. Divide out any numerator factor with any like denominator factor. 2. Multiply numerator by numerator and denominator by denominator. 3. Simplify. **2.** numerator; denominator **3.** Write them as improper fractions; then follow the same process as for multiplying fractions. **4.** The formula is $A = \frac{1}{2}bh$, where A represents the area, b represents the base, and h represents the height. **5.** Radius is the distance from the center to any point on the circle. Diameter is the distance across a circle through its center. **6.** The radius is half the diameter, or the diameter is twice the radius. **7.** The circumference of a circle is the distance around the circle. **8.** An irrational number is a real number that cannot be expressed in the form $\frac{a}{b}$, where a and b are integers and $b \neq 0$. **9.** π represents an irrational number that is the ratio of the circumference of a circle to its diameter. It can be approximated using the fraction $\frac{22}{7}$.

Exercises 1. $\dfrac{8}{35}$ **3.** $-\dfrac{1}{54}$ **5.** $\dfrac{15}{28}$ **7.** $\dfrac{49}{1000}$ **9.** $\dfrac{5}{9}$ **11.** $\dfrac{4}{5}$ **13.** $-\dfrac{1}{3}$ **15.** $\dfrac{13}{20}$ **17.** $-\dfrac{3}{10}$ **19.** $\dfrac{2}{5}$ **21.** $\dfrac{21}{50}$ **23.** $\dfrac{1}{2}$ **25.** Estimate: 6; Actual: $5\dfrac{3}{5}$ **27.** Estimate: 28; Actual: $23\dfrac{1}{3}$ **29.** Estimate: 80; Actual: 74 **31.** Estimate: -3; Actual: $-3\dfrac{1}{2}$

33. Estimate: 42; Actual: $40\dfrac{4}{5}$ **35.** Estimate: -15; Actual: $-13\dfrac{1}{2}$ **37.** $\dfrac{2x^2}{15}$ **39.** $\dfrac{x^2}{6}$ **41.** $\dfrac{2xy^4}{45}$ **43.** $-\dfrac{5k^3}{12}$ **45.** $\dfrac{10x^2y}{7}$

47. $-\dfrac{3m}{10n^3}$ **49.** $\dfrac{25}{36}$ **51.** $-\dfrac{27}{64}$ **53.** $\dfrac{1}{64}$ **55.** $\dfrac{x^3}{8}$ **57.** $\dfrac{8x^6}{27}$ **59.** $\dfrac{m^{12}n^4}{81p^8}$ **61. a.** \$656; **b.** \$574; **c.** \$246; **d.** \$164 **63.** 1560 pieces

65. $\frac{5}{8}$ of the employees **67.** $\frac{2}{3}$ of her long-distance calls; 20 calls **69.** $12\frac{3}{4}$ in. **71.** $42\frac{2}{3}$ mi. **73.** $51\ \text{ft.}^2$ **75.** $9\frac{1}{6}\ \text{ft.}^2$ **77. a.** $4\frac{1}{2}$ m; **b.** $\approx 28\frac{2}{7}$ m **79. a.** $1\frac{53}{220}$ mi.; **b.** $\approx 3\frac{9}{10}$ mi.

Review Exercises 1. -204 **2.** 16 **3.** $3x(3x - 4)$ **4.** $x = -5$ **5.** 2 ft.

Exercise Set 5.4

Prep Exercises 1. Reciprocals are two numbers whose product is 1. Examples may vary. One pair is $\frac{2}{3}$ and $\frac{3}{2}$. **2.** Write an equivalent multiplication by changing the operation from division to multiplication and changing the divisor to its reciprocal. **3.** Write each mixed number as an improper fraction; then follow the process for dividing fractions. **4.** Method 1: Find the square root of the numerator and denominator separately. Method 2: Simplify the fraction; then find the square root of the quotient. **5.** Multiply both sides of the equation by the reciprocal of the coefficient.

Exercises 1. $\dfrac{3}{2}$ **3.** 6 **5.** $-\dfrac{1}{15}$ **7.** $-\dfrac{4}{5}$ **9.** 3 **11.** $\dfrac{9}{25}$ **13.** $-\dfrac{8}{5}$, or $-1\dfrac{3}{5}$ **15.** $\dfrac{1}{24}$ **17.** $\dfrac{3}{4}$ **19.** -18 **21.** Estimate: 10; Actual: $11\dfrac{5}{8}$;

23. Estimate: $2\dfrac{1}{2}$; Actual: $2\dfrac{8}{15}$ **25.** Estimate: $-\dfrac{3}{10}$; Actual: $-\dfrac{1}{4}$ **27.** Estimate: $1\dfrac{6}{7}$; Actual: $1\dfrac{23}{27}$ **29.** $\dfrac{14x^2}{5}$

31. $\dfrac{15m^3}{16n^2}$ **33.** $-\dfrac{9x^4}{8y^5}$ **35.** $\dfrac{8}{9}$ **37.** $\dfrac{11}{6}$ **39.** 6 **41.** 9 **43.** $x = 16$ **45.** $y = -3\dfrac{1}{3}$ **47.** $n = -1\dfrac{4}{5}$ **49.** $m = \dfrac{2}{7}$ **51.** 30 doses; no, there should be 28 doses. **53.** $\frac{7}{8}$ cup **55.** $91\frac{60}{61}$ ft. **57.** $4\frac{6}{13}$ ft. **59.** $989\frac{109}{110}$ ft. **61.** $45\frac{3}{4}$ mph

Review Exercises 1. $9t^3 + 9t^2 - 13t - 16$ **2.** $60x^3y$ **3.** $2 \cdot 3^3 \cdot 7$ **4.** $16x(2x + 1)$ **5.** $=$

Exercise Set 5.5

Prep Exercises 1. The LCM of a set of numbers is the smallest natural number that is divisible by all of the given numbers. **2.** List multiples of the greatest given number until a multiple that is divisible by all of the other given numbers is found. **3.** 1. Find the prime factorization of each given number. 2. Write a factorization that contains each prime factor the greatest number of times it occurs in the factorizations. 3. Multiply to get the LCM **4.** The least common denominator is the least common multiple of the denominators. **5.** 4 **6.** $6x$

Exercises 1. 30 **3.** 36 **5.** 60 **7.** 90 **9.** 72 **11.** 252 **13.** 364 **15.** 1800 **17.** 390 **19.** 3360 **21.** $24xy$ **23.** $30y^3$ **25.** $16mn$

27. $36x^2y^3$ **29.** $\dfrac{9}{30}$ and $\dfrac{25}{30}$ **31.** $\dfrac{21}{36}$ and $\dfrac{11}{36}$ **33.** $\dfrac{3}{60}$ and $\dfrac{34}{60}$ **35.** $\dfrac{27}{36}, \dfrac{6}{36}$, and $\dfrac{28}{36}$ **37.** $\dfrac{7}{12x}$ and $\dfrac{9x}{12x}$ **39.** $\dfrac{9}{16mn}$ and $\dfrac{6n^2}{16mn}$

41. $\dfrac{21}{30y^3z}$ and $\dfrac{25y^2z^2}{30y^3z}$ **43.** $\dfrac{2y^2z}{36x^2y^3}$ and $\dfrac{-15x}{36x^2y^3}$

Review Exercises 1. -12 **2.** 11 **3.** $4x^3 + 39x^2 - 16x - 3$ **4.** $m = -8$ **5.** $x = 53$ **6.** $a = -\dfrac{4}{5}$

Exercise Set 5.6

Prep Exercises 1. 1. Add or subtract numerators and keep the same denominator. 2. Simplify. **2.** 1. Rewrite the fractions with their LCD. 2. Add or subtract numerators and keep the LCD. 3. Simplify. **3.** Multiply the numerator and denominator of $\frac{3}{4}$ by 3 and multiply the numerator and denominator of $\frac{5}{6}$ by 2. **4.** Method 1: Write them as improper fractions; then follow the procedure for adding or subtracting fractions. Method 2: Add or subtract the integer parts and fraction parts separately.

Exercises 1. $\frac{5}{7}$ **3.** $-\frac{2}{3}$ **5.** $\frac{6}{17}$ **7.** $-\frac{1}{5}$ **9.** $\frac{12}{x}$ **11.** $\frac{2x^2}{3}$ **13.** $-\frac{2m}{n}$ **15.** $\frac{4x^2 - 3x + 1}{7y}$ **17.** $\frac{5n^2 + 9}{5m}$ **19.** $1\frac{2}{15}$ **21.** $\frac{5}{18}$ **23.** $\frac{37}{60}$

25. $1\frac{25}{36}$ **27.** $\frac{23x}{24}$ **29.** $\frac{3}{16m}$ **31.** $\frac{13h + 16}{20h}$ **33.** $\frac{6 - 7n}{9n^2}$ **35.** Estimate: 10; Actual: $10\frac{4}{9}$ **37.** Estimate: 7; Actual: $7\frac{1}{2}$

39. Estimate: 8; Actual: $7\frac{1}{2}$ **41.** Estimate: 11; Actual: $10\frac{5}{24}$ **43.** Estimate: 10; Actual: $9\frac{3}{4}$ **45.** Estimate: 10;

Actual: $9\frac{1}{6}$ **47.** Estimate: 3; Actual: $3\frac{7}{24}$ **49.** Estimate: 0; Actual: $\frac{1}{2}$ **51.** $4\frac{5}{12}$ **53.** $-10\frac{1}{2}$ **55.** $-3\frac{5}{8}$ **57.** $-6\frac{5}{8}$

59. $x = \frac{1}{10}$ **61.** $n = 1\frac{1}{20}$ **63.** $b = -\frac{11}{12}$ **65.** $t = -\frac{7}{10}$ **67.** $\frac{7}{8}$ in. **69.** $\frac{11}{16}$ in. **71. a.** $\frac{17}{24}$; **b.** $\frac{1}{8}$ **73.** $68\frac{1}{2}$ in. **75.** 4 ft.

Review Exercises 1. 51 **2.** 10 **3.** 3 **4.** $10x^2 + 3x - 5$ **5.** $-y^3 - y^2 - 4y - 5$ **6.** $-54b^4$ **7.** $2x^2 + x - 15$ **8.** $-4x^2$
9. $3y^3 + 3y^2$

Exercise Set 5.7

Prep Exercises 1. Multiplication **2.** Divide the sum of the numbers by the number of numbers. **3.** Replace d with $5\frac{3}{4}$.
4. A trapezoid is a four-sided figure with one pair of parallel sides. **5.** $A = \frac{1}{2}h(a + b)$, where A represents the area; h, the height, and a and b, the lengths of the parallel sides. **6.** $A = \pi r^2$, where A represents the area, π represents an irrational number with an approximate value of $\frac{22}{7}$, and r represents the radius of the circle.

Exercises 1. $1\frac{1}{2}$ **3.** $2\frac{3}{5}$ **5.** $1\frac{15}{16}$ **7.** $-2\frac{3}{4}$ **9.** $-2\frac{1}{4}$ **11.** $-\frac{1}{8}$ **13.** $-24\frac{5}{6}$ **15. a.** length: $9\frac{1}{4}$ in.; weight: $1\frac{11}{12}$ lb.; **b.** length: $9\frac{1}{4}$ in.;

weight: $1\frac{7}{8}$ lb.; **c.** length: $9\frac{1}{4}$ in.; weight: $1\frac{3}{9}$ lb. **17. a.** $1\frac{1}{4}$ mi.; **b.** $119\frac{2}{5}$ sec.; **c.** $23\frac{22}{25}$ sec.; **d.** $23\frac{4}{5}$; **e.** no mode **19.** 10 **21.** $\frac{45}{64}$

23. $-1\frac{9}{40}$ **25.** $\frac{2}{3}$ **27.** $52\frac{1}{4}$ m^2 **29.** $30\frac{1}{4}$ ft.^{2} **31.** 616 in.^{2} **33.** $4\frac{51}{56}$ m^2 **35.** $240\frac{9}{10}$ m^2 **37.** $\frac{1}{6}x^2 - \frac{27}{4}x$

39. $6y^3 + \frac{11}{5}y^2 + y - \frac{5}{6}$ **41.** $\frac{1}{10}t^3 + \frac{5}{12}t^2 - \frac{2}{3}$ **43.** $-\frac{1}{10}m^4$ **45.** $\frac{1}{2}t^2 - \frac{5}{12}t$ **47.** $x^2 - \frac{1}{4}x - \frac{1}{8}$

Review Exercises 1. $x = 5$ **2.** $n = 11$ **3.** 4 hr. **4.** $6(n + 2) = n - 3$; $n = -3$ **5.** 6 ten-dollar bills and 9 five-dollar bills

Exercise Set 5.8

Prep Exercises 1. Multiply both sides of the equation by the LCD of the fractions. **2.** multiplication **3.** Mistake: $\frac{2}{3}$ of the sum means that the sum must be found before multiplying by $\frac{2}{3}$; so parentheses are needed. Correct: $\frac{2}{3}(n + 6) = 8\frac{1}{2}$ **4.** triangle: $= \frac{1}{2}bh$; trapezoid: $A = \frac{1}{2}h(a + b)$ **5.** Divide the sum of the scores by the number of scores. **6.** Relationship 1: Mary alone raises $2\frac{1}{2}$ times what the rest of the team does. Relationship 2: The total amount raised was \$31,542. **7.** The amount Mary raised alone and the amount the rest of the team raised **8.** Let the variable represent the amount raised by the rest of the team. It is the amount multiplied by $2\frac{1}{2}$ ("acted on") when the first relationship is translated, whereas Mary's amount is isolated.

Exercises 1. $x = -\frac{1}{4}$ **3.** $p = \frac{1}{2}$ **5.** $c = -\frac{7}{30}$ **7.** $a = \frac{2}{3}$ **9.** $f = 3\frac{1}{9}$ **11.** $b = \frac{1}{4}$ **13.** $x = \frac{4}{9}$ **15.** $a = 2\frac{1}{10}$ **17.** $n = -4\frac{3}{4}$ **19.** $x = 10\frac{2}{5}$

21. $\frac{3}{8}n = 4\frac{5}{6}$; $n = 12\frac{8}{9}$ **23.** $y - 8\frac{7}{10} = -2\frac{1}{2}$; $y = 6\frac{1}{5}$ **25.** $3\frac{1}{4} + 2n = -\frac{1}{6}$; $n = -1\frac{17}{24}$ **27.** $\frac{3}{4}(b + 10) = 1\frac{1}{6} + b$; $b = 25\frac{1}{3}$ **29.** $b = 4\frac{3}{5}$ cm

31. 90 **33.** 106 **35.** Many raised \$22,530. The rest of the team raised \$9012. **37.** 1200 gal. each **39.** $\frac{3}{5}$ **41.** 45° and 135°

Review Exercises 1. $<$ **2.** $\frac{3}{8x}$ **3.** $-\frac{14}{45}$ **4.** $\frac{10a}{9b^3}$ **5.** $6\frac{13}{24}$

Chapter 5 Summary and Review Exercises

5.1 Definitions/Rules/Procedures $\frac{a}{b}$; top; bottom; greater than or equal to; less than   Exercises **1.** $\frac{1}{2}$ **2.** $\frac{1}{4}$ **3.** 18 **4.** undefined

Definitions/Rules/Procedures   integer; fraction; integers; quotient; remainder; numerator; numerator   Exercises **5.** $4\frac{4}{9}$ **6.** $-7\frac{1}{4}$

7. $\frac{20}{3}$ **8.** $-\frac{11}{2}$ Definitions/Rules/Procedures   0 and 1; 0 and -1; mixed number; next greater integer; next lesser integer

Exercises **9.**   **10.**

Definitions/Rules/Procedures   name the same number; multiply or divide   Exercises   **11.** 5   **12.** 36   Definitions/Rules/Procedures   evenly divisible; common denominator; numerators   Exercises   **13.** >   **14.** <

5.2 Definitions/Rules/Procedures   1; greatest common factor; all of the common prime factors; polynomials

Exercises   **15.** $\dfrac{3}{7}$ **16.** $-\dfrac{4}{5}$ **17.** $-\dfrac{4m^3}{13n}$ **18.** $\dfrac{3x}{10y^3}$

5.3 Definitions/Rules/Procedures   Multiply; multiply; Divide out; Multiply; improper fractions   Exercises   **19.** $\dfrac{4}{15}$ **20.** $\dfrac{35n}{12p^2}$ **21.** $-\dfrac{h^2}{5}$ **22.** -12 Definitions/Rules/Procedures $\dfrac{a^n}{b^n}$ Exercises   **23.** $\dfrac{1}{64}$ **24.** $\dfrac{9}{25}x^2y^6$

5.4 Definitions/Rules/Procedures   the divisor to its reciprocal; changing the operation symbol from division to multiplication and changing the divisor to its reciprocal   Exercises   **25.** $1\dfrac{5}{9}$ **26.** $\dfrac{4}{3b^2}$ **27.** $-\dfrac{4x^2z^2}{5}$ **28.** $2\dfrac{4}{7}$ Definitions/Rules/Procedures   numerator; denominator; Simplify   Exercises   **29.** $\dfrac{5}{3}$ **30.** 5

5.5 Definitions/Rules/Procedures   the smallest natural number that is divisible by all of the given numbers; list multiples of the greatest given number until you find a multiple that is divisible by all of the other given numbers; each prime factor the greatest number of times it occurs in the factorizations   Exercises   **31.** 168   **32.** $60x^2y$

5.6 Definitions/Rules/Procedures   numerators; denominator   Exercises   **33.** $\dfrac{2}{3}$ **34.** $\dfrac{1}{5}$ **35.** $\dfrac{n}{2}$ **36.** $\dfrac{5x-4}{y}$ Definitions/Rules/Procedures   equivalent fractions with a common denominator   Exercises   **37.** $1\dfrac{1}{30}$ **38.** $-\dfrac{1}{5}$ **39.** $\dfrac{5}{12h}$ **40.** $\dfrac{15-14a}{24a}$

Definitions/Rules/Procedures   improper fractions; improper fractions; integers and fractions   Exercises   **41.** $11\dfrac{7}{24}$ **42.** $-2\dfrac{5}{6}$

5.7 Definitions/Rules/Procedures   Grouping symbols; Exponents or roots; Multiplication or division; Addition or subtraction   Exercises   **43.** $3\dfrac{5}{8}$ **44.** $-6\dfrac{7}{12}$ Definitions/Rules/Procedures   Replace; Calculate; coefficients; variables; exponents; combine like terms; changing the sign of each of its terms; coefficients; Add the exponents; each term; each term; Combine   Exercises   **45.** $-3\dfrac{11}{12}$ **46.** $\dfrac{1}{10}x^2-2x-2$ **47.** $\dfrac{5}{8}n^2-\dfrac{2}{3}n-2$ **48.** $\dfrac{23}{2}y^3-2y^2-\dfrac{5}{6}y+\dfrac{1}{15}$ **49.** $-\dfrac{1}{4}b^4$ **50.** $\dfrac{5}{4}x^2+\dfrac{77}{8}x-3$

5.8 Definitions/Rules/Procedures   multiplying; LCD   Exercises   **51.** $y=4\dfrac{2}{15}$ **52.** $n=-3\dfrac{5}{9}$ **53.** $m=11\dfrac{1}{2}$ **54.** $n=-9\dfrac{1}{3}$

Geometry   equally distant; center; across; distance around; cannot; circumference; diameter; 2; $\dfrac{1}{2}$; $2\pi r$; πd; πr^2; $\dfrac{1}{2}bh$; one; $\dfrac{1}{2}h(a+b)$ Exercises   **55.** $\dfrac{1}{6}$ **56.** 9 ft.^{2} **57.** $1\dfrac{3}{4}$ in.   **58.** $28\dfrac{2}{5}$ cm   **59.** $7\dfrac{6}{7}$ ft.   **60.** 18 servings   **61.** $5\dfrac{1}{4}$ in.   **62.** $\dfrac{7}{20}$ **63.** mean: $18\dfrac{5}{24}$ in.; median: $18\dfrac{3}{8}$ in.   **64.** $110\dfrac{1}{4}$ cm^2 **65.** $12\dfrac{4}{7}$ ft.^{2} **66.** 220 cm^2 **67.** $\dfrac{1}{4}(n+2)=\dfrac{3}{4}n-\dfrac{3}{5}$; $n=2\dfrac{1}{5}$ **68.** $b=2\dfrac{2}{9}$ m   **69.** $24\dfrac{1}{2}$ ft. and $10\dfrac{1}{2}$ ft.   **70.** $77\dfrac{1}{7}°$, $102\dfrac{6}{7}°$

Chapter 5 Practice Test

1. $\dfrac{3}{8}$ **2.**

3. a. = **b.** >   **4. a.** $6\dfrac{1}{6}$; **b.** $-\dfrac{37}{8}$ **5. a.** -16; **b.** 0; **c.** 1   **6. a.** $\dfrac{3}{5}$; **b.** $-\dfrac{3x^2y}{10}$

7. a. $-11\dfrac{5}{8}$; **b.** $\dfrac{4}{3b^2}$ **8. a.** $\dfrac{4}{25}$; **b.** $\dfrac{1}{64}$ **9. a.** $2\dfrac{1}{2}$; **b.** $-\dfrac{8m^3n}{15}$ **10.** $36t^3u$ **11. a.** $\dfrac{11}{20}$; **b.** $-1\dfrac{5}{8}$ **12. a.** $-\dfrac{x}{3}$; **b.** $\dfrac{11}{12a}$ **13.** $2\dfrac{9}{16}$ **14. a.** $-\dfrac{3}{2}n-3$; **b.** $2m^2+m-\dfrac{3}{8}$ **15.** $m=6$ **16.** $\dfrac{5}{8}$ **17.** $22\dfrac{1}{2}$ m^2 **18.** $14\dfrac{13}{14}$ in.   **19.** 8 pieces   **20.** $\dfrac{5}{24}$ **21.** mean: $70\dfrac{7}{16}$ in.; median: $70\dfrac{3}{8}$ in.; no mode   **22.** 62 in.^{2} **23.** $19\dfrac{9}{14}$ in.^{2} **24.** $\dfrac{2}{3}(n+5)=\dfrac{1}{2}n-\dfrac{3}{4}$; $n=-24\dfrac{1}{2}$ **25.** downtown: \$10,400; mall: \$5200

Chapter 1–5 Cumulative Review Exercises

1. false   **2.** false   **3.** false   **4.** true   **5.** 1. Write the prime factorization of each number in exponential form. 2. Create a factorization for the GCF that has only those prime factors common to all factorizations, each raised to the least of its exponents. 3. Multiply.   **6.** 1. Write the fractions with their LCD. 2. Add or subtract the numerators and keep the LCD. 3. Simplify.   **7.** four million, five hundred eighty-two thousand, six hundred one   **8.** 35,000   **9.** coefficient: -1; degree: 1   **10.** 120   **11.** -111 **12.** $20\dfrac{1}{2}$ **13.** $\dfrac{9}{10}$ **14.** $\dfrac{39}{46}$ **15.** $-\dfrac{3x^2}{8}$ **16.** $3\dfrac{2}{3}$ **17.** $\dfrac{12+10x}{15x}$ **18.** $124\dfrac{2}{3}$ **19.** $-b^3+15b^2-\dfrac{11}{12}b$ **20.** $-7x^3-7x+26$ **21.** x^2-49 **22.** $6m(3m^2+4m-5)$ **23.** $y=-1$ **24.** $n=\dfrac{4}{5}$ **25.** mean = \$214,400; median = \$215,000; no mode   **26.** $-\$58,946$ **27. a.** $10y-2$; **b.** 78 ft.   **28.** 2 ft.   **29.** 14 in.   **30.** $71\dfrac{2}{3}°$ and $108\dfrac{1}{3}°$

Exercise Set 6.1
Prep Exercises 1.

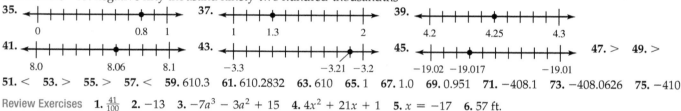

2. 1. Write all digits to the left of the decimal point as the integer part of a mixed number.   2. Write all digits to the right of the decimal point in the numerator of a fraction.   3. Write the denominator indicated by the last place value.   4. Simplify to lowest terms.   **3.** and   **4.** ten   **5.** Compare the digits in the corresponding places from left to right until you find two different digits in the same place. The greater number contains the greater of those two digits.   **6.** eliminated

Exercises   **1.** $\frac{1}{5}$ **3.** $\frac{1}{4}$ **5.** $\frac{3}{8}$ **7.** $\frac{6}{25}$ **9.** $1\frac{1}{2}$ **11.** $18\frac{3}{4}$ **13.** $9\frac{5}{8}$ **15.** $7\frac{9}{25}$ **17.** $-\frac{1}{125}$ **19.** $-13\frac{3}{250}$ **21.** ninety-seven thousandths   **23.** two thousand fifteen millionths   **25.** thirty-one and ninety-eight hundredths   **27.** five hundred twenty-one and six hundred eight thousandths   **29.** four thousand one hundred fifty-nine and six tenths   **31.** negative one hundred seven and ninety-nine hundredths   **33.** negative fifty thousand ninety-two hundred-thousandths

35. [number line: 0, 0.8, 1 with point at 0.8]   **37.** [number line: 1, 1.3, 2 with point at 1.3]   **39.** [number line: 4.2, 4.25, 4.3 with point at 4.25]

41. [number line: 8.0, 8.06, 8.1 with point at 8.06]   **43.** [number line: −3.3, −3.21, −3.2 with point at −3.21]   **45.** [number line: −19.02, −19.017, −19.01 with point at −19.017]   **47.** >   **49.** >

51. <   **53.** >   **55.** >   **57.** <   **59.** 610.3   **61.** 610.2832   **63.** 610   **65.** 1   **67.** 1.0   **69.** 0.951   **71.** −408.1   **73.** −408.0626   **75.** −410

Review Exercises   **1.** $\frac{41}{100}$ **2.** −13   **3.** $-7a^3 - 3a^2 + 15$ **4.** $4x^2 + 21x + 1$ **5.** $x = -17$ **6.** 57 ft.

Exercise Set 6.2
Prep Exercises **1.** align   **2.** 27.31   **3.** The number with the greater absolute value is placed on top.   **4.** 463.4

Exercises   **1.** Estimate: 160; Actual: 159.71   **3.** Estimate: 59.7; Actual: 59.7511   **5.** Estimate: 333; Actual: 333.317   **7.** Estimate: 840; Actual: 841.45   **9.** Estimate: 780 Actual: 784.281   **11.** Estimate: 0.02 Actual: 0.01789   **13.** 10.11   **15.** −4.317   **17.** −0.9985   **19.** −190.02   **21.** −73.25   **23.** 0.9952   **25.** $-12.81y^2 + 5y - 0.93$ **27.** $4.5a^3 + 9a^2 - 1.5a - 10.6$ **29.** $10.2x^2 - 0.9x + 8.4$ **31.** $10.01a^3 - 4.91a^2 + 7.5a + 5.99$ **33.** $0.4n^2 + 10.8n - 3.6$ **35.** $2.8k^3 + 2.5k^2 + 0.06$ **37.** $2.91x^3 - 11.1x^2 + 2.6x - 4.45$ **39.** $m = 10.83$ **41.** $y = -5.81$ **43.** $k = -3.22$ **45.** $46.83 overdrawn   **47.** The client owes $421.94.   **49.** $9.38   **51.** $1919.89   **53.** $35.25 + n = 32.50$; 2.75 decrease   **55.** $18.5 + 36.8 + 18.5 + d = 96.1$; 22.3 m

57. a.

	July 20	July 21	July 22	July 23
Equation	$46.68 + n = 46.61$	$46.61 + n = 45.48$	$45.48 + n = 46.07$	$46.07 + n = 46.15$
Price change	−$0.07	−$1.13	$0.59	$0.08

b. July 21

Review Exercises   **1.** −300   **2.** $\frac{21}{100}$ **3.** 306   **4.** $2\frac{32}{39}$ **5.** 11   **6.** $-18x^7$ **7.** $y^2 - 2y - 15$ **8.** $x = -9$

Exercise Set 6.3
Prep Exercises **1.** Place the decimal point in the product so that the product has the same number of decimal places as the total number of decimal places in the factors.   **2.** 2.12   **3.** Move the decimal point to the right the same number of places as the number of 0s in the power of ten.   **4.** No, because 24.7 is greater than 10. The absolute value of the decimal number needs to be greater than or equal to 1 but less than 10.   **5.** Move the decimal in 4.85 to the right 7 places.   **6.** 5 because there are five places to the right of the new decimal position

Exercises   **1.** Estimate: 0.54; Actual: 054   **3.** Estimate: 12; Actual: 10.585   **5.** Estimate: 700; Actual: 702.5922   **7.** 6.1945   **9.** 196   **11.** −15.08   **13.** −0.152   **15.** 0.00435   **17.** 0.81   **19.** 9.261   **21.** 0.00000081   **23.** −0.064   **25.** 300,000,000 m/sec.; three hundred million m/sec.   **27.** 2,140,000 lt-yr.; two million, one hundred forty thousand lt-yr.   **29.** 307,000,000 people; three hundred seven million people   **31.** 2.3×10^6 blocks   **33.** 2.795×10^9 mi.   **35.** 5.3×10^9 nucleotide pairs   **37.** $5.76x^8$ **39.** $-0.016a^3$ **41.** $0.25t^6$ **43.** $0.9y^2 + 1.3y + 19$ **45.** $-0.136a^3 + 6.8a - 12.92$ **47.** $x^2 + 0.92x - 0.224$ **49.** $6.4x^2 - 18.32x - 3.66$ **51.** $33.64a^2 - 81$ **53.** $14.7k^2 - 4.27k + 0.28$ **55.** $131.52   **57.** $3.84   **59.** $3.75   **61.** $102.96   **63.** $182.40   **65.** −2.8V   **67.** 56.25 ft.   **69.** 23.04 cm²   **71.** 12,551.1 ft.²   **73.** 0.0081 m³

Review Exercises   **1.** 205   **2.** 13   **3.** $6x^4$ **4.** $t = -9$ **5.** $x = 9$ **6.** $a = 3$ **7.** radius: $7\frac{1}{2}$ in.; circumference: $\approx 47\frac{1}{7}$ in.

Exercise Set 6.4
Prep Exercises **1.** 6.2   **2.** two places   **3.** To write a fraction as a decimal number, divide the denominator into the numerator.   **4.** half

Exercises   1. Estimate: 30; Actual: 23.85   **3.** Estimate: 0.025; Actual: 0.024   **5.** Estimate: $666.\overline{6}$ or 600 if 5640 is rounded to 5400; Actual: 600   **7.** Estimate: 5; Actual: 4.8   **9.** Estimate: -30; Actual: -40 **11.** Estimate: 125; Actual: 125   **13.** Estimate: -0.02; Actual: -0.0206 **15.** Estimate: 800; Actual: 814.5   **17.** Estimate: 200; Actual: $178.\overline{18}$ **19.** Estimate: 1; Actual: 1.06   **21.** 0.6   **23.** 0.45   **25.** -0.4375 **27.** $0.4\overline{3}$ **29.** 13.25   **31.** -17.625 **33.** $104.\overline{6}$ **35.** $-216.\overline{571428}$ **37.** 0.04   **39.** 0.5   **41.** ≈ 4.90 **43.** ≈ 14.14 **45.** 1.3   **47.** ≈ 0.09 **49.** $25.5x^2$ **51.** $-6.4m^2n$ **53.** $0.15ac$ **55.** $b = 6.02$ **57.** $h = -1.6$ **59.** $n = 0.6$ **61.** $x = -3.5$ **63.** $613.32   **65.** 40.5 rads   **67.** 400 Ω **69.** ≈ 1.7 hr.   **71.** ≈ 6829.5 mi.   **73.** ≈ 6.69 mi.   **75.** ≈ 1.04 ft.

Review Exercises   1. -5 **2.** 18   **3.** -180 **4.** 56 in.^{2} **5.** 680 cm^2

Exercise Set 6.5

Prep Exercises 1. Writing the decimals as fractions is better in this case because the decimal equivalent of $\frac{2}{3}$ is $0.\overline{6}$, which would have to be rounded for the calculation, thereby yielding an approximation.   **2.** cylinder: $V = \pi r^2 h$; cone: $V = \frac{1}{3}\pi r^2 h$. The volume of a cone is one-third the volume of a cylinder with the same radius and height.   **3.** $V = \frac{1}{3}lwh$; the volume of a pyramid is one-third the volume of a box with the same length, width, and height.   **4.** $V = \frac{4}{3}\pi r^3$, where V represents the volume, π represents the irrational constant with an approximate value of 3.14, and r represents the radius of the sphere.

Exercises   1. 2.64   **3.** 11.73   **5.** 42.99   **7.** -22.24 **9.** -28 **11.** -117.69 **13.** -2.45 **15.** -0.51 **17.** 0.29   **19.** -4.775 **21.** $-0.3\overline{8}$ **23. a.** mean: 22.75 min.; median: 20.5 min; no mode; **b.** $9.17   **25.** mean: $72.\overline{6}$ in.; median: 73 in.; mode: 71.5, 73, and 74 in.   **27.** mean: $50.46; median: $50.38; mode: $50.35   **29.** $327   **31.** $\approx $586.55   **33.** 3.192   **35.** 2.706   **37.** 40   **39.** ≈ 23.6 **41.** 2.25×10^{23} **43.** 9.75   **45.** 1.0609   **47.** 57.6 cm^2 **49.** 1.296 m^2 **51.** ≈ 86.5 in.^{2} **53.** ≈ 0.9 cm^3 **55.** 168,750 ft.^{3} **57.** $\approx 92.10\overline{6}$ in.^{3} **59.** $\approx 2.45 \times 10^{11}$ mi.^{3} **61.** ≈ 2.71 cm^2 **63.** ≈ 134.375 cm^2 **65.** $\approx 151,475.69$ ft.^{3} **67.** $\approx 2,582,957.6$ m^3 **69. a.** 0.000027 in.^{3}; **b.** 37,037 grains

Review Exercises 1. $x = 7$ **2.** $y = 20\frac{1}{4}$ **3.** 300 ft.^{2} **4.** 720 cm^2 **5.** $\frac{1}{2}(x + 12) = \frac{3}{4}x - 9; x = 60$ **6.** 8 five-dollar bills and 12 ten-dollar bills

Exercise Set 6.6

Prep Exercises 1. Multiply both sides of the equation by an appropriate power of 10 as determined by the decimal number with the most decimal places.   **2.** 100   **3.** A right triangle is a triangle that has one angle that measures 90°.   **4.** The legs of a right triangle are the sides that form the right angle.   **5.** The hypotenuse is the longest side in a right triangle.   **6.** $a^2 + b^2 = c^2$, where a and b represent the lengths of the legs and c represents the length of the hypotenuse.

Exercises   1. $n = 0.2$ **3.** $y = 0.6$ **5.** $t = -0.12$ **7.** $k = -6.2$ **9.** $n = -0.58$ **11.** $x = 3.\overline{3}$ **13.** $t = 0.5$ **15.** $x = -11$ **17.** $n = -24$ **19.** $p = 0.8$ **21.** $t = -3.8\overline{3}$ **23.** $y = 2.52$ **25.** ≈ 100 km   **27.** 5 mi.   **29.** 10 in.   **31.** ≈ 91.9 ft.   **33.** 24 ft.   **35.** ≈ 14.1 ft.   **37. a.** ≈ 59.96 ft.; **b.** No, the person is 63 ft. above ground.

39.

Categories	Selling price	Number of books sold	Income
Paperback	0.50	$n + 14$	$0.50(n + 14)$
Hardcover	1.50	n	$1.50n$

41.

Categories	Selling price	Number of T-shirts sold	Income
Children's T-shirts	8.99	$45 - n$	$8.99(45 - n)$
Adult T-shirts	14.99	n	$14.99n$

43. 5 half-dollars and 17 quarters   **45.** 42 of the 12-oz. drinks and 23 of the 16-oz. drinks

Review Exercises   1. twenty-four billion, nine hundred fifteen million, two hundred four   **2.** 46,300,000   **3.** $13\frac{11}{24}$ **4.** $4\frac{9}{20}$ **5.** $4x^6$ **6.** Mistake: divided by 3 rather than by -3 Correct: $x = 2$

Chapter 6 Summary and Review Exercises

6.1 Definitions/Rules/Procedures   a base-10; numerator   Exercises   **1.** $\frac{4}{5}$ **2.** $\frac{3}{125}$ **3.** $-2\frac{13}{20}$ Definitions/Rules/ Procedures   last place value; and   Exercises   **4.** twenty-four and thirty-nine hundredths   **5.** five hundred eighty-one and four hundred fifty-nine thousandths   **6.** two thousand nine hundred seventeen ten-thousandths; compare the digits in the corresponding place values from left to right; closer to zero   Exercises   **7.** $>$ **8.** $<$ Definitions/Rules/Procedures   up; zeros   Exercises   **9.** 32   **10.** 31.8   **11.** 31.81   **12.** 31.805

6.2 Definitions/Rules/Procedures   it aligns with the decimal points in the addends; on top   Exercises   **13.** Estimate: 88; Actual: 88.015   **14.** Estimate: 8; Actual: 7.916   **15.** -6.2 **16.** -6.12 **17.** 23.118   **18.** $11.1y^4 - 12.2y^2 - 8y + 3.8$ **19.** $5.4a^2 + 5.2a + 10.33$ **20.** $-2.7x^3 - 98.1x^2 + 2.1x - 4.2$

6.3 Definitions/Rules/Procedures   the same number of decimal places as the total number of decimal places in the factors   Exercises   **21.** Estimate: 15; Actual: 17.544   **22.** Estimate: -20; Actual: -14.505 **23.** 1.44   **24.** $-3.78x^8y^2$ **25.** $-0.027x^9$ **26.** $0.12n^3 - 0.28n^2 + 0.2n - 3.6$ **27.** $9a^2 - 14.45a + 3.05$ **28.** $1.69x^2 - 17.64$ Definitions/Rules/Procedures   greater than or equal to 1 but less than 10; right; new; original   Exercises   **29.** 3,580,000,000   **30.** -1.03×10^5

6.4 Definitions/Rules/Procedures   directly above its position in the dividend; an integer (multiply by an appropriate power of 10); the same number of places (multiply it by the same power of 10)   Exercises   **31.** Estimate: 5; Actual: 5.25   **32.** Estimate: 4/9 or 0.4; Actual: 0.4   **33.** Estimate: 600 (use $5400 \div 9$ instead of $5000 \div 9$); Actual: 600   **34.** $-0.35a^3c$ Definitions/Rules/Procedures   denomina-tor into the numerator; to the left of a decimal point; Divide the denominator into the numerator   Exercises   **35.** 0.6   **36.** $-4.\overline{18}$ Definitions/Rules/Procedures   half; rational; irrational   Exercises   **37.** 0.8   **38.** ≈ 12.17

6.5 Exercises   **39.** -3.05 **40.** -2.16 **41.** 19.6

6.6 Definitions/Rules/Procedures   decimal places   Exercises   **42.** $y = -0.64$ **43.** $n = -0.02$ **44.** $k = 0.8$ **45.** $x = 8.1$ **46.** $n = 5$

Formulas   each; quantity; $2\pi r$; πd; πr^2; $\frac{1}{2}bh$; $\frac{1}{2}h(a + b)$; $\pi r^2 h$; $\frac{1}{3}lwh$; $\frac{1}{3}\pi r^2 h$; $\frac{4}{3}\pi r^3$; right angle; 90°; directly across; $a^2 + b^2 = c^2$

Exercises   **47.** Radius: 7.5 cm; Circumference: ≈ 47.1 cm; Area: ≈ 176.625 cm^2 **48.** 0.15 m^2 **49.** ≈ 1445.43 cm^2 **50.** ≈ 42.39 in.^{3} **51.** ≈ 20.096 cm^3 **52.** ≈ 904.32 m^3 **53.** 22,696.8 m^3 **54.** ≈ 39.978 in.^{3} **55.** $713.18   **56.** $20.37   **57.** $47.58   **58. a.** $\approx 73.20; **b.** $\approx 72.81; **c.** $71.98   **59.** 1.923   **60.** 11.62 ft.   **61.** 120 min.   **62.** 10 nickels and 19 dimes

Chapter 6 Practice Test

1. fifty-six and seven hundred eighty-nine thousandths   **2.** $\frac{17}{25}$ **3.** $<$ **4. a.** 2; **b.** 2.1; **c.** 2.09; **d.** 2.092   **5.** Estimate: 43; Actual: 41.991   **6.** Estimate: 24; Actual: 23.906   **7.** Estimate: -45 Actual: -38.745 **8.** Estimate: 0.25 Actual: 0.3   **9.** Estimate: 6 Actual: 6.02   **10.** 0.9   **11.** $3.8\overline{3}$ **12.** 2,970,000   **13.** -3.56×10^7 **14.** 5.1344   **15.** 0.256   **16.** $0.9x^3 - 2.3x^2 - 4.7x + 3.8$ **17.** $1.5x^4$ **18.** $1.28n^2 - 18.6n - 9$ **19.** $t = 4.1$ **20.** $k = 0.6$ **21.** $7.03   **22.** $39.95   **23.** $176.96   **24. a.** $8.49; **b.** $8.53; **c.** $8.61;   **25. a.** ≈ 75.36 ft.; **b.** ≈ 452.16 ft.^{2} **26.** $\approx 33.49\overline{3}$ cm^3 **27.** 777.6 ft.^{3} **28.** 785.5 ft.^{2} **29.** ≈ 19.36 ft.   **30.** 8 round-layered and 4 sheet cakes

Chapters 1–6 Cumulative Review Exercises

1. true   **2.** false   **3.** true   **4.** false   **5.** undefined   **6.** coefficients; variables   **7.** 6   **8.**

9. 2.02   **10.** 768,000,000; seven hundred sixty-eight million   **11.** 7   **12.** $-\frac{31}{54}$ **13.** $\frac{8}{5}n$ **14.** $12\frac{1}{8}$ **15.** $\frac{2x + 9}{12}$ **16.** 3.98   **17.** $1.6x^3 + \frac{3}{4}x^2 + 3.55$ **18.** $4.9m^3 - 6.7m^2 + 18.6$ **19.** 2.258   **20.** $4y^2 + 5y - 6$ **21.** $5n(6n^4 + 3n - 5)$ **22.** $k = -6.8$ **23.** $x = \frac{7}{8}$ **24.** $n = 7$ **25.** $1,400,900 profit   **26. a.** $8w^3 - 4w^2$; **b.** 1584 cm^3 **27.** The base is 13 ft., and the equal-length sides are 38 ft. each.   **28.** $\frac{7}{30}$ had no opinion.   **29.** ≈ 81.25 in.^{2} **30.** 9 larger-size and 11 smaller-size bowls

Exercise Set 7.1

Prep Exercises   **1.** A ratio is a comparison between two quantities using a quotient.   **2.** $\frac{a}{b}$ **3.** the ratio of the number of favorable outcomes to the total number of possible outcomes   **4.** A unit ratio is a ratio in which the denominator is 1.   **5.** Divide the denominator into the numerator.   **6.** A rate is a unit ratio comparing two different measurements.   **7.** A unit price is a unit ratio of price to quantity.   **8.** Brand B is the better buy because it costs less per ounce.

Exercises   **1. a.** $\frac{31}{66}$; **b.** $\frac{35}{66}$; **c.** $\frac{31}{35}$; **d.** $\frac{35}{31}$ **3. a.** $\frac{3}{2}$; **b.** $\frac{2}{3}$; **c.** $\frac{3}{5}$; **d.** $\frac{2}{5}$ **5.** $\frac{7}{8}$ **7.** $\frac{14}{9}$ **9. a.** $\frac{7}{30}$; **b.** $\frac{17}{30}$; **c.** 1; **d.** $\frac{1}{2}$ **11. a.** $\frac{5}{8}$; **b.** $\frac{20}{157}$; **c.** $\frac{169}{1054}$; **d.** $\frac{314}{527}$ **13.** $\frac{1}{13}$ **15.** $\frac{2}{13}$ **17.** $\frac{1}{3}$ **19.** 0   **21.** $\frac{7}{13}$ **23. a.** $\frac{1}{5}$; **b.** $\frac{28}{125}$; **c.** $\frac{72}{125}$; **d.** $\frac{97}{125}$ **25.** ≈ 0.16; debt of $0.16 for every $1.00 of income   **27.** ≈ 0.26; $0.26 is paid for the mortgage for every $1.00 of income   **29.** ≈ 16.9; there are about 17 students for every faculty member.   **31.** ≈ 12.08; the stock is selling at $12.08 for every $1 of annual earning.   **33.** 68.2 mi./hr.   **35.** 12¢/min.   **37.** ≈ 7.89 **39.** 16 oz. can   **41.** bag containing 40 diapers   **43.** two 15.5 oz. boxes

45. a.

	Violent crimes	Aggravated assault	Simple assault	Robbery	Rape/Sexual assault
Ratio per 10,000 people	$\frac{171}{10,000}$	$\frac{2}{625}$	$\frac{113}{10,000}$	$\frac{21}{10,000}$	$\frac{1}{2000}$
Ratio per 1000 people	$\frac{17.1}{1000}$	$\frac{3.2}{1000}$	$\frac{11.3}{1000}$	$\frac{2.1}{1000}$	$\frac{0.5}{1000}$
Ratio per 100 people	$\frac{1.71}{100}$	$\frac{0.32}{100}$	$\frac{1.13}{100}$	$\frac{0.21}{100}$	$\frac{0.05}{100}$

b. 2 out of 625 people were victims of aggravated assault.; **c.** 17.1; **d.** 0.21

Review Exercises   **1.** $4\frac{1}{2}$ **2.** 120   **3.** 12.375   **4.** $y = 7\frac{1}{2}$ **5.** $x = 8.6$

Exercise Set 7.2

Prep Exercises   **1.** A proportion is an equation in the form $\frac{a}{b} = \frac{c}{d}$, where $b \neq 0$ and $d \neq 0$.   **2.** cross products   **3.** 1. Find the cross products.   2. Set the cross products equal to each other. 3. Solve the resulting equation.   **4. a.** The units match, pounds with pounds and square feet with square feet.   **b.** The first ratio relates the unknown weight, 5 pounds, to its square footage, 2000 square feet. The second ratio relates the unknown weight, n, to its square footage, 5000 square feet.   **5.** Similar figures are figures with congruent angles and proportional side lengths.

Exercises   **1.** yes   **3.** no   **5.** no   **7.** yes   **9.** no   **11.** yes   **13.** no   **15.** $x = 5$ **17.** $n = -7.5$ **19.** $n = 6$ **21.** $m = -35$ **23.** $h = 25$
25. $b = -25$ **27.** $d = 14\frac{1}{4}$ **29.** $t = 14\frac{2}{3}$ **31.** ≈ 34.3 gal.   **33.** 36 servings   **35.** 1168.8 oz.   **37.** $2\frac{1}{4}$ cups   **39.** $3\frac{1}{4}$ in.   **41.** $1633\frac{1}{3}$ days
43. $c = 8.96$ cm   **45.** $a = 3\frac{11}{18}$ in.; $b = 3\frac{1}{4}$ in.; $c = 4\frac{29}{48}$ in.   **47.** 9.9 m   **49.** $61\frac{3}{5}$ ft.   **51.** 24.8 m

Review Exercises   **1.** $\frac{5}{27}$ **2.** 1344   **3.** $20\frac{1}{4}$ **4.** 5.875   **5.** 3.7

Exercise Set 7.3

Prep Exercises   **1.** A conversion factor is a ratio used to convert from one unit of measurement to another by multiplication.
2. Multiply the given measurement by unit fractions so that the undesired units divide out, leaving the desired units.   **3.** 12; 3;
5280   **4.** 144; 9   **5.** 8; 2; 2; 4   **6.** 16; 2000   **7.** 60; 60; 24; $365\frac{1}{4}$

Exercises   **1.** 156 ft.   **3.** 7.5 ft.   **5.** 34,320 ft.   **7.** 66 in.   **9.** 36,608 yd.   **11.** 2.75 yd.   **13.** 270 ft.^{2} **15.** 13,608 in.^{2} **17.** 450 ft.^{2}
19. 18.75 yd.^{2} **21.** 18 c.   **23.** 20 oz.   **25.** 15 pt.   **27.** 4.25 qt.   **29.** 51.2 oz.   **31.** 5.1 gal.   **33.** 196 oz.   **35.** 11.25 lb.   **37.** 0.6 T
39. 8400 lb.   **41.** 2.5 hr.   **43.** 1.5 min.   **45.** 630 sec.   **47.** 135 min.   **49.** 720 sec.   **51.** ≈ 0.082 yr.   **53.** 1440 min.   **55.** 37,869,120 min.
57. $12.\overline{36}$ mi./min.   **59.** No, the driver and the officer are not in agreement because the officer's estimate converts to $54.\overline{54}$ mi./hr.,
not 40 mi./hr.   **61.** $25,014.\overline{27}$ mi./hr.   **63.** $5.6\overline{81}$ sec.

Review Exercises   **1.** 840   **2.** 950   **3.** 0.329   **4.** 4.56   **5.** 0.009   **6.** 4.8

Exercise Set 7.4

Prep Exercises   **1.** A base unit is a basic unit; other units are named relative to it.   **2.** kilo, hecto, deka, deci, centi, milli   **3.** meter
4. liter   **5.** milliliter   **6.** gram   **7.** 1000   **8.** twice

Exercises   **1.** 450 cm   **3.** 70 m   **5.** 3.8 km   **7.** 9.5 m   **9.** 15.4 dam   **11.** 1.12 hm   **13.** 120 ml   **15.** 5 L   **17.** 0.0095 dal   **19.** 80 cc   **21.** 12 ml
23. 400 cc   **25.** 12,000 mg   **27.** 9 g   **29.** 50 cg   **31.** 3.6 t   **33.** 106 kg   **35.** 6.5 t   **37.** 18,000 cm^2 **39.** 0.22 m^2 **41.** 5 mm^2 **43.** 0.0443 hm^2

Review Exercises   **1.** $21\frac{2}{3}$ **2.** 1.008   **3.** $6.5x^2 - 18.2x - 5.6$ **4.** $481y^2 - 280y + 700$ **5.** $x = 2.8$

Exercise Set 7.5

Prep Exercises   **1.** 2.54 cm   **2.** 2.2 lb.   **3.** $F = \frac{9}{5}C + 32$ **4.** $C = \frac{5}{9}(F - 32)$

Exercises   **1.** ≈ 91.4 m   **3.** ≈ 42.2 km   **5.** 9.525 mm   **7.** ≈ 437.6 yd.   **9.** ≈ 24.84 mi.   **11.** ≈ 17.72 in.   **13.** ≈ 0.473 L   **15.** ≈ 591.25 ml
17. ≈ 0.79 gal.   **19.** ≈ 0.68 oz.   **21.** ≈ 61.29 kg   **23.** ≈ 340.5 g   **25.** ≈ 0.0176 oz.   **27.** ≈ 121 lb.   **29.** 130°F   **31.** 42.8°F
33. −459.67°F   **35.** 37°C   **37.** $38.\overline{8}$°C   **39.** $-20.\overline{5}$°C

Review Exercises   **1.** 0.28   **2.** $0.118\overline{3}$ **3.** $-\frac{4}{3}x - \frac{7}{20}y$ **4.** $t = \frac{2}{3}$ **5.** $m = 10.88$

Exercise Set 7.6

Prep Exercises   **1.** A front-end ratio is the ratio of the total monthly house payment to gross monthly income.   **2.** A back-end ratio is
the ratio of the total monthly debt payments to gross monthly income.   **3.** The PITI payment includes principal, interest, taxes, and
insurance.   **4.** The conversion factor $\frac{1 \text{ ml}}{20 \text{ drops}}$ is inverted. It needs to be $\frac{20 \text{ drops}}{1 \text{ ml}}$ so that milliliters divide and leaving drops/min.

Exercises   **1.** front-end ratio ≈ 0.22; back-end ratio ≈ 0.33; considering only the two ratios and credit score, the Wu family qualifies.
3. back-end ratio ≈ 0.42; considering only the back-end ratio, the Rivers family does not qualify.   **5.** front-end ratio ≈ 0.31; back-end ratio ≈ 0.45; considering only the two ratios, the Bishop family does not qualify.   **7.** $1344   **9.** $533.60
11. $1252.80   **13.** $686.75   **15.** $973.75   **17.** 522.5 mg   **19.** ≈ 5 tablets   **21.** 10 to 11 drops/min.   **23.** 400 units/hr.

Review Exercises   **1.** $\frac{7}{25}$ **2.** $\frac{2}{13}$ **3.** $\approx \$0.14$/oz.   **4.** 25 yd.   **5.** 8500 mg

Chapter 7 Summary and Review Exercises

7.1 Definitions/Rules/Procedures   quotient; $\frac{a}{b}$; divide; an appropriate power of 10   Exercises   **1.** $\frac{11}{16}$ **2.** $\frac{25}{108}$ Definitions/Rules/
Procedures   favorable; possible; favorable; possible   Exercises   **3.** $\frac{2}{13}$ **4.** $\frac{1}{3}$ Definitions/Rules/Procedures   1; the denominator
into the numerator   Exercises   **5.** ≈ 16.4 **6.** ≈ 7.09 Definitions/Rules/Procedures   different; denominator; divide the denominator into the numerator; price to quantity; $\frac{p}{q}$ Exercises   **7.** 70.2 mi./hr.   **8.** 9¢/min.   **9.** 4.5¢/oz.   **10.** 3.9¢/oz.   **11.** 24-oz. cup
12. two 12-oz. cans

7.2 Definitions/Rules/Procedures $\frac{a}{b} = \frac{c}{d}$; cross products; $ad = bc$ Exercises   **13.** no   **14.** yes   Definitions/Rules/
Procedures   Set the cross products equal to each other   Exercises   **15.** $k = -4.5$ **16.** $m = 9\frac{3}{13}$ Definitions/Rules/
Procedures   horizontally; vertically; congruent; proportional   Exercises   **17.** 29.9 m^2 **18.** 280 mi.   **19.** 3.6 cm

20. $a = 6\frac{3}{4}$ ft., $b = 3\frac{3}{8}$ ft., $c = 5\frac{1}{4}$ ft.

7.3 Definitions/Rules/Procedures   ratio; divide out; denominator; numerator; holds; force; 12; 3; 5280; 8; 2; 2; 4; 16; 2000; 60; 60; 24; $365\frac{1}{4}$ Exercises   **21.** 432 in.   **22.** 5.2 mi.   **23.** 2.5 lb.   **24.** 2.4 T   **25.** 192 oz.   **26.** 6.2 gal.   **27.** 135 min.   **28.** 1461 d.   **29.** $24.\overline{1}$ yd.^{2} **30.** 79,200 mi./hr.

7.4 Definitions/Rules/Procedures   basic; meter; liter; gram; kilo-; centi-; places; matter; 1; 1000; twice   Exercises   **31.** 50 dm   **32.** 0.375 m   **33.** 260 g   **34.** 0.95 t   **35.** 28.5 L   **36.** 75 cc

7.5 Definitions/Rules/Procedures   2.54; 3.281; 0.914; 1.094; 1.609; 0.621; 0.946; 1.057; 0.454; 2.2; 0.907; 1.1; $\frac{5}{9}(F-32)$; $\frac{9}{5}C+32$

Exercises   **37.** ≈ 6.10 m   **38.** ≈ 19.31 km   **39.** ≈ 65.83 kg   **40.** ≈ 0.79 gal.   **41.** $32.\overline{2}$°C   **42.** 24.8°F

7.6 Definitions/Rules/Procedures   gross monthly income; debt payments   Exercises   **43.** 12 to 13 drops/min.   **44.** 92.8 mg   **45.** ≈ 0.28 **46.** ≈ 0.40 **47.** $571.30   **48.** $1722

Chapter 7 Practice Test
1. $\frac{19}{26}$ **2.** $\frac{11}{2289}$ **3.** $\frac{1}{5}$ **4.** ≈ 30.8 **5.** $70.4\overline{8}$ mi./hr.   **6.** 4.3¢/oz.   **7.** 20 oz. can   **8.** $n=-5.85$ **9.** 7.84 cm   **10.** $514.50   **11.** 900 mi.
12. 168 in.   **13.** 320 oz.   **14.** 4.5 gal.   **15.** 2.5 hr.   **16.** $27.\overline{7}$ yd.^{2} **17. a.** 5.8 cm; **b.** 0.42 km   **18. a.** 0.024 kg; **b.** 91,000 g   **19. a.** 0.028 m^2; **b.** 80 dm^2 **20. a.** 0.65 L; **b.** 800 cc or cm^3 **21.** ≈ 22.86 m **22.** $-1.\overline{1}$°C   **23.** 10.4° F   **24.** 0.45   **25.** 15 to 16 drops/min.

Chapters 1–7 Cumulative Review Exercises
1. false   **2.** true   **3.** false   **4.** false   **5.** positive   **6.** LCM   **7.** $2^5 \cdot 3 \cdot 5^2$ **8.**

$$\begin{array}{c} \longleftarrow\!\!+\!\!\!-\!\!-\!\!\bullet\!\!-\!\!-\!\!+\!\!-\!\!-\!\!+\!\!\longrightarrow \\ \begin{array}{ccc} -4 & -3\frac{2}{3} & -3 \end{array}\end{array}$$

9. $40k^2$ **10.** 9.14×10^8

11. -91 **12.** $-2\frac{8}{13}$ **13.** $5\frac{9}{10}$ **14.** 3   **15.** -31.2 **16.** 0.0081   **17.** 3.04   **18.** $8.3y^3 - 8.2y^2 - y + 10.2$ **19.** $\frac{1}{2}a^2 - \frac{65}{12}a + 7$ **20.** $\frac{3}{16x^4}$

21. $x = 14\frac{6}{11}$ **22.** $b = -16.32$ **23.** 66,000 ft.   **24.** 480 g   **25.** 96 m^3 **26.** $y^2 - 4y - 45$ **27.** length = 14 m; width = 11 m
28. $41\frac{1}{4}$ ft.^{2} **29.** mean = $5.575 million; median = $5.4 million; no mode   **30.** 16 oz.v

Exercise Set 8.1

Prep Exercises **1.** 100   **2.** $\frac{n}{10}$ **3.** 100%   **4.** 10; 10

Exercises   **1.** $\frac{1}{5}$ **3.** $\frac{3}{20}$ **5.** $\frac{31}{250}$ **7.** $\frac{3}{80}$ **9.** $\frac{91}{200}$ **11.** $\frac{1}{3}$ **13.** 0.75   **15.** 1.35   **17.** 0.129   **19.** 0.0165   **21.** 0.534   **23.** $0.161\overline{6}$ **25.** 25%
27. 40%   **29.** 37.5%   **31.** $16.\overline{6}$%   **33.** $33.\overline{3}$%   **35.** $44.\overline{4}$%   **37.** 96%   **39.** 80%   **41.** 9%   **43.** 120%   **45.** 2.8%   **47.** 0.4%   **49.** $66.\overline{6}$%
51. $163.\overline{63}$%   **53.**

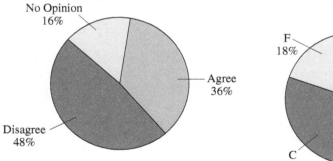

Opinion Poll

55.

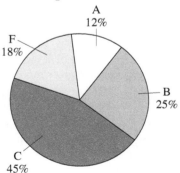

English Grades

Review Exercises   **1.** 14.4   **2.** 1620   **3.** 3.84   **4.** $x = 1333.\overline{3}$ **5.** $y = 38$ **6.** 32

Exercise Set 8.2

Prep Exercises **1.** whole; part   **2.** $0.25 \cdot 60 = n$; $\frac{25}{100} = \frac{n}{60}$ **3.** When the part is unknown, word-for-word translation requires fewer steps because we only need to multiply the whole by the percent, written as a decimal number or fraction. The proportion method requires more steps because we must cross multiply and then divide.   **4.** $0.4 \cdot w = 15$; Proportion: $\frac{40}{100} = \frac{15}{w}$ **5.** $p \cdot 24 = 5$; Proportion: $\frac{P}{100} = \frac{5}{24}$ **6.** When the percent is unknown, using the proportion method requires fewer steps because the answer is the value of the percent. We only need to attach the percent sign. Word-for-word translation requires an extra calculation step because the result is not a percent. We must multiply the result by 100%.

Exercises **1.** part **3.** whole **5.** percent **7.** whole **9.** part **11.** percent **13.** $0.4 \cdot 350 = n$; $\frac{40}{100} = \frac{n}{350}$; 140

15. $n = 0.15 \cdot 78$; $\frac{15}{100} = \frac{n}{78}$; 11.7 **17.** $1.5 \cdot 60 = n$; $\frac{150}{100} = \frac{n}{60}$; 90 **19.** $n = 0.16 \cdot 30\frac{1}{2}$; $\frac{16}{100} = \frac{n}{30\frac{1}{2}}$; 4.88

21. $n = 0.065 \cdot 22{,}800$; $\frac{6.5}{100} = \frac{n}{22{,}800}$; 1482 **23.** $n = 0.055 \cdot 280$; $\frac{5\frac{1}{2}}{100} = \frac{n}{280}$; 15.4 **25.** $0.35 \cdot w = 77$; $\frac{35}{100} = \frac{77}{w}$; 220

27. $4800 = 0.4 \cdot w$; $\frac{40}{100} = \frac{4800}{w}$; 12,000 **29.** $605 = 0.025 \cdot w$; $\frac{2.5}{100} = \frac{605}{w}$; 24,200 **31.** $1.05 \cdot w = 49.14$; $\frac{105}{100} = \frac{49.14}{w}$; 46.8

33. $47.25 = 0.105 \cdot w$; $\frac{10\frac{1}{2}}{100} = \frac{47.25}{w}$; 450 **35.** $p \cdot 68 = 17$; $\frac{P}{100} = \frac{17}{68}$; 25% **37.** $p \cdot 75 = 80$; $\frac{P}{100} = \frac{80}{75}$; 106.$\overline{6}$%

39. $2.142 = p \cdot 35.7$; $\frac{P}{100} = \frac{2.142}{35.7}$; 6% **41.** $7\frac{7}{20}p \cdot 24\frac{1}{2}$; $\frac{P}{100} = \frac{7\frac{7}{20}}{24\frac{1}{2}}$; 30% **43.** $p = \frac{21}{60}$; $\frac{P}{100} = \frac{21}{60}$; 35% **45.** $\frac{43}{65} = p$; $\frac{P}{100} = \frac{43}{65}$;

$\approx 66.15\%$ or $66\frac{2}{13}\%$ **47.** $0.65w = 52$; $\frac{65}{100} = \frac{52}{w}$; 80 **49.** $n = 0.15 \cdot 9\frac{1}{2}$; $\frac{15}{100} = \frac{n}{9\frac{1}{2}}$; 1.425 **51.** $p \cdot 30\frac{1}{3} = 18\frac{1}{5}$; $\frac{P}{100} = \frac{18\frac{1}{5}}{30\frac{1}{3}}$; 60%

53. $n = 0.125 \cdot 440$; $\frac{12\frac{1}{2}}{100} = \frac{n}{440}$; 55 **55.** $1.2 \cdot w = 65.52$; $\frac{120}{100} = \frac{65.52}{w}$; 54.6 **57.** $0.169 \cdot 2450 = n$; $\frac{16.9}{100} = \frac{n}{2450}$; 414.05

59. $\frac{17}{19} = p$; $\frac{P}{100} = \frac{17}{19}$; $\approx 89.47\%$ **61.** Mistake: wrote 2% as 0.2, which is 20% Correct: 1.4 **63.** Mistake: placed the whole, 20, and 100 in the wrong denominators Correct: 40%

Review Exercises **1.** $\frac{5}{12}$ **2.** 268.8 **3.** 1.24 **4.** 250 people disagreed. **5.** $\frac{3}{5}$ were females who agreed.

Exercise Set 8.3
Prep Exercises **1.** 1. Determine whether the percent, the whole, or the part is unknown. 2. Write the problem as a basic percent sentence (if needed). 3. Translate to an equation (word for word or proportion). 4. Solve for the unknown. **2.** 20%; 24; unknown **3.** 12%; ukown; $4800 **4.** unknown; 4000; 2400

Exercises **1.** $0.6 \cdot 800 = n$ or $\frac{60}{100} = \frac{n}{800}$; 480 ml **3. a.** $0.9 \cdot 30 = n$ or $\frac{90}{100} = \frac{n}{30}$; 27 questions; **b.** 10%; **c.** 3 questions

5. $0.1 \cdot 3106 = n$ or $\frac{10}{100} = \frac{n}{3106}$; $310.60 **7.** $0.25 \cdot 16{,}850 = n$ or $\frac{25}{100} = \frac{n}{16{,}850}$; $4212.50

9.

Expense	Percent	Equation	Amount spent
Taxes	18%	$0.18 \cdot 2.34 = n$ or $\frac{18}{100} = \frac{n}{2.34}$	$0.4212
Distribution & Marketing	10%	$0.1 \cdot 2.34 = n$ or $\frac{10}{100} = \frac{n}{2.34}$	$0.2340
Refining	11%	$0.11 \cdot 2.34 = n$ or $\frac{11}{100} = \frac{n}{2.34}$	$0.2574
Crude Oil	61%	$0.61 \cdot 2.34 = n$ or $\frac{61}{100} = \frac{n}{2.34}$	$1.4274

11. $750 = 0.06 \cdot w$ or $\frac{6}{100} = \frac{750}{w}$; $12,500 **13.** $38 = 0.42 \cdot w$ or $\frac{42}{100} = \frac{38}{w}$; 91 shot attempts **15.** $245{,}000 = 0.49 \cdot w$

or $\frac{49}{100} = \frac{245{,}000}{w}$; $500,000 **17.** $p = \frac{48}{54}$ or $\frac{P}{100} = \frac{48}{54}$; 88.$\overline{8}$% **19.** $p = \frac{36}{110}$ or $\frac{P}{100} = \frac{36}{110}$; 32.$\overline{72}$% **21.** $p = \frac{7}{24}$ or $\frac{P}{100} = \frac{7}{24}$; 29.1$\overline{6}$%

23. $p = \frac{31.5}{168}$ or $\frac{P}{100} = \frac{31.5}{168}$; 18.75%

25. a.

	Agreed	Disagreed	No opinion
Percent of Respondents	20%	65%	15%

b.

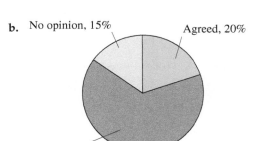

No opinion, 15%
Agreed, 20%
Disagreed, 65%

c. 85%; **d.** 80%

27. a.

	Mortgage	Car loan 1	Car loan 2	Credit card payments	Utilities	Groceries
Percent of total monthly income	≈29.5%	≈7.8%	≈6.4%	≈5.2%	≈10.1%	≈11.2%

b. ≈70.2%   **c.** ≈29.8%   **d.**

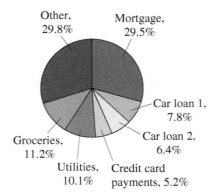

Other, 29.8%
Mortgage, 29.5%
Car loan 1, 7.8%
Car loan 2, 6.4%
Credit card payments, 5.2%
Utilities, 10.1%
Groceries, 11.2%

29. $0.6 \cdot w = 3.5$ or $\dfrac{6}{100} = \dfrac{3.5}{w}$; $58.\overline{3}$ g

31. $0.35 \cdot 119 = n$; $\dfrac{35}{100} = \dfrac{n}{119}$; 41.65 million visits   **33.** $p = \dfrac{2.06 \times 10^{16}}{1.41 \times 10^{18}}$ or $\dfrac{P}{100} = \dfrac{2.06 \times 10^{16}}{1.41 \times 10^{18}}$; about 1.46%

35. $0.08 \cdot 4249 = n$ or $\dfrac{8}{100} = \dfrac{n}{4249}$; commission: $339.92; gross pay: $699.92   **37.** $p = \dfrac{1.4}{6.9}$ or $\dfrac{P}{100} = \dfrac{1.4}{6.9}$; about 20.3%

39. $0.18 \cdot w = 45{,}000$ or $\dfrac{18}{100} = \dfrac{45{,}000}{w}$; $250,000

Review Exercises   **1.** $6.1x - 2.6$ **2.** $x = 14$ **3.** $y = 30.6$ **4.** $m = 1.6$ **5.** $2550

Exercise Set 8.4
Prep Exercises **1.** 3%; $34,000; unknown   **2.** 1. Find the increase amount. 2. Add the increase amount to the initial amount.
3. 1. Find the decrease amount. 2. Subtract the decrease amount from the initial amount.   **4.** 25%; unknown; $80
5. a. initial amount; **b.** increase amount; **c.** No, subtraction is not needed because the increase amount, $5000, is given.
d. $\dfrac{P}{100} = \dfrac{5000}{40{,}000}$ **6. a.** initial amount; **b.** final amount; **c.** Yes, subtraction is needed to find the decrease amount; **d.** $\dfrac{P}{100} = \dfrac{40}{220}$

Exercises   **1.** Sales tax is $9.30; total amount is $195.25.   **3.** The tip is $9.75; total paid is $58.50.   **5.** $30,636   **7.** Discount amount is
$19.80; final price is $46.19.   **9.** $1977.08   **11.** $490; $11,760   **13.** $25,200   **15.** $42.95   **17.** $2596   **19.** $146.97   **21.** 7%   **23.** 3.5%
25. 8%   **27.** 17.6%   **29.** 15%   **31.** ≈9.1%   **33.** ≈2.6%   **35.** ≈7.8%   **37.** ≈23.5%   **39.** ≈35.8%   **41.** ≈174.9%   **43.** Discount
amount is $245; final price is $454.99.   **45.** 22.3%   **47.** Sales tax is $1587; total amount is $28,037.   **49.** $2570   **51.** 33.6%
53. $29,250   **55.** $379.51

Review Exercises   **1.** $\dfrac{1}{125}$ **2.** 1.1664   **3.** $y = \dfrac{1}{3}$ **4.** $x = 3750$ **5.** five $10 bills, eleven $5 bills

Exercise Set 8.5
Prep Exercises **1.** Principal is an initial amount of money borrowed or invested.   **2.** Interest is an amount of money that is a percent
of the principal.   **3.** $I = Prt$, where I represents the simple interest, P represents the principal, r represents the simple interest rate,
and t represents the time in years.   **4.** Add the interest to the principal.   **5.** $B = P\left(1 + \dfrac{r}{n}\right)^{nt}$, where B represents the final balance,
P represents the principal, r represents the APR, n represents the number of compound periods per year, and t represents the time
in years.   **6.** 2   **7.** Subtract the principal from the final balance.   **8.** To amortize means to pay off a loan or debt in installments.

Exercises   **1.** $120; $4120   **3.** $38.50; $388.50   **5.** $7901.25; $20,151.25   **7.** $96; $2496   **9.** $22.68; $2022.68   **11.** $4798.84
13. $1946.25   **15.** $5832; $832   **17.** $943.82; $103.82   **19.** $518.01; $118.01   **21.** $15,142.40; $1142.40   **23.** $1910.48; $310.48
25. $412.41; $122.41   **27.** $1646.80; $346.80   **29.** $571.04; $121.04   **31.** $872.23   **33.** $706.15   **35.** $145.31   **37.** $423.05   **39.** $744.27
41. $853   **43.** $926.92

Review Exercises   **1.** (number line from -7 to beyond -6, with marks at -7, $-6\frac{3}{4}$, $-6\frac{1}{2}$, $-6\frac{1}{4}$, -6 and a point at $-6\frac{1}{2}$)   **2.** $0.8\overline{3}$ **3.** 77.75   **4.** 6   **5.** $x = -1\dfrac{13}{15}$

Chapter 8 Summary and Review Exercises

8.1 Definitions/Rules/Procedures $100; \frac{n}{100}$ Exercises **1.** $\frac{2}{5}$ **2.** $\frac{13}{50}$ **3.** $\frac{13}{200}$ **4.** $\frac{49}{200}$ **5.** 0.16 **6.** 1.5 **7.** 0.032

8. $0.40\overline{3}$ Definitions/Rules/Procedures 100% Exercises **9.** 37.5% **10.** $44.\overline{4}\%$ **11.** 54% **12.** 130%

8.2 Definitions/Rules/Procedures multiplication; equal sign; $\frac{part}{whole}$ Exercises **13.** $n = 15\% \cdot 90; \frac{15}{100} = \frac{n}{90}; 13.5$

14. $12.8\% \cdot w = 5.12; \frac{12.8}{100} = \frac{5.12}{w}; 40$ **15.** $12.5 = p \cdot 20; \frac{P}{100} = \frac{12.5}{20}; 62.5\%$ **16.** $p = \frac{40}{150}, \frac{P}{100} = \frac{40}{150}; 26.\overline{6}\%$

8.3 Definitions/Rules/Procedures percent; whole; part; sales Exercises **17. a.** $65\% \cdot 680 = n$ or $\frac{65}{100} = \frac{n}{680}; 442$ **b.** $12.9\% \cdot 680 = n$

or $\frac{12.9}{100} = \frac{n}{680}; 88$ **18.** \$1170   **19.** $180 = 35\% \cdot w$ or $\frac{35}{100} = \frac{180}{w}; 514.29$ mi.   **20.** $p = \frac{321}{554}$ or $\frac{P}{100} = \frac{321}{554}; \approx 57.9\%$

8.4 Definitions/Rules/Procedures increase; initial; decrease; initial Exercises **21.** $n = 6\% \cdot 495.95$ or $\frac{6}{100} = \frac{n}{495.95}$; sales tax is

\$29.76; total amount is \$525.71.   **22.** $75\% \cdot 56.95 = n$ or $\frac{75}{100} = \frac{n}{56.95}; \$42.71$ **23.** $1102.5 = 3.5\% \cdot w$ or $\frac{3.5}{100} = \frac{1102.5}{w}; \$31,500$

24. 175 lb.   Definitions/Rules/Procedures the initial amount from the final amount; the final amount from the initial amount

Exercises **25.** $p = \frac{27.68}{86.50}$ or $\frac{P}{100} = \frac{27.68}{86.50}; 32\%$ **26.** $p = \frac{3.50}{12}$ or $\frac{P}{100} = \frac{3.50}{12}; \approx 29.2\%$

8.5 Definitions/Rules/Procedures percent; initial; principal; original principal; interest; principal; rate; time; interest; principal; principal; interest Exercises **27.** \$28.80; \$508.80 **28.** \$104; \$8104 **29.** \$785 **30.** \$600 Definitions/Rules/Procedures final balance; principal; annual percentage rate; compound periods; time; principal; final balance; final balance; principal Exercises **31.** \$6298.56; \$1298.56 **32.** \$2033.79; \$233.79 Definitions/Rules/Procedures pay off Exercises **33.** \$917.91 **34.** \$1303.48 or \$1303.49

Chapter 8 Practice Test

1. $\frac{6}{25}$ **2.** $\frac{163}{400}$ **3.** 0.042 **4.** 0.125 **5.** 40% **6.** $55.\overline{5}\%$ **7.** 26% **8.** 120% **9.** $n = 15\% \cdot 76; \frac{15}{100} = \frac{n}{76}; 11.4$ **10.** $6.5\% \cdot w = 8.32;$

$\frac{6.5}{100} = \frac{8.32}{w}; 128$ **11.** $14 = p \cdot 60; \frac{P}{100} = \frac{14}{60}; 23.\overline{3}\%$ **12.** $p = \frac{12}{32}, \frac{P}{100} = \frac{12}{32}; 37.5\%$ **13.** $12\% \cdot 50 = n$ or $\frac{12}{100} = \frac{n}{50}; 6$ students

14. $25\% \cdot 1218 = n$ or $\frac{25}{100} = \frac{n}{1218}; \$304.50$ **15.** $p = \frac{345.75}{2786.92}$ or $\frac{P}{100} = \frac{345.75}{2786.92};$ about 12.4%   **16.** $n = 5\% \cdot 295.75$ or $\frac{5}{100} = \frac{n}{295.75};$

sales tax: \$14.79; total amount: \$310.54   **17.** $n = 70\% \cdot 84.95$ or $\frac{70}{100} = \frac{n}{84.95}; \$59.47$ **18.** $2.5\% \cdot w = 586.25$ or $\frac{2.5}{100} = \frac{586.25}{w};$

\$23,450   **19.** $p = \frac{1.75}{8.75}$ or $\frac{P}{100} = \frac{1.75}{8.75}; 20\%$ **20.** \$300; \$5300   **21.** \$16; \$816   **22.** \$2400   **23.** \$2590.06; \$590.06   **24.** \$1248.72;

\$48.72   **25.** \$901.17

Chapter 1–8 Cumulative Review Exercises

1. true **2.** true **3.** true **4.** false **5.** Find a common denominator. Rewrite. Add numerators and keep the common denominator. Simplify. **6.** Multiply both sides by an appropriate power of 10 as determined by the decimal number with the most decimal

places. **7.** 1,560,000 **8.** 1 **9.** $120xy^2$ **10.** **11.** 20 **12.** $-57\frac{1}{6}$ **13.** 64.975

14. 0.39 **15.** $\frac{8m}{3n}$ **16.** $\frac{12 - 5y}{20}$ **17.** $x^4 - 4x^3 - \frac{14}{15}x^2 + 24.2$ **18.** $24.8x^2 + 27x - 5$ **19.** 0.125 **20.** 71% **21.** $y = 6$ **22.** $x = -4.6$

22. ≈ 8.49 **23.** 152 oz. **24.** 0.08 L **25.** mean: \$41,300,000; median: \$36,000,000; mode: \$32,000,000 **26.** $A = 12x^2 - 17x - 5;$
52 in.^{2} **27.** 40 and 83 **28.** ≈ 81.76 cm^2 **29.** $x \approx 16.1$ cm; $y \approx 19.3$ cm **30.** $85.\overline{3}\%$

Exercises 9.1

Prep Exercises **1.** A point is a position in space having no length or width.   **2.** Parallel lines are two lines that never intersect.
3. The measure of an acute angle is between 0° and 90°, whereas the measure of an obtuse angle is between 90° and 180°.   **4.** The alternate interior angles are congruent.   The consecutive interior angles are supplementary. **6.** 180° **7.** The angles are congruent, and side lengths are equal.   **8.** ASA

Exercises **1.** line $\overleftrightarrow{XY}$ **3.** ray $\overrightarrow{MN}$ **5.** point J **7.** plane R **9.** line segment $\overline{TU}$ **11.** parallel lines a and b, or $a \| b$ **13.** obtuse
15. straight **17.** right **19.** acute **21.** congruent; they are vertical angles. **23.** congruent; they are alternate interior angles.
25. supplementary; they are consecutive exterior angles. **27.** supplementary; they form a straight angle. **29.** congruent; they are alternate exterior angles. **31.** supplementary; they are consecutive interior angles. **33.** congruent; they are corresponding angles.
35. a. 128°; **b.** 52°; **c.** 52°; **d.** 128° **37. a.** 70°; **b.** 70°; **c.** 110°; **d.** 70° **39.** 37°, 143°, 37°, 143° **41.** 126°, 54°, 126°, 54°

43. 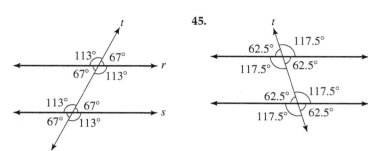   **45.**   **47.** acute triangle; the measure of each angle is less

than 90°.   **49.** obtuse triangle; the measure of one angle is greater than? 90°.   **51.** right triangle; the measure of one angle is 90°.
53. 84°, 58°, 38°   **55.** 128°, 25°, 27°   **57.** congruent; SAS   **59.** not congruent   **61.** congruent; SSS   **63.** congruent; AAS
65. not congruent   **67.** congruent; ASA   **69.** 68; 76°; 36°   **71.** 22; 100°; 58°

Review Exercises   **1.** mean: 81; median: 86   **2.**   **3.**

4. −7   **5.** −5

Exercise Set 9.2
Prep Exercises 1.
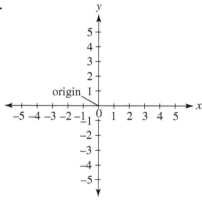
2. horizontal-axis coordinate   **3.** Beginning at the origin, move to the

right or left along the *x*-axis the amount indicated by the first coordinate. From that position on the *x*-axis, move up or down by the
amount indicated by the second coordinate.   **4.** II   **5.** Both coordinates are negative.   **6.** $\left(\dfrac{x_1 + x_2}{2}, \dfrac{y_1 + y_2}{2}\right)$

Exercises   **1.** $A\,(2,3)$; $B\,(-3,0)$; $C\,(-1,-3)$; $D\,(3,-4)$ **3.** $A\,(-4,1)$; $B\,(0,3)$; $C\,(-3,-3)$; $D\,(3,-5)$
5.   **7.**   **9.** II   **11.** I   **13.** IV   **15.** III

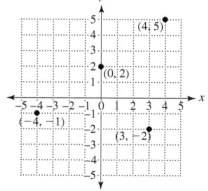   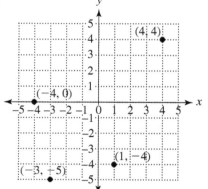

17. $(0, -34)$ is not in a quadrant; it is on the *y*-axis.   **19.** $(39, 0)$ is not in a quadrant; it is on the *x*-axis.   **21.** $(4, 3)$ **23.** $\left(5, \dfrac{1}{2}\right)$

25. $\left(-\dfrac{1}{2}, 1\right)$ **27.** $\left(-\dfrac{7}{2}, \dfrac{3}{2}\right)$ **29.** $\left(-7, -\dfrac{5}{2}\right)$ **31.** $(-1.5, -3.4)$ **33.** $\left(4\dfrac{1}{8}, 5\dfrac{1}{4}\right)$

Review Exercises   **1.**   **2.** 2   **3.** −36   **4.** $x = 9$ **5.** $y = -\dfrac{9}{10}$

Exercise Set 9.3

Prep Exercises 1. 1. Replace each variable in the equation with its corresponding coordinate. 2. Simplify both sides of the equation as needed. If the resulting equation is true, the ordered pair is a solution. **2.** 1. Replace one of the corresponding variables with a value (any value). 2. Solve the equation for the other variable. **3.** a straight line **4.** The third solution acts as a check. It should align with the other two points in a straight line. If it does not, something is wrong. **5.** An x-intercept is a point where a graph intersects the x-axis. **6.** A y-intercept is a point where a graph intersects the y-axis. **7.** 0 **8.** 0

Exercises 1. yes **3.** no **5.** yes **7.** no **9.** no **11.** no

13. $(0, -8)$, $(4, -4)$, and $(8, 0)$

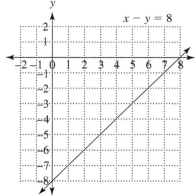

15. $(1, 4)$, $(2, 2)$, and $(3, 0)$

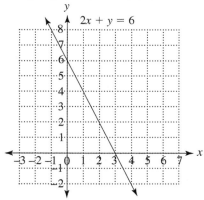

17. $(0, 0)$, $(1, 1)$, and $(2, 2)$

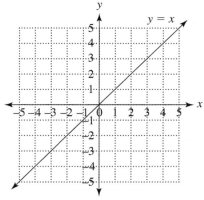

19. $(0, 0)$, $(1, 2)$, and $(2, 4)$

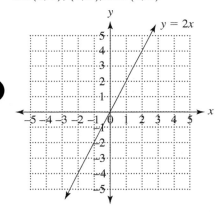

21. $(-1, 5)$, $(0, 0)$, and $(1, -5)$

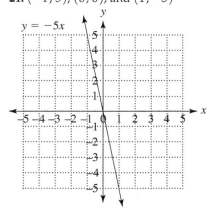

23. $(0, -3)$, $(1, -2)$, and $(2, -1)$

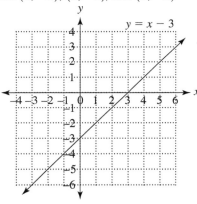

25. $(0, 4)$, $(1, 2)$, and $(2, 0)$

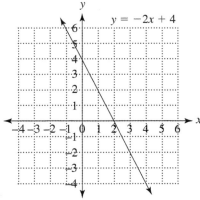

27. $(-1, -1)$, $(0, 2)$, and $(1, 5)$

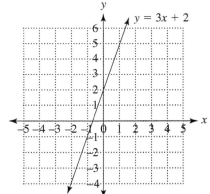

29. $(0, 0)$, $(2, 1)$, and $(4, 2)$

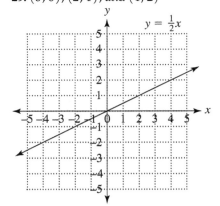

31. $(0, 4)$, $(3, 2)$, and $(6, 0)$

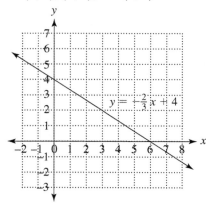

$y = -\frac{2}{3}x + 4$

33. $(0, -8)$, $(1, -4)$, and $(2, 0)$

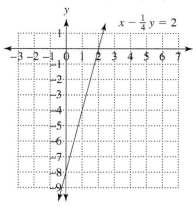

$x - \frac{1}{4}y = 2$

35. $(0, -2.5)$, $(3, -1.3)$, and $(5, -0.5)$

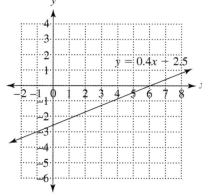

$y = 0.4x + 2.5$

37. $(0, -5)$, $(1, -5)$, and $(4, -5)$

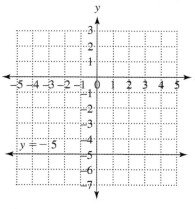

$y = -5$

39. $(7, 0)$, $(7, 2)$, and $(7, 5)$

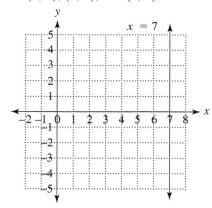

$x = 7$

41. $(12, 0)$, $(0, 6)$ **43.** $(4, 0)$, $(0, -5)$

45. $(-8, 0)$, $(0, 6)$ **47.** $(0.2, 0)$, $(0, 0.65)$ **49.** $(0, 0)$ for both   **51.** $\left(\frac{1}{4}, 0\right)$, $(0, 1)$ **53.** $(25, 0)$, $(0, -5)$ **55.** $(-9, 0)$, no y-intercept
57. no x-intercept, $(0, 4)$ **59.** no x-intercept, $(0, -3)$

Review Exercises   **1.** -64.4 **2.** 40 m   **3.** $45\frac{1}{2}$ ft.^{2} **4.** 5.67 in.^{2} **5.** $55.$ ft.^{2}

Exercises 9.4
Prep Exercises **1.** When the object stops, its velocity is 0. Therefore, we replace v with 0 and solve for t.   **2.** A vertex is a point where two lines join to form an angle.   **3.** a square   **4.** $\left(\dfrac{x_1 + x_2 + x_3}{3}, \dfrac{y_1 + y_2 + y_3}{3}\right)$

Exercises   **1. a.** 600 ft./sec.; **b.** 503.4 ft./sec.; **c.** ≈ 18.6 sec.; **d.**

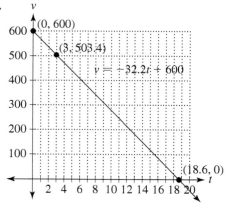

$(0, 600)$
$(3, 503.4)$
$v = -32.2t + 600$
$(18.6, 0)$

3. a. $15,000; **b.** $187,500; **c.**

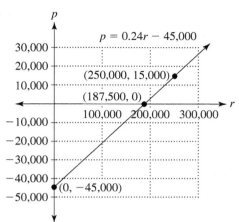

5. a. $300; **b.** $367.50; **c.** $570; **d.**

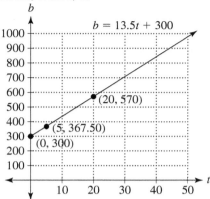

7. a. $(0, 32)$; **b.** $\left(-17\frac{7}{9}, 0\right)$; **c.** 104°F; **d.**

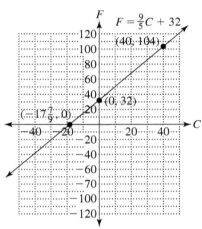

9. a. $T = 60h + 100$; **b.** $250; **c.** $4\frac{1}{4}$ hr.; **d.**

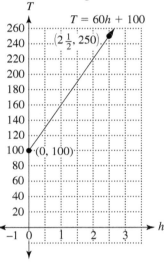

11. $(3, 2)$; 15 square units   **13.** $(1, -2)$; 24 square units   **15.** $(3, 3)$; 36 square units   **17.** $\left(\frac{1}{2}, -2\right)$; 42 square units   **19.** $\left(-\frac{7}{2}, -1\right)$;

30 square units   **21.** $(4, 3)$; 18 square units   **23.** $\left(\frac{3}{2}, -\frac{5}{2}\right)$; 63 square units   **25.** $(1, 4)$; 64 square units   **27.** $\left(\frac{3}{2}, 1\right)$; 24 square units

29. $\left(-\frac{5}{4}, -\frac{5}{2}\right)$; $28\frac{1}{2}$ square units

Review Exercises   **1.** The integers are 25 and 81.   **2.** $\frac{1}{2}$ **3.** mean ≈ 79.2; median $= 83$; mode $= 70$ and 88 **4.** 9 people
5. $440.05

Chapter 9 Summary and Review Exercises

9.1 Definitions/Rules/Procedures   position; forever; end points; one; surface; do not intersect   Exercises   **1. a.** point P; **b.** ray $\overrightarrow{RT}$;
c. parallel lines h and k, or $h \parallel k$ Definitions/Rules/Procedures   intersection; between 0° and 90°; 90°; between 90° and 180°; 180°
Exercises   **2. a.** right; **b.** obtuse; **c.** straight; **d.** acute   Definitions/Rules/Procedures   same; 90°; 180°; congruent; 90°; 5;6; 7; 8; 6;
5; 8; 7; 5; 6; 7; 8   Exercises   **3. a.** congruent; they are vertical angles. **b.** congruent; they are alternate interior angles. **c.** supplementary; they are consecutive exterior angles. **d.** supplementary; they form a straight angle.   **4. a.** 123°; **b.** 57°; **c.** 57°; **d.** 123°
5. 31°, 149°, 31°, 149°   **6.**

Definitions/Rules/Procedures   90°; less than; greater than; 180°

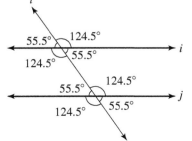

Exercises **7. a.** acute triangle; the measure of each angle is less than 90°; **b.** obtuse triangle; the measure of one angle is greater than 90° **8.** 82°, 58°, 40° **9. a.** not congruent; **b.** congruent; SAS **10.** 32°, 77°, 71°

9.2 Definitions/Rules/Procedures number line; first; second Exercises **11.** $A\,(3,4)$; $B\,(-3,2)$; $C\,(-2,0)$; $D\,(-2,-4)$; $E\,(0,-5)$; $F\,(4,-2)$ Definitions/Rules/Procedures first; second Exercises **12.**

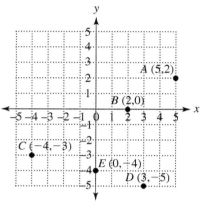

four; $(+,+)$; $(-,+)$; $(-,-)$; $(+,-)$ Exercises **13. a.** II; **b.** III; **c.** I; **d.** on the x-axis

Definitions/Rules/Procedures $\left(\dfrac{x_1+x_2}{2},\dfrac{y_1+y_2}{2}\right)$ Exercises **14.** $(4,5)$ **15.** $\left(1,-\dfrac{1}{2}\right)$

9.3 Definitions/Rules/Procedures true Exercises **16.** no **17.** yes **18.** yes Definitions/Rules/Procedures the other variable; two; straight

Exercises **19.** $(0,6)$, $(2,2)$, and $(3,0)$ **20.** $(0,-2)$, $(2,-1)$, and $(4,0)$ **21.** $(-1,2)$, $(0,3)$, and $(1,4)$

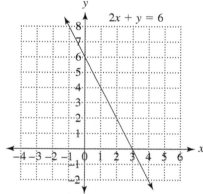
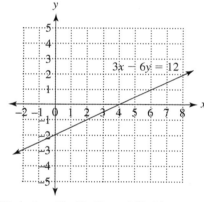
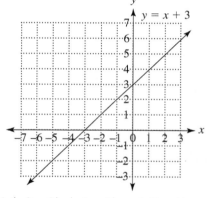

22. $(-1,3)$, $(0,0)$, and $(1,-3)$ **23.** $(-3,-2)$, $(0,0)$, and $(3,2)$ **24.** $(-2,-2)$, $(-1,-3)$, and $(0,-4)$

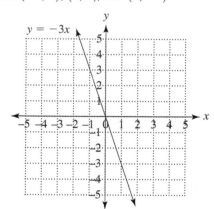
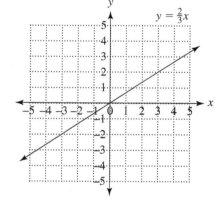
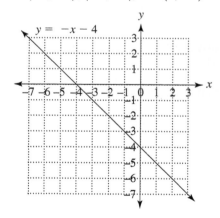

Definitions/Rules/Procedures horizontal; vertical

Exercises   **25.** $(0,7), (2,7),$ and $(5,7)$ **26.** $(-4,0), (-4,4),$ and $(-4,5)$ Definitions/Rules/Procedures $y; x; x; y$

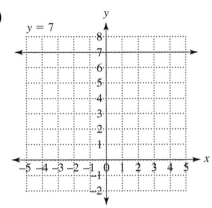
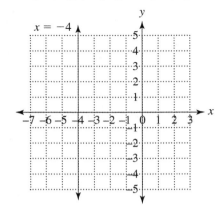

Exercises   **27.** $(2,0); (0,10)$ **28.** $\left(\dfrac{1}{4},0\right), (0,-1)$ **29.** $(0,0)$ for both   **30.** $(6,0)$; no y-intercept   **31.** no x-intercept; $(0,-2)$

9.4 Definitions/Rules/Procedures   point; mean; mean   Exercises   **32.** $(2.5, 1.5)$ **33. a.** \$28,000; **b.** \$30,000; **c.** See below.
34. 40 square units

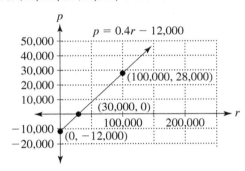

Chaper 9 Practice Test

1. a. line $\overleftrightarrow{AB}$; **b.** ray $\overrightarrow{RS}$ **2. a.** right; **b.** obtuse   **3.** $57°, 123°, 57°, 123°$ **4.**   **5.** congruent; ASA

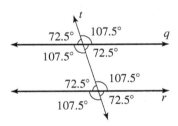

6. $97°, 37°, 46°$ **7.** $A(1,2); B(-3,0); C(-2,-5); D(4,-4)$ **8.** II   **9.** $(0,-4)$ is not in a quadrant; it is on the y-axis.   **11.** $(3,2)$
12. yes   **13.** no   **14.** $(2,0); \left(0,-\dfrac{6}{5}\right)$ **15.** $\left(\dfrac{4}{3},0\right); (0,-4)$ **16.** no x-intercept; $(0,3)$ **17.**

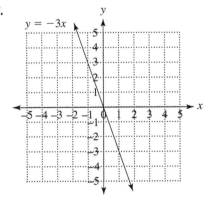

18.

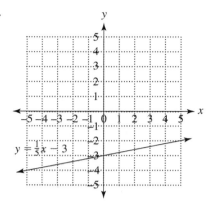

$y = \frac{1}{5}x - 3$

19.

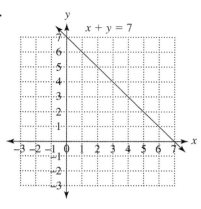

$x + y = 7$

20.

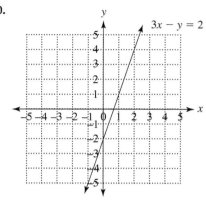

$3x - y = 2$

21.

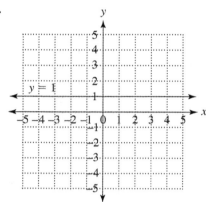

$y = 1$

22.

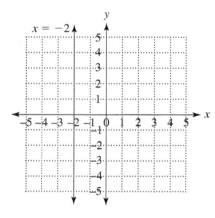

$x = -2$

23. a. 40 m/sec.; **b.** 10.6 m/sec.; **c.** ≈4.1 sec.

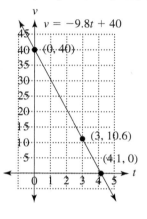

$v = -9.8t + 40$

(0, 40)

(3, 10.6)

(4.1, 0)

24. $(1, -1)$ **25.** 56 square units

Chaper 1–9 Cumulative Review Exercises

1. false   **2.** true   **3.** true   **4.** true   **5.** true   **6.** true   **7.** (1) grouping symbols; (2) exponents or roots from left to right; (3) multiplication or division from left to right; (4) addition or subtraction from left to right   **8.** Add the absolute values and keep the same sign. **9.** Subtract the absolute values and keep the sign of the number with the greater absolute value.   **10.** Change the operation from − to +. Change the subtrahend to its additive inverse.   **11.** positive   **12.** negative   **13. a.** 28,000; **b.** 724,000; **c.** 1800; **d.** 22

14 a.

b.

c.

15. a. 1; **b.** 2; **c.** 4   **16.** four thousand six hundred seven and nine hundredths

17. 3,530,000   **18.** acute   **19.** 35   **20.** $-33\frac{5}{12}$ **21.** -4.725 **22. a.** 0.375; **b.** $0.\overline{3}$; **c.** 0.08; **d.** 0.244; e. 0.045   **23. a.** $\frac{31}{6}$; **b.** $\frac{29}{200}$; **c.** $\frac{3}{50}$;

d. $\frac{11}{200}$; **e.** $\frac{41}{250}$ **24. a.** 25%; **b.** $44.\overline{4}\%$; **c.** 65%; **d.** 3.5%; **e.** 230%   **25. a.** -44; **b.** 88   **26.** $x^4 - 4x^3 - \frac{14}{15}x^2 + 24.2$ **27. a.** $-\frac{5}{18}a^3$;

b. $m^2 - 81$; **c.** $31.95t^2 + 5.7t - 18$ **28.** $2 \cdot 3^2 \cdot 5 \cdot 7$ **29.** $x = 5$ **30.** $k = 12\frac{1}{12}$ **31.** $n = -12.5$ **32.** $u = -5.25$ **33. a.** 12,672 ft.;

b. 4.5 m; **c.** 36 ft.²; **d.** 4.2 m²; **e.** 2.25 T; **f.** 230 g; **g.** 17 qt.; **h.** 0.4 L; **i.** 25°C; **j.** −4°F   **34.** −$58,460   **35. a.** $4h + 8$; **b.** 28 cm; **c.** $h^2 + 3h$;

d. 70 cm²   **36.** The integers are 40 and 83.   **37.** $\frac{3}{16}$; 6 calls   **38. a.** $\frac{23}{30}$; **b.** $\frac{13}{120}$ **39. a.** $2.60; **b.** mean: $13.10; median: $13.00; no mode

40. ≈47.26 cm²   **41.** ≈ 98.91 in.³   **42.** ≈ 21.54 ft.   **43.** $x = 21.6$ cm; $y = 26.\overline{6}$ cm   **44. a.** 180; **b.** $\frac{3}{20}$; **c.** $35.\overline{5}\%$; **d.**

F, $10.\overline{5}\%$

A, 15%

D, $13.\overline{8}\%$

C, $35.\overline{5}\%$

B, 25%

45. $60.76   **46.** $\approx 8.9\%$ **47.** $14.28   **48.**

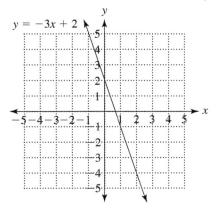

49. $A\!: (5, 1)$; $B\!: (-3, 4)$; $C\!: (-1, -3)$; $D\!: (3, -2)$;

$E\!: (-4, 0)$; $F\!: (0, 3)$ **50.**

$y = -3x + 2$

Photo Credits

Frontmatter
p. i, p. xvi: Pearson Education, Inc.

Chapter 1
p. 1: Brandon Bourdages/Shutterstock; p. 31: EyeWire Collection/Photodisc/Getty Images; p. 70: NASA

Chapter 2
p. 87: SerhioGrey/Shutterstock; p. 103: Radius Images/Alamy

Chapter 3
p. 145: Germanskydiver/Shutterstock; p. 209: National Oceanic and Atmospheric Administration (NOAA)

Chapter 4
p. 224: Maxstockphoto/Shutterstock

Chapter 5
p. 272: Shootz Photography/Shutterstock; p. 305: John Kelly/Photodisc/Getty Images

Chapter 6
p. 376: Image Source/Alamy; p. 44 6: Nagelestock.com/Alamy

Chapter 7
p. 462: Tim Davis/Corbis Super RF/Alamy

Chapter 8
p. 527: Davidwatmough/Dreamstime

Chapter 9
p. 584: Creatas/Thinkstock

Glossary

Bracketed numbers following each definition indicate the section in which the term is covered.

Absolute value: A given number's distance from zero on a number line. **[2.1]**

Acute angle: An angle measuring between 0° and 90°. **[9.1]**

Acute triangle: A triangle in which each angle measures less than 90°. **[9.1]**

Addends: Numbers that are added. **[1.2]**

Addition: The arithmetic operation that combines amounts. **[1.2]**

Addition principle of equality: Adding the same amount to both sides of an equation does not change the equation's solution(s). **[4.2]**

Additive inverses: Two numbers whose sum is zero. **[2.2]**

Algebraic equation: An equation that contains variables. **[3.1]**

Algebraic expression: An expression that contains variables. **[3.1]**

Alternate exterior angles: In a figure formed by a transversal intersecting two lines, a pair of angles outside the lines but on opposite sides of the transversal. **[9.1]**

Alternate interior angles: In a figure formed by a transversal intersecting two lines, a pair of angles between the lines but on opposite sides of the transversal. **[9.1]**

Amortize: To pay off a loan or debt in installments. **[8.5]**

Angle: A figure formed by the intersection of two lines, line segments, rays, or planes. **[9.1]**

Area: The total number of square units that completely fill a shape. **[1.3]**

Arithmetic mean: A quotient of the sum of the given values and the number of values (*see* mean). **[1.5]**

Associative property of addition: $(a + b) + c = a + (b + c)$, where a, b, and c are any numbers. **[1.2]**

Associative property of multiplication: $(ab)c = a(bc)$, where a, b, and c are any numbers. **[1.3]**

Average rate: A measure of the rate at which an object travels a total distance in a total amount of time. **[2.6]**

Axis: A number line used to locate a point in a plane. **[9.2]**

Back-end ratio: The ratio of the total monthly debt payments to gross monthly income. **[7.6]**

Balance method: If we add or remove an amount on one side of an equation, we must add or remove the same amount on the other side to keep the equation balanced. **[4.2]**

Base: In the form b^n, the number b, which is used as a factor n times. **[1.3]**

Base-10 notation: A numeral system that has ten symbols where each place value is a power of 10. **[6.1]**

Base unit: A basic unit; other units are named relative to it. **[7.4]**

Binomial: A polynomial that has exactly two terms. **[3.2]**

Capacity: A measure of the amount a container holds. **[7.3]**

Cartesian coordinate system: Two perpendicular axes that describe the position of a point in a plane (*see* rectangular coordinate system). **[9.2]**

Centroid: The balance point, or center of gravity, of a figure. **[9.4]**

Circle: A collection of points that are equally distant from a central point, called the *center*. **[5.3]**

Circumference: The distance around a circle. **[5.3]**

Coefficient: The numerical factor in a monomial. **[3.2]**

Combinatorics: The branch of mathematics that deals with counting total combinations and arrangements of items. **[1.3]**

Commission: The portion of sales earnings that a salesperson receives. **[8.3]**

Commutative property of addition: $a + b = b + a$, where a and b are any numbers. **[1.2]**

Commutative property of multiplication: $a \cdot b = b \cdot a$, where a and b are any numbers. **[1.3]**

Complementary angles: Two angles whose measurements sum to 90°. **[4.5, 9.1]**

Complex fraction: A fractional expression with a fraction in the numerator and/or denominator. **[5.4]**

Composite form: A geometric form that combines several basic forms. **[6.5]**

Composite number: A natural number that has factors other than 1 and itself. **[3.5]**

Compound interest: Interest calculated using the principal and prior earned interest. **[8.5]**

Congruent angles: Angles that have the same measurement. **[4.5, 9.1]**

Conjugates: Binomials that differ only in the sign of the second term. **[3.4]**

Consecutive exterior angles: In a figure formed by a transversal intersecting two lines, a pair of angles outside the lines on the same side of the transversal. **[9.1]**

Consecutive interior angles: In a figure formed by a transversal intersecting two lines, a pair of angles between the lines on the same side of the transversal. **[9.1]**

Constant: A symbol that does not vary in value. **[1.2]**

Conversion factor: A ratio used to convert from one unit of measurement to another by multiplication. **[7.3]**

Coordinate: A number in an ordered pair that refers to the position on a number line in a coordinate system. **[9.2]**

Corresponding angles: In a figure formed by a transversal intersecting two lines, a pair of angles in the same relative position. **[9.1]**

Cost: Money spent on production, operation, labor, and debts. **[2.3]**

Cubic unit: A $1 \times 1 \times 1$ cube. **[1.6]**

Current: A measure of electricity moving through a wire, measured in amperes, or amps, A. **[1.2, 2.6]**

Decimal notation: A base-10 notation for expressing fractions. **[6.1]**

Deduction: An amount that is subtracted from income. **[8.5]**

Degree: (monomial) The sum of the exponents of all the variables in the monomial; (polynomial) the greatest degree of any of the terms in the polynomial. **[3.2]**

Denominator: The number written in the bottom position in a fraction. **[1.5, 5.1]**

Diameter: The distance across a circle along a straight line through the center. **[5.3]**

Difference: The answer in a subtraction problem. **[1.2]**

Digit: A symbol used to represent a number. **[1.1]**

Distributive property: $a(b + c) = ab + ac$ and $a(b - c) = ab - ac$, where a, b, and c are any numbers. **[1.3]**

Dividend: The number to be divided in a division problem. The dividend is written to the left of a division sign. In fraction form, the dividend is written in the numerator (top). **[1.4]**

Division: Repeated subtraction of the same number. **[1.4]**

Divisor: The number that divides in a division problem. When the division sign is used, the divisor follows the division sign. In fraction form, the divisor is written in the denominator (bottom). **[1.4]**

Ellipsis: Three periods that mean that the pattern continues forever. **[1.1]**

Equation: A mathematical relationship that contains an equal sign. **[1.1, 3.1]**

Equilateral triangle: A triangle with all three sides of equal length. **[4.5]**

Equivalent fractions: Fractions that name the same number. **[5.1]**

Estimate: A quick approximation for a calculation. **[1.2]**

Expanded form: A number written as a sum of all digits multiplied by their place value. **[1.1]**

Exponent: In the form b^n, the number n, which indicates the number of times to use b as a factor. **[1.3]**

Exponential form: A notation having the form b^n, where b is the base and n is the exponent. **[1.3]**

Expression: A constant; a variable; or any combination of constants, variables, and arithmetic symbols that describes a calculation. **[3.1]**

Factored form: A number or an expression written as a product of factors. **[3.6]**

Factorization: A number written as a product of factors. **[3.5]**

Factors: Numbers that are multiplied. **[1.3]**

FOIL: A mnemonic used to describe the order in which two binomials are multiplied: First Outer Inner Last. **[3.4]**

Force: Something that accelerates an object (a push or pull); measured in pounds, lb. **[2.2]**

Formula: An equation that describes a procedure. **[1.3]**

Fraction: A quotient of two numbers or expressions a and b having the form $\frac{a}{b}$, where $b \neq 0$. **[5.1]**

Front-end ratio: The ratio of the total monthly house payment to gross monthly income. **[7.6]**

GCF: *see* greatest common factor

Grade point: A numerical value assigned to a letter grade. **[6.5]**

Grade point average (GPA): The sum of the total grade points earned divided by the total number of credit hours taken. **[6.5]**

Greatest common factor (GCF): The greatest number that divides all given numbers with no remainder. **[3.5]**

Hypotenuse: The side directly across from the 90° angle in a right triangle. **[6.6]**

Improper fraction: A fraction in which the absolute value of the numerator is greater than or equal to the absolute value of the denominator. **[5.1]**

Indeterminate: An expression whose value cannot be determined, for example, $\frac{0}{0}$ or 0^0. **[1.4]**

Inequality: A mathematical relationship that contains an inequality symbol ($<$, $>$). **[1.1]**

Integers: A set of numbers that contains all whole numbers and the negative counting numbers: $\ldots -3, -2, -1, 0, 1, 2, 3, \ldots$ **[2.1]**

Interest: An amount of money that is a percent of the principal. **[8.5]**

Interest rate: A percent used to calculate interest. **[8.5]**

Inverse operations: Operations that undo each other. **[1.2]**

Irrational number: A number that cannot be expressed in the form $\frac{a}{b}$, where a and b are integers and $b \neq 0$. **[5.3, 6.4]**

Isosceles triangle: A triangle with two sides of equal length. **[4.5]**

Key words: Specific words in a sentence that translate to mathematical operations. **[1.2]**

LCD: *see* least common denominator

LCM: *see* least common multiple

Least common denominator (LCD): The least common multiple of the denominators. **[5.5]**

Least common multiple (LCM): The smallest natural number that is divisible by all of the given numbers. **[5.5]**

Legs: The sides that form the 90° angle in a right triangle. **[6.6]**

Like terms: Constant terms or variable terms that have the same variables raised to the same exponents. **[3.2]**

Line: A straight, one-dimensional figure extending forever in both directions. **[9.1]**

Line segment: A straight, one-dimensional figure extending between two endpoints. **[9.1]**

Linear equation: An equation in which each variable term is a monomial of degree 1. **[4.2]**

Loss: A negative net (when revenue is less than cost). **[2.3]**

Lowest terms: A fraction is in lowest terms when the greatest common factor of its numerator and denominator is 1. **[5.2]**

Mass: A measure of the amount of matter that makes up an object. **[7.4]**

Mean: A quotient of the sum of the given values and the number of values (*see* arithmetic mean). **[1.5, 5.7, 6.5]**

Median: The middle number in an ordered list of numbers. **[1.5, 5.7, 6.5]**

Midpoint of two points: The point that lies halfway between the two points on a line segment connecting the two points. **[9.2]**

Minuend: The first number in a subtraction problem. **[1.2]**

Mixed number: An integer combined with a fraction. **[5.1]**

Mode: The value that occurs most often. **[1.5, 5.7, 6.5]**

Monomial: An expression that is a constant or a product of a constant and variables that are raised to whole number powers. **[3.2]**

Multiple: A number that is evenly divisible by a given number. **[5.1]**

Multiplication: Repeated addition of the same number. **[1.3]**

Multiplication principle of equality: Multiplying (or dividing) both sides of an equation by the same nonzero amount does not change the equation's solution(s). **[4.3]**

Multiplicative inverses: Two numbers whose product is 1 (*see* reciprocals). **[5.4]**

Multiplicative property of 0: $0 \cdot n = 0$ and $n \cdot 0 = 0$, where n is any number. **[1.3]**

Multiplicative property of 1: $1 \cdot n = n$ and $n \cdot 1 = n$, where n is any number. **[1.3]**

Multivariable polynomial: A polynomial with more than one variable. **[3.2]**

Natural numbers: $1, 2, 3, \ldots$ **[1.1]**

Net: Money remaining after subtracting costs from revenue (money made minus money spent). **[2.3]**

Net pay: Pay left after all deductions are subtracted from the gross pay. **[8.4]**

Node: A wire connection in a circuit. **[1.2]**

Numbers: Amounts or quantities. **[1.1]**

Numeral: *see* digit **[1.1]**

Numerator: The number written in the top position of a fraction. **[1.5, 5.1]**

Numeric equation: An equation that contains only constants and no variables. **[3.1]**

Obtuse angle: An angle measuring between 90° and 180°. **[9.1]**

Obtuse triangle: A triangle in which one angle measures greater than 90°. **[9.1]**

Ordered pair: A pair of numbers where the order matters, such as those describing a point in a coordinate system. **[9.2]**

Origin: The point where axes of a coordinate plane intersect. **[9.2]**

Parallel lines: Lines lying in the same plane that do not intersect. **[1.6, 9.1]**

Parallelogram: A four-sided figure with two pairs of parallel sides. **[1.6]**

Percent: A ratio representing some part out of 100. **[8.1]**

Perfect square: A number that has a whole-number square root. **[2.4]**

Perimeter: The total distance around a figure. **[1.2]**

Period: A group of places in the place-value system, for example, "billions period." **[1.1]**

Perpendicular lines: Two lines that intersect to form a 90° angle. **[9.1]**

Pi (π): An irrational number that is the ratio of the circumference of a circle to its diameter, usually approximated as $\frac{22}{7}$ or 3.14. **[5.3]**

Place-value system: A system that uses a combination of numerals to represent the numbers greater than 9. **[1.1]**

Plane: A flat, two-dimensional surface extending forever in both dimensions. **[9.1]**

Point: A position in space having no length or width. **[9.1]**

Polynomial: A monomial or a sum of monomials. **[3.2]**

Polynomial in one variable: A polynomial in which every variable term has the same variable. **[3.2]**

Power: *see* exponent. **[1.3]**

Prime factorization: A product written with prime factors only. **[3.5]**

Prime number: A natural number that has exactly two different factors: 1 and the number itself. **[3.5]**

Principal: An initial amount of money invested or borrowed. **[8.5]**

Principal square root: The positive square root of a number. **[2.4]**

Product: The answer in a multiplication problem. **[1.3]**

Profit: A positive net (when revenue is greater than cost). **[2.3]**

Proper fraction: A fraction in which the absolute value of the numerator is less than the absolute value of the denominator. **[5.1]**

Proportion: An equation in the form $\frac{a}{b} = \frac{c}{d}$, where $b \neq 0$ and $d \neq 0$. **[7.2]**

Proportional ratios: Ratios that name the same number and have equal cross products. **[7.2]**

Pythagorean theorem: The sum of the areas of the squares on the legs of a right triangle is the same as the area of the square on the hypotenuse: Given a right triangle, where a and b represent the lengths of the legs and c represents the length of the hypotenuse, $a^2 + b^2 = c^2$. **[6.6]**

Quadrant: One of four regions created by the intersection of the axes in the coordinate plane. **[9.2]**

Quotient: The answer in a division problem. **[1.4]**

Radical sign: $\sqrt{}$ **[2.4]**

Radicand: The number inside the radical sign. **[2.4]**

Radius: The distance from the center to any point on the circle. **[5.3]**

Rate: A ratio comparing two different measurements. **[7.1]**

Ratio: A comparison between two quantities using a quotient. **[7.1]**

Rational expression: A fraction in which the numerator and denominator are polynomials. **[5.2]**

Rational number: A number that can be expressed in the form $\frac{a}{b}$, where a and b are integers and $b \neq 0$. **[5.1, 6.1]**

Ray: A straight, one-dimensional figure extending forever in one direction from a single point. **[9.1]**

Real numbers: The set of all rational and irrational numbers. **[6.4]**

Reciprocals: Two numbers whose product is 1 (*see* multiplicative inverses). **[5.4]**

Rectangular array: A rectangle formed by a pattern of neatly arranged rows and columns. **[1.3]**

Rectangular coordinate system: Two perpendicular axes that describe the position of a point in a plane (*see* Cartesian coordinate system). **[9.2]**

Related equation: A true equation that relates the same pieces of a given equation using the inverse operation. **[1.2]**

Remainder: The amount left over after dividing two whole numbers. **[1.4]**

Resistance: The resistance of a wire to the flow of electricity through it, measured in ohms, Ω. **[2.6]**

Revenue: Income (money made). **[2.3]**

Right angle: An angle measuring 90°. **[1.6, 9.1]**

Right triangle: A triangle with one angle measuring 90°. **[6.6, 9.1]**

Scientific notation: A notation in the form $a \times 10^n$ where a is a decimal number whose absolute value is greater than or equal to 1 but less than 10 and n is an integer. **[6.3]**

Set: A group of elements. **[1.1]**

Similar figures: Figures with the same shape, congruent corresponding angles and proportional side lengths. **[7.2]**

Simple interest: Interest calculated using only the interest rate, the original principal, and the amount of time in which the principal earns interest. **[8.5]**

Simple percent sentence: A percent of a whole amount is a part. **[8.2]**

Simplest form: An equivalent expression written with the fewest symbols and smallest numbers possible. **[3.2, 5.2]**

Solution: A number that makes an equation true when it replaces the variable in the equation. **[1.2, 4.1, 9.3]**

Solve: To find the solution or solutions to an equation. **[4.1]**

Square root: The square root of a given number is a number whose square is the given number. **[2.4]**

Square unit: A 1×1 square. **[1.3]**

Standard form: A number written using the place-value system. **[1.1]**

Statistic: A number used to describe some characteristic of a set of data. **[1.5]**

Straight angle: An angle measuring 180°. **[9.1]**

Subset: A set within a set. **[1.1]**

Subtraction: An operation of arithmetic that can be interpreted as 1. take away, 2. difference, or 3. unknown addend. **[1.2]**

Subtrahend: The number following a minus sign in a subtraction problem. **[1.2]**

Sum: The answer in an addition problem. **[1.2]**

Supplementary angles: Two angles whose measurements sum to 180°. **[4.5, 9.1]**

Surface area: The total number of square units that completely cover the outer shell of an object. **[3.7]**

Term: *see* monomial **[3.2]**

Theoretical probability of equally likely outcomes: The ratio of the number of favorable outcomes to the total number of possible outcomes. **[7.1]**

Trapezoid: A four-sided figure with one pair of parallel sides. **[5.7]**

Trinomial: A polynomial that has exactly three terms. **[3.2]**

Undefined: When no numeric answer exists, as in $n \div 0$ when $n \neq 0$. **[1.4, 3.1]**

Unit fraction: A fraction with a value equivalent to 1. **[7.3]**

Unit price: The price of each unit; a unit rate relating price to quantity. **[6.3, 7.1]**

Unit rate: A rate in which the denominator is 1. **[7.1]**

Unit ratio: A ratio in which the denominator is 1. **[7.1]**

Variable: A symbol that can vary or change in value. **[1.2]**

Vertex: A point where two lines join to form an angle. **[9.1, 9.4]**

Vertical angles: A pair of congruent angles formed by the intersection of two lines. **[9.1]**

Voltage: The electrical pressure created by current, measured in volts, V. **[2.6]**

Volume: The total number of cubic units that completely fill an object. **[1.6]**

Weight: A measure of the downward force due to gravity. **[2.2, 7.3]**

Weighted mean: A mean that takes into account that some scores in the data set have more weight than others. **[6.5]**

Whole numbers: $0, 1, 2, 3, \ldots$ **[1.1]**

x-intercept: A point where a graph intersects the x-axis. **[9.3]**

y-intercept: A point where a graph intersects the y-axis. **[9.3]**

Index

Index of Applications

The Real Number System

Real Numbers

Rational Numbers: Real numbers that can be expressed in the form $\frac{a}{b}$, where a and b are integers and $b \neq 0$, such as $-4\frac{3}{4}$, $-\frac{2}{3}$, 0.018, $0.\overline{3}$, and $\frac{5}{8}$.

Integers: $\dots, -3, -2, -1, 0, 1, 2, 3, \dots$

Whole Numbers: $0, 1, 2, 3, \dots$

Natural Numbers: $1, 2, 3, \dots$

Irrational Numbers: Real numbers that are not rational, such as $-\sqrt{2}$, $-\sqrt{3}$, $\sqrt{0.8}$, and π.

Arithmetic Summary Diagram

Each operation has an inverse operation. In the following diagram, the operations build from the top down. Addition leads to multiplication, which leads to exponents. Subtraction leads to division, which leads to roots.

Properties of arithmetic:

Commutative property of addition
$a + b = b + a$

Associative property of addition
$(a + b) + c = a + (b + c)$

Commutative property of multiplication
$a \cdot b = b \cdot a$

Associative property of multiplication
$(ab)c = a(bc)$

Distributive property
$a(b + c) = ab + ac$

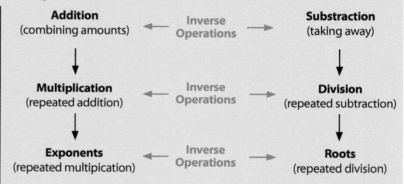

Addition (combining amounts) ← **Inverse Operations** → **Substraction** (taking away)

Multiplication (repeated addition) ← **Inverse Operations** → **Division** (repeated subtraction)

Exponents (repeated multipication) ← **Inverse Operations** → **Roots** (repeated division)

Units of Measurement

Distance	Area	Volume	Capacity
Inches (in.)	Square inches (in.^{2})	Cubic inches (in.^{3})	Ounces (oz.)
Feet (ft.)	Square feet (ft.^{2})	Cubic feet (ft.^{3})	Cups (c.)
Yards (yd.)	Square yards (yd.^{2})	Cubic Yards (yd.^{3})	Pints (pt.)
Miles (mi.)	Square miles (mi.^{2})	Cubic miles (mi.^{3})	Quarts (qt.)
			Gallons (gal.)
Centimeters (cm)	Square centimeters (cm.^{2})	Cubic centimeters (cm^3 or cc)	
Meters (m)	Square meters (m^2)	Cubic meters (km^3)	Milliliters (ml)
Kilometers (km)	Square kilometers (km^2)	Cubic kilometers (km^3)	Liters (l)
Light-years (lt-yr)	Square light-years (lt-yr^2)	Cubic light-years (lt-yr^3)	

Mass	Weight	Temperature	Time	Speed
Slugs	Ounces (oz.)	Degrees Fahreheit (°F)	Years (yr.)	A distance unit over
	Pounds (lb.)	Degrees Celsius (°C)	Days (d.)	a time unit
Milligrams (mg)	Tons (T)	Kelvin (K)	Hours (hr.)	
Grams (g)			Seconds (sec.)	Miles per hour (mi./hr.)
Kilograms (kg)	Newtons (N)			Meters per second (m/sec.)
Metric tons (t)				

Geometry Formulas

Parallelogram
(Includes squares and rectangles)

Area: $A = bh$

Triangle

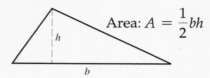

Area: $A = \dfrac{1}{2}bh$

Trapezoid

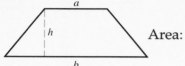

Area: $A = \dfrac{1}{2}h(a + b)$

Box

Volume: $V = lwh$

Surface Area: $SA = 2(lw + wh + lh)$

Pyramid

Volume: $V = \dfrac{1}{3}lwh$

Circle

Circumference: $C = \pi d$ or $C = 2\pi r$

Area: $A = \pi r^2$

Cylinder

Volume: $V = \pi r^2 h$

Cone

Volume: $V = \dfrac{1}{3}\pi r^2 h$

Sphere

Volume: $V = \dfrac{4}{3}\pi r^3$

Pythagorean theorem
(For right triangles)

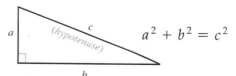

$a^2 + b^2 = c^2$

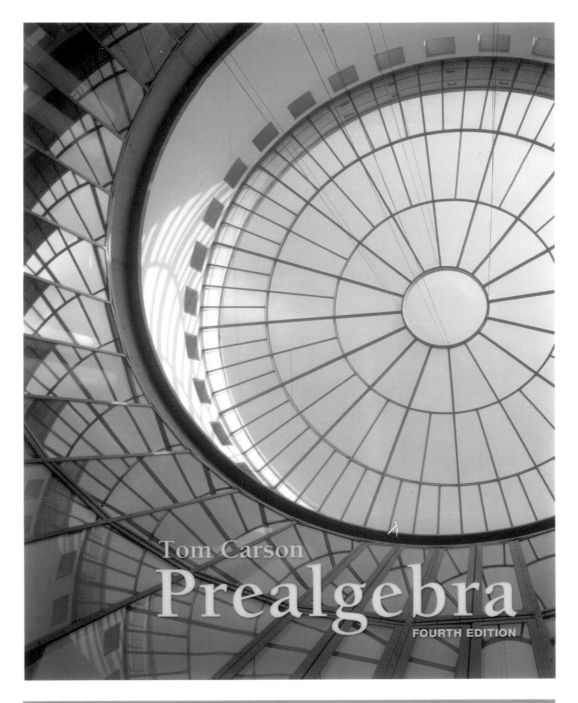

Tom Carson

Prealgebra

FOURTH EDITION

YOUR TEXTBOOK—IN A BINDER-READY EDITION!

This unbound, three-hole punched version of your textbook lets you take only what you need to class and incorporate your own notes—all at an affordable price!

ISBN-13: 978-0-321-78296-0
ISBN-10: 0-321-78296-8

EAN

9 780321 782960

Math Study System at a Glance